T0419708

Plant Virology Protocols

Second Edition

METHODS IN MOLECULAR BIOLOGY

John M. Walker, SERIES EDITOR

Plant Virology Protocols

From Viral Sequence to Protein Function
Second Edition

Gary D. Foster
University of Bristol,
Bristol, UK

I. Elisabeth Johansen
University of Aarhus,
Denmark

Yiguo Hong
University of Warwick,
Warwick, UK

Peter D. Nagy
University of Kentucky,
Kentucky, USA

✲✲ Humana Press

Editors

Gary D. Foster
University of Bristol,
Bristol, UK

Yiguo Hong
University of Warwick,
Warwick, UK

I. Elisabeth Johansen
University of Aarhus,
Denmark

Peter D. Nagy
University of Kentucky,
Kentucky, USA

Series Editor:
John M. Walker
University of Hertfordshire
Hatfield, Hertz
UK

ISBN: 978-1-58829-827-0   e-ISBN: 978-1-59745-102-4

Library of Congress Control Number: 2008920466

Cover Illustration: Background: Derived from Fig.1 of chapter 31
Inset: Derived from Fig. 5 of chapter 37

Printed on acid-free paper

9 8 7 6 5 4 3 2 1

springer.com

Preface

Following the considerable success of the Plant Virology Protocols in the Methods in Molecular Biology volume, Humana Press invited us to produce a second edition of this volume.

The first book *Plant Virology Protocols: From Virus Isolation to Transgenic Resistance* had a trend running through it, which people liked, which was – methods to isolate a virus, clone it, express it, and transform it into plants, and evaluate those plants for transgenic resistance.

For the second edition, we have decided on a different trend running through the book that is – *From Viral Sequence to Protein Function*, which will cover the many new techniques that we now can apply to analyze and understand plant viruses.

This book has been divided into five major parts, containing 44 chapters in total.

Part 1 provides a general introduction to some typical plant viral proteins, and their role in infection and interactions with other viral proteins, with the host, vectors, etc.

Part 2 provides a range of techniques for investigating viral nucleic acid sequence as well functional analysis, with Part 3 covering protein analysis and investigation of protein function.

Part 4 has a wide-ranging remit but centered on techniques for microscopy/GFP visualization and analysis/protein tagging/generation of infectious clones and other such tools.

Part 5 covers the emerging area of genomics, interactions with host factors, and plant-based studies, a theme that will probably expand over the coming years to require an entire book dedicated to this theme alone, perhaps *Plant Virology Protocols Vol 3*!

Plant Virology Protocols is the product of the hard work and major efforts of a large number of individuals who have been supportive and patient during the editing process. The editors would like to thank them all; we hope they and others will find the book useful and informative.

Gary Foster would like to thank, or should it be apologize to, his family (Diana, James, and Kirsty) for agreeing to take on another book, yes I know I promised I would not take another book, but this one was too nice an idea.

Yiguo Hong would like to thank Gary Foster for inviting him to become involved in this project. Thanks also go to Po Tien, Bryan Harrison, John Stanley, and

Michael Wilson who have consistently inspired him to the tiny but extremely exciting world of plant viruses. Yiguo Hong would also like to thank his family (Mei, Elizabeth, and Lucy) for their support.

Elisabeth Johansen would like to thank Gary Foster for the invitation to participate in the challenging process leading to the publication of this book. Thank you for your guidance and encouragement.

Gary D. Foster
I. Elisabeth Johansen
Yiguo Hong
Peter D. Nagy

Contents

Section 3 Protein Analysis and Investigation of Protein Function

Contributors

Padmanaban Annamalai
Department of Plant Pathology, University of California,
Riverside, CA 92521-0122, USA

Špela Baebler
Department of Biotechnology and Systems Biology, National Institute of Biology,
Večna pot 111, 1000 Ljubljana, Slovenia

John F. Bol
Clusius Laboratory, Institute of Biology, Leiden University, Wassenaarseweg 64,
2333 AL Leiden, the Netherlands

Neil Boonham
Central Science Laboratory, Sand Hutton, York YO41 1LZ, UK

Margaret I. Boulton
John Innes Centre, Norwich Research Park, Colney, Norwich NR4 7UH, UK

Véronique Brault
Institut National de la Recherche Agronomique, 28 Rue de Herrlisheim, 68021
Colmar, France

József Burgyán, D.Sc.
Plant Biology Institute, Agricultural Biotechnology Center,
2101 Gödöllö, P.O. Box 411, Hungary

John P. Carr
Department of Plant Sciences, University of Cambridge, Downing Street,
Cambridge CB2 3EA, UK

Romit Chakrabarty
Department of Plant Pathology, 201F Plant Science Building,
University of Kentucky, Lexington, KY 40546, USA

Sean N. Chapman
Plant Pathology Programme, Scottish Crop Research Institute, Scottish Crop
Research Institute, Invergowrie, Dundee DD2 5DA, UK

Vitaly Citovsky
Department of Biochemistry and Cell Biology, State University of New York,
Stony Brook, NY 11794-5215, USA

Tamas Dalmay
School of Biological Sciences, University of East Anglia,
Norwich NR4 7TJ, UK

José-Antonio Daròs
Instituto de Biología Molecular y Celular de Plantas, Universidad Politécnica
de Valencia-Consejo Superior de Investigaciones Científicas, Valencia, Spain

S.P. Dinesh-Kumar
Department of Molecular, Cellular, and Developmental Biology, Yale University,
New Haven, CT 06520-8103, USA

Andrew J. Dingley
Department of Chemistry and School of Biological Sciences, The University
of Auckland, Science Centre, 23 Symonds Street, Auckland, New Zealand

Marc R. Fabian
Department of Biology, York University, 4700 Keele St., Toronto, ON, Canada
M3J 1P3

Lilian H. Florentino
Departamento de Bioquímica e Biologia Molecular/ BIOAGRO-Universidade
Federal de Viçosa-36571.000 Viçosa, MG, Brazil

Ricardo Flores
Instituto de Biología Molecular y Celular de Plantas, Universidad Politécnica
de Valencia-Consejo Superior de Investigaciones Científicas, Valencia, Spain

Elizabeth P.B. Fontes
Departamento de Bioquímica e Biologia Molecular/BIOAGRO-Universidade
Federal de Viçosa-36571.000 Viçosa, MG, Brazil

Gary D. Foster
School of Biological Sciences, University of Bristol, Bristol BS8 1UG, UK

María-Eugenia Gas
Instituto de Biología Molecular y Celular de Plantas, Universidad Politécnica
de Valencia-Consejo Superior de Investigaciones Científicas,
Valencia, Spain

Lee Gehrke
HST Division, MIT E25-545, 77 Massachusetts Avenue, Cambridge,
MA 02139, USA

Raúl C. Gomila
HST Division, MIT E25-545, 77 Massachusetts Avenue, Cambridge,
MA 02139, USA

Michael Goodin
Department of Plant Pathology, 201F Plant Science Building,
University of Kentucky, Lexington, KY 40546, USA

Lisa Gow
School of Biological Sciences, University of Bristol,
Bristol BS8 1UG, UK

Joachim Grötzinger
Biochemisches Institut der Christian-Albrechts-Universität Kiel, Olshausenstr.
40, 24118 Kiel, Germany

Kristina Gruden
Department of Biotechnology and Systems Biology, National Institute of Biology,
Večna pot 111, 1000 Ljubljana, Slovenia

Deyin Guo
National Key Laboratory of Virology and Modern Virology Center, College
of Life Sciences, Wuhan University, Wuhan 430072, P.R. China

John S. Hartung
Molecular Plant Pathology Laboratory, Bldg. 004, BARC-West, ARS/USDA,
10300 Baltimore Ave., Beltsville, MD 20705, USA

Sophie Haupt
University of Dundee at SCRI, Invergowrie, DD2 5DA, UK

Zoltán Havelda
Agricultural Biotechnology Center, Plant Biology Institute, Szent-Györgyi
Albert út 4, Gödöllö H-2001, Hungary

Carmen Hernández
Instituto de Biología Molecular y Celular de Plantas, Universidad Politécnica de
Valencia-Consejo Superior de Investigaciones Científicas, Valencia, Spain

Thien X. Ho
NERC/Centre for Ecology and Hydrology (CEH) Oxford, Mansfield Road,
Oxford OX1 3SR, UK, Department of Biochemistry, University of Oxford, South
Parks Road, Oxford OX1 3QU, UK

Matjaž Hren
Department of Biotechnology and Systems Biology, National Institute of Biology,
Večna pot 111, 1000 Ljubljana, Slovenia

Qi Huang
Floral and Nursery Plants Research Unit, U.S. National Arboretum, U.S.
Department of Agriculture, Agricultural Research Service, Beltsville, MD, USA

Konstantin I. Ivanov
Molecular Cancer Biology Research Program, Biomedicum Helsinki, University
of Helsinki, 00014 Helsinki, Finland

I. Elisabeth Johansen
Department of Genetics and Biotechnology, Faculty of Agricultural Sciences,
University of Aarhus, Thorvaldsensensvej 40, 1871 Frederiksberg C, Denmark

Natalia O. Kalinina
Belozersky Institute of Physico-Chemical Biology, Moscow State University,
Moscow 119992, Russia

C. Cheng Kao
Department of Biochemistry & Biophysics, 103 Biochemistry/Biophysics
Building, Texas A&M University, 2128 TAMU, College Station,
TX 77843-2128, USA

Sakari Kauppinen
Wilhelm Johannsen Center for Functional Genome Research,
Department of Cellular and Molecular Medicine, University of Copenhagen,
DK-2200 Copenhagen N, Denmark and Santaris Pharma,
DK-2970 Hoersholm, Denmark

Young-Chan Kim
Department of Biochemistry & Biophysics, 103 Biochemistry/Biophysics
Building, Texas A&M University, 2128 TAMU, College Station,
TX 77843-2128, USA

Polona Kogovšek
Department of Biotechnology and Systems Biology, National Institute of Biology,
Večna pot 111, 1000 Ljubljana, Slovenia

Dora Chin-Yen Koh
Department of Biological Sciences, National University of Singapore,
14 Science Drive 4, Singapore, Singapore 117543
Department of Neurobiology, The Scripps Research Institute, 10550 Torrey Pines
Road, La Jolla, CA 92037, USA

Hana Krečič-Stres
Department of Biotechnology and Systems Biology, National Institute of Biology,
Večna pot 111, 1000 Ljubljana, Slovenia

Lóránt Lakatos
Plant Biology Institute, Agricultural Biotechnology Center, P.O. Box 411, H-2101
Gödöllö, Hungary

Jung-Youn Lee
Plant and Soil Sciences, Delaware Biotechnology Institute,
University of Delaware, 15 Innovation Way, Newark, DE 19711, USA

Dingxiang Liu
Department of Biological Sciences, National University of Singapore, 14 Science
Drive 4, Singapore, Singapore 117543
Institute of Molecular and Cell Biology, 61 Biopolis Drive, Proteos, Singapore,
Singapore 138673

Huanting Liu
Centre for Biomolecular Science, University of St Andrews, North Haugh,
St. Andrews KY16 9ST, UK

Li Liu
Centre for Infectious Disease, Institute of Cell and Molecular Science,
Barts and The London, Queen Mary's School of Medicine and Dentistry,
The Blizard Building, 4 Newark Street, Whitechapel,
London E1 2AT, UK

George P. Lomonossoff, Ph.D.
John Innes Centre, Norwich NR4 7UH, UK

Inken Lorenzen
Biochemisches Institut der Christian-Albrechts-Universität Kiel,
Olshausenstr. 40, 24118 Kiel, Germany

Ole Søgaard Lund
Department of Plant Biology, Faculty of Life Sciences, University of Copenhagen,
Thorvaldsensvej 40, 1871 Frederiksberg C, Denmark

Stuart A. MacFarlane
Plant Pathology Department, Scottish Crop Research Institute, Invergowrie,
Dundee DD2 5DA, UK

Kristiina M. Mäkinen
Department of Applied Chemistry and Microbiology, P.O. Box 27, University
of Helsinki, 00014 Helsinki, Finland

Andrew J. Maule
John Innes Centre, Norwich Research Park, Colney,
Norwich NR4 7UH, UK

W. Allen Miller
Molecular Cellular and Developmental Biology, Department of Plant Pathology,
Iowa State University, Ames, IA, USA

Diego Molina
Instituto de Biología Molecular y Celular de Plantas, Universidad Politécnica
de Valencia-Consejo Superior de Investigaciones Científicas,
Valencia, Spain

Kirankumar S. Mysore
Plant Biology Division, The Samuel Roberts Noble Foundation,
2510 Sam Noble Pky., Ardmore, OK 73401, USA

Peter D. Nagy
Department of Plant Pathology, University of Kentucky, Lexington,
KY 40546, USA

Peter Palukaitis
Scottish Crop Research Institute, Invergowrie, Dundee DD2 5DA, UK

Tadas Panavas
Department of Plant Pathology, University of Kentucky, Lexington,
KY 40546, USA

Acássia B.L. Pires
Departamento de Ciências Biológicas-Universidade Estadual de
Santa Cruz-45650.000- Ilhéus-BA, Brazil

Judit Pogany
Department of Plant Pathology, University of Kentucky, Lexington,
KY 40546, USA

Maruša Pompe-Novak
Department of Biotechnology and Systems Biology, National Institute of Biology,
Večna pot 111, 1000 Ljubljana, Slovenia

Minna-Liisa Rajamäki
Department of Applied Biology, University of Helsinki, 00014 Helsinki, Finland

K.S. Rajendran
Department of Plant Pathology, University of Kentucky, Lexington, KY 40546,
USA

Aurélie M. Rakotondrafara
Molecular Cellular and Developmental Biology, Department of Plant Pathology,
Iowa State University, Ames, IA, USA

A.L.N. Rao
Department of Plant Pathology, University of California, Riverside,
CA 92521-0122, USA

Maja Ravnikar
Department of Biotechnology and Systems Biology, National Institute of Biology,
Večna pot 111, 1000 Ljubljana, Slovenia

Pia Ruggenthaler
Max F. Perutz Laboratories, Department of Medical Biochemistry, Medical
University of Vienna, Dr. Bohr-Gasse 9, A-1030 Vienna, Austria

Rachel Rusholme
School of Biological Sciences, University of East Anglia,
Norwich NR4 7TJ, UK

Eugene V. Ryabov
Warwick HRI, University of Warwick, Wellesbourne, Warwick CV35 9EF, UK

Hélène Sanfaçon
Pacific Agri-Food Research Centre, P.O. Box 5000, 4200 Highway 97,
Summerland, BC, Canada V0H 1Z0

Anésia A. Santos
Departamento de Bioquímica e Biologia Molecular/ BIOAGRO- Universidade
Federal de Viçosa-36571.000 Viçosa, MG, Brazil

Keith Saunders
Biological Chemistry, John Innes Centre, Norwich NR4 7UH, UK

James E. Schoelz
Division of Plant Sciences, 108 Waters Hall, University of Missouri, Colombia,
MO 65211, USA

Elena Serviene
Department of Plant Pathology, University of Kentucky, Lexington,
KY 40546, USA

William R. Staplin
Molecular Cellular and Developmental Biology, Department of Plant Pathology,
Iowa State University, Ames, IA, USA

Tsubasa Takahashi
Laboratory of Plant Pathology, Iwate University, Ueda 3-18-8,
Morioka 020-8550, Japan

Michael Taliansky
Scottish Crop Research Institute, Invergowrie, Dundee DD2 5DA, UK

Nataša Toplak
Department of Biotechnology and Systems Biology, National Institute of Biology,
Večna pot 111, 1000 Ljubljana, Slovenia

Lesley Torrance
Scottish Crop Research Institute, Invergowrie, Dundee DD2 5DA, UK

Kateryna Trutnyeva
Max F. Perutz Laboratories, Department of Medical Biochemistry,
Medical University of Vienna, Dr. Bohr-Gasse 9, A-1030 Vienna, Austria

Tzvi Tzfira
Department of Molecular, Cellular and Developmental Biology,
University of Michigan, Ann Arbor, MI 48109-1048, USA

Joachim F. Uhrig
Botanical Institute III, University of Cologne, Gyrhofstr. 15, 50931 Cologne,
Germany

Zarir E. Vaghchhipawala
Plant Biology Division, The Samuel Roberts Noble Foundation, 2510 Sam Noble
Pky., Ardmore, OK 73401, USA

Jari Valkonen
Department of Applied Biology, University of Helsinki,
00014 Helsinki, Finland

Elisabeth Waigmann
Max F. Perutz Laboratories, Department of Medical Biochemistry, Medical
University of Vienna, Dr. Bohr-Gasse 9, A-1030 Vienna, Austria

Hui Wang
NERC/Centre for Ecology and Hydrology (CEH) Oxford, Mansfield Road,
Oxford OX1 3SR, UK

K. Andrew White
Department of Biology, York University, 4700 Keele St., Toronto,
ON, Canada M3J 1P3

Sek-Man Wong
Department of Biological Sciences, National University of Singapore,
14 Science Drive 4, Singapore, Singapore 117543
Temasek Life Sciences Laboratory, National University of Singapore, 1 Research
Link, Singapore, Singapore 117604

Sharon Yelton
Department of Plant Pathology, 201F Plant Science Building,
University of Kentucky, Lexington, KY 40546, USA

Nobuyuki Yoshikawa
Laboratory of Plant Pathology, Faculty of Agriculture, Iwate University,
Ueda 3-18-8, Morioka 020-8550, Japan

Francisco M. Zerbini
Departamento de BIOAGRO-Fitopatologia Universidade
Federal de Viçosa-36571.000 Viçosa, MG, Brazil

Guangzhi Zhang
Department of Botany, University of British Columbia, Vancouver,
BC, Canada V6T 1Z4

Xiaohong Zhu
Department of Molecular, Cellular, and Developmental Biology,
Yale University, New Haven, CT 06520-8103, USA

Angelika Ziegler
Scottish Crop Research Institute, Invergowrie, Dundee DD2 5DA, UK

Véronique Ziegler-Graff, Ph.D.
Institut de Biologie Moléculaire des Plantes du Centre National de la Recherche
Scientifique, 12 Rue du Général Zimmer, 67084 Strasbourg, France

Section 1
General Introduction

Chapter 1
Plant–Virus Interactions

Peter Palukaitis, John P. Carr, and James E. Schoelz

Abstract A variety of techniques have been used to examine plant viral genomes, the functions of virus-encoded proteins, plant responses induced by virus infection and plant–virus interactions. This overview considers these technologies and how they have been used to identify novel viral and plant proteins or genes involved in disease and resistance responses, as well as defense signaling. These approaches include analysis of spatial and temporal responses by plants to infection, and techniques that allow the expression of viral genes transiently or transgenically in planta, the expression of plant and foreign genes from virus vectors, the silencing of plants genes, imaging of live, infected cells, and the detection of interactions between viral proteins and plant gene products, both in planta and in various in vitro or in vivo systems. These methods and some of the discoveries made using these approaches are discussed.

Keywords Agroinfection; Agroinfiltration; Green fluorescent protein; Plant gene isolation; Microarrays; PR proteins; Resistance responses; Salicylic acid; Transgenic plants; Virus-induced gene silencing

1 Introduction

Research in plant virology has consisted predominantly of studying diseases induced by plant viruses and characterizing the viruses involved in various diseases of plants (reviewed in ref. 1). While physiological responses in plants to infection by viruses have been measured for many years using biochemical methods, it is only in the last 20 years that plant–virus interactions have been analyzed at the molecular, cellular, and genetic levels (1, 2). A variety of techniques that allowed the characterization of the interactions between plants and viruses has become available. These techniques are described in the following chapters, as are various functions and processes involved in virus–virus, virus–host, and virus–vector interactions.

Prior to the mid 1980s, the technology available allowed only limited molecular analyses of plant–virus interactions. Many of these analyses concerned determining

From: *Methods in Molecular Biology, Vol. 451, Plant Virology Protocols: From Viral Sequence to Protein Function*
Edited by G.D. Foster, I.E. Johansen, Y. Hong, and P.D. Nagy © Humana Press, Totowa, NJ

the genome organization of viruses and the nature of the gene products of viral genomes. In vitro translation was used to assess the number of gene products and the modes of gene expression of plant viruses with plus-sense RNA genomes (reviewed in ref. 3). Because plant viruses do not inhibit the expression of most plant genes, the in vivo detection of viral-encoded gene products was limited largely to the highly expressed capsid protein. Protoplasts prepared from mesophyll cells of leaves were used to study the kinetics of virus replication, as well as whether resistance genes had any effect on virus replication (reviewed in ref. 4); a method still in use today. For viruses with divided genomes, separation of the individual genomic components and reassortment (also referred to as pseudorecombination) were used to map phenotypes, such as symptomatology, and host range to specific genome segments and sometimes to individual genes. However, localization of phenotypic changes to specific nucleotides required the development of cloning techniques for viruses with RNA genomes, in vitro RNA transcription of infectious genomes from cloned cDNAs, sequence determination of the genomes, and mutagenesis methods for modifying specific nucleotides in cloned cDNAs (5–9). These techniques had all been developed by the mid 1980s, although many refinements have been made since then.

2 Approaches to Virus Genome and Gene-Function Analyses

The development of various cDNA cloning techniques has allowed representatives of virtually all viral genera to be cloned and their genome sequences to be determined. This has allowed the number, position, and mode of expression of the various genes to be determined (10). Mutagenesis methods have allowed the roles of various genes and their encoded proteins to be determined, and also whether these genes or sequences support replication, movement, transmission, or act as elicitors of defense functions, and more recently, whether they act as counter defense proteins (e.g., suppressors of RNA silencing) (reviewed in ref. 1). Sequence analysis also has been useful in identifying conserved domains and potential active sites of proteins. Sequencing of strains of the same virus for identifying target sequences combined with mutagenesis has allowed the identification of specific viral sequences involved in eliciting various responses between plants and viruses. These methods have not, however, allowed identification of the host components involved in these interactions. By contrast, various approaches developed and used in the last 20 years have allowed some of these host factors and their functions to be identified.

2.1 Gene Exchange to Localize Determinants

As infectious clones of plant viruses have been developed, gene-exchange experiments have been the preferred method for identification of avirulence (*Avr*) and symptom determinants. The advantage of genetic exchange between two closely related strains is that this allows one to test the effect of viral genes in the context of an infection in a whole plant. All that is required is to have infectious clones of two virus strains that differ in one or more biological properties. In the early days, it was important to locate restriction enzyme sites common to the two virus strains to engineer the exchange, but with the advent of PCR, exchanges can now be made at any point in the virus genome. Reciprocal exchanges are best, because they allow for a more thorough examination of a specific trait: one chimera can be used to prove that a specific viral gene is an *Avr* determinant, whereas the reciprocal chimera shows that no other viral gene product acts as a second *Avr* determinant in a given host.

The first virus *Avr* gene product identified through gene exchange was P6 of cauliflower mosaic virus (CaMV) (11, 12), a CaMV protein that interacts with host ribosomes to reprogram them for expression of the polycistronic CaMV 35 S RNA (13, 14). P6 triggers a hypersensitive response (HR) in *Datura stramonium* and *Nicotiana edwardsonii*, and a nonnecrotic resistance response in *Nicotiana bigelovii*. The same studies also showed that P6 is an important symptom determinant, as it is a primary determinant of chlorosis in crucifers (11). The first *Avr* gene product of an RNA virus to be characterized by gene exchange was the coat protein of tobacco mosaic virus (TMV) (15, 16). As these types of studies have accumulated, it is has been found that virtually any type of viral gene product may trigger a resistance response in plants (17).

2.2   Viral Protein Function Analyses and Localization

The overexpression of viral-encoded proteins in *Escherichia coli*, often tagged with either six histidines (His_6) or glutathione transferase (GST) to aid in their purification, has been very useful in producing proteins that can be used to study various in vitro functions of such proteins. These include binding to other viral-encoded proteins (using GST-pull down assays, coimmunoprecipitation, filter binding assays, or surface-plasmon resonance) or to nucleic acids (18–22). These proteins also have been used to produce antisera that were then used to detect these proteins and localize them in situ by immunogold labeling and electron microscopy (23, 24). *E. coli* expression of viral-encoded proteins has largely superseded expression of such proteins in insect cells, which also has been used.

2.3   Plant Genetic Approaches

Different plant varieties, cultivars, ecotypes, and accessions have been used for identifying viral processes affected by changes in host genes, especially the effects of resistance genes on infection. A number of dominant resistance genes have been isolated, first using insertional mutagenesis and later map-based cloning approaches. These include the resistance genes *N* (against TMV) from tobacco (25), *Rx* [against potato virus X (PVX)] from potato (26, 27), *Tm2* (28) and *Tm2²* (29) (against tomato mosaic virus), as well as *Sw-5* (30) (against tomato spotted wilt virus) from tomato, and *RCY1* (31) (against cucumber mosaic virus), as well as *RTM1* and *2* (32, 33) (against tobacco etch virus) from *Arabidopsis thaliana*. A number of recessive resistance genes also have been identified from *A. thaliana* and various crops species, using candidate gene mapping approaches, demonstrating that the genes encoding translation factors eIF4E, eIF(iso)4E, and eIF4G are involved in resistance to viruses in various groups (reviewed in ref. 34). *A. thaliana* has proven to be particularly useful for isolating genes affecting virus infection, for several reasons, including the small genome size, the large number of ecotypes and mutant genomes available, and especially because the entire genome has been sequenced and annotated.

The study of resistance and host reactions to virus infection has, like most areas of plant virology, been technology driven. Thus, as new techniques become available, dormant topics have become accessible to further investigation. Attempts to better understand host reactions or resistance to plant viruses have in some instances revealed important aspects of viral biology; for example, in the study of recessive genes conferring resistance to potyviruses (reviewed in ref. 34). Paradoxically, the discovery of novel host responses to viruses has sometimes turned out to be more useful for our understanding of resistance to nonviral pathogens, as exemplified by the discovery and subsequent studies of the pathogenesis-related (PR) proteins.

3 Plant Responses Induced by Virus Infection

The work of Ross and Kuć and their respective colleagues in the 1960s and 1970s on induced resistance prompted several research groups to seek new proteins or other factors associated with resistance induction (reviewed in ref. 35). Induced resistance, most commonly referred to as systemic acquired resistance (SAR), is the enhancement of a plant's resistance to disease triggered by previous exposure to an avirulent pathogen.

3.1 "Novel" Proteins: The Discovery of IVR and PR Proteins

Loebenstein and coworkers exploited protoplast techniques to identify an extracellular protein produced by plants following a TMV-induced HR. The HR is a resistance response usually characterized by programmed host-cell death in the vicinity of pathogen entry and often followed by the induction of SAR. This protein inhibited accumulation of several viruses in leaf discs and so was called an inhibitor of virus replication (IVR) (36, 37). Recently, it was shown that constitutive expression of an IVR cDNA in transgenic plants provided some protection against virus infection and, surprisingly, against a fungus (38). The results appear to confirm that IVR may play a role in induced resistance to viruses but they also raise some interesting new questions regarding the mode of action of a factor that can inhibit the life cycles of pathogens as diverse as viruses and fungi.

Probably the most intensively studied inducible gene products associated with the HR and SAR induction are a very diverse set of proteins, the PR proteins. The first PR proteins to be identified were acidic, extracellular proteins belonging to the PR1 and PR2 families (39). PR proteins were discovered independently by van Loon and Gianinazzi, together with their respective colleagues, by comparing the protein compositions of healthy tobacco plants with those of plants infected systemically with TMV or exhibiting the HR in response to the virus (40, 41). Both groups detected "novel", host-encoded proteins accumulating in the leaves of *NN* genotype (TMV-resistant) tobacco plants inoculated with TMV and in leaves expressing SAR. PR proteins are defined now as any plant proteins induced as a result of pathogen infection or attempted infection and many different families have been defined (42). Certain PR proteins, notably members of the PR1 family, and their mRNA transcripts have proved to be very useful molecular indicators for the induction of SAR (43, 44). However, their levels can also increase during systemic infection by certain viruses (45). Furthermore, albeit a correlative relationship between PR1 protein or gene induction and resistance to viruses was suggested in early work (46), constitutive expression of PR1 and several other PR proteins in transgenic plants did not result in enhanced resistance to viruses (47, 48), although it did provide protection against fungal infection (49). Thus far, none of the proteins defined as "PRs" are known to have any antiviral effects. However, it might be argued that a plant RNA-dependent RNA polymerase, RdRp1, which is induced by salicylic acid (SA) and during systemic virus infection and promotes sequence-specific turnover of TMV RNA in tobacco (50), could be considered a PR protein under the current rather wide definition.

Serendipitously, for their pioneering work on PR proteins, Gianinazzi and van Loon both used the native polyacrylamide gel electrophoresis (PAGE) system of Ornstein (51) and Davis (52). Using this system, the highly charged, acidic PR proteins are the most conspicuous bands in the electrophoretic patterns produced on 10% acrylamide gels by proteins extracted from hypersensitively responding tobacco leaves. On one-dimensional SDS-PAGE gels, PR protein bands are completely obscured by a background of polypeptide bands representing constitutive plant proteins. Without the fortuitous use of native PAGE to examine changes in protein composition accompanying the HR, much of the field of molecular plant pathology would have taken a rather different course and our knowledge of plant responses to infection, including mechanisms of resistance to fungal and bacterial infection, the defensive signaling roles of SA, and the regulation of defense genes, would have taken longer to acquire.

3.2 Defensive Signaling: Changes in SA Accumulation

White (53) showed that treatment of plants with aspirin (acetylsalicylic acid) solutions induced SAR and PR protein accumulation in tobacco. This prompted the suggestion that a benzoic acid or salicylic acid-like signal chemical might play a role in the establishment of SAR, following a HR (54). This idea was vindicated through the use of sensitive analytical technologies, which showed that SA levels increase dramatically following a virus or fungus-induced HR (55, 56), as well as the application of plant transformation to create transgenic plants that were unable to accumulate SA (due to constitutive expression of an SA-degrading enzyme) and consequently were unable to exhibit HR-type resistance or SAR against viral and other pathogens (57).

The analytical techniques used for the detection of SA in infected plant tissue were developed from earlier studies on the role of this chemical in triggering heat production in thermogenic plants (58), and most assay methods that have followed take similar approaches. These measurements typically involve extraction of phenolic compounds from plant tissue followed by high performance liquid chromatography (HPLC) and fluorescence detection of the SA peak. In some studies, the presence of SA has been further authenticated by mass spectrometry (55, 56, 58). This type of procedure can be adapted for high-throughput analysis of many samples at once (59).

The HPLC methodology is highly accurate and quantitative, but destructive, and provides no information on the distribution of SA in living tissues. To address this problem, a new method has recently been developed in which genetically engineered, nonpathogenic bacteria (*Acinetobacter sp.* ADP1) harboring an SA-responsive *lux* reporter gene construct are infiltrated into the leaf apoplast or stems of plants following viral challenge. In *NN* genotype tobacco responding to TMV infection, the accumulation of SA around the developing HR lesions could be imaged utilizing the SA-induced bioluminescence from the engineered bacteria (60). By calibrating the imaging system using known amounts of SA into plant leaves, it was possible to make quantitative determinations of SA across the tissues (60).

3.3 Temporal and Spatial Plant Responses to Infection

The technology of the period 1970 to ca.1995 limited the examination of "molecular" markers, that is, proteins and RNA transcripts, for host reactions to virus infection to a relatively hit-or-miss approach. This has changed significantly with the advent of "-omics" approaches in which it is potentially possible to monitor changes in levels of hundreds or thousands of transcripts, proteins, or metabolites in virus-infected plant tissue. For example, using *A. thaliana* DNA microarrays, Whitham and colleagues (45) identified transcripts that increase or decrease in their steady state levels in response to systemic infection by any of five different viruses. In contrast to this group of nonspecifically responding transcripts, levels of certain other host mRNAs are changed only in response to certain specific viruses. For example, the *PR1* gene and several other SA-regulated genes were upregulated by infection with a potyvirus and a cucumovirus but not by infection with viruses belonging to three other genera (45).

Nevertheless, this powerful technology may not be very useful in interpreting the responses of host cells actually infected with the virus. Even with the most concentrated of virus inocula, only about 0.1% of the epidermal cells in directly inoculated leaves initially become infected (61) and although the proportion of virus-harboring cells in systemically infected tissue will be much higher, it will be a mixed population of virus-infected and noninfected cells. Therefore, the results of the current generation of microarray studies using RNA samples taken from systemically infected plants need to be interpreted with caution. In the future, emerging technologies such as single-cell sampling techniques (62), combined with the use of green fluorescent protein (GFP)-expressing viruses and microarray techniques requiring smaller amounts of RNA, are likely allow the very precise assessment of changes in host metabolism and host and viral gene expression in different host cells at different stages of virus infection.

Using currently available technology, virus-induced alterations in metabolism and gene expression are best studied in the earliest phases of infection using directly inoculated tissue because this shows the greatest degree of synchrony. This has been exploited to greatest effect by Maule and collaborators using in situ hybridization techniques to detect host and viral RNAs and microanalytical methods to assess biochemical changes in virus-infected pea seeds and marrow (squash) cotyledons, respectively (63–66; also discussed in ref. 1).

4 Approaches to Study Plant–Virus Interactions

4.1 Development of Transgenic Plants

It has now been over 20 years since the first publications describing the development of transgenic plants, a technique that has truly revolutionized how plant biology research is conducted. In 1983, four groups published their evidence that they had successfully developed transgenic plants. The transgene of choice for three of the four groups was a kanamycin resistance gene, although it was expressed in three different types of plants: petunia, cultivated tobacco (*Nicotiana tabacum*) and a close relative of tobacco (*Nicotiana plumbaginifolia*) (67–69). The fourth group chose to express a bean protein in sunflower (70).

Virologists quickly recognized the practical value of transgenic technology, as it could be used for development of new methods for controlling virus diseases. For example, one of the first applications of transgenic technology involved expression of the TMV coat protein in transgenic tobacco, which protected plants from infection by TMV (71). This technology, called pathogen-derived resistance, has been utilized to protect many types of crop plants from a broad variety of viruses (72). Other transgenic techniques also have been developed; in particular, methods based on homology-dependent gene silencing (now known as RNA silencing). Furthermore, as host virus resistance genes have been cloned, it has been possible to move them across species barriers, as was done in moving the *N* gene from tobacco to tomato (73).

Transgenic plants also have been utilized for basic studies on how viruses cause disease and the strategies plants use to protect themselves from pathogen attack. As virus genes were converted into transgenes for expression in plants, it was discovered that some could elicit symptoms. The first example was P6 of CaMV, which elicited virus-like symptoms in tobacco when expressed as a transgene (74). Viral transgenes also have been used for the development of elegant complementation systems. For instance, transgenic tobacco that expressed the movement protein of TMV could complement movement-defective strains of TMV (75). In addition, attempts to express both wild type and mutant forms of viral transgenes also contributed to the discovery of RNA silencing mechanisms in plants (76, 77) and the discovery of virus-encoded silencing suppressors (78–80). It is now known that several viral genes that function as suppressors of RNA silencing also have the capacity to induce symptoms in plants when expressed as a transgene.

4.2 Virus Gene Expression and Silencing Vectors

Viral-derived gene expression vectors have been used with great success to probe the functions of genes involved in plant defenses. Over the past 10 years, it has become apparent that viruses themselves can be the targets of the plant's gene silencing apparatus. By inserting host nucleotide sequences into a virus vector, researchers can trick the plant into degrading its own mRNAs. Host nucleotide sequences carried in a virus vector will be targeted for degradation, but most importantly, plant mRNAs homologous to the sequence in the virus also will be degraded. This technique is called virus-induced gene silencing (or VIGS). In a classic experiment, Ruiz et al. (81) created a PVX vector that carried portions of the plant gene, phytoene desaturase. When this PVX vector was inoculated to plants, the endogenous phytoene desaturase mRNA was targeted for degradation along with the virus, and this resulted in photobleaching in the leaves. VIGS has been used to silence host resistance genes (82) and also has been used in high-throughput assays to characterize plant genes required for the plant defense response (83).

In addition to characterization of host genes necessary for plant defenses, virus vectors also have been useful for identification and characterization of viral *Avr* genes. In particular, they can be used when no known resistance-breaking strain has been found. For example, all known tomato bushy stunt virus (TBSV) strains elicit HR in *N. tabacum* and *N. edwardsonii*; consequently, gene exchanges are not an option. Scholthof et al. (84) inserted TBSV genes individually into a PVX vector and found that TBSV proteins P19 and P22 were *Avr* determinants. P19 triggered HR in *N. tabacum*, whereas P22 triggered HR in *N. edwardsonii*. Furthermore, this system could be used to dissect the functions of Avr proteins from other functions associated with either P19 or P22; experiments that could not be done in the context of an infectious virus clone. For example, a PVX vector was used to show that the cell-to-cell movement function of P22 could be separated from its capacity to elicit HR in *N. edwardsonii* (85).

4.3 Imaging of Infected Cells

Nondestructive imaging of virus movement through plant tissues at the micro- and macroscopic levels has been possible since the mid-1990s, owing to the development of genetically modified viruses expressing GFP and other fluorophores (86). Experiments with GFP-expressing viruses and viral protein–GFP fusions, especially when combined with imaging using the confocal scanning laser microscope, have revolutionized our understanding of virus movement in susceptible hosts and the mechanisms of intercellular communication in plants (reviewed in ref. 87). Application of these technologies to the investigation of viral movement and distribution around HR lesions and in SA-treated tissues has raised new questions regarding the involvement of mechanisms such as RNA silencing in induced resistance and how antiviral mechanisms may differ between different cell types in the same plant (88, 89).

4.4 Agroinfection and Agroinfiltration

Agroinfection (also known in the literature as agroinoculation) is a technique in which an infectious clone of a virus is inserted into the T-DNA present on the Ti plasmid of *Agrobacterium tumefaciens*. *A. tumefaciens* is subsequently used to deliver the infectious viral DNA into a plant cell, where it is released from the T-DNA and the infection is initiated. Agroinfection was first demonstrated with the caulimoviruses and geminiviruses (90, 91) and is now the primary method for initiating infections with geminivirus and luteoviruses (92, 93). Agroinfection is also increasingly being used for inoculation of infectious clones of RNA viruses (82, 83, 94–97). In this variation, the viral cDNA is expressed from a constitutive promoter such as the CaMV 35 S promoter. Once the viral cDNA is delivered to cells by *A. tumefaciens*, the host RNA polymerase II will utilize the plant promoter to initiate synthesis of an infectious RNA. As with the DNA viruses, agroinfection of infectious cDNAs based on RNA viruses is much more efficient and cost effective, as it eliminates the need for in vitro transcription.

Agroinfiltration is a variation in which individual viral genes are expressed in a transient fashion (98). Agroinfiltration provides a rapid alternative for screening *Avr* genes compared to expression in viral vectors or gene swaps between infectious virus clones. In this technique, a putative *Avr* gene is placed under the control of a promoter such as the 35 S promoter and the cassette inserted into the T-DNA of the Ti plasmid of *A. tumefaciens*. *A. tumefaciens* containing this Ti plasmid is treated with acetosyringone, which mobilizes the transfer of the T-DNA, and after 24 h the cells are infiltrated into the leaf. Plant tissues infiltrated with an *Avr* gene will develop the HR at a rate comparable to a virus-inoculated plant. Agroinfiltration has been used successfully to illustrate the function of *Avr* genes from several viruses. For example, the TMV replicase elicits HR when agroinfiltrated into *N*-gene tobacco (99, 100) and the coat protein of PVX elicits HR when agroinfiltrated into potatoes containing *Rx* gene (26).

As with virus vectors, agroinfiltration has two advantages over gene exchange experiments with infectious virus clones. First, it allows for the identification of an *Avr* gene when no resistance breaking strain of the virus is available. Second, it is possible to isolate domains of Avr proteins capable of eliciting HR. For example, agroinfiltration was used to show that the helicase domain of the TMV replicase protein was responsible for eliciting the HR in *N*-gene tobacco (99, 100). Agroinfiltration also has been developed as a tool for initiation of gene silencing in plants (98) and in the identification of virus suppressors of gene silencing (101).

4.5 Interactions between Host and Viral Molecules

The outcome of many interactions between viruses and their hosts (susceptibility, resistance, or no interaction) is likely to be influenced at some point by the intermolecular interactions of host and viral proteins. Such protein–protein interactions are difficult or impossible to identify *in planta* and so potentially interacting plant and viral proteins have been investigated in vitro or under physiological conditions using the yeast two-hybrid system. In the yeast two-hybrid system, sequences from pairs of candidate interacting proteins are, respectively, engineered to form translational fusions with two separate parts of an artificial transcription factor. If the candidate interacting protein sequences do indeed bind to each other, this will bring together a functional transcription factor that will activate reporter gene activity in the yeast cell (102). The yeast two-hybrid approach is fraught with technical problems but can be extremely informative when its conclusions are backed up by other approaches, for example, in vitro interaction assays using coimmunoprecipitation or analysis of interactions between mutant plants or viruses (see Sects. 2.2 and 2.3).

The best case study in which this combination of approaches has yielded important biological data is in the interaction between the potyviral VpG proteins and the translation factors eukaryotic initiation factor eIF4E and eIF(iso)4E in various hosts (reviewed in ref. 34). The interaction is required for successful infection by potyviruses and underpins most examples of recessive resistance to potyviruses in a wide range of plants including Arabidopsis, pea, tobacco, tomato, pepper, lettuce, and various Brassica species, resistance results in each example from expression of a modified translation factor, with which the potyviral VpG is unable to interact (34, 103–106).

5 Concluding Remarks: Future Directions

Over the past 25 years, infectious clones and virus genome sequences have become a prerequisite for the study of plant viruses. For most viruses, the primary functions for each of their genes have been determined (i.e., coat protein, movement protein, replicase) and secondary and tertiary functions (i.e., silencing suppressor, symptom determinant, Avr determinant) are being explored and characterized. Increased efforts will be made toward understanding differences in interactions of viruses with different cell/tissue types. This latter objective will be aided by the continual development of new techniques; for example, laser capture microscopy/microdissection that will allow genomic, proteomic, and metabolomic analysis of individual virus-infected cells (107). As virologists sort out the mechanisms that lead to symptom development and activation of plant defenses, increasingly there will be a shift toward the host side of the host–virus interaction. The development of plant genome sequences and EST libraries is one step that will contribute to our understanding of how plants respond to virus infection. Another key will be a refinement of techniques to selectively silence or "knockout" plant genes to test their involvement in symptom development or host defenses. Furthermore, new techniques will be developed to facilitate the identification of host proteins that interact with viruses and to visualize where these interactions occur in the cell. For example, fusion of the genes expressing various viral-encoded proteins to sequences encoding TAP (tandem affinity purification) tags and transient expression of the TAP-tagged viral proteins can be used to isolate the viral-encoded proteins and associated plant proteins, with an aim to identifying what plant proteins bind to the various virus-encoded proteins (108, 109). Genome resources, coupled with powerful imaging techniques, hold the potential for a deeper understanding for where host–virus interactions occur in the cell, how viruses cause disease, and how plants defend themselves from pathogen attack.

Acknowledgments JES acknowledges support from the U.S. Department of Agriculture/National Research Initiative Competitive Grant No. 2003–35319–13778. PP was supported by a grant-in-aid from the Scottish Executive Environment and Rural Affairs Department to the SCRI.

References

1. Hull, R. (2002) *Matthews' Plant Virology*, 4th ed. Academic, San Diego.
2. Zaitlin, M. and Palukaitis, P. (2000) Advances in understanding plant viruses and virus disease. *Annu. Rev. Phytopathol.* **38,** 117–143.
3. Davies, J.W. and Hull, R. (1982) Genome expression of plant positive-strand RNA viruses. *J. Gen. Virol.* **61,** 1–14.
4. Takebe, I. (1983) Protoplasts in plant-virus research. *Intl. Rev. Cytol.* **Suppl. 16,** 89–111.
5. Boyer, J.C. and Haenni, A.L. (1994) Infectious transcripts and cDNA clones of RNA viruses. *Virology* **198,** 415–426.
6. Sanger, F., Nicklen, S., and Coulson, A.R. (1977) DNA sequencing with chain-terminating inhibitors. *Proc. Natl. Acad. Sci. USA* **74,** 5463–5467.
7. Sanger, F., Coulson, A.R., Barrell, B.G., Smith, A.J.H., and Roe, B.A. (1980) Cloning in single-stranded bacteriophage as an aid to rapid DNA sequencing. *J. Mol. Biol.* **143,** 161–178.
8. Zoller, M.J. and Smith M. (1982) Oligonucleotide-directed mutagenesis using M13-derived vectors – an efficient and general procedure for production of point mutations in any fragment of DNA. *Nucleic Acids Res.* **10,** 6487–6500.
9. Norris, K., Norris, F., Christiansen, L., and Fiil, N. (1983) Efficient site-directed mutagenesis by simultaneous use of 2 primers. *Nucleic Acids Res.* **11,** 5103–5112.

10. Fauquet, C.M., Mayo, M.A., Maniloff, J., Desselberger, U., and Ball, L.A., eds. (2005) *Virus Taxonomy: Classification and Nomenclature of Viruses. Eighth Report of the International Committee of the Taxonomy of Viruses*. Elsevier, Amsterdam.

11. Daubert, S.D., Schoelz, J.E., Debao, L., and Shepherd, R.J. (1984) Expression of disease symptoms in cauliflower mosaic virus genomic hybrids. *J. Mol. Appl. Genet.* **2,** 537–547.

12. Schoelz, J.E., Shepherd, R.J., and Daubert, S. (1986) Region VI of cauliflower mosaic virus encodes a host range determinant. *Mol. Cell Biol.* **6,** 2632–2637.

13. Bonneville, J.M., Sanfacon, H., Fütterer, J., and Hohn, T. (1989) Posttranscriptional trans-activation in cauliflower mosaic virus. *Cell* **59,** 1135–1143.

14. Gowda, S., Wu, F.C., Scholthof, H.B., and Shepherd, R.J. (1989) Gene VI of figwort mosaic virus (caulimovirus group) functions in posttranscriptional expression of genes on the full-length RNA transcript. *Proc. Natl. Acad. Sci. USA* **86,** 9203–9207.

15. Saito, T., Meshi, T., Takamatsu, N., and Okada, Y. (1987) Coat gene sequence of tobacco mosaic virus encodes host response determinant. *Proc. Natl. Acad. Sci. USA* **84,** 6074–6077.

16. Knorr, D.A. and Dawson, W.O. (1988) A point mutation in the tobacco mosaic capsid protein gene induces hypersensitivity in *Nicotiana sylvestris*. *Proc. Natl. Acad. Sci. USA* **85,** 170–174.

17. Culver, J.N. (1997) Viral avirulence genes, in *Plant–Microbe Interactions, Vol. 2* (Stacey G. and Keen, N., eds.), Chapman and Hall, New York, pp. 196–219.

18. Liu, Y., Burch-Smith, T., Schiff, M., Feng, S., and Dinesh-Kumar, S.P. (2004) Molecular chaperone Hsp90 associates with resistance protein N and its signalling proteins SGT1 and Rar1 to modulate an innate immune response in plants. *J. Biol. Chem.* **279,** 2101–2108.

19. Schmidt, I., Blanc, S., Esperandieu, P., Kuhl, G., Devauchelle, G., Louis, C., and Cerutti, M. (1994) Interaction between the aphid transmission factor and virus particles is a part of the molecular mechanism of cauliflower mosaic virus aphid transmission. *Proc. Natl. Acad. Sci. USA* **91,** 8885–8889.

20. Leh, V., Jacquot, E., Geldreich, A., Hermann, T., Leclerc, D., Cerutti, M., et al. (1999) Aphid transmission of cauliflower mosaic virus requires the viral PIII protein. *EMBO J.* **18,** 7077–7085.

21. Drucker, M., Froissart, R., Hebrard, E., Uzest, M., Esperandieu, P., Mani, J.C., et al. (2002) Intracellular distribution of viral gene products regulate a complex mechanism of cauliflower mosaic virus acquisition by its aphid vector. *Proc. Natl. Acad. Sci. USA* **99,** 2422–2427.

22. Citovsky, V., Knorr, D., Schuster, G., and Zambryski, P. (1990) The P30 movement protein of tobacco mosaic virus is a single-stranded nucleic acid binding protein. *Cell* **60,** 637–647.

23. Wieczorek, A. and Sanfacon, H. (1993) Characterization and subcellular localization of tomato ringspot nepovirus putative movement protein. *Virology* **194,** 734–742.

24. Cillo, F., Roberts, I.M., and Palukaitis, P. (2002) In situ localization and tissue distribution of the replication-associated proteins of *Cucumber mosaic virus* in tobacco and cucumber. *J. Virol.* **76,** 10654–10664.

25. Whitham, S., Dinesh-Kumar, S.P., Choi, D., Hehl, R., Corr, C., and Baker, B. (1994) The product of the tobacco mosaic virus resistance gene *N*: similarity to Toll and the interleukin-1 receptor. *Cell* **78,** 1101–1115.

26. Bendahmane, A., Kanyuka, K., and Baulcombe, D.C. (1999) The *Rx* gene from potato controls separate virus resistance and cell death responses. *Plant Cell* **11,** 781–792.

27. Bendamahne, A., Querci, M., Kanyuka, K., and Baulcombe, D.C. (2000) *Agrobacterium* transient expression system as a tool for the isolation of disease resistance genes: application to the *Rx2* locus in potato. *Plant J.* **21,** 73–81.

28. Lanfermeijer, F.C., Warmink, J., and Hille, J. (2005) The products of the broken TM-2 and the durable Tm-2^2 resistance genes from tomato differ in four amino acids. *J. Exper. Bot.* **56,** 2925–2933.

29. Lanfermeijer, F.C., Dijkhuis, J., Sturre, M.J.G., de Haan, P., Hille, J. (2003) Cloning and characterization of the durable *Tomato mosaic virus* resistance gene *Tm-2^2* from *Lycopersicon esculentum*. *Plant Mol. Biol.* **52,** 1037–1049.

30. Brommonschenkel, S.H., Frary, A., and Tanksley, S.D. (2000) The broad-spectrum tospovirus resistance gene *Sw-5* of tomato is a homolog of the root-knot nematode resistance gene *Mi*. *Mol. Plant-Microbe Interact.* **13,** 1130–1138.

31. Takahashi, H., Miller, J., Nozaki, Y., Sukamto, Takeda, M., Shah, J., et al. (2002) *RCY1*, and *Arabidopsis thaliana RPP8/HRT* family resistance gene, conferring resistance to cucumber mosaic virus requires salicylic acid, ethylene and a novel signal transduction mechanism.

32. Chisholm, S.T., Mahajan, S.K., Whitham, S.A., Yamamoto, M.L., and Carrington, J.C. (2000) Cloning of the Arabidopsis *RTM1* gene, which controls restriction of long-distance movement of *Tobacco etch virus*. *Proc. Natl. Acad. Sci. USA* **97**, 489–494.

33. Whitham, S.A., Anderberg, R.J., Chisholm, S.T. and Carrington, J.C. (2000) Arabidopsis *RTM2* gene is necessary for specific restriction of tobacco etch virus and encodes an unusual small heat shock-like protein. *Plant Cell* **12**, 569–582.

34. Robaglia, C. and Caranta, C. (2006) Translation initiation factors: a weak link in plant RNA virus infection. *Trends Plant Sci.* **11**, 40–45.

35. Gilliland, A., Murphy, A.M., and Carr, J.P. (2006) Induced resistance mechanisms, in *Natural Resistance Mechanisms of Plants to Viruses* (Loebenstein G. and Carr J.P., eds.), Springer, Amsterdam, pp. 125–145.

36. Loebenstein, G. and Gera, A. (1981) Inhibitor or virus replication released from tobacco mosaic virus-infected protoplasts of a local lesion-responding tobacco cultivar. *Virology* **114**, 132–139.

37. Spiegel, S., Gera, A., Salomon, R., Ahl, P., Harlap, S., and Loebenstein, G. (1989) Recovery of an inhibitor of virus replication from the intercellular fluid of hypersensitive tobacco infected with tobacco mosaic virus and from uninfected induced-resistant tissue. *Phytopathology* **79**, 258–262.

38. Akad, A., Teverovsky, E., Gidoni, D., Elad, Y., Kirshner, B., Ray-David, D., et al. (2005) Resistance to *Tobacco mosaic virus* and *Botrytis cinerea* in tobacco transformed with complementary DNA encoding an inhibitor of viral replication-like protein. *Ann. Appl. Biol.* **147**, 89–100.

39. Carr, J.P. and Klessig, D.F. (1989) The pathogenesis-related proteins of plants, in *Genetic Engineering: Principles and Methods, Vol. 11* (Setlow, J.K., ed.), Plenum, New York, pp. 65–100.

40. Gianinazzi, S., Martin, C., and Vallée, J.-C. (1970) Hypersensibilité aux virus, température et protiénes solubles chez le *Nicotiana* 'Xanthi nc'. Apparition de nouvelles macromolécules lors de la répression de la synthèse virale. *C.R. Acad. Sci. Paris D* **270**, 2383–2386.

41. van Loon, L.C. and van Kammen, A. (1970) Polyacrylamide disc electrophoresis of soluble leaf proteins from Nicotiana tabacum var. Samsun and Samsun NN.2. Changes in protein constitution after infection with tobacco mosaic virus. *Virology* **40**, 199–211.

42. van Loon, L.C. and van Strien, E.A. (1999) The families of pathogenesis-related proteins, their activities, and comparative analysis of PR-1 type proteins. *Physiol. Molec. Plant Pathol.* **55**, 85–97.

43. Ward, E.R., Uknes, S.J., Williams, S.C., Dincher,S.S., Wiederhold, D.L., Alexander, D.C., Ahl-Goy, P., Métraux, J.P., and Ryals, J.A. (1991) Coordinate gene activity in response to agents that induce systemic acquired resistance. *Plant Cell* **3**, 1085–1094.

44. Kessmann, H., Staub, T., Hoffmann, C., Maetzke, T., Herzog, J., Ward, E., et al. (1994) Induction of systemic acquired resistance in plants by chemicals. *Annu. Rev. Phytopathol.* **32**, 439–459.

45. Whitham, S.A., Quan, S., Chang, H.S., Cooper, B., Estes, B., Zhu, T., et al. (2003) Diverse RNA viruses elicit the expression of common sets of genes in susceptible *Arabidopsis thaliana* plants. *Plant J.* **33**, 271–283.

46. Fraser, R.S.S. (1998) Introduction to classical crossprotection, in *Methods in Molecular Biology, Vol. 81: Plant Virology Protocols: From Virus Isolation to Transgenic Resistance* (Foster, G.D. and Taylor, S.C., eds.), Humana Press, Totawa, NJ, pp. 13–24.

47. Cutt, J.R., Harpster, M.H., Dixon, D.C., Carr, J.P., Dunsmuir, P., and Klessig, D.F. (1989) Disease response to tobacco mosaic virus in transgenic tobacco plants that constitutively express the pathogenesis-related PR1b gene. *Virology* **173**, 89–97.

48. Linthorst, H.J.M., Meuwissen, R.L.J., Kauffmann, S., and Bol, J.F. (1989) Constitutive expression of pathogenesis-related proteins PR-1, GRP and PR-S in tobacco has no effect on virus infection. *Plant Cell* **1**, 285–291.
49. Alexander, D., Goodman, R.M., Gut-Rella, M., Glascock, C., Weymann, K., Friedrich, L., et al. (1993) Increased tolerance to two oomycete pathogens in transgenic tobacco expressing pathogenesis-related protein-1a. *Proc. Natl. Acad. Sci. USA* **90**, 7327–7331.
50. Xie, Z.X., Fan, B.F., Chen, C.H., and Chen, Z.X. (2001) An important role of an inducible RNA-dependent RNA polymerase in plant antiviral defense. *Proc. Natl. Acad. Sci. USA* **98**, 6516–6521.
51. Ornstein, L. (1964) Disc electrophoresis. I. Background and theory. *Ann. NY Acad. Sci.* **121**, 321–349.
52. Davis, B.J. (1964) Disc electrophoresis. 2. Method and application to human serum proteins. *Ann. NY Acad. Sci.* **121**, 404–427.
53. White, R.F. (1979) Acetylsalicylic acid (aspirin) induces resistance to tobacco mosaic virus in tobacco. *Virology* **99**, 410–412.
54. van Loon, L.C. (1983) The induction of pathogenesis-related proteins by pathogens and specific chemicals. *Netherlands J. Plant Pathol.* **89**, 265–273.
55. Malamy, J., Carr, J.P., Klessig, D.F., and Raskin, I. (1990) Salicylic acid—a likely endogenous signal in the resistance response of tobacco to viral infection. *Science* **250**, 1002–1004.
56. Métraux, J.-P., Signer, H., Ryals, J., Ward, E., Wyss-Benz, M., Gaudin, J., et al. (1990) Increase in salicylic acid at the onset of systemic acquired resistance in cucumber. *Science* **250**, 1004–1006.
57. Gaffney, T., Friedrich, L., Vernooij, B., Negrotto, D., Nye, G., Uknes, S., et al. (1993) Requirement of salicylic acid for the induction of systemic acquired resistance. *Science* **261**, 754–756.
58. Raskin, I., Ehman, A., Melander, W.R., and Meeuse, B.J.D. (1987) Salicylic acid: a natural inducer of heat production in Arum lilies. *Science* **237**, 1601–1602.
59. Muller, A., Duchting, P. and Weiler, E.W. (2002) A multiplex GC-MS/MS technique for the sensitive and quantitative single-run analysis of acidic phytohormones and related compounds, and its application to *Arabidopsis thaliana. Planta* **216**, 44–56.
60. Huang, W.E., Huang, L., Preston, G., Naylor, M., Carr, J.P., Li, Y., et al. (2006) Quantitative *in situ* assay of salicylic acid in tobacco leaves using a genetically modified biosensor strain of *Acinetobacter* sp. ADP1. *Plant J* **46**, in press.
61. Matthews, R.E.F. (1991) *Plant Virology*, 3rd ed. Academic Press, San Diego.
62. Laval, V., Koroleva, O.A., Murphy, E., Lu, C.G., Milner, J.J., Hooks, M.A. and Tomos, A.D. (2002) Distribution of actin gene isoforms in the Arabidopsis leaf measured in microsamples from intact individual cells. *Planta* **215**, 287–292.
63. Técsi, L.I., Maule, A.J., Smith, A.M., and Leegood, R.C. (1994) Complex, localized changes in CO_2 assimilation and starch content associated with the susceptible interaction between cucumber mosaic virus and a cucurbit host. *Plant J.* **5**, 837–847.
64. Técsi, L.I., Maule, A.J., Smith, A.M., and Leegood, R.C. (1994) Metabolic alterations in cotyledons of *Cucurbita pepo* infected by cucumber mosaic virus. *J. Exp. Bot.* **45**, 1541–1551.
65. Técsi, L.I., Smith, A.M., Maule, A.J., and Leegood, R.C. (1996) A spatial analysis of physiological changes associated with infection of cotyledons of marrow plants with cucumber mosaic virus. *Plant Physiol.* **111**, 975–985.
66. Wang, D. and Maule, A.J. (1995) Inhibition of host gene expression associated with plant virus replication. *Science* **267**, 229–231.
67. Bevan, M.W., Flavell, R.B., and Chilton, M.D. (1983) A chimaeric antibiotic resistance gene as a selectable marker for plant cell transformation. *Nature* **304**, 184–187.
68. Fraley, R.T., Rogers, S.G., Horsch, R.B., Sanders, P.R., Flick, J.S., Adams, S.P., et al. (1983). Expression of bacterial genes in plant cells. *Proc. Natl. Acad. Sci. USA* **80**, 4803–4807.
69. Herrera-Estrella, L., Depicker, A., van Montagu, M., and Schell. J. (1983) Expression of chimaeric genes transferred into plant cells using a Ti-plasmid-derived vector. *Nature* **303**, 209–213.

70. Murai, N., Sutton, D.W., Murray, M.G., Slightom, J.L., Merlo, D.J., Reichert, N.A., et al. (1983) Phaseolin gene from bean is expressed after transfer to sunflower via tumor-inducing plasmid vectors. *Science* **222,** 476–482.

71. Powell-Abel, P., Nelson, R.S., De, B., Hoffmann, N., Rogers, S.G., Fraley, R.T., and Beachy, R.N. (1986) Delay of disease development in transgenic plants that express the tobacco mosaic virus coat protein gene. *Science* **232,** 738–743.

72. Goldbach, R., Bucher, E., and Prins, M. (2003) Resistance mechanisms to plant viruses: an overview. *Virus Res.* **92,** 207–212.

73. Whitham, S. McCormick, S., and Baker, B. (1996) The *N* gene of tobacco confers resistance to tobacco mosaic virus in transgenic tomato. *Proc. Natl. Acad. Sci. USA* **93,** 8776–8781.

74. Baughman, G.A., Jacobs, J.D., and Howell, S.H. (1988) Cauliflower mosaic virus gene VI produces a symptomatic phenotype in transgenic tobacco plants. *Proc. Natl. Acad. Sci. USA* **85,** 733–837.

75. Deom, C.M., Oliver, M.J., and Beachy, R.M. (1987) The 30-kilodalton gene product of tobacco mosaic virus potentiates virus movement. *Science* **237,** 389–394.

76. Lindbo J.A. and Dougherty, W.G. (1992) Pathogen-derived resistance to a potyvirus: immune and resistant phenotypes in transgenic tobacco expressing altered forms of a potyvirus coat protein nucleotide sequence. *Mol. Plant-Microbe Interact.* **5,** 144–153.

77. Lindbo, J.A., Silva-Rosales, L., Proebsting, W.M., and Dougherty, W.G. (1993) Induction of a highly specific antiviral state in transgenic plants: implications for regulation of gene expression and virus resistance. *Plant Cell* **5,** 1743–1759.

78. Anandalakshmi, R., Pruss, G.J., Ge, X., Marathe, R., Mallory, A.C., Smith, T.H., and Vance, V.B. (1998) A viral suppressor of gene silencing in plants. *Proc. Natl. Acad. Sci. USA* **95,** 13079–13084.

79. Brigneti, G., Voinnet, O., Li, W.X., Ji, L.H., Sing, S.W., and Baulcombe, D.C. (1998) Viral pathogenicity determinants are suppressors of transgene silencing in *Nicotiana benthamiana*. *EMBO J.* **17,** 6739–6746.

80. Kasschau, K.D. and Carrington, J.C. (1998) A counterdefense strategy of plant viruses: suppression of posttranscriptional gene silencing. *Cell* **95,** 461–470.

81. Ruiz, M.T., Voinnet, O., and Baulcombe, D.C. (1998) Initiation and maintenance of virus-induced gene silencing. *Plant Cell* **10,** 937–946.

82. Liu, Y., Schiff, M., Maranthe, R., and Dinesh-Kumar, S.P. (2002) Tobacco *Rar1, EDS1* and *NPR1/NIM1* like genes are required for *N*-mediated resistance to tobacco mosaic virus. *Plant J.* **30,** 415–429.

83. Lu, R., Malcuit, I., Moffett, P., Ruiz, M.T., Peart, J., Wu, A.J., et al. (2003) High throughput virus-induced gene silencing implicates heat shock protein 90 in plant disease resistance. *EMBO J.* **22,** 5690–5699.

84. Scholthof, H.B., Scholthof K.-B.G., and Jackson A.O. (1995) Identification of tomato bushy stunt virus host-specific symptom determinants by expression of individual genes from a potato virus X vector. *Plant Cell* **7,** 1157–1172.

85. Chu, M., Park, J.-W., and Scholthof, H.B. (1999) Separate regions on the Tomato bushy stunt virus p22 protein mediate cell-to-cell movement versus elicitation of effective resistance responses. *Mol. Plant Microbe Interact.* **12,** 285–292.

86. Oparka, K.J., Boevink, P., and Santa Cruz, S. (1996) Studying the movement of plant viruses using green fluorescent protein. *Trends Plant Sci.* **1,** 412–418.

87. Heinlein, M. (2002). Plasmodesmata: dynamic regulation and role in macromolecular cell-to-cell signaling. *Curr. Opin. Plant Biol.* **5,** 543–552.

88. Wright, K.M., Duncan, G. H., Pradel, K. S., Carr, F., Wood, S., Oparka, K. J., and Santa Cruz, S. (2000) Analysis of the *N* gene hypersensitive response induced by a fluorescently tagged tobacco mosaic virus. *Plant Physiol.* **123,** 1375–1385.

89. Murphy, A.M. and Carr, J.P. (2002) Salicylic acid has cell-specific effects on *Tobacco mosaic virus* replication and cell-to-cell movement. *Plant Physiol.* **128,** 552–563.

90. Grimsley, N., Hohn, B., Hohn, T., and Walden, R. (1986) 'Agroinfection', an alternative route for viral infection of plants by using the Ti plasmid. *Proc. Natl. Acad. Sci. USA* **83,** 3282–3286.

91. Grimsley, N., Hohn, T., Davies, J.W., and Hohn, B. (1987) *Agrobacterium*-mediated delivery of infectious maize streak virus into maize plants. *Nature* **325,** 177–179.

92. Leiser, R.M., Ziegler-Graff, V., Reutenauer, A., Herrbach, E., Lemaire, H., Guilley, H., et al. (1992) Agroinfection as an alternative to insects for infecting plants with beet western yellows luteovirus. *Proc. Natl. Acad. Sci. USA* **89,** 9136–9140.

93. Prüfer, D., Wipfscheibel, C., Richards, K., Guilley, H., Lecoq, H., and Jonard, G. (1995) Synthesis of a full-length infectious cDNA clone of cucurbit aphid-borne yellows virus and its use in gene exchange experiments with structural proteins from other luteoviruses. *Virology* **214,** 150–158.

94. Turpen, T.H., Turpen, A.M., Weinzettl, N., Kumagai, M.H., and Dawson, W.O. (1993) Transfection of whole plants from wounds inoculated with *Agrobacterium-tumefaciens* containing cDNA of tobacco mosaic virus. *J. Virol. Methods* **42,** 227–240.

95. Lamprecht, S. and Jelkmann, W. (1997) Infectious cDNA clones used to identify strawberry mild yellow edge-associated potexvirus as causal agent of the disease. *J. Gen. Virol.* **78,** 2347–2353.

96. Liu, L. and Lomonossoff, G.P. (2002) Agroinfection as rapid method for propagating Cowpea-mosaic virus-based constructs. *J. Virol. Methods* **105,** 343–348.

97. Chiba, M., Reed, J.C., Prokhnevsky, A.I., Chapman, E.J., Mawassi, M., Koonin, E.V., et al. (2006) Diverse suppressors of RNA silencing enhance agroinfection by a viral replicon. *Virology* **346,** 7–14.

98. Schob, H., Kunc, C., and Meins, F. (1997) Silencing of transgenes introduced into leaves by agroinfiltration: a simple, rapid method for investigating sequence requirements for gene silencing. *Mol. Gen. Genet.* **256,** 581–585.

99. Abbink, T.E.M., Tjernberg, P.A., Bol, J.F., and Linthorst, H.J.M. (1998) Tobacco mosaic virus helicase domain induces necrosis in *N* gene-carrying tobacco in the absence of virus replication. *Mol. Plant-Microbe Interact.* **11,** 1242–1246.

100. Erickson, F., Holzberg, S., Calderon-Urrea, A., Handley, V., Axtell, M., Corr, C., and Baker, B. (1999) The helicase domain of the TMV replicase proteins induces the N-mediated defence response in tobacco. *Plant J.* **18,** 67–75.

101. Johansen, L.K. and Carrington, J.C. (2001) Silencing on the spot. Induction and suppression of RNA silencing in the *Agrobacterium*-mediated transient expression system. *Plant Physiol.* **126,** 930–938.

102. Yocum, R.R., Hanley, S., West, R. Jr., and Ptashne, M. (1984) Use of lacZ fusions to delimit regulatory elements of the inducible divergent GAL1–GAL10 promoter in *Saccharomyces cerevisiae. Mol. Cell. Biol.* **4,** 1985–1998.

103. Leonard, S., Plante, D., Wittmann, S., Daigneault, N., Fortin, M.G., and Laliberte, J.F. (2000) Complex formation between Potyvirus VPg and translation eukaryotic initiation factor 4E correlates with virus infectivity. *J. Virol.* **74,** 7730–7737.

104. Leonard, S., Viel, C., Beauchemin, C., Daigneault, N., Fortin, M.G., and Laliberte, J.F. (2004) Interaction of VPg-Pro of Turnip mosaic virus with the translation initiation factor 4E and the poly(A)-binding protein in planta. *J. Gen. Virol.* **85,** 1055–1063.

105. Moury, B., Morel, C., Johansen, E., Guilbaud, L., Souche, S., Ayme, V., et al. (2004) Mutations in *Potato virus Y* genome-linked protein determine virulence toward recessive resistances in *Capsicum annuum* and *Lycopersicon hirsutum. Mol. Plant-Microbe Interact.* **17,** 322–329.

106. Nicolas, O., Dunnington, S.W., Gotow, L.F., Pirone, T.P. and Hellmann, G.M. (1997) Variations in the VPg protein allow a Potyvirus to overcome *va* gene resistance in tobacco. *Virology* **237,** 452–459.

107. Simone N.L, Bonner R.F., Gillespie J.W., Emmert-Buck M.R., and Liotta L.A. (1998) Laser-capture microdissection: opening the microscopic frontier to molecular analysis. *Trends Genet.* **14,** 272–276.

108. Rubio, V., Shen, Y.P., Saijo, Y., Liu, Y.L., Gusmaroli, G., Dinesh-Kumar, S.P., and Deng, X.W. (2005) An alternative tandem affinity purification strategy applied to Arabidopsis protein complex isolation. *Plant J.* **41,** 767–778.

109. Earley, K.W., Haag, J.R., Pontes, O., Opper, K., Juehne, T., Song, K., and Pikaard, C.S. (2006) Gateway-compatible vectors for plant functional genomics and proteomics. *Plant J.* **45,** 615–629.

Chapter 2
Role of Capsid Proteins

John F. Bol

Abstract Coat proteins (CPs) of all plant viruses have an early function in disassembly of parental virus and a late function in assembly of progeny virus. Depending on the virus, however, CPs may play a role in many steps of the infection cycle in between these early and late functions. It has been shown that CPs can play a role in translation of viral RNA, targeting of the viral genome to its site of replication, cell-to-cell and/or systemic movement of the virus, symptomatology and virulence of the infection, activation of R gene-mediated host defenses, suppression of RNA silencing, interference with suppression of RNA silencing, and determination of the specificity of virus transmission by vectors. These functions are reviewed in this chapter.

Keywords Virus assembly; Virus disassembly; Translation of viral RNA; Replication of viral RNA; Cell-to-cell movement; Long-distance movement; Hypersensitive response; RNA silencing; Vector transmission

1 Introduction

With the exception of umbraviruses, the genomic RNA or DNA of plant viruses is encapsidated by one ore more types of coat (or capsid) protein (CP) molecules. In the classical view, CP protects the viral genome from degradation during virus multiplication in the infected plant and transmission of the virus from plant to plant. In the past decades, however, it has become clear that, depending on the virus, CP can be involved in almost every step of the viral infection cycle, including delivery of the virus into the plant cell, disassembly of virus particles, translation of viral RNA, replication of the viral genome, assembly of progeny virus, virus movement in the plant, activation or suppression of host defenses, and transmission of the virus to healthy plants. Recent data indicate that many steps of the infection cycle are tightly linked.

This chapter will briefly review known functions of CP with reference to the methods used to analyze the role of CP in plant virus infection. A more extensive

From: *Methods in Molecular Biology, Vol. 451, Plant Virology Protocols: From Viral Sequence to Protein Function*
Edited by G.D. Foster, I.E. Johansen, Y. Hong, and P.D. Nagy © Humana Press, Totowa, NJ

review has been published by Callaway et al. (1). Emphasis will be on viruses with a positive-strand RNA genome as these represent the majority of plant viruses.

2   Virus Entry and Translation of Viral RNA

Initiation of infection by plus-strand RNA viruses requires uncoating of virus particles and translation of genomic RNA into viral proteins including the RNA-dependent RNA polymerase (replicase) required for viral minus-strand RNA synthesis. It has been proposed that the rigid rod-shaped *Tobacco mosaic virus* (TMV) particles are destabilized after entry into the plant cell by interaction with lipid containing structures, by interaction with a hypothetical subcellular receptor-like component, or by exposing the virus to a low calcium concentration and raised pH. This latter condition would negatively charge carboxylate groups in CP, affecting carboxyl–carboxylate interactions between CP subunits and carboxyl–phosphate interactions between CP and RNA. Elimination of these interactions by mutagenesis of participating Glu and Asp residues to Gln and Asn affects TMV disassembly (2). In vitro, exposure of TMV particles to pH 8.0 results in dissociation of CP from the 5′-terminal 200 nucleotides of the viral RNA and the partially uncoated particle acts as a messenger for translation of the 126 kDa and 183 kDa replicase proteins in a cell-free system in a process called cotranslational disassembly. Electron micrographs revealed "striposomes" consisting of ribosomes attached to one end of less-than-full-length virus particles. After electroporation of protoplasts with TMV particles, the 5′-terminal region of the viral RNA, including most or all of the 183 kDa open reading frame (ORF), became susceptible to ribonuclease within 2–3 min. Uncoating of the 3′ region of the RNA began between 2 and 5 min after electroporation and occurred in the 3′–5′ direction. These observations are compatible with the hypothesis that TMV RNA is cotranslationally uncoated from the 5′ terminus by ribosomes, whereas the 3′ terminus is coreplicationally uncoated by traversing replicase proteins (2). However, a fundamental difficulty of in vivo experiments is that plant cells are exposed to large numbers of virus particles, which may obscure the minor fraction of the inoculum that establishes the infection.

After uncoating, viral plus-strand RNA has to compete with a vast excess of cellular mRNAs for the translational machinery of the host. The translational efficiency of cellular messengers is greatly enhanced by the formation of a closed-loop structure, because of an interaction between the poly(A)-binding protein (PABP), bound to the 3′ poly(A) tail, and the eIF4G subunit of the initiation factor complex eIF4F, bound to the 5′ cap structure (Fig. 1a). Viral messengers without a cap or poly(A) tail use alternative strategies to form a closed-loop structure (3). Viruses from the genera *Alfamovirus* (type species *Alfalfa mosaic virus*, AMV) and *Ilarvirus* in the family Bromoviridae require viral CP for efficient translation of the viral RNAs. The 3′ end of the three genomic RNAs and subgenomic CP messenger, RNA 4, of these viruses can adopt two mutually exclusive conformations: a strong CP-binding site (CPB) or a tRNA-like structure (TLS) resembling the TLS of other viruses in the family Bromoviridae. The 5′ termini of the RNAs are capped. A mixture of the three genomic RNAs of AMV has a low intrinsic infectivity (Fig. 1b, panel AMV wt), which is increased 1,000-fold by binding of CP to the 3′

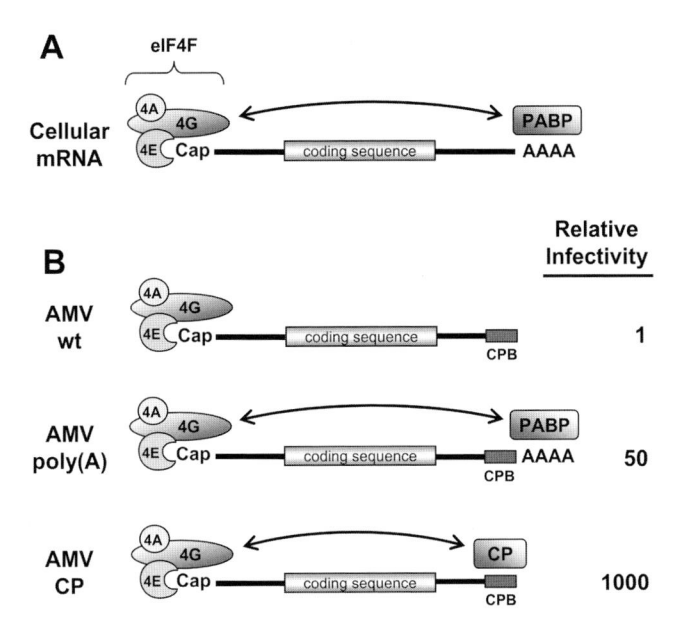

Fig. 1 Coat protein (CP) initiates *Alfalfa mosaic virus* (AMV) infection by mimicking the function of the poly(A) binding protein (PABP). (A) Translational efficiency of cellular mRNAs is strongly enhanced by the formation of a closed-loop structure, because of an interaction between PABP, bound to the 3′ poly(A) tail, and the eIF4G subunit of the initiation factor complex eIF4F, bound to the 5′ cap structure. (B) The tripartite AMV genome is represented by a single RNA molecule with the 3′ terminus folded into the CP-binding (CPB) structure. In the absence of CP the genomic RNAs have a low intrinsic infectivity (panel AMV wt), which is stimulated 50-fold by extension of the RNAs with an artificial 3′ poly(A) tail (panel AMV poly(A)) and 1,000-fold by binding of CP to the 3′ termini of the RNAs (panel AMV CP). It has been shown that, like PABP, CP specifically interacts with eIF4G and stimulates translation in vivo of AMV RNAs 40-fold (4)

end of the RNAs (Fig. 1b, panel AMV CP). Extension of the 3′ termini of the viral RNAs with an artificial poly(A) tail, to allow binding of PABP, increased infectivity 50-fold (Fig. 1b, panel AMV poly(A)) when compared with the CP-free inoculum (4). This suggested that CP mimics the function of PABP in translation of the viral RNAs. Transfection of carrot protoplasts with a transcript containing the luciferase ORF fused with a 3′ sequence consisting of the AMV 3′ UTR revealed that binding of CP to this UTR enhanced translational efficiency of the reporter construct 40-fold. In GST pull-down assays, a CP-GST fusion specifically pulled down the eIF4F (and eIFiso4F) complex from a wheat germ extract. Far Western analysis of protein blots run with recombinant wheat germ initiation factors revealed that AMV CP specifically interacted with the eIF4G and eIFiso4G subunits of eIF4F and eIFiso4F, respectively (4). These results support the notion that, by analogy to PABP, CP increases translational efficiency of AMV RNAs by the formation of a closed-loop structure through its simultaneous interactions with the 3′ end of the viral RNAs and the eIF(iso)4G subunit present in the cap-bound eIF(iso)4F complex. It has been proposed that CP in the inoculum initiates infection by promoting translation of RNAs 1 and 2 of alfamo- and ilarviruses into the replicase proteins required for viral minus-strand RNA synthesis. Such a mechanism explains why AMV CP is no longer required to initiate infection when RNAs 1 and 2 in the inoculum are extended with an artificial 3′ poly(A) tail or when polyadenylated RNAs 1 and 2 are expressed from nuclear genes in transgenic tobacco plants (4).

The genome of DNA viruses has to move to the nucleus of the plant cell to initiate transcription of mRNAs encoding the replicase proteins. Geminiviruses with a monopartite single-stranded DNA genome in the genera *Mastrevirus* and *Begomovirus* encode CPs that act as nuclear shuttles to traffic viral DNA into and out of the nucleus. Trafficking of CP/DNA complexes could be monitored in these experiments by microinjection of protoplasts with *E. coli* expressed GFP-tagged CP or DNA labeled with the fluorescent TOTO-1 dye (5, 6). A similar role of CP in nuclear transport of the double-stranded DNA genome of pararetroviruses from the family Caulimoviridae has been studied by expression of GFP-tagged mutant CP in plasmid-transfected plant protoplasts (7). Thus, CP may promote early events in the initiation of infection by plant DNA viruses.

3   Replication of the Viral Genome

There is growing evidence that translation and replication of positive-strand RNA viruses are tightly coupled. The genomic RNA has to be cleared from ribosomes before initiation of minus-strand RNA synthesis occurs. After translation of AMV RNAs, CP has to dissociate from the 3′ termini to allow the formation of the TLS-structure that is required for minus-strand promoter activity. One possibility is that this dissociation is induced by the binding of the newly synthesized replicase proteins to a minus-strand promoter hairpin upstream of the CPB/TLS sequence (4). As dissociation of CP strongly reduces translational efficiency of the viral RNAs, the replicase proteins could trigger the switch from translation to replication. So far, however, a role of CP in the replication of plant viral RNAs or DNAs remains to be demonstrated (4).

4   Virus Assembly

Encapsidation of newly synthesized plant viral RNA has been proposed to occur upon exit of the RNA from vesicles that contain viral replication complexes. A tight link between replication and encapsidation has been suggested for both DNA and RNA viruses. In the yeast two-hybrid system and by using Far-Western assays, CP of the pararetrovirus *Cauliflower mosaic virus* (CaMV) was shown to interact with the viral transactivator protein (TAV), supporting the notion that translation of viral RNA on the surface of cytoplasmic inclusion bodies (viroplasm) and its packaging and reverse transcription in proviral capsids are linked (8). TAV is the main component of the inclusion body matrix and mediates reinitiation of translation of the polycistronic CaMV RNA through interactions with eIF3 and the 60S ribosomal subunit. Transient expression of BMV (bromovirus) RNAs and CP from a T-DNA vector in agroin-filtrated leaves results in encapsidation of viral RNAs as well as host RNAs. Only upon coexpression of functional replicase proteins, the encapsidation of host RNAs was excluded (9). Probably, a link with replication increases the specificity of the encapsidation process. In the family Bromoviridae, encapsidation of RNAs 1 and 2 by the RNA 3 encoded CP occurs (by definition) in *trans*. However, encap-sidation in protoplasts of AMV RNA 3 with a knock-out mutation in the *CP* gene could not be complemented by coreplicating wild-type RNA 3 (4). This observa-tion points to a coupling between RNA 3 replication, synthesis of subgenomic RNA 4, translation of RNA 4 into CP, and encapsidation of RNA 3 (and possibly RNA 4). In view of the evidence that various steps in the viral replication cycle are linked, results from in vitro encapsidation studies should be confirmed by experi-ments done in vivo.

For a few RNA viruses, the RNA sequence that acts as the origin of assembly (oas) in in vitro packaging assays has been identified. Some of these oas sequences have been inserted into hetrologous RNAs to confirm that they direct encapsidation of the RNA by CP in vivo. Most detailed studies have focused on the assembly of the rigid rod-shaped particles of TMV (*vulgare* strain) (2). The TMV oas is com-posed of one essential hairpin structure and two accessory hairpins located in the movement protein (*MP*) gene between bases 5,290 and 5,527. According to the most widely accepted model, a 20S disk of two layers of 17 CP subunits each binds to the oas and converts to a protohelical form. This RNA–protein complex initiates helical rod elongation in the 5′ direction of the RNA by using 20S disks and in the 3′ direction by using CP monomers or trimers. Potex- and potyviruses have parti-cles with flexuous rod-shaped morphology. In the RNAs of the potexviruses *Papaya mosaic virus* and *Potato virus X* (PVX) and the potyvirus *Tobacco vein mottling virus*, oas sequences have been mapped near the 5′ end in in vitro packag-ing assays (10, 11). The flexuous rod-shaped particles of closteroviruses contain five viral proteins. The 5′ terminal ~630 nucleotides of the RNA are associated with the minor CP (CPm) to form the tail structure, whereas the remainder of the RNA is associated with the major CP. The tail is extended with segments consisting of the virus-encoded homolog of cellular Hsp70 (Hsp70h) and viral proteins p64 and p20 (12). CP, CPm, Hsp70, and p64 are required for virion assembly. Sequences in the 5′ UTR of closterovirus RNA have been implicated in the formation of virions.

Viruses in the family Bromoviridae have icosahedral symmetry. The 3′ end of the RNAs of bromo- and cucumoviruses contains a tRNA-like structure (TLS) whereas the 3′ termini of alfamo- and ilarvirus RNAs can be folded either in a TLS-structure or in a structure with a high affinity for CP. Surprisingly, this CP-binding structure was found to be dispensable for encapsidation of RNAs 1 and 2 of the alfamovirus AMV. Transient expression of 3′-terminally truncated AMV RNAs 1 and 2 from a T-DNA vector in agroinfiltrated leaves supported replication of RNA 3, and the truncated RNAs were encapsidated by the RNA 3 encoded CP (4).

In RNA 3 of the bromovirus (BMV), the signal for in vitro packaging was found to consist of a 69-nucleotide sequence in the 3′ region of the MP ORF and the 3′ TLS of 200 nucleotides. The TLS could perform its function in either *cis* or *trans*. When added in *trans* as a 200 nucleotide fragment to 3′ terminally truncated RNA 3, the TLS fragment was not copackaged with the truncated RNA 3 in an in vitro assay. Expression of BMV CP and TLS-defective viral RNAs from a T-DNA vector in agroinfiltrated leaves revealed that the TLS was not required for encapsidation of BMV RNAs in vivo, and it was proposed that its function in encapsidation could be taken over by cellular tRNAs (9). Accumulation of nonreplicating AMV and BMV RNAs in protoplasts was increased 20-fold or more by expression of the cognate CP. This illustrates that encapsidation protects the viral RNAs from degradation (4, 9). For another isometric plant virus, *Turnip crinkle virus* (TCV, genus *Carmovirus*), studies done in vivo revealed that a 186-nucleotide region at the 3′ end of the *CP* gene was indispensable for viral RNA encapsidation (13).

5   Virus Cell-to-Cell and Systemic Movement

From primary infected cells, plant viruses move cell-to-cell through plasmodesmata, are transported from mesophyll cells into phloem tissue, and exit from the vasculature to enter the healthy upper leaves of the plant. The role of CP in this process has been recently reviewed (refs. 1, 14; *see also* Chap. 3). Generally, virus movement in plant tissue is monitored by insertion of the *GFP* reporter gene in the viral genome and the effect of mutations in viral genes is analyzed. At the level of cell-to-cell movement, a subdivision can be made into viruses with CP-independent and CP-dependent movement. CP-independent viruses include tobamo-, diantho-, carmo-, hordei- and umbraviruses. Viruses that do require CP for cell-to-cell movement can be further subdivided into those moving as virus particles and those moving by other mechanisms that do not necessarily involve virion formation. Transport of virus particles through plasmodesmata-penetrating tubules made up of viral MP has been observed in plant tissues infected with *Cowpea mosaic virus* (CPMV, *Comovirus*), *Grapevine fanleaf virus* (*Nepovirus*), and the pararetroviruses CaMV (*Caulimovirus*) and *Commelina yellow mottle virus* (*Badnavirus*). By blot overlay assays, a specific interaction between CPMV MP and virions was shown. The interaction involved the large CP subunit in the virion and the C-terminus of MP. CaMV virions may interact with MP through the virion associated protein (VAP) (15). Closteroviruses do not form tubules, yet they are transported as viral particles. The structural proteins CP, CPm, p64, and Hsp70h are required for virion formation and cell-to-cell transport; the p20 protein is dispensable for virion formation and cell-to-cell movement, but is necessary for transport through the vascular system. Flexuous rod-shaped potex- and potyviruses also require CP for cell-to-cell movement, but it is not fully clear whether these viruses are transported as virions or VNP complexes.

In the family Bromoviridae, AMV, BMV, and *Cucumber mosaic virus* (CMV) require CP for cell-to-cell movement, but *Cowpea chlorotic mottle virus* (CCMV) does not. The MPs of AMV, BMV, and CMV assemble into virion-containing tubular structures at the surface of infected protoplasts, but such structures have not been observed in plasmodesmata in leaf tissue infected with these viruses. Movement of BMV strain M1 requires CP that is encapsidation competent but BMV strain M2 does not require CP for cell-to-cell movement. AMV and CMV require CP for cell-to-cell movement but movement is observed for some CP mutants that are unable to form virions. Moreover, C-terminal point mutations or deletions in the MP of BMV and CMV result in movement of these viruses that is no longer CP-dependent. Probably, viruses in the family Bromoviridae move cell-to-cell as VNP complexes. With the exception of CCMV, CP of these viruses may play an auxiliary role in MP-mediated virus transport, such as suppression of host defense mechanisms. A differential requirement for CP in virus movement is also observed in the family Geminiviridae of viruses with a single-stranded DNA genome. Geminiviruses with a monopartite genome of the genus *Mastrevirus* require CP for cell-to-cell movement whereas bipartite viruses from the genus *Begomovirus* do not. The mastrevirus CP has a functional analogy with the begomovirus BV1 protein. Note that the genus *Begomovirus* contains both monopartite and bipartite geminiviruses (5, 6).

Most viruses require CP for systemic movement through the phloem either as virions or viral nucleoprotein (VNP) complexes. In specific host plants, CP is dispensable for systemic spread of the tombusvirus *Tomato bushy stunt virus*, the hordeivirus *Barley stripe mosaic virus* and for tobraviruses. Umbraviruses do not encode a CP and move systemically as VNPs consisting of viral RNA and the ORF3-encoded protein. Although TMV generally requires CP for systemic movement, CP deletion mutants can move long distances in *N. benthamiana*. The mechanism of systemic movement is poorly understood.

6 Vector Transmission

In addition to mechanical transmission, plant viruses are transmitted from plant to plant by vectors such as nematodes, fungi, or insects (including leafhoppers, planthoppers, whiteflies, aphids, mealybugs, thrips, beetles, and mites). Generally, transmission requires virion formation in the source plant, and CP is a major determinant of the specificity of the virus-vector interaction (ref. 1; *see also* Chap. 6). CP subunits in the viral capsid may interact directly with putative receptors in the vector or via accessory viral proteins. In transmission of, for instance, the cucumovirus (CMV) by aphids or the tombusvirus *Cucumber necrosis virus* (CNV) by zoospores of the fungus *Olpidium bornovanus*, CP is believed to be the sole virus-encoded determinant. Interaction of CNV particles with the zoospores in vitro results in a conformational change of the virus that renders CP in the viral capsids susceptible to digestion with trypsin (16). Luteoviruses are transmitted by aphids in a circulative, nonpropagative manner that requires virions to traverse the aphid hindgut epithelial cells into the body cavity (hemocoel) and then traverse accessory salivary gland cells into the salivary canal. Transmission can be studied by feeding aphids on purified virus or homogenates of protoplasts infected with wild-type or mutant virus to which sucrose has been added. Efficiency of virus transmission to oat plants can be measured and virus can be detected in various organelles of the aphid by electron microscopy and RT-PCR. In addition to CP, transmission of luteoviruses was shown to be dependent on the presence in virions of a few copies of a readthrough protein (RTP) consisting of the CP sequence fused to a C-terminal extension. The RTP is not required for uptake of virions by the aphid or their trafficking to the hemocoel, but appears to be required for transport of virus through membranes of the aphid salivary gland (1). Umbraviruses do not encode CP and are transmitted by aphids only when encapsidated by CP and RTP of a helper luteovirus. To this goal, the seven definitive umbravirus species are each associated with a specific luteovirus.

Aphid transmission of potyviruses requires the viral helper component, protease (HC-Pro) as an accessory protein. By site-directed mutagenesis, it has been shown that interaction between HC-Pro and potyvirus CP involves a PKT-motif in HC-Pro and a DAG-motif near the N-terminus of CP. Retention of HC-Pro on the aphid's stylet involves a KITC-motif in HC-Pro. Electron microscopic observations revealed an association of potyvirus particles and HC-Pro with the cuticle lining of the mouth parts of aphid vectors. These data support the hypothesis that HC-Pro forms a bridge between virus particles and the aphid food canal (1). It has been proposed that non-structural protein 2b encoded by RNA 2 of tobraviruses act as accessory proteins in transmission of these viruses by trichodorid nematodes (genera *Trichodorus* and *Paratrichodorus*) in a way that resembles the role of HC-Pro in virus transmission by aphids. *Tobacco rattle virus* (TRV) particles ingested by root-feeding nematodes are retained as clumps associated with the oesophageal cuticle and are released during subsequent feeding on roots of healthy plants. In yeast two-hybrid assays, a specific interaction between TRV CP and its cognate 2b protein was observed and in thin sections of tobravirus-infected plants the 2b protein colocalized with virus particles (17).

Transmission by aphids of the caulimovirus (CaMV) involves two viral accessory proteins: VAP and the aphid transmission factor (ATF). VAP is bound to virions and associates with MP to permit cell-to-cell movement or with ATF to facilitate aphid transmission of the virus. The interactions between these viral proteins were mapped by Far Western and GST pull-down assays. In transmission of CaMV by aphids, ATF is believed to bridge virion–VAP complexes with the inner lining of the aphid stylet (*see* ref. 15).

7 Plant Response to Virus Infection

Successful infection of a plant requires the virus to overcome host defense mechanisms. Two major defense mechanisms are mediated by plant resistance genes (*R* genes) and the mechanism of RNA interference (RNAi). The gene-for-gene hypothesis predicted that defense mechanisms mediated by *R* genes are activated by an interaction between the product of a viral avirulence (*Avr*) gene, termed as effector, and the product of a plant *R* gene (*see* ref. 18). However, with a few exceptions, such interactions were not found and the gene-for-gene model was modified into the "guard hypothesis." This hypothesis predicts that the viral effector targets a key component (guardee) of the basal defense system of the plant in order to invade successfully. A virus-induced change in the structure of the guardee is recognized by an R protein (guard), which subsequently activates defense mechanisms leading to a hypersensitive response of the plant to virus infection. In a susceptible host that lacks the *R* gene, the viral effectors act as virulence factors (18).

The interaction between the carmovirus TCV and *A. thaliana* ecotypes containing the resistance gene *HRT* (the guard) lends support to the guard hypothesis. A yeast two-hybrid screen and in vitro GST pull-down assays revealed that TCV CP interacts with the host transcription factor TIP (the guardee). Confocal microscopy of leaves agroinfiltrated with GFP-tagged TIP showed that TIP localizes to the nucleus. However, coexpression of GFP-TIP and TCV CP prevented the nuclear localization of TIP. The interaction between TIP and CP is required for *HRT*-mediated defense responses (19). Agroinfiltration of *N. benthamiana* transformed with the potato resistance gene *Rx1* with a construct expressing CP of the potexvirus PVX or coexpression of the potato resistance gene *Rx2* and PVX CP in agroinfiltrated *N. tabacum* confirmed that PVX CP is the effector in resistance mediated by resistance genes *Rx1* and *Rx2* in potato. CP of cucumovirus CMV strain Y mediates resistance conferred by the *RCY1* gene of *A. thaliana* (18). Structural studies using site-directed mutagenesis of TMV CP revealed that maintenance of the three-dimensional fold of this CP is essential for elicitation of the *N'*-mediated hypersensitive response in *Nicotiana sylvestris* (2).

Defense mediated by RNA silencing (RNAi) is triggered in virus-infected plants by double-stranded RNA (dsRNA) derived from viral replication intermediates or in the case of plant DNA viruses from annealing of overlapping complementary viral transcripts (ref. 20; *see also* Chap. 5 for details on RNAi). To overcome this plant defense mechanism, many viruses have evolved suppressors of gene silencing, which interfere with the RNA silencing pathway at different levels. CPs of several plant viruses have been identified as suppressors of gene silencing (18, 20). CP of the carmovirus TCV suppresses RNA silencing possibly by interfering the function of a Dicer-like ribonuclease. This function of TCV CP is not related to its role in *HRT*-mediated resistance. CP of the closterovirus *Citrus tristeza virus* suppresses intercellular silencing. The small CP subunit of the comovirus CPMV has been reported to act as a weak suppressor of gene silencing. CP of the Satellite of *Panicum mosaic virus* (family *Tombusviridae*) acts as a pathogenicity factor. This CP did not suppress gene silencing but interfered with the suppressor activity of the PVX (potexvirus) p25 protein (21).

8   Future Directions

In addition to their structural roles, many novel and unexpected functions of viral CPs have been discovered in the past decades. Further research will undoubtedly shed new light on the role of these multifunctional proteins in virus replication and their interactions with viral and host components. Plant viral-based vectors have a high potential for the production of safe and cheap vaccines by directing the synthesis of virions that display foreign peptides fused to CP on the surface of viral particles (1). During evolution, CPs have been adapted to the strategy of the virus to evade the activation of host defense mechanisms and almost every man-made change in the CP sequence affects symptomatology of the infection. Further studies on the roles of CP will provide insight in virus–plant interactions.

References

1. Callaway, A., Giesman-Cookmeyer, D., Gillock, E.T., Sit, T.L., and Lommel, S.A. (2001) The multifunctional capsid proteins of plant RNA viruses. *Annu. Rev. Phytopathol.* **39**, 419–460.
2. Culver, J.N. (2002) Tobacco mosaic virus assembly and disassembly: Determinants in pathogenicity and resistance. *Annu. Rev. Phytopathol.* **40**, 287–308.
3. Dreher, T.W. and Miller, W.A. (2006) Translational control in positive strand RNA plant viruses. *Virology* **344**, 185–197.
4. Bol, J.F. (2005) Replication of Alfamo- and Ilarviruses: Role of the coat protein. *Annu. Rev. Phytopathol.* **43**, 39–62.
5. Rojas, M.R., Jiang, H., Salati, R., Xoconostle-Cázares, B., Sudarshana, M.R., Lucas, W.J., and Gilbertson, R.L. (2001) Functional analysis of proteins involved in the movement of the monopartite begomovirus, *Tomato yellow leaf curl virus. Virology* **291**, 110–125.
6. Liu, H., Boulton, M.I., Oparka, K.J., and Davies, J.W. (2001) Interaction of the movement and coat proteins of maize streak virus: Implications for the transport of viral DNA. J. *Gen. Virol.* **82**, 35–44.
7. Guerra-Peraza, O., Kirk, D., Seltzer, V., Veluthambi, K., Schmit, A.C., Hohn, T., and Herzog, E. (2005) Coat proteins of Rice tungro bacilliform virus and Mungbean yellow mosaic virus contain multiple nuclear-localization signals and interact with importin α. J. Gen. Virol. 86, 1815–1826.
8. Himmelbach, A., Chapdelaine, Y., and Hohn, T. (1996) Interaction between *Cauliflower mosaic virus* inclusion body protein and capsid protein: Implications for viral assembly. *Virology* **217**, 147–157.
9. Annamalai, P. and Rao, A.L.N. (2005) Replication-independent expression of genome components and capsid protein of brome mosaic virus in planta: A functional role for viral replicase in RNA packaging. *Virology* **338**, 96–111.
10. Kwon, S-J., Park, M-R., Kim, K-W., Plante, C.A., Hemenway, C.L. and Kim, K-H. (2005) *cis*-Acting sequences required for coat protein binding and in vitro assembly of *Potato virus X. Virology* **334**, 83–97.
11. Wu, X. and Shaw, J.G. (1998) Evidence that assembly of a potyvirus begins near the 5′ terminus of the viral RNA. J. *Gen. Virol.* **79**, 1525–1529.
12. Peremyslov, V.V., Andreev, I.A., Prokhnevsky, A.I., Duncan, G.H., Taliansky, M.E., and Dolja, V.V. (2004) Complex molecular architecture of beet yellows virus particles. *Proc. Natl. Acad. Sci. USA* **101**, 5030–5035.
13. Qu, F. and Morris, T.J. (1997) Encapsidation of turnip crinkle virus is defined by a specific packaging signal and RNA size. J. *Virol.* **71**, 1428–1435.
14. Scholthof, H.B. (2005) Plant virus transport: motions of functional equivalence. *Trends Plant Sci.* **10**, 376–382.
15. Stavolone, L., Villani, M.E., Leclerc, D., and Hohn, T. (2005). A coiled-coil interaction mediates cauliflower mosaic virus cell-to-cell movement. *Proc. Natl. Acad. Sci. USA* **102**, 6219–6224.

16. Kakani, K., Reade, R., and Rochon, D. (2004) Evidence that vector transmission of a plant virus requires conformational change in virus particles. J. *Mol. Biol.* **338**, 507–517.

17. Vellios, E., Duncan, G., Brown, D., and MacFarlane, S. (2002) Immunogold localization of tobravirus 2b nematode transmission helper protein associated with virus particles. *Virology* **300**, 118–124.

18. Soosaar, J.L.M., Burch-Smith, T.M., and Dinesh-Kumar, S. (2005) Mechanisms of plant resistance to viruses. *Nat. Rev. Microbiol.* **3**, 789–798.

19. Ren, T., Qu, F., and Morris, T.J. (2005) The nuclear localization of the Arabidopsis transcription factor TIP is blocked by its interaction with the coat protein of *Turnip crinkle virus.* *Virology* **331**, 316–324.

20. Roth, B.M., Pruss, G.J., and Vance, V.B. (2004) Plant viral suppressors of RNA silencing. *Virus Res.* **102**, 97–108.

21. Qiu, W. and Scholthof, K.G. (2004) Satellite panicum mosaic capsid protein elicits symptoms on a nonhost plant and interferes with a suppressor of virus-induced gene silencing. *Mol. Plant-Microbe Int.* **17**, 263–271.

Chapter 3
Role of Plant Virus Movement Proteins

Michael Taliansky, Lesley Torrance, and Natalia O. Kalinina

Abstract Plant viruses spread from the initially infected cells to the rest of the plant in several distinct stages. First, the virus (in the form of virions or nucleic acid protein complexes) moves intracellularly from the sites of replication to plasmodesmata (PD, plant-specific intercellular membranous channels), the virus then transverses the PD to spread intercellularly (cell-to-cell movement). Long-distance movement of virus occurs through phloem sieve tubes. The processes of plant virus movement are controlled by specific viral movement proteins (MPs). No extensive sequence similarity has been found in MPs belonging to different plant virus taxonomic groups. Moreover, different MPs were shown to use different pathways and mechanisms for virus transport. Some viral transport systems require a single MP while others require additional virus-encoded proteins to transport viral genomes. In this review, we focus on the functions and properties of different classes of MPs encoded by RNA containing plant viruses.

Keywords cell-to-cell movement, long-distance movement, movement protein, plasmodesmata, phloem

1 Introduction

To induce disease, plant viruses must spread from the initially infected cells to the rest of the plant. The systemic spread of plant viruses proceeds in several distinct stages. First, the virus (in the form of virions or nucleic acid protein complexes) moves intracellularly from the sites of replication to plasmodesmata (PD), which are unique intercellular membranous channels that span cell walls linking the cytoplasm of contiguous cells. The virus then transverses the PD to spread intercellularly (cell-to-cell movement). Virus systemic movement between organs (long-distance movement) occurs through vascular tissue, usually phloem sieve tubes. Similar pathways are employed by the plant host to traffic endogenous macromolecules suggesting that viruses hijack host transport systems for their own movement.

The first evidence suggesting that the processes of plant virus movement are controlled by specific viral proteins came from early studies of temperature-sensitive

From: *Methods in Molecular Biology, Vol. 451, Plant Virology Protocols:*
From Viral Sequence to Protein Function
Edited by G.D. Foster, I.E. Johansen, Y. Hong, and P.D. Nagy © Humana Press, Totowa, NJ

mutants of *Tobacco mosaic virus* (TMV) indicating that the TMV genome encodes a nonstructural 30 kDa transport or "movement" protein (MP) (see 1–3 for reviews). These studies gave rise to similar investigations on other plant viruses and it soon became clear that MPs are general feature of both RNA and DNA-containing plant viruses (see 4–6 for reviews). However, no extensive sequence similarity has been described in MPs belonging to different plant virus taxonomic groups. Moreover, different MPs were shown to use different pathways and mechanisms for virus transport: some MPs transport viral genomes in the form of virus particles, others such as TMV traffic viral RNA as ribonucleoprotein (RNP) complexes (4, 7–10). The concept of functional diversity of movement mechanisms was strengthened by the fact that some viral transport systems require a single MP while others require additional virus-encoded proteins to transport viral genomes. Such diversity possibly reflects the fact that different viruses have usurped different aspects of the host cellular machinery for movement.

Rather than attempt to cover all aspects of virus movement, in this short review, we will focus on functions and properties of different classes of MPs encoded by RNA containing plant viruses. The reader is referred to recent excellent reviews on the movement of DNA-containing viruses (7, 11) and different aspects related to virus spread, such as the cell biology of virus movement, the role of plant host defense responses, and RNA silencing (see 12–16 for reviews). Numerous original papers have been devoted to this topic but because of space limitation we have been unable to cite all of them and mainly refer readers to reviews.

2 Viruses That Require a Single MP for Cell-to-Cell Movement

2.1 RNA Binding

TMV 30 kDa MP. It is well established that TMV employs only one virus-encoded MP, the 30 kDa protein, to mediate cell-to-cell spread (17) Moreover, intercellular spread is coat protein (CP) independent (18), showing that the virus moves in a nonvirion form. Experiments in vitro showed that the TMV MP was able to bind, in a cooperative but sequence non-specific manner, to single-stranded nucleic acids (RNA or DNA) and two RNA-binding domains were identified, located at residues 112–185 and 186–268 (19, 20).

Electron microscopy (20) and atomic force microscopy (AFM) (21) showed that the TMV MP and viral RNA form thin, elongated RNP particles having diameter of approximately 2–3 nm which is close to the dimensions of the dilated PD channels being of the order 3–4 nm (22, see below). These data supported the idea that the TMV MP form viral RNP (vRNP) particles that move to and through PD. Recently, it has been shown that TMV replicase may also be involved in cell-to-cell movement and both nonconserved region and RNA helicase domain are important (23, 24). In addition, the TMV MP is able to bind GTP and has a conserved motif, responsible for this activity (25, 26).

Umbraviruses and dianthoviruses. Umbraviruses (see 27 for review) and dianthoviruses (28) are two other virus groups encoding single MPs that move intercellularly independently of the CP. MPs of *Groundnut rosette virus* (GRV; umbravirus), the ORF4 protein, and *Red clover necrotic mosaic virus* (RCNMV; dianthovirus), 35 kDa MP, were shown to bind to ssRNA (28, 29), though the mechanisms of interaction with RNA for these proteins were different from that of the TMV MP. The RCNMV MP bound to RNA in a cooperative manner (30) but, in contrast to the TMV MP, did not unfold the RNA, instead it formed compact RNAase-resistant vRNP complexes (28). The GRV MP bound to RNA noncooperatively and only to a limited extent leaving extensive, unpacked RNA sequences in a loose globular form (29).

2.2 Interaction with Plasmodesmata

Early microinjection experiments with fluorescently labeled dextran probes showed that only molecules of less than about 1 kDa could diffuse through PD (see 31 for review). Therefore, to allow movement of vRNP, the TMV MP must increase PD permeability. Immunogold-labeling experiments showed that this MP was able to accumulate in secondary PD (32–34). Interaction of the transgenically expressed or microinjected TMV MP with PD led to a significant increase in PD size exclusion limit (SEL) (35, 36) allowing the transport through PD of large dextrans of 10–20 kDa. Such a SEL corresponds to a dilated channel diameter, potentially allowing passage of 2.0–3.0 nm wide TMV vRNP complexes. TMV MP amino acid residues 3–5 and 195–213 were shown to be responsible for PD localization (37) and residues 126–224 were essential for gating PD. A strong correlation was found between the ability of the TMV MP to accumulate in and increase SEL of PD and the capacity to move vRNP from cell to cell. However, during native TMV infection the ability of MP to gate PD was restricted only to the leading edge of infection (33, see below).

Interaction with PD and modification of their SEL appears to be a property of many viruses. For example, the GRV ORF4 MP was shown to accumulate in PD (38) and the RCNMV MP increased PD SEL (26).

2.3 Role of Phosphorylation

As mentioned above, an ability of the TMV MP to gate PD is exhibited only at the infection front and is inactivated in the centre of infectious foci (33). Thus, the MP may be suggested to exist in two forms within PD of infected cells: (1) an active form that transiently increases the SEL during early stages of infection and (2) an inactive form that still localizes to the PD channels but no longer increases the SEL. Since MP has been shown to be phosphorylated in vivo (39, 40), it can be suggested that the ability of the TMV MP to modify the SEL may depend on phosphorylation. In agreement with this idea, it was shown that substitution of each of three potential amino acids that are phosphorylation sites in the TMV MP C terminus with asparagine, a negatively charged amino acid that mimics the phosphorylated MP status, inactivated the ability of MP to increase the SEL in tobacco plants (41). However, it should be noted that the C-terminal region of the MP is not essential for TMV movement (42), but it may be important in regulation of this process.

It has been demonstrated that TMV RNA coated by the TMV MP is not translatable in vitro and is noninfective in protoplasts. In contrast, translation and replication of TMV vRNP occurred *in planta*, suggesting that the vRNP undergoes modification upon passage through PD (43). Phosphorylation of the TMV MP within the vRNP complex resulted in conversion of the nontranslatable form of vRNP into one that was translatable in vitro and infectious in protoplasts and plants (44).

Thus, the TMV MP phosphorylation during movement through PD may play dual roles: (1) to inactivate ability of the MP to modify the SEL and (2) to destabilize the vRNP complex thereby allowing the release of RNA for further translation and replication.

2.4   Intracellular Movement of TMV vRNP

The TMV MP contains a transmembrane domain (45) and when fused to fluorescent proteins was found to label the cortical endoplasmic reticulum (ER) early in infection (46, 47). The ER is implicated in the infection process, as sites for viral replication and protein synthesis (48). TMV vRNP complex is also likely to be assembled on the ER. The ER passes through PD in the form of desmotubules and is tightly entwined with the actin cytoskeleton (49), and hence the simplest way for the vRNP complex to get to PD is via the ER/actin network. Consistent with this idea, the TMV MP was found to be associated with actin filaments (50). For trafficking along the ER/actin network, the TMV vRNP complexes may specifically interact with myosin motors. Interestingly, both actin and myosin have been detected within PD (51).

Although association of the TMV MP with microtubules was also found, their involvement in vRNP delivery to PD seems unlikely. Indeed, agents disrupting microtubules did not affect TMV movement (47). Moreover, tagging of the microtubules by the TMV MP was only detected during the late stages of infection, suggesting that this mechanism is used for entry of the TMV MP into the 26S proteosome for protein turnover (10, 47).

3   Viruses That Require a Single MP and CP but Not Virus Particles for Cell-to-Cell Movement

In addition to an MP, a number of RNA viruses require a CP for effective cell-to-cell movement. For those viruses which move in the form of virions (e.g., comoviruses), the obvious role of the CP is in the assembly of virions and will be discussed below. *Cucumber mosaic virus* (CMV, a member of the *Cucumovirus* genus) represents another situation where the CP, but not virus particles, is essential for the cell-to-cell movement of CMV (52).

Several lines of evidence have indicated that both the 3a MP and CP are necessary for cell-to-cell movement (see 53 for review). However, microinjection experiments showed that similar to the TMV MP, the CMV 3a MP alone was able to increase PD SEL and traffic the coinjected CMV RNA from cell to cell (54). These findings were consistent with virus movement being in the form of vRNP complex. In support of this idea, experiments using CMV MP mutants showed that virion assembly was not a prerequisite for virus movement. Furthermore, the 3a MP was demonstrated to bind ssRNA (30) and form filamentous vRNP particles in vitro (55) but they were not as stable as those formed by the TMV MP. It seems that such particles formed in the absence of a CP were unable to move from cell to cell. A model was proposed in which it was suggested that the role of the CMV CP in virus movement was to induce a conformational change in the 3a protein allowing it to form movement-competent vRNP complexes (53, 55, 56). In support of this model, CMV expressing the 3a protein containing a deletion of the C-terminal 33 amino acids (3aΔC33 MP mutant) was shown to be able to move from cell to cell in the absence of the CMV CP (57). It was postulated that the RNP complex formed by the 3aΔC33 MP alone was movement competent, but for the wt CMV MP, the CP may be required to modify its interaction with viral RNA. Using AFM-based approaches, it was shown that the 3aΔC33 MP bound RNA more strongly than did the wt MP (55, 56). Such an increase in binding affinity would likely lead to the formation of stabilized and hence movement-competent vRNP complexes without involvement of the CP. It is conceivable that the CMV CP could alter the wt MP conformation to increase its binding efficiency. Thus, deleting the C terminus of the MP might have the same effect on the overall conformation as adding CP. The results of the experiments with the MP and CP chimeras of two related cucumoviruses CMV and *Tomato aspermy virus* (TAV) showed that the C-terminal two-thirds of the CP and the 29 aa long C-terminal region of the MP need to be compatible to enable virus movement (58).

Similar to CMV the C-terminal parts of the MPs of *Alfalfa mosaic virus* (AMV; the genus *Alfamovirus*) (59) and *Brome mosaic virus* (BMV; the genus Bromovirus) are also involved in the requirement for CP in cell-to-cell movement. In contrast, the viral CP was found to be dispensable for cell-to-cell movement of another bromovirus, *Cowpea chlorotic mottle virus* (CCMV; type I) (60, 61). Thus, with these viruses there are two types of MPs: MPs that do not require CP (CCMV) and MPs that require CP but not virions (CMV, AMV, and BMV; type II). Point or deletion mutations into the C terminus of these MPs change type II to type I.

4   Viruses That Require a Triple Gene Block for Movement

Viruses in about eight different genera carry a specialized, evolutionarily conserved genetic element of three partially overlapping ORFs called the triple gene block (TGB) that encodes three proteins (named according to their position in the module),

TGB1, TGB2, and TGB3. All three TGB proteins are essential for virus cell-to-cell and long-distance movement. Although the genome position varies, comparisons of primary structure, genome organization, and biological properties allow the classification of TGB-containing viruses into two groups represented by the hordei-like and the potex-like viruses. The hordei-like group includes rod-shaped viruses with segmented ssRNA genomes in the genera *Hordeivirus, Benyvirus, Pomovirus*, and *Pecluvirus*; whereas the potex-like group contains filamentous viruses with monopartite ssRNA genomes in the genera *Potexvirus, Carlavirus, Foveavirus*, and *Allexivirus* (see 62 for review).

The TGB1 proteins encoded by viruses of both groups contain an NTPase/helicase domain with seven conserved motifs of superfamily I (SF-I) helicase closely related to the helicases involved in replication of alpha-like viruses (63–65). The TGB1's display NTPase activity and ATP-dependent helicase activity (66–73). In addition to the NTPase/helicase domain, the hordei-like (but not potex-like) TGB1 proteins contain an additional domain of variable mass at the N terminus (62). The two smaller proteins (TGB2 and TGB3) possess predicted hydrophobic transmembrane segments (62, 74–77). The TGB2 proteins are more conserved in molecular organization in viruses of different genera (78–80) than the TGB3 (81, 82).

A crucial difference between hordei-like and potex-like transport systems is the requirement of CP for movement (83). The CP is dispensable for cell-to-cell and long-distance movement mediated by Barley stripe mosaic hordeivirus (BSMV), Peanut clump pecluvirus (PCV), and Potato mop-top pomovirus (PMTV) TGB (84–88). In contrast, the potexvirus CPs of *Potato virus X* (PVX), *White clover mosaic virus* (WClMV), and *Papaya mosaic virus* (PMV) are essential for cell-to-cell movement (89–91). These observations indicate that the structure and composition of the movement forms of hordei-like and potex-like viruses may be different.

4.1 Hordei-Like Viruses

The hordei-like TGB1 proteins display sequence-nonspecific RNA binding of two types: noncooperative and cooperative (67, 92–94). Mutations in the N-terminal extension domain of *Poa semilatent hordeivirus* (PSLV) TGB1 abolishing noncooperative RNA binding do not affect cell-to-cell movement, but block long-distance movement (94). The C-terminal NTPase/helicase domain of PSLV and potex-like viruses has been shown to be responsible for cooperative RNA binding but it binds to RNA more weakly than the extension domain (66, 68, 69, 71, 72, 94). RNA-binding activity of hordei-like TGB1 is suggested to be responsible for formation of vRNP particles. Taking into account that viruses of this group do not require the CP for movement, it seems conceivable that they move in the form of such vRNP. Consistent with this idea, RNP complexes composed of viral RNA and TGB1 were isolated from BSMV-infected plants (95).

Individually expressed hordei-like TGB1 proteins [PCV, *Beet necrotic yellow vein virus* (BNYVV), PSLV, BSMV, PMTV) were unable to target and modify PD and to move from cell to cell, indicating that the protein depends on TGB2/TGB3 for intracellular movement (74, 96–101). The hordei-like and potex-like TGB2 and TGB3 apparently function as membrane anchors for delivery of vRNP complex to PD. The PSLV, BSMV, PMTV, and PVX TGB2s expressed alone as fluorescent fusion proteins were distributed throughout the cell endomembrane network, mostly the cortical ER and PMTV TGB2 were observed in granules moving on the actin/ER network. TGB3 contains the PD targeting signal and accumulates in peripheral membrane bodies, which are derivatives of ER structures associated with the PD neck regions (93, 102–104); when coexpressed in the same cell, TGB2 and TGB3 colocalize and are targeted to the PD by TGB3 (75, 102, 103). The hydrophobic sequences of TGB2 and TGB3 are necessary for interaction with cellular membranes and their function in cell-to-cell movement (77, 81, 105, 106). In some hosts, PVX TGB2 and BSMV TGB2 may increase PD SEL and move through PD (76, 107–109). However, in general, TGB2 and TGB3 are not thought to exit the infected cell (62, 71, 75, 110). Recently, Haupt et al. (75) also showed that PMTV TGB proteins associate with components of the plant endocytic pathway. TGB2 colocalizes in vesicles derived from the plasma membrane and containing markers of the early endosome such as the Rab GTPase Ara7 (*At*RabF2b), TGB2 also interacts with a tobacco DNAJ protein belonging to the RME family of J-domain chaperones. They hypothesize that after delivery of vRNP complex to PD for intracellular transport, the TGB proteins are incorporated into the membranes of endocytic vesicles and are recycled through the endocytic pathway (75).

Importantly, heterologous TGB3s could facilitate intracellular trafficking of TGB1 to the PD, however, they are not competent to mediate subsequent transport of TGB1 through PD. This agrees with previous data showing that only a complete set of TGB proteins encoded by the same virus is competent for the cell-to-cell movement functions (96, 98, 100). Although transport of TGB1 to the cell periphery is independent of the NTPase and helicase activities (69, 71, 74), they are crucial when TGB1 or vRNP move through PD to the next cell (74, 85, 99). One can speculate that TGB1 NTPase/helicase activity is required for vRNP unfolding (disassembly) for genome translocation or for microchannel dilation during trafficking of complexes through PD additionally to or independent from cellular helicase activity (9).

Recent studies suggest that there are some subtle difference in function between TGB proteins and that BSMV and PMTV TGB2 functions in supporting virus replication as well as movement (109; L.T., unpublished results).

4.2 Potex-Like Viruses

As mentioned above, in addition to TGB proteins potexviruses require the CP for cell-to-cell movement. In contrast to hordei-like counterpart, the potex-like (PVX,

WClMV) TGB1 protein is able to target PD, to increase PD SEL as well as to mediate its own cell-to-cell movement (69, 71, 110–114). Microinjection experiments showed that the WClMV TGB1 protein can mediate intercellular transport of potexviral RNA and the CP (71). Collectively these data led to the suggestion that potexvirus cell-to-cell movement occurs in the form of nonvirion vRNP represented by TGB1–RNA–CP complex (71, 110).

However, recent structural analysis of the products assembled from PVX RNA, CP and TGB1 in vitro detected only partially assembled virions with the TGB1 bound to the terminal head-like structure (115). Thus, it appears conceivable that potex-like viruses can move in the form of fully or partially assembled virions containing one or more TGB1 molecules at one end of the particles presumably associated with 5′ end of viral RNA (116). It should be noted that full or partially assembled PVX particles are entirely untranslatable; however, they become translatable after interaction of TGB1 to the end the helical particles (115, 116). This interaction induces a linear destabilization of the whole virion particle (117, 118). Thus, fast and reversible conformational changes of virion particles induced by TGB1, could support the model in which virions are unwound during transport through PD, presenting translatable viral RNA in the adjacent cells (76, 117, 118). These data support the virion model of cell-to-cell potexvirus movement suggested by Santa-Cruz *et al.* (119) based on the detection of fibrillar material, in PD of PVX-infected tissues, that was immunoreactive with virion-specific antisera.

However, the experiments with PVX CP mutants show that virion formation is not sufficient for effective cell-to-cell and long-distance transport and suggest additional movement-related activities for the C-terminal region of the CP (120, 121). Importantly, transient complementation experiments show that the CPs of potex- and potyviruses have common (yet unknown) function(s) that is (are) necessary for the transport process and not connected with genome transencapsidation. The recent data show that the CP of both types of filamentous viruses display NTPase activity in vitro (122).

5 Viruses That Require Multiple Proteins for Cell-to-Cell Movement

5.1 Potyviruses

A very different system for intercellular movement is used by another group of filamentous viruses, the potyviruses. Potyviruses do not encode a dedicated MP, but movement functions have been allocated to several proteins. Genetic studies showed that cell-to-cell movement requires an assembly competent CP of *Tobacco etch virus* (TEV) (123), suggesting that intercellular transport involves virion formation. Moreover, the potyviral CP has been shown to induce an increase in PD SEL, traffic through PD and facilitate cell-to-cell movement of viral RNA during microinjection experiments (124). Another potyviral protein that displayed the same activities in microinjected tissues is the helper component – proteinase (HC-Pro, which also acts as a suppressor of RNA silencing) (see 10 for review).

Several reports implicated the cylindrical inclusion (CI) protein, an RNA helicase also required for genome replication in potyvirus cell-to-cell movement (see 125 for review). Mutational analysis of the TEV CI protein (126) identified mutants defective in cell-to-cell movement but not in accumulation in single cells. Ultrastructural studies (127, 128) supported the role of CI and CP in cell-to-cell movement by detecting continuous channels through the centre of the CIs and the plasmodesmata containing viral RNA and CP in the form of fibrillar material similar to potyvirus particles (128).

Finally, mutational analysis demonstrated that the genome linked protein (VPg) is also involved in cell-to-cell movement (129). VPg forms a covalent linkage to the 5′ end of the viral RNA being therefore an integral part of potyvirus particles. Although an exact mechanism of the involvement of VPg in virus movement is unclear, it has recently been shown that VPg can potentiate cell-to-cell movement through interaction with two host proteins, the translation initiation factor eIF4E (130) and PVIP (for potyvirus VPg interacting protein) (131). Thus, in addition to the obvious role in virus translation eIF4E may also assist in intracellular movement of virus particles providing the mechanism for binding to microtubules through its strong affinity with eIF4G (130).

AFM analysis of potyviruses *Potato virus Y* and *Potato virus A* revealed the presence of protrusions (tips) at one (presumably 5′) end of the particles that are attached to the virions at least at some stages of the virus infection process (132). The tips contain virus-encoded VPg, HC-Pro and possibly host proteins interacting with them such as eIF4E and PVIP. Collectively these results suggest a model for potyvirus movement in which these tips play the role of a guide device for directional trafficking of virions to and through PD (132), perhaps in conjunction with eIF4G (130). In adjacent cells, eIF4E can initiate translation of the virus through a mechanism of cotranslational disassembly which may facilitate the process of trafficking through PD as well as release of viral RNA for further replication/translation. Directional trafficking and cotranslational disassembly mediated by specific structures at one end (associated with 5′-end of viral RNA) of virus particles may be general mechanism used by different groups of filamentous viruses (see below).

5.2   Closteroviruses

Another example of filamentous viruses that contain additional proteins at one of the end of the virus particles is closteroviruses. Closteroviruses possess exceptionally long (approx. 1,300 nm) filamentous virus particles composed of a main "body" and short (70–100 nm) "tail" whose principal components are the major and minor capsid proteins (CP and CPm), respectively (see 133 for review). AFM studies of *Beet yellows virus* (BYV), a closterovirus, demonstrated that the tail possesses a striking, segmented morphology encapsidating the 5′-terminal ~650 nt long part of the viral RNA (134, 135). In addition to the CPm, three other virus-encoded proteins are incorporated into the tail: Hsp-70 homologue of cellular Hsp70 molecular chaperones (Hsp70h), 64 kDa protein (p64), and 20 kDa protein (p20) (135). Genetic analysis showed that CPm, Hsp-70h, and p64 are each required for virion assembly and subsequent cell-to-cell movement suggesting that the formation of the "tailed" virions is a prerequisite for intercellular trafficking (136, 137). In contrast, p20 is dispensable for virion assembly and cell-to-cell movement, but is necessary for BYV transport through the plant vascular system (138).

One of most remarkable features of closteroviruses is that the Hsp70h is an integral component of the virion tail. So far, only closteroviruses have been found to harbor an Hsp70 gene. BYV Hsp70h has been localized in PD confirming its specific role in intercellular movement. Although the specific roles of the other closteroviral tail components in cell-to-cell movement are unclear, the whole tail structure was proposed to be a viral transport device (135, 136).

What could be the reason for the attachment of a complex transport device to the virions? One possibility is that the closterovirus virions are exceptionally long and simply cannot pass from cell to cell without an extra mechanism powered by the ATPase activity of Hsp70h. Plasmodesmal localization of Hsp70h is also suggestive of a specific role played by the thin tails in the entry of virions into the narrow channels of plasmodesmata (139). A complementary possibility is a need for 5′ to 3′ directional transport of the viral RNA. Because the large BYV genome needs to be protected from degradation by the host machinery used in RNA silencing, particle disassembly upon arrival in the adjacent cells should be tightly regulated and coupled to primary translation (139). Thus, one of the tail functions could be control of virion disassembly.

In addition to specific tail-based transport device, BYV cell-to-cell movement was shown to require another virus-encoded protein, the 6 kDa protein (p6) (139). P6 is a single span transmembrane protein that resides and functions in the ER. However, it is unclear how this protein aids in viral cell-to-cell movement.

5.3 Terminal Transport Devices of Filamentous Viruses

A novel implication from the information presented above is that filamentous plant viruses such as potexviruses (116, 117), potyviruses (132) and closteroviruses (135) use a novel type of transport device attached to one of the ends of the virus particles (associated with 5′-end of viral RNA). Such devices can provide specific mechanism for directional virus intercellular movement. These devices can also be involved in polar destabilization of virus particles after their delivery to adjacent cells to release viral RNA for further translation and replication. Interestingly, not only filamentous plant viruses but also filamentous bacteriophages containing circular ssDNA genomes have a polar structure with virions composed of thousands of helically arranged copies of a single major CP with a few minor proteins at the tips essential for various bacteriophage functions (140).

6   Viruses That Form Tubules: Comoviruses and Nepoviruses

Another completely different strategy for intercellular movement is used by comoviruses and nepoviruses (e.g., *Cowpea mosaic virus*, CPMV, a comovirus). These are small spherical viruses that spread from cell to cell as virions through specific tubular structures in extensively modified PD (see 141 for review). The tubule structures, which have a diameter of 30 nm and are filled with a single row of spherical virions, penetrate the cell wall, and protrude into the cytoplasm of neighboring cells (142). Using mutational analysis, it was shown that that the 48 kDa MP is the only virus protein required for tubule formation (143). Experiments with metabolic inhibitors indicated that targeting of the CPMV MP to peripheral punctate structures representing the potential origins of tubules involves neither the cytoskeleton, nor secretory pathway (144). It is possible that the CPMV MP moves along the ER or simply arrives at the plasma membrane by diffusion. Interestingly, the MP of another virus using the same (tubular) strategy for virus movement, *Grapevine fanleaf virus*, a nepovirus, may use the secretory pathway and the cytoskeleton for intracellular trafficking (145).

Mutational analysis revealed several functional domains within the CPMV MP. The tubule-forming domain is essential for targeting the MP in the form of dimers/multimers to the plasma membrane (146). The C-terminal domain of the MP is involved in incorporation of the virions in the tubules, probably during assembly of tubules. Thus, one possible mechanism of comovirus/nepovirus movement is that the tubule containing virus particles traverses the PD channel where it is degraded releasing virions. Alternatively, virions may flow through the tubules (see 141 for review).

7   Complementation of Cell-to-Cell Movement

In spite of the striking diversity of virus transport systems and lack of similarity between MPs of different virus groups, complementation of plant virus movement is a very common phenomenon (2). It has been demonstrated using different experimental approaches including (1) mixed infections with movement-deficient (dependent) virus and helper virus (2, 147), (2) infections with recombinant viral genomes bearing a heterologous MP genes, and (3) complementation of a movement-deficient virus in transgenic plants expressing the MP of a helper virus (see 147 for review).

8 Long-Distance Movement Proteins

It is generally accepted now that virus cell-to-cell movement and long-distance transport are distinct processes in which different virus-encoded factors are involved. It is not clear how viruses enter, move through, and exit from the vascular system, which is usually surrounded by bundle sheath cells and contains various cell types including vascular parenchyma cells, companion cells, and enucleate sieve elements (148, 149). Thus, transport of a virus into and within vascular tissue implies movement from mesophyll cells to bundle sheath cells, from bundle sheath cells to vascular parenchyma and/or companion cells, and entry into sieve elements. Virus exit from vascular tissue presumably involves the same steps in reverse order. With only a few exceptions (150), CP is essential for efficient long-distance transport of plant viruses; in the rare instances in which the CP gene is partially or wholly dispensable for systemic spread, the time required for systemic invasion is often increased (151, 152). Several viruses also encode proteins that provide additional functions needed for the systemic spread of infection. For example, genetic analyses showed that two proteins encoded by BYV, L-Pro, and p20 play specific roles in long-distance movement of this virus (see 139 for review).

An interesting situation is represented by the phloem-limited viruses such as *Potato leafroll virus* (PLRV, polerovirus) where movement in parenchyma tissues is limited and therefore systemic spread is based on the long-distance movement through the phloem. It is interesting, however, that the PLRV dedicated 17 kDa MP has properties strikingly similar to those of the TMV 30 kDa MP which functions in cell-to-cell movement, for example, ssRNA binding, protein dimerization, phosphorylation by a membrane-bound protein kinase, plasmodesmal localization in both virus-infected and transgenic plants and ability to increase plasmodesmal SELs (153–156). Nevertheless, in spite of the similarities in properties between the TMV and PLRV MPs, PLRV, unlike TMV, is unable to spread out of the plant vasculature and it is assumed that PLRV MP mediates virus movement only between cells within the phloem tissues. It should be noted that PLRV can also use another strategy for phloem movement independent of PLRV MP where the CP and its translational readthrough product (RT protein), a minor structural protein of PLRV virions, are involved in virus systemic spread (see 157 for review).

Members of the genus *Umbravirus* represent a special situation because they do not encode a CP, but nonetheless accumulate and spread systemically very efficiently within infected plants (see 27 for review). Genetic analysis of GRV and another umbravirus, *Pea enation mosaic virus*-2, showed that the proteins encoded by ORF3 of these viruses are essential for umbravirus long-distance movement and can functionally replace the CP of TMV for long distance movement (158, 159, M.T., unpublished results). Localization studies showed that the GRV ORF3 protein accumulated in cytoplasmic granules of filamentous RNP particles that contained viral RNA and the ORF3 protein (160). The granules were detected in all types of cells and were abundant in phloem-associated cells. It was suggested that these RNP particles serve to protect viral RNA, and may be the form in which it moves through the phloem.

Another quite unexpected finding was that in addition to the cytoplasmic granules containing RNP particles, the ORF3 protein was also found in nuclei, preferentially targeting nucleoli (38, 161). Functional analysis of ORF3 protein mutants revealed a correlation between the ORF3 protein nucleolar localization and its ability to form the RNP particles and transport viral RNA long distances. It was also shown that the ORF3 protein interacts with a nucleolar protein, fibrillarin, redistributing it from the nucleolus to cytoplasm and such an interaction is absolutely essential for umbravirus long-distance movement (M.T., N.K., unpublished results). The study of the fibrillarin involvement in long-distance virus movement is ongoing and will certainly clarify this interesting phenomenon.

9 Concluding Remarks

As illustrated in some of the examples given above, studies investigating the interaction or association of plant proteins with viral MPs have provided new insights into the mechanisms of virus movement. Viral MPs have been shown to interact with numerous cellular proteins, such as pectin methylesterases (162, 163), protein kinases (164), homeodomain proteins (165), DNAJ-like proteins (75, 166), Rab 5 ortholog Ara 7 (*At*RabF2b) which functions in the endocytic pathway (76), transcriptional coactivators (167), and some others (see 51 for review); long-distance umbraviral MP interacts with nucleolar protein fibrillarin. The role of the most of such interactions in virus movement remains obscure awaiting future studies.

The ability of viruses to spread also depends on their capacity to combat host defense mechanisms such as RNA interference (RNAi). Interestingly, some viral MPs such as PVX TGB1 also play a role of RNAi suppressors (168) suggesting possible involvement of viral MPs in the battle with RNAi-mediated host defense response and potential cross-links between virus movement and RNAi pathways in plants. This suggestion also raises issues to consider for future work in this area.

Acknowledgments Scottish Crop Research Institute is grant-aided by the Scottish Executive Environment and Rural Affairs Department. NOK is supported in part by the Grant of Russian Foundation for Basic Research.

References

1. Atabekov, J.G. and Dorokhov, Yu. L. (1984) Plant virus-specific transport function and resistance of plants to viruses. *Adv. Virus Res.* **29**, 313–364.
2. Atabekov, J.G. and Taliansky, M.E. (1990) Expression of a plant virus-coded transport function by different viral genomes. *Adv. Virus Res.* **38**, 201–248.
3. Hull, R. (1989) The movement of viruses in plant. *Annu. Rev. Phytopathol.* **27**, 213–240.
4. Carrington, J.C., Kasschau, K.D., Mahajan, S.K., and Schaad, M.C. (1996) Cell-to-cell and long-distance transport of viruses in plants. *Plant Cell* **8**, 1669–1681.
5. Lucas, W.J. and Gilbertson, R.L. (1994) Plasmodesmata in relation to viral movement within leaf tissue. *Annu. Rev. Phytopathol.* **32**, 387–411.

6. Maule, A.J. (1991) Virus movement in infected plants. *Crit. Rev. Plant Sci.* **9**, 457–473.
7. Gilbertson, R.L. and Lucas, W.J. (1996) How do viruses traffic on the 'vascular highway'? *Trends Plant Sci.* **1**, 260–268.
8. Ghoshroy, S., Lartey, R., Sheng, J., and Citovsky, V. (1997) Transport of proteins and nucleic acids through plasmodesmata. *Ann. Rev. Plant Physiol. Plant Mol. Biol.* **48**, 27–49.
9. Lazarowitz, S.G. and Beachy, R.N. (1999) Viral movement proteins as probes for intracellular and intercellular trafficking in plants. *Plant Cell* **11**, 535–548.
10. Lucas, W.J. (2006) Plant viral movement proteins: Agents for cell-to-cell trafficking of viral genomes. *Virology* **344**, 169–184.
11. Rojas, M.R., Hagen, C., Lucas, W.J., and Gilbertson, R.L. (2005) Exploiting chinks in the plant's armor: evolution and emergence of geminiviruses. *Annu. Rev. Phytopathol.* **43**, 361–394.
12. Roberts, A.G. and Oparka, K.J. (2003) Plasmodesmata and the control of symplastic transport. *Plant Cell Environ.* **26**, 103–124.
13. Dangl, J.L. and Jones, J.D., (2001) Plant pathogens and integrated defence responses to infection. *Nature* **411**, 826–833.
14. Carrington, J.C., Kasschau, K.D., and Johansen, L.K. (2001) Activation and suppression of RNA silencing by plant viruses. *Virology* **281**, 1–5.
15. Baulcombe, D. (2004) RNA silencing in plants. *Nature* **431**, 356–363.
16. Vionnet, O. (2005) Non-cell autonomous RNA silencing. *FEBS Lett.* **579**, 5858–5871.
17. Deom, C.M, Oliver, M.J., and Beachy, R.N. (1987) The 30-kilodalton gene product of tobacco mosaic virus potentiates virus movement. *Science* **237**, 389–394.
18. Dawson, W.O., Bubrick, P., and Grantham, G.L. (1988) Modifications of the tobacco mosaic virus coat protein gene affecting replication, movement and symptomatology. *Phytopathology* **78**, 783–789.
19. Citovsky, V., Knorr, D., Schuster, G., and Zambryski, P. (1990) The p-30 movement protein of tobacco mosaic-virus is a single-stranded nucleic-acid binding-protein. *Cell* **60**, 637–647.
20. Citovsky, V., Wong, M.L., Shaw. A.L., Prasad, B.V.V. and Zambryski, P. (1992) Visualization and characterization of tobacco mosaic-virus movement protein-binding to single-stranded nucleic-acids. *Plant Cell* **4**, 397–411.
21. Kiselyova, O.I., Yaminsky, I.V., Karger, E.M., Frolova, O.Y., Dorokhov, Y.L. and Atabekov, J.G. (2001) Visualization by atomic force microscopy of tobacco mosaic virus movement protein-RNA complexes formed in vitro. *J. Gen. Virol.* **82**, 1503–1508.
22. Lucas, W.J. (1995) Plasmodesmata-intercellular channels for macromolecular transport in plants. *Curr. Opin. Cell Biol.* **7**, 673–680.
23. Hirashima, K. and Watanabe,Y. (2001) Tobamovirus replicase coding region is involved in cell-to-cell movement. *J. Virol.* **75**, 8831–8836.
24. Hirashima, K. and Watanabe,Y. (2003) RNA helicase domain of tobamovirus replicase executes cell-to-cell movement possibly through collaboration with its nonconserved region. *J. Virol.* **77**, 12357–12362.
25. Mushegian, A.R. and Koonin, E.V. (1993) Cell-to-cell-movement of plant viruses. Insights from amino acid sequence comparisons of movement proteins and from analogies with cellular transport system. *Arch. Virol.* **133**, 239–257.
26. Li, Q.B. and Palukaitis, P. (1996) Comparison of the nucleic acid- and NTP-binding properties of the movement protein of cucumber mosaic cucumovirus and tobacco mosaic tobamovirus. *Virology* **216**, 71–79.
27. Taliansky, M.E. and Robinson, D.J. (2003) Molecular biology of umbraviruses: phantom warriors. *J. Gen. Virol.* **84**, 1951–1960.
28. Fujiwara, T., Giesman-Cookmeyer, D., Ding, B., Lommel, S.A., and Lucas, W.J. (1993) Cell-to-cell trafficking of macromolecules through plasmodesmata potentiated by the red-clover necrotic mosaic-virus movement protein. *Plant Cell* **5**, 1783–1794.
29. Nurkiyanova, K.M., Ryabov, E.V., Kalinina, N.O., Fan, Y.C., Andreev, I., Fitzgerald, A.G., Palukaitis, P., and Taliansky, M. (2001) Umbravirus-encoded movement protein induces tubule

formation on the surface of protoplasts and binds RNA incompletely and non-cooperatively. *J. Gen. Virol.* **82**, 2579–2588.

30. Osman, T.A., Hayes, R.J., and Buck, K.W. (1992) Cooperative binding of the red clover necrotic mosaic virus movement protein to single-stranded nucleic acids. *J. Gen. Virol.* **73**, 223–227.

31. Heinlein, M. and Epel, B.L. (2004) Macromolecular transport and signaling through plasmodesmata. *Int. Rev. Cytol.* **235**, 93–164.

32. Atkins D., Hull, R., Wells, B., Roberts, K., Moore, P., and Beachy, R.N. (1991) The tobacco mosaic virus 30 K movement protein in transgenic tobacco plants is localized to plasmodesmata. *J. Gen. Virol.* **72**, 209–211.

33. Oparka, K.J., Prior, D.A.M., Santa Cruz, S., Padgett, H.S., and Beachy, R.N. (1997) Gating of epidermal plasmodesmata is restricted to the leading edge of expanding infection sites of tobacco mosaic virus (TMV). *Plant J.* **12**, 781–789.

34. Tomenius, K., Clapham, D., and Meshi, T., (1987) Localization by immunogold cytochemistry of the virus-coded 30 K protein in plasmodesmata of leaves infected with tobacco mosaic-virus. *Virology* **160**, 363–370.

35. Wolf S., Deom, C.M., Beachy, R., and Lucas, W.J. (1989) Movement protein of tobacco mosaic-virus modifies plasmodesmatal size exclusion limit. *Science* **246**, 377–379.

36. Waigmann, E. and Zambryski, P. (1995) Tobacco mosaic virus movement protein-mediated protein transport between trichome cells. *Plant Cell* **7**, 2069–2079.

37. Ding, B., Haudenshield, J.S., Hull, R.J., Wolf, S., Beachy, R.N., and Lucas, W.J. (1992) Secondary plasmodesmata are specific sites of localization of the tobacco mosaic-virus movement protein in transgenic tobacco plants. *Plant Cell* **4**, 915–928.

38. Ryabov, E.V., Oparka, K.J., Santa Cruz, S., Robinson, D.J., and Taliansky, M.E. (1998) Intracellular location of two groundnut rosette umbravirus proteins delivered by PVX and TMV vectors. *Virology* **242**, 303–313.

39. Watanabe, Y., Ogawa T., and Okada, Y. (1992) In vivo phosphorylation of the 30-kDa protein of tobacco mosaic virus. *FEBS Lett.* **313**, 181–184.

40. Haley, A., Hunter, T., Kiberstis, P., and Zimmern, D. (1995) Multiple serine phosphorylation sites on the 30 kDa TMV cell-to-cell movement protein synthesized in tobacco protoplast. *Plant J.* **8**, 715–724.

41. Waigmann, E., Chen, M.H., Bachmaier, R., Ghoshroy, S., and Citovsky, V. (2000) Regulation of plasmodesmal transport by phosphorylation of tobacco mosaic virus cell-to-cell movement protein. *EMBO J.* **19**, 4875–4884.

42. Berna, A., Gafny, R., Wolf, S., Lucas, W.J., Holt, C.A., and Beachy, R.N. (1991) The TMV movement protein-role of the C-terminal 73 amino acids in subcellular localization and function. *Virology* **182**, 682–689.

43. Karpova, O.V., Ivanov, K.I., Rodionova, N.P, Dorokhov, Y.L., and Atabekov, J.G. (1997) Nontranslatability and dissimilar behavior in plants and protoplasts of viral RNA and movement protein complexes formed in vitro. *Virology* **230**, 11–21.

44. Karpova, O.V., Rodionova, N.P, Ivanov, K.I., Kozlovsky, S.V., Dorokhov, Y.L., and Atabekov, J.G. (1999) Phosphorylation of tobacco mosaic virus movement protein abolishes its translation repressing ability. *Virology* **261**, 20–24.

45. Brill, L.M., Nunn, R.S., Kahn, T.W., Yeager, M., and Beachy, R.N. (2000) Recombinant tobacco mosaic virus movement protein is an RNA-binding, alpha-helical membrane protein. *Proc. Natl. Acad. Sci.* **97**, 7112–7117.

46. Heinlein M., Padgett, H.S., Gens, J.S., Pickard, B.G., Casper S.J., Epel, B.L., and Beachy, R.N. (1998) Changing patterns of localization of the tobacco mosaic virus movement protein and replicase to the endoplasmatic reticulum and microtubules during infection. *Plant Cell* **10**, 1107–1120.

47. Gillespie, T., Boevink, P., Haupt, S., Roberts, A.G., Toth, R., Valentine, T., Chapman, S., and Oparka, K.J. (2002) Functional analysis of a DNA-shuffled movement protein reveals that microtubules are dispensable for the cell-to-cell movement of Tobacco mosaic virus. *Plant Cell* **14**, 1207–1222.

48. Noueiry, A.O. and Ahlquist, P. (2003) Brome mosaic virus RNA replication: revealing the role of the host in RNA virus replication. *Annu. Rev. Phytopathol.* **41**, 77–98.
49. Boevink, P., Oparka, K.J., Santa Cruz, S., Martin, B., Betteridge, A., and Hawes, C. (1998) Stacks on tracks: the plant Golgi apparatus traffics on an actin/ER network. *Plant J.* **15**, 441–447.
50. McLean, B.G., Zupan, J., and Zambryski, P.C. (1995) Tobacco mosaic virus movement protein associates with the cytoskeleton in tobacco cells. *Plant Cell* **7**, 2101–2114.
51. Oparka, K.J. (2004) Getting the message across: how do plant cells exchange macromolecular complexes? *Trends Plant Sci.* **9**, 33–41.
52. Kaplan, I.B., Zhang, L., and Palukaitis, P. (1998) Characterization of cucumber mosaic virus. V. Cell-to-cell movement requires capsid protein but not virions. *Virology* **246**, 221–231.
53. Palukaitis, P. and Garcia-Arenal, F. (2003) Cucumoviruses. *Adv. Virus Res.* **62**, 241–323.
54. Ding, B., Li, Q., Nguyen, L. Palukaitis, P., and Lucas, W.J. (1995) Cucumber mosaic virus 3a protein potentiates cell-to-cell trafficking of CMV RNA in tobacco plants. *Virology* **207**, 345–353.
55. Kim, S.H., Kalinina, N.O., Andreev, I., Ryabov, E.V., Fitzgerald, A.G., Taliansky, M.E., and Palukaitis, P. (2004) The C-terminal 33 amino acids of the cucumber mosaic virus 3a protein affect virus movement, RNA binding and inhibition of infection and translation. *J. Gen. Virol.* **85**, 221–230.
56. Andreev, I.A, Kim, S.H., Kalinina, N.O., Rakitina, D.V., Fitzgerald, A.G., Palukaitis, P., and Taliansky, M.E. (2004) Molecular interactions between a plant virus movement protein and RNA: force spectroscopy investigation. *J. Mol. Biol.* **339**, 1041–1047.
57. Nagano, H., Mise, K., Furusawa, I., and Okuno, T. (2001) Conversion in the requirement of coat protein in cell-to-cell movement mediated by the cucumber mosaic virus movement protein. *J. Virol.* **75**, 8045–8053.
58. Salanki, K., Gellert, A., Huppert, E., Naray-Szabo, and G., Balazs, E. (2004) Compatibility of the movement protein and the coat protein of cucumoviruses is required for cell-to-cell movement. *J. Gen. Virol.* **85**, 1039–1048.
59. Sanchez-Navarro, J.A. and Bol, J.F., (2001) Role of the alfalfa mosaic virus movement protein and coat protein in virus transport. *Mol. Plant Microbe Interact.* **14**, 1051–1062.
60. Rao, A.L.N. (1997) Molecular studies on bromovirus capsid protein: III. Analysis of cell-to-cell movement competence of coat protein defective variants of cowpea chlorotic mottle virus. *Virology* **232**, 385–395.
61. Osman, F., Schmitz, I., and Rao, A.L.N. (1999) Effect of C-terminal deletion in the movement protein of cowpea chlorotic mottle virus on cell-to-cell and long-distance movement. *J. Gen. Virol.* **80**, 1357–1365.
62. Morozov, S.Yu. and Solovyev, A.G., (2003) Triple gene block: modular design of a multifunctional machine for plant virus movement. *J. Gen. Virol.* **84**, 1351–1366.
63. Gorbalenya, A.E., Koonin, E.V., Donchenko, A.P., and Blinov, V.M. (1989). Two related superfamilies of putative helicases involved in replication, recombination, repair and expression of DNA and RNA genomes. *Nucleic Acids Res.* **17**, 4713–4730.
64. Gorbalenya, A.E. and Koonin, E.V. (1993) Helicases: amino acid sequence comparisons and structure-function relationships. *Curr. Opin. Struct. Biol.* **3**, 419–429.
65. Koonin, E.V. and Dolja, V.V. (1993) Evolution and taxonomy of positive-strand RNA viruses: implications of comparative analysis of amino acid sequences. *Crit. Rev. Biochem. Mol. Biol.* **28**, 375–430.
66. Rouleau, M., Smith, R.J., Bancroft, J.B., and Mackie, G.A. (1994) Purification, properties, and subcellular localization of foxtail mosaic potexvirus 26-kDa protein. *Virology* **204**, 254–265.
67. Donald, R.G., Lawrence, D.M., and Jackson, A.O. (1997). The barley stripe mosaic virus 58-kilodalton beta(b) protein is a multifunctional RNA binding protein. *J. Virol.* **71**, 1538–1546.
68. Kalinina, N.O., Fedorkin, O.N., Samuilova, O.V., Maiss, E., Korpela, T., Morozov, S.Yu., and Atabekov, J.G. (1996) Expression and biochemical analyses of the recombinant potato virus X 25 K movement protein. *FEBS Lett.* **397**, 75–78.

69. Morozov, S.Yu., Solovyev, A.G., Kalinina, N.O., Fedorkin, O.N., Samuilova, O.V., Schiemann, J., and Atabekov, J.G. (1999) Evidence for two nonoverlapping functional domains in the potato virus X 25 K movement protein. *Virology* **260**, 55–63.

70. Kalinina, N.O., Rakitina, D.V., Solovyev, A.G., Schiemann, J., and Morozov, S.Yu. (2002) RNA helicase activity of the plant virus movement proteins encoded by the first gene of the triple gene block. *Virology* **296**, 321–329.

71. Lough, T.J., Shash, K., Xoconostle-Cazares, B., Hofstra, K.R., Beck, D.L., Balmori, E., Forster, R.L., and Lucas, W.J. (1998) Molecular dissection of the mechanism by which potexvirus triple gene block proteins mediate cell-to-cell transport of infectious RNA. *Mol. Plant Microbe Interact.* **11**, 801–814.

72. Wung, C.H., Hsu, Y.H., Liou, D.Y., Huang, W.C., Lin, N.S., and Chang, B.Y. (1999) Identification of the RNA-binding sites of the triple gene block protein 1 of bamboo mosaic potexvirus. *J. Gen. Virol.* **80**, 1119–1126.

73. Bayne, E.H., Rakitina, D.V., Morozov, S.Yu., and Baulcombe, D.C. (2005) Cell-to-cell movement of potato potexvirus X is dependent on suppression of RNA silencing. *Plant J.* **44**, 471–482.

74. Zamyatnin, A.A., Jr, Solovyev, A.G., Savenkov, E.I., Germundsson, A., Sandgren, M., Valkonen, J.P.T., and Morozov, S.Yu. (2004) Transient coexpression of individual genes encoded by the triple gene block of Potato mop-top virus reveals requirements for TGBp1 trafficking. *Mol. Plant Microbe Interact.* **17**, 921–930.

75. Haupt, S., Cowan, G.H., Ziegler, A., Roberts, A.G., Oparka, K.J., and Torrance, L. (2005) Two plant-viral movement proteins traffic in the endocytic recycling pathway. *Plant Cell* **17**, 164–181.

76. Verchot-Lubicz, J. (2005) A new cell-to-cell transport model for potexviruses. *Mol. Plant Microbe Interact.* **18**, 283–290.

77. Schepetilnikov, M.V., Manske, U., Solovyev A.G., Zamyatnin, A.A. Jr., Schiemann, J., and Morozov, S.Yu. (2005) The hydrophobic segment of Potato virus X TGBp3 is a major determinant of the protein intracellular trafficking. *J. Gen. Virol.* **86**, 2379–2391.

78. Morozov, S.Yu., Lukasheva, L.I., Chernov, B.K., Skryabin, K.G., and Atabekov, J.G. (1987) Nucleotide sequence of the open reading frames adjacent to the coat protein cistron in potato virus X genome. *FEBS Lett.* **213**, 438–442.

79. Skryabin, K.G., Morozov, S.Yu., Kraev, A.S., Rozanov, M.N., Chernov, B.K., Lukasheva, L.I., and Atabekov, J.G. (1988) Conserved and variable elements in RNA genomes of potexviruses. *FEBS Lett.* **240**, 33–40

80. Solovyev, A.G., Savenkov, E.I., Agranovsky, A.A., and Morozov, S.Yu. (1996). Comparisons of the genomic cis-elements and coding regions in RNA beta components of the hordeiviruses barley stripe mosaic virus, lychnis ringspot virus, and poa semilatent virus. *Virology* **219**, 9–18.

81. Morozov, S.Yu., Miroshnichenko, N.A., Solovyev, A.G., Zelenina, D.A., Fedorkin, O.N., Lukasheva, L.I., Grachev, S A., and Chernov, B.K. (1991) In vitro membrane binding of the translation products of the carlavirus 7-kDa protein genes. *Virology* **183**, 782–785.

82. Koenig, R., Pleij, C.W., Beier, C., and Commandeur, U. (1998) Genome properties of beet virus Q, a new furo-like virus from sugarbeet, determined from unpurified virus. *J. Gen. Virol.* **79**, 2027–2036.

83. Callaway, A., Giesman-Cookmeyer, D., Gillock, E.T., Sit, T.L., and Lommel, S.A. (2001). The multifunctional capsid proteins of plant RNA viruses. *Annu. Rev. Phytopathol.* **39**, 419–460.

84. Petty, I.T., French, R., Jones, R.W., and Jackson, A.O. (1990) Identification of barley stripe mosaic virus genes involved in viral RNA replication and systemic movement. *EMBO J.* **9**, 3453–3457.

85. Lawrence, D.M. and Jackson, A.O. (2001b) Requirements for cell-to-cell movement of barley stripe mosaic virus in monocot and dicot hosts. *Mol. Plant Pathol.* **2**, 65–75.

86. Herzog, E., Hemmer, O., Hauser, S., Meyer, G., Bouzoubaa, S., and Fritsch, C. (1998) Identification of genes involved in replication and movement of peanut clump virus. *Virology* **248**, 312–322.

87. McGeachy, K.D. and Barker, H. (2000) Potato mop-top virus RNA can move long distance in the absence of coat protein: evidence from resistant, transgenic plants. *Mol. Plant Microbe Interact.* **13**, 125–128.

88. Savenkov, E.I., Germundsson, A., Zamyatnin, A.A.,Jr., Sandgren, M., and Valkonen, J.P.T. (2003) Potato mop-top virus: the coat protein-encoding RNA and the gene for cystein-rich protein are dispensable for systemic virus movement in Nicotiana benthamiana. *J. Gen. Virol.* **84**, 1001–1005.

89. Chapman, S., Hills, G., Watts, J., and Baulcombe, D. (1992) Mutational analysis of the coat protein gene of potato virus X: effects on virion morphology and viral pathogenicity. *Virology* **191**, 223–230.

90. Forster, R.L., Beck, D.L., Guilford, P.J., Voot, D.M., Van Dolleweerd, C.J., and Andersen, M.T. (1992) The coat protein of white clover mosaic potexvirus has a role in facilitating cell-to-cell transport in plants. *Virology* **191**, 480–484.

91. Sit, T.L. and AbouHaidar, M.G. (1993) Infectious RNA transcripts derived from cloned cDNA of papaya mosaic virus: effect of mutations to the capsid and polymerase proteins. *J. Gen. Virol.* **74**, 1133–1140.

92. Bleykasten, C., Gilmer, D., Guilley, H., Richards, K.E., and Jonard, G. (1996) Beet necrotic yellow vein virus 42 kDa triple gene block protein binds nucleic acid in vitro. *J. Gen. Virol.* **77**, 889–897.

93. Cowan, G.H., Lioliopoulou, F., Ziegler, A., and Torrance, L. (2002). Subcellular localisation, protein interactions, and RNA binding of potato mop-top virus triple gene block proteins. *Virology* **298**, 106–115.

94. Kalinina, N.O., Rakitina, D.A., Yelina, N.E., Zamyatnin, A.A., Jr., Stroganova, T.A., Klinov, D.V., Prokhorov, V.V., Ustinova, S.V., Chernov, B.K., Schiemann, J., Solovyev, A.G., and Morozov, S.Yu. (2001) RNA-binding properties of the 63 kDa protein encoded by the triple gene block of poa semilatent hordeivirus. *J. Gen. Virol.* **82**, 2569–2578.

95. Brakke, M.K., Ball, E.M., and Langenberg, W.G. (1988) A non-capsid protein associated with unencapsidated virus RNA in barley infected with barley stripe mosaic virus. *J. Gen. Virol.* **69**, 481–491.

96. Lauber, E., Bleykasten-Grosshans, C., Erhardt, M., Bouzoubaa, S., Jonard, G., Richards, K.E., and Guilley, H. (1998) Cell-to-cell movement of beet necrotic yellow vein virus. I. Heterologous complementation experiments provide evidence for specific interactions among the triple gene block proteins. *Mol. Plant Microbe Interact.* **11**, 618–625.

97. Erhardt, M., Herzog, E., Lauber, E., Fritsch, C., Guilley, H., Jonard, G., Richards, K., and Bouzoubaa, S. (1999) Transgenic plants expressing the TGB1 protein of peanut clump virus complement movement of TGB1-defective peanut clump virus but not of TGB1-defective beet necrotic yellow vein virus. *Plant Cell Rep.* **18**, 614–619.

98. Erhardt, M., Stussi-Garaud, C., Guilley, H., Richards, K.E., Jonard, G., and Bouzoubaa, S. (1999) The first triple gene block protein of peanut clump virus localizes to the plasmodesmata during virus infection. *Virology* **264**, 220–229.

99. Erhardt, M., Morant, M., Ritzenthaler, C., Stussi-Garaud, C., Guilley, H., Richards, K.E., Jonard, G., Bouzoubaa, S., and Gilmer, D. (2000) P42 movement protein of beet necrotic yellow vein virus is targeted by the movement proteins p13 and p15 to punctate bodies associated with plasmodesmata. *Mol. Plant Microbe Interact.* **13**, 520–528.

100. Solovyev, A.G., Savenkov, E.I., Grdzelishvili, V.Z., Kalinina, N.O., Morozov, S.Yu., Schiemann, J., and Atabekov, J.G. (1999) Movement of hordeivirus hybrids with exchanges in the triple gene block. *Virology* **253**, 278–287.

101. Lawrence, D.M. and Jackson, A.O. (2001). Interactions of the TGB1 protein during cell-to-cell movement of barley stripe mosaic virus. *J. Virol.* **75**, 8712–8723.

102. Solovyev, A.G., Stroganova, T.A., Zamyatnin, A.A., Jr., Fedorkin, O. N., Schiemann, J., and Morozov, S.Yu. (2000) Subcellular sorting of small membrane-associated triple gene block proteins: TGBp3-assisted targeting of TGBp2. *Virology* **269**, 113–127.

103. Zamyatnin, A.A., Jr., Solovyev, A.G., Sablina, A.A., Agranovsky, A.A., Katul, L., Vetten, H.J., Schiemann, J., Hinkkanen, A.E., Lehto, K., and Morozov, S.Yu. (2002) Dual-colour

imaging of membrane protein targeting directed by poa semilatent virus movement protein TGBp3 in plant and mammalian cells. *J. Gen. Virol.* **83**, 651–662.

104. Gorshkova, E.N., Erokhina, T.N., Stroganova, T.A., Yelina, N.E., Zamyatnin, A.A., Jr., Kalinina, N.O., Schiemann, J., Solovyev, A.G., and Morozov, S.Yu. (2003) Immunodetection and fluorescent microscopy of transgenically expressed hordeivirus TGBp3 movement protein reveals its association with endoplasmic reticulum elements in close proximity to plasmodesmata. *J. Gen. Virol.* **84**, 985–994.

105. Krishnamurthy, K., Heppler, M., Mitra, R., Blancaflor, E., Payton, M., Nelson, R.S., and Verchot-Lubicz, J. (2003) The Potato virus X TGBp3 protein associates with the ER network for virus cell-to-cell movement. *Virology* **309**, 135–151.

106. Mitra, R., Krishnamurthy, K., Blancaflor, E., Payton, M., Nelson, R.S., and Verchot-Lubicz, J. (2003) The Potato virus X TGBp2 protein association with the endoplasmic reticulum plays a role in but is not sufficient for viral cell-to-cell movement. *Virology* **312**, 35–48.

107. Tamai, A. and Meshi, T. (2001) Cell-to-cell movement of potato virus X: the role of p12 and p8 encoded by the second and third open reading frames of the triple gene block. *Mol. Plant Microbe Interact.* **14**, 1158–1167.

108. Krishnamurthy, K., Mitra, R., Payton, M. E., and Verchot-Lubicz, J. (2002) Cell-to-cell movement of the PVX 12 K, 8 K, or coat proteins may depend on the host, leaf developmental stage, and the PVX 25 K protein. *Virology* **300**, 269–281.

109. Torrance, L., Cowan, G.H., Gillespie, T., Ziegler, A., and Lacomme, C. (2006) Barley stripe mosaic virus encoded triple gene block 2 and γb localize to chloroplasts in virus infected monocot and dicot plants revealing hitherto unknown roles in virus replication. *J. Gen. Virol. In press.*

110. Lough, T.J., Netzler, N.E., Emerson, S.J., Sutherland, P., Carr, F., Beck, D.L., Lucas, W.J., and Forster, R.L. (2000) Cell-to-cell movement of potexviruses: Evidence for a ribonucleoprotein complex involving the coat protein and first triple gene block protein. *Mol. Plant Microbe Interact.* **13**, 962–974.

111. Angell, S.M., Davies, C., and Baulcombe, D.C. (1996) Cell-to-cell movement of Potato virus X is associated with a change in the size-exclusion limit of plasmodesmata in trichome cells of *Nicotiana clevelandii. Virology* **216**, 197–201.

112. Malcuit, I., Marano, M.R., Kavanagh, T.A., De Jong, W., Forsyth, A., and Baulcombe, D.C. (1999) The 25-kDa movement protein of PVX elicits *Nb*-mediated hypersensitive cell death in potato. *Mol. Plant Microbe Interact.* **12**, 536–543.

113. Yang, Y., Ding, B., Baulcombe, D.C., and Verchot, J. (2000) Cell-to-cell movement of the 25 K protein of potato virus X is regulated by three other viral proteins. *Mol. Plant-Microbe Interact.* **13**, 599–605.

114. Howard, A.R., Heppler, M.L., Ju, H.-J., Krishnamurthy, K., Payton, M.E., and Verchot-Lubicz, J. (2004) Potato virus X TGBp1 induces plasmodesmata gating and moves between cells in several host species whereas CP moves only in Nicotiana benthamiana leaves. *Virology* **328**, 185–197.

115. Karpova, O.V., Zayakina, O.V., Arkhipenko, M.A., Sheval, E.V., Kiselyova, O.I., Poljakov, V.Yu., Yaminsky, I.V., Rodionova, N.P., and Atabekov, J.G. (2006) Potato virus RNA-mediated assembly of single-tailed ternary complexes "coat protein-RNA-movement protein" *J.Gen.Virol.* **87**, 2731–2740.

116. Atabekov, J.G., Rodionova, N.P., Karpova, O.V., Kozlovsky, S.V., and Poljakov, V.Y. (2000) The movement protein-triggered in situ conversion of Potato virus X virion RNA from a nontranslatable into a translatable form. *Virology* **271**, 259–263.

117. Kiselyova, O.I., Yaminsky, I.V., Karpova, O.V., Rodionova, N.P., Kozlovsky, S.V., Arkhipenko, M.V., and Atabekov, J.G. (2003) AFM study of Potato virus X disassembly induced by movement protein. *J. Mol. Biol.* **332**, 321–325.

118. Rodionova, N.P., Karpova, O.V., Kozlovsky, S.V., Zayakina, O.V., Arkhipenko, M.V., and Atabekov, J.G. (2003) Linear remodeling of helical virus by movement protein binding. *J. Mol. Biol.* **333**, 565–572.

119. Santa Cruz, S., Roberts, A.G., Prior, D.A.M., Chapman, S., and Oparka, K.J. (1998) Cell-to-cell and phloem-mediated transport of potato virus X: the role of virions. *Plant Cell* **10**, 495–510.

120. Fedorkin, O.N., Merits, A., Lucchesi, J., Solovyev, A.G., Saarma, M., Morozov, S.Yu., and Makinen, K. (2000) Complementation of the movement-deficient mutations in Potato virus X: Potyvirus coat protein mediates cell-to-cell trafficking of C-terminal truncation but not deletion mutant of potexvirus coat protein. *Virology* **270**, 31–42.
121. Fedorkin, O.N., Solovyev, A.G., Yelina, N. E., Zamyatnin, A.A., Jr, Zinovkin, R.A., Makinen, K., Schiemann, J., and Morozov, S.Yu. (2001) Cell-to-cell movement of Potato virus X involves distinct functions of the coat protein. *J. Gen. Virol.* **82**, 449–458.
122. Rakitina, D.V., Kantidze, O.L., Leshchiner, A.D., Solovyev, A.G., Novikov, V.K., Morozov, S.Yu., and Kalinina, N.O. (2005) Coat proteins of two filamentous viruses display NTPase activity in vitro *FEBS Lett.* **579**, 4955–4960.
123. Dolja, V.V., Halderman-Cahill, R., Montgomery, A.E., Vandenbosch, K.A., and Carrington, J.C., (1995) Capsid protein determinants involved in cell-to-cell and long distance movement of tobacco etch potyvirus. *Virology* **206**, 1007–1016.
124. Rojas, M.R., Zerbini, F.M., Allison, R.F., Gilbertson, R.L., and Lucas, W.J. (1997) Capsid protein and helper compoment-proteinase function as potyvirus cell-to-cell movement proteins. *Virology* **237**, 283–295.
125. Revers, F., Le Gall, O., Candresse, T., and Maule, A.J., (1999) New advances in understanding the molecular biology of plant/potyvirus interactions. *Mol. Plant Microbe Interact.* **12**, 367–376.
126. Carrington, J.C., Jensen, P.E., and Schaad, M.C. (1998) Genetic evidence for an essential role for potyvirus CI protein in cell-to-cell movement. *Plant J.* **14**, 393–400.
127. Rodriguez-Cerezo, E., Findlay, K.,Shaw, J.G., Lomonossoff, G.P., Qiu, S.G., Linstead, P., Shanks, M., and Risco, C. (1997) The coat and cylindrical inclusion proteins of a potyvirus are associated with connections between plant cells. *Virology* **236**, 296–306.
128. Roberts, I.M., Wang, D., Findlay, K., and Maule, A.J. (1998) Ultrastuctural and temporal observations of the potyvirus cylindrical inclusions (CIs) show that the CI protein acts transiently in aiding virus movement. *Virology* **245**, 173–181.
129. Nicolas, O., Dunnington, S.W., Gotow, L.F., Pirone, T.P., and Hellmann, G.M. (1997) Variations in the VPg protein allow a potyvirus to overcome va gene resistance in tobacco. *Virology* **237**, 452–459.
130. Gao, Z., Johansen, E., Eyers, S., Thomas, C.L., Ellis, T.H.N., and Maule A.J. (2004). The potyvirus recessive resistance gene, sbm1, identifies a novel role for translation inititiation factor eIF4E in cell-to-cell trafficking. *Plant J.* **40**, 376–385.
131. Dunoyer, P., Thomas, C., Harrison, S., Revers, F., and Maule, A. (2004) A cysteine-rich plant protein potentiates potyvirus movement through an interaction with the virus genome-linked protein VPg. *J. Virol.* **78**, 2301–2309.
132. Torrance, L., Andreev, I.A., Gabrenaite-Verhovskaya, R., Cowan, G., Mäkinen, K., Taliansky, M.E. (2006). An unusual structure at one end of potato potyvirus particles. *J. Mol. Biol.* **357**, 1–8.
133. Dolja, V.V. (2003) Beet yellow virus: the importance of being different. *Mol. Plant Pathol.* **4**, 91–98.
134. Napuli, A.J., Falk, B.W., and Dolja, V.V. (2000) Interaction between HSP70 homolog and filamentous virions of the Beet yellows virus. *Virology* **274**, 232–239.
135. Peremyslov, V.V., Andreev, I.A., Prokhnevsky, A.I., Duncan, G. H., Taliansky, M.E., and Dolja, V.V. (2004). Complex molecular architecture of beet yellows virus particles. *Proc. Natl. Acad. Sci. USA* **101,** 5030–5035.
136. Alzhanova, D.V., Napuli, A.J., Creamer, R., and Dolja, V.V. (2001) Cell-to-cell movement and assembly of a plant closterovirus: roles for the capsid proteins and Hsp70 homolog. *EMBO J.* **20**, 6997–7007.
137. Peremyslov, V.V., Hagiwara, Y., and Dolja, V.V. (1999) HSP70 homolog functions in cell-to-cell movement of a plant virus. *Proc. Natl. Acad. Sci.* **96**, 14771–14776.
138. Prokhnevsky, A.I., Peremyslov, V.V., Napuli, A.J., and Dolja, V.V. (2002) Interaction between long-distance transport factor and Hsp70-related movement protein of Beet yellows virus. *J. Virol.* **76**, 11003–11011.

139. Dolja, V.V., Kreuze, J.F., and Valkonen J.P. (2006) Comparative and functional genomics of closteroviruses. *Virus Res.* **117**, 38–51.
140. Marvin, D.A. (1998). Filamentous phage structure, infection and assembly. *Curr. Opin. Struct. Biol.* **8**, 150–158.
141. Pouwels, J., Carette, J.E., Van Lent, J., and Wellink, J. (2002) Cowpea mosaic virus: effect on cell host processes. *Mol. Plant Pathol.* **3**, 411–418.
142. Van Lent, J., Storms, M., van der Meer, F., Wellink, J., and Goldbach, R. (1991) Tubular structures involved in movement of cowpea mosaic virus are also formed in infected cowpea protoplasts. *J. Gen. Virol.* **72**, 2615–2623.
143. Kasteel, D.T., Perbal, M.C., Boyer, J.C., Wellink, J., Goldbach, R.W., Maule, A.J., and Van Lent, J.W. (1996) The movement proteins of cowpea mosaic virus and cauliflower mosaic virus induce tubular structures in plant and insect cells. *J. Gen. Virol.* **77**, 2857–2864.
144. Pouwels, J., Van der Krogt, G.N., Van Lent, J., Bisseling, T., and Wellink, J. (2002) The cytoskeleton and the secretory pathway are not involved in targeting the cowpea mosaic virus movement protein to the cell periphery. *Virology* **297**, 48–56.
145. Laporte, C., Vetter, G., Loudes, A.M., Robinson, D.G., Hillmer, S., Stussi-Garaud, C., and Ritzenthaler, C. (2003) Involvement of the secretory pathway and the cytoskeleton in intracellular targeting and tubule assembly of Grapevine fanleaf virus movement protein in tobacco BY-2 cells. *Plant Cell* **15**, 2058–2075.
146. Bertens, P., Wellink, J., Goldbach, R., and van Kammen, A. (2000) Mutational analysis of the cowpea mosaic virus movement protein. *Virology* **267**, 199–208.
147. Atabekov, J.G., Malyshenko, S.I., Morozov, S.Yu., Taliansky, M.E., Solovyev, A.G., Agranovsky, A.A., and Shapka, N.A. (1999) Identification and study of tobacco mosaic virus movement function by complementation tests. *Philos. Trans. R. Soc. Lond., B, Biol. Sci.* **354**, 629–635.
148. Nelson, R.S. and van Bel, A.J.E. (1998) The mystery of virus trafficking into, through and out of the vascular tissue. *Progr. Bot.* **59**, 476–533; Nelson RS, Citovsky V.Plant viruses. Invaders of cells and pirates of cellular pathways. Plant Physiol. 2005 Aug;138(4):1809–14. Review.
149. Oparka, K.J. and Turgeon, R. (1999) Sieve elements and companion cells-traffic control centers of the phloem. *Plant Cell* **11**, 739–750; Oparka KJ, Cruz SS. The great escepe: Phloem Transport and Unloading of Macromolecules Annu Rev Plant Physiol Plant Mol Biol. 2000 51:323–347.
150. Swanson, M., Barker, H., and MacFarlane, S.A. (2002). Rapid vascular movement of tobraviruses does not require coat protein: evidence from mutated and wild-type viruses. *Ann. Appl. Biol.* **141,** 259–266.
151. Cadman, C.H. (1962) Evidence for association of tobacco rattle virus nucleic acid with a cell component. *Nature* **193**, 49–52.
152. Scholthof, H.B., Scholthof, K.B., Kikkert, M., and Jackson, A.O. (1995) Tomato bushy stunt virus spread is regulated by two nested genes that function in cell-to-cell movement and host-dependent systemic invasion. *Virology* **213**, 425–438.
153. Tacke, E., Prufer, D., Schmitz, J. and Rohde, W. (1991) The potato leafroll luteovirus 17 K protein is a single-stranded nucleic acid-binding protein. *J. Gen. Virol.* **72**, 2035–2038.
154. Tacke, E., Schmitz, J. Prufer, D. and Rohde, W. (1993) Mutational analysis of the nucleic acid-binding 17 kDa phosphoprotein of potato leafroll luteovirus identifies an amphipathic alpha-helix as the domain for protein/protein interactions. *Virology* **197**, 274–282.
155. Sokolova, M., Prufer, D., Tacke, E., and Rohde, W. (1997) The potato leafroll virus 17 K movement protein is phosphorylated by a membrane-associated protein kinase from potato with biochemical features of protein kinase C. *FEBS Lett.* **400**, 201–205.
156. Schmitz, J., Stussi-Garaud, C., Tacke, E., Prufer, D., Rohde, W., and Rohfritsch, O. (1997) In situ localization of the putative movement protein (pr17) from potato leafroll luteovirus (PLRV) in infected and transgenic potato plants. *Virology* **235**, 311–322.
157. Taliansky, M., Mayo, M.A., and Barker, H. (2003) Potato leafroll virus: a classic pathogen shows some new tricks. *Mol. Plant Pathology* **4**, 81–89.

158. Ryabov, E.V., Robinson, D. J., and Taliansky, M. E. (1999) A plant virus-encoded protein facilitates long-distance movement of heterologous viral RNA. *Proc. Natl. Acad. Sci. USA* **96**, 1212–1217.

159. Ryabov, E.V., Robinson, D.J., and Taliansky, M.E. (2001) Umbravirus-encoded proteins both stabilize heterologous viral RNA and mediate its systemic movement in some plant species. *Virology* **288**, 391–400.

160. Taliansky, M., Roberts, I.M., Kalinina, N., Ryabov, E.V., Raj, S.R., Robinson, D.J., and Oparka, K.J. (2003) An umbraviral protein, involved in long-distance RNA movement, binds viral RNA and forms unique, protective ribonucleoprotein complexes. *J. Virol.* **77**, 3031–3040.

161. Ryabov, E.V., Kim, S.-H., and Taliansky, M. E. (2004) Identification of a nuclear localization signal and nuclear export signal of the umbraviral long-distance RNA movement protein. *J. Gen. Virol.* **85**, 1329–1333.

162. Dorokhov Y.L., Makinen, K., Frolova, O.Y., Merits, A., Saarinen, J., Kalkkinen, N., Atabekov, J.G., and Saarma, M. (1999). A novel function for a ubiquitous plant enzyme pectin methylesterase: the host-cell receptor for the tobacco mosaic virus movement protein. *FEBS Lett.* **461**, 223–228.

163. Chen M.H., Sheng, J.S., Hind G., Handa, A.K., and Citovsky, V. (2000) Interaction between the tobacco mosaic virus movement protein and host cell pectin methylesterases is required for viral cell-to-cell movement. *EMBO J.* **19**, 913–920.

164. Yoshioka, K., Matsushita, Y., Kasahara, M., Konagaya, K., and Nyunoya, H. (2004) Interaction of tomato mosaic virus movement protein with tobacco RIO kinase. *Mol. Cells* **17**, 223–229.

165. Desvoyes, B., Faure-Rabasse, S., Chen, M.H., Park, J.W., and Scholthof, H.B. (2002) A novel plant homeodomain protein interacts in a functionally relevant manner with a virus movement protein. *Plant Physiol.* **129**, 1521–1532.

166. Soellick, T., Uhrig, J.F., Bucher, G. L., Kellmann, J.W., and Schreier, P.H. (2000) The movement protein NSm of tomato spotted wilt tospovirus (TSWV): RNA binding, interaction with the TSWV N protein, and identification of interacting plant proteins. *Proc. Natl. Acad. Sci. USA* **97**, 2373–2378.

167. Matsushita, Y., Miyakawa, O., Deguchi, M., Nishiguchi, M., and Nyunoya, H. (2002) Cloning of a tobacco cDNA coding for a putative transcriptional coactivator MBF1 that interacts with the tomato mosaic virus movement protein. *J. Exp. Bot.* **53**, 1531–1532.

168. Voinnet, O., Lederer, C., and Baulcombe, D.C. (2000) A viral movement protein prevents spread of the gene silencing signal in *Nicotiana benthamiana*. *Cell* **103**, 157–167.

Chapter 4
Multiple Roles of Viral Replication Proteins in Plant RNA Virus Replication

Peter D. Nagy and Judit Pogany

Abstract Identification of the roles of replication factors represents one of the major frontiers in current virus research. Among plant viruses, the positive-stranded (+)RNA viruses are the largest group and the most widespread. The central step in the infection cycles of (+)RNA viruses is RNA replication, which leads to rapid production of huge number of viral (+)RNA progeny in the infected plant cells. The RNA replication process is carried out by the virus-specific replicase complex consisting of viral RNA-dependent RNA polymerase, one or more auxiliary viral replication proteins, and host factors, which assemble in specialized membranous compartments in infected cells. Replication is followed by cell-to-cell and long-distance movement to invade the entire plant and/or encapsidation to facilitate transmission to new plants. This chapter provides an overview of our current understanding of the role of viral replication proteins during genome replication. The recent significant progress in this research area is based on development of powerful in vivo and in vitro approaches, including replicase assays, reverse genetic approaches, intracelular localization studies and the use of plant or yeast model hosts.

Keywords Brome mosaic virus, Tomato bushy stunt virus, Tobacco mosaic virus, RNA-dependent RNA polymerase, replicase, helicase, chaperone

1 Introduction

Plant viruses with positive-stranded RNA genomes are the largest group and the most widespread among plant viruses, causing numerous diseases of economically important crops. Plant RNA viruses have also been developed as efficient gene-expression vectors in biotechnological applications as well as in nanotechnology. The central step in the infection cycles of positive-stranded RNA viruses is RNA replication, which leads to rapid and robust production of large number of viral genomic (g)RNA progenies in the infected plant cells. The RNA replication process is carried out by viral and host-coded proteins in specialized membranous compartments in infected cells via the production of minus-stranded RNA intermediates,

From: *Methods in Molecular Biology, Vol. 451, Plant Virology Protocols:*
From Viral Sequence to Protein Function
Edited by G.D. Foster, I.E. Johansen, Y. Hong, and P.D. Nagy © Humana Press, Totowa, NJ

followed by synthesis of the plus-stranded (+)RNA progeny. The newly synthesized (+)RNA can then participate in cell-to-cell and long-distance movement to invade the entire plant or in encapsidation to facilitate transmission to new plants.

This chapter focuses on replication of (+)RNA viruses of plants with the goal of providing an overview of our current understanding of the role of viral replication proteins during genome replication. The recent significant progress in this research area is based on development of powerful in vivo and in vitro approaches, including in vitro replicase assays, with either purified recombinant replicase proteins or with partially purified replicase complexes; reverse genetic approaches; intracellular localization studies; and the use of plant or yeast model hosts. Many of these approaches are presented in more details in the following chapters of this book.

Identification of the roles of various replication-associated or replication-modulating viral and host factors represents one of the major frontiers in current virus research. The emerging picture is that in spite of sequence diversity among viral replication proteins, the mechanism of RNA genome replication and the functions of viral replication factors might be analogous to some extent among various (+)RNA viruses. Indeed, all (+)RNA viruses replicate their genomes through minus-stranded replication intermediates, which are less abundant than the new (+)gRNA progeny due to asymmetrical RNA strand replication. Moreover, the genomes of all known (+)RNA viruses of plants code for an RNA-dependent RNA polymerase (RdRp) and one or more auxiliary replication proteins, which, likely together with some host factors, assemble the virus-specific replicase complexes (1–5). In addition, RNA replication takes place on the cytoplasmic surfaces of membranous compartments derived from intracellular organelles, such as endoplasmic reticulum, mitochondrium, chloroplast, vacuole, or peroxisome. Our current view is that viral replication proteins perform multiple functions to complete the whole replication process. These functions include (1) selective recognition of the viral RNA, (2) template recruitment to the site of replication, (3) assembly of the specialized replication complexes, and (4) synthesis of the viral RNA progeny.

Most of our current knowledge on plant virus RNA replication comes from studies performed with advanced model viruses. Therefore, this chapter will provide deeper insight into the known roles of viral RNA replication proteins based on three most intensively studied RNA viruses, namely bromo-, tobamo-, and tombusviruses. Detailed description on RNA replication of these and other plant viruses and the roles of replication proteins can be found in several recent publications (2, 4–6).

2   General Features of Replication Proteins

2.1   *Production of Viral Replicase Proteins in Infected Cells*

All positive-stranded RNA viruses of plants generate their replication proteins via direct translation of their gRNAs, which can serve as mRNAs, by the host ribosomes after their entry to the host cells. Different virus groups utilize various mechanisms to allow the production of two or more replication proteins, which are essential for replication (7, 8). Regardless of the gene expression strategies, the replication proteins are produced early in the infection process and usually in lesser amounts than the structural proteins, such as the coat protein. We provide some details for the expression strategies of the three most intensively studied virus groups as examples (see below).

Tombusviruses have a single ~4,800 nt gRNA, which is uncapped and nonpolyadenylated (6). They use a 3′ located translational enhancer, which compensate for the lack of the above translational elements, to make the gRNA efficient in translation in plant cells (9). The two replication proteins, denoted p33 and p92^{pol}, are directly translated from the gRNA via sharing the same initiation codon. The translation stop codon of p33, however, can be "read through" by the ribosome with 5% efficiency, resulting in misincorporation of mostly a tyrosine residue, followed by continuation of translation to the end of p92^{pol} open reading frame (ORF) (6, 10). This expression strategy guarantees that p33 sequence overlaps with the N-terminal region of p92^{pol} and that p92^{pol} is produced ~20-fold lesser amount than p33 (10).

Tobamoviruses, similar to tombusviruses, use ribosomal read-through strategy to produce two partly overlapping replication proteins from the same initiation codon located 5′ proximally in the 6.3 kb one-component gRNA, which is capped at the 5′ end and carries a tRNA-like structure at the 3′ end (11). The two replication proteins, denoted 126 K and 183 K, are expressed in 10:1 ratio (12, 13). The 126 K protein has an N-terminal capping domain, an RNA-binding site in the intervening region (14) and a C-terminal helicase domain (15), whereas the 183 K protein has an additional RdRp domain in its unique C terminus (11).

Bromoviruses use another gene expression strategy that is based on a tri-component RNA genome (i.e., the whole genomic sequence is divided among three separate gRNAs), which are used as separate mRNAs by the host translational apparatus (2, 7). RNA1 and RNA2, which are capped at the 5′ end and carry tRNA-like structures at their 3′ ends, code for the two essential replication proteins, denoted 1a and 2a^{pol}. These proteins are produced in different amounts due to selective downregulation of translation of RNA2 by the host translation factor Ded1p via specific interaction with a unique sequence at the 5′ noncoding region of RNA2 (16). The auxiliary 1a protein is present 25-fold excess in comparison with 2a^{pol} at the sites of replication (17).

2.2 *Comparison of Replication Proteins*

Among the two or more replication proteins translated from the invading viral RNAs, one of the proteins is the RdRp, whereas the usually more abundant protein(s) has regulatory and auxiliary functions (1–3). Although 3D structure is not yet available for a plant virus RdRp, bioinformatics based comparison between RdRps of animal viruses (hepatitis C virus, poliovirus, bovine virus diarrhea virus), a bacteriophage (Fi6), and plant virus RdRps reveals significant similarities (18, 19). Therefore, functional domains in plant viral RdRp, including the catalytic site and domains involved in binding to ribonucleotides and to the viral RNA template, can

A   TBSV

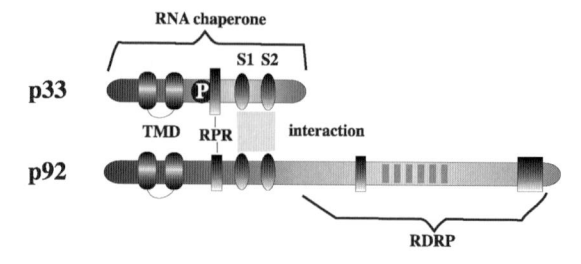

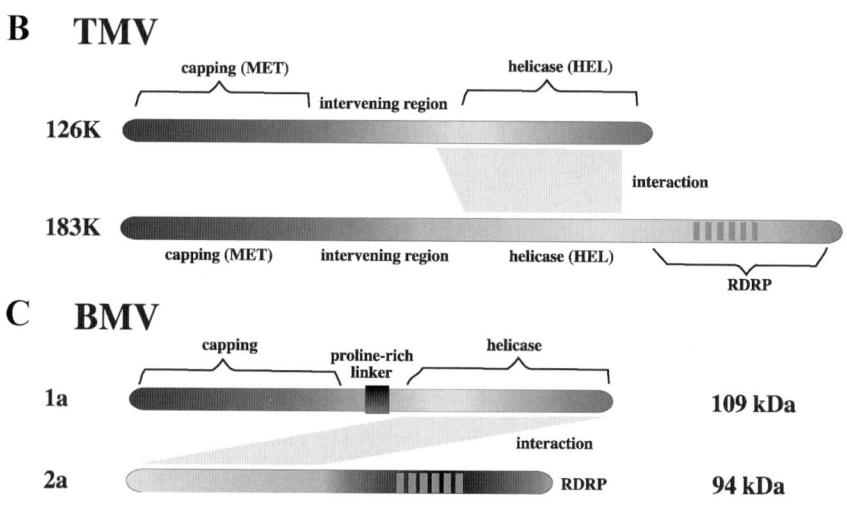

Fig. 1 The known domains in the replication proteins of (**a**) *Tomato bushy stunt virus*, (**b**) *Tobacco mosaic virus*, and (**c**) *Brome mosaic virus*. The transmembrane (TMD), the arginine-proline-rich (RPR) RNA binding, the p33–p33/p92 interaction (S1 and S2) and the RNA-dependent RNA polymerase (RdRp) domains in the TBSV proteins are depicted. The phosphorylation site is marked with "P." Additional mapped RNA binding sequences are shown with *gray boxes*, whereas the interacting protein sequences are connected with *shaded rectangular boxes*

be predicted with high certainty (19). In addition, mutagenesis of the predicted domains performed with either purified recombinant viral RdRps, with partially purified replicase complexes or in vivo using infectious transcripts confirmed the essential nature of these domains in viral RNA replication (1, 3, 6, 11). The known and predicted functional domains for the tombusvirus p92[pol], the tobamovirus 183 K, the bromovirus 2a[pol] RdRp proteins re shown in Fig. 1.

The possible functions of the auxiliary replication proteins and the presence of functional domains in these proteins are less conserved. For example, some of the auxiliary replication proteins have capping and helicase domains, while others lack these domains (3, 20). In spite of these major differences in conserved sequences and biochemical functions, the various auxiliary replication proteins might perform several similar functions as discussed below. The characterized domains in the tombusvirus p33, the tobamovirus 126 K, the bromovirus 1a auxiliary proteins are shown in Fig. 1.

3 Composition of the Viral Replicase Complex

The viral replicase is the key enzyme in virus replication. The replicase complex has to perform many functions during replication, including recognition of minus and plus-strand initiation promoters located at the 3′ terminus of either the (+) or (−)RNA, de novo (primer-independent) or primer-dependent initiation followed by synthesis of full-length complementary RNA strands, strand separation, and possibly repair of viral RNAs with damaged termini. In addition, for some viruses, the viral replicase has to recognize additional regulatory elements, such as replication silencer and replication enhancer, which either down- or upregulate RNA synthesis (5). Also, the viral replicase can synthesize subgenomic RNAs in selected viruses (3, 11, 21). Moreover, the replicase plays a major role in RNA virus evolution by creating mutations or supporting RNA recombination via template switching during RNA synthesis (22–28).

3.1 Proteomics-Based Analysis of the Viral Replicase Complex

Although it is currently unknown how the viral replicase could perform many functions, it is assumed that the multifunctionality is due to the complex nature/composition of the viral replicase complex. To analyze the protein composition of a viral replicase complex, the tombusvirus replicase was solubilized using nonionic detergents from a membrane-enriched fraction of the infected cells, followed by two-dimensional (2D) gel electrophoresis to separate the individual proteins. Mass spectrometry analysis of the 2D separated proteins identified the two viral replication proteins (p33 and p92^{pol}) and four host proteins, including the heat shock protein 70 (Hsp70, called Ssa1/2p in yeast), glyceraldehyde-3-phosphate dehydrogenase (GAPDH, called Tdh2/3p in yeast, which is an RNA-binding protein), pyruvate decarboxylase (Pdc1p), and an unidentified acidic protein (29). It is likely that additional host proteins, which might be lost during purification, also associated with the viral replicase in a temporal fashion.

At least four host proteins have been detected in the purified TMV replicase preparation by silver staining of SDS-PAGE gel (30). Western blotting led to the identification of an RNA-binding protein, GCD10, which is one of the subunits of the 10-component eIF-3 complex (30). An additional host protein in the TMV replicase might be translation elongation factor 1A, which was found to bind to the methyltransferase (capping) domain of TMV 126 K protein based on coimmunoprecipitation (31).

A highly purified replicase preparation for BMV contained the 1a and 2a^{pol} replicase proteins and ~10 host proteins based on silver-stained SDS-PAGE analysis (32). One of the host proteins identified was the p41 subunit of the eIF-3 complex, based on immunodetection. In addition, p41 is bound to the 2a^{pol} affinity resin in vitro (32). The function of p41 in the BMV replicase is currently unknown.

3.2   Structure of the Viral Replicase Complex

Although high-resolution images of viral replication complexes are currently not yet available for (+)RNA viruses of plants, low-resolution images of the sites of BMV replication, containing the active viral replicase complexes, have been published by the Ahlquist group (17). Using electron microscopy (EM) and immuno-EM, they found that the sites of BMV replication consist of 50–60 nm spherule-like structures with cellular membranes surrounding the replication proteins and the viral RNA. Interestingly, the EM images revealed the existence of a small opening (membranous neck) from individual spherules that likely serve as a gate for communication/transportation of molecules between the spherules and the cytoplasm (17). In a broad sense, we can regard one separate spherule as one active/matured replicase complex. Studies on the molecular composition of a single spherule based on immunolabeled images revealed that one spherule could contain 25-fold more 1a than 2a^{pol}, whereas the actual number and nature of host molecules within single spherules are currently unknown.

High-resolution EM studies on recombinant TMV replication proteins revealed ring-like structures, which likely consist of hexameric 126 K/183 K complexes (15). Oligomerization of the helicase subunits in 126 K/183 K might facilitate efficient positioning of the RdRp domain over the initiation site in the (+)RNA template (15). The ratio of 126 K versus 183 K in the TMV replicase complex is currently undecided with reports favoring 1:1 or 5:1 (14, 15, 33).

Altogether, the current models on the replicase complexes of plant (+)RNA viruses predict that highly structured protein–protein and protein–RNA complexes with the help of cellular membranes facilitate the formation of the active replicase complex. The nature of the known protein–protein and protein–RNA interactions will be discussed in some detail below.

4   Interactions between Replication Proteins and Viral RNA during Replication

Selection of the viral RNA for replication from the vast pool of mRNAs that is actively translated at any given time point in the infected cells is one of the most important steps during replication. The emerging picture is that the viral auxiliary protein could be responsible for the selective binding to the viral RNA, likely via specific recognition of a cis-acting RNA element (5, 34). Moreover, the specific auxiliary protein–viral RNA interaction is also important for recruitment of the viral RNA to the site of replication.

The interaction between replication protein and the viral RNA has been studied in vitro using various methods, such as gel mobility shift assay with purified protein and RNA components, cross-linking studies, template competition in replicase assay, and surface plasmon resonance (SPR) assay; and with in vivo approaches, such as copurification, colocalization in the same intracellular compartments, and the yeast three-hybrid assay.

In the case of tombusviruses, the purified recombinant p33 replication cofactor can bind to various RNAs in vitro based on gel mobility shift assay and SPR analysis (35). However, in the presence of TBSV (+)RNA, p33 was found to show strong binding specificity in vitro to an internal recognition element, termed p33RE (36). Detailed mutagenesis/binding studies revealed that p33 recognizes a C·C mismatch present within the long stem portion of a stem-loop structure [RII(+)-SL]. Mutations within the C·C mismatch (except for a C to U mutation in the 3′ proximal C) was found to render the TBSV genomic RNA as well as the defective interfering RNA replication incompatible in plant cells, whole plants, and yeast, demonstrating the significance of p33–TBSV (+)RNA interaction (36, 37). P33 binds to the RNA within its RPR (arginine/poline-rich) domain (Fig. 1), in which a central arginine is especially crucial for binding based on in vivo and in vitro studies. Interestingly, mutations within the RPR motif not only affected replication, but recombination as well (38, 39).

Based on gel mobility shift experiments with purified recombinant 126 K or its truncated derivatives, the TMV 126 K replication protein has been shown to bind to the viral RNA in vitro (15). In addition, UV cross-linking of purified 126 K and the TMV 3′ UTR revealed specific interaction between tyrosine residues in the intervening region of 126 K and small stem-loops within the 3′ sequence of the viral (+)RNA (14).

Another auxiliary protein, the 1a protein of BMV, was also found involved in template selection/recruitment based on cell fractionation assay (40, 41). Mutations within the helicase domain of 1a inhibited the recruitment of viral RNA into membrane-associated, nuclease resistant state, without affecting the ER localization of 1a protein (41). Specific recognition of the viral RNA depends on the 1a responsive element (RE), which is present at the 5′ end of RNA2 and the intergenic noncoding region in RNA3 (40, 42).

5 Replication Protein–Protein Interactions

In addition to the protein–RNA interactions, the multimolecular replicase complex is likely held together by number of protein–protein interactions as well. These interactions have been studied in some detail for a number of plant viruses as discussed below.

The tombusvirus p33 protein has been shown to interact with itself and with p92^{pol} based on yeast two-hybrid and pull-down experiments (43, 44). The interaction involves two short amino acid stretches, denoted S1 and S2, both containing large aromatic and positively charged amino acid residues. Kinetics studies with surface plasmon resonance assay defined stronger interaction between p33 molecules involving the S1 subdomain (in the nanomolar range) than with the S2 subdomain (in the micromolar range) (44). Mutagenesis studies revealed essential roles for the p33:p33/p92^{pol} domain in both p33 and p92^{pol} replication proteins during tombusvirus replication (43). In vitro template assays with affinity-purified tombusvirus replicase complexes from yeast cells expressing the mutated p33 and the wt p92^{pol} revealed that the p33–p33/p92^{pol} interaction is critical for the assembly of the tombusvirus replicase in vivo (44). Moreover, the p33–p33/p92^{pol} interaction domain was also important for p33 to bind selectively to the tombusvirus RNA template carrying the p33RE. The need for the functional p33–p33/p92^{pol} domain in RNA binding suggests that dimer or multimer formation between p33 molecules is likely needed to create a complex with defined structure for selective RNA recognition/binding (36). The p33–p33/p92^{pol} interaction domain also affected the intracellular localization of p33 and p92^{pol} (45). Altogether, the above data suggest that tombusvirus replication depends on the formation of multimolecular p33–p33 and p33–p92^{pol} complexes that affect the assembly of the viral replicase complex, recruitment of the viral RNA for replication, and the intracellular localization of p33 replication cofactor (5).

Yeast two-hybrid assay revealed that 126 K of TMV interacted with 126 K and 183 K proteins (46). Gel-filtration and EM studies confirmed that 126 K proteins interact and form hexamer-like structures (15). These protein–protein interactions play important roles in TMV replication, because mutations in the helicase domain, which disrupted 126 K self-interaction in a temperature-sensitive manner, also inhibited virus replication when present in the viral gRNA (46).

The multifunctional BMV 1a protein has been shown to self-interact and can also bind to the N-terminal region of 2a^{pol} (47, 48). The 1a–1a interaction involves the capping domain and is likely important in the assembly of spherules formed in the ER (the site of BMV RNA replication), whereas interaction between the helicase domain of 1a and the N-terminal region of 2a^{pol} (47, 49) is needed for the recruitment of 2a^{pol} (which in the absence of 1a is a cytoplasmic protein) to the spherules (50).

6 Subcellular Localization of Viral Replication Proteins

Production of the replication proteins via translation of the viral gRNAs and then the selection of the viral gRNA for replication likely takes place in the cytoplasm, whereas replication of plant RNA viruses occurs on the cytoplasmic surfaces of various organelle-derived membrane surfaces (2, 3). Therefore, the viral replication proteins and the viral gRNA, possibly together with host factors, must be transported/recruited to the sites of replication. The current models involve the recruitment of replication proteins and the viral gRNA as complexes, which are likely formed during template selection (specific binding of viral and/or host proteins to the template RNA, see above). Intracellular localization of viral replication proteins and the viral RNA has been intensively studied for many plant viruses, using epifluorescent or confocal microscopy, EM, and immuno-EM as well as cell-fractionation methods.

Tombusviruses have been shown to replicate either on peroxisomal (such as TBSV and several other tombusviruses) or mitochondrial (*Carnation Italian ringspot virus*) membranes (45, 51–54). Therefore, the viral RNA and the replication proteins must be targeted to these compartments. p33 replication cofactor plays a major role in this process via its peroxisomal targeting sequences (45, 52, 54). Colocalization data also suggest that p92^{pol} can be targeted to the peroxisomal membranes even in the absence of the peroxisomal targeting domain when coexpressed with wt p33 (45), suggesting that the mutant p92^{pol} might be "piggybacking" on p33 to the site of replication. Interestingly, this mutant p92^{pol} protein is still partly functional in the absence of peroxisomal targeting or the RPR RNA-binding domains, but it is nonfunctional if the p33–p33/p92^{pol} interaction domain is missing (45).

The tombusviral RNA is colocalized with peroxisomal marker proteins when expressed together with p33 (45). The viral gRNA is likely recruited in "cis" by the newly produced p33 proteins after translation by binding to the same gRNA that was used for their synthesis ("cis-preferential" binding) (38, 55). Altogether, it seems that p33 replication protein plays a master role in intracellular targeting of other p33 and p92^{pol} proteins as well as the TBSV RNA, likely in the form of multimolecular complexes, to the site of replication.

TMV infections lead to intensive membrane rearrangements in cells. Both replication proteins were found associated with ER membranes, although they lack recognizable ER retention motifs or membrane spanning domains (56, 57). Two Arabidopsis genes, *TOM1* and *TOM3*, with several membrane-spanning domains, however, interact with 126/183 K proteins and likely act as tethers for the TMV replication proteins (58, 59). Interestingly, a nuclear localization signal in combination with the membrane-binding domain in 126 K is required for the formation of characteristic spot-like bodies that contain high concentration of replication proteins (56). These structures and/or the cellular environment might facilitate the efficient assembly of the TMV replicase complex.

Intracellular localization of BMV replicase has been studied in detail, confirming the formation of spherules (the sites of viral RNA replication, see above) from perinuclear ER membranes using double-label immunofluorescence, in vivo labeling with bromo-UTP, and confocal microscopy (60, 61). While the 1a protein localized to the ER membrane in the absence of additional viral factors, recruitment of both 2a^{pol} and the viral (+)RNAs to the ER depended on 1a (50, 62). Although 1a lacks typical membrane spanning domains, the binding of 1a to the membrane is resistant to high salt or high pH treatment, but not to treatment with nonionic detergents (63). Membrane flotation gradient analysis revealed that the N-terminal capping domain is responsible for 1a localization to ER (63).

7 Biochemical Functions of Viral Replicase Proteins

Because the viral replicase has multiple functions performed in association with membranes in infected cells, as described above, generally, it is difficult to develop assays to characterize all the activity of the replicase in vitro. Nevertheless, three different approaches have been used successfully to obtain functional viral replicases or RdRps: (1) partially or highly purified replicase preparations that can initiate complementary RNA synthesis on added RNA templates have been obtained from infected plant tissues for a number of plant RNA viruses, such as BMV (64), *Cucumber mosaic virus* (CMV) (65), *Turnip yellow mosaic virus* (66, 67), *Alfalfa mosaic virus* (68), tombusviruses (TBSV and *Cucumber necrosis virus*, CNV) (69), and others (3, 70). (2) Purified recombinant replicase expressed in yeast has been obtained for BMV (71) and a tombusvirus (CNV) (72, 73). (3) Purified recombinant RdRp from *Escherichia coli* without host factors and additional viral proteins, were obtained for *Tobacco vein mottling virus* (74), *Bamboo mosaic virus* (75), and *Turnip crinkle virus* (TCV) (76). These preparations were active in RNA synthesis on exogenous templates in vitro. Most of our current knowledge on the polymerase function of viral replicases is based on these preparations.

7.1 Polymerase Activity of the Viral Replicase or Viral RdRp

In vitro assays with a number of plant virus replicases/RdRps revealed that initiation of minus-strand (complementary-) RNA synthesis occurs on simple promoter sequences, often consisting of a defined secondary structure (tRNA-like or other stem-loop structures) with a short single-stranded tail containing C bases (70). The promoter likely serves as a binding site for the replicase, whereas the initiation site allows the replicase to start complementary RNA synthesis de novo at a defined position. Multiple initiation sites, denoted CCA boxes, can also serve as simple promoters, suggesting that template recognition by the replicase is not a stringent process in vitro (67, 77).

Initiation on minus-strand intermediate RNA can be performed by the same viral replicase/RdRp that also recognizes the (+)RNA (70). The defined plus-strand initiation promoters and subgenomic promoters consist of short unstructured sequences or simple hairpins with a single-stranded tails. Interestingly, none of the characterized plus-strand initiation promoters are similar in sequence/structure with the minus-strand initiation promoters, suggesting the viral replicase/RdRp can recognize more than one sequences/structures (70). The recognition of more than one promoter by the viral replicase/RdRp might be an important regulatory step to allow asymmetrical RNA synthesis, which results in more abundant plus-strand viral RNA progeny than minus-strand intermediate. In addition, regulatory RNA elements, such as replication enhancers, could also contribute to asymmetrical strand synthesis (78–80).

7.2 Helicase Activity of Auxiliary Viral Replication Proteins

Most cellular processes involving RNA use helicases, which are capable of modifying RNA structures and protein–RNA interactions by unwinding RNAs in nucleotide triphosphate (NTP)-dependent manner. These helicases contain signature motifs, which are involved in NTP hydrolysis and RNA binding. Based on the presence of the helicase motifs, a large number of viral-coded helicases have been predicted. These helicases are often auxiliary replication proteins, such as TMV 126 K and BMV 1a protein. In vitro assay with purified recombinant 126 K, which included only the helicase domain (HEL) (Fig. 1) showed that HEL could unwind partially double-stranded RNA in the presence of NTP in vitro (15). Similar to other helicases, the recombinant 126 K protein and the BMV 1a protein has NTPase function in vitro (15, 41). Interestingly, Goregaoker and Culver found antagonistic relationship between ATP binding and RNA binding, suggesting that ATP binding/hydrolysis might induce structural changes allowing the helicase domain of 126 K to bind/release RNA (15). The helicase function of 126 K was predicted to play a role in unwinding of dsRNA replication intermediates or destabilizing strong secondary structures during TMV replication (15).

7.3 RNA Chaperone Activity of Auxiliary Viral Replication Proteins

Unlike larger RNA viruses, small RNA viruses of plants do not code for RNA helicase-like proteins (20). To regulate the structure of viral RNA during replication, these RNA viruses might recruit host helicases. Alternatively, or in addition, small RNA viruses might produce viral auxiliary replication proteins with RNA chaperone activity that could alter RNA structure without the use of NTP hydrolysis. Indeed, the tombusvirus p33 replication co-factor has been shown to have RNA chaperone activity by facilitating initiation of RNA synthesis by the viral RdRp in vitro (Stork and Nagy, submitted). It was proposed that binding of the p33 cofactor to the AU-rich portion of a double-stranded RNA could lead to more efficient loading of the viral RdRp onto the template (Stork and Nagy, submitted).

7.4 Capping Activity of Auxiliary Viral Replication Proteins

Similar to cellular mRNAs, many plus-stranded viral RNAs are capped at the 5′ end, which facilitates their translation. However, the plant plus-stranded viral RNAs replicate in the cytoplasm, whereas capping of the host mRNAs takes place in the nucleus. Therefore, many plant plus-stranded viral RNAs code for an auxiliary protein with capping function. For example, the recombinant 1a helicase-like protein of BMV (81, 82) and the 126 K of TMV (83) have capping-related activities, such as methyltransferase and guanylyl transferase activity in vitro, that introduce a cap-structure to the 5′ end of the viral (+)RNA during replication.

8 Posttranslational Modification

Most proteins produced in host cells are modified posttranslationally, via phosphorylation, acetylation, ubiquitination, etc., to regulate their functions, alter intracellular distribution, or affect the stability of the given protein. Viral replication proteins are also likely modified in infected cells as shown for TYMV replication proteins, which are both phosphorylated and ubiquitinated (84). Similarly, CMV 1a helicase-like protein (85–87) and the tombusvirus p33 and the TCV p28 replication cofactors have been found to be phosphorylated in infected plants (86).

The posttranslational modifications have been identified by using metabolic labeling of infected cells, followed by immunopurification of viral proteins (84–86). Specific antibodies recognizing phosphorylated serine or threonine residues (86) or mass spectrometry can also be used if purified replicase proteins are available. The relevance of posttranslational modification can be studied by mutational modification of the modified residues, which can inhibit posttranslational modification. For example, replacing serines or threonines with alanines would inhibit phosphorylation at selected positions. Another useful modification is replacement of serine/threonine with aspartic acid. The aspartic acid mutation introduces negative charge, similar to phosphorylation, thus serving as phosphorylation mimicking mutation (87, 88). However, none of these mutations can "mimic" the reversible nature of phosphorylation.

Posttranslational modification of viral replication proteins might affect protein–protein interactions. For example, phosphorylation of CMV 1a helicase-like protein was found to inhibit interaction with the 2a RdRp protein in vitro (85). The inhibition of 1a–2a interaction could prevent the assembly of new replicase complexes at late time points, which could be beneficial for "shutting down" replication and for facilitating the use of the viral RNA for additional functions, such as cell-to-cell spread and encapsidation (85).

Another example is the role of posttranslational modification in replicase protein–viral RNA interaction. For example, serine and threonine residues located in the vicinity of the RPR RNA-binding domain in TBSV p33 replication cofactor could be phosphorylated in vitro (86), and phosphorylation was shown to reduce RNA binding by p33 in vitro (87). Phosphorylation of the tombusvirus p33 cofactor also inhibited the assembly of the viral replicase complex based on the use of phosphorylation mimicking mutations (87). The authors proposed that if phosphorylation takes place reversibly, then the same replicase complex would be able to release the viral RNA progeny, followed by new rounds of RNA synthesis and release.

9 Future Directions

In spite of the recent introduction of a large number of methods into plant virus replication studies, our knowledge in many areas is still far from complete. For example, proteomics-based analysis of the viral replicase complexes is expected to lead to further identification of novel host proteins recruited into RNA virus replication. Then, combined use of genetics, biochemistry, and cell biology could help in dissecting the detailed functions of viral and subverted host proteins. In addition, determination of three-dimensional structures of plant viral RdRp and the auxiliary replication proteins with bound RNAs as well as low and high-resolution imaging of viral replicase complexes should help unravel the mechanism and the regulation of RNA replication. Proteomics approaches should also accelerate future studies on various posttranslational modifications of viral replication proteins that could affect or regulate their functions during the replication process. These advances will lead to better understanding of virus replication and host–virus interactions, which are key aspects of viral pathogenesis.

Acknowledgment The authors thank Drs. H. Jaag, Z. Li, and R. Wang for discussion. The authors apologize to those colleagues whose works on replication of (+)RNA viruses were not mentioned in this review due to page restrictions. This work was supported by NIH-NIAID.

References

1. Ahlquist, P. (2002) *Science* **296,** 1270–3.
2. Ahlquist, P., Noueiry, A. O., Lee, W. M., Kushner, D. B., and Dye, B. T. (2003) *J Virol* **77,** 8181–6.
3. Buck, K. W. (1996) *Adv Virus Res* **47,** 159–251.
4. Noueiry, A. O., and Ahlquist, P. (2003) *Annu Rev Phytopathol* **41,** 77–98.
5. Nagy, P. D., and Pogany, J. (2006) *Virology* **344,** 211–20.
6. White, K. A., and Nagy, P. D. (2004) *Prog Nucleic Acid Res Mol Biol* **78,** 187–226.
7. Dreher, T. W., and Miller, W. A. (2006) *Virology* **344,** 185–97.
8. Miller, W. A., and White, K. A. (2006) *Annu Rev Phytopathol* **44,** 447–467.
9. Wu, B., and White, K. A. (1999) *J Virol* **73,** 8982–8.
10. Scholthof, K. B., Scholthof, H. B., and Jackson, A. O. (1995) *Virology* **208,** 365–9.
11. Buck, K. W. (1999) *Philos Trans R Soc Lond B Biol Sci* **354,** 613–27.
12. Osman, T. A., and Buck, K. W. (1996) *J Virol* **70,** 6227–34.
13. Young, N., Forney, J., and Zaitlin, M. (1987) *J Cell Sci Suppl* **7,** 277–85.
14. Osman, T. A., and Buck, K. W. (2003) *J Virol* **77,** 8669–75.
15. Goregaoker, S. P., and Culver, J. N. (2003) *J Virol* **77,** 3549–56.
16. Noueiry, A. O., Chen, J., and Ahlquist, P. (2000) *Proc Natl Acad Sci U S A* **97,** 12985–90.
17. Schwartz, M., Chen, J., Janda, M., Sullivan, M., den Boon, J., and Ahlquist, P. (2002) *Mol Cell* **9,** 505–14.
18. van Dijk, A. A., Makeyev, E. V., and Bamford, D. H. (2004) *J Gen Virol* **85,** 1077–93.
19. O'Reilly, E. K., and Kao, C. C. (1998) *Virology* **252,** 287–303.
20. Koonin, E. V., and Dolja, V. V. (1993) *Crit Rev Biochem Mol Biol* **28,** 375–430.
21. White, K. A. (2002) *Virology* **304,** 147–54.
22. Nagy, P. D., Zhang, C., and Simon, A. E. (1998) *Embo J* **17,** 2392–403.
23. Nagy, P. D., Dzianott, A., Ahlquist, P., and Bujarski, J. J. (1995) *J Virol* **69,** 2547–56.
24. Roossinck, M. J. (2003) *Curr Opin Microbiol* **6,** 406–9.
25. Nagy, P. D., and Simon, A. E. (1997) *Virology* **235,** 1–9.
26. Cheng, C. P., and Nagy, P. D. (2003) *J Virol* **77,** 12033–47.
27. Kim, M. J., and Kao, C. (2001) *Proc Natl Acad Sci U S A* **98,** 4972–7.
28. Wierzchoslawski, R., and Bujarski, J. J. (2006) *J Virol* **80,** 6182–7.
29. Serva, S., and Nagy, P. D. (2006) *J Virol* **80,** 2162–9.
30. Osman, T. A., and Buck, K. W. (1997) *J Virol* **71,** 6075–82.
31. Yamaji, Y., Kobayashi, T., Hamada, K., Sakurai, K., Yoshii, A., Suzuki, M., Namba, S., and Hibi, T. (2006) *Virology* **347,** 100–8.
32. Quadt, R., Kao, C. C., Browning, K. S., Hershberger, R. P., and Ahlquist, P. (1993) *Proc Natl Acad Sci U S A* **90,** 1498–502.
33. Watanabe, T., Honda, A., Iwata, A., Ueda, S., Hibi, T., and Ishihama, A. (1999) *J Virol* **73,** 2633–40.
34. Ahlquist, P., Schwartz, M., Chen, J., Kushner, D., Hao, L., and Dye, B. T. (2005) *Vaccine* **23,** 1784–7.
35. Rajendran, K. S., and Nagy, P. D. (2003) *J Virol* **77,** 9244–58.
36. Pogany, J., White, K. A., and Nagy, P. D. (2005) *J Virol* **79,** 4859–69.
37. Monkewich, S., Lin, H. X., Fabian, M. R., Xu, W., Na, H., Ray, D., Chernysheva, O. A., Nagy, P. D., and White, K. A. (2005) *J Virol* **79,** 4848–58.
38. Panaviene, Z., Baker, J. M., and Nagy, P. D. (2003) *Virology* **308,** 191–205.
39. Panaviene, Z., and Nagy, P. D. (2003) *Virology* **317,** 359–72.

40. Sullivan, M. L., and Ahlquist, P. (1999) *J Virol* **73,** 2622–32.
41. Wang, X., Lee, W. M., Watanabe, T., Schwartz, M., Janda, M., and Ahlquist, P. (2005) *J Virol* **79,** 13747–58.
42. Baumstark, T., and Ahlquist, P. (2001) *Rna* **7,** 1652–70.
43. Rajendran, K. S., and Nagy, P. D. (2004) *Virology* **326,** 250–61.
44. Rajendran, K. S., and Nagy, P. D. (2006) *Virology* **345,** 270–9.
45. Panavas, T., Hawkins, C. M., Panaviene, Z., and Nagy, P. D. (2005) *Virology* **338,** 81–95.
46. Goregaoker, S. P., Lewandowski, D. J., and Culver, J. N. (2001) *Virology* **282,** 320–8.
47. O'Reilly, E. K., Wang, Z., French, R., and Kao, C. C. (1998) *J Virol* **72,** 7160–9.
48. Kao, C. C., and Ahlquist, P. (1992) *J Virol* **66,** 7293–302.
49. O'Reilly, E. K., Paul, J. D., and Kao, C. C. (1997) *J Virol* **71,** 7526–32.
50. Chen, J., and Ahlquist, P. (2000) *J Virol* **74,** 4310–8.
51. Burgyan, J., Rubino, L., and Russo, M. (1996) *J Gen Virol* **77,** 1967–74.
52. Navarro, B., Rubino, L., and Russo, M. (2004) *J Virol* **78,** 4744–52.
53. Rubino, L., and Russo, M. (1998) *Virology* **252,** 431–7.
54. McCartney, A. W., Greenwood, J. S., Fabian, M. R., White, K. A., and Mullen, R. T. (2005) *Plant Cell* **17,** 3513–31.
55. Oster, S. K., Wu, B., and White, K. A. (1998) *J Virol* **72,** 5845–51.
56. dos Reis Figueira, A., Golem, S., Goregaoker, S. P., and Culver, J. N. (2002) *Virology* **301,** 81–9.
57. Mas, P., and Beachy, R. N. (1999) *J Cell Biol* **147,** 945–58.
58. Hagiwara, Y., Komoda, K., Yamanaka, T., Tamai, A., Meshi, T., Funada, R., Tsuchiya, T., Naito, S., and Ishikawa, M. (2003) *Embo J* **22,** 344–53.
59. Yamanaka, T., Imai, T., Satoh, R., Kawashima, A., Takahashi, M., Tomita, K., Kubota, K., Meshi, T., Naito, S., and Ishikawa, M. (2002) *J Virol* **76,** 2491–7.
60. Restrepo-Hartwig, M., and Ahlquist, P. (1999) *J Virol* **73,** 10303–9.
61. Restrepo-Hartwig, M. A., and Ahlquist, P. (1996) *J Virol* **70,** 8908–16.
62. Chen, J., Noueiry, A., and Ahlquist, P. (2001) *J Virol* **75,** 3207–19.
63. den Boon, J. A., Chen, J., and Ahlquist, P. (2001) *J Virol* **75,** 12370–81.
64. Miller, W. A., Bujarski, J. J., Dreher, T. W., and Hall, T. C. (1986) *J Mol Biol* **187,** 537–46.
65. Hayes, R. J., and Buck, K. W. (1990) *Cell* **63,** 363–8.
66. Mouches, C., Bove, C., and Bove, J. M. (1974) *Virology* **58,** 409–23.
67. Yoshinari, S., Nagy, P. D., Simon, A. E., and Dreher, T. W. (2000) *Rna* **6,** 698–707.
68. Quadt, R., and Jaspars, E. M. (1991) *FEBS Lett* **278,** 61–2.
69. Nagy, P. D., and Pogany, J. (2000) *Virology* **276,** 279–88.
70. Kao, C. C., Singh, P., and Ecker, D. J. (2001) *Virology* **287,** 251–60.
71. Quadt, R., Ishikawa, M., Janda, M., and Ahlquist, P. (1995) *Proc Natl Acad Sci U S A* **92,** 4892–6.
72. Panaviene, Z., Panavas, T., and Nagy, P. D. (2005) *J Virol* **79,** 10608–18.
73. Panaviene, Z., Panavas, T., Serva, S., and Nagy, P. D. (2004) *J Virol* **78,** 8254–63.
74. Hong, Y., and Hunt, A. G. (1996) *Virology* **226,** 146–51.
75. Li, Y. I., Cheng, Y. M., Huang, Y. L., Tsai, C. H., Hsu, Y. H., and Meng, M. (1998) *J Virol* **72,** 10093–9.
76. Rajendran, K. S., Pogany, J., and Nagy, P. D. (2002) *J Virol* **76,** 1707–17.
77. Deiman, B. A., Verlaan, P. W., and Pleij, C. W. (2000) *J Virol* **74,** 264–71.
78. Panavas, T., and Nagy, P. D. (2005) *J Virol* **79,** 9777–85.
79. Ray, D., and White, K. A. (2003) *J Virol* **77,** 245–57.
80. Nagy, P. D., Pogany, J., and Simon, A. E. (1999) *Embo J* **18,** 5653–65.
81. Ahola, T., and Ahlquist, P. (1999) *J Virol* **73,** 10061–9.
82. Kong, F., Sivakumaran, K., and Kao, C. (1999) *Virology* **259,** 200–10.
83. Merits, A., Kettunen, R., Makinen, K., Lampio, A., Auvinen, P., Kaariainen, L., and Ahola, T. (1999) *FEBS Lett* **455,** 45–8.
84. Hericourt, F., Blanc, S., Redeker, V., and Jupin, I. (2000) *Biochem J* **349,** 417–25.
85. Kim, S. H., Palukaitis, P., and Park, Y. I. (2002) *Embo J* **21,** 2292–300.
86. Shapka, N., Stork, J., and Nagy, P. D. (2005) *Virology* **343,** 65–78.
87. Stork, J., Panaviene, Z., and Nagy, P. D. (2005) *Virology* **343,** 79–92.
88. Waigmann, E., Chen, M. H., Bachmaier, R., Ghoshroy, S., and Citovsky, V. (2000) *Embo J* **19,** 4875–84.

Chapter 5
Role of Silencing Suppressor Proteins

József Burgyán

Abstract RNA silencing suppressors, developed by plant viruses, are potent arms in the arm race between plant and invading viruses. In higher plants, these proteins efficiently inhibit RNA silencing, which has evolved to defend plants against viral infection in addition to regulation of gene expression for growth and development. Virus-encoded RNA-silencing suppressors interfere with various steps of the different silencing pathways and the mechanisms of suppression are being progressively unraveled. Our better understanding of action of silencing suppressors at molecular level dramatically improved our basic knowledge about the intimate plant–virus interactions and also provided valuable tools to unravel the diversity, regulation, and evolution of RNA-silencing pathways.

Keywords RNA silencing; VIGS; Plant virus silencing suppressors; Mechanism of silencing suppression; siRNA; p19; p21; HC-Pro

1 Introduction

RNA-silencing suppressor proteins are effective arms of a counter-defensive strategy of plant viruses, representing a viral adaptation to host antiviral defense. RNA silencing is efficiently triggered by double-stranded RNA structures, and the activation of RNA silencing by viruses leads to sequence-specific degradation of the genome of inducer viral RNAs. Because the majority of known plant viruses have RNA genomes and replicate via dsRNA intermediates and the single-stranded viral genome forms a secondary-structure featured RNA, it is not surprising that plant viruses are strong inducers as well as targets of Virus Induced RNA Silencing (VIGS, Fig. 1). VIGS also operates against DNA viruses, and thereby, dsRNA may be formed by annealing of overlapping complementary transcripts or single-stranded viral RNAs, which may be converted to dsRNA by plant-encoded RNA-dependent RNA polymerase. In plants, virus-induced gene silencing prevents virus accumulation. Consistently, viruses replicating in plant cells have evolved various strategies to counteract this antiviral defense mechanism. The most important

From: *Methods in Molecular Biology, Vol. 451, Plant Virology Protocols:*
From Viral Sequence to Protein Function
Edited by G.D. Foster, I.E. Johansen, Y. Hong, and P.D. Nagy © Humana Press, Totowa, NJ

Virus induced RNA silencing

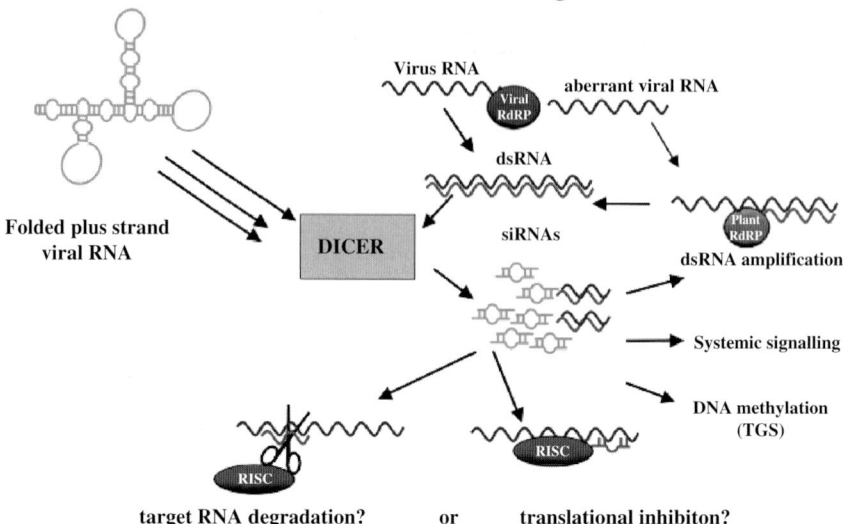

Fig. 1 Simplified model for plant RNA silencing. The virus induced RNA silencing is initiated by dicing of double-stranded (ds) or highly structured viral RNAs into 21–24 nt siRNAs. Ds viral RNAs can be produced by viral RdRp or plant RDR and then diced to small ds siRNAs. Viral siRNAs activate RISC complex for target cleavage or translational arrest and may also guide plant RdRp to amplify dsRNA, which are diced again to siRNAs. These siRNAs are also responsible for systemic signaling and transcriptional silencing (TGS)

counter-defensive strategy of plant viruses involves suppressor proteins of silencing, which are encoded in the genomes of both RNA and DNA viruses (1). Viral proteins having other functions in the virus life cycle demonstrated silencing suppressor activity (Table 1). These suppressor proteins were identified in almost all viral genomes, however, these proteins are structurally diverse without any common sequence motifs. This diversity of silencing suppressor proteins suggests that these proteins probably evolved independently in different virus groups. Accordingly, it was suggested that different suppressors inhibit the antiviral silencing mechanisms at different steps. Although, our knowledge is still limited about the molecular bases of silencing suppressor proteins, it seems that despite of the high diversity of these proteins, the sequestration of small interfering (si) RNA – the conserved elements of silencing machinery – is the most common strategy of silencing suppression (2, 3).

In higher plants, there are at least three RNA-silencing pathways, which are involved in antiviral defense, regulation of plant gene expression, and the condensation of chromatin into heterochromatin (4). Since these RNA-silencing pathways intersect, it is very likely that silencing suppressor proteins expressed by viruses and counteracting with antiviral defense could also impair other gene-silencing pathways involved plant gene regulations (1, 5).

Table 1 RNA silencing suppressor proteins encoded by plant viruses

Viral families	Viruses	Suppressors	Other functions	References
Positive-strand RNA viruses				
Carmovirus	Turnip Crinkle virus	P38	Coat protein	(57)
Cucumovirus	Cucumber mosaic virus; Tomato aspermy virus	2b	Host-specific movement	(40)
Closterovirus	Beet yellows virus	P21	Replication enhancer	(51)
	Citrus tristeza virus	P20	Replication enhancer	(58)
		P23	Nucleic acid binding	
		CP	Coat protein	
Comovirus	Cowpea mosaic virus	S protein	Small coat protein	(59)
Hordeivirus	Barley yellow mosaic virus	γb	Replication enhancer; movement; seed transmission; phatogenicity determinant	(60)
Pecluvirus	Peanut clump virus	P15	Movement	(61)
Polerovirus	Beet western yellows virus; Cucurbit aphid-born yellow virus	P0	Phatogenicity determinant	(52, 62)
Potexvirus	Potato virus X	P25	Movement	(63)
Potyvirus	Potato virus Y; Tobacco etch virus; Turnip yellow virus	Hc-Pro	Movement; polyprotein processing; aphid transmission; phatogenicity determinant	(35, 39, 40, 64)
Sobemovirus	Rice yellow mottle virus	P1	Movement; phatogenicity determinant	(35)
Tombusvirus	Tomato bushy stunt virus; Cymbidium ringspot virus; Carnation Italian ringspot virus	P19	Movement; phatogenicity determinant	(37)
Tobamovirus	Tobacco mosaic virus; Tomato mosaic virus	P30	Replication	(65)
Tymovirus	Turnip yellow mosaic virus	P69	Movement; phatogenicity determinant	(66)
Negative-strand RNA viruses				
Tospovirus	Tomato spotted with virus	NSs	Phatogenicity determinant	(67)
Tenuivirus	Rice hoja blanca virus	NS3	Unknown	(67)
Double-strand RNA viruses				
Phytoreovirus	*Rice dwarf virus*	Pns10	Unknown	(68)
DNA viruses				
Gemini virus	African cassava mosaic virus	AC4 AC2	putative synergistic genes	(55)
	Tomato yellow leaf curl virus	C2	Transcriptional activator protein (TrAP)	(35)

2   RNA Silencing

RNA silencing is a eukaryotic gene regulatory system that inhibits gene expression through RNA-mediated sequence-specific interactions. RNA silencing is conserved across kingdoms and is manifested as quelling in fungi, RNA interference (RNAi) in animals, and cosuppression or posttranscriptional gene silencing (PTGS) in plants. The unifying feature of RNA silencing is the presence of 21–24-nucleotide (nt) siR-NAs (6–8). In addition, biochemical and genetic analyses have shown that the core mechanisms of RNA silencing are shared among different eukaryotes (4, 8–12). RNA silencing is triggered by ds or self-complementary foldback RNA that are processed into 21–24-nt siRNA or microRNA (miRNA) duplexes by the RNase III-type DICER enzymes (13, 14). These siRNAs guide the sequence-specific inactivation of target mRNAs by the RNA-induced silencing complex (RISC) (15). As a general rule, RISC mediates cleavage of target mRNA when there is perfect or near perfect base pairing between mRNA and the short guide RNA and translation repression when there is partial complementarity (16–19). SiRNAs can also guide another effector complex, namely the RNA-induced initiation of transcriptional gene silencing (RITS) complex, to direct the chromatin modification of homologous DNA sequences (20).

Regardless of the origin, the occurrence of dsRNA in the cytoplasm of plant cells induces PTGS. It has been demonstrated previously that an RNA-dependent RNA polymerase (RDR6) is also involved in PTGS (21, 22), presumably by converting target ssRNA into dsRNA, which is then processed by DICER to generate 21-nt siRNAs. These siRNAs program RISC for target cleavage and they are also involved in the spread of PTGS either in short range or long range (23, 24).

miRNA, the other class of small regulatory RNAs, are endogenous RNAs that regulate gene expression in plants and animals. These approximately 21-nt RNAs are processed from stem-loop regions of long primary transcripts by a Dicer-like 1 (DCL1) enzyme. Then miRNAs are loaded into RISC, which generally cleaves the complementary mRNAs modulating the expression of these mRNAs (which control cell differentiation, development, and probably many other cellular functions in a sequence-specific manner) (25). However, the biogenesis of plant miRNAs consists of an additional step, they are methylated on the ribose of the last nucleotide by the miRNA methyltransferase HEN1 (26).

A third class of small RNAs, the so-called "*trans*-acting siRNAs," are processed by DCL4 from long dsRNA produced by a plant RNA-dependent RNA polymerase (RDR6); they mainly target the expression of other genes rather than their own expression (27–31).

2.1   *Virus-Induced Gene Silencing*

Plant viruses are known as strong inducers as well as targets of virus-induced gene silencing (Figure 1). During virus infection, the accumulation of 21-nt ds siRNAs – the hallmark of gene silencing – is observed in virus-infected tissues (7, 32), indicating the activation of VIGS. High siRNA level correlates with the activity of VIGS, resulting in lower viral titer and, in some cases, immunity or recovery in upper noninoculated leaves (32, 33). Thus, VIGS acts as an RNA-mediated defense response to protect plants against viral infection (1, 34). Therefore, to infect plants, viruses have to evade or suppress RNA silencing. One of the most compelling evidence for the antiviral function of VIGS is that many of plant viruses have evolved proteins that suppress various steps of the silencing machinery (1, 5, 35–37).

2.2 Virus-Encoded Silencing Suppressors and Strategies of Suppression Virus Induced Gene Silencing

Theoretically, viruses can counteract with RNA-silencing mediated defense at least at three steps: (1) preventing the generation of siRNAs, (2) inhibiting the incorporation of siRNAs into effector complexes, and (3) interfering with one of the effector complexes. These suppression strategies may have different effects on other RNA-silencing pathways that manifest in their different influence on the accumulation of various short RNAs.

So far more than a dozen silencing suppressors (Table 1) have been identified from different types of viruses, including positive strand RNA, negative strand RNA, and ssDNA viruses. No sequence homology has been detected between distinct silencing suppressors. Consistently, it was suggested that different suppressors inhibit silencing mechanisms at different steps.

Although our knowledge is still limited about the mechanism of silencing suppression of different suppressors, a few silencing suppressor proteins have been studied in detail. Early reports showed that the potyvirus-encoded helper component proteinase (HC-Pro) enhances the replication of unrelated viruses (38). Indeed, the potyvirus-encoded HC-Pro was the first viral protein identified as a suppressor of transgene and virus-induced RNA silencing. Analyses of data from variable experimental systems led to the development of several different models for the mechanism of HC-Pro silencing suppression. HC-Pro was proposed to reverse established RNA silencing by acting on RISC (35, 39, 40) involving rgs-CaM, a calmodulin related protein, which is cellular negative regulator of silencing (41). Other observation suggested that HC-Pro acts downstream of an RNA-dependent RNA polymerase and impairs the DICER activity (42, 43). Recent comparative study of different silencing suppressor proteins predicted that RISC activation was suppressed through interaction between HC-Pro and a protein or complex required for siRNA duplex unwinding (44).

The tombusviral 19 kDa protein (p19) is one of the best-studied silencing suppressors so far. Indeed, recent advances in understanding the molecular mechanism underlying p19 suppressing activity revealed that p19 specifically binds 21-nt ds siRNAs in vitro and in vivo, preventing siRNA incorporation into effector complexes such as RISC (37, 45). This model was well supported by the three-dimensional X-ray crystal structure of a p19–siRNA complex, which revealed that a p19 dimer acts as a ds RNA caliper binding the ends of the siRNA duplex while measuring its length (46, 47). Moreover, Lakatos et al. (2006) showed that p19 can only prevent the assembly of RISC by sequestering of ds siRNA or miRNA duplexes; however, if the RISC complex once assembled p19 has no more effect on it. The reason for this is that p19 is not able to bind ss siRNA or miRNA which guide the assembled RISC for target cleavage.

The third well-studied suppressor is p21 of *Beet yellows virus* (BYV). The molecular base of the mechanism of p21 silencing suppression was found to be very similar to p19. It was shown that p21 inhibits silencing pathways by binding siRNAs or ds miRNA intermediates (1, 44).

Our very recent in vivo and in vitro studies further clarified the mode of action these three well-studied silencing suppressors (HC-Pro, p19, and p21); it was demonstrated undoubtedly that HC-Pro, similar to p19 and p21, impairs RNA silencing by si- and miRNA sequestration, which results in the inhibition of siRNA guided RISC assembly (2).

Since these suppressors bind si and miRNA duplexes, it might be expected that they interfere with the 3′ methylation of si- and miRNAs. Indeed, recent results support this hypothesis. Transgenic expression of the HC-Pro results in a marked decrease in the 3′ terminal modification of viral siRNAs but does not significantly affect the modification of endogenous miRNAs (48). Surprisingly, the effects of p19 and HC-Pro on the 3′ terminal methylation were clearly different and may reflect the fine differences in how and perhaps where these suppressors bind and sequester small RNA duplexes (Lozsa et al., unpublished data).

2.2.1 Is siRNA Duplex Sequestration a Widely Used Strategy to Suppress RNA Silencing?

Inhibition of antiviral RNA silencing is a critical prerequisite for the successful systemic invasion for most plant viruses. Silencing inhibition through siRNA sequestration seems advantageous, as production of siRNAs is a conserved element of the antiviral silencing in any host. p19, p21, and HC-Pro are structurally and evolutionarily unrelated proteins, each representing a small protein family specific to a respective viral taxon (46, 47, 49–51). Although only a limited number of silencing suppressors have been proved unequivocally to bind small RNA duplexes (2, 43–45), there are several other silencing suppressor proteins that are suggested to be siRNA-binding proteins, such as p14 of *Pothos latent virus*, 2b protein of *Cucumber mosaic virus*, p38 of *Turnip crinkle virus*, p15 *Peanut clump virus*, γB3 of *Barley stripe mosaic virus* (3). Thus, siRNA duplex-binding mechanism represents a recurring mechanism that has evolved independently in several virus families (e.g., *Tombusviridae, Potyviridae, Bromoviridae, Closteroviridae*) within the positive-strand RNA viruses.

Although siRNA sequestration is a widely used strategy to suppress RNA silencing, there are also other known or predicted mechanisms of silencing suppression. P0 RNA silencing suppressor protein of poleroviruses *Beet western yellows polerovirus* (BWYV) acts as an F-box protein that targets an essential component of silencing machinery. It was suggested that P0 interacts with its substrate protein to assign it for ubiquitination and finally for degradation (52).

The coat protein of Turnip crinkle virus (TCV) was suggested to inhibit DICER activities (53), this was confirmed by Merai et al. (2006) who showed that CP of TCV is dsRNA binding protein, which likely interacts with either short or long viral-derived dsRNA. A more recent report demonstrated that p38 specifically inhibits DCL4 activity (54).

The A-AC4 protein of a Gemini virus was shown to bind to mature miRNAs and it was predicted that A-AC4 recruits the matured miRNAs by interacting with one or more cellular factors that are associated with RISC-loading complex or RISC (55). These alternative ways of silencing suppression extend dimension to how viruses have evolved with distinct mechanisms to modify the cell system to host virus replication in plants.

2.2.2 Link between Silencing Suppressors and Viral Symptoms

Although many viral suppressors (Table 1) were previously identified as pathogenic determinants that are largely responsible for virus-induced symptoms, the molecular basis for virus-induced disease in plants has been a long-standing mystery. The recent development in our understanding of the mechanism of silencing suppression provides a better insight into the molecular mechanism of virus-induced symptoms. It is well established that the antiviral and endogenous silencing pathways share common elements (e.g., endogenous small regulatory RNAs such as si-, tasi-, and ds miRNA intermediates) and silencing suppressors often interact with these common elements. Thus, it is reasonable to suggest that many of the virus-induced symptoms are the consequence of the interaction of silencing suppressors and endogenous RNA silencing mediated developmental pathways that share components with the antiviral RNA silencing. Indeed, TuMV HC-Pro was shown to inhibit both endogenous mi- and siRNA mediated gene regulations, which resulted in the overexpression of miRNA targeted genes in HC-Pro expressing transgenic plants leading to phenotype resembling to TuMV infected plants (43, 56).

3 Conclusion and Further Directions

During the last few years, dramatic progress has been made in the understanding of biological roles and pathways involved in RNA silencing. A large number of new silencing suppressor proteins were described, and the discovery of the molecular bases of silencing suppression inspired new concepts about the existence of cellular negative regulators of RNA silencing, such as silencing suppressors, and the discovery that many of the symptoms caused by viruses are the consequence of different interactions between the silencing suppressors and the regulatory pathways of endogenous RNA silencing.

The function of RNA silencing is to protect plants against viral invasion, but surprisingly it seems that viruses may exploit this defense response to keep the virus titer at a tolerable level in plant tissues preventing the detrimental effects of virus over-accumulation. Thus, the action of RNA silencing ensures both the virus and the plants survive. Silencing suppressors is likely involved in this fine-tuning of plant–virus interplay for joint survival; however, our knowledge about the orchestration of this intim*r*te plant–virus interplay is very limited and remains for future exploration.

Silencing suppressors (p19 and HC-Pro), which target the most conserved elements of silencing pathways, could also be used as a powerful tool to dissect the RNA-silencing pathways not only in plants but also in animals.

References

1. Voinnet, O., Induction and suppression of RNA silencing: insights from viral infections. Nat Rev Genet, 2005. **6**(3): p. 206–20.
2. Lakatos L., C.T., Pantaleo V., Chapman E.J., Carrington J.C., Liu Y.P., Dolja V.V., Fernández Calvino L., López-Moya J.J., Burgyán J., Comparative study of viral encoded silencing suppressors: small RNA binding is a common strategy to suppress RNA silencing. EMBO J, 2006. **25**: 2768–2780.
3. Mérai, Z., Kerényi, Z., Kertész, S., Magna, M., Lakatos, L., and Silhavy, D., Double-stranded RNA binding could be a general plant RNA viral strategy to suppress RNA silencing J. Virol. 2006. **80**(12): 5747–56.
4. Baulcombe, D., RNA silencing in plants. Nature, 2004. **431**(7006): p. 356–63.
5. Silhavy, D. and J. Burgyan, Effects and side-effects of viral RNA silencing suppressors on short RNAs. Trends Plant Sci, 2004. **9**(2): p. 76–83.
6. Hamilton, A., et al., Two classes of short interfering RNA in RNA silencing. Embo J, 2002. **21**(17): p. 4671–4679.
7. Hamilton, A.J. and D.C. Baulcombe, A species of small antisense RNA in posttranscriptional gene silencing in plants. Science, 1999. **286**(5441): p. 950–2.
8. Plasterk, R.H., RNA silencing: the genome's immune system. Science, 2002. **296**(5571): p. 1263–5.
9. Hannon, G.J. and D.S. Conklin, RNA interference by short hairpin RNAs expressed in vertebrate cells. Methods Mol Biol. Vol. 257. 2004. 255–66.
10. Meister, G. and T. Tuschl, Mechanisms of gene silencing by double-stranded RNA. Nature, 2004. **431**(7006): p. 343–9.
11. Voinnet, O., RNA silencing: small RNAs as ubiquitous regulators of gene expression. Curr Opin Plant Biol, 2002. **5**(5): p. 444.
12. Zamore, P.D., Ancient pathways programmed by small RNAs. Science, 2002. **296**(5571): p. 1265–9.
13. Bernstein, E., et al., Role for a bidentate ribonuclease in the initiation step of RNA interference. Nature, 2001. **409**(6818): p. 363–6.
14. Nykanen, A., B. Haley, and P.D. Zamore, ATP requirements and small interfering RNA structure in the RNA interference pathway. Cell, 2001. **107**(3): p. 309–21.
15. Hammond, S.M., et al., An RNA-directed nuclease mediates post-transcriptional gene silencing in Drosophila cells. Nature, 2000. **404**(6775): p. 293–6.
16. Doench, J.G., C.P. Petersen, and P.A. Sharp, siRNAs can function as miRNAs. Genes Dev, 2003. **17**(4): p. 438–42.

17. Hutvagner, G. and P.D. Zamore, RNAi: nature abhors a double-strand. Curr Opin Genet Dev, 2002. **12**(2): p. 225–32.
18. Chen, X., A microRNA as a translational repressor of APETALA2 in Arabidopsis flower development. Science, 2003. **11**: p. 11.
19. Aukerman, M.J. and H. Sakai, Regulation of flowering time and floral organ identity by a microRNA and its APETALA2-like target genes. Plant Cell, 2003. **10**: p. 10.
20. Verdel, A., et al., RNAi-mediated targeting of heterochromatin by the RITS complex. Science, 2004. **303**(5658): p. 672–6.
21. Dalmay, T., et al., An RNA-dependent RNA polymerase gene in Arabidopsis is required for posttranscriptional gene silencing mediated by a transgene but not by a virus. Cell, 2000. **101**(5): p. 543–53.
22. Mourrain, P., et al., Arabidopsis SGS2 and SGS3 genes are required for posttranscriptional gene silencing and natural virus resistance. Cell, 2000. **101**(5): p. 533–42.
23. Himber, C., et al., Transitivity-dependent and -independent cell-to-cell movement of RNA silencing. Embo J, 2003. **22**(17): p. 4523–33.
24. Voinnet, O., Non-cell autonomous RNA silencing. FEBS Lett, 2005. **579**(26): p. 5858–71.
25. Bartel, D.P., MicroRNAs: genomics, biogenesis, mechanism, and function. Cell, 2004. **116**(2): p. 281–97.
26. Chen, X., MicroRNA biogenesis and function in plants. FEBS Lett, 2005. **579**(26): p. 5923–31.
27. Vazquez, F., et al., Endogenous trans-Acting siRNAs Regulate the Accumulation of Arabidopsis mRNAs. Mol Cell, 2004. **16**(1): p. 69–79.
28. Peragine, A., et al., SGS3 and SGS2/SDE1/RDR6 are required for juvenile development and the production of trans-acting siRNAs in Arabidopsis. Genes Dev, 2004. **18**(19): p. 2368–79.
29. Allen, E., et al., microRNA-directed phasing during trans-acting siRNA biogenesis in plants. Cell, 2005. **121**(2): p. 207–21.
30. Gasciolli, V., et al., Partially redundant functions of Arabidopsis DICER-like enzymes and a role for DCL4 in producing trans-acting siRNAs. Curr Biol, 2005. **15**(16): p. 1494–500.
31. Dunoyer, P., C. Himber, and O. Voinnet, DICER-LIKE 4 is required for RNA interference and produces the 21-nucleotide small interfering RNA component of the plant cell-to-cell silencing signal. Nat Genet, 2005. **37**(12): p. 1356–60.
32. Szittya, G., et al., Short defective interfering RNAs of tombusviruses are not targeted but trigger post-transcriptional gene silencing against their helper virus. Plant Cell, 2002. **14**(2): p. 359–72.
33. Ratcliff, F., B.D. Harrison, and D.C. Baulcombe, A Similarity Between Viral Defense and Gene Silencing in Plants. Science, 1997. **276**(5318): p. 1558–60.
34. Moissiard, G. and O. Voinnet, Viral suppression of RNA silencing in plants. Molecular plant pathology, 2004. **5**(1): p. 71–82.
35. Voinnet, O., Y.M. Pinto, and D.C. Baulcombe, Suppression of gene silencing: a general strategy used by diverse DNA and RNA viruses of plants. Proc Natl Acad Sci U S A, 1999. **96**(24): p. 14147–52.
36. Li, W.X. and S.W. Ding, Viral suppressors of RNA silencing. Curr Opin Biotechnol, 2001. **12**(2): p. 150–4.
37. Silhavy, D., et al., A viral protein suppresses RNA silencing and binds silencing-generated, 21- to 25-nucleotide double-stranded RNAs. Embo J, 2002. **21**(12): p. 3070–80.
38. Pruss, G., et al., Plant viral synergism: the potyviral genome encodes a broad-range pathogenicity enhancer that transactivates replication of heterologous viruses. Plant Cell, 1997. **9**(6): p. 859–68.
39. Anandalakshmi, R., et al., A viral suppressor of gene silencing in plants. Proc Natl Acad Sci U S A, 1998. **95**(22): p. 13079–84.
40. Brigneti, G., et al., Viral pathogenicity determinants are suppressors of transgene silencing in *Nicotiana benthamiana*. Embo J, 1998. **17**(22): p. 6739–46.

41. Anandalakshmi, R., et al., A calmodulin-related protein that suppresses posttranscriptional gene silencing in plants. Science, 2000. **290**(5489): p. 142–4.
42. Mallory, A.C., et al., HC-Pro suppression of transgene silencing eliminates the small RNAs but not transgene methylation or the mobile signal. Plant Cell, 2001. **13**(3): p. 571–83.
43. Dunoyer, P., et al., Probing the microRNA and small interfering RNA pathways with virus-encoded suppressors of RNA silencing. Plant Cell, 2004. **16**(5): p. 1235–50.
44. Chapman, E.J., et al., Viral RNA silencing suppressors inhibit the microRNA pathway at an intermediate step. Genes Dev, 2004. **18**(10): p. 1179–86.
45. Lakatos, L., et al., Molecular mechanism of RNA silencing suppression mediated by p19 protein of tombusviruses. Embo J, 2004. **23**(4): p. 876–84. Epub 2004 Feb 19.
46. Vargason, J., et al., Size selective recognition of siRNA by an RNA silencing suppressor. Cell, 2003. **115**(7): p. 799–811.
47. Ye, K., L. Malinina, and D.J. Patel, Recognition of small interfering RNA by a viral suppressor of RNA silencing. Nature, 2003. **3**: p. 3.
48. Ebhardt, H.A., et al., Extensive 3′ modification of plant small RNAs is modulated by helper component-proteinase expression. Proc Natl Acad Sci U S A, 2005. **102**(38): p. 13398–403.
49. Dolja, V.V., J.F. Kreuze, and J.P. Valkonen, Comparative and functional genomics of closteroviruses. Virus Res, 2006. **117**(1): p. 38–51.
50. Koonin, E.V., et al., Evidence for common ancestry of a chestnut blight hypovirulence-associated double-stranded RNA and a group of positive-strand RNA plant viruses. Proc Natl Acad Sci U S A, 1991. **88**(23): p. 10647–51.
51. Reed, J.C., et al., Suppressor of RNA silencing encoded by Beet yellows virus. Virology, 2003. **306**(2): p. 203–9.
52. Pazhouhandeh, M., et al., F-box-like domain in the polerovirus protein P0 is required for silencing suppressor function. Proc Natl Acad Sci U S A, 2006. **103**(6): p. 1994–9.
53. Qu, F., T. Ren, and T.J. Morris, The coat protein of turnip crinkle virus suppresses posttranscriptional gene silencing at an early initiation step. J Virol, 2003. **77**(1): p. 511–22.
54. Deleris et al., A molecular framework for induction and suppression of antiviral RNA silencing in plants. Science, in press
55. Chellappan, P., R. Vanitharani, and C.M. Fauquet, MicroRNA-binding viral protein interferes with Arabidopsis development. Proc Natl Acad Sci U S A, 2005. **102**(29): p. 10381–6.
56. Kasschau, K.D., et al., P1/HC-Pro, a viral suppressor of RNA silencing, interferes with Arabidopsis development and miRNA unction. Dev Cell, 2003. **4**(2): p. 205–17.
57. Thomas, C.L., et al., Turnip crinkle virus coat protein mediates suppression of RNA silencing in Nicotiana benthamiana. Virology, 2003. **306**(1): p. 33–41.
58. Lu, R., et al., Three distinct suppressors of RNA silencing encoded by a 20-kb viral RNA genome. Proc Natl Acad Sci U S A, 2004. **101**(44): p. 15742–7.
59. Liu, L., et al., Cowpea mosaic virus RNA-1 acts as an amplicon whose effects can be counteracted by a RNA-2-encoded suppressor of silencing. Virology, 2004. **323**(1): p. 37–48.
60. Yelina, N.E., et al., Long-distance movement, virulence, and RNA silencing suppression controlled by a single protein in hordei- and potyviruses: complementary functions between virus families. J Virol, 2002. **76**(24): p. 12981–91.
61. Dunoyer, P., et al., Identification, subcellular localization and some properties of a cysteine-rich suppressor of gene silencing encoded by peanut clump virus. Plant J, 2002. **29**(5): p. 555–67.
62. Pfeffer, S., et al., P0 of beet Western yellows virus is a suppressor of posttranscriptional gene silencing. J Virol, 2002. **76**(13): p. 6815–24.
63. Voinnet, O., C. Lederer, and D.C. Baulcombe, A viral movement protein prevents spread of the gene silencing signal in *Nicotiana benthamiana*. Cell, 2000. **103**(1): p. 157–67.
64. Kasschau, K.D. and J.C. Carrington, A counterdefensive strategy of plant viruses: suppression of posttranscriptional gene silencing. Cell, 1998. **95**(4): p. 461–70.
65. Kubota, K., et al., Tomato mosaic virus replication protein suppresses virus-targeted posttranscriptional gene silencing. J Virol, 2003. **77**(20): p. 11016–26.

66. Chen, J., et al., Viral virulence protein suppresses RNA silencing-mediated defense but upregulates the role of microrna in host gene expression. Plant Cell, 2004. **16**(5): p. 1302–13.
67. Bucher, E., et al., Negative-strand tospoviruses and tenuiviruses carry a gene for a suppressor of gene silencing at analogous genomic positions. J Virol, 2003. **77**(2): p. 1329–36.
68. Cao, X., et al., Identification of an RNA silencing suppressor from a plant double-stranded RNA virus. J Virol, 2005. **79**(20): p. 13018–27.

Chapter 6
Role of Vector-Transmission Proteins

Véronique Ziegler-Graff and Véronique Brault

Abstract Most phytoviruses rely on vectors for their spread and survival. Although a great variety of virus vectors have been described, there are relatively few different mechanisms mediating virus transmission by vectors: virions can either be internalized into vector cells where replication may or may not take place or they can simply be adsorbed on the vector's surface or cuticle. Virus transmission by vectors requires tight associations between viral proteins, generally capsid proteins, and vector compounds, usually referred to as receptors. This review will focus on the viral determinants involved in virus transmission. Only the best-known models for which molecular data are available are described.

1 Introduction

To survive in nature, plant viruses face two main problems: the need to spread from one plant to another and to penetrate the physical barrier that represents the plant cuticle and the cellulose-containing cell wall. Plant viruses have therefore evolved specific associations with various vectors which can carry out these essential functions of dissemination and introduction into the plant. The vast majority of these vectors are insects, such as aphids, leafhoppers, planthoppers, beetles, and whiteflies (1), although examples of transmission by mites, nematodes and fungi are also known (2). These associations rely on specific interactions between virus-encoded proteins and vector constituents. As early as 1918, scientists became aware of the relationship existing between aphid vectors and viruses (3). In the early 1970s, it was discovered that potyvirus transmission requires a supplementary factor in addition to purified virus. This factor is a nonstructural protein called the "helper component" (HC). Studies of viral determinants governing transmissibility have since benefited from molecular technology. The domains involved can usually be identified by comparative studies of the protein sequences of transmissible and nontransmissible forms. Mutating the genes in question in infectious viral cDNA clones can then be used to confirm function in transmission tests.

This chapter will focus on the best-studied models of viruses transmitted by vectors, and in particular, those for which molecular data are available. These viruses

From: *Methods in Molecular Biology, Vol. 451, Plant Virology Protocols:*
From Viral Sequence to Protein Function
Edited by G.D. Foster, I.E. Johansen, Y. Hong, and P.D. Nagy © Humana Press, Totowa, NJ

are mostly transmitted by insects. We will nevertheless present current knowledge for a few examples of nematode and fungi-transmitted viruses (4).

1.1   Two Main Types of Transmission

Historically, virus transmission has mostly been studied on aphid-transmitted viruses. The relationship established by such viruses with their vectors is of two major types depending upon whether they are transmitted in a "nonpersistent" or a "persistent" fashion (2), although a third intermediate group of viruses, which is transmitted in a "semipersistent" fashion, was added later. These terms are based on criteria which are relatively easy to evaluate: the time required by the vector to acquire the virus and the time the vector retains its inoculation capacity (persistence). The nonpersistent viruses have a half-life in their vectors of minutes, semipersistent viruses of several hours and persistent viruses of days and even weeks or months. An important step in understanding the fate of the virus in the vector was provided when the concept of the "stylet-borne" virus was introduced to describe the nonpersistent relationship, and the "circulative" virus for viruses that enter the vector's body. Circulative viruses can be recovered from the vector's hemolymph and also be transmitted after injection into the vector's body. Moreover, they are retained after molting. Among circulative viruses, a further distinction is made between "propagative" and "nonpropagative" viruses depending on whether they replicate in the vector. Multiplication can be shown by an increase of virus titer in vectors after they have been removed from the viral source.

2   Virus Transmitted by Insects

2.1   Noncirculative Viruses

Transmission of noncirculative viruses is based on retention of virions along the epicuticle of the vector's mouthparts or foregut, followed by their release and inoculation into a plant. This association has a very short life time. Transmission electron microscopy studies associated with immunogold labeling localized the main attachment site for potyviruses to the food canal in the rostrum (5). Subsequent studies correlated this association with transmissible viruses (6). Virus release from the tip of the stylets occurs during the salivation phase when aphids sample the content of epidermal cells (7, 8). The interaction between vector components and virus particles can be either direct (capsid strategy) or indirect, involving one or more virus-encoded proteins called HCs (helper strategy). The following section will give an overview of the best-studied models for both strategies, concentrating on aphid-transmitted viruses.

2.1.1 Capsid Strategy

2.1.1.1 Cucumoviruses

Cucumoviruses (*Cucumovirus, Bromoviridae*) have isometric particles separately encapsidating three genomic positive-sense RNAs. Experiments based on in vitro reassembled particles with the genomes and the capsid proteins of two strains of *Cucumber mosaic virus* (CMV), a highly and a poorly aphid-transmitted strain, demonstrated that efficiency of transmission segregated with the source of the coat protein (CP) (9). The CP is the only viral product involved in transmission (Fig. 1). A reverse genetics approach identified on the CMV CP amino acids important for transmission by *Aphis gossypii* or *Myzus persicae* (10). Interestingly, these amino acids differed for the two aphid species, suggesting that virus-binding sites are different in each vector or that virion stability is differently affected in the two insect environments. These amino acids in question were in some cases exposed on the outer surface but in other cases buried in the virion. Loss of transmissibility could therefore be related to lack of interaction with the aphid cuticular receptors or to altered particle stability (11). A negatively charged loop structure on the virion surface was also shown to be involved in transmissibility without affecting infectivity or virion formation (10).

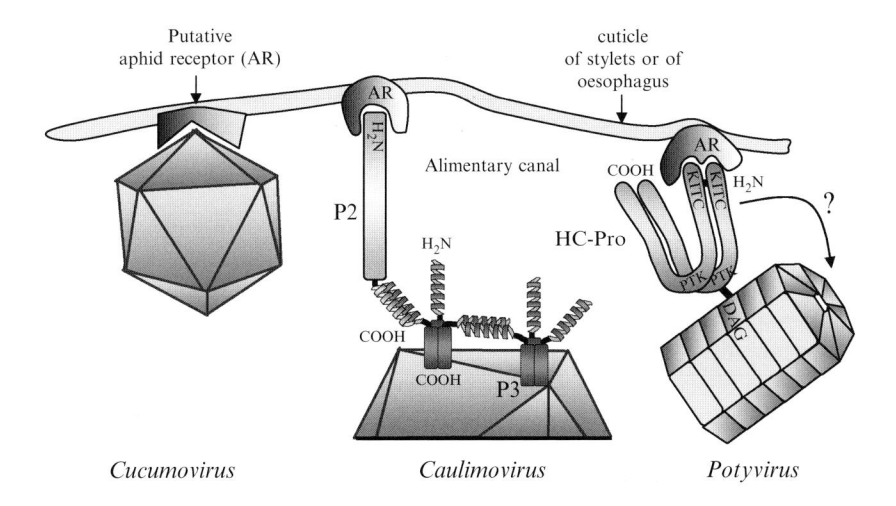

Cucumovirus   *Caulimovirus*   *Potyvirus*

Fig. 1 Hypothetical model of retention on the cuticle of the stylet or the esophagus of viruses aphid transmitted by the noncirculative mode. Cucumoviruses are retained by the putative aphid receptor(s) (AR) without the involvement of any other viral proteins while caulimoviruses and potyviruses need additional viral components to be adsorbed to the alimentary tract. Although HC-Pro of potyviruses has been observed at the extremity of the virus particle, it is not known if this location controls aphid transmission of the virion

2.1.2 Helper Strategy

To be successfully transmitted, some viruses require an extra component, other than the virus particle, that is believed to act as a "bridge" between aphid cuticle and virion (12). This strategy has been adopted by two very different types of viruses, the potyviruses and the caulimoviruses (for a detailed review (13)).

2.1.2.1 Potyviruses

Potyviruses (*Potyvirus, Potyviridae*) have flexuous particles containing a monopartite single-stranded RNA molecule of about 10 kb. Their genome encodes a single polyprotein that is cleaved into nine or ten products by virus-encoded proteases (14). Two viral proteins are required for aphid transmission: the CP and the above mentioned HC. The latter also contains a C-terminal domain with proteolytic activity and has therefore been renamed HC-Pro. HC-Pro is a multifunctional protein involved in aphid transmission, genome amplification, long-distance movement, and suppression of posttranscriptional gene silencing (13).

The CP is organized in three domains, the N and C-terminal domains being exposed on the surface of the particle. Comparison of aphid-transmissible isolates and nonaphid-transmissible isolates identified a highly conserved DAG (Asp-Ala-Gly) motif in the CP N-terminal domain of transmissible isolates. Evidence for implication of this motif in aphid transmission has been obtained by site-directed mutagenesis on full-length infectious clones (13). When the DAG triplet of a transmissible isolate of *Tobacco vein mottling virus* (TVMV) was changed to DAE, transmission was lost. Conversely, modification of the DTG motif of a *Zucchini yellow mosaic virus* (ZYMV) nontransmissible isolate to DAG restored transmission. Further substitutions in the DAG motif or its flanking sequence also affected potyvirus transmission by aphids. In vitro binding overlay assays between CP and HC-Pro have defined a seven amino acid motif, containing the DAG triplet, as the minimal CP sequence needed for interaction. However, in vivo recognition requires a larger domain located in the N terminal part of the CP as demonstrated by site-directed mutagenesis or sequence exchange between transmissible and nontransmissible isolates.

Comparing the HC-Pro sequences of transmissible and nontransmissible potyviruses allowed the identification of two conserved motifs, KITC (Lys-Ile-Thr-Cys) and PTK (Pro-Thr-Lys) in the N terminal and the central region of the protein, respectively. By site-directed mutagenesis on full-length cDNA clones, their role in aphid transmission has been confirmed. In particular, the crucial importance of the positively charged Lys of the KITC motif and of the Pro or Thr residues in the PTK triplet has been highlighted. The bridge hypothesis has been further clarified by testing the interaction between CP and HC-Pro of several nontransmissible ZYMV isolates altered either in the KITC or PTK motif: the KITC motif is involved in HC-Pro binding to the aphid's mouthparts, whereas the PTK motif interacts with virions (Fig. 1). These findings were extended by examining the retention capacity of PVY in the aphid stylet. A mutation in the KITC motif, responsible for the loss of transmission, was sufficient to impede virus retention while CP binding to HC-Pro was unaffected. The KITC motif is located in a cysteine-rich domain of the HC-Pro N terminus which could be organized as a zinc finger-like motif. Deletion mutants lacking this region were fully infectious but have lost their aphid transmissibility.

Earlier studies on HC-Pro of PVY (*potato virus Y*), TVMV, and TuMV (*Turnip mosaic virus*) showed that the biologically active form in transmission could be an oligomer (probably a dimer or trimer). Indeed Plisson et al. (15) showed that a biochemically active His-tagged HC-Pro purified from *Lettuce mosaic virus* infected plants, behaves as a dimer in solution (Fig. 1). The existence of multimers of dimers of HC-Pro in infected plants was recently confirmed by Ruiz-Ferrer et al. (16) by sedimentation velocity of highly purified HC-Pro of *Tobacco etch virus*. Structural analysis of HC-Pro fused to a His-tag revealed two independent structural domains: an N-terminal domain required for aphid transmission connected to a C-terminal domain by a highly structured hinge. All functions except aphid transmission were associated with the hinge and the C-terminal domain (15).

Unexpectedly, HC-Pro was also reported to interact with the genome-linked protein (VPg) in a yeast two-hybrid system. More recently, atomic force microscopy and immunogold-labeling electron microscopy showed that the filamentous PVY virions were associated at one end to the VPg and HC-Pro (17). The association of additional virus-associated proteins to the virion extremity has already been described for several rod-shaped virus groups (clostero-, potex-, beny- pomoviruses), but the relevance of this interaction in aphid transmission or other viral functions remains to be elucidated.

2.1.2.2 Caulimoviruses

Cauliflower mosaic virus (CaMV) is the type member of the genus *Caulimovirus* (*Caulimoviridae*). The isometric particles of about 50 nm of diameter contain the circular double stranded DNA genome. Caulimoviruses differ from RNA viruses in numerous aspects concerning their gene expression strategy and their infection cycle (18).

Like potyviruses, CaMV transmissibility is lost upon virus purification. Lack of transmission can be restored when plants are coinfected with a transmissible isolate, showing that CaMV requires a HC for aphid transmission. This HC (or aphid transmission factor, ATF) was identified by sequence comparison between transmissible and nontransmissible strains as the P2 protein. Interestingly, soluble P2 expressed in baculoviruses could assist a nontransmissible CaMV strain when acquired from infected plants but not from a purified virus preparation. The significance of this observation became apparent with the discovery of a second helper factor, P3 protein. P3 expressed in bacteria can interact directly with purified particles and mediates association of the virion–P3 complex with P2 to promote transmission. Therefore, the bridge hypothesis formulated for potyvirus transmission also applies to CaMV but with two viral components, rather than one, required for linking the virus particle to the aphid stylet (Fig. 1).

The transmission mechanism of CaMV involves an unusual sequential acquisition of the viral components forming the transmissible complex. In infected plant cells, P2 and virions are physically separated: P2 is localized within electron-lucent inclusion bodies (elIB) and most of the virions are sequestered, as P3–virion complexes, into electron-dense inclusion bodies (edIB). This spatial separation controls transmission of CaMV, as the vector must first take up P2 from elIB before it can acquire virions from edIB (19) or from phloem sap that is deprived of P2 (20).

P3 has a bipolar structure with the N-terminal domain interacting directly with the C-terminal domain of P2 via a coiled-coil interaction, and the C-terminal domain responsible for direct association with the virions. Based on low-resolution three-dimensional structures of native and P3-decorated virions, a model has been proposed in which P3 is deeply anchored via its C terminus in the inner shell of the virion in a trimeric structure. This allows the N terminus of P3 located on the surface of the virion to interact by coiled-coil recognition either with other P3 or P2 molecules (21). Interestingly, free P3 does not bind P2 and it is possible that a conformational change of P3, occurring during the oligomerization step, could expose a P2-binding site (19, 21). More recently, it has been suggested that P3 oligomerization could involve a third viral partner, the CaMV movement protein (P1), which is known to interact with P3 via a coiled-coil interaction (22). By a simple structural switch, the multifunctional protein P3 might thus link and regulate two important steps in the viral infection cycle, cell-to-cell movement and aphid transmission.

2.2   Circulative Nonpropagative Viruses

2.2.1   Luteoviruses

Members of the *Luteoviridae* family (genera *Luteovirus, Polerovirus, Enamovirus*) have icosaedric particles of about 25 nm of diameter containing a single-stranded RNA. Luteoviruses are exclusively transmitted by aphids by a circulative and nonpropagative mode. In the vector, virions are always enclosed in characteristic vesicles and are transported through the gut and accessory salivary gland epithelia by a transcytosis mechanism based on the presence of specific virus receptors in these locations (23, 24). Virus particles are composed of two structural proteins: the major CP and a minor readthrough protein (RT), with a C-terminal domain (RTD) exposed on virion surface. The RT is produced by translational expression of the CP cistron leaky termination codon.

Both CP and RT are involved in luteovirus transmission (23, 25, 26). Structural proteins of *Beet western yellows virus* (BWYV, *Polerovirus*) are known to be glycosylated and, although the nature of the sugar has not been precisely identified, glycosylation is suspected to be involved in aphid transmission of BWYV (27).

The crucial role of RTD in aphid transmission has been demonstrated using engineered RTD mutants. RTD-deletion mutants, while infectious in plants, are nontransmissible even after microinjection in the vector's body. However, the RTD is not strictly required for virus attachment to the gut epithelium and transport into the hemolymph, although it appears to enhance this process. Specific point mutations in putatively surface-exposed RTD domains resulted in low transport of mutated virions through gut cells without affecting transcytosis of virions across the accessory salivary glands (ASG) (24), suggesting that recognition at the gut and the ASG involves different viral domains. The RTD could also be required for stabilization of virions as it contains domains involved in binding to the GroEL homolog symbionin present in the hemolymph (4). A role for the CP in aphid transmission has also been established using virus mutants with point mutations targeting amino acids potentially exposed on the virion surface. Some of these mutants were found to be nontransmissible or poorly transmissible, highlighting in particular the importance of amino acids near the C terminus of the CP (28).

Overlay assays with total aphid extracts identified several proteins that exhibit affinity for BWYV in vitro and could therefore be virus receptors or factors involved in intermediate steps of the endocytotic pathway. Among these proteins, Rack-1 (receptor for activated C kinase) displayed differential in vitro binding capacity between two BWYV RTD mutants (both inefficiently transported through gut cells) and wild type virus. Thus, Rack-1 could enhance polarized transport of virions through the gut epithelium by binding to an RTD motif. Another aphid protein, glyceraldehyde-3-phosphate dehydrogenase, bound the aforesaid BWYV mutants as efficiently as the wild type virus and could be involved in the capture of virions from the food bolus by a CP motif interaction (29). Finally, an important characteristic of luteovirus transmission is its high specificity; each member in this family is efficiently transmitted by only one or two aphid species (30). Recently, chimeras obtained by exchanging the RTD of two viruses with different vectors have unequivocally demonstrated that the origin of the RTD determines vector specificity (31).

2.2.2 Geminiviruses

Members of the *Geminiviridae* family have single-stranded DNA genomes contained in geminate particles. These viruses have been split into four genera (*Mastrevirus, Curtovirus, Begomovirus*, and *Topocuvirus*) based on their genome organization and vectors (14). Their genomes are composed of one or two components with the latter referred to as DNA A and DNA B. These viruses are transmitted in a circulative manner by leafhopper, treehopper, or whitefly, with whitefly transmission being confined to begomoviruses.

The CP of geminiviruses is essential for transmission and vector specificity: exchange of the CP gene between a leafhopper-transmitted geminivirus and a whitefly transmitted geminivirus modified vector specificity of the chimeric viruses (32). Mutations targeting amino acids predicted to be exposed on the surface of leafhopper-transmitted geminiviruses revealed the importance of the C-terminal part of the CP in virion formation and the involvement of the N terminus in virus movement in the leafhopper. Indeed, a mutant carrying a two amino acid change in the CP N terminus formed virus particles, was infectious to plants and was acquired, but was not inoculated, by the leafhopper (33). It should be mentioned, however, that electron cryomicroscopy and image reconstruction have pinpointed another region of the *Maize streak virus* CP crucial for leafhopper transmission (34).

Amino acids involved in whitefly transmission of *Tomato yellow leaf curl virus* (TYLCV, begomovirus) have been localized in the CP central region and were correlated with virion formation (35). For the bipartite begomoviruses, the critical residues were identified in the central and C-terminal regions of the CP and may be involved in posttranslational modification and virion stability instead of, or in addition to, playing a role in recognition by putative receptors in the whitefly (36, 37). Finally, in the transmission of TYLCV by whiteflies, symbionin has been suggested to play a key role by its interaction with the CP in the hemolymph (38).

Geminiviruses are generally considered to be transmitted in a nonpropagative fashion but the absence of replication of geminiviruses in their vector has been brought into question for two species (TYLCV and *Squash leaf curl virus*, SCLV). Although the presence of viral DNA replicative forms has not been reported, TYLCV transcripts were recently detected in the vector, which would favour the idea that TYLCV replicates in the insect (39).

2.2.3 Nanoviruses

Nanoviruses (*Nanovirus, Circoviridae*) have multiple small icosahedral particles of about 17–22 nm diameter that contain a circular single-stranded DNA molecule. They are transmitted by aphids in a circulative and nonpropagative manner. As purified *Faba bean necrotic yellows virus* could not be aphid transmitted, the involvement of a putative assistant factor has been suggested. This HC could participate in the transport of virions across the hemocoel – accessory salivary gland interface of the aphid (40).

2.3   *Propagative Viruses*

2.3.1   Tospoviruses

The genus *Tospovirus* is the only plant-pathogenic genus in the family *Bunyaviridae*. The type member, *Tomato spotted wilt virus* (TSWV), is transmitted by several thrips species, *Frankliniella occidentalis* being the most prevalent and efficient vector (41). Virus particles of 80–120 nm diameter are pleiomorphic and roughly spherical. The three negative or ambisense genomic single-stranded RNAs are individually encapsidated by the N protein (nucleocapsid protein) and enclosed in a lipid-bilayer membrane of host origin (virus envelope). Two structural glycoproteins (G_N and G_C) are embedded in the envelope, producing spikes on the surface of virions. Evidence for the role of G_N and G_C in transmission by thrips has come from envelope-deficient mutants, which are infectious to plants when mechanically inoculated, but are not acquired by the vector (42). It is, however, unknown if their function in transmission involves the sugar moiety bound to the proteins or the proteins themselves.

TSWV transmission by thrips is circulative and propagative. Once acquired from an infected plant, virions are transported into the vector through the gut cells and are internalized in the cells of visceral muscle surrounding the gut. Virions are then addressed toward the primary salivary gland cells from which they are excreted into a new host plant together with saliva. Virus replication is thought to proceed in these different organs. The most striking and unique feature of tospovirus transmission is the strict dependence of the developmental stage of the thrips for successful transmission. Thus, adult thrips can only transmit the virus when it is acquired during larval stage (41). The close proximity of the different organs sustaining virus replication during the larval stage could account for the relationship between larval development and virus acquisition (43).

Electron microscopy observations, as well as immunolocalization using antiidiotypic antibodies mimicking the viral glycoproteins, have revealed binding of TSWV envelope proteins to the brush border apical plasmalemma, suggesting a role for these proteins in the first attachment step and in the interaction with putative virus receptors (41). This is in agreement with the function of glycoproteins in animalinfecting members of *Bunyaviridae*, where the glycoproteins on the particle surface are known to be involved in the attachment and fusion events leading to virus entry (44).

The G_N protein contains a specific motif characteristic of a cell-adhesion molecule and direct evidence for such a function in thrips has been obtained after expressing the protein as a recombinant form. Once acquired by the vector, the recombinant protein not only binds specifically to the midgut epithelium of thrip nymphs without the assistance of any other viral protein, but it can also inhibit TSWV acquisition when mixed with purified virus (45). The G_C protein, on the other hand, may act as the fusion protein mediating virus entry in insect cells. This role has been attributed on the basis of sequence homology with other bunyavirus proteins known to fulfill such a function. Moreover the pH-dependent cleavage of G_C is also consistent with its putative role in pH-dependent endocytosis (46). The G_N and G_C glycoproteins have been used in screens to identify thrips proteins exhibiting virus-binding capacity. Two thrip proteins have been identified as candidate TSWV receptors by gel overlay assays, a 50 kDa protein found in the midgut of larvae, and a 94 kDa protein, which may be involved in invasion of other organs, is absent from this organ (41).

2.3.2 Rhabdoviruses

Members in the *Rhabdoviridae* family (genera *Nucleorhabdovirus* and *Cytorhabdovirus*) are transmitted in a circulative and propagative manner mostly by leafhoppers or planthoppers although a few members are vectored by aphids or lacebugs (14). Most plant rhabdovirus particles have a bacilliform shape consisting of a lipid layer surrounding a ribonucleoprotein core composed of the N protein bound to the genomic single-stranded negative RNA. The membrane envelope, of host origin, contains the matrix (M) protein and glycoproteins (G) which form surface spikes. After ingestion by the vector, rhabdoviruses move through the midgut into the hemolymph. From there, they can invade a wide range of cells of various organs (brain, nerve ganglia, ovary, fat body, muscle, salivary glands, etc.). They are finally introduced into new host plants with saliva after their transport through the salivary gland (47).

Viral glycoproteins appear to have an important function in receptor recognition on insect cells but they are not essential for plant infectivity. Blocking of glycoproteins by specific antibodies or removal of glycoprotein spikes drastically reduced infectivity for insect cells. These viral determinants could participate in the interaction of the virus with host cell receptors as demonstrated for rabies virus, a vertebrate rhabdovirus. Plant rhabdovirus glycoprotein sequences do not share homology with their counterparts in vertebrates but they share similar structural features thought to be important for receptor recognition (48).

2.3.3 Reoviruses

Members of the *Reoviridae* family (genera *Fijivirus, Oryzavirus, Phytoreovirus*) have icosahedral double-shelled particles of 70–80 nm diameter. Their genome is composed of 10–12 segments of double-stranded RNA encapsidated in structural proteins. Reoviruses are transmitted in a circulative and propagative manner by planthoppers or leafhoppers. Most of the members in this family have spikes on the outer and inner shells (14). Evidence for a role of virus surface proteins in the attachment step of reoviruses in vectors has been obtained by expressing the spike protein of *Rice ragged stunt virus* (RRSV, *Oryzavirus*) in *Escherichia coli*. When this protein is delivered to the planthopper, it can block subsequent transmission of RRSV. This protein is believed to be the viral determinant recognized by a putative membranous receptor in planthopper (49).

Rice dwarf virus (RDV, *Phytoreovirus*) particles do not possess spikes on the particle surface. The role of the outer capsid protein (protein P2) in RDV transmission by leafhoppers has been demonstrated using insect cell cultures, which did not sustain internalization of a P2-deprived virus whose surface protein has been removed chemically. When such particles were microinjected in the vector hemolymph, transmission was nevertheless possible, suggesting a role for P2 in gut internalization. Additionally, an RDV mutant unable to synthesize and incorporate the protein in the particle was also unable to proliferate in insects and to be vectored whatever the mode of virus acquisition (50). This observation suggests that P2 is also required for additional steps in the virus cycle in the insect.

3 Virus Transmitted by Vectors Other than Insects

3.1 Transmission by Nematodes

Some viruses are transmitted from plant to plant by soil-inhabiting plant parasitic nematodes (51). Known vector nematodes belong to the families *Longidoridae* (longidorids) and *Trichodoridae* (trichodorids) and transmit nepoviruses and tobraviruses, respectively. These nematodes are ectoparasites that remain outside the roots while feeding on epidermal cells located just behind the root tip. Transmission is a noncirculative process in which virus particles are retained at specific sites on the surface of the esophagus. Nematodes can transmit virus even after serial feeds on noninfected tissues and retain virus for periods of months. However, virions are lost during moulting, together with the cuticle lining the esophagus.

Nepo- and tobraviruses are both single-stranded RNA viruses with two genomic components (RNA1 and RNA2) encapsidated into separate spherical or rod-shaped particles, respectively. Production of pseudorecombinants in which the two genomic RNAs of different isolates of the same virus were mixed allowed the vector specificity and transmissibility to be assigned to RNA2 for both nepo- and tobraviruses (52).

3.1.1 Nepoviruses

Nepovirus particles are isometric and about 30 nm diameter. Most of the work on nepoviruses has been conducted on *Grapevine fanleaf virus* (GFLV), which is transmitted specifically by *Xiphinema index*. The CP, encoded by the C-terminal gene of RNA2, is the only viral determinant governing nematode specificity of GFLV, as shown by exchanging RNA2 sequences of GFLV by analogous sequences from *Arabis mosaic virus*, another nepovirus transmitted by a different nematode species (53, 54).

3.1.2 Tobraviruses

The situation is more complex for tobraviruses as in addition to the CP gene, RNA2 encodes two or three other proteins (2b, 2c, 9 K). Replacement experiments between transmissible and poorly transmissible isolates of *Tobacco rattle virus* (TRV) or *Pea early browning virus* (PEBV), suggested that additional proteins to the CP could be involved in tobravirus transmission. Involvement of 2b protein was demonstrated by correlating deletion of the 2b gene with an absence of nematode transmission of tobraviruses. Protein 2b can act *in trans* to assist a nontransmissible virus and direct interaction between CP and 2b has been demonstrated using the yeast two-hybrid system. The tobravirus 2b protein can therefore be considered as a true HC, bridging the virion to the vector mouthparts. Moreover, immunogold-localization studies showed binding of TRV and PEBV 2b to viral particles. A role for the 2c protein in enhancing transmission efficiency has been observed for PEBV but not for TRV. This difference may reflect differences in transmission specificity (52).

3.2 Transmission by Fungi

Fungus-borne viruses can be split in two categories: (1) viruses belonging to the *Tombusviridae* family, which have isometric particles and are transmitted by *Olpidium* spp. and (2) rod-shaped viruses mainly belonging to the *Potyviridae* or to an unassigned family that are transmitted by plasmodiophorids (55). These soil-inhabiting vectors are obligate parasites living within root cells. They survive in the soil in the form of resting spores which can remain viruliferous for many years and produce motile zoospores as a means of dispersal. Two types of fungal transmission are described based on the mode of virus acquisition. In nonpersistent (also called in vitro) transmission, virions are adsorbed to the surface of the zoospores and are not present in resting spores, while in persistent (or in vivo) transmission, virions are acquired by the fungus as it develops in the plant and resting spores harbour virions internally. All viruses transmitted in a nonpersistent manner are *Olpidium* transmitted while viruses transmitted in a persistent manner are plasmodiophorid vectored. There is no evidence of virus replication in the vectors.

3.2.1   Tombusviruses

Purified particles alone are sufficient to initiate fungal transmission, which rules out the existence of a HC. Reciprocal exchanges carried out between the CP gene of *Cucumber necrosis virus* (CNV, *Tombusvirus*) and the related but nontransmissible *Tomato bushy stunt virus* (*Tombusvirus*) have shown that virion uptake by the zoospore is dependent on the CP and a single amino acid change in the CNV CP resulted in loss of transmissibility (55). Analysis of several naturally occurring transmission mutants of CNV showed that all the CP mutations occurred at the surface of the particle, raising the possibility that the critical amino acids may serve as an attachment site for interaction of CNV with a putative zoospore receptor. Recently, Kakani et al. (56) have reported that treatment of zoospores with periodate and trypsin reduced CNV binding, suggesting the involvement of glycoproteins in virus attachment. CNV binding to zoospores was also inhibited by preincubating purified virus with specific sugars before contact with zoospores, confirming the involvement of glucidic moieties in CNV transmission. Kakani et al. (57) provided evidence that zoospore-bound CNV particles are in a modified conformation compared to native CNV particles and exhibit a different proteolytic pattern after trypsin digestion. This conformational change seems to be related to the virus transmission process since a virus mutant incapable of such modification is only poorly transmitted.

3.2.2   Bymoviruses, Furoviruses, Benyviruses, and Pomoviruses

All these viruses have multipartite genomes consisting of two to five RNAs. The viral determinant involved in fungal transmission has been shown in most cases to be a CP-fusion protein (RT) synthesized by a translational readthrough. An exception is for bymoviruses (*Potyviridae*) for which the transmission determinant (P2 protein) is expressed independently from the CP but liberated from a polyprotein by internal cleavage (55).

After repeated mechanical transmission, the genomes of the bymoviruses, furoviruses, and benyviruses frequently undergo natural deletions in P2 or the RT gene which render the virus nontransmissible. A KTER motif in the RT of *Beet necrotic yellow vein virus* (BNYVV, *Benyvirus*) has been shown to be required for transmission, but this motif seems to be restricted to benyviruses. Whereas only limited sequence similarity exists between the RT domains of beny-, furo-, pomoviruses, and the P2 protein of bymoviruses, these proteins nevertheless share structural similarities. In particular, all contain two complementary transmembrane domains, which are thought to be involved in virion acquisition from plant cytoplasm across the fungal membrane. Electron microscopy studies have revealed that the RT proteins of BNYVV and *Potato mop top virus* (*Pomovirus*) are associated with one extremity of the particle, suggesting a potential interaction with vector receptors (58, 59). Finally, involvement of another viral product in fungal transmission is suspected for BNYVV because the presence of RNA4 is also associated with efficient fungal transmission of this virus (60).

References

1. Herrbach, E. (2005). Arthropod transmission. In *Viruses and Virus Diseases of Poaceae (Graminae)*, Lapierre H and A. Signoret P, eds. (INRA editions, Versailles), pp. 114–124.
2. Nault, L.R. (1997). Arthropod transmission of plant viruses: a new synthesis. In *Ann Entomol Soc Am*, **90**, pp. 521–541.
3. McClintock, J.A., and Smith, L.B. (1918). True nature of spinach-blight and relation of insects to its transmission. *J Agric Res Washington DC* **14**, 1–59.
4. van den Heuvel, J.F.J.M., Franz, A.W.E., and van der Wilk, F. (1999). Molecular basis of plant virus transmission. *Trends Microbiol* **7**, 71–76.
5. Ammar, E.D, Järlfors, U., and Pirone, T.P. (1994). Association of potyvirus helper component protein with virions and the cuticle lining the maxillary food canal and foregut of an aphid vector. *Phytopathology* **84**, 1054–1060.
6. Wang, R.Y., Ammar, E.D., Thornbury, D.W., Lopez-Moya, J.J., and Pirone, T.P. (1996). Loss of potyvirus transmissibility and helper-component activity correlate with non-retention of virions in aphid stylets. *J Gen Virol* **77**, 861–867.
7. Martin, B., Collar, J.L., Tjallingii, W.F., and Fereres, A. (1997). Intracellular ingestion and salivation by aphids may cause the acquisition and inoculation of non-persistently transmitted plant viruses. *J Gen Virol* **78**, 2701–2705.
8. Powell, G. (2005). Intracellular salivation is the aphid activity associated with inoculation of non-persistently transmitted viruses. *J Gen Virol* **86**, 469–472.
9. Gera, A., Loebenstein, G., and Raccah, B. (1979). Protein coats of two strains of cucumber mosaic affect transmission by *Aphis gossypii*. *Phytopathology* **69**, 396–399.
10. Palukaitis, P., and Garcia-Arenal, F. (2003). Cucumoviruses. *Adv Virus Res* **62**, 241–323.
11. Ng, J.C., Josefsson, C., Clark, A.J., Franz, A.W., and Perry, K.L. (2005). Virion stability and aphid vector transmissibility of Cucumber mosaic virus mutants. *Virology* **332**, 397–405.
12. Pirone, T.P., and Blanc, S. (1996). Helper-dependent vector transmission of plant viruses. *Annu Rev Phytopathol* **34**, 227–247.
13. Syller, J. (2006). The roles and mechanisms of helper component proteins encoded by potyviruses and caulimoviruses. *Physiol Mol Plant Pathol*, 67, 119–130.
14. Hull, R. (2002). Matthew's Plant Virology (Academic Press, San Diego, USA).
15. Plisson, C., Drucker, M., Blanc, S., German-Retana, S., Le Gall, O., Thomas, D., and Bron, P. (2003). Structural characterization of HC-Pro, a plant virus multifunctional protein. *J Biol Chem* **278**, 23753–23761.

16. Ruiz-Ferrer, V., Boskovic, J., Alfonso, C., Rivas, G., Llorca, O., Lopez-Abella, D., and Lopez-Moya, J.J. (2005). Structural analysis of tobacco etch potyvirus HC-pro oligomers involved in aphid transmission. *J Virol* **79**, 3758–3765.

17. Torrance, L., Andreev, I.A., Gabrenaite-Verhovskaya, R., Cowan, G., Makinen, K., and Taliansky, M.E. (2006). An unusual structure at one end of potato potyvirus particles. *J Mol Biol* **357**, 1–8.

18. Blanc, S., Hebrard, E., Drucker, M., and Froissart, R. (2001). Molecular aspects of virus-vector interactions. In virus-insect-plant interactions, K.F. Harris, ed. (Academic press Inc, USA), pp. 143–166.

19. Drucker, M., Froissart, R., Hebrard, E., Uzest, M., Ravallec, M., Esperandieu, P., Mani, J.C., Pugniere, M., Roquet, F., Fereres, A., and Blanc, S. (2002). Intracellular distribution of viral gene products regulates a complex mechanism of cauliflower mosaic virus acquisition by its aphid vector. *Proc Natl Acad Sci U S A* **99**, 2422–2427.

20. Palacios, I., Drucker, M., Blanc, S., Leite, S., Moreno, A., and Fereres, A. (2002). Cauliflower mosaic virus is preferentially acquired from the phloem by its aphid vectors. *J Gen Virol* **83**, 3163–3171.

21. Plisson, C., Uzest, M., Drucker, M., Froissart, R., Dumas, C., Conway, J., Thomas, D., Blanc, S., and Bron, P. (2005). Structure of the mature P3-virus particle complex of cauliflower mosaic virus revealed by cryo-electron microscopy. *J Mol Biol* **346**, 267–277.

22. Stavolone, L., Villani, M.E., Leclerc, D., and Hohn, T. (2005). A coiled-coil interaction mediates cauliflower mosaic virus cell-to-cell movement. *Proc Natl Acad Sci U S A* **102**, 6219–6224.

23. Gildow, F.E. (1999). Luteovirus transmission and mechanisms regulating vector specificity. In *The Luteoviridae*, G.H. Smith and H. Baker, eds. (CAB International, Oxon, UK), pp. 88–113.

24. Brault, V., Herrbach, E., and Reinbold, C. (2007). Electron microscopy studies on luteovirid transmission by aphids. *Micron* **38**, 302–312.

25. Brault, V., Ziegler-Graff, V., and Richards, K. (2001). Viral determinants involved in luteovirus-aphid interactions, K.F. Harris, ed. (Academic press, Inc., USA), pp. 207–232.

26. Gray, S., and Gildow, F.E. (2003). Luteovirus-aphid interactions. *Annu Rev Phytopathol* **41**, 539–566.

27. Seddas, P., and Boissinot, S. (2006). Glycosylation of beet western yellows virus proteins is implicated in the aphid transmission of the virus. *Arch Virol* **151**, 967–984.

28. Brault, V., Bergdoll, M., Mutterer, J., Prasad, V., Pfeffer, S., Erdinger, M., Richards, K.E., and Ziegler-Graff, V. (2003). Effects of point mutations in the major capsid protein of beet western yellows virus on capsid formation, virus accumulation, and aphid transmission. *J Virol* **77**, 3247–3256.

29. Seddas, P., Boissinot, S., Strub, J.M., Van Dorsselaer, A., Van Regenmortel, M.H., and Pattus, F. (2004). Rack-1, GAPDH3, and actin: proteins of Myzus persicae potentially involved in the transcytosis of beet western yellows virus particles in the aphid. *Virology* **325**, 399–412.

30. Herrbach, E. (1999). Virus-vector interactions, Introduction. In *The Luteoviridae*, G.H. Smith and H. Barker, eds. (CAB International, Oxon, UK), pp. 85–88.

31. Brault, V., Perigon, S., Reinbold, C., Erdinger, M., Scheidecker, D., Herrbach, E., Richards, K., and Ziegler-Graff, V. (2005). The polerovirus minor capsid protein determines vector specificity and intestinal tropism in the aphid. *J Virol* **79**, 9685–9693.

32. Gray, S.M., and Banerjee, N. (1999). Mechanisms of arthropod transmission of plant and animal viruses. *Microbiol Mol Biol Rev* **63**, 128–148.

33. Soto, M.J., Chen, L.F., Seo, Y.S., and Gilbertson, R.L. (2005). Identification of regions of the Beet mild curly top virus (family Geminiviridae) capsid protein involved in systemic infection, virion formation and leafhopper transmission. *Virology* **341**, 257–270.

34. Böttcher, B., Unseld, S., Ceulemans, H., Russell, R.B., and Jeske, H. (2004). Geminate structures of African cassava mosaic virus. *J Virol* **78**, 6758–6765.

35. Noris, E., Vaira, A.M., Caciagli, P., Masenga, V., Gronenborn, B., and Accotto, G.P. (1998). Amino acids in the capsid protein of tomato yellow leaf curl virus that are crucial for systemic infection, particle formation, and insect transmission. *J Virol* **72**, 10050–10057.

36. Kheyr-Pour, A., Bananej, K., Dafalla, G., A, Caciagli, P., Noris, E., Ahoonmanesh, A., Lecoq, H., and Gronenborn, B. (2000). Watermelon chlorotic stunt virus from the Sudan and Iran: sequence comparisons and identification of a whitefly-transmission determinant. *Phytopathology* **90**, 629–635.
37. Höhnle, M., Höfer, P., Bedford, I.D., Briddon, R.W., Markham, P.G., and Frischmuth, T. (2001). Exchange of three amino acids in the coat protein results in efficient whitefly transmission of a nontransmissible Abutilon mosaic virus isolate. *Virology* **290**, 164–171.
38. Morin, S., Ghanim, M., Sobol, I., and Czosnek, H. (2000). The GroEL protein of the whitefly Bemisia tabaci interacts with the coat protein of transmissible and nontransmissible begomoviruses in the yeast two-hybrid system. *Virology* **276**, 404–416.
39. Sinisterra, X.H., McKenzie, C.L., Hunter, W.B., Powell, C.A., and Shatters, Jr. R.G. (2005). Differential transcriptional activity of plant-pathogenic begomoviruses in their whitefly vector (Bemisia tabaci, Gennadius: Hemiptera Aleyrodidae). *J Gen Virol* **86**, 1525–1532.
40. Franz, A.W.E., van der Wilk, F., Verbeek, M., Dullemans, A.M., and van den Heuvel, J.F. (1999). Faba bean necrotic yellows virus (genus Nanovirus) requires a helper factor for its aphid transmission. *Virology* **262**, 210–219.
41. Whitfield, A.E., Ullman, D.E., and German, T.L. (2005). Tospovirus-thrips interactions. *Annu Rev Phytopathol* **43,** 459–489.
42. Nagata, T., Nagata-Inoue, A.K., Prins, M., Goldbach, R., and Peters, D. (2000). Impeded thrips transmission of defective tomato spotted wilt virus isolates. *Phytopathology* **90**, 454–459.
43. Moritz, G., Kumm, S., and Mound, L. (2004). Tospovirus transmission depends on thrips ontogeny. *Virus Res* **100**, 143–149.
44. Ludwig, G.V., Israel, B.A., Christensen, B.M., Yuill, T.M., and Schultz, K.T. (1991). Role of La Crosse virus glycoproteins in attachment of virus to host cells. *Virology* **181**, 564–571.
45. Whitfield, A.E., Ullman, D.E., and German, T.L. (2004). Expression and characterization of a soluble form of tomato spotted wilt virus glycoprotein GN. *J Virol* **78**, 13197–13206.
46. Whitfield, A.E., Ullman, D.E., and German, T.L. (2005). Tomato spotted wilt virus glycoprotein G(C) is cleaved at acidic pH. *Virus Res* **110**, 183–186.
47. Ammar, E.D., and Nault, L.R. (1985). Assembly and accumulation sites of maize mosaic virus in its planthopper vector. *Intervirology* **24**, 33–41.
48. Hogenhout, S., A, Redinbaugh, M., G, and Ammar, E., D, (2003). Plant and animal rhabdovirus host range: a bug's view. *Trends Microbiol* **11**, 264–271.
49. Zhou, G.Y., Lu, X.B., Lu, H.J., Lei, J.L., Chen, S.X., and Gong, Z.X. (1999). Rice ragged stunt Oryzavirus: role of the viral spike protein in transmission by the insect vector. *Ann Appl Biol* **135**, 573–578.
50. Omura, T., Yan, J., Zhong, B., Wada, M., Zhu, Y., Tomaru, M., Maruyama, W., Kikuchi, A., Watanabe, Y., Kimura, I., and Hibino, H. (1998). The P2 protein of rice dwarf phytoreovirus is required for adsorption of the virus to cells of the insect vector. *J Virol* **72**, 9370–9373.
51. Taylor, C.E., and Brown, D.J.F. (1997). Nematode vectors of plant viruses. (Wallingford,UK: CAB International), p. 286.
52. MacFarlane, S.A. (2003). Molecular determinants of the transmission of plant viruses by nematodes. *Mol Plant Pathol.* **4**, 211–215.
53. Belin, C., Schmitt, C., Demangeat, G., Komar, V., Pinck, L., and Fuchs, M. (2001). Involvement of RNA2-encoded proteins in the specific transmission of Grapevine fanleaf virus by its nematode vector Xiphinema index. *Virology* **291**, 161–171.
54. Andret-Link, P., Schmitt-Keichinger, C., Demangeat, G., Komar, V., and Fuchs, M. (2004). The specific transmission of Grapevine fanleaf virus by its nematode vector Xiphinema index is solely determined by the viral coat protein. *Virology* **320**, 12–22.
55. Rochon, D., Kakani, K., Robbins, M., and Reade, R. (2004). Molecular aspects of plant virus transmission by olpidium and plasmodiophorid vectors. *Annu Rev Phytopathol* **42**, 211–241.
56. Kakani, K., Robbins, M., and Rochon, D. (2003). Evidence that binding of cucumber necrosis virus to vector zoospores involves recognition of oligosaccharides. *J Virol* **77**, 3922–3928.
57. Kakani, K., Reade, R., and Rochon, D. (2004). Evidence that vector transmission of a plant virus requires conformational change in virus particles. *J Mol Biol* **338**, 507–517.

58. Haeberle, A.M., Stussi-Garaud, C., Schmitt, C., Garaud, J.C., Richards, K.E., Guilley, H., and Jonard, G. (1994). Detection by immunogold labelling of P75 readthrough protein near an extremity of beet necrotic yellow vein virus particles. *Arch Virol* **134**, 195–203.

59. Cowan, G.H., Torrance, L., and Reavy, B. (1997). Detection of potato mop-top virus capsid readthrough protein in virus particles. *J Gen Virol* **78**, 1779–1783.

60. Richards, K.E., and Tamada, T. (1992). Mapping functions on the multipartite genome of beet necrotic yellow vein virus. *Annu Rev Phytopathol* **42**, 211–241.

Section 2
Viral Nucleic Acid Sequence/Function Analysis

Chapter 7
In Vivo Analyses of Viral RNA Translation

William R. Staplin and W. Allen Miller

Abstract Positive-strand RNA viruses often use noncanonical strategies to usurp the host translational machinery for their own benefit. These strategies have been analyzed using transient expression assays in the absence of replication, with reporter genes replacing viral genes. A sensitive and convenient reporter assay is the dual luciferase system using *Renilla* (*Renilla reniformis*) and firefly (*Photinus pyralis*) reporter genes. Use of recombinant viral constructs containing the reporter luciferase gene allows us to discern whether a particular RNA sequence or secondary structure elicits an effect on initiation of translation or recoding. This chapter describes a standard luciferase protocol that can be molded to fit any viral sequence, in order to detect *cis*-acting regulatory elements in viral RNA.

Keywords Cap-independent translation; Electroporation; Oat protoplast; Plant cell suspension culture; Recoding; Ribosomal frameshift

1 Introduction

For viral replication to occur, viral messenger RNA must compete effectively with host mRNAs for the translational machinery of the cell. In most cases the viral 5′ and/or 3′ untranslated regions (UTR) control translation, and regulate RNA stability (1–3). Reporter constructs have demonstrated the utility of a gutted virus, with a reporter gene replacing the coding regions, while retaining the flanking viral UTRs (4). Replacing the coding region with a reporter gene allows rapid, quantitative assessment of viral sequences that direct such events as cap-dependent translation, cap-independent translation, poly(A) tail-independent translation, ribosomal frameshifting, and stop codon readthrough (4–7). The firefly luciferase (LUC) reporter system is advantageous in that it is simple to quantify, extremely sensitive (10^{-20} mole detection limit), and reproducible (5). The following protocol describes in detail, a transient in vivo assay with the firefly luciferase gene flanked by the 5′ and 3′ UTRs of a model virus, Barley yellow dwarf virus (BYDV), using a commercial kit. This system can be used to assess 22 samples per 50 mL of protoplasts in 1 day and to quantify LUC activity in order to map viral RNA structures that control translation.

From: *Methods in Molecular Biology, Vol. 451, Plant Virology Protocols:*
From Viral Sequence to Protein Function
Edited by G.D. Foster, I.E. Johansen, Y. Hong, and P.D. Nagy © Humana Press, Totowa, NJ

1.1 Advantages of the In Vivo Luciferase Reporter Translation System for a Range of Viral Translational Requirements

Protoplasts are cells from which the cell wall has been removed. In vivo assays utilizing oat protoplasts provide a unique environment that takes advantage of the assay being in plant cells, while being an expedient way to assess viral gene translation through transfected recombinant viral constructs (6). While in vitro (cell-free) translation assays provide useful information on translational control signals in a highly defined environment, they may not accurately reflect the conditions in the cell. For example, commonly used wheat germ or rabbit reticulocyte translation extracts are insensitive to the presence of a poly(A) tail on an mRNA, whereas a poly(A) tail is required for efficient translation, in vivo. This is an important consideration when studying poly(A) tail-dependent viral RNAs, such as the Potyviridae family (e.g., Potato virus Y) and the Comoviridae family (e.g., Cowpea mosaic virus). In vivo assays can also provide useful information for assessing cap-dependent and cap-independent translations of plant viral RNAs. 5′ cap (m^7GpppN)-dependent viral RNAs that follow traditional eukaryotic translation requirements include Potato virus X, Tobacco mosaic virus (7), and Alfalfa mosaic virus. Cap-independent viral RNAs include those in the Tombusviridae (e.g., Tomato bushy stunt virus), Luteoviridae (e.g., BYDV (1)), and Potyviridae families, among others.

Recoding signals such as ribosomal frameshift or in-frame stop codon readthrough elements can be detected using a dual luciferase system that encodes the Renilla luciferase in the 0 frame and firefly reporter genes in a −1 or +1 frame for frameshifting (8–10) or in the 0 frame for readthrough (11, 12). In contrast to the single reporter constructs, these constructs are useful in that the upstream ORF (Renilla luciferase) serves as an internal control for transformation efficiency and RNA stability.

1.2 Advantages of Using Protoplasts and Recombinant RNA for Transfections

Protoplast assays take place in living cells (hence they are in vivo). Thus, they provide a more realistic system for assessing gene expression in plant cells. Oat protoplasts readily take up nucleic acid, upon electroporation (13). Thus, they can be easily transfected with experimental luciferase-encoding RNA constructs and yield reproducible data. We use this system to show how viral RNA secondary structures have multifaceted roles in facilitating viral translation initiation, readthrough, and frameshift. While experimental RNA constructs may be affected by stability, independent of translation, RNA stability can be assayed easily (9, 10). Whereas, when DNA is introduced, many ill-defined processing events may affect gene expression, including promoter activity, splicing, and selective nuclear export (14). Also, RNA expressed from DNA in vivo would be capped and polyadenylated assuming a pol II promoter is used. Hence, we advise introducing RNA directly into cells when investigating RNA virus gene expression (2).

2 Materials

2.1 Oat Protoplast Cell Culture: Transport and Shipping of Oat Protoplasts

Oat protoplasts can be easily transported or restored following shipping by using the following protocol (13):

1. Fresh Murashige and Skoog (MS) media is prepared with a solidifying agent (e.g., phytagel, phytoblend) and is aseptically transferred to 50 mL polypropylene sterile tubes.
2. With the conical tubes angled, the media will solidify, creating a large surface area. Aseptically insert a sterile, circular piece of Whatman 3 MM filter paper on top of the solidified media in the conical tube. A portion of pelleted oat protoplast culture is then applied carefully to the top of the paper filter. The cells are now ready for shipping. In order to re-establish a suspension culture, cells from the paper disk are transferred aseptically into 40 mL of fresh MS media and grown under continuous gyration, rotating on a gyrotory shaker at 160–220 rpm (New Brunswick Scientific Classic Series C24 incubator/shaker). See note 1. Oat suspension culture may be grown indefinitely this way, with subculturing by a 1:5 dilution into MS media without phytagel, every 7 days, at 20–25 °C.

2.2 Solutions and Reagents

MS media with phytagel (1.0 L): One packet of MS media without agar (MPBio Cat. Number 2633020), 87.6 mM Sucrose, 10 mL of 100X vitamin solution, 2.5 g/L phytagel (Sigma Cat Number P8169), and adjust to pH 5.7, and bring volume up to 1 L. Autoclave, cool to room temperature, and store at 4 °C.

100X Vitamin Solution: Pyridoxine-HCl (50 mg), thiamine-HCl (50 mg), nicotinic acid (50 mg), myoinositol (10 g). Resuspend components into 100 mL of double distilled water (ddH$_2$0), dispense into 1 mL aliquots, and store at −20 °C.

MS media (1 L): One packet of MS media without agar (MPBio Cat. No. 2633020), 87.6 mM sucrose, 10 mL of 100X vitamin solution, and adjust pH to 5.7 and bring volume up to 1 L. Autoclave, cool to room temperature, and store at 4 °C.

MS Media with 0.4 M Mannitol (1 L): One packet of MS media without agar (MPBio Cat No. 2633020), 0.4 M Mannitol, 87.6 mM Sucrose, and adjust the pH to 5.7, and bring volume up to 1 L. Autoclave, cool to room temperature, and store at 4 °C.

Enzyme solution (50 mL): Always prepare a fresh batch of enzyme reagent in Artificial Sea Water (ASW):0.6 M Manitol: 0.1% (w/v) Driselase (Sigma), 0.175% cellulase (Onozuka RS, Yakult Pharmaceuticals), 0.8% (w/v) hemicellulase (Sigma). Stir for 15 min, while sealed with Parafilm, until all of the reagents have gone into solution, adjust the pH to 5.6–5.7, and bring enzyme volume up to 50 mL. Filter through a 0.2 μm filter. 50 mL of enzyme solution is satisfactory for digesting 10 mL of packed cells, or approximately 20–22 mL samples. Enzyme solution may be scaled up linearly.

ASW:0.6 M mannitol: Working reagent is used at a 1:1 ratio. 311 mM NaCl, 18.8 mM MgSO$_4$, 6.8 mM CaCl$_2$ (1.0 g CaCl·2H$_2$0), 10 mM MES, 6.9 mM KCl, 16.7 mM MgCl, 1.75 mM NaHC0$_3$. Adjust pH to 6.0 and bring volume up to 1 L. Prepare separately 1 L of 0.6 M Mannitol. Add equal volumes of ASW and 0.6 M Mannitol, for a total of 2 L. Autoclave, cool to room temperature (22 °C), and store at 4 °C.

Electroporation Buffer: 10 mg KH$_2$P0$_4$, 57.5 mg NaHP0$_4$·7H$_2$0, 130 mM NaCl, 0.2 M mannitol. Adjust the volume to 495 mL with ddH$_2$0. Adjust pH to 7.2 and bring volume up to 1,996 mL. Once the solution is cooled to room temperature (22 °C), store at 4 °C. On the day of the electroporation, add CaCl$_2$ (Final concentration = 3.2 mM) to the aliquot of electroporation buffer, as well as filter-sterilized spermidine 0.45 μM, prior to resuspending the washed oat protoplasts in the electroporation buffer.

2.3   Equipment

Sterile 150 mL glass flasks exclusively used for oat-cell subculture.
Sterile HEPA filtered hood.
Gyrotory shaker (e.g., Model G2 New Brunswick Scientific Co., Inc.).
Motorized pipetter (e.g., Eppendorf Easy Pet).
Centrifuge (e.g., Sorvall® RC-5C Plus) with SH-3000 or compatible bucket rotor
Electroporator square wave pulse (e.g., BTX® T820 ElectroSquare Porator or
  BioRad GenePulser XCell™ with CE module).

2.4   Disposable Sterile Materials

50 mL conical centrifuge tubes
10 mL serological glass pipettes
100 × 15 mm sterile plastic petri dishes
Wide orifice tips for P-1000 Pipetter
4 mm electroporation cuvettes (available from BTX or BioRad)
Six-well cell culture plates

3   Methods

3.1   Preparation of Oat Protoplasts From Suspension Culture for Electroporation of Viral Constructs

1. Using a glass pipette, aseptically remove 40–50 mL of oat suspension culture from the maintenance flask and dispense into a disposable 50 mL conical tube. After

15 min, the cells will settle to the bottom of the conical tube. Meanwhile, transfer the remaining amount (5–10 mL) to a new maintenance flask (See note 2).

2. Decant the cell-free growth media with the pipetter, and resuspend settled cells, with slow pipetting action to avoid cell damage, in 20 mL of cell-wall digestion solution. The enzyme solution must be freshly prepared the day of the digestion. Using a sterile glass pipette, very gently pipette up and down the settled cells, or alternatively, gently add the enzyme digestion solution to the conical tube's side, and turn the recapped conical tube on its side, tilting the tube from side to side for 5 min, until the pellet is fully resuspended.

3. Transfer the resuspended cells into a 145 × 22 mm large Petri dish, or four small Petri dishes (100 × 15 mm), and add equal volumes of the remaining enzyme solution (30 mL) to the granular slurry of undigested cells and enzyme solution. Secure the Petri dish lid with Parafilm, carefully wrap with aluminum foil so as to keep the resuspended cells in the dark, and shake at 25 °C, at 42 rpm, for 16–19 h, using a rotary shaker (e.g., Model G2 New Brunswick Scientific Company, Inc.).

3.2 Preparing Cells for Recombinant RNA Electroporation

1. Following the 16–19 h gentle digestion of the cell membrane, oats cells will have a homogenous, milky, and translucent appearance, devoid of cell clumps.

2. Using a drop of digested cells, dilute digested cells 1:10 with ASW:0.6 M mannitol, and inspect the prepared oat protoplasts under a light microscope under 100X magnification, and count the free floating protoplasts using a hemocytometer. After two washes with the 0.6 M ASW:0.6 M mannitol solution, the cells should be completely devoid of any cell aggregates (see note 3).

3. Collect the cells by gently suspending g the membrane bound protoplasts with a 10 mL serological pipette, transferring them into two sterile 50 mL conical tubes. Centrifuge the cells at 100 × g for 5 min in a 4 °C Sorvall centrifuge, using a SH-3000 Sorvall® swinging bucket rotor (see note 4).

4. Wash and pellet the cells again by removing the digestion solution and cellular debris by aspiration. Resuspend the cells in 10 mL of buffered wash solution (ASW:0.6 M mannitol). Cells should appear as a pale yellowish pellet in the transparent ASW:0.6 M mannitol wash solution (see note 5).

5. Repeat Step 4 and resuspend cells in 10 mL of ASW:0.6 M mannitol. After the second wash, the cells should be completely free of debris. While the cells are being washed, RNA samples designated to be electroporated into the oat protoplast cells should be resuspended in an appropriate concentration and volume (RNase free/DNase-free water, RNA Storage Solution®, or RNA Resuspension Solution®) (see note 6).

6. It is always advisable to resuspend the washed oat protoplasts in slightly more electroporation volume, (e.g., 3 mL) more than is necessary for the total volume (mL) of cells, for the total number of cuvettes, if for some reason a sample needs to be redone (see note 7).

7. Following the aspiration of the ASW:0.6 M mannitol wash solution, wash cells in the electroporation buffer one final time by gently resuspending both oat cell pellets into 10 mL of electroporation buffer with 0.2 mM spermidine, and combine the pelleted oat protoplasts into a new 50 mL conical tube. Repeat the centrifugation step, $100 \times g$ for 5 min at 4 °C, aspirate off the electroporation buffer wash. Resuspend cells to the concentration of 6×10^6 cells/mL/sample. They are ready for aliquoting into the prechilled cuvettes (see note 8).

8. Generally, an initial harvest of 10 mL of packed cells from the previous day (prior to the overnight cell wall digestion) will yield 6×10^7 protoplasts. For transient expression of reporter genes on the experimental RNA, cells may be resuspended in up to 20–25 mL of electroporation buffer, for roughly 2.5×10^6 cells/mL/sample.

9. Aliquot 4 mL of MS + 0.4 M mannitol into each well of a six-well plate, such that one RNA sample, electroporated into cells in triplicate, can be incubated in three wells of a six-well plate, or two electroporated samples per six-well plate. At least two additional wells of a six-well plate are necessary for the negative "No RNA" control samples.

3.3 Electroporation of RNA into Cells

1. The process of preparing to add the RNA, electroporating the RNA into the cells, and transferring the cells to one well of a six-well plate demands a methodical and efficient performance (13). The RNA samples should be at a similar concentration (e.g., 0.4 μg/μL) that will allow for a consistent volume, (3–20 μL) to be added to each cuvette, for each representative sample.

2. Keep all components on ice (protoplast cells in electroporation buffer with 0.2 M spermidine, RNA samples, and cuvettes), and transfer 1 mL of protoplast suspension into the 4 mm electroporation cuvettes. Use either the P-1000 pipette with a wide-bore 1 mL pipet tip to aliquot the protoplasts, or the motorized pipet filler/dispenser, with a 10 mL serological pipette. Gently suspend the cells, prior to dispensing 1 mL aliquots into the cuvettes.

3. When ready to electroporate, prime a mechanical P-1000 pipette with 1 mL of the MS media from one of wells, and set aside for rinsing the cuvette after electroporation. Add a total of 0.1–1.5 pmol of an experimental mRNA (e.g., 1–1.5 μg) containing the firefly luciferase reporter and 0.1 pmol of an internal control renilla luciferase RNA construct to the washed cells within the 4 mm cuvette. After adding the RNA, the metal contacts on the cuvette are briefly wiped with a Kimwipe tissue and immediately placed in the cuvette holder of the electroporator.

4. Electroporate the cells at a defined voltage and capacitance setting (see note 9). Immediately pour the electroporated cells from the cuvette into a well of the six-well plate, which also contains 3 mL of MS + 0.4 M mannitol.

5. Using the P-1000 pipetter primed with the 1 mL of MS media, gently rinse the cuvette chamber, and pour the contents back into its respective well on the six-well plate. At this juncture, change the P-1000 tip, primed again with 1 mL of MS media from the next adjacent well, and the whole electroporation process is repeated. Different types of protoplasts require different electroporation settings. Therefore, optimization of the appropriate mass of RNA electroporated with the protoplast type needs to be done along with the wild type positive control construct, prior to experimental analyses (see note 10). For example, the optimal pulse for Cowpea protoplasts is 90 V with exponential decay and a 50 ms time constant, which gives 900–950 µF (2).

6. The six-well plates are subsequently covered with their respective lids and incubated at room temperature (22 °C) for 3½–4 h; however, cells may be harvested as soon as 30 min postelectroporation, for RNA stability time course experiments, described below.

3.4 Harvesting Cells, Lysing Cells, and Monitoring Renilla and Firefly Luciferase Activity on a Luminometer

1. After the 30 min to 4 h posttransfection incubation, inspect the electroporated cells briefly under the light microscope, then transfer to separate 15 mL conical centrifuge tubes. Centrifuge the cells, as described above, and remove the MS media (see note 11).

2. For the dual luciferase protocol, the cells are resuspended with a P-200 Pipette tip, in 2X Passive Lysis Buffer (PLB). PLB is a patented cell-lysis reagent from Promega that rapidly disrupts protoplast membranes. PLB is also formulated to minimize Renilla luciferase substrate, coelenterazine autoluminescence, once the Stop N Glo® is added to the sample, later in the procedure. A suggested volume of 50 µL PLB will be sufficient to detect Renilla and firefly activity. This volume can be scaled up to 100 µl, while the Luciferase A Reagent II (LAR II) firefly reagent volume needs to match the LAR II added after the cells are lysed.

3. Transfer the resuspended cells to a prelabeled microcentrifuge tube, secure it on a vibratory shaker, and shake at room temperature for 15 min. PLB reagent will actively release the Renilla and firefly luciferase reporter enzymes into the cell lysate. PLB exhibits minimal coelenterazine autoluminescence and serves as an ideal lysing reagent for assays involving the dual luciferase reporter genes.

4. While the protoplasts are being lysed, aliquot 50 µL of LAR II into prelabeled 1.7 mL microcentrifuge tubes with a tube for each sample as well as extra microcentrifuge tubes for calibration of the positive control.

5. Turn on the luminometer for the instrument's lamp to warm up (approximately 20 min).

6. Add 2–20 µL of lysate to the 50 µL of LAR II, firefly Beetle Luciferin substrate, and monitor luciferase expression in a luminometer (e.g., Turner Designs TD 20/20). The LAR II substrate can typically be prealiquoted into designated

microcentrifuge tubes or disposable plastic vials, for each of the samples and extra tubes for setting sensitivity readings. The luminometer can be preprogrammed to allow a 2 s preread delay, followed by a 10 s luminescent measurement period. The positive control sample will set the upper luciferase limit for each experimental sample, by determining the standard cell lysate volume that will be added to the 50 μL of LAR II substrate reagent. For example, if 5 μL of positive-control cell lysate is shown to provide adequate luciferase expression, without compromising the sensitivity of the reading, set the sensitivity to the positive control's upper luciferase activity and 5 μL becomes the designated volume for each of the experimental and the no RNA negative control samples.

7. After the firefly reading, the luminometer provides a 2 s delay for 50 μL of Stop N Glo® to be added, which quenches the firefly luciferase activity, and contains Renilla substrate, coelenterazine, for selective luminance of the Renilla luciferase over the 10 s luminescent period. The Renilla luciferase activity can then be used to normalize the firefly readings, by dividing each firefly reading by the subsequent Renilla luminance reading.

3.5   *Experimental Design to Detect Translational Control*

1. To measure cap-independent translation, measure translation of a reporter RNA with and without a 5′ cap. Deletions and point mutations can be used to identify the sequence required for cap-independent translation. Luciferase expression should be reduced sharply from the uncapped RNA when the cap-independent translation element is disrupted. Expression should be restored fully when that same transcript contains a 5′ cap, but not by the presence of a poly(A) tail. Capped transcripts can be prepared by in vitro transcription in the presence of an 8:1 ratio of m^7GTP:GTP. The mMessage mMachine® kit from Ambion is efficient for producing capped transcripts.

2. In all known cap-independently translated animal viral mRNAs, and in some, but not all, plant viral mRNAs (1), cap-independent translation signals serve as internal ribosome entry sites (IRESes). To determine if a sequence is an IRES, it should facilitate translation of a second ORF in a bicistronic construct when placed between the two ORFs (Fig. 1a). Controls must be performed to verify that the translation of the second ORF does not involve reinitiation by ribosomes that had translated the first ORF. To confirm this, translation of the second ORF should be independent of the translation efficiency of the first ORF. To test this, introduction of a highly stable ($\Delta G \leq -60$ kcal/mol) stem-loop in the 5′ UTR should prevent translation of the first ORF, but not affect translation of the second ORF located downstream of the IRES. Another approach would be to compare translation of capped vs. uncapped forms of the dicistronic reporter. Absence of a 5′ cap should nearly abolish translation of the first ORF, but not affect the one located downstream of the IRES. In this case, care must be taken to

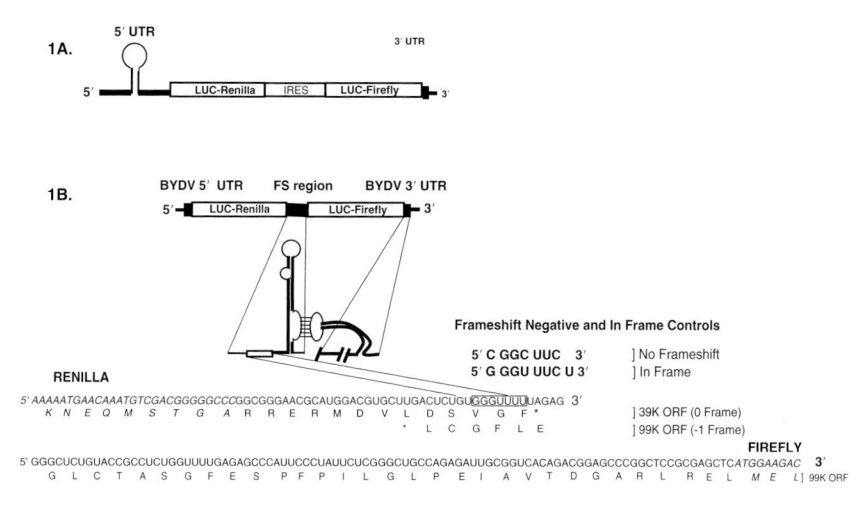

Fig. 1 (**a**) Bicistronic construct for detection of IRES function. The suspected IRES region is positioned in the intergenic region upstream of the firefly reporter gene (the second ORF). The firefly luciferase translation should be independent of translation efficiency of the Renilla luciferase (the first ORF). Thus, presence, in the 5′ UTR, of a highly stable stable stem-loop structure ($\Delta G \leq -60$ kcal/mol) that prevents ribosomes from translating the Renilla luciferase gene, should not affect IRES-mediated translation of the firefly luciferase ORF. (**b**) A bicistronic reporter used to detect −1 frameshift (FS) activity in BYDV RNA. The BYDV frameshift site, including the overlapping portions of the viral 39 K and 99 K ORFs, is fused to the upstream (Renilla) and downstream (firefly) luciferase ORFs so that −1 frameshifting is required for translation of the firefly luciferase. Unlike other frameshift elements, BYDV also requires a stem-loop in the 3′ UTR that base pairs to a bulge in the stem-loop adjacent to the slippery site (19) (shown below map). The sequence of the end of the Renilla luciferase ORF, the BYDV sequence, and the beginning of the firefly luciferase ORF is shown with luciferase-derived sequences in italics. The two lines represent one continuous sequence. Three point mutations to the heptanucleotide FS site (5′-C GGC UUC -3′), block −1 frameshifting, allowing translation only of the Renilla luciferase gene for the no frameshift control. Insertion of an extra base (C) in the slippery heptanucleotide site (5′-G GGU UUC U-3′) serves as the in-frame construct, enabling full translation of both the Renilla and firefly luciferase genes

measure the effects of lacking a cap on RNA stability (see below). To stabilize mRNA lacking a functional cap, RNA can be transcribed with an A(5′)ppp(5′)A modification instead of a cap. This blocks 5′ to 3′ exonucleases that attack uncapped mRNAs, but does not facilitate recruitment of the translational machinery to the cap (see refs. 15, 16).

3. To measure poly(A) tail-independent translation, translation of a reporter RNA with and without a 3′ poly(A) tail is measured. Deletions and point mutations can be used to identify the sequence required for poly(A) tail-independent translation. Luciferase expression should be reduced in the nonpolyadenylated RNA when the cap-independent translation element is disrupted. Expression should be restored fully when that same transcript contains a poly(A) tail, but not by addition of a

5′ cap, if the transcript lacks a cap. Polyadenylation is often less stimulatory to translation than presence of a 5′ cap. Hence, poly(A) tail "mimic" sequences can be more difficult to map than cap-independent translation elements.

Polyadenylated transcripts can be prepared by in vitro transcription of the reporter construct from a vector that harbors a poly(A) tract immediately downstream of the viral 3′ UTR, and immediately upstream of the restriction site used to linearize the plasmid prior to transcription. It is convenient to have unique restriction sites flanking the poly(A) tract, which should be at least 60 nt long, so that the same plasmid can be linearized at either of the two unique restriction sites to make transcripts that differ only by presence or absence of the poly(A) tail.

4. To measure ribosomal frameshifting, we use a system that places the putative frameshift sequence between the Renilla (upstream) and firefly (downstream) luciferase ORFs. The 3′ end of the Renilla ORF is fused in-frame with the zero frame portion of the frameshift site, and the −1 (or + 1) frame of the frameshift site is fused with the 5′ end of the firefly luciferase ORF (Fig. 1b). To calculate the frameshift efficiency, the ratio of firefly/Renilla luciferase activity obtained from the wild type frameshift signal is divided by the ratio of firefly/Renilla activity obtained with a version of the frameshift region that is altered by a one-base insertion (for −1 frameshifting, a one-base deletion for + 1) in the frameshift site which places both Renilla and firefly ORFs in the same reading frame. For example, the BYDV slippery site sequence is G GGU UUU, and an insertion that places both ORFs in the same frame alters the sequence to G GGU UUC U. The normalized frameshift values are then taken times 100 for the percent frameshift. All samples must be tested at least in triplicate.

3.6 RNA Stability Assays

Mutations that affect reporter gene expression may affect RNA stability rather than translation efficiency (17). To distinguish between these possibilities, the half-life of the RNA must be determined. Simply measuring physical half-life by northern blot hybridization, for example, may not reflect the true half-life of the mRNA potentially available for translation. Usually the electroporated mRNA is present in a million-fold excess over the protoplasts. After electroporation, the exact cellular location of much of this mRNA is unknown. It may adhere to the cell membrane or end up in compartments where it is unavailable to the translational machinery. Hence, a more reliable estimate of stability is obtained by measuring functional half-life of the reporter mRNA. This is done by tracking luciferase activity over a time course (e.g., 30 min–8 h posttransfection), to quantify protein accumulation, as a function of time (Fig. 2). The levels of luciferase level off sooner if the mRNA has a short half-life, and thus ceases to program the synthesis of more luciferase.

1. Using the equation described by Danthinne et al. (18), $t_{1/2} = P(\infty) \times \ln 2/aR_0$, functional luciferase protein accumulation can be recorded from mutant and wild type constructs and the functional RNA half-lives can be compared.

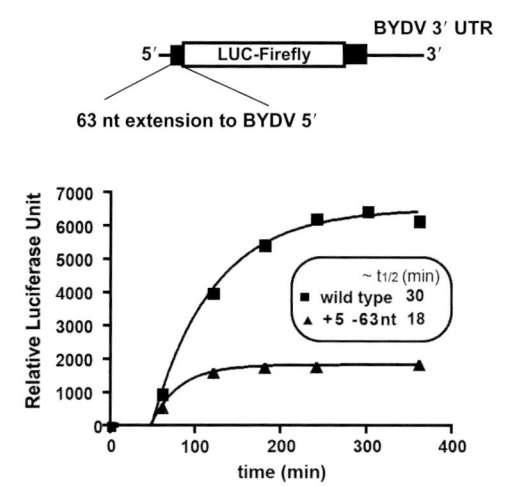

Fig. 2 Example of an RNA stability assay of BYDV showing a time course of firefly luciferase accumulation for wild type (wt) and a mutant construct with a 63 nt extension to the 5′ UTR (+5′ −63 nt), with the approximate functional half-lives of 30 and 18 min, respectively, calculated as described in Sect. 3.5. Reprinted from ref. 20, Fig. 5c

Transfected protoplasts should be harvested in triplicate, lysed, and analyzed for relative light unit (RLU) luciferase activity over six to eight time points post-transfection. A best-fit curve is calculated from these RLU time course values using GraphPad 4.0 software or another statistical program. After plotting a best-fit curve of the RLU luciferase activity for experimental and wild type constructs, the slope (RLU/min) of the best fit line (a) is obtained, from the time point following the lag phase at which the first luciferase protein is synthesized. The RNA half-life ($t_{1/2}$) can then be calculated using RLU activity saturation point as P (∞), the initial RNA input (R_0), and the best fit curve slope (a).

2. If there is no significant change in experimental RNA half-life ($t_{1/2}$) of a mutant mRNA, then a significant change in luciferase activity generated by this mRNA relative to wild type must be attributed to a change in translation efficiency.

4 Notes

1. Another option for establishing an oat protoplast cell suspension is to remove the entire disk from the shipping tube, and aseptically transfer it to a flask with MS media (13). In order to save time for weekly passages and to prepare for a protoplast electroporation experiment 1 week prior to the experiment date, 1 L of media can be aliquoted into 25 flasks (i.e., 40 mL per flask), with the flasks individually plugged with cotton, and wrapped with a 20 × 20 cm square of aluminum foil, autoclaved, and stored at 4 °C.

2. Typically one flask with 50 mL of culture is adequate for growing 10 mL of packed cells (i.e., cell culture transferred to a 50 mL conical tube and allowed 15–20 min time for cells to

settle). Digestion medium is prepared fresh in ASW:0.6 M Mannitol Digestion Solution. Initially combine the components into a beaker with 50 mL of ASW:Mannitol solution while stirring on a stir plate. A standard digestion solution volume is 50 mL for one 50 mL flask of suspended cells. Cellulase is an essential reagent for the gradual digestion of cell walls, while retaining the integrity of the cell membranes.

3. Adequately grown suspension culture shows a fine granulated consistency with minimal cell aggregations. One 50 mL culture will yield approximately 10 mL of packed cells. If exceeding 10 mL, increase the total amount of enzyme solution to maintain a 5:1 ratio of enzyme solution to packed cells (13).

4. Washing the protoplasts free of cellular debris: In order to minimize cell rupturing and membrane damage, use a motorized pipet filler/dispenser (e.g., Eppendorf EasyPet) to gently resuspend the cells. The protoplasts should be carefully brought up into the dispenser and gradually released against the side of the conical tube.

5. Protoplasts do not retain their cell membrane integrity with dramatic changes in osmotic pressure. Therefore, it is necessary to ensure that the concentrations of components in the buffer wash solution (ASW:0.6 M mannitol) are correct, to prevent continual presence of disrupted cell membranes due to unfavorable osmotic wash conditions. Electroporation of RNA into cells with damaged membranes will yield erroneous results.

6. To verify the concentration and integrity of the RNA, we analyze it by agarose gel electrophoresis. Internal controls include electroporating cells in triplicate with the same RNA construct derived from the same RNA preparation. At least three separate RNA preparations need to be done in triplicate to assess how the viral mutation will affect the luciferase activity.

7. There are variations on how many, and at what time frame, RNA samples can be electroporated into 1 mL of oat protoplasts within one cuvette and it is up to the investigator to set up the relevant number of transfections. We constructed a recombinant viral vector with the 5′ and 3′ UTRs flanking a reporter fluorescent marker (e.g., firefly gene) and a second RNA control construct, which encodes another fluorescent marker (e.g., Renilla gene) that is added simultaneously, to normalize for the firefly readings. RNA samples may also be electroporated as a two-step electroporation with the experimental RNA electroporated a second time 4-h post primary transfection.

8. For assays designed for IRES detection (19), frameshift and readthrough constructs, the positive and negative RNA constructs are treated as separate constructs and typically are electroporated as separate triplicate samples. Whatever the configuration, it is critical that the investigator tailor an experiment with a statistically relevant number of samples that will be electroporated, and that at least two different recombinant RNA preparations are prepared per sample, and tested for luciferase activity in triplicate. Three electroporations per sample translates into three mL of resuspended oat protoplasts (1 mL per cuvette), plus two samples reserved for "No RNA" control samples. For example, to assess one point mutation on the 5′ and 3′ sides of a stem-loop structure, as well as compensatory mutations for a proposed secondary structure totals 17 cuvettes: nine cuvettes (three samples per mutant construct), three negative control cuvettes, three positive control constructs per experimental construct, and two only controls.

9. If using a BTX® Electro Squareporator T820, the optimal setting is one pulse for 6 ms at 300 V. If using the Biorad GenePulser Xcell™, the optimal setting is at 300 V/500 μF. The first time one uses electroporation, we advise optimizing the voltage by electroporating a positive control mRNA over a range of voltages between 200 and 500 V. Also the capacitance (μF) should be optimized. Prior to the luciferase assay, examine cells under a microscope immediately after electroporation to determine the amount of damage. There is a trade-off between electroporation efficiency and cell damage caused by the electroporation.

10. Before the electroporation process, position the P-1000 pipetter, with large bore pipette tips, and the cuvettes with 1 mL of resuspended oat protoplasts in close proximity to one another to help cycle through all the cuvettes quickly. It also helps to have the cuvettes, designated RNA, and six-well plates prelabeled, to maintain the respective RNAs, their designated cuvettes, and the six-well plates in a quickly recognizable format. The electroporation steps need to be done in a synchronous and consistent pattern, so that all of the respective RNA, electroporated cells, and plates are handled in a consistent manner.

11. The cells can be frozen at this point in the protocol. However, slightly stronger luminescent signals are detected when the Renilla and firefly luciferase reporter activities are measured on the same day as the electroporation.
12. RNA stability assays are not usually warranted for dual luciferase frameshift and readthrough constructs because the insertion of both reporter genes on the same construct serves as its own internal control. However, if Renilla is the internal control and shows variability between samples, variable RNA stability is possible, and RNA stability should be tested.

References

1. Kneller, E.P., Rakotondrafara, A.M., Miller, W.A., *Cap-independent translation of plant vial RNAs*. Virus Res, 2005. **119**: 63–75.
2. Matsuda, D., Dreher, T., *In Vivo Translation Studies of Plant Viral RNAs Using Reporter Genes*. Current Protocols in Microbiology. 2005. 16 K.2.1–16 K.2.11.
3. Gallie, D.R., *The cap and poly(A) tail function synergistically to regulate mRNA translational efficiency*. Cold Spring Harbor Laboratory, 1991: p. 2108–2116.
4. Gallie, D.R., Kobayashi, M., *The role of the 3′ Untranslated region of nonpolyadenylated plant mRNAs in regulating translational efficiency*. Gene, 1994. **142**: 159–165.
5. Brasier, A.R.a.F., J.J., *Nonisotopic assays for reporter gene activity*. In: Current Protocols in Molecular Biology, R. Brent. F.M. Ausubel, R.E. Kingston, D.D. Moore, J.G. Seidman, J.A. Smith, K. Struhl (eds.). 1995, Hoboken, N.J.: John Wiley & Sons. 9.7.12–9.7–21.
6. Matsuda, D., Bauer, L., Tinnesand, K., Dreher, T.W., *Expression of the two nested overlapping reading frames of TYMV RNA is enhanced by a 5′ cap and by 5′ and 3′ viral sequences*. J Virol, 2004. **78**: 9325–9335.
7. Yamaji, Y., Kobayashi, T., Hamada, K., Sakurai, K, Yoshii, A., Suzuki, M., Namba, S., Hibi, T. et al., *In vivo interaction between Tobacco mosaic virus RNA-dependent RNA polymerase and host translation elongation factor 1A*. Virology, 2006. **347**, 100–108.
8. Brierley, I., Dos Ramos, F.J., *Programmed ribosomal frameshifting in HIV-1 and the SARS-CoV*. Virus Res, 2006. **119**(1): 29–42.
9. Baril, M., Brakier-Gingras, L., *Translation of the F protein of hepatitis C virus is initiated at a non-AUG codon in a + 1 reading frame relative to the polyprotein*. Nucleic Acids Res, 2005. **33**(5): 1474–1486.
10. Plant, E.P., Perez-Alvarado, G.C., Jacobs, J.L., Mukhopadhyay, B., Hennig, M., Dinman, J.D. *A three-stemmed mRNA pseudoknot in the SARS coronavirus frameshift signal*. PLoS Biol, 2005. **3**(6): e172.
11. Dreher, T.W., Miller, W.A., *Translational control in positive strand RNA plant viruses*. Virology, 2006. **344**(1): 185–197.
12. McInerney, P., T. Mizutani, T. Shiba, *Inorganic polyphosphate interacts with ribosomes and promotes translation fidelity in vitro and in vivo*. Mol Microbiol, 2006. **60**(2): 438–447.
13. Rakotondrafara, A.M., Jackson, J.R., Pettit Kneller, E., Miller, W.A. Preparation and electroporation of oat protoplasts from cell suspension culture. Curr. Protocols Micro., 2007. 16D.3.1–16D.3.12.
14. Kozak, M., *New ways of initiating translation in eukaryotes?* Mol Cell Biol, 2001. **21**: 1899–1907.
15. Thoma, C., et al., *Enhancement of IRES-mediated translation of the c-myc and BiP mRNAs by the poly(A) tail is independent of intact eIF4G and PABP*. Mol Cell, 2004. **15**(6): 925–935.
16. Sanchez, M., et al., *Iron Regulation and the Cell Cycle: Identification of an iron-responsive element in the 3′ -untranslated region of human cell division cycle 14A mRNA by a refined microarray-based screening strategy* J Biol Chem, 2006. **281**(32): 22865–22874.

17. Meulewaeter, F., Van Montagu, M., Cornelissen, M., *Features of the autonomous function of the translational enhancer domain of satellite tobacco necrosis virus.* RNA, 1998. **4**(11): 1347–1356.

18. Qin, X., Sarnow, P., *Preferential translation of internal ribosome entry site-containing mRNAs during the mitotic cycle in mammalian cells.* J Biol Chem, 2004. **279**: 13721–13728.

19. Barry, J.K., Miller, W.A., *A-1 ribosomal frameshift element that requires base pairing across four kilobases suggests a mechanism of regulating ribosome and replicase traffic on a viral RNA.* Proc Natl Acad Sci U S A, 2002. **99**(17): 11133–11138.

20. Rakotondrafara, A.M., et al., *Oscillating kissing stem-loop interactions mediate 5′ scanning-dependent translation by a viral 3′-cap-independent translation element.* RNA, 2006. **12**: 1893–1906.

Chapter 8
In Vitro Analysis of Translation Enhancers

Aurélie M. Rakotondrafara and W. Allen Miller

Abstract The genomes of many plant viruses contain translation-enhancing sequences that allow them to compete successfully with host messenger RNAs for the translation machinery. Identification of translation enhancer elements is valuable, both to gain understanding of virus gene expression control and to apply them as tools for engineering gene expression in plant cells. Here, we describe experiments designed to detect viral elements that enhance translation, focusing on cap-independent translation activity, using a high fidelity cell-free wheat germ translation extract.

Keywords Plant RNA viruses; Translation enhancer; Cap-independent translation; In vitro translation; Wheat germ extract; Cap analogs; Ribosome scanning; Internal ribosome entry site

1 Introduction

Translation is a key step in gene expression of most plant viruses because most have an RNA genome. Initiation is the rate limiting and most regulated step in translation. The first step of translation initiation involves recognition of the messenger RNA by the protein synthesis machinery (1). For classical capped and polyadenylated mRNAs, this recognition is achieved by binding of the 5' cap (m^7GpppN) on the mRNA to the cap-binding translation initiation factor complex, eIF4F, which then recruits the 40S ribosomal subunit. This process is enhanced by the 3' poly(A) tail, which is brought to close proximity of the 5' cap by interacting with eIF4F via poly(A) binding protein. Thus, the poly(A) tail and the 5' cap synergistically stimulate translation initiation. Once positioned at the 5' end of the RNA, the 40S ribosomal subunit progresses in a 5' to 3' direction in search of the initiation codon.

Many plant viral RNAs harbor elements that enhance the recruitment of the host translation machinery in the presence or absence of a 5' cap and/or a 3' poly(A) tail. Translation enhancers can be categorized in two groups: (1) elements that substantially stimulate translation in combination with the 5' cap and (2) elements that

From: *Methods in Molecular Biology, Vol. 451, Plant Virology Protocols:*
From Viral Sequence to Protein Function
Edited by G.D. Foster, I.E. Johansen, Y. Hong, and P.D. Nagy © Humana Press, Totowa, NJ

promote translation cap-independently. The former elements include the Potato virus X 5' leader motifs which are required for efficient translation of the capped, polyadenylated RNA (2), the Tobacco mosaic virus (TMV) Ω sequence (3) and the Alfalfa mosaic virus (AMV) RNA 4 5' UTR (4) which stimulate cap-dependent translation by an order of magnitude relative to that of the RNA with a cap only. The Ω sequence stimulates translation of uncapped RNA, but addition of a cap stimulates it even more (3). The 3' end of Turnip yellow mosaic virus (TYMV) (5), which consists of a tRNA-like structure, and the nonpolyadenylated AMV 3' noncoding region (6) substitute for the role of a poly(A) tail on capped mRNAs. The cap-independent translation elements include the internal ribosome entry sites in the 5' UTRs of potyviral RNAs and the cap-independent translation elements in the 3' UTRs of uncapped, nonpolyadenylated viral RNAs of luteoviruses and members of the Tombusviridae family (for review see ref. 7). It is likely that many other plant viruses have translational enhancers. Here, we describe an approach to identify cap-independent translation elements in plant viral RNAs.

1.1   Advantages of the In Vitro Translation System

In vitro cell-free translation systems consist of crude cell extracts containing active components for protein synthesis, including tRNAs, ribosomes, and translation factors, supplemented with an energy regenerating system (8). They require only the addition of amino acids and the mRNA template (see note 1) and optimization of the ionic conditions to initiate protein expression. The most commonly used systems include the S30 supernatant from wheat germ extract (WGE) (9) or a rabbit reticulocyte lysate (RRL) (10). Both systems are commercially available (e.g., Promega, Madison, WI, Ambion, Austin, TX). The WGE is a higher fidelity system than RRL for translation of plant or animal mRNAs (11). The RRL is less cap-dependent and is prone to initiate translation at internal sites and to read through stop codons.

Usually, in vitro translation extracts are pretreated with micrococcal nuclease to remove endogenous host mRNAs. This results in a less competitive environment when compared to cells in terms of availability of the translation machinery to any input mRNA. In this state, the presence of a poly(A) tail on the mRNA does not provide a translational advantage. The synergistic stimulation of translation by the cap and the poly(A) tail, which is normally observed in cells, can be reproduced in the presence of competing mRNAs, or in a system partially depleted of ribosomes and/or translation factors (12).

The cell-free translation system provides an easy-to-use, well-defined environment that can be tuned for efficient translation of any particular mRNA. Moreover, it allows rapid, sensitive detection of any translation product, by the incorporation of radiolabeled amino acids added to the reaction mixture or by direct measurement of the enzymatic activity of the encoded protein.

In vitro translation is used widely to study plant viral translation independent of viral replication or other steps of the viral life cycle. It is an ideal system to identify the minimal functional sequence sufficient for translatability of an mRNA. However, it is possible that the signals defined in vitro may not be sufficient for full translational control in cells. Moreover, some translation enhancers are not readily detected in vitro, including the poly(A) tail on most mRNAs. Thus, it is important to complement in vitro data with experimental measurements in cells.

This protocol focuses not on the preparation of in vitro translation extracts (see ref. 8) but on experiments designed to study cap-independent translation enhancers using the wheat germ extract, which include mapping of the translation element within the viral genome, determining the effect of free cap-analog on cap-independent translation, investigating the requirement for 5' ribosome scanning, and testing for an internal ribosome entry site (IRES).

2 Materials

2.1 Wheat Germ Translation

1. Commercial wheat germ in vitro translation system (e.g., Promega, cat no. L4370, L4380, L4390, L4400), which includes the wheat germ extract, 1 M potassium acetate, 1 mM amino acids mixture (minus methionine, minus cysteine, or minus leucine) and BMV RNAs, provided as a positive control and molecular weight marker when synthesizing radiolabeled proteins
2. [^{35}S]-labeled methionine (1,200 Ci mM^{-1}) at 10 mCi mL^{-1} (e.g., Amersham, Arlington Heights, IL) (see note 2)
3. Ribonuclease inhibitor (e.g., Promega RNAsin supplied at 33 U µL^{-1}, Ambion SuperRNAsin supplied at 20 U µL^{-1})
4. Nuclease-free water or DEPC-treated water
5. In vitro transcribed RNAs (see note 3). Always include a positive control (e.g., wild type RNA) for any preparation

2.2 SDS-PAGE Gel and Autoradiography

1. SDS-polyacrylamide gel electrophoresis system (refer to ref. 13)
2. Gel-drying apparatus
3. phosphorimager screen
4. phosphorimager apparatus (e.g., Typhoon, GE Healthcare)

2.3 Quantification of Luciferase Activity

1. Luminometer (e.g., Turner Designs TD-20/20 Luminometer)
2. Luciferase assay system reagent (Promega, E1500)

3 Methods

3.1 In Vitro Translation Assay

The in vitro translation standard reaction using radioactive detection is carried out according to manufacturer's instructions, subject to alteration of conditions discussed in Sect. 3.2, in a final small volume of 12.5 μL for each sample (see notes 4–6). Caution must be taken to avoid RNase contamination. Carry out the preparation of the translation mixture on ice. This minimizes possible inactivation of the wheat germ extract components and helps to synchronize translation initiation for all samples at a similar time point. Rapidly freeze the unused extract to −80 °C.

1. Assemble the following reaction:
   - 2 μL amino acid minus mixture (minus methionine) (see note 2).
   - 1 μL 1 M potassium acetate for a final concentration of 130 mM (see Sect. 3.2.1).
   - Ribonuclease inhibitor at a final concentration of 1 U μL⁻¹.
   - 0.65 μL [³⁵S]-labeled methionine, 1150 Ci/mmole, 8.3 μM.
   - 6.5 μL WGE.
   - 0.1 pmol in vitro transcribed RNA (see Sect. 3.2.2). Use 0.25 μg of the provided BMV RNA control as a molecular weight marker.
   - Bring the final volume to 12.5 μL with nuclease-free water.
2. Incubate the reaction at 25 °C for 60 min (see Sect. 3.2.3).
3. Stop the reaction by placing samples on ice and by adding gel loading buffer.
4. Run samples by standard SDS polyacrylamide gel electrophoresis (refer to ref. 13) (see note 7).
5. Dry gel and expose it to a phosphorimager screen.
6. Quantify the radioactivity of the band migrating at the expected size of your protein of interest using image analysis software (e.g., ImageQuant). The intensity of the band reflects the level of protein expression based on the amount of labeled amino acids incorporated into the protein during synthesis. It is important to subtract background counts, which is determined by the quantification of an equivalent area on a negative reaction control lane (translation reaction without mRNA template added). For an accurate quantification of different reaction products, the number of methionine codons (or other labeled amino acids) present in each protein of interest must to be taken into account.

3.2 Essential Optimizations of the In Vitro Translation Reaction

The translation efficiency of any particular mRNA depends on the condition used in the in vitro reaction. Prior to any definitive assay, it is imperative to determine the appropriate conditions (potassium, magnesium, mRNA concentrations, incubation time) to optimize the translation reaction for sensitivity to changes in the translation efficiency of the tested mRNA.

3.2.1 Optimization of the Magnesium and Potassium Concentrations

The optimal ionic conditions for translation vary for each mRNA transcript. The Promega WGE contains an endogenous level of 53 mM potassium acetate and 2.1 mM magnesium acetate. To optimize ionic conditions for the mRNA of interest, in separate reactions, test concentrations ranging from 2 to 5 mM magnesium acetate and 50 to 200 mM of potassium acetate.

3.2.2 Optimization of mRNA Concentration

In order to measure the changes in translation efficiency of intact mRNAs, the translation system should not be saturated by the amount of mRNA template added to the reaction. Thus, it is necessary to generate an mRNA concentration curve.

Test an increasing concentration of mRNA (0.05–5 pmol) in 12.5 μL translation reactions as described in Sect. 3.1.

1. Plot the level of the protein synthesized for the given mRNA vs. concentration. The optimal concentration of mRNA is that in which there is a linear response of the protein product to the amount input mRNA.

3.2.3 Optimization of the Reaction Incubation Time

The reaction should be run long enough to obtain sufficient product, but should be terminated before the products reach saturation.

1. Perform a time course with a subsaturating concentration of mRNA template in a 50 μL translation reaction. Adjust the recommended volume of each component to the wheat germ mixture accordingly.
2. Remove 5 μL of the translation mix from the on-going translation reaction at different time points (from 0 to 3 h) taking more samples at early time intervals.
3. Rapidly freeze each aliquot until the analysis of the results of the translation products for all samples is performed.
4. Plot the level of the protein synthesized from the given mRNA vs. time. The optimal incubation time is around the time point at which 50% of maximum protein accumulates, where the response of protein produced over time is still linear.

3.3   In Vitro Translation Mediated by a Translation Enhancer on a Reporter mRNA

A conventional approach to study viral translation is to place a reporter gene in between the viral UTRs in place of part or all viral coding regions. In some cases, the coding region of the gene of interest contributes to the translation efficiency. To ensure that the replacement of the coding region with a reporter does not interfere with the normal activity of the putative translation element, it is important to check that the reporter RNA construct behaves similarly to an all-virus based mRNA construct in a standard in vitro translation reaction. If the viral ORF does appear to affect translation, it can then be included as an extension of the UTR in the reporter construct, or the reporter gene can be fused in-frame to the virus ORF. The reporter construct can reveal the minimal sequence required for full activity of the translation element in a heterologous mRNA. The translatability of the reporter mRNA can be easily detected both in vitro and in vivo via enzymatic assay, which is an alternative to the standard radioactive translation product detection (see note 8).

Different reporters, which are not endogenous to the cell-free extract, are commonly used to study translation efficiency. These include the chloramphenicol acetyl-transferase (CAT), β-glucuronidase (GUS), and luciferase reporter genes. The luciferase assay, which measures the light intensity emitted during hydrolysis of ATP by the enzyme in presence of the substrate luciferin, is (1) fast, allowing immediate reading of the enzyme activity upon addition of the substrate, (2) does not require additional sample preparation, (3) is more sensitive than other reporter genes, and (4) has no background activity in the in vitro translation system.

1. Carry out the in vitro translation reaction as described in Sect. 3.1, excluding the addition of the radiolabeled amino acid (see notes 5 and 6). Each sample should be tested at least in triplicate.
2. Stop the in vitro translation reaction by placing the samples on ice for about 10 min.
3. Thaw the luciferase assay reagent (see note 9) and aliquot 50 μL in fresh eppendorf tubes.
4. Add 1–5 μL of the positive control in vitro translation reaction (the positive control is an mRNA, such as one with wild type viral UTRs known to translate efficiently) to the luciferase reagent and set the sensitivity of the luminometer to ensure no signal saturation. Adjust the volume of the sample to use, accordingly.
5. Start reading all samples. When analyzing the data, the percentage of luciferase activity of each RNA sample can be normalized relative to the positive control, which is then defined as 100%. (see note 10)

3.4   In Vitro Analysis of Cap-Independent Translation Elements

Several in vitro assays can be performed to test the presence of a cap-independent translation element within a plant viral RNA.

3.4.1 Mapping the Cap-Independent Translation Element Region

The first step in the identification of a translation enhancer is to define the region within the viral genome that contributes most to the translatability of the uncapped RNA in vitro. The translation elements can reside within an UTR or be part of the viral coding region. Map the 5' and 3' boundaries of the translation element by testing the effect of progressive truncation, and internal deletion, on the translation of the transcript of the viral RNA genome or subgenomic RNA. Compare the translation of all RNA transcripts in both capped and uncapped forms (see note 11). A cap-independent translation element is defined as the minimal region that is necessary and sufficient for the translation of an uncapped mRNA, and can be replaced functionally by the addition of a 5' cap.

1. Progressively truncate the mRNA transcript from the 3' end by linearization of the cDNA clone with various restriction enzymes (see note 12).
2. Test the efficiency of translation of each truncated mRNA transcript in both uncapped and capped forms in the in vitro translation reaction optimized as under Sect. 3.2.
3. Compare the efficiency of translation of each truncated mRNA to that of the wild type, full-length viral RNA (in both capped and uncapped forms). In the presence of the cap-independent translation element, the addition of a cap generally does not provide more than two or threefold stimulation of translation to that of the uncapped form. The loss of the cap-independent translation element should cause at least a fivefold drop in translation efficiency of the truncated mRNA when compared to that of the full-length RNA, although this may vary depending on the system. The addition of a cap should fully restore translation of the RNA, from which the translation element has been deleted, to wild type level (see notes 13–15). Test the putative cap-independent translation element in the context of a nonviral RNA (e.g., reporter construct, see Sect. 3.3) and see whether it confers translation at a similar level as in the natural context viral RNA when comparing capped vs. uncapped RNAs. Keep in mind that additional sequences elsewhere in the genome could contribute to full cap-independent translation activity, particularly in cells.

3.4.2 Effect of Free Cap Analog on Cap-Independent Translation

A functional assay to test the cap-independent translation mechanism is to determine the effect of free cap analog (m^7GpppG) on the efficiency of translation of the mRNA of interest. Free cap analog inhibits cap-mediated translation by competing for the cap-binding pocket in the cap-binding factor eIF4E. If the translatability of the particular mRNA of interest does not depend on eIF4E, but relies simply on recruitment of the ribosomes in the absence of a cap-binding protein, the presence of free cap analog should have no or little effect on its translation.

1. Carry out the in vitro translation reaction as described in Sect. 3.1 with the optimal translation condition for the particular mRNA of interest. In separate

reactions, add increasing concentrations of free cap analogs (m^7GpppG, ranging from 0 to 0.4 mM) to the translation reaction.

2. As control, in a separate reaction, add similar increasing concentrations of GTP to the translation reaction. Because GTP lacks the 7-methyl group, it does not bind the cap-binding pocket of eIF4E and thus does not inhibit the cap-dependent translation. This confirms the specificity of the effect of the cap analog on the RNA translatability (see note 16).

3. Compare the efficiency of translation of the uncapped mRNA to the capped form with a functional translation element, and a capped mRNA lacking a translation element, in the presence of increasing concentration of either m^7GpppG or GTP. Determine the concentration of cap analog required for each mRNA to inhibit translation by 50%. The difference in translation behavior of both capped and uncapped RNAs in the presence of cap analog indicates their requirement for the cap-binding protein, and hence cap-dependence.

3.4.3   Testing for the Requirement for Ribosome Scanning from the 5' End

The cap-independent translation elements can either bind the ribosomal subunits internally near the start codon, regardless of its position relative to the 5' end, or recruit the translation initiation complex cap-independently, but requires a free 5' end to initiate ribosome scanning through the 5' UTR to reach the first initiation codon. The latter mechanism thus far appears to be limited to plant viruses.

The standard assay for 5' ribosome scanning is to (1) place a stem-loop structure ($\Delta G = -30$ kcal mol^{-1}) within 12 nucleotides of the 5' end in one construct, and at a much more 5' distal position in the 5' UTR of another construct (14) (see note 17), or (2) insert an upstream AUG within the 5' UTR of the uncapped mRNA, out of frame relative to the main open reading frame (ORF).

1. Perform the in vitro translation assay as described in Sect. 3.1.

2. Compare the efficiency of translation of the mRNA containing the above stem-loops or the out-of-frame upstream AUG to that of the wild type mRNA. If the cap-independent translation mechanism is 5' end dependent, the presence of the stable stem-loop at the very 5' end of the RNA will inhibit translation, by blocking access of the 5' end to the ribosome (see note 18). When located at a distal position from the 5' end, the same stem-loop structure will have no effect on the binding and scanning of the ribosomes through the 5' UTR (see note 17). Also, if the cap-independent translation mechanism relies on ribosome scanning, it will favor the recognition of the upstream AUG, which would drastically reduce initiation from the start codon of the main ORF located further downstream (see notes 19 and 20).

3.4.4   Testing for an IRES

A conventional assay to determine whether cap-independent translation mechanism occurs via internal ribosome entry is to test the translation element in the context of a dicistronic mRNA (15) (see note 21). In an internal ribosome entry mechanism, the 40S ribosomal subunit binds directly to the IRES element, independent of the 5' end, and initiates translation at an internal AUG. The assay from the dicistronic construct determines whether the putative IRES element in the intercistronic region will support translation of the downstream ORF, which normally remains untranslated (or at a background level). If the translation element is an IRES, translation of the downstream ORF is independent of the translatability of the first ORF and is controlled solely by the translation element placed in the intergenic region.

1. Insert the putative IRES element in between two standard reporter genes (Fig 1a).
2. Test the dicistronic construct in both capped and uncapped forms in a standard in vitro translation reaction and measure the translation efficiency of each ORF. IRES-mediated translation may require higher salt concentration for optimal expression. Further optimization of the ionic condition of the in vitro translation may be needed (see Sect. 3.2.1). Include all control constructs to demonstrate specificity of translation (Fig 1b). By removing the 5' cap or by blocking ribosome entry from the 5' end with the addition of a stem-loop (see Sect. 3.4.3), the translation of the first ORF is abolished. However, the translatability of the second ORF should not be affected if the sequence inserted in between the two ORFs is an IRES. An IRES element is demonstrated by its ability to sustain a level of expression of the second ORF that is many fold higher than is obtained in the presence of a nonIRES sequence between the two ORFs. To confirm the IRES activity of the translation element, it is necessary to test the same constructs in vivo, and ensure that the transcript remains intact because it is possible that the translation of second ORF results from a translation of a truncated form of the RNA transcript.

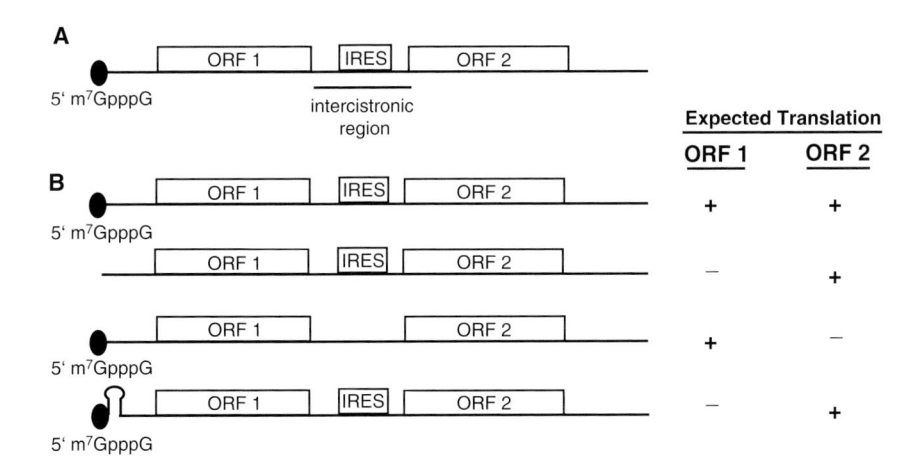

Fig. 1 (a) Schematic representation of a dicistronic construct to test the IRES activity of a translation element from an intercistronic position. (b) Various dicistronic RNA constructs used to demonstrate internal initiation of translation at the IRES. The expected translational activity of each open reading frame (ORF) in the different contexts is shown

4   Notes

1. Some in vitro translation systems can be programmed with a DNA template because they are coupled transcription and translation systems (e.g., Promega, Ambion). The mRNA is transcribed from the DNA template driven by a SP6, T7 RNA polymerase promoter, and without further purification, it is translated. While convenient, this system may not be appropriate to compare quantitatively the translation efficiency among different RNA samples. It is difficult to estimate the amount of RNA present at a given time and some mutations may affection transcription rate.

2. The WGE lacks endogenous amino acids, which permits addition of radiolabeled residues in the reaction. ^{35}S-labeled methionine is most frequently used to label proteins, unless the protein of interest does not bear any methionine in its sequence. ^{3}H-labeled leucine can also be used. When adding ^{35}S-labeled methionine to the reaction mix, use the amino acids mixture minus methionine.

3. Check all in vitro transcribed RNAs by agarose gel electrophoresis prior to assay to ensure that they are intact, with the expected size and concentration. It is important to be accurate and consistent with the concentration of the RNA used throughout the assay, to ensure that the amount of protein synthesized reflects translation efficiency of each mRNA and not variations in RNA concentration.

4. While the user can follow the recommended final reaction volume of 50 µL per sample as instructed by the manufacturer, the authors found that a 12.5 µL reaction volume is more than sufficient for an in vitro translation assay and minimizes the use of the expensive reagents. If a larger volume reaction is needed, increase the recommended volumes described in Sect. 3.1 accordingly.

5. It is important to include for each preparation, an in vitro translation reaction without mRNA template added. This helps to determine background level of expression of endogenous mRNAs present in the WGE, and the specificity of any translation products observed in the assay to mRNA template added to the reaction.

6. It is recommended to prepare a master mix, from which aliquots are removed for each sample, prior to the addition of the RNA transcripts. This provides accuracy in component concentration and decreases deviation in between samples.

7. The quantification of the synthesized proteins can also be performed by trichloroacetic acid precipitation of labeled proteins followed by scintillation counting to determine the approximate percentage of incorporation of the radiolabeled amino acids. The main advantage of separating the protein product on a SDS-PAGE is that it determines the size of the translation products and reveals products of premature termination, proteolysis, or internal initiation.

8. The enzymatic assay measures indirectly the translation efficiency of the RNA, as we assume that there is a linear correlation between enzyme activity and amount of enzyme expressed. This is in contrast with the quantification of protein expression by radiolabeling, which measures direct accumulation of the translation products.

9. When testing a large number of samples, it is advisable to use the Steady-glo® Luciferase assay system (Promega, E2510). The Steady-glo® luciferase has a signal half-life of more than 1 h (compared to the 1 min signal half-life of the standard Luciferase assay system). However, be aware of the lower sensitivity of the Steady-glo® luciferase system.

10. If no luciferase activity is observed in any of the samples, test your reporter constructs in presence of radiolabeled amino acids for a standard quantification of the translation products as described in Sect. 3.1. While the luciferase gene tolerates some fusions to its N-terminal domain, it remains possible that the fusions to the reporter construct may have altered the proper folding or activity of the luciferase enzyme.

11. Capped transcripts are synthesized during in vitro transcription in the presence of cap analog (m7GpppG) and normal GTP at a 4:1 ratio (e.g., mMESSAGE mMachine, Ambion). The cap analog is incorporated only as the 5' terminal G of the transcript.

12. The secondary structures of all mutants should be predicted using a program such as MFOLD (16) prior to construction.

13. If translation of a truncated mRNA transcript is not restored in the presence of a cap, it is worth verifying the capping efficiency during RNA transcription. The efficiency of capping for each transcription can be a subject of variation. Thus, it is suggested to repeat each translation assay in replicate with different batches of in vitro transcribed RNAs.

14. Uncapped mRNAs are more susceptible to degradation than the capped forms, especially in cells (17). It is advisable to confirm that instability of the mRNA does not account for the loss of translation of the uncapped transcripts lacking a putative translation enhancer. For this purpose, perform (1) a functional mRNA stability assay, by comparing the translation efficiency of each mRNA over a time course and (2) physical mRNA stability assay, by extracting total RNA from the translation mixture at different time points and analyzing RNA accumulation over time by a northern blot hybridization (17). If RNA degradation is a concern, test translation of the mRNA transcript in the presence of a nonfunctional-ApppG cap as the 5' terminal nucleotide. The m^7ApppG cap increases the stability of the transcript but has no stimulatory effect on translation, as it is unable to recruit the translation factors.

15. It is important to test each mRNA in vivo to determine whether the boundaries of the putative translation element are the same in vivo and in vitro. Additional sequences have been observed to be necessary for full expression of the viral RNA in vivo (7).

16. If translation is inhibited by the presence of GTP, it may be necessary to optimize the magnesium concentration. The excess of nucleotide may chelate the magnesium present in the translation reaction, which results in nonspecific inhibition of translation.

17. The presence of an extremely stable stem-loop structure ($\Delta G = -61$ kcal mole^{-1}) inhibits translation regardless of its position within the 5' UTR, as the stem structure is too stable to be disrupted by the scanning ribosome complex (14, 18).

18. Predict secondary structure of the 5' UTR of the particular mRNA of interest in the presence of the stable stem-loop using a program such as MFOLD (16), to ensure that there are no other alterations of the secondary structure, which may result in artifactual translation inhibition.

19. It remains possible that some detectable level of translation can be measured from the downstream start codon, which results from leaky scanning or reinitiation. The efficiency of translation at the upstream AUG is also influenced by the sequence context surrounding the start codon. The optimal translation initiation context in plant system is the consensus: A(C/A)AAUGG (19).

20. Discontinuous scanning or shunting of the ribosome remains a possibility to be tested. Such a mechanism involves specific sequences that act as "take-off" and "landing" sites for the ribosome, and secondary structures within the 5' UTR that block the linear progression of the 40S ribosomal subunit (20).

21. Circular mRNA can also be used to test IRES activity (21). In this context, the translation efficiency of the IRES is measured in the absence of a free 5' end to demonstrate the direct binding of the 40S ribosomal subunit to the IRES element independently of the 5' end. However, construction of homogeneous circular mRNAs is technically difficult.

References

1. Kapp, L. D., and Lorsch, J. R. (2004) The molecular mechanics of eukaryotic translation. *Annu. Rev. Biochem.* **73,** 657–704.
2. Zelenina, D. A., Kulaeva, O. I., Smirnyagina, E. V., Solovyev, A. G., Miroshnichenko, N. A., Fedorkin, O. N., et al. (1992) Translation enhancing properties of the 5'-leader of potato virus X genomic RNA. *FEBS Lett.* **296,** 267–270.
3. Gallie, D. R. (2002) The 5'-leader of tobacco mosaic virus promotes translation through enhanced recruitment of eIF4F. *Nucleic Acids Res.* **30,** 3401–3411.

4. Neeleman, L., Olsthoorn, R. C., Linthorst, H. J., and Bol, J. F. (2001) Translation of a non-polyadenylated viral RNA is enhanced by binding of viral coat protein or polyadenylation of the RNA. *Proc. Natl. Acad. Sci. USA* **98**, 14286–14291.

5. Matsuda, D., and Dreher, T. W. (2004) The tRNA-like structure of Turnip yellow mosaic virus RNA is a 3'-translational enhancer. *Virology* **321**, 36–46.

6. Neeleman, L., Linthorst, H. J., and Bol, J. F. (2004) Efficient translation of alfamovirus RNAs requires the binding of coat protein dimers to the 3' termini of the viral RNAs. *J. Gen. Virol.* **85**, 231–240.

7. Kneller Pettit, E. L., Rakotondrafara, A. M., and Miller, W. A. (2006) Cap-independent translation of plant viral RNAs. *Virus Res.* **119**, 63–75.

8. Lax, S. R., Lauer, S. J., Browning, K. S., and Ravel, J. M. (1986) Purification and properties of protein synthesis initiation and elongation factors from wheat germ. *Methods Enzymol.* **118**, 109–128.

9. Roberts, B. E., and Paterson, B. M. (1973) Efficient translation of TMV RNA and rabbit globin 9S RNA in a cell-free system from commercial wheat germ. *Proc. Natl. Acad. Sci. USA* **70**, 2330–2334.

10. Pelham, H. R., and Jackson, R. J. (1976) An efficient mRNA-dependent translation system from reticulocyte lysates. Eur. J. Biochem. 67, 247–256.

11. Kozak, M. (1989) Context effects and inefficient initiation at non-AUG codons in eucaryotic cell-free translation systems. *Mol. Cell. Biol.* **9**, 5073–5080.

12. Michel, Y. M., Poncet, D., Piron, M., Kean, K. M., and Borman, A. M. (2000) Cap-poly(A) synergy in mammalian cell-free extracts. Investigation of the requirements for poly(A)-mediated stimulation of translation initiation. *J. Biol. Chem.* **275**, 32268–32276.

13. Turner, R., and Foster, G. D. (1998) In vitro transcription and translation. *Methods Mol. Biol.* **81**, 293–299.

14. Kozak, M. (1989) Circumstances and mechanisms of inhibition of translation by secondary structure in eucaryotic mRNAs. *Mol. Cell. Biol.* **9**, 5134–5142.

15. Jackson, R. J., Hunt, S. L., Reynolds, J. E., and Kaminski, A. (1995) *in* "Cap-independent translation" (Sarnow, P., Ed.), Vol. 203, pp. 1–29, Springer-Verlag, Berlin Heidelberg.

16. Zuker, M. (2003) Mfold web server for nucleic acid folding and hybridization prediction. *Nucleic Acids Res.* **31**, 3406–3415.

17. Gallie, D. R. (1991) The cap and poly(A) tail function synergistically to regulate mRNA translational efficiency. *Genes Dev.* **5**, 2108–2116.

18. Guo, L., Allen, E., and Miller, W. A. (2001) Base-pairing between untranslated regions facilitates translation of uncapped, nonpolyadenylated viral RNA. *Mol. Cell* **7**, 1103–1109.

19. Joshi, C. P., Zhou, H., Huang, X., and Chiang, V. L. (1997) Context sequences of translation initiation codon in plants. *Plant Mol. Biol.* **35**, 993–1001.

20. Ryabova, L. A., and Hohn, T. (2000) Ribosome shunting in the cauliflower mosaic virus 35S RNA leader is a special case of reinitiation of translation functioning in plant and animal systems. *Genes Dev.* **14**, 817–829.

21. Chen, C. Y., and Sarnow, P. (1998) Internal ribosome entry sites tests with circular mRNAs. *Methods Mol. Biol.* **77**, 355–363.

Chapter 9
Identification of Plant Virus IRES

Sek-Man Wong, Dora Chin-Yen Koh, and Dingxiang Liu

Abstrat Plant RNA viruses exploit nonorthodox strategies, such as the use of internal ribosomal entry sites (IRES), to express multiple genes from a single RNA species. IRES elements have been reported in tobacco etch virus (TEV), crucifer infecting tobamovirus (crTMV), hibiscus chlorotic ringspot virus (HCRSV), and many other animal and plant RNA viruses. In this chapter, the methodology used to identify and characterize a plant virus IRES element, including construction of a translation reporter vector for testing the IRES activity, testing the IRES activity in coupled in vitro transcription and translation systems and mammalian cells analysis of RNA stability, and sucrose gradient analysis and polysome profiling, is presented.

Keywords IRES; Plant virus; Translation; Gene expression

1   Introduction

In eukaryotes, translation initiation involves recruitment of 40S ribosomal subunits at either the 5′ m7G cap structure or internal ribosome entry sites (IRES). The ribosomal subunit together with other factors locates a start codon and protein synthesis begins following binding of the 60S ribosomal subunit, terminating at a stop codon (1, 2). To fully use their compact genomes, viruses have evolved various mechanisms either to redirect the translational machinery to favor viral transcripts or to regulate the expression of internal genes. Genome partitioning and the use of sgRNAs are common mechanisms used by many plant viruses to make their internal genes accessible for the ribosome (3). In addition, nonorthodox strategies such as the use of internal ribosomal entry sites (IRES) elements have been exploited by viruses to express multiple genes from a single RNA species. IRES elements are initially discovered in picornavirus RNAs to confer internal initiation independent of the 5' end (4, 5). Since then, functional IRES elements have been reported in various viral and cellular messenger RNAs (mRNAs). Cellular mRNAs that contain IRES elements encode a wide variety of

From: *Methods in Molecular Biology, Vol. 451, Plant Virology Protocols:*
From Viral Sequence to Protein Function
Edited by G.D. Foster, I.E. Johansen, Y. Hong, and P.D. Nagy © Humana Press, Totowa, NJ

proteins such as translation initiation factors, transcription factors, oncogenes, homoeotic genes, and growth factors (6). IRES elements have also been reported in plant viruses such as *Tobacco etch virus* (TEV) (7) and Crucifer infecting tobamovirus (crTMV) (8, 9).

In *Tobacco mosaic virus* (TMV) U1, only the 5' proximal gene of the gRNA is accessible to the ribosomes. An uncapped dicistronic sgRNA1 directs the expression of only the MP, while a capped monocistronic sgRNA2 is responsible for the expression of the CP (10). The gRNA of crTMV is able to direct the synthesis of CP in vitro. A 148-nt region preceding the CP of crTMV was tested in a bicistronic construct to contain an IRES element with relatively short and simple structure (8). IRES elements identified in crTMV have been reported to be active in rabbit reticulocyte lysate system (8) and have recently been demonstrated to function in yeast and HeLa cells (11).

IRES elements, most well studied in picornaviruses, share common features that are responsible for activity. Most of the known viral IRESs are located in 5' UTRs, highly structured and contain multiple conserved AUGs. Parts of the secondary structure of IRES elements associated with activity include sequences that form part of double-stranded regions (12, 13) or sequences located in apical or internal loops (14, 15). Disruptions of these regions have been associated with the modification of essential RNA–protein interactions (16). The GNRA tetraloop, an example of a conserved motif located at a distal loop in the central domain of *Foot and mouth disease virus* (FMDV) IRES, has been indicated to be involved in long range RNA interactions (14). Such RNA–RNA interactions, dependent on RNA concentration, ionic conditions, and temperature (17, 18), suggest dynamism in the tertiary structure of IRES that may play an important role in the IRES activity. Identification and characterization of a putative IRES element can be tested in wheat germ extract, rabbit reticulocyte lysate, and mammalian cells.

2   Materials

1. Acetic acid.
2. pBluescript®.
3. Coommasie blue R-250 (Sigma).
4. T7 RNA polymerase.
5. A bicistronic construct containing GFP gene as the 5' cistron and the envelope (E) protein gene of coronavirus infectious bronchitis virus (IBV) as the 3' cistron.
6. [α-^{32}P]UTP.
7. [^{35}S]methionine.
8. Phenol (pH 8).
9. 0.5 M NH$_4$OAc.
10. tRNA (10 mg mL^{-1}).

11. Chloroform.
12. 10–30% (w/v) linear sucrose gradient.
13. 25 mM Tris-HCl (pH 7.6).
14. 100 mM KCl.
15. 5 mM $MgCl_2$.
16. Ultrahigh speed centrifuge and rotors.
17. Absolute ethanol.
18. 1% formaldehyde agarose gel.
19. Cos-7 cells.
20. β-tubulin antibody.
21. Coupled in vitro transcription translation(TnT) kit (Promega).
22. 30 °C heat block.
23. Complete Dulbecco's modified Eagles medium (Gibco Life Technologies).
24. 10% bovine calf serum (Hyclone).
25. 1% penicillin/streptomycin (Gibco Life Technologies).
26. X-ray film and cassette.
27. X-ray film developer.
28. 0.1% SDS (v/v).
29. Recombinant vaccinia virus (vTF7–3).
30. DOTAP transfection reagent (Roche).
31. Humidified 5% CO_2 incubator.
32. GS-710 calibrated imaging densitometer (Bio-Rad).
33. Molecular Analyst computer software (Bio-Rad).
34. 1X Laemmli's sample buffer; 24 mM Tris-HCl pH 6.8, 100 mM DTT, 2% SDS, 0.1% bromophenol blue, 20% glycerol.
35. Protein gel running buffer: 2.88% glycine (w/v), 0.6% Tris-HCl, and 0.1% SDS (v/v).
36. Amplify™ (Amersham Pharmacia Biotech).
37. Nitrocellulose membrane (Stratagene).
38. Semidry transfer cell (Bio-Rad Trans-Blot SD).
39. Western blot blocking buffer: 5% skim milk powder in TBST including 20 mM Tris-HCl (pH 7.4), 150 mM NaCl, 0.1% Tween 20.
40. IgG conjugated with horseradish peroxidase (Dako).
41. Chemiluminiscence detection kit (ECL, Amersham Pharmacia Biotech).

3   Methods (see notes 1–4)

3.1   *Coupled In Vitro Transcription and Translation (see note 5)*

1. Plasmid DNAs (5 μg) are linearized and extracted by phenol/chloroform and ethanol precipitated before adding to the 50 μL reaction mix containing the following components:

1. Linearized DNA 2 μg
2. 5X transcription buffer (Promega) 10 μL
3. 100 mM DTT 5 μL
4. NTPs (2.5 mM each of ATP, GTP, CTP, TTP) 10 μL
5. RNase inhibitor (40 U μL^{-1}) 0.5 μL
6. T7 RNA polymerase 1 μL
7. Sterile water to 50 μL

Incubate the reaction mix at 37 °C for 1 h.
Check the integrity of the transcripts by agarose gel electrophoresis.
2. In vitro translation reaction includes the following components:

8. 1 μg RNA template (after in vitro transcription) 2 μL
9. Wheat germ extracts 35 μL
10. Minus methionine amino acid mixture 1 μL
11. 10 mM methionine 1 μL
12. RNase inhibitor (40 U μL^{-1}) 1 μL
13. 50 μCi of [^{35}S] methionine 1 μL
14. Add sterile water to 50 μL

Incubate the reaction at 30 °C for 1 h.
3. Reaction products are separated by sodium dodecyl sulfate-polyacrylamide gel electrophoresis (SDS-PAGE) and detected by autoradiography.

3.2   Transient Expression of Constructs in Cos-7 Cells

1. Culture Cos-7 cells (ATCC-CRL-1651) in complete Dulbecco's modified Eagles medium (Gibco Life Technologies) supplemented with 10% bovine calf serum (Hyclone) and 1% penicillin/streptomycin (Gibco Life Technologies) at 37 °C in humidified 5% CO_2 incubator.
2. Infect Cos-7 cells with 10 PFU/cell of recombinant vaccinia virus (vTF7–3) for 2 h at 37 °C prior to transfection.
3. Transfect the vTF7–3-infected cells with the plasmids (see note 6) using DOTAP transfection reagent (Roche).
4. After 20–24 h post-infection, harvest the cells.
5. Lyse transfected Cos-7 cells
6. Mix total proteins with 2X SDS loading buffer and subjected to SDS-PAGE.

3.3   Analyses of Protein Products using SDS-PAGE

1. Discontinuous polyacrylamide gel electrophoresis system (19) is used. Separating gels of different concentrations (12.5, 15, or 17.5%) and 3% stacking gels are cast between two glass plates.

2. Typically, an aliquot of the translation reaction is added to the 1X Laemmli's sample buffer and boiled at 100 °C for 2 min and cooled on ice before loading on a SDS-polyacrylamide gel.
3. In a reservoir of protein gel running buffer, gels are run vertically at 20 mA until the bromophenol blue dye reached the bottom of the gel.
4. After electrophoresis, gels containing [^{35}S]-methionine are fixed in 50% methanol and 10% acetic acid for 30 min.
5. The signal is enhanced with the use of Amplify™ (Amersham Pharmacia Biotech) for 15 min.
6. Gels are dried under vacuum at 80 °C for 1 h and later exposed to X-ray film (Biomax, Kodak) for autoradiography at −80 °C overnight.
7. Gels containing unlabeled polypeptides are stained in 50% methanol, 10% acetic acid and 0.05% coommasie blue R-250 (Sigma) for 30 min at RT°C and destained in 50% methanol, 10% acetic acid before drying.

3.4 Western Blotting

1. After SDS-PAGE, proteins are transferred to nitrocellulose membrane (Stratagene) by a semidry transfer cell (Bio-Rad Trans-Blot SD) and blocked overnight at 4 °C in blocking buffer.
2. The membrane is incubated for 2 h at room temperature in a dilution (1:1000) of a specific antiserum against the reporter gene in blocking buffer.
3. After three washes for 15 min each with TBST, the membrane was blocked for 20 min before it is incubated with either antirabbit or antimouse IgG conjugated with horseradish peroxidase (Dako) diluted 1:2500 in blocking buffer for 1 h at room temperature.
4. After three washes with TBST, the membrane was treated using a chemiluminiscence detection kit (ECL, Amersham Pharmacia Biotech) accordingly.

3.5 Densitometry

1. The intensities of the protein bands are measured by a GS-710 calibrated imaging densitometer (Bio-Rad) and analyzed using Molecular Analyst computer software (Bio-Rad).
2. The protein band of interest is normalized accordingly to the intensities of the corresponding p28 band or a consistent background band.

3.6 RNA Stability Test (see note 7)

1. Equal amounts of templates are linearized for in vitro transcription in the presence of [α-^{32}P]UTP.

2. Equal amounts of the labeled RNA (2 μL of the 25 μL reaction) are added directly into the in vitro translation reaction mixture. At specific time intervals, a portion (5 μL) is withdrawn from each translation reaction mixture.
3. Prior to adding 100 μL of phenol (pH 8), 94 μL 0.5 M NH$_4$OAc and 1 μL tRNA (10 mg mL^{-1}) are added to the reaction mixture. The sample is vortexed and centrifuged at 20,000 g for 5 min.
4. The supernatant is added to 100 μL chloroform, vortexed and centrifuged at 5,000 g for 5 min.
5. The RNA is precipitated by transferring the supernatant into 300 μL absolute ethanol and placed in −80 °C for 30 min before obtaining the pellet.
6. The extracted RNAs are resolved in a 1% agarose gel containing 0.1% SDS before autoradiography.

3.7 Sucrose Gradient Analysis and Polysome Profiling

The integrity of mRNA derived from a test construct is analyzed by sucrose gradient analysis as described by Pelletier and Sonenberg (5).

1. The DNA template is linearized for in vitro transcription in the presence of [α-^{32}P]UTP.
2. The in vitro synthesized RNA is extracted with phenol/chloroform, precipitated with ethanol, and incubated in a 50 μL translation reaction containing 35 μL of wheat germ extract at 30 °C for 10 min.
3. The translation mixture is cooled on ice and layered onto a 10–30% (w/v) linear sucrose gradient (2 mL) in buffer containing 25 mM Tris-HCl (pH 7.6), 100 mM KCl, and 5 mM MgCl$_2$. The gradient was then subjected to centrifugation at 259,000 g in an appropriate rotor at 4 °C for 60 min.
4. Fractions of 200 μL are collected from the top of the gradient and measured at absorbance 260 nm to obtain the polysome profile.
5. Based on the polysome profile, fractions are pooled and RNA extracted by phenol/chloroform and ethanol precipitation.
6. The transcripts are resolved on a 1% formaldehyde agarose gel and analyzed by autoradiography.

3.8 Functionality of the IRES Element in Mammalian Cells

To test if the plant virus IRES element is functional in mammalian cells, the GFP gene is cloned upstream of the IRES region in a bicistronic construct.

1. In a transient expression system (20), Cos-7 cells are infected with vaccinia virus, which possesses a T7 polymerase gene, and transfected with T7-promoter driven plasmid DNA.

2. The expression of GFP is detected in transfected cells by viewing under ultraviolet light.
3. The cells are harvested and western blot detection is performed using the expression of β-tubulin gene as a control to normalize the densitometry readings.
4. The expression of both GFP and reporter gene positioned after the IRES indicates that the IRES element s active in mammalian cells.
5. The IRES activity in Cos-7 cells is further analyzed by expression of deletion constructs.

4 Notes

1. Construction of a translation reporter vector. The vector can be a pBluescript® or any suitable plasmid that contains a T7-RNA polymerase promoter so that a full-length sequence containing a putative IRES element can be cloned into it for transcription. The transcribed RNA template can then be used for in vitro translation. A reporter gene (such as green fluorescent protein, GFP) at the 5' end is used as an internal control for the internal initiation event and the stability of RNA templates during translation.
2. The region containing the IRES element is inserted into a bicistronic construct containing GFP gene as the 5' cistron and the envelope (E) protein gene of coronavirus *Infectious bronchitis virus* (IBV) (21) as the 3' cistron, giving rise to construct pGFP-IRES-E.
3. A series of sequential deletions of the full-length sequence containing a putative IRES element is carried out to delineate the region responsible for internal initiation. When the expression of the reporter gene is observed to be reduced by several folds, it indicates that the IRES element has been deleted. Construct pGFP-E, containing the GFP gene and the E protein gene in different reading frames, are used as controls (22).
4. Expression of pGFP-E resulted in the detection of GFP. With the insertion of the putative IRES element inserted between the 5' and 3' cistrons, the GFP encoded by the 5' cistron and the E protein encoded by the 3' cistron were expressed.
5. To avoid repeated freezing and thawing of the coupled transcription and translation kit. It is advisable to aliquot the wheat germ extracts after the first usage.
6. Constructs containing deletions under the control of a T7 promoter are expressed transiently in semiconfluent monolayers of vTF7–3-infected Cos-7 cells.
7. To handle RNA with care so that it will not be degraded by RNase and affect the results.

Acknowledgments The authors would like to thank the National University of Singapore for financial support of our research on Hibiscus latent Singapore virus (RP 154–000–295–112) and Hibiscus chlorotic ringspot virus (RP 154–000–252–112); and A*Star for research support in Institute of Molecular and Cell Biology.

References

1. Kozak, M. (1989) Circumstances and mechanisms of inhibition of translation by secondary structure in eukaryotic mRNAs. *Mol. Cell. Biol.* **9,** 5134–5142.
2. Hershey, J. W. B., and Merrick, W. C. (2000) The pathway and mechanism of initiation of protein synthesis. *In* "Translational Control of Gene Expression" (N. Sonenberg, et al. eds), pp. 33–88. Cold Spring Harbor Laboratory Press, Cold Spring Harbor, NY.

3. Agranovsky, A., and Morozov, S. (1999) Gene expression in positive strand RNA viruses: Conventional and aberrant strategies, pp 99–119. *In* "Molecular biology of plant viruses" (C. L. Mandahar, ed.). Kluwer Academic Press, Norwell, MA.

4. Jang, S. K., Krausslic, H. G., Nicklin, M. J., Duke, G. M., Palmenberg, A. C., and Wimmer, E. (1988) A segment of the 5' non-translated region of encephalomyocarditis virus RNA directs internal entry of ribosomes during in vitro translation. *J. Virol.* **62**, 2636–2643.

5. Pelletier, J., and Sonenberg, N. (1988) Internal initiation of translation of eukaryotic mRNA directed by a sequence derived from poliovirus RNA. *Nature* **33**, 320–325.

6. Martinez-Salas, E., Ramos, R., Lafuente, E., and Lopez de Quinto, S. (2001) Functional interactions in internal translation initiation directed by viral and cellular IRES elements. *J. Gen. Virol.* **82**, 931–984.

7. Niepel, M., and Gallie, D. R. (1999) Identification and characterization of functional elements within tobacco etch virus 5' leader required for cap-independent translation. *J. Virol.* **73**, 9080–9088.

8. Ivanov, P. A., Karpova, O. V., Skulachev, M. V., Tomashevskaya, O. L., Rodionova, N. P., Dorokhov, Y. L., et al. (1997) A tobamovirus genome that contains an internal ribosome entry site functional in vitro. *Virology* **232**, 32–43.

9. Skulachev, M. V., Ivanov, P. A., Karpova, O. V., Korpela, T., Rodionova, N. P., Dorokhov, Yu. L., et al. (1999) Internal initiation of translation directed by the 5'-untranslated region of the tobamovirus subgenomic RNA I$_2$. *Virology* **263**, 139–154.

10. Palukaitis, P., and Zaitlin, M. (1986) Tobacco mosaic virus. Infectivity and replication. *In* "The Plant Viruses" (M. H. V. van Regenmortel and M. Fraenkel-Conrat, eds). Vol. 2, pp. 105–131. Plenum Press, NY.

11. Dorokhov, Y. L., Skulachev, M. X., Ivanov, P. A., Zvereva, S. D., Tjulkina, L. G., Merits, A., et al. (2002) Polypurine (A)-rich sequences promote cross-kingdom conservation of internal ribosome entry. *Proc. Natl. Acad. Sci. USA* **99**, 5301–5306.

12. Jang, S. K., and Wimmer, E. (1990) Cap-independent translation of encephalomyocarditis virus RNA: structural elements of the internal ribosomal entry site and involvement of a cellular 57-kD RNA-binding protein. *Genes Dev.* **4**, 1560–1572.

13. Hoffman, M. A., and Palemberg, A. C. (1996) Revertant analysis of J-K mutations in the encephalomyocarditis virus internal ribosomal entry site detects an altered leader protein. *J. Virol.* **70**, 6425–6430.

14. Lopez de Quinto, S., and Martinez-Salas, E. (1997) Conserved structural motifs located in distal loops of aphthovirus internal ribosome entry site domain 3 are required for internal initiation of translation. *J. Virol.* **71**, 4171–4175.

15. Jubin, R., Vantuno, N. E., Kieft J. S., Murray, M. G., Doudna, J. A., Lau, J. Y., et al. (2000) Hepatitis C virus internal ribosome entry site (IRES) stem loop IIId contains a phylogenetically conserved GGG triplet essential for translation and IRES folding. *J. Virol.* **74**, 10430–10437.

16. Lopez de Quinto, S., and Martinez-Salas, E. (2000) Interaction of the eIF4G initiation factor with the aphthovirus IRES is essential for internal translation initiation *in vivo*. *RNA* **6**, 1380–1392.

17. Ramos, R., and Martinez-Salas, E. (1999) Long-range RNA interactions between structural domains of the aphthovirus internal ribosome entry site (IRES). *RNA* **5**, 1374–1383.

18. Kieft, J. S., Zhou, K., Jubin, R., Murray, M. G., Lau, J. Y., and Doudna, J. A. (1999) The hepatitis C virus internal ribosome entry site adopts an ion-dependent tertiary fold. *J. Mol. Biol.* **292**, 513–529.

19. Laemmli, U. K. (1970) Cleavage of structural proteins during the assembly of the head of bacteriophage T4. *Nature* **227**, 680–685.

20. Fuerst, T. R., Niles, E. G., Studier, F. W., and Moss, B. (1986) Eukaryotic transient-expression system based on recombinant vaccinia virus that synthesizes bacteriophage T7 RNA polymerase. *Proc. Natl. Acad. Sci. USA* **83**, 8122–8127.

21. Lim, K. P., Xiu, H. Y. and Liu, D. X. (2001) Physical interaction between the membrane (M) and envelope (E) proteins of the coronavirus avian infectious bronchitis virus (IBV). *Adv Exp Med Biol.* **494**, 595–602.

22. Koh D. C., Wong, S. M., and Liu, D. X. (2003) Synergism of the 3'-untranslated region and an internal ribosome entry site differentially enhances the translation of a plant virus coat protein. *J. Biol. Chem.* **278**, 20565–20573.

Chapter 10
Analysis of Geminivirus DNA Replication by 2-D Gel

Keith Saunders

Abstract The technique described was developed for the separation of begomovirus DNA. DNA products resulting from and during geminiviral replication are characterized by the application of strand-specific separation and identification by strand-specific DNA probing of Southern blots. The mapping of the initiation site of complementary-strand DNA synthesis, by this technique is also presented.

Keywords Geminivirus; ssDNA; dsDNA; Rolling circle replication

1 Introduction

Geminiviruses have a single-stranded DNA genome composed either of approximately 2.6 kbp for a single genomic component virus, for example, *Tomato yellow leaf curl virus* or approximately 5.2 kbp for bipartite members such as *African cassava mosaic virus* (ACMV) (for a review see ref. 1). Satellite DNA molecules are associated with some geminivirus infections and are approximately half the size (1.4 kbp) of the single genomic component (2). Geminivirus DNAs are encapsidated in twinned quasi-isometric particles. In addition to the single-stranded genomic DNA, infected plant material also contains several other distinct geminiviral DNA species and heterogeneous forms (H DNA) generated as a consequence of viral rolling circle replication. From the encapsidated genomic virus single-stranded DNA, two predominant double-stranded DNA forms are produced. Supercoiled DNA (SC DNA) is generated for transcription purposes for the expression of geminiviral genes, and open circular DNA (OC DNA) for the provision as a template for replication, resulting in the eventual production of progeny single-stranded viral DNA (3). Linear DNA (L DNA) forms are also generated during replication. Besides monomeric genomic DNA, dimeric and trimeric DNA forms are also prevalent in infected plants.

The two-dimensional (2-D) technique described in this paper was developed with the whitefly transmitted begomoviruses (ACMV) but it could equally be applied to other members of the geminivirus family, such as mastriviruses and curtoviruses, to any associated satellite DNAs (4), or to any other single-stranded DNA viruses. The separation of the various DNAs by 2-D agarose gel electrophoresis coupled with their prior separation into single-stranded or double-stranded forms combined with strand-specific hybridization of resulting Southern blots has allowed for the detailed characterization of begomovirus replication. Mapping the initiation site of DNA synthesis, primed by RNA, is also possible by modification of the denaturant used in the second dimension.

2   Materials

2.1   Inoculation of Plants

1. Infectious clones of a begomovirus
2. Celite abrasive or fine carborundum

2.2   Isolation of Nucleic Acids from Plants

1. Extraction buffer (EB): 1% triisopropylnaphthalene sulfonate, 6% 4-amino salycilate, 5% phenol, 50 mM Tris-HCl pH 8.0
2. Ribonuclease free H_2O
3. Phenol: Buffered to pH 8.0. Best obtained as a commercially prepared solution
4. Phenol–Chloroform mixture: phenol:chloroform:isoamyl alcohol (25:24:1). This is extremely corrosive and toxic and is best bought as a prepared solution.
5. Ethanol
6. 70% ethanol in H_2O
7. 3 M Na acetate pH 5.2

2.3   Purification of geminivirus DNA, Separation of dsDNA and ssDNA by Benzoylated-Naphthoylated DEAE (BND)-Cellulose Chromatography

1. Nucleic acid digestion buffer: 50 mM Tris-HCl, 5 mM $MgCl_2$ pH 7.9
2. Ribonuclease A: 50 µg mL^{-1}
3. Ribonuclease T_1: 100 U mL^{-1}
4. Chloroform
5. Isopropanol
6. BND running buffer (BND-RB): 10 mM Tris-HCl, 800 mM NaCl, 1 mM EDTA, pH 8.0
7. 2X BND-RB: 20 mM Tris-HCl, 1.6 M NaCl, 2 mM EDTA, pH 8.0

8. BND elution buffer (BND-EB): 10 mM Tris-HCl, 1 M NaCl, 1 mM EDTA 1.8% Caffeine, pH 8.0
9. 3 M Na acetate pH 5.2

2.4 2-D Agarose Gel Electrophoresis

2.4.1 Neutral/Alkaline Denaturing Gels

1. Agarose
2. First dimension buffer (neutral dimension) (TNE): 40 mM Tris, 20 mM Sodium acetate, 2 mM EDTA pH 7.5
3. Second dimension alkaline denaturing and running buffer: 30 mM NaOH, 2 mM EDTA
4. Loading buffer: 2% Orange G, 25% Ficoll, 5 mM EDTA in TNE
5. Depurination solution: 100 mM HCl
6. Denaturing solution: 0.5 M NaOH, 1.5 M NaCl
7. Neutralizing solution: 1 M Tris-HCl, 1.5 M NaCl pH 7.4
8. Transfer solution 3 M NaCl, 300 mM tri-sodium citrate

2.4.2 Neutral/Formamide Gels

1. Agarose
2. Loading buffer: 2% Orange G, 25% Ficoll, 5 mM EDTA in TNE
3. First dimension buffer (neutral dimension) (TNE): 40 mM Tris-acetate, 20 mM Sodium acetate, 2 mM EDTA pH 7.5
4. Second dimension formamide equilibration buffer: 50% (v:v) solution of formamide in 20 mM MOPS, 8 mM sodium acetate, 1 mM EDTA (pH 7.0)
5. Second dimension formamide running buffer: 20 mM MOPS, 50 mM Na acetate, 10 mM EDTA pH 7.0

3 Methods

3.1 Inoculation of Plants (see note 3)

1. ACMV is derived from infectious clones (pJS092 and pJS094) of a Kenyan isolate (5).
2. *Nicotiana benthamiana* is mechanically inoculated with 1 μg of each genomic DNA component following excision of the cloned viral insert using the appropriate restriction enzymes. The second and third true leaves, not fully expanded, are dusted with celite abrasive or fine carborundum. The solution containing

the geminiviral components $100\,ng\,\mu L^{-1}$ is applied to the leaf ($5\,\mu L$ per leaf) and the mixture is gently rubbed between the thumb and first finger of a "gloved hand" to disrupt the leaf surface.

3. Systemically infected leaves are harvest 10 days post inoculation.

3.2 Isolation of Nucleic Acids from Plants (see notes 1 and 2)

1. Harvest infected plant leaves, wash, and blot dry with a paper towel.
2. Grind leaves, up to 1 g in mortar with 1 mL of EB. For less material use approximately $100\,\mu L$ of EB for each $100\,\mu g$ leaf material.
3. Transfer to a sealable tube and extract the homogenate three times with phenol: chloroform.
4. Precipitate the total nucleic acids from the aqueous phase by the addition of 0.1 volume of 3 M Na acetate pH 5.2 and 2 volumes of ethanol.
5. Recover the nucleic acids by centrifugation at $15,000\,g$ and wash the nucleic acid pellet with 70% ethanol. Centrifuge again.
6. Briefly dry the nucleic acid pellet under vacuum and resuspend in ribonuclease free H_2O.
7. Determine the concentration of the nucleic acid and adjust the concentration to $5\,mg\,mL^{-1}$.

3.3 BND Cellulose Chromatography, Separation of ssDNA and dsDNA (see note 5)

1. Between 1 and 5 mg of total nucleic acids are digested with ribonucleases A ($50\,\mu g\,mL^{-1}$) and T_1 ($100\,U\,mL^{-1}$) in nucleic acid digestion buffer in a total volume of 5 mL. Nucleic acids are recovered by sequential phenol, phenol–chloroform, and chloroform extractions, followed by precipitation with ethanol. Nucleic acids are resuspended in BND-RB buffer with the final concentration adjusted to that of $5\,mg\,mL^{-1}$.
2. Preparation of BND-cellulose: Allow the BND-cellulose to swell by mixing 1 g of BND-cellulose with 2 mL BND-RB. Transfer to a suitable disposable plastic column for chromatography. The column volume containing the BND-cellulose should be no more than 2 mL. Equilibrate the column with the addition of 2 column volumes of BND-RB, that is, 4 mL. Gravity flow is sufficient to run the column.
3. Mix the DNA sample with an equal volume of 2X BND-RB. The total amount of DNA should not exceed 2 mg and the total volume should not exceed 2 mL.
4. Apply to the column and collect the solution that has passed down the column.
5. Reapply this solution and repeat this three more times.

6. Wash the column with 3 column volumes of BND-RB collecting the flow through (FT-DS).
7. Repeat the previous step with fresh BND-RB again collecting the FT-DS.
8. The FT-DS fraction contains double-stranded DNA. To recover this DNA by precipitation add 0.7 volumes of isopropanol to the combined FT-DS fractions.
9. Wash the column with 4 column volumes of BND-EB and collect the flow through (FT-SS).
10. The FT-SS fractions contain single-stranded nucleic acids and are recovered by the addition of 0.7 volumes of isopropanol.
11. Wash pellets with 70% ethanol and dissolve in $100\,\mu L$ H_2O.

3.4   2-D Agarose Gel Electrophoresis (see note 4)

3.4.1   Neutral/Alkaline Denaturing Gels

1. The sample well is formed by placing a 2-mm round rivet (or a sealed Pasteur pipette) suspended approximately 2 mm from the bottom, in a vertical position approximately 1 cm from the corner of a 20×20 cm gel apparatus. Do not allow the rivet to touch the bottom of the apparatus, thereby allowing for a layer of agarose to be formed below the sample well. For a 1.2% agarose gel 0.5 cm deep, melt 2.4 g agarose in 200 mL TNE buffer and pour gel. For convenience, the gel is best cast on a glass plate that will fit inside the electrophoresis apparatus.
2. Assemble the gel in its running apparatus and remove the well former. Pour in TNE buffer and submerge the gel to a depth of approximately 0.5 cm. Set up a pump to circulate the buffer.
3. Run $10\,\mu g$ of nucleic acids in a total volume of $10\,\mu L$ including the loading buffer solution. Apply to the sample well.
4. Electrophoresis is performed at $1\,V\,cm^{-1}$ for 24 h.
5. Incubate the entire gel in second dimension alkaline denaturing and running buffer for 45 min at room temperature on a rotating platform.
6. Electrophoresis is continued for 24 h at $90°$ orientation to the first dimension in second dimension alkaline denaturing and running buffer.
7. Resolved nucleic acids are transferred to Hybond N by overnight capillary action in transfer solution following sequential incubations on a rotating platform in depurination solution (20 min); denaturing solution (45 min) and neutralizing solution (45 min). The gel should be submerged to a depth of approximately 1 cm during the washing steps.

3.4.2   Neutral/Formamide Gels

1. The sample well is formed as described in the previous section.
2. If required for control purposes, the nucleic acid samples are incubated for 2 h in the solution of ribonucleases. Nucleic acids are recovered by phenol–chloroform

extraction followed by ethanol precipitation and are suspended in 10 µL loading buffer.

3. Assemble the gel as described above for alkaline/neutral gels. Apply the sample to the well.

4. After electrophoresis, performed at 1.25 V cm^{-1} for 24 h, the length of the gel containing the separated nucleic acids is excised from the remainder of the gel and the resulting "gel strip" is incubated in second dimension formamide equilibration buffer for 15 min at 65 °C.

5. Melt 1.2% agarose in second dimension formamide equilibration buffer. Do not heat formamide in a microwave oven and allow the agarose solution to cool before adding the formamide. Set the first dimension gel strip in the casting tray at a right angle to the first dimension and pour the liquid second dimension gel solution around the first dimension strip. Allow to set.

6. Electrophoresis is continued for 24 h at 1.25 V cm^{-1} in second dimension formamide running buffer. Set up a pump to circulate the running buffer.

7. Nucleic acids are blotted onto Hybond N in transfer solution by overnight capillary action without prior treatment of the gel.

3.5 Detection of Geminivirus DNA and RNA Resolved by 2-D Electrophoresis (see note 6)

Following the transfer of the nucleic acid to Hybond N, standard hybridization techniques are employed to determine the identity of the differently resolved geminiviral DNA or RNAs. Both double-stranded DNA and strand-specific RNA probes maybe used in this analysis. Double-stranded probes are best generated by using PCR with primers designed to selected geminiviral sequences. Alternatively DNA fragments suitable for labeling maybe obtained by restriction endonuclease digestion of cloned geminiviral DNAs. Strand-specific RNA probes can be obtained by cloning geminiviral DNA sequences downstream of a T7 RNA polymerase promoter and consequently generating labeled single-stranded RNA by the use of T7 RNA polymerase.

4 Notes

1. To preserve any RNA bound to DNA, samples resolved by neutral/formamide electrophoresis are not subjected to ribonuclease digestion.

2. The extraction of total nucleic acids from plants is based on a method described by Covey and Hull (6). The extraction of nucleic acids from plants may now be achieved, following the manufacturers instructions, with the purchase of a suitable commercially available plant DNA extraction kit.

3. The genomic components of many begomoviruses have been molecularly cloned, transferred to binary vectors, and can be readily introduced into *N. benthamiana* or the natural host plant of the virus in question by agroinoculation with transformed *Agrobacterium tumefaciens*. Begomovirus cloned genomic DNA components may also be coated onto gold particles and applied to leaf surfaces by biolistic methods. The ACMV isolate used in the original studies could be propagated by mechanical inoculation.

4. 2-D agarose gel electrophoresis is a powerful method for the resolving a complex population of DNA molecules. Fig. 1a shows a typical 2-D gel with 1-D migration in neutral gel at the top and 1-D migration under alkaline conditions aligned at the side. Thus due to their size and conformation, the separation of DNAs composed of single or double strands is possible under neutral electrophoretic conditions. Retardation of dsDNA relative to ssDNA in the neutral dimension is achieved by the high salt concentration in the agarose gel. In contrast, under denaturing conditions double-stranded DNA is separated into its single strands and consequently any molecules of similar mass migrate together. If a radioactive probe is used, by overexposing the filter to X ray film, heterogeneous DNA (H1–H5) (Fig. 1a) are readily observed. Figure 1b shows the theoretical migration of various DNA forms. The identity of the heterogeneous DNA can be made by tracing their migration with respect to that of the homogeneous DNAs. Dimeric DNA forms, DNAs whose mass is double that of the genomic size (DSS and DSC), can also be identified.

5. Further differentiation of the viral DNAs can be made by separation of complex DNA mixture into ssDNA and dsDNA prior to 2-D agarose gel electrophoresis. BND-cellulose readily binds ssDNA and any ssDNA possessing any dsDNA regions. DNAs that are entirely double stranded do not interact with BND-cellulose and are present in the flow through fraction. Thus, Fig. 2 shows the 2-D separation of geminivirus dsDNA confirming that OC, L, and SC DNAs are double stranded (Fig. 2a). In contrast, these specific DNA species are absent in DNA fractions

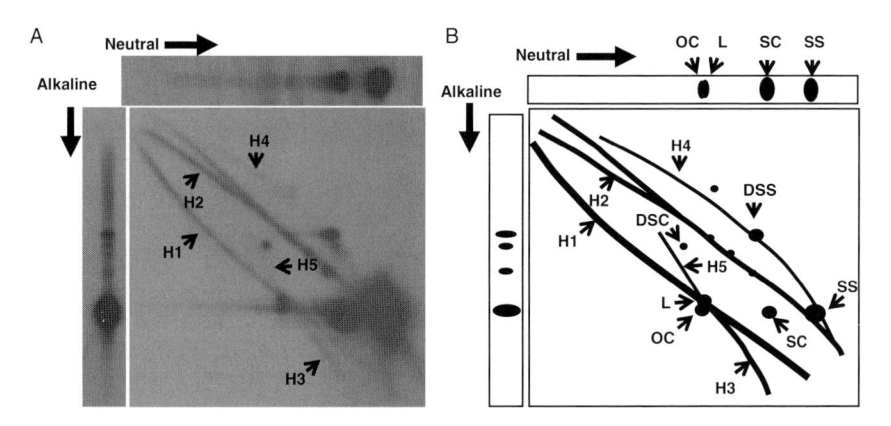

Fig. 1 2-D electrophoresis of geminivirus intracellular DNA. Total cellular DNA was isolated and treated with ribonuclease prior to resolving on a two-dimensional gel. (**a**) Within the complex DNA population, double-stranded, single-stranded, and DNAs composed of double strands with single-stranded components are separated in the first neutral dimension according to their size and conformation. In the second dimension, under alkaline denaturing conditions, separation is achieved according to their single-stranded size. (**b**) Theoretical migration of the various forms. Sample well is located at the top left. A double-stranded geminivirus-specific probe was used to detect the various DNA forms. H, heterogeneous; D, dimeric; L, linear; SC, supercoiled; SS, single-stranded DNA forms are indicated. For a detailed characterization of the various DNA species see ref. (7)

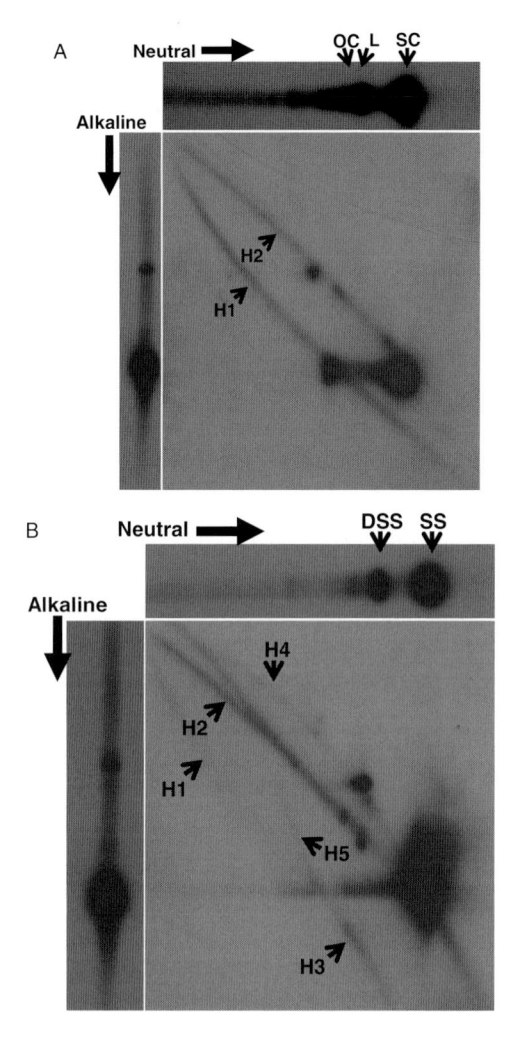

Fig. 2 2-D electrophoresis of geminivirus DNAs separated by BND-cellulose chromatography prior to electrophoresis. The electrophoretic and probe conditions were as described in Fig. 1. (**a**) Separation of totally double-stranded DNA forms, open circular (OC), linear (L), and supercoiled (SC) DNAs. These DNAs are excluded from BND cellulose. (**b**) DNA forms possessing single-stranded DNA (SS and DSS) eluted from BND cellulose. The composition of the various heterogeneous forms (H1–H5) is described in the notes

that have the ability to bind to BND-cellulose (Fig. 2b). Clearly, heterogeneous DNAs H1 and H2 are formed of both ssDNA and dsDNA (Fig. 2). Overlaying the X-ray films of the 2D separations in Fig. 2a, b results in the separations seen in Fig. 1a.

6. The strand specificity of the ssDNA species maybe characterized further by probing with a strand-specific probe (for further details see ref. 7). With modification to the denaturant used in the second dimension, that is, replacing alkaline with formamide, it is also possible to characterized RNA/DNA hybrids and to map priming sites on the geminivirus genome (8).

References

1. Rojas, M. R., Hagen, C., Lucas, W. J., and Gilbertson, R. L. (2005) Exploiting chinks in the plant's armor: Evolution and emergence of geminiviruses. *Annu. Rev. Phytopathol.* 43, 361–394.
2. Saunders, K., Bedford, I. D., Briddon, R. W., Markham, P. G., Wong, S. M., and Stanley, J. (2000) A unique virus complex causes *Ageratum* yellow vein disease. *Proc. Natl. Acad. Sci. USA* 97, 6890–6895.
3. Hanley-Bowdoin, L., Settlage, S. B., Orozco, B. M., Nagar, S., and Robertson, D. (1999) Geminiviruses: Models for plant replication, transcription and cell cycle regulation. *Crit. Rev. Plant Sci.* 18, 71–106.
4. Briddon, R. W., and Stanley, J. (2006) Subviral agents associated with plant single-stranded DNA viruses. *Virology* 344, 198–210.
5. Stanley, J. (1983) Infectivity of the cloned geminivirus genome requires sequences from both DNAs. *Nature* 305, 643–645.
6. Covey, S. N., and Hull, R. (1981) Transcription of cauliflower mosaic virus DNA. Detection of transcripts, properties, and location of the gene encoding the virus inclusion body protein. *Virology* 111, 463–474.
7. Saunders, K., Lucy, A. P., and Stanley, J. (1991) DNA forms of the geminivirus African cassava mosiac virus consistent with a rolling circle mechanism of replication. *Nucleic acids Res.* 19, 2325–2330.
8. Saunders, K., Lucy, A., and Stanley, J. (1992) RNA-primed complementary-sense DNA synthesis of the geminivirus African cassava mosaic virus. *Nucleic Acids Res.* 20, 6311–6315.

Chapter 11
Begomoviruses: Molecular Cloning and Identification of Replication Origin

Lilian H. Florentino, Anésia A. Santos, Francisco M. Zerbini, and Elizabeth P.B. Fontes

Abstract The *Begomovirus* genus is the largest genus of the *Geminiviridae* family and comprises the whitefly transmitted geminiviruses that infect dicotyledonous plants. They can be either mono or bipartite. In this chapter, we describe the cloning of begomovirus replication modules and the subsequent functional characterization of geminivirus replication origins.

Keywords Geminivirus; Begomovirus; Origin of replication; Replication assay; Replicon module

1 Introduction

In the field, mixed virus infections occur in the same plant with biological and epidemiological implications. In the case of the *Geminiviridae* family, the frequent occurrence of multiple infections has provided the means for interspecies recombination that has been shown to play a significant role in geminivirus diversity and their emergence as agriculturally important pathogens. However, diagnosis of these mixed infections is often complicated by the synergistic effect of virus interactions that cause a more than additive effect in the severity of symptoms. Thus, very frequently the correct discrimination of the set of symptoms caused by each virus does not reproduce faithfully the scenario caused by a simultaneous infection of two or more geminiviruses. In these cases, it is really necessary to isolate or separate biologically the different viruses in order to proceed with the characterization of the etiological agents of the disease (see chapter "Geminivirus: Biolistic Inoculation and Molecular Diagnosis"). To this end, host plants are bombarded with the geminivirus replicative forms (RF) extracted from infected plants and, after successive inoculations, they are biologically isolated in the permissive host.

The biological isolation of a geminivirus allows the cloning of the viral genomic components and subsequent sequencing, phylogenetic analysis, and functional characterization of the viral genome. Currently more than 130 species and strains of the *Geminiviridae* family have been cloned and the molecular variability of some

From: *Methods in Molecular Biology, Vol. 451, Plant Virology Protocols: From Viral Sequence to Protein Function*
Edited by G.D. Foster, I.E. Johansen, Y. Hong, and P.D. Nagy © Humana Press, Totowa, NJ

of these viruses has been assayed by sequencing of polymerase chain reaction (PCR) products. The current taxonomic criteria for distinction of species and strains of the family *Geminiviridae* rely on the sequence identity of the viral genomes and on their biological properties. For the case of the bipartite begomoviruses, the transreplication properties between heterologous DNA-A and DNA-B genomic components have also been taken into consideration for taxonomic classification of new species or strains. In general, the DNA-A-mediated transreplication of heterogenomic DNA-B components is limited to isolate/strains of a particular virus, due to Rep specificity for its cognate replication origin (1–5). Several exceptions in the literature indicate, however, that incompatibility between heterologous DNA-A and DNA-B can not be considered as absolute for the taxonomic classification of distinct species of begomoviruses (6, 7). This is particularly true for the case of naturally occurring recombinant progenies whose replication module (Rep and replication origin) was originated as a unit from a parental virus (8). Frequently, the enhanced fitness of recombinant progenies is associated with a more perfect fit between the Rep DNA-binding domain and Rep-binding motifs in the origin of replication. Thus, in the current agricultural scenario of rapidly emerging new geminiviruses, the identification and characterization of replication modules (Rep protein and origin of replication) of novel geminivirus isolates may provide insights into the evolutionary selection pressure toward enhanced fitness.

The geminivirus replication origins lie in the intergenic region of the viral genome (see chapter "Geminivirus: Biolistic Inoculation and Molecular Diagnosis") and they are structurally organized in at least three functional modules: (1) the signature stem-loop structure containing an invariant nonanucleotide motif that corresponds to the DNA cleavage site to initiate the rolling circle replication, (2) a specific high-affinity binding site for the Rep protein, and (3) an AG motif located between the Rep-binding site and the hairpin structure (Fig. 1a) (2, 9). By sequence comparison among geminivirus intergenic regions one may predict the putative Rep binding motif and identify the invariant hairpin structure of recently isolated new geminivirus genomes (Fig. 1b). Nevertheless, the functional characterization of the plus-strand minimal origin requires an analysis of the capacity of intergenic region sequences to support episomal replication in plant protoplasts in the presence of the cognate Rep protein. In this chapter, we describe some cloning strategies of the geminiviral genome and, subsequently, a replication assay for geminiviral replicons in tobacco or tomato protoplasts. These assays allow the structural and functional characterization of a plus-strand minimal origin of novel geminivirus isolates.

Due to the small size of the viral genome and the large number of geminiviral sequences in the GeneBank, the molecular cloning of known geminiviral genomic components has progressively become a fairly easy task, based on the design of specific and overlapping primers that allow the amplification of the full-length circular genome. However, the cloning of a new geminivirus under investigation requires a more elaborated strategy. Several protocols for the molecular cloning of DNA viruses have been described, which are based primarily on the previous biological isolation of the viral agent from symptomatic plants, as described in chapter "Geminivirus: Biolistic Inoculation and Molecular Diagnosis." Here we describe

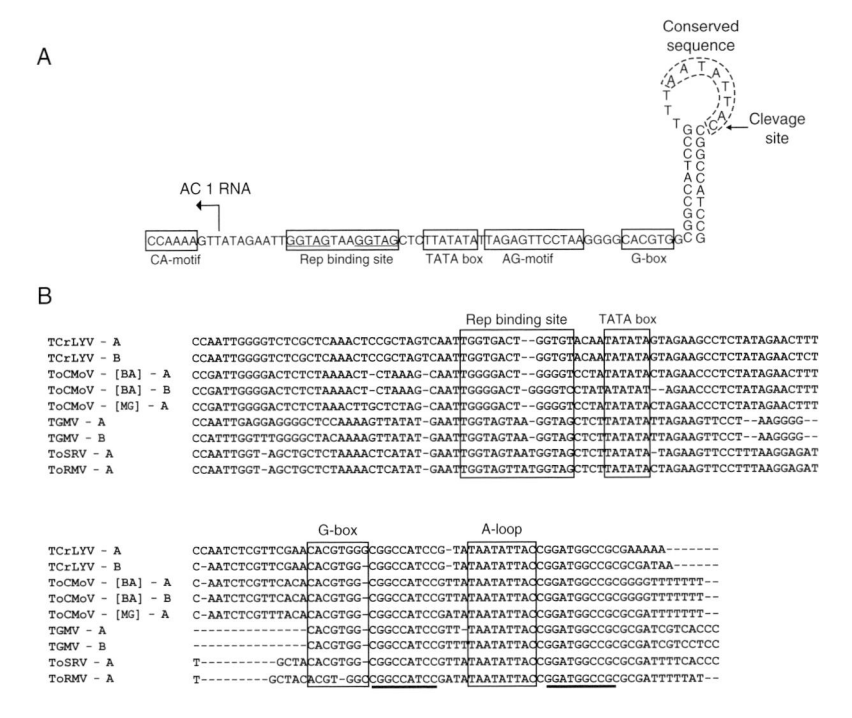

Fig. 1 The geminivirus replication origin. (**a**) The minimal origin of begomovirus replication. The DNA sequence corresponds to the TGMV DNA-A positions 54– 153. The hairpin structure is indicated and the conserved nonananucleotide loop sequence is marked. The initiation site and direction of synthesis for (+) strand DNA replication and AC1 transcription are also indicated. The Rep binding site and other functional elements are boxed as determined in ref. 9. AG motif and CA motif are essential for virus replication; TATA box and G-box correspond to binding sites for transcription factors. (**b**) Common region of DNA-A and DNA-B from Brazilian tomato-infecting begomoviruses. A multiple sequence alignment of common region sequences from the indicated begomoviruses was obtained with the CLUSTAL-W program. The nucleotide sequences have been aligned by introducing gaps (shown as dashes) to maximize identity. The putative AC1-binding site (repeats), the TATA and G-box of the leftward promoter and the conserved nonanucleotide (A-loop) are marked by open boxes. The conserved stem–loop structure is underlined

two cloning strategies: (1) PCR-based cloning of the full-length viral genome and (2) shotgun cloning of the viral DNA-A and DNA-B, which have been successfully used in our laboratory.

The cloning and sequencing of a new virus allow the prediction of the coding capacity of the viral genome and the subsequent identification of the viral genes by sequence comparisons. For the functional characterization of its origin of replication, the deduced replication module (Rep gene and origin of replication) has to be isolated, such that the Rep gene is cloned in a plant expression vector and the putative origin sequences into a pUC-based vector (2, 9). This strategy has been used with success to delimit the minimal geminivirus origin of replication and its functional modules (2, 9).

2   Materials

2.1   Common Materials Needed for all Procedures Described in this Chapter

1. Liquid nitrogen
2. Ice
3. Micropipettes
4. Tips
5. 1.7 mL microcentrifuge tubes
6. Vortex
7. Microcentrifuge
8. Sterile distilled water

2.2   Isolation of Total DNA from Infected Plants

See chapter "Geminivirus: Biolistic Inoculation and Molecular Diagnosis," Sect. 2.2.

2.3   Polymerase Chain Reaction (see notes 1 and 2)

1. High fidelity, thermostable DNA-dependent DNA polymerase (1–3 units)
2. Reaction buffer 10X (according to the manufacturer's instruction)
3. Template DNA (0.1–1 μg infected plant DNA for a 50 μL reaction)
4. Primers (10 μM) 18–25 deoxynucleotides
5. Divalent ion (1–4 mM per reaction)
6. 10 mM dNTPs (0.2 mM per reaction)
7. Sterile 0.5 mL or 0.2 mL microcentrifuge tubes
8. Thermocycler

2.4   Cloning of the amplified fragment

Several commercially available kits for cloning PCR products have been developed, such as TOPO cloning of PCR products (Invitrogen) and pGEMT cloning (Promega). They are quite appropriate for cloning partial amplified sequences of viral DNA, but they rely on an Adenosine extension added at the 5' end of the PCR product by the *Taq* DNA polymerase that complements a

Thymidine overhang in the vector. However, when the use of high-fidelity DNA-dependent DNA polymerase is required, the resulting amplified fragment is blunt ended due to the 3' exonuclease proofreading activity of the polymerase. To overcome this problem, we describe here a classical protocol for cloning the specific amplified fragments of begomovirus DNA-A and DNA-B, taking advantage of the presence of a *Pst*I site in the begomovirus degenerate primers (11). The amplified fragment is desalted and then digested with *Pst*I prior to cloning.

2.4.1 Digestion of the Amplified Fragment with *Pst*I

1. Restriction enzyme *Pst*I
2. Enzyme reaction buffer (according to manufacturer's instructions)
3. Water bath at 37 °C

2.4.2 Vector Preparation for Cloning (Digestion and Dephosphorylation)

This procedure includes the isolation of plasmid DNA, digestion with the appropriate enzyme (*Pst*I), desalting and dephosphorylation of the vector (pUC118, for instance).

1. Purified pUC118, as described in ref. 10
2. Restriction enzyme *Pst*I
3. Enzyme reaction buffer (according to manufacturer's instruction)
4. Water bath at 37 °C
5. Alkaline Phosphatase
6. Dephosphorylation buffer: 50 mM Tris-HCl pH 9.0, 1 mM $MgCl_2$

2.4.3 Ligation Reaction for Transformation of Escherichia coli and Diagnosis of Transformed Colonies

The ligation reaction and transformation of *E. coli* are performed according to standard techniques of molecular cloning into bacterial plasmids (10).

1. Previously purified, digested, and dephosphorylated vector (Sect. 2.4.2)
2. Purified and digested amplified viral DNA fragment
3. T4 DNA ligase
4. T4 DNA ligase buffer (10X)
5. Water bath at 14 °C or 4 °C
6. *E. coli* competent cells prepared as described in ref. 10
7. LB medium
8. Solid LB medium containing the appropriate antibiotic for selection (see note 3)

2.5 Designing of Overlapping Primers Harboring a Restriction Site for Cloning of the Full-Length Viral Genome

WebCUTTER or NetCUTTER software.

2.6 Shotgun Cloning of Full-Length Viral DNA

2.6.1 Isolation of *Geminivirus* RF from Infected Plants (*12*) (*see* note 4)

1. Lysis buffer: 10 mM Tris-HCl pH 7.5; 10 mM EDTA; 1% (w/v) SDS; 0.14 mM β-mercaptoethanol. β-mercaptoethanol must be added immediately before use
2. 5 M NaCl
3. Phenol:Chloroform 1:1 (v/v)
4. 3 M NaOAc
5. Isopropanol
6. 70% (v/v) ethanol
7. TE Buffer: 10 mM Tris-HCl pH 7.6; 1 mM EDTA; 50 µg mL^{-1} RNase. It is important to add RNase immediately before use

2.6.2 Digestion of Total DNA from Infected Plants

1. Total DNA from infected plants
2. Restriction enzymes (from the multiple cloning site (MCS) of the vector)
3. Restriction enzymes buffers (according to the manufacturer's instructions)
4. Water bath

2.6.3 Agarose Gel Electrophoresis

1. Electrophoresis apparatus
2. Power supply
3. 10 mg mL^{-1} ethidium bromide stock (see note 5)
4. Agarose 1% agarose and 0.2 µg mL^{-1} ethidium bromide in 0.5X TBE
5. Sample loading buffer (10 X): 0.1 M Tris-HCl, pH 8.0, 50% (v/v) glycerol, 0.1% (w/v) bromophenol blue
6. 10X TBE (Tris-Borate-EDTA) buffer: 0.89 M Tris-borate, pH 8.3, 0.02 M EDTA (see note 6)

2.6.4 Probe Preparation (Random Labeling)

1. 25 ng DNA heterologous probe (DNA-A and DNA-B of known, cloned begomoviruses)
2. Random primers (10 µM)

3. dCTP or dATP reaction buffer (according to the manufacturer's instruction)
4. 50 µCuries of α-^{32}P-dCTP or α-^{32}P-dATP
5. 1 U of T7 polymerase
6. Spin columns
7. Water bath at 37 °C

2.6.5 Southern Blot

1. Nylon membrane
2. Whatman 3MM Chr Blotting Papers
3. Radiolabeled probes (DNA-A and DNA-B of known begomoviruses)
4. 0.25 M HCl depurination solution
5. Denaturing solution: 0.5 M NaOH, 1.5 M NaCl
6. Neutralization solution: 1 M Tris-HCl, pH 7.0, 1.5 M NaCl
7. UV cross-linker
8. 20X SSC (see note 7)
9. Prehybridization solution: 6X SSC, 0.2% (w/v) Ficoll 400, 0.2% (w/v) polyvi-nylpyrrolidine (PVP), 0.2% (w/v) bovine serum albumin (BSA), 0.5% (w/v) SDS (sodium dodecyl sulfate), 1% (w/v) sodium pirophosphate, and 100 µg mL^{-1} denatured fragmented salmon sperm DNA. The salmon sperm DNA is added to the solution immediately before use. Boil salmon sperm DNA before add-ing to the solution.
10. Hybridization solution: 6X SSC, 0.06% (w/v) Ficoll 400; 0.06% (w/v) PVP, 0.06% (w/v) BSA, 0.7% (w/v) SDS, 1.3 mM EDTA, 13 mM Tris-HCl, pH 7.5, 50 ng denatured radiolabeled probe, and 100 µg mL^{-1} denaturated fragmented salmon sperm DNA. The probe and salmon sperm DNA are added to the solution immediately before use. Boil probe and salmon sperm DNA for 5 min and incubate on ice for 5 min before adding to the hybridi-zation solution.
11. Washing solution I: 1X SSC, 0.1% (w/v) SDS, 0.02% (w/v) sodium pyrophosphate
12. Washing solution II: 0.1X SSC, 0.1% (w/v) SDS, 0.02% (w/v) sodium pyrophosphate
13. Intensifying screen
14. Pressure blotter apparatus

2.6.6 Cloning of Full-Length Viral DNA (see note 8)

1. Total DNA from infected plants digested with the appropriate restriction enzyme
2. Digested and dephosphorylated plasmid DNA
3. T4 DNA ligase
4. *E. coli* competent cells

2.6.7   Diagnosis of Transformed Colonies by Dot Blot

Except for the pressure blotter that here is replaced by the dot blotter apparatus with a vacuum pump, all the other materials and solutions are the same as in Sect. 2.6.5. Alternatively to the dot blotter, the colonies can be spotted directly on the nylon membrane. The size of the insert of a positive clone must be confirmed by electrophoresis.

2.6.8   PCR-Based Diagnostic of Transformed Colonies

1. Sterile 0.5 mL or 0.2 mL microcentrifuge tubes
2. Overnight grown colonies in solid selective media
3. Primers (10 μM) – 18–25 nucleotides
4. 25 mM $MgCl_2$ (1–4 mM per reaction)
5. 10 mM dNTPs (0.2 mM per reaction)
6. *Taq* DNA polymerase (1–3 U)
7. Thermocycler
8. 10X *Taq* buffer: 0.2 M Tris-HCl, pH 8.3, 0.5 M KCl

2.7   Replication Assays in Protoplasts and Identification of Origin of Replication

2.7.1   Tissue Culture Stocks and Solutions

1. B1 Inositol stock: 10 g of Inositol, 0.1 g of Thiamine-HCl. Add dH_2O to 1 L and autoclave.
2. Miller's I: 60 g of KH_2PO_4, add dH_2O to 1 L. Autoclave.
3. Tobacco suspension cell media: 4.3 g of MS salts, 10 mL of B1 Inositol stock, 3 mL of Miller's I, 30 g of Sucrose, 20 μL of 10 mg mL^{-1} 2,4-D, dH_2O to 1 L. Adjust pH to 5.5–5.7 and autoclave.
4. Tobacco protoplast media: Tobacco suspension cell media plus 0.4 M mannitol. Adjust pH 5.5–5.7 and autoclave.

2.7.2   Protoplast Preparation

1. Protoplasting basic solution (Enzyme solution): 0.4 M mannitol, 20 mM MES pH 5.5, 1% (w/v) cellulose, 0.1% (w/v) pectolyase. Stir for 1 h at room temperature (RT). Spin at 10,000 rpm for 10 min. Filter sterilize with 0.2 μ filter.
2. *Nicotiana tabacum* suspension cells, 4–5 days after subculture
3. 100 × 25 mm Petri dishes
4. Orbital shaker

2.7.3   Electroporation of Viral DNA Replicons into Protoplasts

1. 15 μg DNA cassette A: pUC-based plasmid containing viral DNA sequences (potential replication origin sequences)
2. 15 μg DNA cassette B: Plant expression vector (pMON921, ref. 2) harboring the cognate Rep coding region under the control of the 35S promoter and the 3' end of the pea E9 *rbcS* gene.
3. 40 μg of salmon sperm DNA
4. Sterile electroporation buffer (EB): 0.8% NaCl, 0.02% KCl, 0.02% KH_2PO_4, 0.11% Na_2HPO_4, 0.4 M mannitol, pH 6.5
5. Electroporation cuvettes: 0.4 cm gap cuvettes
6. Electroporator

2.7.4   Analysis of Nascent Viral Replicon in Protoplasts

1. Lysis buffer: 50 mM Tris-HCl pH 7.6, 100 mM NaCL, 50 mM EDTA, 0.5% (w/v) SDS, 1.0 mM β-mercaptoethanol. β-mercaptoethanol must be added immediately before use.
2. Buffered Phenol
3. Phenol:Chloroform 1:1 (v/v)
4. Chloroform
5. Isopropanol
6. Absolute ethanol
7. 70% (v/v) ethanol
8. TNE Buffer: 10 mM Tris-HCl pH 7.6, 100 mM NaCl, 1 mM EDTA, 100 μg mL^{-1} RNase. It is important to add RNase immediately before use.
9. The detection of newly synthesized DNA is performed by Southern blot as described in Sect. 2.6.5 (see note 10).

3   Methods

3.1   *PCR-Based Cloning of Full-Length Viral DNA (Fig. 2)*

This strategy requires the partial sequencing of a genome fragment of the new geminivirus, as the basis for primer design. In the case of begomoviruses, this is commonly achieved by amplification of a genome fragment with the DNA-A and DNA-B-specific degenerate primers, as described in the accompanying chapter, and subsequent cloning and sequencing of the amplified fragment. The partial sequence of the new begomovirus is then used to design overlapping forward and reverse primers that allow the amplification of the full-length genomic component. In addition, for cloning purposes, both primers must contain at the 5' end a restriction site that is unique in the viral genome (12). Thus, the primers are designed such

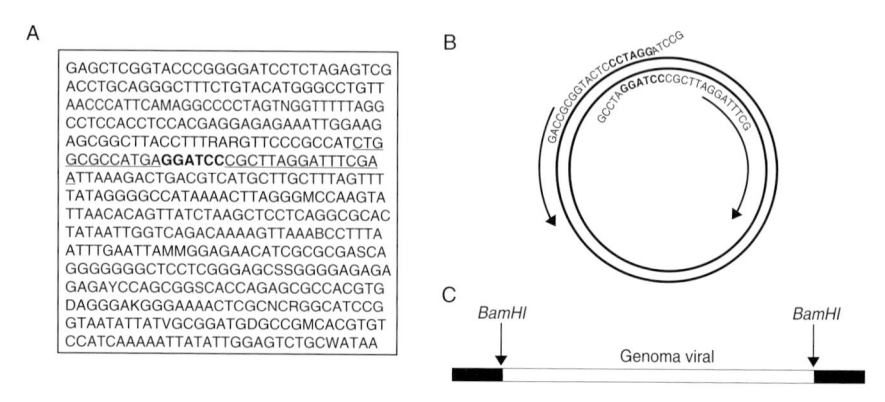

Fig. 2 PCR-based cloning of full-length viral DNA (**a**) Partial sequence of a begomovirus-amplified fragment by the degenerate primers. The *Bam*HI restriction site, GGATCC, is indicated in bold. The underlined sequences correspond to primers for cloning the full-length DNA-A. (**b**) Annealing position and direction of primer extension on the viral genome. The sequences in bold correspond to the *Bam*HI site and the nonannealing sequences at the ends of the primers correspond to random sequences used as adaptors. (**c**) PCR amplification of full-length circular DNA-A. The amplified fragment illustrates the full-length DNA-A flanked by the *Bam*HI site. The *black boxes* correspond to the random sequences inserted by the primers. The intact viral DNA-A is recovered by digestion with *Bam*HI and recircularization of the amplified fragment. The digested *Bam*HI amplified fragment may be cloned into a previously *Bam*HI digested pUC-derived vector

that the amplified fragment is flanked by the selected restriction site that constitutes the cloning site. To increase the efficiency of restriction enzyme digestion of the amplified fragment, random nonannealing sequences are included as 5' extensions of each primer, which serve as adaptors.

The replicative form of the viral DNA, isolated from infected plants, is enriched by purification from an agarose gel and used as template in PCR with the specific primers and a high-fidelity DNA-dependent DNA polymerase. The resulting amplified fragment consists of the full-length genomic component flanked by the restriction site. The amplified viral genome is digested with the restriction site-specific endonuclease, separated by electrophoresis, purified from the agarose gel and then cloned into the same site of an appropriate bacterial vector. The procedures for desalting, dephosphorylation of the restricted vector and ligation of the DNA fragment into the vector are conducted essentially as described (10).

3.1.1 Amplification of a Fragment from the Begomovirus DNA-A or DNA-B

The conditions of PCR for amplification of begomovirus DNA have been described in chapter "Geminivirus: Biolistic Inoculation and Molecular Diagnosis," Sects.

4–8, 3.3.3. While those conditions have been optimized for diagnostic assays using *Taq* DNA polymerase, they may vary substantially depending on the DNA polymerase to be used. Several high-fidelity DNA-dependent DNA polymerases are commercially available, such as *Taq* platinum high fidelity, *Pfx* or *Pfu* (Invitrogen) and *Tli* (Promega). Among these enzymes, the requirements for optimal enzyme activity differ with respect to the divalent ion, concentration of DNA template, annealing and extension temperatures. Therefore, the PCR protocol must be established according to the manufacturer's instructions.

3.1.2 Cloning the Amplified Fragment into a Plasmid

After amplification of a viral DNA fragment by PCR using degenerated primers associated with high-fidelity DNA-dependent DNA polymerase, the amplified fragment should be desalted and then digested with *Pst*I, using the protocol provided by the manufacturer. The cloning vector should replicate to a high copy number in *E. coli* and contain a *Pst*I restriction site in its MCS. Both pBS-based and pUC-based plasmids have been largely used for viral DNA cloning. The steps for vector preparation are as follows:

1. Isolation of plasmid DNA
2. *Pst*I digestion according to the manufacturer's instructions
3. Desalting of the linearized vector
4. Dephosphorylation of the vector using CIAP (calf intestinal alkaline phosphatase)
5. Purification of the linearized vector from 1% agarose gel

The ligation reactions should be performed according to standard techniques of molecular cloning into bacterial plasmids (10).

The success of the reaction depends on the concentration of the DNA fragments to be ligated, and a 3:1 molar ratio of fragment to vector is considered optimal. To figure out the correct molar concentration of the vector and fragment ends, we have used the equation

$$X = (Z \times B)/(A \times 3,148),$$

where

$X = \mu L$ of DNA;
$Z = f$mol wanted in the reaction;
$B =$ size of the fragment (kb);
$A =$ DNA concentration (ng μL^{-1}).

Furthermore, in a ligation reaction of cohesive-ended fragments, the amount of insert and vector ends should be 9–90 *f*mol and 3–30 *f*mol, respectively. For blunt-ended fragment ligation, while the 3:1 molar ratio of fragment to vector should be maintained, the amount of insert should be increased to 45–180 *f*mol and the vector to 15–60 *f*mol.

After the ligation reaction, the transformation of *E. coli* and diagnostic of positive clones are performed as described (10). The sequencing of the positive clones will allow the design of primers for amplification of the full-length genomic components (Fig. 2a).

3.1.3   Design of Partially Overlapping Primers Containing a Unique Restriction Site for Cloning of the Full-Length Viral Genome

The sequencing of the cloned fragment allows the identification of a possible restriction site in the viral DNA also present in the MCS of the vector (Fig. 2a). The specific primers are designed in such a way that the amplified fragment is flanked by this unique restriction site (Fig. 2b). Both the forward and reverse primers will contain the restriction site that will be extended at the 3'direction with viral sequences and at the 5' end with a random sequence of six deoxynucleotides, which works as an adaptor to improve the efficiency of enzyme digestion (Fig. 2c). The final size of the primers may vary from 20 to 30 mer nucleotides. The partial overlap of the primers at the restriction site will allow the recircularization of the viral DNA inside the host cells. A schematic representation of the primer is shown in Fig. 2.

3.1.4   Cloning of Full-Length Viral DNA-A and DNA-B (see note 10)

This cloning strategy is based on the amplification of full-length DNA-A and DNA-B of begomoviruses and subsequent cloning into an appropriate vector. The success of this approach depends on the efficiency of the primers to be designed based on the sequences of the cloned viral fragments, as obtained in Sects. 3.1.1–3.1.3.

1. Isolate total DNA from infected plants as described in chapter "Geminivirus: Biolistic Inoculation and Molecular Diagnosis" (see Sects. 4–8, 3.1.1).
2. Assemble a 50 μL reaction for PCR as described in Sects. 4–8, 3.3.3, with the designed DNA-A or DNA-B-specific primers.
3. Check the amplified fragment by electrophoresing a 5 μL aliquot of the PCR products in an agarose gel. The size of the amplified fragment should be approximately 2.6 kb.
4. Desalt the rest of the reaction and digest the purified fragment with the restriction endonuclease that corresponds to the selected site for primer designing.
5. Purify the full-length digested viral DNA from an agarose gel using commercially available kits.
6. Prepare the vector for cloning through digestion with the same enzyme (see Step 4) and followed by dephosphorylation as described in Sect. 3.1.2 (Fig. 4a).
7. Assemble a 10–15 μL ligation reaction as described in Sect. 3.1.2, and incubate at 14–16 °C for 4 h or at 4 °C for 12 h.
8. Use half of the ligation reaction for transformation of *E. coli* (strain JM109 or DH5α or others) competent cells, as described (10).

9. Proceed to the diagnosis of positive clones by PCR-specific amplification or by sizing the insert through plasmid DNA digestion and gel electrophoresis (Fig. 4c).

3.2 Shotgun Cloning of Full-Length Viral DNA (Fig. 4)

This approach is based on the previous knowledge of a restriction map of the genomic components for the selection of the cloning site. For new viruses whose DNA sequences are unknown the restriction pattern is deduced from Southern blotting of previously digested DNA from infected plants, using begomovirus heterologous probe. For high-quality results, the protocol for DNA isolation from infected plants should render an enriched fraction of the viral replicative form (dsDNA), as described below. The isolated DNA is digested separately with different restriction enzymes, which are present in the MCS of the cloning vector and, then, analyzed by Southern Blot using DNA-A and DNA-B-specific probes (Fig. 3). After identification of unique restriction sites in the viral DNA, the isolated DNA is cleaved with the corresponding restriction enzyme, cloned into a previously digested and dephosphorylated vector and screened by dot blot.

3.2.1 Isolation of *Geminivirus* RF from Infected Plants (*12*)

1. Harvest young leaves (2 g) from infected plants and immediately freeze them in liquid nitrogen.
2. Grind to a fine powder in liquid nitrogen with a mortar and pestle.
3. Homogenize with 10 mL of Lysis buffer until the mixture is frozen.
4. Incubate at 4 °C for 10 min.
5. Centrifuge at 15,000 g during 15 min.
6. Transfer the supernatant to a clean, properly labeled tube.
7. Add 1 M NaCl to 1 mM final concentration and mix by inverting the tube.
8. Incubate for 12–14 h at 37 °C, under gentle agitation.
9. Centrifuge at 75,000 g for 1 h.
10. Transfer the supernatant to a clean, properly labeled tube and add equal volume of phenol: chloroform (1:1 v/v) and 3 M NaOAc to 0.3 M final concentration. Mix gently (do not vortex).
11. Centrifuge at 16,000 g for 10 min.
12. Carefully remove the supernatant with a pipette and transfer it to a clean, properly labeled microtube.
13. Precipitate the DNA with 1 volume of isopropanol. Mix well and incubate at RT for 10–30 min.
14. Centrifuge at 16,000 g for 15 min.
15. Carefully discard the supernatant with a pipette.
16. Wash the precipitate with 1 mL of 70% ethanol.
17. Centrifuge at 16,000 g for 10 min.
18. Discard the supernatant.

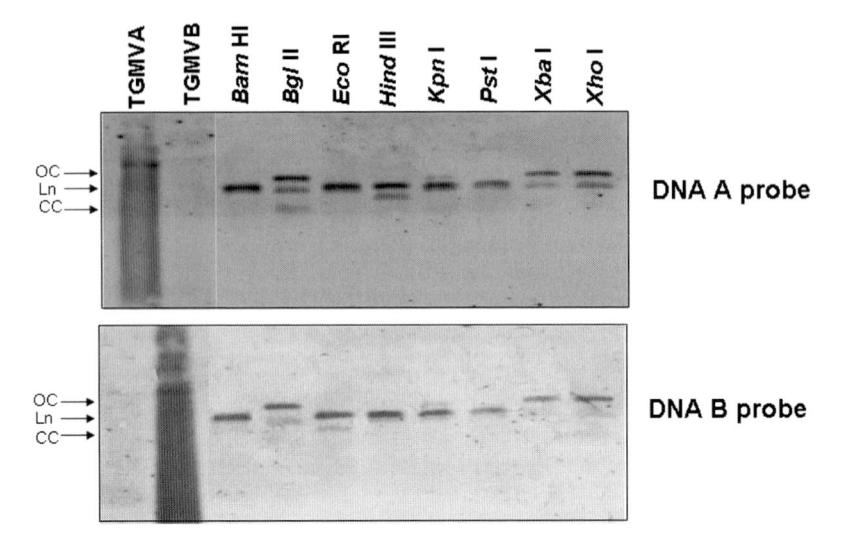

Fig. 3 Restriction mapping of viral DNA-A and DNA-B. Viral DNA was purified from infected tomato leaves and digested by several restriction enzymes, resolved on an agarose gel, transferred to nylon membranes, fixed on the membrane by UV, and hybridized with a ^{32}P-labeled probe specific for begomovirus DNA-A or DNA-B. The positions of open circular (OC) and supercoiled (CC) dsDNA as well as the linearized form (Ln) of viral DNA are indicated on the left. TGMV DNA-A and DNA-B were used as control

19. Invert the microtube on absorbent paper and allow the pellet to air dry or dry the pellet by Speedvac.
20. Resuspend the pellet in 100–250 μL of sterile dH$_2$0 or TE buffer, containing 50 μg mL^{-1} RNase.

3.2.2   Restriction Mapping of A and B Genomic Components (Fig. 3)

The goal is to identify a unique restriction site in the DNA-A and DNA-B to be used as the cloning site. Thus, the chosen enzymes should have restriction sites in the MCS of the vector. The single digestion reactions of plant DNA must be assembled according to the manufacturer's instructions.

3.2.2.1   Random Labeling Protocol for Probe Preparation (see note 11)

Thaw primers, DNA, buffers, and radioisotope on ice at least 30 min prior to use.

1. To prepare DNA probe, mix 25 ng DNA in 23 μL of dH$_2$O and 10 μL random primers.
2. Boil for 5 min and leave at RT.

3. Add 10 μL of 5X dCTP or dATP buffer, 5 μL (50 μcuries) of [α-^{32}P]-dCTP or [α-^{32}P]-dATP, and 1 μL of T7 polymerase (2 U μL^{-1}).
4. Incubate at 37 °C for 10 min.
5. Add 50 μL of dH$_2$O.
6. Eliminate nonincorporated radioisotope with spin columns.

3.2.2.2 DNA Digestion, Electrophoresis, and Southern Blot (see note 12)

1. Digest 10–20 μg of total DNA from infected plants with each one of the restriction enzymes present in the MCS of the vector to be used for cloning.
2. Separate the digested DNA by electrophoresis in 1% agarose gel.
3. After electrophoresis, treat the agarose gel with depurination, denaturing, and neutralization solutions as described (10).
4. Transfer the digested DNA to a nylon membrane either by capillarity (see ref. 10) or by using a pressure blotter.
5. UV crosslink the DNA to the membrane.
6. Add 15–25 mL of prehybridization solution to the membrane in a hybridization bag or in hybridizarion tubes.
7. Incubate at 65 °C in a hybridization oven or shaker for 2 h or overnight.
8. Remove prehybridization solution and add 10–20 mL of hybridization solution. Keep the volume as low as possible to maximize hybridization.
9. Hybridize at 65 °C overnight shaking at constant speed.
10. Wash the membrane three times with 300–400 mL of 1X SSC, 0.1% SDS for 30 min each at 65 °C.
11. Wash the membrane —one to two times with 300–400 mL of 0.1X SSC, 0.1% SDS for 30 min each at 65 °C.
12. Transfer the membrane to a new hybridization bag and seal it.
13. Place the membrane in a cassette with intensifying screen, put film on, at −80 °C overnight or up to 3 days depending on the exposure needed.
14. Develop the film by autoradiography.

3.2.3 Cloning of Full-Length Viral DNA

Based on the Southern of restricted viral DNA it is possible to identify a unique restriction site for cloning (Fig. 3).

1. Digest 10–20 μg of total DNA from infected plants with the restriction enzyme identified.
2. Separate the digested DNA by electrophoresis in a 1% agarose gel (Fig. 4a).
3. Cut the region of the gel around the 2.6-kb fragments and purify the digested DNA from the gel slice.
4. Prepare the vector for cloning (digestion with the same enzyme and dephosphorylation), as described in Sect. 3.1.2.

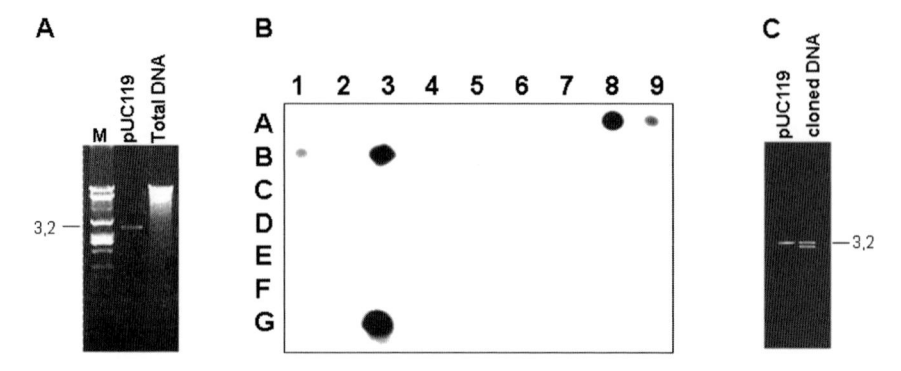

Fig. 4 Shotgun cloning of full-length viral DNA. (**a**) Digestion of DNA for cloning of viral DNA-A. Total DNA from infected tomato and the pUC118 vector were digested with *Bam*HI, and visualized on an ethidium bromide-stained agarose gel after electrophoresis. M corresponds to digested lambda DNA markers and the number to the left corresponds to the size in kb. (**b**) Screening of positive viral DNA-A clones. Plasmid DNA isolated from transformed colonies was blotted onto a nylon membrane and hybridized with a ^{32}P labeled begomovirus-specific probe. (**c**) Diagnostic of positive clones from B by restriction digestion. Plasmid DNA was isolated from an overnight culture of a positive colony, digested with *Bam*HI, separated by electrophoresis and visualized by ethidium bromide. The size of the insert was 2.6 kb and corresponds to the size of a full-length DNA-A

5. Assemble a 10–15 μL ligation reaction as described in Sect. 3.1.2, and incubate at 14–16 °C for 4 h or at 4 °C for 12 h.
6. Use half of the ligation reaction for transformation of *E. coli* (strain JM109 or DH5α or others) competent cells, as described (10).
7. Proceed to the screening of the transformed bacterial colonies by Dot-Blot (Fig. 4b) or by PCR of colonies using specific primers.

3.2.4 Dot Blot Screening

1. Cut the nylon membrane and the Whatman 3MM filter with the same dimensions as the transfer region of the dot blotter to be used.
2. Submerge the membrane in 6X SSC solution for 5 min.
3. Assemble the dot blot apparatus with the membrane.
4. Apply the samples into the wells under vacuum (2 μL of DNA miniprep, 30 μL of 20X SSC, 68 μL of dH2O).
5. Incubate the membrane over a Whatman 3MM filter that has been saturated with denaturing solution for 5 min.
6. Incubate for 5 min over a filter saturated with neutralization solution.
7. Dry the membrane over the filter paper.

8. Immobilize the nucleic acids through UV crosslinking.
9. Prehybridize, hybridize, and wash the membrane as described in Sect. 3.2.2.2.

3.2.5 Colony PCR-Based Screening (see note 13)

In this screening, an overnight-grown colony is directly used in a PCR-based diagnostic assay as the source of template DNA.

1. With a sterile pipette tip, pick a portion of an overnight grown colony from the Petri dish and suspend in $20\,\mu L$ of sterile dH_2O.
2. Use $5\,\mu L$ of the mixture for PCR and reserve the remaining $15\,\mu L$ to inoculate $2\,mL$ of LB medium supplemented with the proper antibiotic for plasmid isolation.
3. Assemble the reaction as described in chapter "Geminivirus: Biolistic Inoculation and Molecular Diagnosis."
4. The conditions of PCR are as described in chapter "Geminivirus: Biolistic Inoculation and Molecular Diagnosis," except that the period for the initial step at $94\,^{\circ}C$ should be extended to $10\,min$ to lyse the cells and expose the plasmid DNA.
5. Analyze the PCR product by electrophoresis.

3.3 Replication Assays in Protoplasts and Identification of Replication Origin

Due to the specificity of the Rep protein of a given begomovirus to its cognate replication origin (ori), any replication assay for functional characterization of replication origins requires that the replication module (Rep + ori) be treated as an unit (2, 13). Thus, it is necessary to clone the Rep ORF of the new virus under characterization in a plant expression vector to mediate its cognate origin-supported episomal replication. The cloning and sequencing of viral DNA-A allow the prediction of the origin of replication by sequence homology as well as the identification of the putative Rep ORF.

Knowing the DNA-A sequence of the new geminivirus, both the putative replication origin and the Rep ORF can be isolated by PCR using specific primers, and transferred to the appropriate plasmids. The replication origin is often cloned into a pUC vector. The Rep gene has to be cloned into a plant expression cassette, for instance pMON921 (2), which contains the Cauliflower mosaic virus 35S promoter with a duplicated enhancer region (E35S) and the 3' end of the pea E9 *rbcS* gene.

3.3.1   Replicon Construct (see note 14)

1. Design specific primers for amplification of the putative origin sequences. Strategically insert a restriction site into the primer sequence to facilitate cloning into a pUC vector.
2. Isolate the putative replication origin by PCR using the cloned viral DNA as a template and a high fidelity DNA-dependent DNA polymerase.
3. Desalt the amplified fragment, digest it with the appropriate restriction enzyme and purify it from an agarose gel.
4. Prepare the vector (pUC118, pUC119, or pBSKS) for cloning (digestion with the same enzyme and dephosphorylation), as described in Sect. 3.1.2.
5. Assemble a 10–15 µL ligase reaction, as described in Sect. 3.1.2, and incubate at 14–16 °C for 4 h or at 4 °C for 12 h.
6. Use half of the ligation reaction for transformation of *E. coli* (strain JM109 or DH5α or others) competent cells, as described (10).
7. Proceed to the diagnostic of the positive clones by electrophoresis of digested DNA.

3.3.2   Cloning of the Rep ORF into a Plant Expression Vector

1. Design specific primers for amplification of the putative Rep ORF. Strategically insert a restriction site into the primer sequence to facilitate cloning into a plant expression cassette.
2. Isolate the putative Rep ORF by PCR using the cloned viral DNA as a template and a high fidelity DNA-dependent DNA polymerase and prepare the amplified fragment for cloning as described in Sect. 3.3.1.
3. Prepare the plant expression vector, such as pMON921, as described in Sect. 3.1.2.
4. Follow the standard procedures for molecular cloning into bacterial vectors.

3.3.3   Protoplast Preparation and Electroporation of Viral Constructs

1. Subculture the *N. tabacum* suspension cells by diluting 5 mL of cultured cells into 45 mL of tobacco suspension cell medium.
2. 4–5 days after subculture, pour cells into 50 mL tubes.
3. Spin cells at 500 rpm for 2 min.
4. Wash cells with 0.4 M mannitol.
5. Spin cells at 500 rpm for 2 min.
6. Take fresh weights. Subtract out tube weight.
7. Add enzyme solution based on fresh weight. Usually 5–10 mL is sufficient.
8. Plate out in 100 × 25 mm Petri dishes.
9. Shake on platform shaker at 60 rpm for 30 min. Check under microscope to examine protoplast formation.

10. Shake extra 15–30 min (see note 15).
11. Pour cells into 50 mL tubes. Spin 500 rpm for 2 min.
12. Wash twice with 0.4 M mannitol.
13. Wash twice with EB.
14. Screen cells through 100 μ mesh.
15. Count cells using hemacytometer.
16. Resuspend cells in appropriate amount of EB to a final concentration of 5×10^6 cells mL^{-1} and keep on ice until ready to electroporate.
17. Electroporate at 500 μF, 250 Volts (see note 16).
18. Keep cells on ice after electroporation and prior to plating out.
19. Plate out cells into 7–9 mL of protoplast media.
20. Incubate protoplasts for 48 h in 25 °C growth chamber for DNA extraction.

3.3.4   DNA Extraction from Electroporated Protoplasts

1. Wash cells twice in 7–10 mL of tobacco protoplast media.
2. Resuspend washed cells in 400 μL of lysis buffer.
3. Sonicate cells
4. Incubate at RT for 15 min.
5. Add 150 μL of buffered phenol, shake vigorously. Let sit at RT for 2 min.
6. Add 150 μL of CHCl$_3$, vortex fast.
7. Spin for 2 min.
8. Re-extract with 1 volume of CHCl$_3$.
9. Precipitate with 1 volume of isopropanol for 5 min at RT or leave at −20 °C overnight.
10. Centrifuge at 16,000 g for 5 min.
11. Wash with 500 μL of 70% (v/v) ethanol.
12. Resuspend in 200 μL of TNE buffer and incubate at 37 °C for 30 min.
13. Extract with 1 volume of phenol/CHCl$_3$.
14. Extract with 1 volume of CHCl$_3$.
15. Reprecipitate with 500 μL of 100% ethanol. Incubate at −20 °C for 10 min or overnight.
16. Centrifuge at 16,000 g for 5 min.
17. Wash with 500 μL of 70% (v/v) ethanol.
18. Centrifuge at 16,000 g for 5 min, dry in speedvac.
19. Resuspend in 40 μL of TE.
20. Quantitate 1 μL on fluorometer.

3.3.5   Probe Preparation

The DNA used as probe can be any DNA fragment of the constructed replicon and frequently is a 1–1.5 kb fragment of the vector (pUC or pBS vectors). For DNA labeling, follow the protocol as described in Sect. 3.2.2.1.

3.3.6 DNA Digestion, Electrophoresis, and Southern Blot

1. Double digest 5 µg of total DNA from protoplasts with a restriction enzyme that has a unique site in the replicon and the enzyme *Dpn*I (see note 17).
2. Separate the digested DNA by electrophoresis in 1% agarose gel and follow the procedures as described in the Sect. 3.2.2.2 for DNA transferring, UV crosslinking and Southern hybridization.

4 Notes

1. The use of a high fidelity, thermostable DNA-dependent DNA polymerase for cloning of PCR product is crucial to avoid misincorporation of deoxynucleotides in the amplified fragment. Several high-fidelity, thermostable DNA-dependent DNA polymerases are commercially available (*Taq* platinun high fidelity, *Pfx, Pfu, Tli* etc.). It is important that the reactions are performed according to the manufacturer's instructions.
2. The universal and degenerate primers for begomoviruses DNA-A or DNA-B specific amplification are presented in chapter "Geminivirus: Biolistic Inoculation and Molecular Diagnosis" (Table 2).
3. 100 µg mL^{-1} ampicillin in the case of pUC-derived or pBS-derived vectors, such as pUC118, pUC119, pBSII KS (+/−), pBSII SK (+/−).
4. The efficiency of each DNA extraction protocol depends on the plant species. For instance, quick methods do not work properly in the case of *Sida aurens* and *Sida rhombifolia* due to the large amounts of phenolic compounds and polysaccharide contaminants. Therefore, frequently the described protocols of DNA extraction from infected plants should be adjusted for the species under investigation. In general, the DNA isolation protocols described in the chapter 39 are well suited as the starting material for viral DNA cloning. Nevertheless, we describe in the present chapter a distinct protocol for isolation of Geminivirus RF, which produces high quality RF-enriched fractions and, thus, is more appropriate for Southern blot-based restriction mapping of unknown viral genomes.
5. Dissolve 0.2 g of ethidium bromide in 20 mL H$_2$O. Mix, wrap in aluminum foil, and store at 4 °C. CAUTION: Ethidium bromide is a mutagen and must be handled carefully. Wear gloves.
6. Add 108 g Tris base, 55 g boric acid, 40 mL 0.5 M EDTA, to make 1 L of 10X TBE (pH 8.0) with sterile water.
7. Dissolve 175.3 g NaCl and 88.2 g sodium citrate in 800 mL of distilled H$_2$O. Adjust the pH to 7.0 with a few drops of 1 M HCl. Adjust the volume to 1 L with additional distilled H$_2$O. Sterilize by autoclaving.
8. To enrich for the viral fragment to be cloned, the digested DNA should be fractionated by electrophoresis and 2.5–3.0 kb sized fragments should be purified directly from the gel.
9. For Southern blot analysis, an additional *Dpn*I digestion of the total DNA is required to distinguish newly synthesized DNA from input DNA. Thus, total DNA from protoplasts must be double digested with *Dpn*I and a restriction enzyme that has a unique site in the DNA construct for replicon linearization, prior to electrophoresis. While the newly synthesized DNA in plant cells is resistant to *Dpn*I digestion, the input DNA which has been replicated in *E. coli* will be degraded.
10. Because the primer was designed on the basis of partial sequence of the viral DNA, it is possible that the selected restriction site is also present in the remaining sequence of the viral genome. This possibility will be confirmed with the determination of the insert size of the resulting clone. If this is the case, one may still clone a full-length viral DNA by partial digestion of the amplified fragment.

11. The DNA used as probe corresponds to DNA restricted fragments from cloned viral DNA-A and/or DNA-B and must cover the entire genome.

12. The efficiency of the heterologous probe may vary among begomoviruses and, therefore, the temperature of hybridization and washing conditions may have to be adjusted for any particular case.

13. One may use an overnight liquid culture of transformed bacterial colonies for PCR diagnosis of positive clones. In this case, the reaction is assembled with 3 µL of LB cultured bacterial cells for a 50 µL reaction. After PCR (or DNA digestion by restriction enzymes), it is important to visualize the reaction product by electrophoresis in ethidium bromide-stained gels to confirm the size of the fragment. The positive clones are grown in LB media (with antibiotic) up to the exponential phase of growth to make glycerol stocks and be stored at −80 °C.

14. The viral sequences to be cloned may be mutated to precisely characterize the cis-regulatory elements for origin function. Deletion mutants are often used to determine the minimal origin of replication (9).

15. Generally, formation of protoplast by enzyme digestion is complete in 45 min. Do not go longer than 1 h.

16. Use 400 µL cells + 400 µL EB containing viral replicon (15 µg) + Rep expression cassette (15 µg) + DNA carrier (40 µg of sheared salmon sperm DNA).

17. The inclusion of the *Dpn*I enzyme in the digestion is required to distinguish the newly synthesized replicon in the protoplast from the input DNA. While the newly synthesized DNA in plant cells is resistant to *Dpn*I digestion, the input DNA, which has been replicated in *E. coli*, will be degraded.

References

1. Fontes, E. P. B., Luckowand, V. A., and Hanley-Bowdoin, L. (1992) A geminivirus replication protein is a sequence-specific DNA binding protein. *Plant Cell* **4**, 597–608.

2. Fontes, E. P. B., Eagle, P. A., Sipe, P. S., Luckow, V. A., and Hanley-Bowdoin, L. (1994) Interaction between a geminivirus replication protein and origin DNA is essential for viral replication. *J. Biol. Chem.* **269**, 8459–8465.

3. Fontes, E. P. B., Gladfelter, H. J., Schaffer, R. L., Petty, I. T. D., and Hanley-Bowdoin, L. (1994) Geminivirus replication origins have a modular organization. *Plant Cell* **6**, 405–416.

4. Lazarowitz, S. G., Wu, L. C., Rogers, S. G., and Elmer, J. S. (1992) Sequence-specific interaction with the viral AL1 protein identifies a geminivirus DNA replication origin. *Plant Cell* **4**, 799–809.

5. Chatterji, A., Chatterji, U., Beachy, R. N., and Fauquet, C. M. (2000) Sequence parameters that determine specificity of binding of the replication-associated protein to its cognate site in two strains of Tomato leaf curl virus-New Delhi. *Virology* **273**, 341–350.

6. Frischmuth, T., Engel, M., Lauster, S., and Jeske, H. (1997) Nucleotide sequence evidence for the occurrence of three distinct whitefly transmitted, Sida-infecting bipartite geminiviruses in Central America. *J. Gen. Virol.* **78**, 2675–2682.

7. Gilbertson, R. L., Hidayat, S. H., Paplomatas, E. J., Rojas, M. R., Hou, Y-M., and Maxwell, D. P. (1993) Pseudorecombination between infectious cloned DNA components to tomato mottle and bean dwarf mosaic geminiviruses. *J. Gen. Virol.* **74**, 23–31.

8. Galvão, R. M., Mariano, A. C., Luz D. F., Alfenas, P. F., Andrade, E. C., Zerbini, F. M., Almeida, M. R., and Fontes, E. P. B. (2003) A naturally occurring recombinant DNA-A of a typical bipartite begomovirus does not require the cognate DNA-B to infect *Nicotiana benthamiana* systemically. *J. Gen. Virol.* **84**, 715–726.

9. Orozco B. M., Gladfelter, H. J., Settlage, S. B., Eagle, P. A., Gentry, R. N. and Hanley-Bowdoin L. (1998). Multiple cis-elements contribute to geminivirus origin function. *Virology* **242**, 346–356.

10. Sambrook, J., Fritsch, E. F., and Maniatis, T. (1989) Molecular Cloning: A laboratory manual. Cold Spring Harbor, NY Cold Spring Harbor Laboratory Press.

11. Rojas, M. R., Gilbertson, R. L., Russel, D. R., and Maxwell, D. P. (1993) Use of degenerate primers in the polymerase chain reaction to detect whitefly-transmitted geminiviruses. *Plant Dis.* **77,** 340–347.

12. Patel, V. P., Rojas, M. R., Paplomatas, E. J., and Gilbertson, R. L. (1993) Cloning biologically active geminivirus DNA using PCR and partially overlapping primers. *Nucleic Acids Res.* **21,** 1325–1326.

13. Gladfelter, H. J., Eagle, P. A., Fontes, E. P. B., and Hanley-Bowdoin, L. (1997) Two domains of the AL1 protein mediate geminivirus origin recognition. *Virology* **239,** 186–197.

Chapter 12
Analysis of Viroid Replication

Ricardo Flores, María-Eugenia Gas, Diego Molina, Carmen Hernández, and José-Antonio Daròs

Abstract Viroids, as a consequence of not encoding any protein, are extremely dependent on their hosts. Replication of these minimal genomes, composed exclusively by a circular RNA of 246–401 nt, occurs in the nucleus (family Pospiviroidae) or in the chloroplast (family Avsunviroidae) by an RNA-based rolling-circle mechanism with three steps: (1) synthesis of longer-than-unit strands catalyzed by host DNA-dependent RNA polymerases recruited and redirected to transcribe RNA templates, (2) cleavage to unit-length, which in family Avsunviroidae is mediated by hammerhead ribozymes, and (3) circularization through an RNA ligase or autocatalytically. This consistent but still fragmentary picture has emerged from a combination of studies with in vitro systems (analysis of RNA preparations from infected plants, transcription assays with nuclear and chloroplastic fractions, characterization of enzymes and ribozymes mediating cleavage and ligation of viroid strands, dissection of 5′ terminal groups of viroid strands, and in situ hybridization and microscopy of subcellular fractions and tissues), and in vivo systems (tissue infiltration studies, protoplasts, studies in planta and use of transgenic plants expressing viroid RNAs).

Keywords Viroids; Catalytic RNAs; Rolling-circle replication; Hammerhead ribozymes

1   Introduction

There is profuse experimental evidence supporting the notion that RNA viruses, because of the limited information contained in their genomes, rely on diverse host proteins that are detracted from their normal functions to assist in the replication – as well as in the transcription, translation, and movement – of the RNA from these pathogens (see for reviews refs. 1–3). The situation is even more extreme in viroids, which despite being exclusively composed by a small circular RNA of 246–401 nucleotide residues (nt) without any apparent protein-coding capacity, are able to replicate autonomously in certain plants (see for reviews refs. 4–7). Viroid-based

From: *Methods in Molecular Biology, Vol. 451, Plant Virology Protocols: From Viral Sequence to Protein Function*
Edited by G.D. Foster, I.E. Johansen, Y. Hong, and P.D. Nagy © Humana Press, Totowa, NJ

systems, therefore, offer a simplified context for identifying host proteins assisting the replication of a RNA without interference by switches between replication and transcription, or between transcription and translation, as when dealing with viral RNAs.

The approximately 30 sequenced viroid species have been classified into two distinct families (7). Most viroids have a characteristic central conserved region (CCR) in their predicted rod-like or quasi-rod-like secondary structure (8) and belong to the family Pospiviroidae, type member *Potato spindle tuber viroid* (PSTVd), whereas four viroids that lack the CCR but are able to form self-cleaving hammerhead ribozymes are grouped in the family Avsunviroidae, type member *Avocado sunblotch viroid* (ASBVd) (see for reviews refs. 6, 9). Moreover, PSTVd replicates and accumulates in the nucleus in contrast to ASBVd that replicates and accumulates in the chloroplast (see later), with the other members of both families in which this question has been examined behaving as their respective type species. Analysis of viroid replication is deeply influenced by the different subcellular sites in which this process takes place.

In the following sections, we will review the main in vitro and in vivo methods that have been used for dissecting viroid replication. In most instances, these methods have been developed using either PSTVd-infected tomato (an experimental host easy to propagate in which this viroid reaches high titers and incites symptoms in a relatively short time) or ASBVd-infected avocado (this viroid and other members of the family Avsunviroidae have a host range essentially restricted to their natural hosts and closely-related species), and then extended to other viroid–host combinations.

2   Methods In Vitro

2.1   *Analysis of RNA Preparations from Viroid-Infected Plants*

Total RNAs from viroid-infected tissue were obtained by extraction with buffer-saturated phenol. The aqueous phase is then made to contain STE (50 mM Tris-HCl, 100 mM NaCl, 1 mM EDTA, pH 7.2) and 35% ethanol, and mixed with moderate shaking for some minutes with nonionic cellulose equilibrated with STE/35% ethanol. Under these conditions, viroid RNAs (genomic and replicative intermediates), because of their high content in secondary structure, are bound to the cellulose that is sedimented by low-speed centrifugation (the procedure can also be performed in chromatographic columns) and washed several times with fresh STE/35% ethanol. Viroid RNAs are released by washing with STE/0% ethanol and recovered by ethanol precipitation (10) (see note 1). The resulting preparations can be analyzed by electrophoresis, although, in certain cases, an additional clarification with methoxyethanol is needed to remove polysaccharides (11) (see note 2).

Because of their small size, viroid RNAs are electrophoretically fractionated in 5% polyacrylamide gels under native or denaturing conditions, or under a combination of both (an approach termed double or sequential PAGE). Native conditions allow analysis of genomic and subgenomic viroid RNAs together with some double-stranded intermediates. The resolution is increased using denaturing conditions, although double-stranded complexes are disassembled. However, double PAGE is the best option for detecting the characteristic viroid circular RNAs: following the first separation under native conditions and staining with ethidium bromide, the segment of the gel delimited by appropriate markers is excised and applied on top of a second denaturing gel (containing urea and a buffer of low ionic strength), which is run with an electric intensity suitable to heat the gel at 50–60 °C (12) (see note 3). Under these conditions, the circular viroid RNAs are characteristically retarded, whereas the linear viroid and host RNAs of similar size migrate more rapidly (Fig. 1). In addition to PAGE, electrophoresis in agarose gels has been used for detecting multimeric viroid RNA intermediates (13). After electroblotting to nylon membranes, identification of viroid RNAs is performed by hybridization with cDNA or cRNA probes labeled with radioactivity or chemically (see note 4). The same methodologies have been also applied to RNA preparations obtained from subcellular fractions.

Studies of this kind showed that in PSTVd-infected tomato, the most abundant viroid circular RNA, arbitrarily assigned as having (+) polarity, is accompanied by oligomeric (–) RNAs, and led to propose that the latter were replicative intermediates resulting from reiterative transcription of the former through a single rolling circle mechanism (13, 14) (Fig. 2). Previously, detection in infected tissues of (–) RNA sequences of a viroid closely related to PSTVd had indicated that viroid replication was an RNA-based process (15), and differential centrifugation studies had

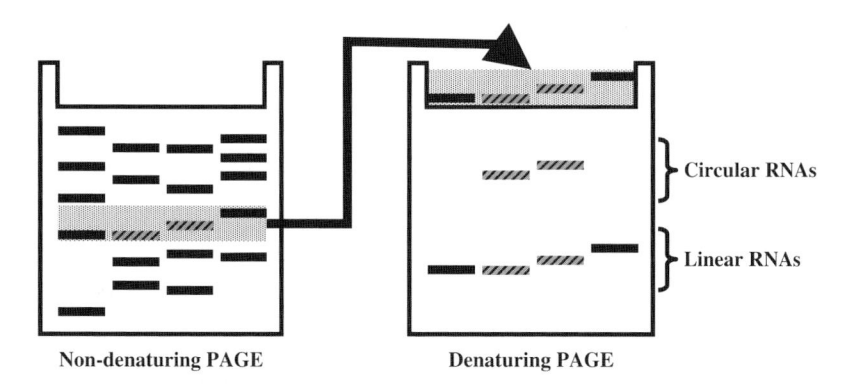

Non-denaturing PAGE   **Denaturing PAGE**

Fig. 1 Analysis of viroid circular RNAs by double (or sequential) PAGE. RNA preparations are first fractionated under nondenaturing conditions (*left panel*). After staining with ethidium bromide, the segment of the gel delimited by appropriate markers is cut and directly applied on top of a second denaturing gel (*right panel*). In this second gel, viroid circular RNAs (hatched) display mobilities significantly slower than their linear counterparts or cellular RNAs of similar size (*solid*)

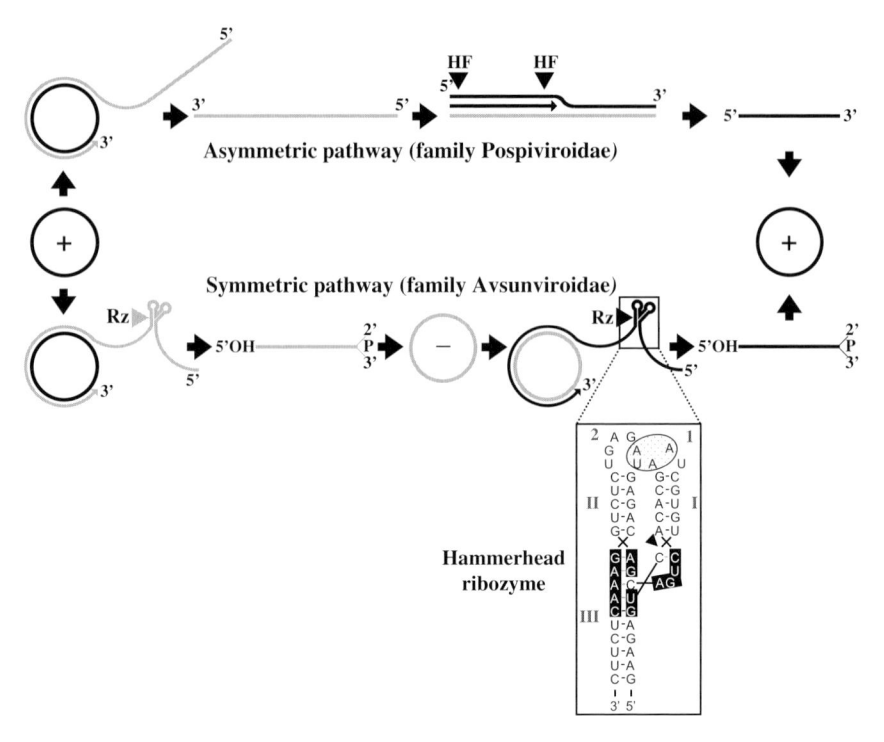

Fig. 2 Asymmetric and symmetric pathways of the rolling-circle mechanism proposed for replication of members of the families Pospiviroidae and Avsunviroidae, respectively. *Black* and *gray lines* refer to (+) and (−) strands, respectively. *Arrowheads* mark cleavage sites of a host factor (HF) or ribozyme (Rz), with the resulting terminal 5′ and 3′ groups being indicated (those corresponding to the family Pospiviroidae have not yet been conclusively determined). The structure of a natural hammerhead ribozyme is displayed in the boxed inset, with Roman and Arabic numerals denoting helices I, II, and III, and loops 1 and 2, respectively, and the arrowhead the self-cleavage site. The nucleotides conserved in most natural hammerhead structures are on a black background. *Short black* and *gray lines* indicate canonical and noncanonical base pairs, respectively, and the oval a tertiary interaction between loops 1 and 2 that enhances catalytic activity

showed that PSTVd (16) and its complementary strands (17) accumulate in the nucleus, suggesting the involvement of a nuclear RNA polymerase in replication. In contrast, similar experiments identified first ASBVd (18) and then multistranded complexes containing viroid (+) and (−) circular RNAs in the chloroplast (19, 20). These results indicate that replication of ASBVd occurs in this organelle through a symmetric mechanism with two rolling circles, in which the two circular RNAs – and not only the (+) circular RNA as in the asymmetric mechanism operating in PSTVd – are the templates, and the participation of a chloroplastic RNA polymerase (Fig. 2).

2.2 In Vitro Transcription Assays with Nuclear and Chloroplastic Preparations from Viroid-Infected Plants

Partially purified nuclei from *Gynura aurantiaca* infected with *Citrus exocortis viroid* (CEVd), a member of the family Pospiviroidae, were obtained by gentle homogenization of the tissue, treatment with Triton X-100, and differential centrifugation. Incubation in the presence of the four NTPs, one of them labeled at $[\alpha^{32}P]$, followed by phenol extraction and PAGE analysis under native and denaturing conditions showed synthesis of CEVd-specific RNAs. This synthesis was unaffected by pretreatments of the nuclei-rich preparations with actinomycin D or DNase that severely reduced synthesis of cellular RNAs. However, when α-amanitin was included in the in vitro system, synthesis of CEVd RNAs was markedly reduced by concentrations of 10 nM or greater (see note 5). Collectively, these data support that viroid RNA elongation is most likely catalyzed by the α-amanitin sensitive DNA-dependent RNA polymerase II acting on a RNA template (21). A similar study – using nuclei from PSTVd-infected *Solanum demissum* cells and analysis of the in vitro-synthesized RNAs by dot-blot hybridization with cDNA probes specific for viroid strands and for three cellular RNAs transcribed by the DNA-dependent RNA polymerases I, II, and III – showed that synthesis of (+) and (−) PSTVd RNAs was inhibited by the same α-amanitin concentration that inhibited synthesis of the cellular RNA transcribed by RNA polymerase II. This toxin did not influence transcription of the two other cellular RNAs mediated by RNA polymerases I and III. Together with the observation that synthesis of PSTVd RNAs was unaffected by actinomycin D, these results support the involvement of RNA polymerase II in elongation of (+) and (−) PSTVd RNAs (22).

Alternatively, a CEVd RNA elongation activity isolated as a chromatin-enriched fraction from infected tomato leaves was solubilized with ammonium sulfate. The nucleoprotein complexes in the soluble fraction, which bound to a monoclonal antibody to the carboxy-terminal domain of the largest subunit of RNA polymerase II, were affinity-purified and examined by dot-blot hybridization with cDNA probes specific for both viroid strands. Detection of (+) and (−) CEVd RNAs supports a role for RNA polymerase II in viroid replication and provides direct evidence of an association in vivo between host RNA polymerase II and viroid RNAs (23).

On the other hand, chloroplast-enriched preparations were obtained by gentle disruption of protoplasts from ASBVd-infected young avocado leaves followed by differential centrifugation and a final centrifugation step through a Percoll gradient. When these preparations were assayed for their in vitro ability to transcribe ASBVd RNAs, as well as representative genes of the three classes of chloroplastic genes according to their promoter structure, high concentrations of α-amanitin had no effect on gene or on viroid transcription, but tagetitoxin (5–10 μM) prevented transcription of all the genes without affecting synthesis of ASBVd strands; only at higher tagetitoxin concentrations (50–100 μM) a 25% inhibition was observed. Because previous studies on the sensitivity to tagetitoxin of the two main choroplastic RNA polymerases – the plastid-encoded polymerase (PEP) with a multisubunit structure similar to the *Escherichia coli* enzyme and a single-subunit nuclear-encoded polymerase (NEP) resembling phage RNA polymerases – have identified PEP and NEP as the most plausible candidates for the tagetitoxin-sensitive and -resistant RNA polymerases, respectively, these results suggest that NEP is the RNA polymerase required in ASBVd replication (24) (see note 6).

2.3 *In Vitro Approaches for Analysis of Cleavage and Ligation of Viroid Strands: Enzymes and Ribozymes*

Longer-than-unit-length PSTVd RNAs were produced by in vitro transcription of recombinant plasmids containing appropriate cDNA inserts. These RNAs, mimicking the replicative intermediates generated in vivo, were incubated with a nuclear extract from potato cell suspensions that were enzymatically converted into protoplasts and then mechanically disrupted. The nuclei, obtained by centrifugation through a two-step gradient, were subsequently used to prepare the nuclear extract (25). This study showed that the PSTVd transcript is correctly processed only if the CCR is folded into a multi-helix junction containing a hairpin capped by a GNRA tetraloop. The cleavage-ligation site was mapped with S1 nuclease and primer extension at a specific position in the upper strand of the CCR, and the structural motifs involved in the processing mechanism were analyzed by UV cross-linking, chemical mapping, phylogenetic comparisons, and thermodynamic calculations. The first cleavage occurs within the stem capped by the GNRA tetraloop, with a local conformational change switching the tetraloop motif into a loop E motif that stabilizes a base-paired 5′ terminus. The second cleavage yields a unit-length linear intermediate with its 3′ terminus also base-paired and most probably juxtaposed to the 5′ terminus. This particular folding facilitates ligation to mature circles autocatalytically or enzymatically (with low and high efficiency, respectively) (25). Additional fractionation of the nuclear extract should help to identify the host RNase and RNA ligase involved. The proposed mechanism, however, may not apply to other members of the family Pospiviroidae that are unable to form the GNRA tetraloop and the loop E.

In vitro transcription of recombinant plasmids containing tandem dimeric ASBVd-cDNA inserts showed that (+) and (−) ASBVd RNAs self-cleave at two specific sites in each transcript generating exact unit-length strands. Self-cleavage occurs during in vitro transcription and after incubation of the purified primary transcripts in a protein-free medium at pH 8 and in the presence of 6 mM magnesium ions, producing 5′-OH and 2′,3′-cyclic phosphodiester termini (26). Similar results have been obtained with the other members of the family Avsunviroidae, indicating that the cleavage activity does not reside in a host RNase but in a hammerhead ribozyme, a small motif embedded in the viroid strands of both polarities (see for a review ref. 27). Moreover, primer-extension experiments of the monomeric linear viroid RNAs extracted from infected tissues have mapped their most abundant 5′ termini at those predicted by the hammerhead structures, thus showing that they are also operative in vivo (27, 28). Hammerhead ribozymes have a core of conserved nucleotides flanked by three double-stranded regions with loose sequence requirements that are capped by loops, and X-ray crystallography has revealed that the actual shape does not resemble a hammerhead but rather a Y wherein the stems III and II are almost colinear (Fig. 2, inset). In vitro transcriptions of viroid RNA regions corresponding to the hammerhead domains (in the presence of antisense oligonucleotides to prevent self-cleavage during transcription) have generated sufficient RNA amounts for determination of the self-cleavage kinetics of distinct natural hammerhead in their *cis* context (see note 7). In these experiments, the disappearance with time of the radioactive primary transcript and the concomitant emergence of its self-cleavage products are assessed by denaturing PAGE (see note 8). The data obtained show that modifications of loops 1 and 2 of natural hammerheads (initially regarded as catalytically irrelevant) induce a severe reduction in their catalytic activity, indicating that these peripheral regions play a critical role through tertiary interactions that may favor the active site at the low magnesium concentration existing in vivo (29, 30) (Fig. 2, inset). These interactions could also be stabilized by chloroplast proteins behaving as RNA chaperones, thus explaining why they facilitate the hammerhead-mediated self-cleavage of a viroid RNA (see later). On the other hand, in vitro incubation of certain unit-length viroid RNAs resulting from self-cleavage led to self-ligation through a 2′,5′-phosphodiester bond (31). If such an atypical bond indeed exists in natural viroids, replication in the family Avsunviroidae would be an RNA-based mechanism only demanding a host RNA polymerase. However, the alternative involvement of an RNA ligase or hammerhead-mediated ligation, leading in both instances to 3′,5′-phosphodiester bonds, cannot be excluded.

2.4 Characterization of 5′-Terminal Groups for Mapping Initiation Sites of Viroid Strands

Because RNA folding occurs during transcription, the initiation sites of nascent viroid strands may determine the adoption of transient metastable structures functionally relevant in replication. Moreover, the region adjacent to the initiation sites may provide hints about the nature of the promoters involved. Data on the initiation sites for PSTVd, obtained by in vitro transcription assays (see earlier) of the PSTVd monomeric (+) circular RNA with either a potato nuclear extract or with purified RNA polymerase II from wheat germ and tomato, are restricted to the minus polarity strand and not coincidental (5, 32). This discrepancy may in part reflect the difficulties to reconstitute in vitro an initiation complex reproducing the in vivo situation. A way to circumvent this problem is to map the 5′ termini of viroid primary transcripts. In chloroplasts, the 5′ termini of primary transcripts, but not those resulting from their processing, have a free triphosphate group that can be specifically capped in vitro with $[\alpha^{32}P]GTP$ and guanylyltransferase. This labeling, combined with RNase protection assays (RPA), has been used to map the transcription start sites of ASBVd (+) and (−) RNAs isolated from infected avocado at similar A + U-rich terminal loops in their predicted quasi-rod-like secondary structures (33). Attempts to extend the same methodology to a second chloroplastic viroid, *Peach latent mosaic viroid* (PLMVd) (34), failed because the RNAs of this viroid accumulate in vivo to considerably lower levels than those of ASBVd. However, a combination of in vitro capping and an RNA ligase-mediated rapid amplification of cDNA ends methodology developed for identifying the genuine capped 5′ termini of eukaryotic messenger RNAs has mapped the PLMVd (+) and (−) initiation sites at a similar double-stranded motif of 6–7 bp. This motif, which also includes the conserved GUC triplet preceding the self-cleavage site in both polarity strands, is located at the base of the hammerhead arm that presumably contains the promoters for a chloroplastic RNA polymerase (28). Returning to members of the family Pospiviroidae with nuclear replication, it is possible that the 5′ triphosphate of their primary transcripts could be capped in vivo. If so, this would mark unambiguously the transcription initiation sites in this viroid family.

2.5 Identification of Viroid RNA-Binding Proteins by RNA-Ligand Screening of a cDNA Expression Library

To set up the screening protocol, two components of the human U1 snRNP were used as a model system. The RNA was the U1-RNA stem-loop II, and the protein the N-terminal recognition motif of the U1A protein fused to beta-galactosidase and expressed by a recombinant lambda phage. Following binding of the fusion protein to nitrocellulose membranes, hybridization with a ^{32}P-labeled U1-RNA ligand was performed for detecting specific RNA–protein interactions. Parameters influencing the specificity and sensitivity were investigated, with processing the membranes in the presence of transition metals greatly increasing the signal-to-background ratio. Specific RNA–protein interactions could be observed in the presence of a large excess of recombinant phages from a cDNA library, and only moderate binding affinities were required (35). Application of this methodology led to the identification of a viroid RNA-binding protein 1 (VIRP1) from tomato with specificity for monomeric and oligomeric PSTVd (+) RNAs. The specificity of this interaction was examined by different in vitro methodologies that included Northwestern blotting, plaque lift, and electrophoretic mobility shift assays, as well as by immunoprecipitation from extracts of PSTVd-infected tomato leaves. Sequence analysis revealed that VIRP1 is a member of a family of bromodomain-containing transcriptional regulators associated with chromatin remodeling. VIRP1 is the first member of this family for which a specific RNA-binding activity is shown, and it could be involved in viroid replication and in RNA-mediated chromatin remodeling.

2.6 Analysis by *In Situ* Hybridization and Microscopy of Subcellular Fractions and Tissues from Viroid-Infected Plants

To investigate the intracellular localization of PSTVd, isolated nuclei from viroid-infected tomato plants were bound to microscope slides, fixed with formaldehyde, and hybridized with biotinylated strand-specific cDNA probes, with the bound probe being detected with lissamine–rhodamine conjugated streptavidin. The highest fluorescence corresponding to PSTVd (+) and (−) strands was found in the nucleoli, and examination of the distribution of the fluorescence signals by confocal laser scanning microscopy and three-dimensional reconstructions showed that both viroid strands were homogeneously distributed throughout the nucleolus (36). This work confirmed results of previous fractionation studies, but the localization of (+) and (−) PSTVd RNAs in the nucleolus was in apparent contradiction with that of the RNA polymerase II catalyzing their synthesis, which is a nucleoplasmatic enzyme. To solve this discrepancy, the possibility was advanced that synthesis by RNA polymerase II could occur in the nucleoplasm, with the oligomeric (+) strands being then transferred to the nucleolus where processing to the mature monomeric circular RNAs would take place. More recent studies using fluorescence in situ hybridization have revealed that in PSTVd-infected cultured cells and plants, the (−) strand was localized in the nucleoplasm, whereas the (+) strand was localized in the nucleolus as well as in the nucleoplasm with distinct spatial patterns. Interestingly, the presence of the PSTVd (+) RNA in the nucleolus caused the redistribution of a small nucleolar RNA (37). These results support a model in which the synthesis of the (−) and (+) PSTVd RNAs occurs in the nucleoplasm, with the (−) strand remaining anchored in the nucleoplasm and the (+) strand RNA being transported selectively into the nucleolus where it is processed. On the other hand, confocal laser scanning microscopy and transmission electron microscopy in conjunction with in situ hybridization have been used to determine the subnuclear (ultrastructural) and tissue (histological) localizations of two other members of the family Pospiviroidae: CEVd in tomato and *Coconut cadang cadang viroid* (CCCVd) in palm. Both viroids were found in the vascular tissues as well as in the nuclei of mesophyll cells of infected host plants. At the subnuclear level, however, CEVd was distributed across the entire nucleus, in contrast to CCCVd that was mostly concentrated in the nucleolus with the remainder distributed throughout the nucleoplasm (38).

Regarding the family Avsunviroidae, biotin- and digoxigenin-labeled RNA probes with subsequent detection by a gold-labeled anti-biotin and anti-digoxigenin antibodies were used to localize ASBVd in chloroplasts, mostly on the thylakoid membranes, of infected avocado leaves (39, 40). A predominant chloroplast localization has also been observed for PLMVd, another member of this family (41). Therefore, subcellular localization has emerged as a criterion for discriminating members of both viroid families, with deep implications in their replication and evolutionary origin.

3   Methods In Vivo

3.1   *Infiltration Studies*

To assess whether inhibition of viroid synthesis by actinomycin D reflected a direct or indirect effect (see later), foliar tissue of *G. aurantiaca* infected with CEVd or potato tuber sprouts infected with PSTVd were vacuum infiltrated with actinomycin D and ^{32}P in a Barth's solution, and incubated overnight before nucleic acid extraction and analysis. Using the incorporation of ^{32}P into DNA, determined by the acid-precipitable radioactivity after treatment with RNase A, it was possible to discriminate between sublethal doses of actinomycin D (0–10 μg ml^{-1}) at which DNA synthesis was not affected, and higher doses (10–40 μg ml^{-1}) at which inhibition of DNA synthesis could reflect general cellular impairment. The incorporation of ^{32}P into total RNA, determined by the acid-precipitable radioactivity after treatment with DNase I, decreased over the actinomycin D dose range as predicted by the mode of action of actinomycin D, which by binding to DNA inhibits DNA-directed RNA synthesis. The incorporation of ^{32}P into the 5S ribosomal RNA, estimated by autoradiography of nondenaturing polyacrylamide gels, followed the same trend. However, synthesis of CEVd in *G. aurantiaca* and of PSTVd in potato, estimated also by autoradiography, was not affected by actinomycin D, supporting a primary role of viroid complementary RNAs as intermediates of viroid replication (42).

3.2   *Protoplasts*

Initial studies used tomato protoplasts to examine the effects of different inhibitors on synthesis of *Cucumber pale fruit viroid* (CPFVd), a member of the family Pospiviroidae. Freshly isolated protoplasts were inoculated and viroid (and host RNA) synthesis was followed by ^{3}H-uridine incorporation. Because viroid replication increased to a detectable level 48–72 h after inoculation, as revealed by counting the radioactivity of the corresponding band in a denaturing polyacrylamide gel, the inhibitors were applied during this period. Actinomycin D (20 μg ml^{-1}) inhibited total and viroid RNA synthesis to a similar extent (80–85%), suggesting a nonspecific toxic effect of this drug and leaving open the possibility that a host RNA polymerase could catalyze elongation of viroid strands. A similar general toxic effect on RNA synthesis was observed when cycloheximide (1–50 μg ml^{-1}) was applied, precluding any conclusion on possible requirements of newly synthesized proteins for viroid replication. However, application of α-amanitin (50 μg ml^{-1}) inhibited viroid replication to 75%, whereas synthesis of total and certain specific cellular RNAs (including 5S and 7S RNAs) was not appreciably affected. Moreover, incubation of protoplasts with ^{3}H-labelled α-amanitin indicated that a concentration of 50 μg ml^{-1} (10^{-5} M) in the medium led to an intracellular concentration of 10^{-8} M. Since this concentration is known to inhibit RNA polymerase II from different sources – inhibition of RNA polymerase III requires 1,000-fold higher concentrations – this enzyme was proposed to mediate replication of CPFVd. Additional control experiments showed that tomato protoplasts inoculated with tobacco mosaic virus (TMV) were able to replicate and accumulate the virus in the presence of α-amanitin (50 μg ml^{-1}), further discarding that the marked inhibition of viroid replication could result from an indirect toxic effect of this drug (43).

More recently, an efficient electroporation protocol to inoculate protoplasts from cultured cells of tobacco (BY2) and *Nicotiana benthamiana* with in vitro transcripts of PSTVd has been developed. This protocol has permitted to characterize viroid structural features that influence replication efficiency at the cellular level. In situ hybridization showed that 60–70% of the cells were infected, and Northern blot hybridization revealed that PSTVd (+) and (−) strands could be detected by as early as 6-h postinoculation (h.p.i.). Interestingly, the predominant (−) strands were multimers and accumulated to higher levels between 6 and 24 h.p.i., whereas (+) strands were present mostly as circular monomers and dominated the population of viroid RNAs during the second part of the experimental period (that ended 144 h.p.i.). These results add further support for an asymmetrical rolling-circle mechanism for replication of PSTVd. Moreover, replication assays of certain PSTVd variants showed that specific substitutions in loop E enhanced accumulation in tobacco (but not in *N. benthamiana)* cells, indicating that in addition to its role in processing, loop E also seems to have a role in modulating replication efficiency when adapting to a specific host (44).

3.3 In Planta Studies

3.3.1 Bioassays with Natural and Artificial Variants

The possibility of synthesizing full-length viroid cDNAs, first with classical approaches and then with RT-PCR amplification, opened the door to reverse genetics studies with viroids (45). Remarkably, recombinant bacterial plasmids containing monomeric and, particularly, multimeric viroid cDNAs of representative members of the two families, or their corresponding inserts released with restriction enzymes, are infectious when inoculated to appropriate host plants. This entails that the alien DNA is transcribed by a host RNA polymerase, with the resulting multimeric viroid RNA – which mimics the natural replicative intermediates of the rolling-circle mechanism – then initiating an RNA–RNA amplification cascade. There are many studies of this kind and here we will only present some illustrative examples.

To investigate the role in viroid replication (and pathogenesis) of the secondary structure domains previously advanced for PSTVd and related viroids (8), a series of chimeras was constructed by exchanging certain domains between CEVd and *Tomato apical stunt viroid* (TASVd), also of the family Pospiviroidae. The chimeras tested were replicated stably in tomato, but those containing the right side of TASVd accumulated to higher levels early in infection, and the infected plants developed more severe symptoms than those whose right halves were derived from

CEVd. These results support the proposed modular structure of these viroids and indicate that replication levels are particularly dependent on certain domains (46).

In contrast to the rod-like (or quasi-rod-like) structure of most members of the family Pospiviroidae, the predicted conformations of PLMVd (34) and *Chrysanthemum chlorotic mottle viroid* (CChMVd) RNAs, of the family Avsunviroidae, are branched. The covariations found in a number of natural CChMVd variants support that the same, or a closely related conformation, exists in vivo. Besides, the CChMVd natural variability also supports that the branched conformation is additionally stabilized by a kissing-loop interaction resembling another one proposed in PLMVd from in vitro assays (47). More specifically, site-directed mutagenesis combined with bioassays and progeny analysis showed that single CChMVd mutants affecting the kissing loops had low or no infectivity (with the infectivity being recovered in double mutants restoring the interaction), that mutations affecting the structure of the regions adjacent to the kissing loops reverted to wild-type or led to rearranged stems supporting also their interaction, and that interchange between four nucleotides of each of the two kissing loops generated a viable CChMVd variant with eight mutations. Preservation of a similar kissing-loop interaction in two hammerhead viroids with an overall low sequence similarity suggests that it facilitates in vivo the adoption and stabilization of a compact folding critical for viroid viability (48).

3.3.2   Ultraviolet-Irradiaton of Viroid-Infected Tissues

The search for cellular factors that assist viroid replication, based on in vitro analysis of subcellular fractions or in vitro binding between viroid RNAs and host proteins expressed from cDNA libraries (see above), have met with partial success and left unanswered whether the observed interactions occur also in vivo and their possible functional significance. To circumvent these limitations, a more direct approach has been developed. Ultraviolet (UV) cross-linking is a powerful methodology for characterizing RNA–protein interactions in ribonucleoprotein complexes. UV light is a "zero-length" cross-linking agent able to induce formation of covalent bonds between nucleic acids and proteins at their contact points, thereby "freezing" the interaction between the two molecules. To screen for host proteins directly interacting with viroid RNAs in vivo, ASBVd-infected avocado leaves were UV-irradiated (see note 9). This led to the identification of several ASBVd-host protein adducts and to the characterization, via mass-spectroscopy, of the protein component of the most abundant cross-linked species. This component is formed by two closely-related chloroplast RNA-binding proteins that belong to a family whose members have been previously shown to be involved in different steps of RNA metabolism in this organelle. At least one of these avocado proteins behaves as an RNA chaperone and stimulates in vitro, and possibly in vivo, the hammerhead-mediated self-cleavage of multimeric ASBVd transcripts (49). This methodology is restricted to viroids that, like ASBVd, accumulate in vivo to high concentrations but with the increasing sensitivity of proteomics tools should be soon applicable to other members of the group.

3.4 Transgenic Plants

To reexamine the question of whether the monomeric (−) circular RNA of PSTVd could serve as a template for synthesis of (+) strand progeny – infected plants appear to contain only multimeric linear (−) viroid RNAs (see earlier) – a ribozyme-based expression system for the production of precisely full-length (−) PSTVd RNA whose termini are capable of undergoing circularization in vitro has been developed. Mechanical inoculation of tomato seedlings with electrophoretically purified (−) circular PSTVd RNA led to a small fraction of plants becoming infected. Ribozyme-mediated production of (−) PSTVd RNA in transgenic *N. benthamiana* plants, obtained by conventional protocols based in *Agrobacterium tumefaciens*, resulted in the appearance of monomeric circular (−) PSTVd RNA and large amounts of (+) PSTVd progeny. However, no monomeric circular (−) PSTVd RNA could be detected in naturally infected plants by using either PAGE or more sensitive approaches like RPA. Therefore, although not a component of the normal replicative asymmetric pathway, precisely full-length (−) PSTVd RNA appears to contain all of the structural and regulatory elements necessary for initiation of viroid replication (50).

On the other hand, since *Arabidopsis thaliana* was adopted as the model organism for higher plants, multiple tools, resources, and experimental approaches have been developed, prominent among which is the availability of the complete sequence of its genome. Research on plant viruses that naturally or experimentally infect *A. thaliana* has benefited from the use of such a versatile system. However, no viroid has been shown to infect *A. thaliana*. To circumvent this problem and to explore whether this model plant can be used to tackle questions on viroid replication, *A. thaliana* was transformed with cDNAs expressing dimeric (+) transcripts of representative species of the families Pospiviroidae and Avsunviroidae, which, as indicated previously, replicate in the nucleus and the chloroplast, respectively. Correct processing to the circular (+) monomers was always observed, demonstrating that *Arabidopsis* has the appropriate RNase and RNA ligase. Northern-blot hybridization also revealed the multimeric (−) RNAs of CEVd and *Hop stunt viroid* (HSVd) of the family Pospiviroidae, but not of ASBVd of the family Avsunviroidae, showing that the first RNA–RNA transcription of the rolling circle mechanism occurs in *A. thaliana* for the two nuclear viroids, and that their multimeric (−) RNAs remain unprocessed as in typical hosts. Moreover, transgenic *A. thaliana* expressing HSVd dimeric (−) transcripts accumulated the circular (+) monomers, although at low levels, together with the unprocessed primary transcript that served as the template for the second RNA–RNA transcription. Therefore, processing of HSVd dimeric transcripts, and most likely of other members of the family Pospiviroidae, appears to be a polarity intrinsic property, which dictates the susceptibility to and the specificity of the reactions mediated by the host enzymes (51). The recent finding by fluorescence in situ hybridization of PSTVd (−) strands accumulating in the nucleoplasm, and of PSTVd (+) strands accumulating in the nucleolus as well as in nucleoplasm (see earlier), provides an explanation for this different behavior and suggests that processing of the (+) strands occurs in the nucleolus, where processing of the precursors of rRNAs and tRNAs also occurs. Which factors determine this differential traffic of viroid (+) and (−) strands remain an intriguing issue. In any case, these results show that *A. thaliana* transformed with cDNAs expressing viroid transcripts can be used to address certain steps, particularly cleavage and ligation, of the replication cycle.

4 Notes

1. Phenol extraction and fractionation with nonionic cellulose. Tissue type: usually leaves but occasionally fruits or young bark.

   a. Homogenate 10 g of plant tissue in a Polytron for 3 min with 40 ml of water-saturated phenol (neutralized), 10 ml of 0.2 M Tris-HCl pH 8.9, 2.5 ml 0.1 M EDTA pH 7.0, 2.5 ml 5% SDS and 1.25 ml β-mercaptoethanol. (To prepare water-saturated phenol mix 500 g phenol and 200 ml water; keep the solution in a dark bottle at 4 °C).
   b. Centrifuge at 8,000 rpm for 15 min.
   c. Remove aqueous phase and reextract with 0.5 volumes of water-saturated phenol.
   d. Centrifuge at 8,000 rpm for 15 min.
   e. Remove aqueous phase, bring it to a final volume of 20 ml with water, and add 3.7 ml of 10× STE (1 M NaCl, 0.5 M Tris, 10 mM EDTA, pH 7.2 with HCl), 13.4 ml of ethanol little by little while shaking and 1.25 g of nonionic cellulose (CF11, Whatman). Shake it at room temperature for at least 1 h. (To prepare 1 l 10× STE, dissolve 60.57 g Tris, 58.44 g NaCl, and 3.72 g EDTA, and bring it to pH 7.2 with HCl).
   f. Centrifuge at 3,000 rpm for 5 min and discard the supernatant.
   g. Wash the pellet three times with 30 ml of 35% ethanol in 1× STE; each time the mixture is centrifuged at 3,000 rpm for 5 min at room temperature.
   h. To elute the nucleic acids, resuspend the CF11 pellet in 3.3 ml 1× STE, centrifuge the mixture at 3,000 rpm for 5 min at room temperature, and recover the supernatant. Repeat this step two more times.
   i. Add to the combined supernatant 2.5 volumes of ethanol and mix. Keep the sample for at least 2 h at −20 °C.
   j. Centrifuge at 8,000 rpm for 30 min and discard the supernatant.
   k. Dry the pellet for 30 min, and resuspend it in water (0.25 ml of water per 10 g of fresh tissue).

2. Clarification with methoxyethanol for removing polysaccharides.

   a. Mix on ice in a 15 ml Corex tube one volume of extract, one volume of K2HPO4 2.5 M, pH 8.0 (for 2 ml of extract, add 2 ml of K2HPO4 2.5 M and 40 μl of H3PO4 85%), and one volume of methoxyethanol.
   b. Shake and keep on ice for 5 min.
   c. Centrifuge at 3,000 rpm for 3 min. Recover the aqueous phase and transfer it to a new tube.
   d. Add per volume of the aqueous phase, 0.05 volumes of sodium acetate 3 M, pH 5.5 and 0.5 volumes of CTAB 1% (CTAB is cetyl trimethyl ammonium bromide).
   e. Shake and keep on ice for 5 min.
   f. Centrifuge at 8,000 rpm for 20 min at 4 °C.
   g. Discard the supernatant and dry the pellet.
   h. Resuspend the pellet in 2 ml of 1 M NaCl.
   i. Add 6 ml of cold ethanol and mix.
   j. Keep the sample at −20 °C for at least 2 h.
   k. Centrifuge at 8,000 rpm for 20 min and discard the supernatant.
   l. Dry the pellet and resuspend it in water (0.25 ml of water per 10 g of fresh tissue).

3. Although bands from each lane of the native gel can be cut individually, it is more convenient to cut a single segment of this gel delimited by appropriate markers and to apply it on top of a second denaturing gel with a single well. Accommodating the single segment into the single well requires some practice and it is facilitated by lubricating the well with a small volume of buffer. Avoid bubbles between both contacting gel surfaces.

4. When the preparations analyzed contain sufficient amounts of viroid RNA, the band corresponding to the circular form can be directly visualized with ethidium bromide or silver staining. Only the first staining is compatible with subsequent examination by Northern blot hybridization.

5. Actinomycin D and α-amanitin are highly toxic even at very low concentrations. They should be handled with care.

6. Two notes of caution in this respect. First, because RNA polymerases form part in vivo of large complexes, in vitro replication experiments with highly purified RNA polymerases may be of limited relevance for reproducing the physiological context. Second, despite being in vitro transcriptions with nuclear and chloroplastic preparations more appropriate for this purpose, most of these systems fail to reinitiate and they just elongate viroid strands that were initiated in vivo.

7. During gel elution of the uncleaved primary transcript of certain hammerheads, as the PLMVd (+) hammerhead, extensive self-cleavage is observed even when the buffer contains EDTA at high concentration. Gel elution in the presence of 40% formamide decreases considerably self-cleavage.

8. The presence in the denaturing gel of 7–8 M urea may not be sufficient to fully denature certain RNAs, which will display an electrophoretic mobility inconsistent with their size. Incorporation of 40% formamide to the gel usually solves this problem.

9. Exposure to UV light should be avoided. Use appropriate masks for eye protection.

References

1. Buck, K. W. (1996) Comparison of the replication of positive-stranded RNA viruses of plants and animals. *Adv. Virus Res.* **47**, 159–251.

2. Lai, M. M. C. (1998) Cellular factors in the transcription and replication of viral RNA genomes: A parallel to DNA-dependent RNA transcription. *Virology* **244**, 1–12.

3. Ahlquist, P., Noueiry, A. O., Lee, W. M., Kushner, D. B., and Dye, B. T. (2003) Host factors in positive-strand RNA virus genome replication. *J. Virol.* **77**, 8181–8186.

4. Diener, T. O. (2003) Discovering viroids – a personal perspective. *Nature Rev. Microbiol.* **1**, 75–80.

5. Tabler, M. and Tsagris, M. (2004) Viroids: Petite RNA pathogens with distinguished talents. *Trends Plant Sci.* **9**, 339–348.

6. Flores, R., Hernández, C., Martínez de Alba, E., Daròs, J. A., and Di Serio, F. (2005) Viroids and viroid-host interactions. *Ann. Rev. Phytopathol.* **43**, 117–139.

7. Flores, R., Randles, J. W., Owens, R. A., Bar-Joseph, M., and Diener, T. O. (2005) Viroids. In Virus Taxonomy. Eight Report of the International Committee on Taxonomy of Viruses (Fauquet, C. M., Mayo, M. A., Maniloff, J., Desselberger U. and Ball A. L. eds). London: Elsevier/Academic Press, pp. 1145–1159.

8. Keese, P. and Symons, R. H. (1985) Domains in viroids: Evidence of intermolecular RNA rearrangements and their contribution to viroid evolution. *Proc. Natl. Acad. Sci. USA* **82**, 4582–4586.

9. Flores, R., Daròs, J. A., and Hernández, C. (2000) The *Avsunviroidae* family: Viroids with hammerhead ribozymes. *Adv. Virus Res.* **55**, 271–323.

10. Semancik, J. S., Morris, T. J., Weathers, L. G., Rodorf, B. F., and Kearns, D. R. (1975) Physical properties of a minimal infectious RNA (viroid) associated with the exocortis disease. *Virology* **63**, 160–167.

11. Bellamy, A. R. and Ralph, R. K. (1968) Recovery and purification of nucleic acids by means of cetyltrimethylammonium bromide. *Methods Enzymol.* **XII**, 156–160.

12. Flores, R., Durán-Vila, N., Pallás, V., and Semancik, J. S. (1985) Detection of viroid and viroid-like RNAs from grapevine. *J. Gen. Virol.* **66**, 2095–2102.

13. Branch, A. D. and Robertson, H. D. (1984) A replication cycle for viroids and other small infectious RNAs. *Science* **223**, 450–454.

14. Branch, A. D., Benenfeld, B. J., and Robertson, H. D. (1988) Evidence for a single rolling circle in the replication of potato spindle tuber viroid. *Proc. Natl. Acad. Sci. USA* **85**, 9128–9132.

15. Grill, L. K. and Semancik, J. S. (1978) RNA sequences complementary to citrus exocortis viroid in nucleic acid preparations from infected *Gynura aurantiaca. Proc. Natl. Acad. Sci. USA* **75**, 896–900.

16. Diener, T. O. (1971) Potato spindle tuber "virus": A plant virus with properties of a free nucleic acid. III. Subcellular location of PSTV-RNA and the question of whether virions exist in extracts or *in situ. Virology* **43**, 75–89.

17. Spiesmacher, E., Mühlbach, H. P., Schnölzer, M., Haas, B., and Sänger, H. L. (1983) Oligomeric forms of potato spindle tuber viroid (PSTV) and of its complementary RNA are present in nuclei isolated from viroid-infected potato cells. *Biosci. Rep.* **3**, 767–774.

18. Mohamed, N. A. and Thomas W. (1980) Viroid-like properties of an RNA species associated with the sunblotch disease of avocados. *J. Gen. Virol.* **46**, 157–167.

19. Daròs, J. A., Marcos, J. F., Hernández, C., and Flores, R. (1994) Replication of avocado sunblotch viroid: Evidence for a symmetric pathway with two rolling circles and hammerhead ribozyme processing. *Proc. Natl. Acad. Sci. USA* **91**, 12813–12817.

20. Navarro, J. A., Daròs, J. A., and Flores, R. (1999) Complexes containing both polarity strands of avocado sunblotch viroid: Identification in chloroplasts and characterization. *Virology* **253**, 77–85.

21. Flores, R. and Semancik, J. S. (1982) Properties of a cell-free system for synthesis of citrus exocortis viroid. *Proc. Natl. Acad. Sci. USA* **79**, 6285–6288.

22. Schindler, I. M., and Mühlbach, H. P. (1992) Involvement of nuclear DNA-dependent RNA polymerases in potato spindle tuber viroid replication: A reevaluation. *Plant Sci.* **84**, 221–229.

23. Warrilow, D. and Symons, R. H. (1999) Citrus exocortis viroid RNA is associated with the largest subunit of RNA polymerase II in tomato *in vivo. Arch. Virol.* **144**, 2367–2375.

24. Navarro, J. A., Vera, A., and Flores, R. (2000) A chloroplastic RNA polymerase resistant to tagetitoxin is involved in replication of avocado sunblotch viroid. *Virology* **268**, 218–225.

25. Baumstark, T., Schröder, A. R. W., and Riesner, D. (1997) Viroid processing: Switch from cleavage to ligation is driven by a change from a tetraloop to a loop E conformation. *EMBO J.* **16**, 599–610.

26. Hutchins, C., Rathjen, P. D., Forster, A. C., and Symons, R. H. (1986) Self-cleavage of plus and minus RNA transcripts of avocado sunblotch viroid. *Nucleic Acids Res.* **14**, 3627–3640.

27. Flores, R., Hernández, C., de la Peña, M., Vera, A., and Daròs, J. A. (2001) Hammerhead ribozyme structure and function in plant RNA replication. *Methods Enzymol.* **341**, 540–552.

28. Delgado, S., Martínez de Alba, E., Hernández, C., and Flores, R. (2005) A short double-stranded RNA motif of peach latent mosaic viroid contains the initiation and the self-cleavage sites of both polarity strands. *J. Virol.* **79**, 12934–12943.

29. De la Peña, M., Gago, S., and Flores, R. (2003) Peripheral regions of natural hammerhead ribozymes greatly increase their self-cleavage activity. *EMBO J.* **22**, 5561–5570.

30. Khvorova, A., Lescoute, A., Westhof, E., and Jayasena, S. D. (2003) Sequence elements outside the hammerhead ribozyme catalytic core enable intracellular activity. *Nature Struct. Biol.* **10**, 708–712.

31. Côté, F. and Perreault, J. P. (1997) Peach latent mosaic viroid is locked by a 2′,5′-phosphodiester bond produced by *in vitro* self-ligation. *J. Mol. Biol.* **273**, 533–543.

32. Kolonko, N., Bannach, O., Aschermann, K., Hu, K. H., Moors, M., Schmitz, M., Steger, G., and Riesner, D. (2006) Transcription of potato spindle tuber viroid by RNA polymerase II starts in the left terminal loop. *Virology* **347**, 392–404.

33. Navarro, J. A. and Flores, R. (2000) Characterization of the initiation sites of both polarity strands of a viroid RNA reveals a motif conserved in sequence and structure. *EMBO J.* **19**, 2662–2670.
34. Hernández, C. and Flores, R. (1992) Plus and minus RNAs of peach latent mosaic viroid self-cleave *in vitro* via hammerhead structures. *Proc. Natl. Acad. Sci. USA* **89**, 3711–3715.
35. Sagesser, R., Martínez, E., Tsagris, M., and Tabler, M. (1997) Detection and isolation of RNA-binding proteins by RNA-ligand screening of a cDNA expression library. *Nucleic Acids Res.* **25**, 3816–3822.
36. Harders, J., Lukacs, N., Robert-Nicoud, M., Jovin, J. M., and Riesner, D. (1989) Imaging of viroids in nuclei from tomato leaf tissue by *in situ* hybridization and confocal laser scanning microscopy. *EMBO J.* **8**, 3941–3949.
37. Qi, Y. and Ding, B. (2003) Differential subnuclear localization of RNA strands of opposite polarity derived from an autonomously replicating viroid. *Plant Cell* **15**, 2566–2577.
38. Bonfiglioli, R. G., Webb, D. R., and Symons, R. H. (1996) Tissue and intra-cellular distribution of coconut cadang cadang viroid and citrus exocortis viroid determined by *in situ* hybridization and confocal laser scanning and transmission electron microscopy. *Plant J.* **9**, 457–465.
39. Bonfiglioli, R. G., McFadden, G. I., and Symons, R. H. (1994) *In situ* hybridization localizes avocado sunblotch viroid on chloroplast thylakoid membranes and coconut cadang cadang viroid in the nucleus. *Plant J.* **6**, 99–103.
40. Lima, M. I., Fonseca, M. E. N., Flores, R., and Kitajima, E. W. (1994) Detection of avocado sunblotch viroid in chloroplasts of avocado leaves by *in situ* hybridization. *Arch. Virol.* **138**, 385–390.
41. Bussière, F., Lehoux, J., Thompson, D. A., Skrzeczkowski, L. J., and Perreault, J. P. (1999) Subcellular localization and rolling circle replication of peach latent mosaic viroid: Hallmarks of group A viroids. *J. Virol.* **73**, 6353–6360.
42. Grill, L. K. and Semancik, J. S. (1980) Viroid synthesis: The question of inhibition by actinomycin D. *Nature* **283**, 399–400.
43. Mühlbach, H. P. and Sänger, H. L. (1979) Viroid replication is inhibited by α-amanitin. *Nature* **278**, 185–188.
44. Qi, Y. and Ding, B. (2002) Replication of potato spindle tuber viroid in cultured cells of tobacco and *Nicotiana benthamiana:* the role of specific nucleotides in determining replication levels for host adaptation. *Virology* **302**, 445–456.
45. Cress, D. E., Kiefer, M. C., and Owens, R. A. (1983) Construction of infectious potato spindle tuber viroid cDNA clones. *Nucleic Acids Res.* **11**, 6821–6835.
46. Sano, T., Candresse, T., Hammond, R. W., Diener, T. O., and Owens, R. A. (1992) Identification of multiple structural domains regulating viroid pathogenicity. *Proc. Natl. Acad. Sci. USA* **89**, 10104–10108.
47. Bussière, F., Ouellet, J., Côté, F., Lévesque, D., and Perreault, J. P. (2000) Mapping in solution shows the peach latent mosaic viroid to possess a new pseudoknot in a complex, branched secondary structure. *J. Virol.* **74**, 2647–2654.
48. Gago, S., De la Peña, M., and Flores, R. (2005) A kissing-loop interaction in a hammerhead viroid RNA critical for its *in vitro* folding and *in vivo* viability. *RNA* **11**, 1073–1083.
49. Daròs, J. A. and Flores, R. (2002) A chloroplast protein binds a viroid RNA *in vivo* and facilitates its hammerhead-mediated self-cleavage. *EMBO J.* **21**, 749–759.
50. Feldstein, P. A., Hu, Y., and Owens, R. A. (1998) Precisely full length, circularizable, complementary RNA: An infectious form of potato spindle tuber viroid. *Proc. Natl. Acad. Sci. USA* **95**, 6560–6565.
51. Daròs, J. A. and Flores, R. (2004) *Arabidopsis thaliana* has the enzymatic machinery for replicating representative viroid species of the family Pospiviroidae. *Proc. Natl. Acad. Sci. USA* **101**, 6792–6797.

Chapter 13
Biochemical Analyses of the Interactions Between Viral Polymerases and RNAs

Young-Chan Kim and C. Cheng Kao

Abstract The interaction between viral polymerases and their cognate RNAs is vital to regulate the timing and abundance of viral replication products. Despite this, only minimal detailed information is available for the interaction between viral polymerases and cognate RNAs. We study the biochemical interactions using two viral polymerases that could serve as models for other plus-strand RNA viruses: the replicase from the tripartite brome mosaic virus (BMV), and the recombinant RNA-dependent RNA polymerase (RdRp) from hepatitis C virus (HCV). Replicase binding sites in the BMV RNAs were mapped using a template competition assay. The minimal length of RNA required for RNA binding by the HCV RdRp was determined using fluorescence spectroscopy. Lastly, regions of the HCV RdRp that contact the RNA were determined by a method coupling reversible protein-RNA crosslinking, affinity purification, and mass spectrometry. These analyses of RdRp–RNA interaction will be presented as three topics in this chapter.

Keywords Brome mosaic virus; Hepatitis C virus; RNA replication; RNA-dependent; RNA polymerase; Template competition assay; Fluorescence spectroscopy; Reversible crosslinking; Mass spectrometry

Topic I. A Template Competition Assay to Identify RNA Elements that Bind the BMV Replicase

1 Introduction

The interaction between RNA elements with viral and cellular proteins regulates all aspects of RNA virus infection, from RNA encapsidation to intracellular trafficking, translation, and replication. Therefore, to understand RNA virus infection, it is important to identify the *cis*-acting RNA elements.

Three general steps are routinely used to identify viral *cis*-acting elements: (1) predict the folding of the RNAs using a program such as Mfold, (2) probe the structure

From: *Methods in Molecular Biology, Vol. 451, Plant Virology Protocols: From Viral Sequence to Protein Function*
Edited by G.D. Foster, I.E. Johansen, Y. Hong, and P.D. Nagy © Humana Press, Totowa, NJ

of the RNA in solution using various enzymes or chemicals, and (3) construct specific mutations in key predicted elements and assess the effects of the mutations on viral infection and/or RNA synthesis in vitro. While this approach has led to many important advances in defining viral RNA elements, it relies on prior knowledge that a particular region contains an activity of interest. To identify viral replicase-binding elements more systematically, we used a template competition assay.

The template competition assay uses two RNAs, a template RNA that will direct the synthesis of a product of known length and a competitor RNA that does not give rise to a product but could titrate the replicase away from the functional template. By varying the concentration of competitor, the IC_{50} value (the concentration of the competitor required to reduce synthesis from the reference RNA template to 50%) can be determined for comparison between competitors.

It is worthwhile to note that regulatory RNA elements can lie within the RNAs that initiate the infection and also in the complementary RNAs synthesized as replication intermediates during infection. To produce both plus- and minus-strand RNAs, the template for in vitro transcription consists of DNA fragments generated by PCR to contain T7 and SP6 promoters at the two termini (Fig. 1a). These promoters are part of the oligonucleotide primers used for PCR. Furthermore, each DNA fragment was overlapped by 40-nt with its neighboring fragment to minimize the chance that an RNA motif was cleaved in designing the primers. A review of functional templates that can direct RNA synthesis in vitro by the BMV replicase is previously described (1). The template used in this work directs the synthesis of the BMV subgenomic RNA and is named –20/13. The original manuscript describing this work was published in (2).

2 Materials

2.1 RNAs

RNA–20/13 (33-nt) is the template for the competition assays. It was chemically synthesized by Dharmacon, Inc. (Boulder, CO) and stored frozen in aliquots to ensure consistent quality and concentration of the RNA used throughout the experiment. Transcripts that serve as competitors are made with commercially available kits (Epicentre Technologies, Madison, WI), treated with DNaseI, extracted with phenol–chloroform, and the RNAs precipitated with LiCl, according to the manufacturer's protocol. The concentrations of all RNAs were determined by spectrometry and its quality was examined following electrophoresis and staining with 0.05% Toluidine Blue.

2.2 BMV Replicase

BMV replicase was prepared from BMV-infected barley as previously described (3).

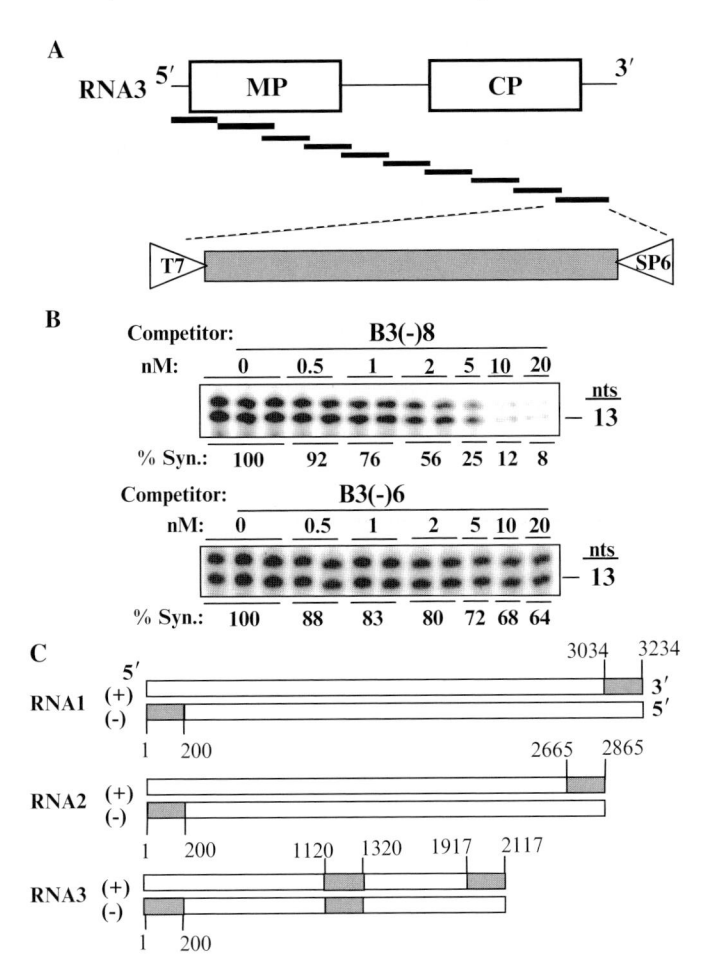

Fig. 1 Mapping of the replicase binding sites in the BMV RNA using a template competition assay. (**A**) A schematic of BMV RNA3 and the DNA fragments representing the templates for in vitro transcription. Each DNA fragment contained a T7 promoter on the left end and a SP6 promoter on the right end to allow for the synthesis of both the plus-and minus-strand RNAs. (**B**) A representation of the results from a high efficiency binder and one that does not bind the replicase well. (**C**) Locations of the eight high affinity binders within the plus- and minus-strand BMV RNAs. RNA segments containing high affinity binding sites are shown in grey

3   Methods

3.1   Template Competition Assay

1. Template competition assays were performed in an RNA synthesis reaction programmed with a 2 nM concentration of template RNA and increasing concentrations of competitors. The initial and final ratios of the template to competitor are

1:0.25 and 1:10, respectively. The template RNA, −20/13, directs the synthesis of 13-nt and 14-nt products, with the latter being the result of nontemplated nucleotide addition to the product RNA.

2. RNA synthesis assay. Each reaction is synthesized by mixing

   a. 3.33 μL of 220 mM Na-glutamate and 44 mM $MgCl_2$, pH [8.2]
   b. 3.33 μL NTPs [2 mM GTP, 1 mM ATP, 1 mM UTP]
   c. 1 μL of 20% TX-100
   d. 1 μL of 0.5 M dithiothreitol
   e. 0.4 μL of [α-^{32}P] CTP (400 Ci mmol^{-1}, 10 mCi mL^{-1})
   f. RNA(s) and water to make the final volume 32 μL

The number of reactions can be scaled up according to the need. We usually make a minimum of 12 reactions per experiment in a master mix. The use of the master mix will increase the precision of the reaction.

3. The reaction is started by the addition of the BMV replicase and incubated for 60 min at 25 °C. Our standard reaction will contain 8 μL of replicase. The amount of enzyme needed for each reaction must be determined empirically since it is currently not possible to make biochemically pure replicases from plus-strand RNA viruses. Any adjustments in enzyme volume can be compensated by adjusting the volume of water in item "f" above.

4. After the 1 h incubation, add 60 μL of water to increase the volume for better RNA recovery in subsequent steps. The reaction products are extracted with phenol/chloroform (1:1, vol/vol) and precipitated with six volumes of ethanol, 10 μg of glycogen, and 10 μL of 5 M ammonium acetate. The samples are incubated on ice for 30 min prior to precipitating the RNAs by centrifugation for 20 min at 14,000 RPM at 4 °C.

5. Following removal of the ethanol, the pellet is washed briefly with ice cold 70% ethanol and then dried in a vacuum dryer.

6. The dried pellet is solubilized by adding 8 μL of loading buffer (45% deionized formamide, 1.5% glycerol, 0.04% bromophenol blue [BPB], and 0.004% xylene cyanol), vortexed briefly and centrifuged for 2 s to pull the sample to the bottom of the tube. The sample is then heated to 90 °C for 3 min prior to electrophoresis.

7. Gel electrophoresis is performed in 20% PAGE with 7.5 M urea. The percentage of polyacrylamide should be adjusted to optimize the separation of the replicase products. The electrophoresis was performed until the BPB migrated to the bottom of the gel. This will remove the unincorporated radiolabel and make the quantification of replicase products more accurate.

8. The 13- and 14-nt bands should be quantified together by the use of a PhosporImager and Molecular Dynamics software.

3.2   Analysis of Results

The competition assay can be used to determine whether an RNA element can bind the BMV replicase. In our effort to identify replicase-binding elements systematically, we identified three classes of RNAs. The majority of the RNA fragments had IC_{50}s greater than 20 nM. An example of an RNA in this class is B3(-)6 (Fig. 1b). There were two competitors that had IC_{50}s between 17 and 19 nM that could possibly contain weak binding sites. Eight competitors had IC_{50} values between 3 and 5 nM, which we designated as high affinity binders. The locations of the eight binding sites from all three of the BMV RNAs are shown in Fig. 1c. Several of the eight RNA fragments include sequences known to contain replicase-binding motifs. For example, within the 3′ 200-nt of the plus-strand BMV RNAs resides Stem-loop C (SLC), which is the site for binding of the replicase for minus-strand initiation (4–6). Another element lies within the intercistronic region of minus-strand RNA3, the location of the subgenomic promoter (7).

It is interesting that the numbers of binding elements within the three BMV RNAs are different: BMV RNA3 has two elements in the plus- and minus-strand RNAs, while RNA1 and RNA2 have one primary replicase binding site, a piece in both the plus- and minus-strand RNAs. Since RNA1 and RNA2 encode replication proteins, it is possible that the proteins act in *cis* of the RNA to confer some degree of specificity. In contrast, RNA3 may require additional elements to recruit the replicase. Additional evidence for the roles of each of the eight RNA elements in BMV RNA synthesis is in ref. (2).

4 Notes

The eight binding sites represent high affinity sites that we identified with our replicase preparation. Additional replicase-binding sites in BMV RNAs could exist if some form of the replicase is not well represented in our replicase preparations; the BMV replicase is known to have different capabilities for minus- and subgenomic modes of RNA synthesis over the course of infection (6).

Topic II: Analysis of HCV RdRp–RNA Interaction Using Fluorescence Anisotropy

1 Introduction

RdRp is the catalytic subunit of the viral replicase. Hence, its interaction with RNA should exert a primary influence on all aspects of viral RNA synthesis. RdRp interaction with RNA will also be of interest in terms of the mechanism of polymerase action and can be compared with interactions of other template-dependent RNA polymerases, such as the better studied T7 RNA polymerase (8). Several assays can be used to assess RdRp–RNA binding, such as the retention of the RNA–RdRp complex in a filter, alteration in the electrophoretic mobility of the RNA in a nondenaturing gel, or the resistance of the RNA to digestion by RNAse or chemicals that would digest unprotected RNAs (9–13). We used fluorescence spectroscopy to determine the equilibrium binding of RNA template by recombinant HCV RdRp.

This method can provide quantitative results and ability to calculate the affinity and cooperativity of RdRp–RNA interaction.

The specific protocol used is the measurement of global anisotropy of RNA. Briefly, anisotropy is the measurement of the emission polarization from a fluorescent sample on excitation with fully polarized light. The changes in the polarization of the fluorescent sample depend on the rotational motion of the molecule, which is inversely proportional to the molecular size (14). When a protein binds a RNA, the increase in mass will make the protein–RNA complex remain relatively polarized because of a much slower rate of rotation, thus giving a higher anisotropy value. This relative change of anisotropy values can be used in standard models for molecular interactions to derive the properties of a particular protein–RNA interaction.

One consideration for this analysis is that the intrinsic fluorescence of protein would interfere with measurements of the RNA. Therefore, to eliminate this, we used extrinsically labeled RNA that contained a 5′ fluorescein. In this section, we measured the differences in the binding of the HCV RdRp to RNA as a function of RNA length.

2   Materials

2.1   5′-Fluorescein-Labeled-RNAs

RNAs were synthesized chemically by Dharmacon Inc to contain a 5′ fluorescein and purified after electrophoresis in denaturing polyacrylamide gels. Each RNA contains the nucleotides needed for specific de novo initiation of RNA synthesis by the HCV RdRp in vitro and can direct RNA synthesis in a radio-labeled gel-based assay (15). The names and sequences of the RNAs are shown in Fig. 2a.

2.2   Recombinant RdRp

The RdRp, named $\Delta21$, is derived from the hepatitis C virus type 1b, BK strain. This protein lacks the hydrophobic C-terminal 21-residues of the HCV NS5B protein and was expressed in *E. coli* BL21 (DE3) using from a pET21 vector. The protein contains a C-terminal hexa-histidine tag and was purified by immobilized metal affinity chromatography and poly-U ion exchange chromatography as previously described (16).

2.3   Binding Buffer

The basic buffer contained 50 mM HEPEs (pH 7.5), 5 mM $MgCl_2$, and 0.002% Tween 20. This buffer could be made in 10× stock to facilitate the addition of various concentrations of NaCl.

A

RNA	Sequence	50mM NaCl		K_{d} (μM) with NaCl at:		
		K_{d} (μM)	n	75mM	100mM	125mM
F-5	5′-Fl-UAUAC-3′	2.06±0.39	1.06±0.13	3.50±0.74	7.31±1.23	12.±2.70
F-7	5′-Fl-CGUAUAC-3′	0.35±0.02	1.23±0.11	0.59±0.05	0.73±0.07	1.32±0.19
F-9	5′-Fl-CUCGUAUAC-3′	0.33±0.01	1.47±0.15	ND	ND	ND
F-14	5′-Fl-UAAUUCUC GUAUAC-3′	0.42±0.03	1.73±0.16	0.54±0.03	0.61±0.05	0.97±0.10

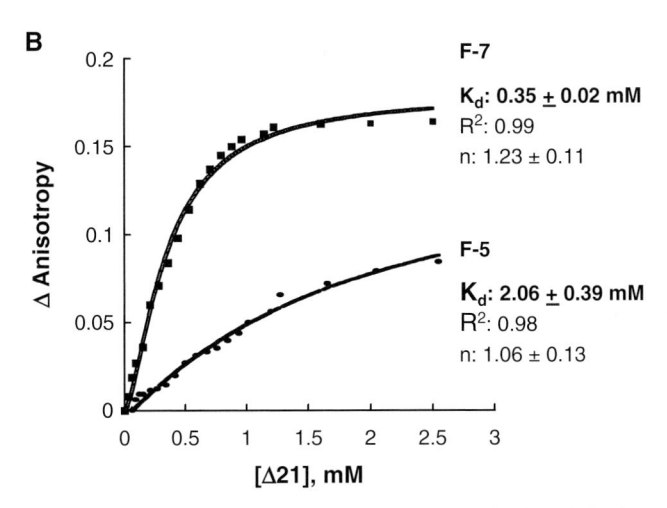

Fig. 2 Analysis of the RNA length required for stable interaction with the HCV RdRp using a fluorescence assay. (**A**) Sequences of RNAs used and the affinities with the HCV RdRp Δ21. Fl denotes a fluorescein and ND denotes not determine. (**B**) The binding isotherms for the RNAs of 7- and 5-nt

2.4   Fluorometer and Cuvette

A Perkin-Elmer luminescence spectrometer LS55 and cuvettes with an optical path length of 0.4 cm were used.

3   Methods

3.1   Fluorescence Measurements

1. All measurements can be made at 22–23 °C temperature.
2. The optimal emission wavelength is determined by doing a scan of the fluorescence at an excitation wavelength of 490 nm. The expected emission optimum for fluorescein is 518~525 nm. We obtained an optimum at 520 nm.

3. The experiment is performed with an integration time of 1 s and excitation and emission slit width of 5 nm.
4. The sample lacking Δ21 is loaded into the cuvette containing a magnetic stir bar set to stir constantly at a moderate speed.
5. After the sample was equilibrated for 60 s, ten independent measurements were recorded and an average value calculated.
6. An aliquot of Δ21 (at a volume less than 1% of the total sample volume) can be added to a concentration at least 15-fold lower than the RNA, allowed to equilibrate for 60 s, and ten anisotropy values were recorded and averaged.
7. The changes in quantum yield of F-RNAs should be measured throughout the titration to monitor photobleaching. There should be no significant change of fluorescence during the titration.
8. The process in steps 6 and 7 is repeated for all concentrations of Δ21 added. The titration should continue until the change in the anisotropy values reaches saturation.
9. The titration could also be done with set concentrations of Δ21 and RNA, but with increasing concentrations of NaCl or an inhibitor.

3.2 Data Analysis

1. The binding isotherms for each titration are constructed by plotting the gain of anisotropy vs. final concentration of added protein for each different concentration of NaCl.
2. A nonlinear least square fitting of the binding isotherm can be derived using a program such as KaleidaGraph (Synergy Software, Reading, PA), or GraphPad (San Diego, CA).
3. The equilibrium dissociation constants (K_d) can be determined by nonlinear regression analysis using a Hill equation as a binding model ($\Delta A = B_{max} x^n/[x^n + K_d^n]$). In this equation, ΔA is the value of anisotropy change by the ligand binding, B_{max} is the value of maximum anisotropy change, x is the total concentration of the added protein and the exponential term (n) is the Hill coefficient, which can be used to estimate the extent of cooperative binding.
4. If no cooperativity was involved in protein–RNA binding and the initial concentration of fluorescently-labeled molecule should be considered, one can use the quadratic equation as a binding model ($\Delta A = B_{max} [(x + K_d + P) - \{(x + K_d + P)^2 - 4xp\}^{1/2}]/(2P)$). In this equation, ΔA is the value of anisotropy change by the ligand binding, B_{max} is the value of maximum anisotropy change, P is the initial concentration of fluorescently-labeled molecule, which was used for titration, and x is the total concentration of added protein.

3.3 Results

After monitoring the fluorescence anisotropy changes of the RNAs in response to Δ21 titration, the K_ds and Hill coefficients of the four RNAs in Fig. 2a were determined (Fig. 2a). Representative binding isotherms and curve fittings are presented in

Fig. 2b. At 50 mM NaCl, the K_d of $\Delta 21$ to the RNAs of F-14, F-9, and F-7 were all between 0.29 and 0.33 μM. However, the K_d of $\Delta 21$ to the F-5 was approximately fivefold higher (Fig. 2a), suggesting that an RNA length longer than 5-nt is minimally required for stable binding under the conditions used.

To address whether $\Delta 21$ binding to RNAs of different lengths was qualitatively different, we examined the K_ds for F-5, F-7, and F-14 in the presence of NaCl concentrations of up to 125 mM. The K_d of F-5 increased by 5.8-fold in a buffer containing 125 mM NaCl, while the K_ds with RNAs of 7- and 14-nt increased by 3.8- and 2.3-fold, respectively, over this range of salt concentrations (Fig. 2a). These results indicate that nonspecific ionic interaction contributed more to the binding affinity with the 5-nt RNA than to that with longer RNAs. Therefore, 7-nt RNA appears to be a minimal length of RNA that can confer stable binding to the HCV RdRp during de novo initiation of RNA synthesis and the elongation stage of RNA synthesis.

Topic III: Identification of RdRp Residues that Contact RNA Using Reversible Cross-Linking, RNA Affinity Chromatography, and Mass Spectrometry

1 Introduction

Identification of RNA contact sites in a protein is important for both structural and mechanistic understanding of protein–RNA interaction. We developed a method combining reversible formaldehyde cross-linking, RNA affinity chromatography, and mass spectrometric analyses to map the residues in the HCV RdRp that can contact an initiation competent RNA. Reversible formaldehyde cross-linking is widely used to study DNA–protein and RNA–protein interactions, especially in vivo (17, 18). Cross-linking of RNA to protein using formaldehyde occurs through the linkage between the side-chains of Lys, Arg, His, Cys, aromatic residues and bases (A, G, and C) of RNA by the formation of methylene bridges via dehydration and Schiff base formation (19). These cross-links can be reversed by heating or acidification of the cross-linked conjugates to release formaldehyde (17). The overall scheme of these approaches is summarized in Fig. 3a.

2 Materials

2.1 7-nt RNA

The RNA used for cross-linking to the HCV RdRp was synthesized by Dharmacon, Inc. and purified after electrophoresis in denaturing polyacrylamide gels. This RNA, 7C, was characterized in topic 2 of this chapter and contains the sequence: 5′ CGUAUAC 3′. A 5′ biotinylated form of the RNA is named 7CB.

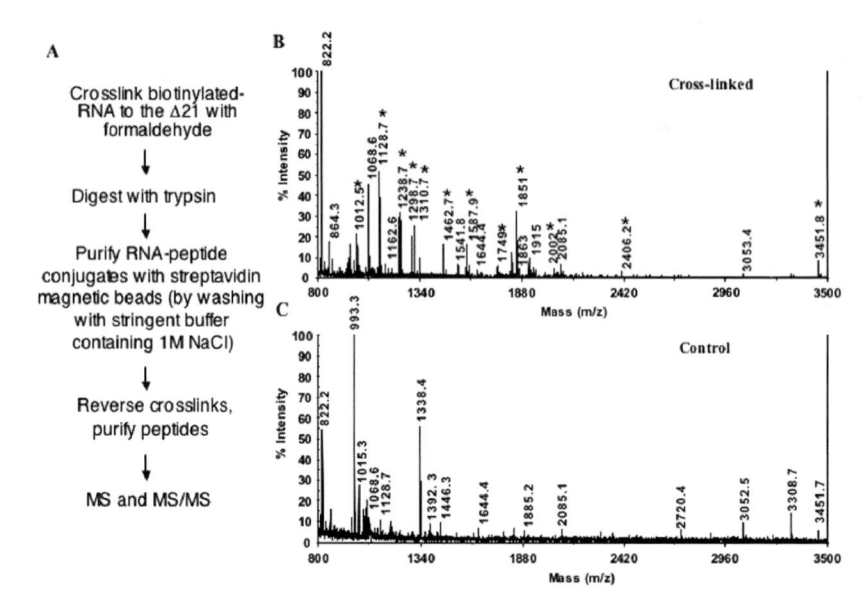

Fig. 3 A method to rapidly map the interaction between the HCV RdRp and RNA. (**A**) the schema used in the experiment. (**B**) A representative mass spectra for a cross-linked reaction and a background control. Peptide masses are indicated in Daltons above the ions, and the ones identified with asterisks are unique in the sample of interest

2.2 Reagents for Cross-Linking Reactions, Digestion, and Enrichment Process

1. Formaldehyde (38% stock solution)
2. 2× binding buffer: 40 mM HEPES (pH 7.5), 8 mM $MgCl_2$, and 2 mM DTT
3. 2 M glycine
4. 4–12% gradient Nu-PAGE gel (Invitrogen, Carlsbad, CA)
5. Proteomics grade trypsin (Trypsin Gold™, Promega, Madison, WI)
6. Streptavidin magnetic beads (New England Biolab, Beverly, MA)
7. Ziptip™ (Millipore, Bedford, MA)

2.3 Other Items

1. PhosphorImager and Molecular Dynamics software (Molecular Dynamics)
2. SpeedVac™ concentrator

3   Methods

3.1   *Cross-Linking of 7-nt RNA to Δ21 Using Formaldehyde*

3.1.1   Analytical Scale Cross-Linking Reaction

1. An analytical crosslinking reaction will help define the conditions for a preparative reaction. The first parameter to test is the formaldehyde concentration. Set up as many reactions as to be tested to contain 1 μM (final concentration) each of Δ21 and 5′-radio-labeled 7C in 20 mM HEPES (pH 7.5), 4 mM $MgCl_2$, 1 mM DTT in 20 μL reaction, and incubated at room temperature for 5 min (see note 1).
2. Formaldehyde is added to the reaction to final concentrations of between 0.01% and 0.5%. All of the reactions are incubated for 5 min at room temperature to allow the cross-linking (see note 2).
3. The cross-linking reaction is quenched by the addition of glycine to a final concentration of 0.2 M.
4. Mix 10 μL of the reaction products with 4× SDS sample buffer and electrophorese the products on the appropriate SDS-PAGE.
5. The cross-linked products are visualized and quantified using the PhosphoImager and Molecular Dynamics software (Molecular Dynamics).
6. The optimal concentration of formaldehyde for the preparative analysis can be selected from this analytical experiment. It is also a good idea to check the time required for RNA–protein crosslinking. We found that Δ21 crosslinked to the RNA 7CB best at 0.1% formaldehyde for 5 min.

3.2   *A Preparative Scale Cross-Linking Reaction*

1. The preparative cross-linking reaction (100 μL) contains a final concentration of 2 μM Δ21 and 4 μM 7CB RNA in Buffer H (20 mM HEPES (pH 7.5), 4 mM $MgCl_2$, and 1 mM DTT).
2. Formaldehyde is added to 0.1% final concentration for 5 min at room temperature before the addition of 0.2 M glycine to quench the reaction.

3.3   *Digestion of the Cross-Linked RNA–Protein Conjugates Using Trypsin*

1. The digestion reaction is adjusted to be 100 mM NH_4HCO_3 (pH 7.8) by adding 5 μL of a 0.5 M NH_4HCO_3.

2. Proteomics grade trypsin (Trypsin Gold, Promega) is added to 1/50th of the concentration of the Δ21 in the cross-linking reaction and incubated at 37 °C overnight.

3.4 Enrichment of Cross-Linked RNA-Peptide Conjugates with Streptavidin Beads

1. Streptavidin magnetic beads (New England Biolab, Beverly, MA) are used to capture the 7CB RNA and 7CB RNA-peptide conjugates.
2. Packed beads (50 μL) are washed thrice with buffer H, discarding the wash buffer each time.
3. The Streptavidin beads are added to the trypsinized cross-linked reaction products to bind the biotinylated RNA and RNA-peptide conjugates. The slurry is allowed to mix for 30 min.
4. The sample is washed three times with a buffer containing 20 mM Hepes, 1 M NaCl, 1 mM EDTA and 1 mM DTT and twice with 25 mM ammonium bicarbonate (pH 7.8) to remove unbound RNA and peptides.

3.5 Reversal of Cross-Linked RNA-Peptide Conjugates and Retrieval of Cross-Linked Peptides

1. The RNA-peptide conjugates are reversed by incubating the samples at 70 °C for 1 h (see note 3).
2. The retrieved cross-linked peptide samples are prepared for mass spectrometry by pelleting the beads at 2,000 g for 3 min, and transfer the supernatant to a Ziptip (Millipore, Bedford, MA) by pipetting the supernatant up and down for a minimum of eight times. The Ziptip is then centrifuged at 3 K for 3 min to desalt and concentrate the sample.
3. Wash the Ziptip twice with 20 μL of 0.1% trifluoroacetic acid.
4. The final peptide preps are eluted with 2.5 μL of 70% acetonitrile/0.1% trifluoroacetic acid.

3.6 Identification and Mapping of the Cross-Linked Peptides Using Mass Spectrometry

1. All MALDI MS analyses were performed on an ABI 4700 Proteomics Analyzer (Applied Biosystems, Framingham, MA) equipped with a 200 Hz Nd:YAG laser (PowerChip, JDS Uniphase, San Jose, CA) and controlled by the Applied Biosystems 4000 series Explorer V3.0 software package.

2. All MS analyses were performed using the dried droplet method and 5 mg mL^{-1} α-cyano-4-hydroxycinnamic acid (Sigma, St. Louis, MO) in 60% acetonitrile as the matrix.
3. The laser intensity was set just above the threshold required to ionize the peptides. For the tandem MS experiments, the acceleration was 1 kV in all cases, the collision gas was air, and the laser intensity was increased by 10% over the MS mode experiment. The number of laser shots used to obtain a spectrum varied from 500–5,000, depending on signal quality.
4. The fragmentation data obtained in these experiments were analyzed with the Applied Biosystems Data Explorer software package. In some cases, internal calibrants were used to maximize mass accuracy in MS mode (+/− 10 ppm). Accuracy in MS/MS mode is routinely less than 0.2 Da.
5. Peaks were selected with a signal to noise ratio (S/N) greater than 5 and for having the appropriate isotope ratios.
6. Peptide mass fingerprinting was performed by comparing calculated and observed peptide masses using an established database and algorithms, such as Mascot, PeptideMass, and MS-Fit (http://www.matrixscience.com/, http://ca.expasy.org/tools/peptide-mass.html, http://prospector.ucsf.edu/ucsfhtml4.0/msfit.htm).

3.7 Analysis of Results

RNA 7CB can direct the de novo initiation of RNA synthesis by HCV RdRp, hence it is a bona fide template for the HCV RdRp. The MALDI-MS spectrum of the peptides cross-linked to 7CB is shown in Fig. 3b. The experiment was repeated thrice, and identical results were observed each time. As a control, a sample was treated identically to the cross-linked sample with the exception that formaldehyde was not used (Fig. 3b). Peptides unique to the cross-linked sample are indicated with an asterisk (Fig. 3b, top panel). Each of the identified peptide peaks showed unique isotope distributed peptides (data not shown). A number of low intensity peptide peaks were also observed. These corresponded to derivatives of higher intensity peptide peaks except that they had missed one or two potential trypsin cleavage sites (Fig. 4a). The presence of overlapping peptides was helpful in confirming the assignment of the peaks. Despite the stringent washing steps included in the protocol, several peptides were observed in the control sample, although in low abundance compared with corresponding peaks from cross-linked sample, indicating that some uncross-linked peptides were not completely eliminated by this highly sensitive analysis. There also were low intensity peaks in both cross-linked and control samples that could not be assigned. These peaks could be attributed to several factors, including the matrix, multiply cross-linked peptides, other possible modified peptides, or peptides that interacted with the streptavidin resin. Nonetheless, by comparing the peaks from the non-cross-linked control sample, the unassignable peaks did not affect the identification of the cross-linked peptides from the spectra. The amino acid position, observed and calculated masses, and sequences of the identified peptides by peptide mass fingerprinting results are summarized in Fig. 4a.

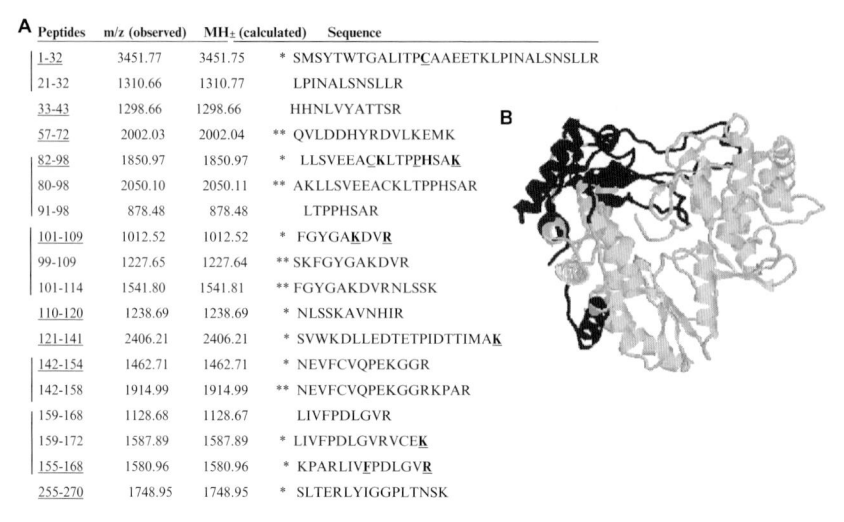

Peptides	m/z (observed)	MH± (calculated)		Sequence
1-32	3451.77	3451.75	*	SMSYTWTGALITPCAAEETKLPINALSNSLLR
21-32	1310.66	1310.77		LPINALSNSLLR
33-43	1298.66	1298.66		HHNLVYATTSR
57-72	2002.03	2002.04	**	QVLDDHYRDVLKEMK
82-98	1850.97	1850.97	*	LLSVEEACKLTPPHSAK
80-98	2050.10	2050.11	**	AKLLSVEEACKLTPPHSAR
91-98	878.48	878.48		LTPPHSAR
101-109	1012.52	1012.52	*	FGYGAKDVR
99-109	1227.65	1227.64	**	SKFGYGAKDVR
101-114	1541.80	1541.81	**	FGYGAKDVRNLSSK
110-120	1238.69	1238.69	*	NLSSKAVNHIR
121-141	2406.21	2406.21	*	SVWKDLLEDTETPIDTTIMAK
142-154	1462.71	1462.71	*	NEVFCVQPEKGGR
142-158	1914.99	1914.99	**	NEVFCVQPEKGGRKPAR
159-168	1128.68	1128.67		LIVFPDLGVR
159-172	1587.89	1587.89	*	LIVFPDLGVRVCEK
155-168	1580.96	1580.96	*	KPARLIVFPDLGVR
255-270	1748.95	1748.95	*	SLTERLYIGGPLTNSK

Fig. 4 Peptides in the HCV RdRp that were crosslinked to RNA and their location in RdRp structure. (**A**) A summary of the peptides identified to interact with RNA 7CB. The bold and underlined residues were predicted to contact RNA by Bressanelli et al. (20). The number of asterisks in front of the peptide sequences indicates the number of missed cleavage sites by trypsin (internal K or R) in the identified peptides. The underlined peptides are represented as darker portions in the model of the HCV RdRp (Panel **B**). The structure of HCV NS5B Δ21 was constructed from coordinates in PDB ID code 1QUV. Where there are overlapping peptides, only the representative peptide in that overlapped region is shown

The identified peptides are colored in the three-dimensional structures of HCV RdRp (PDB ID code: 1QUV). An asymmetric distribution of the cross-linked peptides was observed, with no peptides from the thumb subdomain, and only one from the palm subdomain of the HCV RdRp (Fig. 4b). Instead, the peptides mostly mapped in the fingers and the connecting loops between fingers and thumb subdomains. Overall, the cross-linked peptides are located along the putative RNA-binding channel predicted from the crystal structure (20, Fig. 4b). Two of the identified peptides (amino acids 142–154 and 155–168) mapped within the active site cavity. Residues R158 contained with the peptide 155–168 was previously determined to interact with the α-phosphate of GTP and shown to be important for RdRp activity (21, 22). The biological relevance of the binding residues could be determined by assessing the effects of mutations on RNA synthesis in purified HCV RdRp and on the replication of subgenomic HCV replicons in cultured cells. This analysis was performed in the work by Kim et al. (23).

4 Notes

1. Binding buffer: Cross-linking buffer components should not contain free amino groups, since these will compete for cross-linking to the specific amino acids in protein or to bases in RNA with formaldehyde. Therefore, Hepes, phosphate, or acetate buffers are recommended for the cross-linking reaction.

2. Higher concentrations of formaldehyde will increase protein–protein cross-linking. Therefore, optimization of cross-linking time and formaldehyde concentration should be performed before a preparative reaction.
3. Reversal of the cross-linked RNA-peptide: Cross-linked RNA-peptide conjugate was routinely reversed by incubating for 1 h at 70 °C. Acidification of the RNA–peptide conjugate can also be used to reverse the conjugate.

References

1. Kao, C.C., Singh, P., and Ecker, D.J. (2001) De novo initiation of viral RNA-dependent RNA synthesis. *Virology* 287, 251–260.
2. Choi, S.K., Hema, M., Gopinath, K., Santos, J., and Kao, C. (2004) Replicase-binding site on plus- and minus-strand brome mosaic virus RNAs and their roles in RNA replication in plant cells. *J. Virol.* 78, 13420–13429.
3. Sun, J.H., Adkins, S., Faurote, G., and Kao, C.C. (1996) Initiation of (–)-strand RNA synthesis catalyzed by the BMV RNA-dependent RNA polymerase: synthesis of oligonucleotides. *Virology* 226, 1–12.
4. Kim, C.H., Kao, C.C., and Tinoco, I. (2000) RNA motifs that determine specificity between a viral replicase and its promoter. *Nat. Struct. Biol.* 7, 415–423.
5. Chapman, M.R., and Kao, C.C. (1999) A minimal RNA promoter for minus-strand RNA synthesis by the brome mosaic virus polymerase complex. *J. Mol. Biol.* 286, 709–720.
6. Sivakumaran, K., Hema, M., and Kao, C.C. (2003) Brome mosaic virus RNA syntheses in vitro and in barley protoplasts. *J. Virol.* 77, 5703–5711.
7. Siegel, R.W., Adkins, S., and Kao, C.C. (1997) Sequence-specific recognition of a subgenomic promoter by a viral RNA polymerase. *PNAS* 94, 11238–11243.
8. Tahirov, T.H., Temiakov, D., Anikin, M., Parlan, V., McAllister, W.T., Vassylyev, D.G., and Yokoyama, S. (2002) Structure of a T7 RNA polymerase elongation complex at 2.9 A resolution. *Nature* 420, 43–50.
9. Beckman, S.E., and Kirkegaard, K. (1998) Site size of cooperative single-stranded RNA binding by poliovirus RNA-dependent RNA polymerase. *J. Biol. Chem.* 273, 6724–6730.
10. Kao, C.C., Yang, X., Kline, A., Wang, Q.M., Barket, D., and Heinz, B.A. (2000) Template requirements for RNA synthesis by a recombinant hepatitis C virus RNA-dependent RNA polymerase. *J. Virol.* 74, 11121–11128.
11. Sun, X.L., Johnson, R.B., Hockman, M.A., and Wang, Q.M. (2000) De novo RNA synthesis catalyzed by HCV RNA-dependent RNA polymerase. *Biochem. Biophys. Res. Comm.* 268, 798–803.
12. Kim, M. J., Zhong, W., Hong, Z., and Kao, C.C. (2000) Template nucleotide moieties required for de novo initiation of RNA synthesis by a recombinant viral RNA-dependent RNA polymerase. *J. Virol.* 74, 10312–10322.
13. Cheng, J.C., Chang, M.F., and Chang, S.C. (1999) Specific interaction between the hepatitis C virus NS5B RNA polymerase and the 3′ end of the viral RNA. *J. Virol.* 73, 7044–7049.
14. Hill, J.J., and Royer, C.A. (1997) Fluoresence approaches to study of protein-nucleic acid complexation. *Meth. Enzymol.* 278, 390–416.
15. Ranjith-Kumar, C.T., Gutshall, L.L., Kim, M.J., Sarisky, R.T., and Kao, C.C. (2002) Requirement for de novo initiation of RNA synthesis by recombinant flaviviral RNA-dependent RNA polymerases. *J. Virol.* 76, 12526–12536.
16. Ranjith-Kumar, C.T., Kim, Y.C., Gutshall, L.L., Silvermann, C., Khandekar, S., Sarisky, R.T., and Kao, C.C. (2002) Mechanism of de novo initiation by the hepatitis C virus RNA-dependent RNA polymerase: role of divalent metals. *J. Virol.* 76, 12513–12525.
17. Orlando, V., Strutt, H., and Paro, R. (1997) Analysis of chromatin structure by in vivo formaldehyde crosslinking. *Methods* 11, 205–214.

18. Niranjanakumari, S., Lasda, E., Brazas, R., and Garcia-Blanco, M.A. (2002) Reversible crosslinking combined with immunoprecipitation to study RNA–protein interaction in vivo. *Methods* 26, 182–190.

19. Metz, B., Kersten, G.F.A., Hoogerhout, P., Brugghe, H.F., Timmermans, H.A.M., Jong, A., Meiring, H., Hove, J., Hennink, W.E., Crommelins, D.J.A., and Jiskoot, W. (2004) Identification of formaldehyde-induced modifications in proteins. *J. Biol. Chem.* 279, 6235–6243.

20. Bressanelli, S., Tomei, L., Roussel, A., Incitti, I., Vitale, R.L., Mathieu, M., and De Francesco, R. (1999) Crystal structure of the RNA-dependent RNA polymerase of hepatitis C virus. *PNAS* 96, 13034–13039.

21. Bressanelli, S., Tomei, L., Rey, F.A., and De Francesco, R. (2002) Structural analysis of the hepatitis C virus RNA polymerase in complex with ribonucleotides. *J. Virol.* 76, 3482–3492.

22. Ranjith-Kumar, C.T., Sarisky, R.T., Gutshall, L.L., Thomson, M., and Kao, C.C. (2004) De novo initiation pocket mutations have multiple effects on hepatitis C virus RNA-dependent RNA polymerase activities. *J. Virol.* 78, 12207–12217.

23. Kim, Y.C., Russell, W.K., Ranjith-Kumar, C.T., Thomson, M., Russell, D.H., and Kao, C.C. (2005) Functional analysis of RNA binding by the hepatitis C virus RNA-dependent RNA polymerase. *J. Biol. Chem.* 280, 38011–38019.

Chapter 14
In Situ Detection of Plant Viruses and Virus-Specific Products

Andrew J. Maule and Zoltán Havelda

Abstract The ability to combine nucleic acid hybridisation or immunospecific reactions with structural and ultrastructural analysis of virus-infected tissues has provided the opportunity to resolve the spatial details of infection with respect to the production of virus-specific products and the nature of the host response. These technologies may seem lengthy and complex but offer high rewards in terms of revealing the details of host–virus interactions not otherwise accessible.

Keywords Plant virus; In situ hybridisation; Immunocytochemistry; Spatial resolution

1 Introduction

Plant viruses infect susceptible host tissues in a progressive mode whereby multiplication at the primary infection site is extended through cell-to-cell spread to become dispersed in a single tissue, and by long distance spread in the vasculature to achieve a systemic infection. Within the vasculature, virus movement occurs passively according to sink-source demands in the host plant. Symptoms also differ between inoculated tissues and distal tissues infected via the vasculature. Hence, both virus multiplication and the host response to virus invasion are highly spatially regulated. For example, cells at the edge of an expanding lesion reveal very different properties with respect to host gene expression from cells either outside or in the centre of the lesion (1, 2). Hence, virus infection should be viewed as a dynamic interaction with the host where the details of that interaction can only be understood when spatial relationships are taken into account. Since the properties of the interaction can vary across single cell boundaries technologies that provide an equivalent level of resolution are required. In situ hybridisation and immunocytochemistry provide such resolution

In situ hybridisation involves the hybridisation of tagged RNA probes to target viral nucleic acids within thin sections of tissue and the detection of the positive reaction with an antibody recognising the substituted nucleotide used to tag the RNA probe. An example of the power of in situ hybridisation is shown in Fig. 1

From: *Methods in Molecular Biology, Vol. 451, Plant Virology Protocols: From Viral Sequence to Protein Function*
Edited by G.D. Foster, I.E. Johansen, Y. Hong, and P.D. Nagy © Humana Press, Totowa, NJ

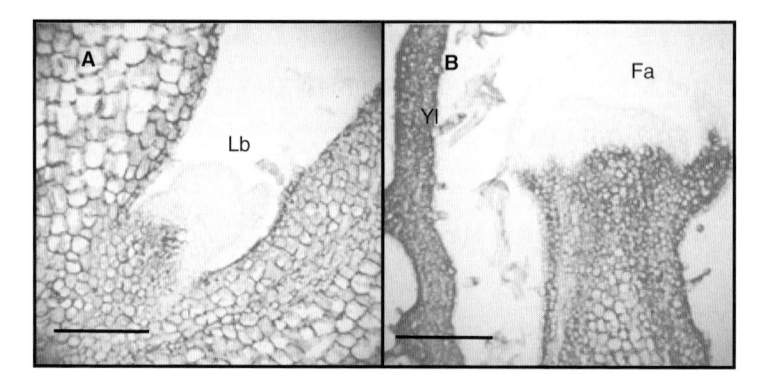

Fig. 1 Meristematic exclusion of *Cymbidium ringspot tombusvirus (CymRSV)* after 7-day post-infection. Longitudinal sections of apical region of *Nicotiana benthamina* were hybridised with digoxigenin-11-UTP-labeled minus-sense RNA probe corresponding to the *CymRSV* coat protein encoding region and detected with an alkaline phosphatase-conjugated antidigoxigenin antibody. (**a**) Lateral bud (Lb); (**b**) Floral apex (Fa); Young leaf (Yl). Bars = 200 μm

where the tissue localization of *Cymbidium ringspot tombusvirus* relative to plant meristems is illustrated. The most effective probes for in situ hybridisation are short (150–300 nucleotides) single stranded (ss) RNAs synthesised in vitro, from a cloned DNA template, with a proportion of UTP substituted with digoxigenin-UTP (DIG). The presence of DIG is detected with an anti-digoxigenin antibody. Microscopic examination of the section reveals the precise spatial distribution of the viral RNA at the tissue, cellular, or subcellular levels. Immunocytochemistry is the equivalent technique for the detection of viral proteins using specific antibodies. Immunocytochemistry is a less demanding technique that should be used preferentially when appropriate antibodies are available and when knowledge about virus location is the primary quest. In situ hybridisation used in parallel with immunocytochemistry is very powerful allowing the discriminative localisation of specific viral nucleic acids and proteins. This can be best achieved by processing consecutive tissue sections in parallel allowing near absolute spatial comparison between the techniques.

The elegance of both techniques lies in their specificity for the target molecules (see note 1). The value of specific antibodies is well known but in situ hybridisation can equally be used to differentiate particular virus nucleic acids. Hence, for example, by separately using complementary ss RNA probes information about different phases of the virus replication cycle may be explored. For example, analysis of the −ve strand of + ve ss RNA viruses reveals aspects of the viral RNA replication process (3). The resolution of the technology is limited by the fixation, embedding, and sectioning for light microscopy, although we have applied the same principles to sections for electron microscopy.

In situ hybridisation is a demanding and time-consuming technique. The methods described are mostly based on published manual methods (4–6) with modification and specific adaptations. The best results come from careful preparation and a rigorous attention to cleanliness and detail. Initially, this all seems very daunting but usually after two or three runs through the complete process a routine is established and good quality data begin to emerge. Relatively recent developments have provided the opportunities to automate and accelerate many of the processes, but at a high cost. These costs may be appropriate for organizations where demand is high or high through-put is required but for many labs where answers to specific research questions are required the cost may not be justified. Examples of automated technology include machines for automated dehydration and wax infiltration of fixed materials (e.g. Sakura Tissue-Tek VIP processor), the use of temperature controlled wax embedding work stations (e.g. GeneQ Wax Embedding Center) and automated slide processors for in situ hybridisation (e.g. Intavis AG, InsituPro VS). Particularly with the manual procedures but also to some degree with automation, some optimization is likely to be necessary. Patience is important. Good luck!

2 Materials

2.1 Tissue Fixation and Embedding

1. Solution of paraformaldehyde (*Sigma*; 4% w/v) and 0.1% Triton-X 100 in phosphate-buffered saline (PBS; 0.13 M NaCl, 7 mM Na_2HPO_4, 3 mM NaH_2PO_4; pH 6.7) (PFA). (See note 2). Stored in aliquots at $-20\,°C$. An alternative fixative is 3.7% formaldehyde, 5% acetic acid, 50% ethanol (FAA). (See note 6)
2. Vacuum chamber and incubation chamber
3. 10× saline: 8.5% w/v NaCl in water
4. Ethanol (100%)
5. Wax (eg. Paraplast plus (Sigma-Aldrich))
6. Eosin Y disodium salt (Fluka)
7. Histoclear (National Diagnostics, Atlanta, Georgia) or Roti-Clear (Roth, Germany)

2.2 Sectioning and Slide Preparation

1. Wax: Polyester wax (BDH), Paraplast extra (Sigma) or Paraplast plus (Sigma). The choice of wax influences the ease of handling and the quality of tissue preservation (see note 3)
2. Rotary microtome (e.g. HM335E (Microm, Germany) and hotplate
3. Histoclear
4. Commercially prepared poly-L-lysine slides (eg. Poly-Prep Slides, Sigma)
5. Protease from Streptomyces griseus (Pronase) (Sigma), 40 mg mL^{-1} in water and self-digested at $37\,°C$ for 2 h to remove contaminant nuclease activities. Store in aliquots at $-20\,°C$
6. Acetic anhydride (Sigma)
7. Triethanolamine (Sigma)

2.3 Hybridisation Probe Preparation

1. Commercially available RNA polymerase and buffer such as T7 or SP6 RNA polymerase (Fermentas)
2. dNTPs set (Fermentas)
3. RiboLock RNase Inhibitor (Fermentas)
4. Digoxygenin-11-UTP (Roche)
5. DNase I (RNase free) (Fermentas)
6. Ribonucleic acid, transfer from *E. coli* Strain W
7. Blocking Reagent (for nucleic acid hybridisation) (Roche)
8. SB buffer: 0.1 M Tris, 0.1 M NaCl, 0.05 M $MgCl_2$, pH 9.5. Prepare 10x stock (1M Tris, 1M NaCl, pH 9.5) and add $MgCl_2$ to 1' buffer just before use.
9. NBT: 50 mg mL^{-1}, BCIP: 50 mg mL^{-1}
10. Anti-Digoxigenin-AP, F'ab fragments (Roche)
11. 10x TBE buffer: 0.9 M Tris, 0.9 M boric acid, 0.02 M EDTA, pH 8.0
12. FDE: 10 mL deionised formamide, 200 μL 0.5 M EDTA (pH 8.0), 10 mg xylene cyanol, 10 mg bromphenol blue
13. 10x TBS buffer: 1 M Tris 1.5 M NaCl, pH 7.2

2.4 Hybridisation and Washing

1. Hybridisation solution: 0.3 M NaCl, 10 mM Tris-HCl pH 6.8, 10 mM $NaHPO_4$ pH 6.8, 5 mM EDTA, 50% formamide, 10% dextran sulphate, 1x Denhardt's solution, 1 mg mL^{-1} tRNA (see note 4)
2. Hybridisation washing solution: 50% formamide in 2x SSC
3. 10x NTE buffer: 5 M NaCl, 100 mM Tris HCl pH 7.5, 10 mM EDTA
4. RNaseA (Sigma)
5. Coverslips

2.5 Hybridisation Detection

1. Alcian Blue solution in 3% acetic acid
2. *Or* Calcofluor (0.1% w/v in water)
3. DPX Mountant for histology (Fluka)

2.6 Primary and Secondary Antibodies for Immunolocalisation

1. Pimary (specific) antibody for target detection
2. Secondary antibody for detection of primary antibody (eg. monoclonal Anti-Rabbit immunoglobulins Clone RG-16)

2.7   Immunoreaction, Washing and Detection

Data recording: Standard light microscope with camera, or fluorescence microscope if using calcofluor as a counterstain for the cell wall structure of the section

3   Methods

3.1   Tissue Fixation and Embedding

1. Tissue fixation in PFA can be a long and slow process taking up to 12 days; careful planning and preparation is advised. Table 1 systematically lays out the steps involved. (see note 5). Fixation with FAA is usually quicker.
2. The tissue is fixed in PFA by incubating overnight at 0 °C with gentle shaking (see note 6). Two-three hours is usually sufficient for fixation in FAA, again at 0 °C. It is advisable to use an excess volume of fixative with respect to the mass of tissue to be fixed, e.g. 20 mL for 10×25 mm^2 tissue pieces. To facilitate rapid fixation, the fixative solution should be fully vacuum infiltrated into the tissue. Submerge the tissue under the liquid surface with a sterile blue tip and place in a vacuum chamber. Apply a vacuum for 20–30 s or until air bubbles are visible on the surface of the tissue. Hold the vacuum for 5–10 min. Release the vacuum very slowly; tapping the chamber gently will help dislodge the bubbles from the tissue surface. Repeat the treatment. Change the fixative for fresh solution and repeat the vacuum treatment until the infiltration is complete (see note 7).
3. For embedding in wax the tissue must be completely dehydrated and the water replaced with molten wax before being solidified in the same wax. This involves a series of stepwise treatments (Table 2). Briefly, the fixative is washed away with 0.85% w/v NaCl (saline), displaced with increasing concentrations of ethanol (see note 8), displaced again with increasing concentrations of histoclear (source; see note 9) and finally substituted with molten wax. All these treatments should be carried out with gentle shaking at the specified temperatures (Table 2). The final stages must be carried out above the melting temperature of the wax. The handling of the tissue pieces after dehydration when they become white/translucent is facilitated by the addition of a few grains of the stain EosinY to

Table 1 Timetable for tissue fixation and embedding

Day	Treatment
1	Tissue fixation
2	Dehydation in EtOH graded series - 1 × 30 min and 2 × 3 h steps
3	Complete dehydration to 100% EtOH – 2 × 4 h steps
4	Graded exchange of EtOH with Histoclear; start wax infiltration: 1–3 h steps all day
5	Gradual increase in wax to Histoclear proportions: 3–4 h steps all day
6	Gradual increase in wax to Histoclear proportions: 3–4 h steps all day
7–9	Additions of fresh molten wax: 4 h steps all day
10–11	Additions of fresh molten wax: 4 h steps all day
12	Mount the tissue in blocks

Table 2 Tissue Dehydration

Day	Treatment	Stage	Time	Temperature
2	Dehydration to 85% EtOH			
	1× saline (0.85% NaCl)		30 min	Ice
	50% EtOH: 50% 1× saline		3 h	Ice
	70% EtOH: 30% 1× saline		3 h	Ice
	85% EtOH: 15% 1× saline		O/N	4 °C
3	Dehydration to 100% EtOH			
	95% EtOH: 5% water		4 h	4 °C
	100% EtOH		4 h	4 °C
	100% EtOH		O/N	4 °C
4	Exchange EtOH with Histoclear			
	100% EtOH		2 h	RT
	3:1 EtOH:histoclear		1–3 h	RT
	1:1 EtOH:histoclear		1–3 h	RT (add Eosin Y at this satge)
	1:3 EtOH:histoclear		1–3 h	RT
	100% histoclear		1 h	RT
	100% histoclear		1 h	RT
	100% histoclear + approx 10% wax chips		O/N	RT
5	Increase wax to histoclear proportions			
	Add approx 10% wax chips		Every 3–4 h	RT
	Add approx 10% wax chips		O/N	42 °C
6	Increase wax to histoclear proprtions			
	Add approx 10% wax chips		Every 3–4 h	42 °C
	Put approx. 200 g wax to melt O/N at 58 °C			
7–9	Additions of fresh molten wax			
	Change half the molten wax with fresh wax (try not to disturb the tissues at the bottom of the container; it is easier at this stage to leave the caps off the container)		Every 3–4 h	58 °C
10–11	Additions of fresh molten wax			
	Change all the molten wax with fresh wax		Every 3–4 h	58 °C
12	Block the tissues			

the 100% ethanol and ethanol/histoclear stages, which gives the tissue a red colour. For FAA fixed materials, this rigorous but lengthy procedure has been successfully replaced with brief (30 min) washes in 50%, 60% and 70% ethanol to substitute for the steps on Day 2 (Table 2).

4. Prior to sectioning the tissue must be solidified in a wax block. This can be done in any convenient receptacle (see note 10). To allow orientation of the tissue in the block, it is best to work on a heated block.

3.2 Sectioning and Slide Preparation

1. Tissue sections (10–20 μm) are obtained using a conventional rotary microtome. If very thin sections are required, then a retracting rotary microtome (e.g. HM335E (Microm, Germany) should be used to avoid the compression of the

tissue block by the up-stroke of the knife (see note 11). Sections should be mounted onto poly-L-lysine-coated pre-prepared slides (see note 12).

2. Before hybridisation the sections must be de-waxed, rehydrated and buffer equilibrated (see note 13). Again this involves a series of graded treatments. Briefly, these involve removing the wax with Histoclear (see note 14), washing with ethanol and rehydration in a graded ethanol/saline series (Table 3).

3. After hydration and buffer equilibration, it is possible to stop if the aim is to subject the section to immunohistochemistry. (*see* Sect. 3.6).

4. To maximise the availability of the target nucleic acids for hybridisation, the section needs to be de-proteinised. Equilibrate the slides in pronase buffer (50 mM Tris-HCl pH 7.5, 5 mM EDTA) for 2 min, and then incubate in pronase solution (0.125 µg mL^{-1}) for 10 min at room temp. Wash the slides in PBS for 2 min and postfix in PFA (in a fume hood) for 30 min.

5. To eliminate background reaction due to electrostatic binding of the hybridisation probe, amino groups on the section should be acetylated using an acetic anhydride treatment. Rinse the slides twice in PBS for 2 min. In a fume hood, incubate the slides in buffered acetic anhydride (add 0.5 mL acetic anhydride to 100 mL 0.1 M triethanolamine-HCL, pH 8) for 10 min at room temp with gentle agitation (see note 15). Rinse the slides twice more in fresh PBS for 2 min.

Table 3 Dewaxing the section in preparation for hybridisation

Treatment	Time	Notes
Histoclear 1	10 min	
Histoclear 2	10 min	This histoclear can be reused as histoclear 1
100% EtOH	5 min	
100% EtOH	3 min	
95% EtOH/5% 1× saline	2 min	
85% EtOH/15% 1× saline	2 min	
50% EtOH/50% 1× saline	2 min	
30% EtOH/70% 1× saline	2 min	
100% 1× saline	2 min	Slides can be used for immunocytochemistry after this stage
Pronase buffer	2 min	
Pronase treatment	10 min	
PBS	2 min	
Post-fixation	30 min	In the fume hood
PBS	2 min	In the fume hood
PBS	2 min	
Acetylation	10 min	In the fume hood
PBS	2 min	
1× saline (fresh)	2 min	
30% EtOH/70% 1× saline	2 min	These solutions can the same as those used in reverse earlier
50% EtOH/50% 1× saline	2 min	
85% EtOH/15% 1× saline	2 min	
95% EtOH/5% 1× saline	2 min	
100% EtOH	2 min	

6. Dip slides in fresh saline solution for 2 min and dehydrate through a graded etha-nol/saline series (see Table 3). Repeat the 100% EtOH treatment. Now the slides are ready for hybridisation. You can stop here and keep the slides safely in EtOH for few hours.

3.3 *Hybridisation Probe Preparation*

1. The RNA probe is prepared in an in vitro transcription reaction using a DNA template cloned into an appropriate vector possessing RNA polymer-ase promoters (e.g., Bluescript, pGEM) (see note 16). Map the plasmid to determine the orientation of the DNA template. Linearise the plasmid with the appropriate restriction enzyme (generating a blunt or 5′-protruding end) and check that the digestion is complete by agarose gel electrophore-sis. Purify the DNA by phenol extraction, ethanol-precipitate, wash with 75% ethanol, dry and resuspend in a suitable volume of nuclease-free water (see note 17).
2. Set up the transcription labelling reaction (30 μL) by adding linearized DNA template (ca.0.5 μg), 3 μL 10× transcription buffer, 3 μL 10× DIG/NTP mix, 0.5 μL RNase inhibitor (5 U), required RNA polymerase enzyme (ca. 10 U) and add RNase free water to 30 μL. Incubate for at 37 °C for 60 min. Various com-mercial kits are also available for preparing DIG-labelled RNA probes. These are costly alternatives but can work well.
3. To assess the quality of the probe analyse 3 μL of labelled RNA transcript by electrophoresis in 1× TBE buffer using nuclease-free 1.2% agarose gel contain-ing ethidium bromide (see note 18). The RNA product should be observed as a single band with little or no degradation products and about 5–20 times more intense than the template DNA band.
4. To remove the DNA template, add 7 μL 10× DNase buffer, 5 U RNase-free DNase, 2 μL tRNA (100 μg μL^{-1}) and water to100 μL. Incubate 15–20 min at 37 °C. Extract with phenol/chloroform then chloroform and precipitate the remaining RNA by adding 5 μL 4 M Na-acetate and 500 μL EtOH and cooling on dry ice or at −70 °C for 30 min. Collect the RNA pellet by centrifugation, wash twice with 70% EtOH for 5 min, dry and resuspend in 30–50 μL sterile water. Alkaline-hydrolysis might be necessary if your probe is significantly longer than 300 bases (see note 19).
5. Before starting the in situ hybridisation, it is advisable to test the probe in a "spot test". Spot 0.5 μL probe onto a piece of Hybond N membrane and UV-crosslink. Wet the membrane in TBS and incubate in blocking reagent/TBS for 30 min. Add anti-DIG-alkaline phosphatase, Fab fragments (Roche) (1:5,000) in TBS and incubate the membrane for 30 min at 37 °C. Wash three times with TBS for 5 min then equilibrate in SB buffer. Develop the colour reaction by adding SB buffer containing NBT and BCIP (add 30 μL NBT and 33 μL BCIP solution to 10 mL SB buffer). Dark colour spots should emerge after a few hour (over night)

incubation of the membrane. Stop the reaction by rinsing the membrane with water and dry the membrane.

6. Add equal amount of deionized formamide to the probe and store at −20 °C.

3.4 Hybridisation and Washing

1. Prepare hybridisation solution. You need 100–200 µL hybridisation solution per slide depending on the number and size of the sections. Prepare a little more than needed to account for losses.
2. Denature 2–4 µL probe (containing 50% formamide) per slide at 80 °C for 5 min. Cool down on ice and centrifuge briefly to minimise losses due to evaporation (see note 20). Add the probe to the hybridisation solution and mix well and keep at room temperature.
3. Apply the hybridisation solution with the probe directly onto the slide and cover it with a sterilized coverslip (see note 21).
4. Hybridisation is carried out in a closed environment saturated with 50% formamide/2× SSC. Prepare a plastic box with blotting paper in the bottom soaked with 50% formamide/2× SSC. Place the slides on a horizontal support (e.g. glass plate) inside the plastic box. Make sure that the slides do not touch the blotting paper. Close the box and seal with clingfilm or Parafilm. Incubate the slides at 50 °C overnight. Prepare the washing solution, 50% formamide/2× SSC, for the next day and put it also at 50 °C.
5. Prepare enough washing solution for four washes.
6. After hybridisation, place the slides into a holder (or a Coplin jar) and wash them with the washing solution for 30 min at 50 °C. Be sure that the coverslips or Parafilm have fallen off after the first wash. Immerse the slides in fresh washing solution and incubate at 50 °C for 60 min. Repeat the procedure. (see Table 4).
7. Immerse the slides in NTE buffer, pre-warmed to 37 °C. Repeat in fresh buffer. Incubate slides in NTE containing 20 µg mL^{-1} RNase A at 37 °C for 30 min (see note 22). Rinse the slides in NTE for 5 min and transfer the rack to washing solution (50% formamide/2× SSC) for 60 min at 50 °C. Dip the slides into 1× SSC for 2 min then into TBS twice for 5 min each time. Slides are now ready for the detection step.

3.5 Hybridisation Detection

1. Incubate the slides in their rack in 0.5% Blocking reagent (Roche) in TBS (see note 23) for 30 min. Incubate the slides for 30 min in the second blocking solution containing also 1% BSA and 0.3% Triton-X 100.
2. Add anti-DIG-alkaline phosphatase, F'ab fragments (Roche) (1:2,000) to the required amount of second blocking solution; you need 0.5 mL per slide. Place

Table 4 Post-hybridisation protocol

Treatment	Time	Temperature
50% formamide/2× SSC	30 min	50 °C
50% formamide/2× SSC	60 min	50 °C
50% formamide/2× SSC	60 min	50 °C
NTE	5 min	37 °C
NTE	5 min	37 °C
NTE/RnaseA	30 min	37 °C
50% formamide/2× SSC	60 min	50 °C
1× SSC	2 min	RT
TBS	5 min	RT
TBS	5 min	RT
TBS/blocking reagent	30 min	RT
TBS/1% BSA/0.3% Triton X 100	30 min	RT
TBS/1% BSA/0.3% Triton X 100/anti-DIG antibody	90 min	RT
TBS	5 min	RT
TBS	5 min	RT
TBS	5 min	RT
TBS	5 min	RT
SB	5 min	RT
SB	5 min	RT
SB/NBT/BCIP	Undefined	RT

the slides on a support and put them into moist plastic or glass chamber (see note 24) on a tray. Apply the antibody solution onto the slides. Cover the chamber and incubate for 90 min at room temperature (see note 25).

3. Stop the reaction by washing the slides (transferred to a rack) at least four times in TBS for 5 min.

4. Equilibrate the slides in SB buffer for few minutes.

5. Develop the colour reaction by adding SB buffer containing NBT and BCIP (add 30 μL NBT and 33 μL BCIP solution to 10 mL SB buffer) (see note 26). Observe the slides after 2–24 h. Stop the reaction by rinsing the slides with water and air dry.

6. If a strong background develops, it may be possible to reduce it by washing the slides in a graded EtOH series (see note 27).

7. To make the tissue structure visible, either the slides can be viewed with Nomarski optics or it is necessary to counterstain the sections. For the latter, dip the slides for <1 min in 0.25% Alcian blue in 3% acetic acid. The slides should be monitored for the intensity of staining. Rinse the slides for <1 min in water. Tissue having no hybridisation signals should show a faint blue staining. Alternatively, the structure of the tissue can be revealed by staining with 0.1% Calcofluor. Air-dry the slides.

8. Cover the section with a coverslip using mounting solution (DPX), about 100–200 μL per slide. Leave the slides to dry for a few hours. Now the sections are ready for data recording. Until that time, store the mounted slides in the dark.

3.6 Primary and Secondary Antibodies for Immunolocalisation

1. In situ detection of viral proteins by immunohistochemistry can be a quicker and simpler assay for the location of the virus infection. This is achieved by first incubating the sections with a specific antibody raised against the target virus protein. After washing, binding of the specific antibody is detected indirectly using a reporter molecule attached to a secondary antibody, e.g. commercially available alkaline phosphatase-conjugated anti-rabbit immunoglobulins).
2. To carry out immunohistochemical detection follow the in situ hybridisation protocol from the beginning to step 2 of Sect. 3.2. Transfer the slides from saline to TBS incubate for 5 min.
3. Move the slides in blocking solution (5% powder milk (see note 28), 0.1% Tween 20 in TBS) and incubate at room temperature for 30 min with gentle agitation.

3.7 Immunoreaction, Washing and Detection

1. Prepare an incubation chamber as described in step 2 of Sect. 3.5. Add the specific antibody to the blocking solution (see note 29). Spread about 500 µL of the antibody solution on each slide and incubate in the covered box > 90 min.
2. Wash the slides for four times in TBS, 5 min each.
3. Add the secondary antibody (see note 30) to TBS (1:2,000), spread about 500 µL per slide and incubate in the covered box > 90 min.
4. Wash the slides for four times in TBS 5 min each and transfer to SB buffer. Develop the colour reaction, wash, counter stain and cover slides as described for in situ hybridisation (from step 4 of Sect. 3.5).

3.8 Data Recording

1. Images at the required magnification (see note 31) viewed through a microscope are best recorded with a digital camera.
2. Sections stained with Alcian Blue can be viewed under bright field microscopy. Sections stained with Calcofluor should be viewed using a combination of fluorescence (calcofluor staining of the cell walls) and bright field (blue in situ staining) microscopy. These images often have an orange background.

3.9 Ultrastructural In Situ Hybridisation

1. We (AJM and K. Findlay, unpublished data) have used in situ hybridisation as a technology to detect pea seed borne mosaic virus RNA at the ultrastructural

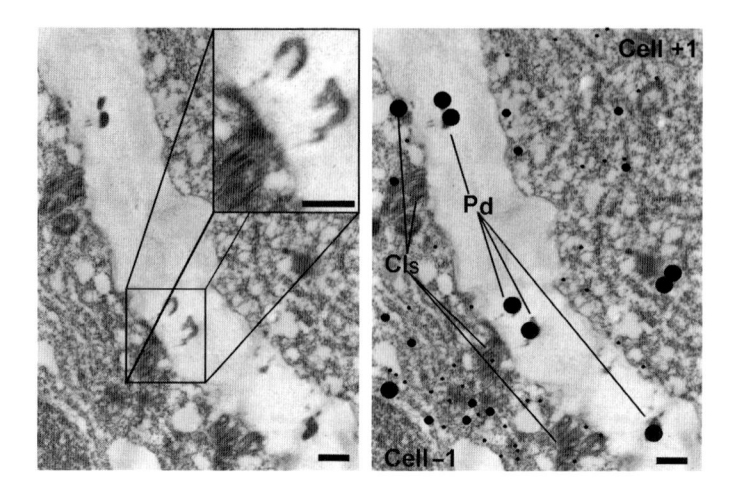

Fig. 2 EM micrograph of cells precisely at the extreme boundary of infection of a pea cotyledon with *Pea seed borne mosaic virus*. The section was probed with a negative sense RNA hybridisation probe and the reaction developed use anti-digoxigenin tagged with 10 nm gold particles. The *left hand image* is the unannotated micrograph of the interface between the last heavily infected cell (−1), identified by the presence of potyvirus cylindrical inclusions on the plasma membrane adjacent to plasmodesmata, the newly infected cell (+1), identified by positive labelling for + ve sense RNA even though virus specific inclusions are not yet visible. *Inset*: Magnified view of this micrograph to show gold particles associated with plasmodesmata and CIs. The *right hand image* has been annotated to show the relative location of the respective cells, location of CIs attached to the cell wall and plasmodesmata (Pd), and the location of viral + ve strand RNA (*large dots* – >10 gold particles; *intermediate dots* – 5–10 gold particles; *small dots* – <5 gold particles). Bars = 500 nm

level in pea cotyledonary tissues (Fig. 2). The technology is based upon the fixation and embedding of material for electron microscopy and the processing of thin sections by subjecting them to the procedures for in situ hybridisation. The brief methods described below assume a familiarity with the procedures for fixation, sectioning and viewing tissues and sections for electron microscopy.

2. Pea cotyledons were fixed (see note 32) in 2% (v/v) glutaraldehyde/0.2% (v/v) saturated aqueous picric acid and embedded in the acrylic resin, LR White, according to the protocol described by Leitch et al. (5) and modified from the protocol by McFadden et al. (6). After UV polymerisation of the resin, ultrathin sections (100 nm) were collected on carbon and pyroxylin-coated gold grids.

3. RNA probes for hybridisation were prepared as in Sect. 3.3. (see note 33)

4. Sections supported on pyroxylin and carbon-coated, gold grids were pre-hybridised for 1 h on drops of the hybridisation solution minus the probe. Grids were then hybridised overnight with a final concentration of approximately 0.5 ng μL^{-1} kb^{-1} of probe in 50% formamide, 10% dextran sulphate, 3 mM NaCl, 100 mM PIPES, 10 mM EDTA, 2.5 mg mL^{-1} blocking tRNA (modified from the protocol described in Leitch et al. (5)).

5. After hybridisation, the grids were washed as follows: two washer for 20 min in 2× SSC then one 20 min wash in 0.1× SSC followed by a further 45 min wash in 0.1× SSC. All the hybridisation and washing steps to this point were carried out at 42 °C.

6. For detecting the hybridised RNA, grids were equilibrated in PBS for 15 min then blocked in 1% acetylated bovine serum albumin (BSA-C) (Aurion, The Netherlands)/PBS/0.1% Tween 20 for 1 h before being incubated with anti-DIG antibody conjugated to gold particles (British BioCell International; see note 34) at a 1:100 dilution in 0.1% BSA-C/PBS/0.01% Tween 20 for 4 h.

7. After two washes in PBS for 20 min the grids were rinsed in water for 10 min before staining with 2% uranyl acetate and 1% lead citrate. All these steps were carried out at room temperature.

8. Grids were viewed using a transmission electron microscope.

4 Notes

1. One factor worthy of comment is the possibility that the in situ hybridisation techniques as described may detect small interfering RNAs (siRNAs) produced during all virus infections as a consequence of the action of RNA silencing in viral double stranded RNAs. Although we have not tested this formally, the similarity of the hybridisation conditions for sections (described here) and those used for the detection of siRNAs on filters means that the detection of siRNAs remains a possibility. This is significant as it has been shown that siRNAs can act in a non-cell-autonomous fashion and can move 10–15 cells away from their site of production.

2. This is best prepared in a screw-top bottle (eg. Duran type) in a fume cupboard. Take 50 mL of water and using a solution of 5 M KOH adjust to pH > 12. Add 4 g paraformaldehyde and heat gently to 60 °C on a heating plate. Shake vigorously for about 30 s, release the pressure every 5–10 s. The paraformaldehyde should dissolve completely, although very slight cloudiness is acceptable. Cool it on ice. Bring the pH back down to 7 using H_2SO_4 (Do not use HCl as this releases a carcinogen). Then add 2× or 10× PBS stock and water to a final composition of 4% paraformaldehyde in 1× PBS in 100 mL. Add 0.1 mL of Triton-X 100.

3. Different commercial waxes have different properties, most obviously in their different melting temperatures. For example, Polyester wax (BDH), Paraplast X-tra (Sigma) or Paraplast plus (Sigma) have MPs of 37, 50–54 and 56–57 °C, respectively. The lower MP waxes can make embedding and block preparation easier. They also have the potential advantage of helping to preserve antigenicity if the material is to be used also for immunocytochemistry. For polyester wax, the temperatures for embedding and section drying should also be lowered correspondingly. In our view, the major deciding factor is the quality of tissue preservation. Compact tissues (e.g. pea seeds; 7) are successful with polyester wax, whereas better tissue preservation of thin leaf pieces is achieved with Paraplast plus. Two disadvantages of polyester wax are that it removes Eosin Y from stained tissues making them less visible in wax blocks and sectioning can be difficult if the ambient temperature is high.

4. Prepare the desired amount of hybridisation solution. For 1 mL add 100 µL 10× salts buffer (3 M NaCl, 0.1 M Tris-HCl pH 6.8, 0.1 M NaHPO4 buffer pH 6.8, 50 mM EDTA), 500 µL deionized formamide, 200 µL 50% dextran sulphate, 10 µL 100 mg tRNA, 10 µL 100× Denhardt's solution, water. The volume of the probe usually does not alter significantly the concentration of hybridisation solution. If you wish to use higher volume of probe you should take account of the dilution factor in the preparation of hybridisation solution.

5. Since in situ hybridisation depends upon the detection of RNA using RNA hybridisation probes, every effort should be made to avoid contamination with RNases. Hence, the working environment should be clean and tidy and an adequate supply of clean, nuclease free tubes,

bottles, etc. should be established at the beginning of the experiment. All aqueous solutions should be sterilised and preferably aliquotted for single usage. Because of its hazardous nature, we do not favour the use of diethyl pyrocarbonate (DEPC) for the removal of RNases, although DEPC-treated water can be purchased commercially.

6. The choice of plant tissue critically influences the subsequent steps in the process for fixation and embedding. Logically, the larger the tissue piece the longer is the period required for fixation and embedding. However, for fixation, incubation at 4 °C overnight in PFA has proved to be good for small (<2 mm^2) and large (~25 mm^2) leaf pieces and even for half pea seeds (7). FAA fixative is attractive for the speed of fixation and ease of tissue infiltration. However, although widely suitable it has not proven to give optimal results in all cases. This should be determined comparitively. For reasons that remain unclear we do not know of anyone successfully performing in situ hybridisation on the lamina tissues from mature *Arabidopsis* leaves. Reports of unsuccessful attempts point to difficulties in tissue preservation (although we have found this not to be a major problem) or excessive background signals. We are confident that these problems should be surmountable.

7. When the tissues are fully infiltrated they should sink below the surface and appear 'water-soaked'.

8. Air bubbles appear when ethanol is mixed with water. This can reduce the efficiency of the embedding procedure. It is advisable, therefore, to de-gas (under vacuum) all water/ethanol mixtures before use.

9. Histoclear ((National Diagnostics, Atlanta, Georgia) or Roti-clear (Roth, Germany) is a non-toxic organic oil extracted from oranges. Nevertheless, it has as strong odour and is best handled in a fume hood. Since the oil can dissolve plastic, only glass containers should be used for stages including Histoclear/ Roti-clear.

10. If large numbers of replicate leaf samples are to be processed, it is possible to 'block' a stack of leaf pieces. Sections taken at right angles to the stack result in many leaf cross-sections for each section from the block.

11. For successful sectioning, the block (on a suitable mount for the microtome) should be trimmed so that the upper and lower faces are parallel. Traditionally, a trapezoid shape was recommended. Practically, it probably makes little difference and a square or rectangular block works as well. Some practice is required to perfect the sectioning technique. Repeated sectioning should lead to the formation of ribbons of sections. These are useful because they provide the potential to identify and process consecutive sections with parallel hybridisation or hybridisation and immunodetection treatments. In this case, the trapezoid shape is advantageous as it demarcates the serial sections. To maximize the numbers of sections on each slide, it is preferable to work with as small a block face as is practicable.

12. Wax sections need to be 'stretched' before adhesion to the glass slide. Sections should be lifted onto a layer of de-gassed water on a slide held on a warmed flat plate (40–42 °C). Once the section has stretched, drain away the excess water and leave the slide until the section has dried onto the slide.

13. The use of stainless steel or solvent-resistant plastic racks for moving slides between the solvents during in situ hybridisation is advisable since this significantly facilitates the handling of the slides. The size of the rack determines the volume of working solutions.

14. For economy, the second Histoclear wash can be retained and used as the first wash in the next experiment.

15. For the acetylation step work under fume hood. To prepare acetylation buffer add 1.25 mL triethanolamine and 0.5 mL HCl to 98.25 mL water and stir. Add 0.5 mL acetic anhydride to triethanolamine buffer and stir vigorously. Since the acetic anhydride is very unstable in water, it has to be added just before using. Fresh triethanolamine buffer and acetic anhydride must be used if you have a second rack of slides.

16. Alternatively, a PCR product can also be used directly as the template DNA. In this case, the PCR anti-sense primer includes the T7 RNA polymerase promoter sequence. This approach allows the easy preparation of template DNA.

17. It is advisable to check the quality and amount of linearized template DNA in a pilot in vitro RNA transcription reaction, replacing the DIG-labelled nucleotide with UTP. Check the RNA product on a RNase-free agarose gel (see note 18).

18. For agarose gel electrophoresis, denature the RNA sample by adding an equal vol FDE and incubate for 5 min at 65 °C, cool on ice and load onto the gel; the gel contains ethidium bromide. The gel is run between 80 V and 100 V and the quality of the RNA samples can be visualized using a UV trans-illuminator.

19. The optimal size for facilitating the efficient penetration of the probe into the tissue is between 150 and 300 bases. The following equation describes the general rule for alkaline-hydrolysis: time of hydrolysis = $(L_0-L_f)/K(L_0)(L_f)$, where L_0 = transcript length (Kb), L_f = desired length (kb) and K = ~0.11 breaks min^{-1} kb^{-1}. Add 50 μL 2× carbonate buffer ((80 mM $NaHCO_3$, 120 mM Na_2CO_3, pH 10.2) to 50 μL in vitro transcription reaction mix and incubate at 60 °C for the desired length of time. Precipitate the RNA using 250 μL EtOH in presence of 10 μL 4 M Na-Acetate and 5 μL 10% acetic acid by cooling in dry ice or −70 °C freezer for 30 min. Collect the RNA pellet by centrifugation, wash twice with 70% EtOH for 5 min, dry and resuspend the pellet in 30–50 μL sterile water.

20. The precise amount of probe needed for any specific target needs to be determined empirically. Since viruses usually accumulate to high level in the tissue, less probes may be needed to get satisfactory results.

21. If the size of sections is relatively large it can be difficult to cover the slide without having bubbles around and on the sections. To avoid this problem, cut Parafilm similar to the coverslip in size and use it to cover the sections.

22. This is to remove unhybridised RNA probe.

23. Blocking reagent should be made fresh. Dissolve at 60 °C with continuous stirring. The solution will remain turbid.

24. A dampened paper towel in the bottom of the box works well.

25. To avoid the antibody solution draining from the slides, level the container using a spirit level.

26. To develop the colour reactions put the slides into a chamber similar to that used for the antibody incubation, and cover them individually with about 1 mL substrate solution. Remove slides one by one from equilibrating SB buffer and immediately apply the substrate solution since after drying it can be difficult to spread the liquid.

27. Wet slides in water. Pass through them an EtOH series of 70, 95 and 100% EtOH series for 5 min and repeat the process in reverse direction. The washing time depends on the intensity of the signal and background.

28. Commercially available skimmed milk powder is perfect for this purpose.

29. The specific antibody should be diluted depending on its quality (between 1:100 and 1:2,000). The usage of recombinant antigens, instead of purified proteins from virus-infected plants, for production of specific antibodies will help to eliminate the undesired background binding.

30. Dilute secondary antibody according to the manufacturer's instruction. The specificity of the secondary antibody depends on the organism used for the production of the primary specific antibody (eg. rabbit).

31. To assist in calculating the final magnification of the image, it is useful to take a picture of a scaled graticule or graduated slide (e.g. a Bürker cell used for counting cell number (protoplast, etc.) at the same magnification.

32. These conditions were found to be satisfactory for pea cotyledons but have not been tested for a range of other tissues.

33. Control probes should also be prepared. For RNA viruses, strongest hybridisation would be expected using a negative sense probe, while a positive sense probe should detect the location of virus replication. To distinguish between specific hybridisation and 'background' a probe solution lacking any DIG-labelled RNA or, preferably, a completely heterologous probe (e.g. non-plant sequences) should be used.

34. Figure 2 shows results using a secondary antibody tagged with 10 nm gold particles. These small particles are difficult to visualise in light micrographs. The use of 15 or 20 nm gold particles is recommended.

Acknowledgements   Kim Findlay is acknowledged for advice on the methodology for in situ hybridisation for EM. Z. H. was supported by Hungarian Scientific Research Fund (OTKA; K61461) and Bolyai Janos Fellowship. The John Innes Centre is grant-aided by the United Kingdom Biotechnology and Biological Research Council.

References

1. Havelda, Z. and Maule, A. J. (2000) Complex spatial responses to cucumber mosaic virus infection in susceptible *Cucurbita pepo* cotyledons. *Plant Cell* **12**, 1975–86.
2. Escaler, M. Aranda, M. A. Thomas, C. L. and Maule, A. J. (2000) Pea embryonic tissues show common responses to the replication of a wide range of viruses. *Virology* **267**, 318–25.
3. Wang, D. and Maule, A. J. (1995) Inhibition of host gene expression associated with plant virus replication. *Science* **267**, 229–31.
4. Jackson, D. P. (1992) *In situ* hybridisation in plants, *in* Molecular Plant Pathology: A Practical Approach, (Bowles, D.J., Gurr, S.J. and McPhereson, M., eds.), Oxford University Press, Oxford, England,
5. Leitch, A. R. Schwarzacher, T. Jackson, D. and Leitch, I. J. (1994). *In Situ* Hybridisation. RMS Microscopy Handbooks 27. Bios Scientific Publishers.
6. McFadden, G. I. Bonig, I. Cornish, E. C. and Clarke, A. E. (1988). A simple fixation and embedding method for use in hybridisation histochemistry of plant tissues. *Histochem. J.* **20**, 575–85.
7. Wang, D. and Maule, A. J. (1994) A model for seed transmission of a plant virus: genetic and structural analyses of pea embryo invasion by pea seed-borne mosaic virus. *Plant Cell* **6**, 777–87.

Chapter 15
Detection of siRNAs and miRNAs

Sakari Kauppinen and Zoltán Havelda

Abstract Small RNAs such as small interfering RNAs (siRNAs) and microRNAs (miRNAs) play crucial roles in establishing general host defense mechanisms against viral infections in plants and the development of disease symptoms. Understanding these fundamental processes requires the sensitive and specific detection of small RNA species. However, because of the small size of miRNAs and siRNAs, their detection is technically demanding. Here, we describe methods for robust and sensitive detection of small RNAs by Northern blot analysis and in situ hybridization.

Keywords siRNA; miRNA; Detection; Northern blot; In situ hybridization; LNA

1 Introduction

Post-transcriptional gene silencing (PTGS) has been identified as an antiviral defence system in plants (1). Double-stranded (ds) RNA or highly structured RNA, which is generated during viral replication, serve as inducers of PTGS (1, 2), thereby mediating the processing of these RNA precursors to ca. 21–26 nucleotide small interfering RNAs (siRNAs). In turn, the siRNAs guide the cleavage of their cognate target RNAs (3). To evade this robust host defense mechanism, plant viruses generally encode silencing suppressors (4). Regulation of endogenous gene expression is mediated by another class of 21 nucleotide small RNAs, called microRNAs (miRNAs) (5). Recent data show that virus-induced disease development can be associated, at least partly, by interference with miRNA-mediated regulation (6). Investigation of small RNA-mediated processes in virus-infected plants is the prerequisite for understanding the relationship between viruses and their host plants. Northern blot analysis combined with polyacrylamide gel electrophoresis is widely used to examine the expression of (si)miRNAs, since it allows both quantitation of the expression levels as well as determination of the RNA size. A major drawback of using conventional DNA oligonucleotide probes in miRNA detection is their poor sensitivity, especially when monitoring expression

From: *Methods in Molecular Biology, Vol. 451, Plant Virology Protocols:*
From Viral Sequence to Protein Function
Edited by G.D. Foster, I.E. Johansen, Y. Hong, and P.D. Nagy © Humana Press, Totowa, NJ

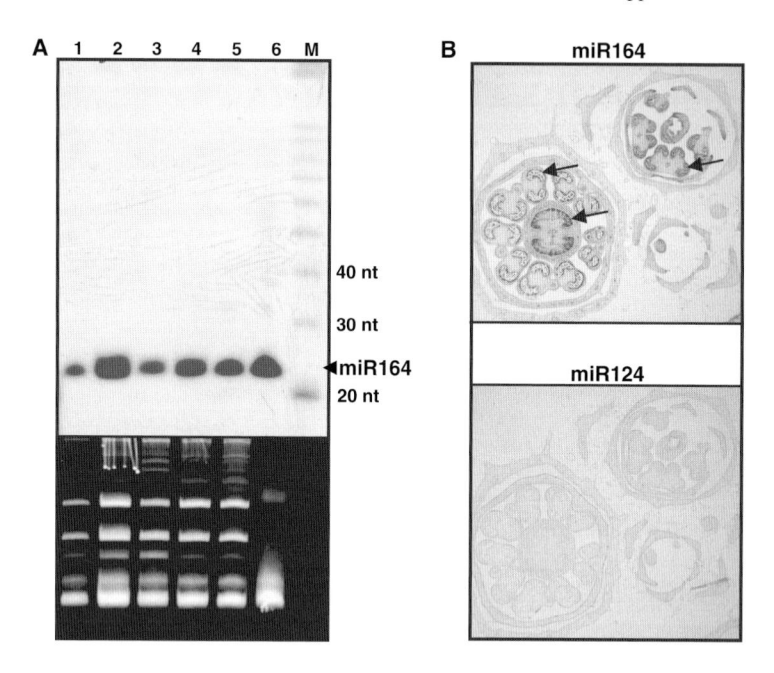

Fig. 1 Detection of miR164 by Northern blot analysis and in situ hybridization. (**a**) Total RNA samples (15 μg per sample) from *N. benthamiana* flowers (1), *A. thaliana* young leaves (2), fully developed leaves (3), young flowers (4), mature flowers (5), siliques (6) were electrophoresed in 12% polyacrylamide gels under denaturing conditions, blotted and hybridized with ³²P-labelled LNA oligonucleotide probe detecting miR164 (*upper panel*). After 5 h of hybridization at 50 °C, the membrane was washed in 2× SSC at 65 °C for 10 min, then 0.1 SSC 65 °C for 5 min. To eliminate the unspecific background, the membrane was treated with RNase A digestion and then washed 0.1 SSC 65 °C for 5 min and exposed overnight. M: Decade™ Marker (Ambion). The gel loading controls are shown from ethidium bromide staining (*bottom panels*). (**b**) Cross sections of *N. benthamiana* young flowers were hybridized with an LNA-modified oligonucleotide probe detecting miR164 (*upper panel*). *Arrows* indicate the sites of miR164 accumulation. An LNA-modified oligonucleotide probe specific for mouse miR-124 was used as negative control (*bottom panel*)

of low-abundant miRNAs. Consequently, a large amount of total RNA per sample is required for Northern analysis, which is not feasible when the cell or tissue source is limited. Here, we describe detection of viral siRNAs, based on radiolabelled RNA probes. In addition, we report a new method for sensitive and specific detection of miRNAs using locked nucleic acid (LNA)-modified probes. The LNA-based detection method shows dramatically improved sensitivity of at least one order of magnitude, while simultaneously being highly specific (7). Finally, we describe the use of LNA probes for in situ detection of iRNAs in tissue sections Figure 1 (8).

2 Materials

2.1 RNA Extraction

1. Trizol (Invitrogen) or TRI Reagent (Sigma)
2. Liquid nitrogen (optional)
3. Chloroform
4. Isopropyl alcohol
5. 70% ethanol
6. Agarose
7. 10× TBE buffer: 0.9 M Tris, 0.9 M boric acid, 0.02 M EDTA, pH 8.0
8. FDE: 10 mL deionised formamide, 200 µL 0.5 M EDTA (pH 8.0), 10 mg xylene cyanol, 10 mg bromphenol blue
9. Ethidium bromide solution: 10 mg mL^{-1} in water

2.2 Denaturing PAGE and Capillary Transfer

1. Penguin Electrophoresis System or equivalent
2. 12% acrylamide solution: 24 mL 40% acrylamide/bis (19/1) solution, 8 mL 10× TBE, 10 mL water and 40 g urea, water
3. TEMED (Sigma)
4. Ammonium persulphate (APS): Prepare 10% solution in water and immediately freeze in aliquots for single use at −20 °C
5. FLS: 10 mL deionised formamide, 200 µL 0.5 M EDTA (pH 8.0), add xylene cyanol and bromphenol blue to get a faint blue solution. High concentration of dye can interfere with separation of small RNA species
6. Nytran N membrane (Schleicher and Schuell, Germany) or Hybond-N + membrane (Amersham Pharmacia Biotech)
7. 20× SSC: 3 M NaCl 175.3 g, 0.3 M Sodium Citrate-2H$_2$O, pH 7.0

2.3 Marker and Oligonucleotide Probe Preparation

1. Synthetic RNAs
2. LNA-modified oligonucleotide (Exiqon)
3. T4 polynucleotide kinase (Fermentas)
4. [γ -^{32}P]ATP

2.4 RNA Probe Preparation

1. Restriction endonuclease, T4 DNA ligase (Fermentas)
2. Phenol

3. Chloroform
4. T7 RNA polymerase (Fermentas)
5. Ribonuclease inhibitor
6. Ribonucleotid (rNTP) mixture: 5 mM ATP, 5 mM CTP, 5 mM GTP and 0.5 mM UTP
7. α^{32}P-UTP

2.5 Hybridization

1. Hybridization solution: 50% deionised formamide, 0.5% SDS, 5× SSPE, 5× Denhardt's solution and 20 μg mL^{-1} sheared, denatured salmon sperm DNA
2. RNase A stock solution 10 mg mL^{-1}
3. RNase A buffer: 0.5 M NaCl, 10 mM TRIS, pH:7.5, 1 mM EDTA, 20 μg mL^{-1} RNase A

2.6 In Situ Hybridization

1. DIG Oligonucleotide 3 -End Labelling Kit (Roche)
2. TS buffer: 0.15 M NaCl, 0.1 M TRIS, pH 7.5
3. TSB buffer: Dissolve 0.5% Blocking Reagent (Roche) in TS buffer at 65 °C
4. SB buffer: 0.1 M NaCl, 0.05 M MgCl$_2$ 0.1 M TRIS, pH 9.5
5. NBT: 50 mg mL^{-1}, BCIP: 50 mg mL^{-1}
6. Anti-DIG-alkaline phosphatase, Fab fragments (Roche)
7. Solution and reagents for the in situ hybridization process are described previously (9)

3 Methods

3.1 RNA Extraction

1. Intact, high quality RNA is the prerequisite for the reliable detection of small RNAs. Use TRIZOL (Invitrogen) or TRI Reagent (Sigma) for purifying RNA samples. This will ensure that the samples will be essentially free of genomic DNA and have adequate quality. Do not use minicolumn-based purification methods for RNA extraction since these may result in loss of the small RNA fraction. The RNA extraction should be carried out according to the manufacturer's instructions. Here, we briefly describe the RNA extraction method using the TRI Reagent (see note 1).

2. Freeze plant tissues in liquid nitrogen and grind them to a powder using mortar and pestle, add 400 μL of TRI Reagent during the homogenization. Alternatively, to avoid use of liquid nitrogen, ice cold mortar and pestle can be used for homogenization of plant tissue. Add 600 μL of TRI Reagent to the sample and pipette into a 1.5 mL test tube.

3. Shake the tubes for 15 s and incubate at room temperature for 5 min to permit dissociation of nucleoprotein complexes.

4. Add 200 μL of chloroform and shake manually for 15 s.

5. Centrifuge the sample for 15 min at 4 °C.

6. Following centrifugation (12,000 g) the aqueous phase is transferred to a 1.5 mL test tube containing 500 μL of isopropyl alcohol for RNA precipitation.

7. After 15 min of incubation at room temperature, the RNA is pelleted by centrifugation for 10 min at 4 °C. Wash the RNA pellet in 1 mL of 70% ethanol and centrifuge for 5 min. Air-dry the pellet (do not use vacuum dryer) and resuspend in 30–50 μL of sterile RNase-free H2O.

8. To assess the quality of the sample analyse 1–5 μL of purified RNA sample by electrophoresis in 1× TBE buffer using nuclease free 1.2% agarose gel containing ethidium bromide.

9. For agarose gel electrophoresis, agarose and autoclaved 1× TBE are mixed and heated in a microwave oven until boiling. The solution is allowed to cool to approximately 60 °C and ethidium bromide is added. Pour the solution into a gel holder.

10. Place the gel into an appropriate electrophoresis tank containing 1× TBE. Add water to the RNA sample to bring up the volume to 5 μL and mix with 5 μL of FDE. Denature the RNA sample in FDE for 10 min at 65 °C, cool on ice and load onto the gel.

11. The gel is run between 80 and 100 V and the quality of the RNA samples can be visualized in UV light by use of a UV trans-illuminator.

3.2   Preparation of Denaturing PAGE

1. The correct running of denaturing PAGE is critical for separation of small RNA species. The improperly run denaturing PAGE may result in the appearance of artificial bands. Since several types of equipments are available for PAGE, always follow the manufacturers' instructions.

2. Gel setup using Penguin Electrophoresis System, 20 × 20 cm^2 glass plate gel sandwich with 1.5 mm spacers. Clean the apparatus with detergent and rinse it thoroughly with autoclaved sterile distilled water.

3. Set up gel apparatus using 1.5 mm spacer (see note 3)

4. Prepare 80 mL of 12% acrylamide solution (see note 2). Warm up the solution shortly by using microwave oven to dissolve the urea. Cool down the solution to room temperature and bring up the volume to 80 mL with water.

5. After adding 480 μL of 10% APS, 32 μL of TEMED, pour the gel and allow it to polymerize at least for 1 h.

6. Carefully remove the comb (you need flat wells), assemble gel onto the apparatus and rinse wells thoroughly with 1× TBE using a syringe and a needle. No leaks should be observed.
7. Pre-run the gel at 400 V (40 mA) for 60 min. During the pre-run, the gel should warm up to ensure proper denaturing conditions.
8. Rinse wells again immediately before loading sample on the warm gel.

3.3 Sample Preparation and Gel Electrophoresis

1. Total RNA (1–100 μg) can be loaded into the wells depending on the abundance of the small RNA species of interest (see note 4)
2. Add the desired amount of RNA sample in 20 μL to 20 μL FLS (see note 5) and denature the sample at 65 °C for 20 min. Chill on ice for 1 min and spin down.
3. Load 10–20 μL of FDE in empty well; this will help monitor the electrophoresis. Load samples and run them at 200 V while they enter into the gel, then run at 400 V until the dye reaches the bottom of the gel. At this stage, xylene cyanol dye can be detected in the middle of the gel.
4. Dismantle the apparatus and soak the gel in 500 mL of 1× TBE containing ethidium bromide (2–5 μL from ethidium bromide stock solution to 100 mL of 1× TBE) for 10 min to stain tRNAs and 5S RNA for loading control (see note 6). Rinse gel in 1× TBE for 5–10 min to wash away the excess of ethidium bromide for visualisation.

3.4 Capillary Gel Transfer

1. Soak the gel in 20× SSC for 10 min. Transfer the RNAs to the membrane by capillary blotting (10) using 20× SSC (see note 7)
2. After blotting, rinse the membrane with 2× SSC for 2 min and let it dry. Using a pencil, mark the position of the bands on the top of the membrane.
3. Fix RNA on the membrane by ultraviolet cross-linking with the RNA sample side up.

3.5 Preparation of Radio-Labelled RNA Markers

1. It is necessary to run marker RNAs next to the small RNA samples, in order to correctly assess the size of the RNA species. Synthetic RNAs of different lengths (for example; 18, 21 and 25 nucleotides in length, see note 8) should be radio-labelled using T4 polynucleotide kinase (see probe preparation for miRNA detection, 3.7.1).

2. Remove the unincorporated nucleotides by filtration on a spincolumn centrifugation. Dilute to 100 μL and load 1 μL (or less) onto the gel and/or measure the radioactivity to determine the optimal amount of radio-labelled marker. Since RNA is very unstable, it is important to work on ice and use sterile tips and tubes during marker preparation.

3.6   Detection of Viral siRNAs

3.6.1   RNA Probe Preparation for Detecting Viral siRNAs

1. To detect virus-specific siRNAs, clone the full-length or part of the viral genome into a plasmid containing T7 promoter. You can use either labelled sense or antisense RNA transcript for detecting the siRNAs. The antisense probe will result in stronger signals, since it will detect siRNAs deriving from the secondary structures in the positive strand (in the case of plus-strand RNA viruses) (2).
2. Linearise the plasmid with the appropriate restriction enzyme (generating blunt or 5′ protruding end) and check the digestion by agarose gel electrophoresis. Then perform phenol extraction, ethanol precipitation (2.5 volume 100% ethanol and 1/2 volume 7.5 M NH4 OAc), and wash with 75% ethanol, dry, and resuspend in suitable volume of nuclease-free water.
3. Check the quality of DNA template in a 10 μL test in vitro transcription (as described in Step 4 using rUTP insted of α^{32}P-UTP) and analyse 3 μL of the RNA product on an agarose gel (as described for total RNA quality control in 3.1). The RNA product should be observed as a single band with little or no degradation products and about 5–20 times more intense than the template DNA band.
4. Set up the labelling reaction by adding 2 μL of 5× buffer (Fermentas), 0.2 μL of Ribonuclease inhibitor (8 U), the optimal amount of linearised DNA template (about 100–200 ng), 1 μL of rNTP mixture, 0.5 μL of T7 RNA polymerase (10 U), make up to 8 μL with water and 2 μL of α^{32}P-UTP (0.8 MBq). Incubate for 1 h at 37 °C. Place on ice.
5. Add 40 μL of 1× TE, remove unincorporated free nucleotides by microspin column purification and check for the incorporation.

3.6.2   Hybridization of RNA Probe

1. Prehybridize the fixed membrane in hybridization solution at 37 °C for 60 min.
2. Heat the labelled probe for 1 min at 95 °C and cool down in ice before addition to the hybridization solution. Perform hybridizations in the same solution at 37 °C.

3. After overnight hybridization, wash the membrane twice in 2× SSC, 0.1% SDS at 37 °C for 5 min.
4. Expose the membrane to evaluate the signals. Do not allow the membrane to dry (store moist in saranwrap).
5. If stringent wash is necessary, gradually increase the temperature (or lower the salt concentration) and check the membrane by counter and/or by exposure. It is always a good idea to include a negative control sample on the membrane (eg. a RNA sample from mock-inoculated plants).

3.7   MicroRNA Detection

3.7.1   Preparation of Radio-Labelled Oligonucleotide Probe for miRNA Detection

1. Conventional DNA oligonucleotide probes can be used for miRNA detection, but the sensitivity and specificity of radio-labelled DNA probes is often not sufficient. Here, we describe a method, based on the use of LNA-modified probes, which show dramatically improved sensitivity of at least one order of magnitude, while simultaneously being highly specific.
2. Radiolabel LNA-modified oligonucleotide (Exiqon), complementary to the target microRNA, by combining 10 pmol of LNA-modified oligonucleotide probe with 1 µL of 10× T4 polynucleotide kinase buffer and bring up the volume to 8.5 µL with water. Add 0.5 µL of T4 polynucleotide kinase, 1 µL of $[\gamma$-^{32}P]ATP (0.4 MBq) and incubate the reaction at 37 °C for 60 min. Add 40 µL of 1× TE to the reaction.
3. Check the labelling efficiency in spot test by dispensing 0.5 µL labelled probe on a marked piece of membrane, leave it dry and UV cross-link. It is not necessary to remove the unincorporated free nucleotide if you are using this probe checking method.
4. Wash in 0.1× SSC, 0.1% SDS at 65 °C for 15 min. Expose and check the labelling efficiency (see note 9).

3.7.2   Hybridization of LNA-Modified Oligonucleotide Probe

1. Prehybridize as described for RNA probe at 40–50 °C (see note 10) for 30 min.
2. Heat the labelled LNA probe for 1 min at 95 °C and cool down in ice, before addition to the hybridization solution.
3. Hybridize the membrane from 4 h up to over night at 50 °C.
4. Wash the membrane twice in 2× SSC, 0.1% SDS at 40–50 °C for 5 min. If you use more than one probe wash the membranes separately to avoid cross-hybridization of the different LNA probes in the washing solutions.
5. Expose the membrane to check the signals. Do not allow the membrane to dry.
6. If stringent wash is required, wash twice at high stringency in 0.1 SSC, 0.1% SDS at 65 °C for 5 min.

3.7.3 Removal of Unspecific Background

1. The use of some LNA probes may result in undesired background hybridization.
2. To eliminate the background treat the hybridized membrane with 50–100 mL RNase A buffer containing RNAse A, at 37 °C for 30 min. Shake gently. This will remove the imperfect duplexes formed during hybridization. (Never allow the membrane to get dry after hybridization. The RNase A-digested membrane is not reusable.)
3. Wash the membrane again at high stringency in 0.1× SSC, 0.1% SDS at 65 °C for 5 min.

3.8 Stripping the Northern Filters

1. Wash filter in 100 mL of 0.1% SDS, 5 mM EDTA at 95 °C for 30–60 min with gentle shaking. Avoid wrinkling of the filter and do not let it dry out before complete stripping of blot.
2. Expose the filter to ensure that the probe has been stripped completely.
3. Abundant miRNAs require longer treatment at boiling temperature. New, unstripped Northern blots are the best for detecting new miRNAs.

3.9 Detection of miRNAs by In Situ Hybridization

3.9.1 Labelling of LNA-Modified Oligonucleotides for In Situ Hybridization

1. Label the LNA-modified oligonucleotide using the DIG Oligonucleotide 3 -End Labelling Kit (Roche), according to the manufacturer's instructions. It is not necessary to purify the probe after labelling.
2. Remove 0.5 μL (for probe checking) from the 20 μL reaction and add 20 μL deionised formamide.
3. To check the quality of the probe, spot the 0.5 μL aliquot (and the labelled control oligonucleotide provided by the kit) on a piece of membrane and UV cross-link.
4. Put the membrane in TS buffer for 2 min and treat the membrane in TSB buffer for 30 min.
5. Remove TSB and add 5 mL TS buffer containing Anti-DIG-alkaline phosphatase, Fab fragments (Roche) 1:2,000. Hybridize with gentle shaking at 37 °C for 30 min.
6. Wash at least three times in TS buffer and transfer to SB buffer for 2 min.
7. Develop the colour reaction by adding SB buffer containing NBT and BCIP (add 30 μL NBT and 33 μL BCIP solution to 10 mL SB buffer), stop the reaction by rinsing the membrane with water and dry the membrane.
8. Two to six microlitres of formamide-containing probe solution is used per slide in 100–200 μL hybridization solution.

3.9.2 In Situ Hybridization

1. LNA-modified oligonucleotide probes are well suited for detection of miRNAs in tissue sections. miRNA in situ hybridization works well in *Nicotiana benthamiana*. However, in situ detection of miRNAs in *Arabidopsis thaliana* is technically more demanding. We describe here a method, based on the fixation of tissue samples in 4% formaldehyde, which gives good results in *N. benthamiana* and also works in *A. thaliana* but is less efficient. In *A. thaliana*, empirical tests may be necessary to determine the optimal conditions (fixative, hybridization conditions, etc.) for successful in situ hybridization. The in situ hybridization method is based on a previously described protocol (9) and only the critical steps are indicated.
2. Apply 5–10 µm cross and longitudinal sections onto poly-L-lysine microscopic slides.
3. Add 2–6 µL labelled LNA-modified oligonucleotide probe to 100–200 µL hybridization solution and hybridize overnight at 50–60 °C.
4. Perform wash at 50–60 °C (depending on the temperature of hybridization) in three times 1× SSC/50% formamide (Wash the slides having different probe separately in the first wash to avoid cross-contamination of probes.) During the washing process, the slides are RNaseA-treated in RNaseA buffer at 37 °C for 30 min.
5. Depending on the signal intensity (few hours to two days), stop the colour reactions by rinsing the slides with distilled water.
6. The slides are further washed through ethanol dilution series to eliminate unspecific signals and counter stained with Alcian Blue (Fluka).

4 Notes

1. Total RNA can also be extracted using the traditional phenol/chloroform method (11). However, samples prepared in this way contain significant amounts of genomic DNA and sometimes cause problems if you want to load high amounts of RNA for gel electrophoresis.
2. You can adjust the acrylamide content (about 8–15%) of the gel depending on the required resolution.
3. Thinner spacer also can be used. Modify the running condition according the size of the spacer.
4. Usually it is not necessary to further purify the small RNA fraction for Northern blot analyses. However, if the target small RNA is low-abundant, the small RNA fraction can be separated from higher molecular weight RNA species and concentrated. The easiest and most reliable way to purchase small RNA purification kit such as for example the flashPAGE™ Fractionator (Ambion).
5. The volume of the sample can be reduced to about 10 µL (5 µL RNA sample and 5 µL FLS) depending on the concentration of the sample and size of the wells. Smaller volume of loaded sample may result in sharper bands.
6. Alternatively, to check the loading the membrane can be stripped and rehybridized with probes detecting for example 5S rRNA.

7. Alternatively, you can carry out the transfer with a semi-dry transfer cell according the manufacturers' instruction.
8. Alternatively, you can purchase a RNA marker ladder and use it according the manufacturers' instructions (for example Decade™ Markers (Ambion).
9. The probe can be also checked by removal of the unincorporated free nucleotide with spin columns followed by checking the incorporation of the radio-labelled nucleotide.
10. When using LNA-modified probes, the hybridization temperature can be increased up to 60 °C. This can increase the specificity of hybridization.

Acknowledgements Z. H. was supported by Hungarian Scientific Research Fund (OTKA; K61461) and Bolyai Janos Fellowship.

References

1. Baulcombe, D. (2004) *Nature* **431,** 356–63.
2. Szittya, G., Molnar, A., Silhavy, D., Hornyik, C., and Burgyan, J. (2002) *Plant Cell* **14,** 359–72.
3. Herr, A. J. (2005) *FEBS Lett* **579,** 5879–88.
4. Silhavy, D. and Burgyan, J. (2004) *Trends Plant Sci* **9,** 76–83.
5. Chen, X. (2005) *FEBS Lett* **579,** 5923–31.
6. Chapman, E. J., Prokhnevsky, A. I., Gopinath, K., Dolja, V. V., and Carrington, J. C. (2004) *Genes Dev* **18,** 1179–86.
7. Valoczi, A., Hornyik, C., Varga, N., Burgyan, J., Kauppinen, S., and Havelda, Z. (2004) *Nucleic Acids Res* **32,** e175.
8. Valoczi, A., Varallyay, E., Kauppinen, S., Burgyan, J., and Havelda, Z. (2006) *Plant J* **47,** 140–51.
9. Jackson, D., P. (1992) *in* "Molecular Plant Pathology: A Practical Approach", pp. 163–74., Oxford University Press, Oxford, England.
10. Sambrook, J., Fritsch, E. F., and Maniatis, T. (1989) Molecular cloning: a laboratory manual, Cold Spring Harbor Laboratory Press, Cold Spring Harbor, N.Y.
11. White, J. L., and Kaper, J. M. (1989) *J Virol Methods* **23,** 83–93.

Chapter 16
Cloning of Short Interfering RNAs from Virus-Infected Plants

Thien X. Ho, Rachel Rusholme, Tamas Dalmay, and Hui Wang

Abstract During their infection in plants, viruses can form double stranded (ds) RNA structures. These dsRNAs can be recognized by plants as "aberrant" signals and short interfering RNA (siRNA) molecules of 19–25 nt will be produced with sequences derived from the viral source. Knowledge about antiviral siRNA profiles including siRNA size, distribution, polarity, etc. provides valuable insights to plant–virus interactions. In this chapter, we describe a simple method for cloning siRNA from virus-infected plants. This protocol includes isolation of small RNAs, their ligation to a pair of 5′ and 3′ adapters, RT-PCR/PCR amplification, and subsequent concatamerization before pGEM-T cloning and sequencing. Concatamers containing as many as 15 small RNA inserts can be produced. This protocol has successfully been applied to leaf materials of monocots and dicots infected with poty-, carmo-, and sobemo-viruses.

Keywords RNA silencing; siRNA cloning; siRNA profile; Plant–virus interaction

1 Introduction

RNA silencing is an ancient defense mechanism of many eukaryotic organisms triggered by intracellular presence of double stranded RNA (dsRNA). This leads to homology-dependent degradation of the dsRNA sequence to protect the host genomes against aberrant endogenous or exogenous RNA sequences. The phenomenon was discovered in 1990 (1, 2) and termed as posttranscriptional gene silencing (PTGS) in plants. The process involves the recognition and cleavage of dsRNA by a ribonuclease III like enzyme termed DICER into 19–25 nt duplex termed short interfering RNA (siRNA) (3–5). The siRNA duplex represents both polarities and has two nucleotide 3′ overhangs with 5′ phosphate and 3′ hydroxyl groups (6, 7). One strand of the siRNA duplex is then selected to be incorporated into a RNA-induced silencing complex (RISC) (8). The single stranded siRNA guides the complex to search for and degrade perfectly complementary RNA sequences, while mediating translation repression of partially complementary targets (3, 5, 6, 8–13).

From: *Methods in Molecular Biology, Vol. 451, Plant Virology Protocols:*
From Viral Sequence to Protein Function
Edited by G.D. Foster, I.E. Johansen, Y. Hong, and P.D. Nagy © Humana Press, Totowa, NJ

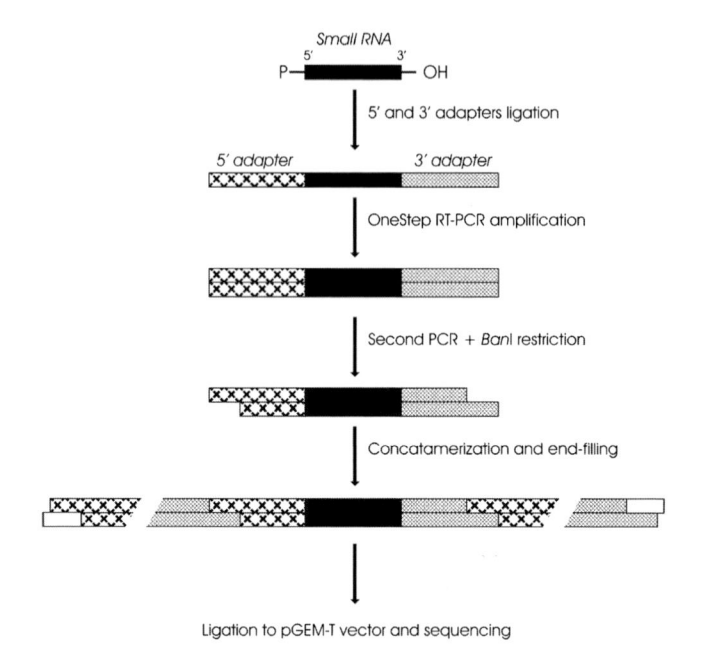

Fig. 1 The siRNA cloning protocol

So far the only reliable approach for obtaining a siRNA profile is through the cloning and sequencing of small RNAs isolated from the total RNA of a virus-infected plant (14). Therefore, an efficient siRNA cloning and sequencing method would provide significant benefit to siRNA research. In this chapter, we describe a simplified method (Fig. 1) to clone siRNA from virus-infected plants. With this method, siRNA profiles of both monocots and dicots can be obtained with relative ease.

2 Materials

Viral infections need to be confirmed by virus detection techniques e.g., RT-PCR/ PCR, ELISA, etc. Samples can be processed directly after harvest or from −70 °C storage. Young symptomatic leaf tissues are preferred because they can be ground easily and yield better RNA quantity and quality than other materials. Old yellow tissues must not be used as they will greatly reduce the RNA quality. As the case of RNA work, tissues must be handled carefully, especially from sample harvest to RNA isolation. If the samples are stored at −70 °C, frozen tissues are not allowed to thaw before RNA isolation. After being removed from the freezer, samples must be kept completely frozen on dry ice and directly transferred to prechilled mortar and pestle containing liquid nitrogen. Normally 2 g of infected leaves are sufficient for total RNA isolation.

Conventional chemicals and equipments for molecular biology work are needed, e.g. absolute and 70% ethanol, RNase-free water, 20 mg mL^{-1} glycogen, Hyperladder I, IV, and V molecular weight markers (Bioline), 1× TAE, 1× TBE, 25:24:1 (v/v/v) phenol:chloroform:isoamyl alcohol, 0.3 M and 5 M NaCl, agarose, agarose gel running apparatus, RNase-free eppendorf tubes, micropipettes and tips, sharp razor blades, homogenizer, thermal cycler, temperature-controlled water bath, UV transilluminator, bench top refrigerated microcentrifuge, etc. More specific chemicals and equipments are listed in details in the following sections (see note 1).

2.1 Small RNA Isolation

2.1.1 Total RNA Isolation

1. Tri Reagent (Sigma-Aldrich), kept at 4 °C
2. Chloroform
3. Isopropanol
4. Mortar and pestle
5. 50 mL polypropylene screw cap tube (Nalgene)
6. Liquid nitrogen
7. Beckman JA-20 rotor
8. Any Beckman centrifuge (e.g., J2-HC), which can accommodate Beckman JA-20 rotor (see note 2)

2.1.2 Isolation of 19–25 nt Small RNA from Total RNA

1. RNA loading buffer: 5 mM EDTA, 0.1% bromophenol blue, 95% formamide
2. Urea
3. 40% polyacrylamide, kept at 4 °C
4. 10% ammonium persulfate (freshly made)
5. 5× TBE
6. TEMED
7. 19 and 24 nt RNA markers (Table 1)
8. Mini-protean 3 cell system (Bio-Rad), for running 10-well 15% denaturing polyacrylamide gel (80 × 73 × 1 mm^3)
9. Gel staining solution: 100 mL of 1× TBE with 10 µL of 10 mg mL^{-1} ethidium bromide

Table 1 Oligonucleotides used in siRNA cloning

Name	Sequence	Concentration	Notes
5′ adapter	5′-N6accctcttggcacccactAAA-3′	100 pmol	Uppercase: RNA; lowercase: DNA; N6: amino 6-carbon spacer, for blocking the 5′ end
3′ adapter	5′-UUUaccaggcacccagcaatgN3-3′	100 pmol	Uppercase: RNA; lowercase: DNA; N3′: amino modifier, for blocking the 3′ end
19 nt RNA size marker	5′-CCUGGCUACCCCAAGCACA-3′	100 pmol	
24 nt RNA size marker	5′- ACUAGCCUAUCCUAGAAGAGAU CC -3′	100 pmol	
Forward primer	5′-ACCCTCTTGGCACCCACTAAA-3′	100 pmol	Underline indicates *Ban*I site
Reverse primer	5′-CATTGCTGGGTGCCTGGTAAA-3′	100 pmol	Underline indicates *Ban*I site

2.2 Amplification of Small RNA Sequences

2.2.1 Ligation of the 5′ and 3′ Adapters to Small RNA

1. 5′ and 3′ adapter oligonucleotides (Table 1)
2. T4 RNA ligase and 10× T4 RNA ligase buffer (Amersham)
3. 50% (v/v) aqueous dimethyl sulfoxide (DMSO)
4. 0.1% (w/v) bovine serum albumin (BSA)

2.2.2 RT-PCR Amplification of the Adapter-Linked Small RNA

1. OneStep RT-PCR Kit (Qiagen)
2. PCR primers (Table 1)
3. Electroelution: molecular porous membrane-tubing (Spectrum Laboratories Inc., Spectra/Por Membrane MWCO: 3,500)

2.2.3 Secondary Large Scale PCR Amplification of the 65 bp Product

1. REDTaq DNA polymerase and 10× REDTaq PCR reaction buffer (Sigma)
2. dNTP Mix (dATP, dCTP, dGTP, dCTP, 25 mM each, pH 7.0) (Bioline)
3. PCR primers (Table 1)

2.3 Concatamerization of the PCR Products

2.3.1 *Ban*I Digestion

1. 20 U µL^{-1} *Ban*I restriction endonuclease and 10× NEB buffer 4 (New England Biolabs)

2.3.2 Concatamerization

1. 400 U µL^{-1} T4 DNA ligase and T4 DNA ligase buffer (New England Biolabs)
2. QIAquick Gel Extraction kit (Qiagen)

2.3.3 End-Filling

1. 1 U µL^{-1} REDTaq DNA polymerase and 10× REDTaq PCR reaction buffer (Sigma)
2. dNTP Mix (dATP, dCTP, dGTP, dCTP, 25 mM each, pH 7.0) (Bioline)
3. QIAquick PCR Purification kit (Qiagen)

2.4 Cloning the DNA Concatamers to pGEM-T Easy Vector and Sequencing

1. pGEM-T Easy vector system I (Promega)
2. JM109 competent cells, > 10 °cfu µg^{-1} (Promega)
3. T7 and SP6 primers
4. REDTaq DNA polymerase and 10× REDTaq PCR reaction buffer (Sigma)
5. dNTP Mix (dATP, dCTP, dGTP, dCTP, 25 mM each, pH 7.0) (Bioline)
6. Chemicals and equipments for standard cloning and sequencing: Luria Broth agar, ampicillin, Petri plates, X-gal, IPTG, ABI Terminator Kit

3 Methods

3.1 Total RNA Isolation

1. If a fresh sample is used as starting material, process them as soon as possible (at no more than 2 h after harvest) to minimize RNA degradation. If −70 °C frozen sample is used, keep it on dry ice before grinding to avoid thawing (see note 3).

2. Cut the sample into small pieces in a prechilled mortar with liquid nitrogen. Grind it carefully in liquid nitrogen.

3. After grinding, immediately add 40 mL of Tri Reagent into the mortar and homogenize the mixture (see note 4). Transfer the mixture to a 50 mL polypropylene tube and leave at room temperature (RT) for 5 min.

4. Centrifuge the homogenate at 12,000 g (or 10,000 rpm using the JA-20 rotor) for 10 min at 4 °C to precipitate the insoluble material.

5. Transfer the clear supernatant to a fresh tube and allow the liquid to stand for 5 min at RT.

6. Add 8 mL of chloroform to the tube, cover it tightly, shake vigorously for 15 s and allow it to stand for 15 min at RT.

7. Centrifuge the tube at 12,000 g for 15 min at 4 °C. Carefully transfer the colorless upper aqueous phase to a fresh tube without disturbing the red phase at the bottom of the tube. Add 20 mL of isopropanol to the aqueous mixture.

8. Vortex the tube briefly and allow it to stand for 10 min at RT, then centrifuge at 12,000 g for 10 min at 4 °C. The RNA precipitate will form a white pellet on the side and the bottom of the tube.

9. Dissolve the pellet in 150 μL of RNase-free water. Measure the quality and quantity of the RNA solution by calculating the A260/A280 ratio, and proceed immediately to Step 3.2 or store it at −70 °C for subsequent use. By this method, 2 g of infected leaves normally yield ~300 μg of total RNA.

3.2 Isolation of 19–25 nt Small RNA from Total RNA

1. Prepare 15% polyacrylamide gel containing urea:

   a. Slightly heat 4.2 g of urea in 3 mL of RNase-free water until the urea is dissolved.

   b. Add 3.74 mL of 40% polyacrylamide, 0.5 mL of 5× TBE, 100 μL of freshly made 10% ammonium persulfate.

   c. Add 5 μL of TEMED and immediately but gently stir the solution. Load the solution onto the Mini-protean 3 cell system (80 × 73 × 1 mm^3) with 10-well comb.

   d. Leave the gel to set at RT for 30–60 min.

2. Dilute the total RNA (see Sect. 3.1) in 150 μL of RNA loading buffer. Denature the mixture at 90 °C for 30 s before loading onto the 15% polyacrylamide-urea gel. The RNA molecular weight markers (19 and 24 nt) are needed in every gel.

3. Run the gel in 1× TBE for 2–3 h at 120 V until the bromophenol blue dye reaches the bottom of the gel.

4. Carefully remove the gel from the apparatus, stain it by shaking in ethidium bromide solution (TBE) at RT for 5 min, and visualize the RNA in the gel using a UV transilluminator (Fig. 2) (see note 5).

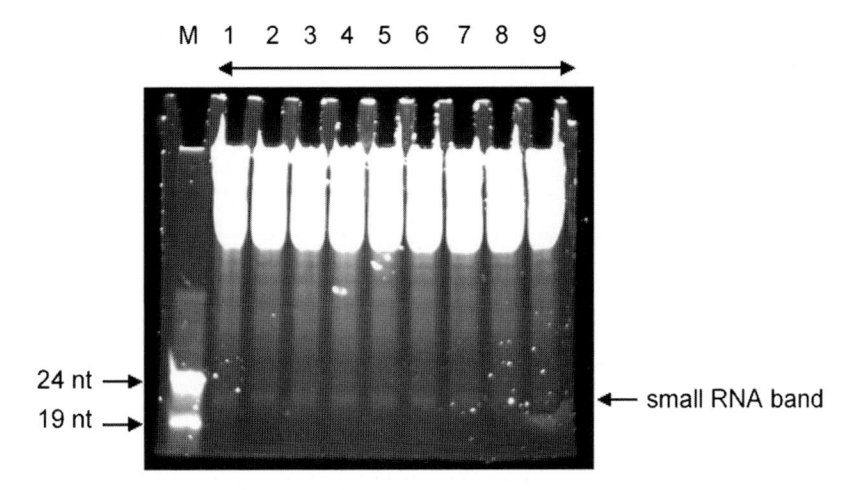

Fig. 2 Polyacrylamide gel electrophoresis of total RNA from a virus-infected plant. The small RNA is visible as a distinct band within 19–24 nt region. *Lane M*: 19 and 24 nt molecular weight markers; *lanes 1–9*: total RNA

5. Excise the section containing 19–25 nt RNA band, cut it to small pieces and elute the small RNA in a 1.5 mL eppendorf tube containing 500 μL of 0.3 M NaCl at 4 °C overnight. Carefully transfer the liquid to a new eppendorf tube (see note 6).
6. Perform ethanol precipitation to the oligonucleotides:
   a. Add three volumes of absolute ethanol and 0.2 μL of 20 mg mL^{-1} glycogen into the tube and centrifuge at 13,000 rpm in a refrigerated microcentrifuge at 4 °C for 15 min.
   b. Discard the liquid, add 1 mL of ice-cold 75% ethanol into the tube and centrifuge at 13,000 rpm at 4 °C for 15 min.
   c. Remove the ethanol and dry the pellet at RT for 3 min.
7. Dissolve the pellet in 7 μL of RNase-free water.

3.3 Amplification of Small RNA Sequences

3.3.1 Ligation of the 5′ and 3′ Adapters to Small RNA

1. Prepare the ligation mix as follows: 7 μL of small RNA preparation (see Sect. 3.2.), 1 μL of 5′ adapter, 1 μL of 3′ adapter (Table 1), 2 μL of 10× RNA ligase buffer, 6 μL of 50% DMSO, and 2 μL of BSA (see note 7).
2. Mix all the components of the reaction and give the tube a heat shock at 90 °C for 30 s and immediately chill on ice for 20 s. Add 1 μL of T4 RNA ligase to the reaction, mix gently and incubate at 37 °C for 2 h.
3. Add 37 μL of RNase-free water and 3 μL of 5 M NaCl to the reaction.

4. Perform phenol:chloroform:isoamyl alcohol extraction:

   a. Add one volume of 25:24:1 phenol:chloroform:isoamyl alcohol to the RNA solution and vortex vigorously for 10 s and centrifuge at 7,000 rpm for 5 min.
   b. Carefully transfer the top aqueous phase containing the RNA to a new tube.
   c. If a white precipitate is visible at the aqueous/organic interface, repeat the extraction again.

5. Ethanol-precipitate the adapted small RNA product (see Sect. 3.2) and dissolve it in 5 μL of RNase-free water.

3.3.2 RT-PCR Amplification of the Adapter-Linked Small RNA

1. Use OneStep RT-PCR Kit for reverse transcription and small scale PCR amplification. Heat shock 5 μL of adapter-linked RNA at 90 °C for 30 s. Set up the RT-PCR reaction as: 5 μL of RNA sample, 29 μL of RNase-free water, 10 μL of 5× Qiagen OneStep RT-PCR buffer, 2 μL of dNTP mix (provided in the kit), 1 μL of forward primer, 1 μL of reverse primer (Table 1), and 2 μL of Qiagen OneStep RT-PCR Enzyme Mix.
2. Perform the reaction at 50 °C for 30 min, 95 °C for 15 min, 30 cycles of PCR (94 °C for 1 min, 50 °C for 1 min, and 72 °C for 1 min), and a final extension period of 10 min at 72 °C.
3. Run the PCR product on a 3.5% agarose gel in 1× TBE with Hyperladder V as molecular weight marker (Fig. 3a). There will be two PCR bands: the lower one of ~45 bp is the product of 5′ and 3′ adapters ligation without small RNA insert, and the upper band of ~65 bp contains the adapted single small RNA insert. Excise the ~65 bp band using a sharp razor blade.
4. Perform electroelution to the RT-PCR product:

   a. Soak 10 cm of molecular porous membrane-tubing (see Sect. 2.2.2) for 30 min in distilled water.
   b. Tightly tie off one end of the tubing with a plastic clip.
   c. Slide the gel slice into the tubing and fill the tubing with 1× TBE buffer, use another clip to seal the top of the tubing.
   d. Place the sealed tubing in a horizontal gel tank and run in 1× TBE buffer for 1 h at 110 V.
   e. After electroelution is complete, reverse the polarity of the electrodes and run at 110 V for 20 s, carefully open one end of the tubing and collect TBE buffer that contains the 65 bp DNA.

5. Perform one phenol:chloroform:isoamyl alcohol extraction.
6. Ethanol-precipitate the DNA, dissolve it in 30 μL of distilled water, and store it at −20 °C. If any subsequent steps fail, reamplification can be carried out from this 65 bp DNA (see note 8).

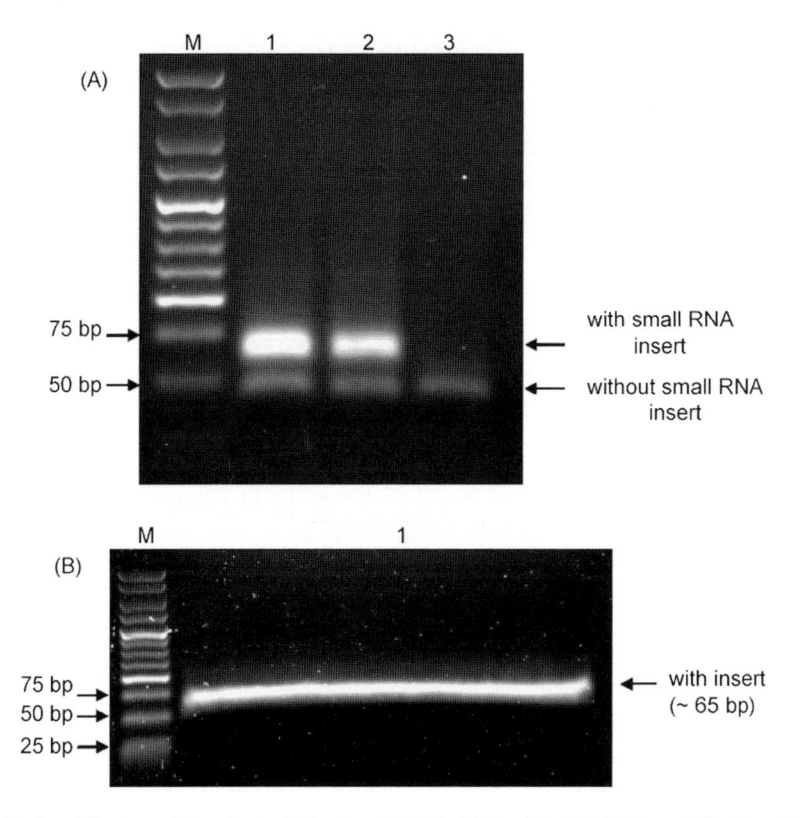

Fig. 3 Amplification of the adapter-linked small RNA. (**a**) OneStep RT-PCR amplification of the small RNA. The desired band having small RNA insert is approximately 65 bp and gel-purified from the lower band (~45 bp) that is amplified as the adapters only without a small RNA insert. *Lane M*: Hyperladder V molecular weight marker; *lanes 1, 2*: OneStep RT-PCR products of small RNA; *lane 3*: negative control (adapters only). (**b**) Secondary PCR amplification of the ~65 bp DNA product. *Lane M*: Hyperladder V molecular weight marker; *lane 1*: PCR product containing small RNA inserts

3.3.3 Secondary Large Scale PCR Amplification of the 65 bp Product

1. Set up the secondary PCR reaction: 5 µL of ~65 bp product, 100 µL of 10× REDTaq PCR reaction buffer, 20 µL of dNTP mix, 10 µL of forward primer, 10 µL of reverse primer, 830 µL of distilled water, and 25 µL of REDTaq DNA polymerase.
2. Perform the reaction at 94 °C for 2 min, followed by 30 cycles of PCR (94 °C for 1 min, 50 °C for 1 min, and 72 °C for 1 min), and a final extension at 72 °C for 10 min.

3. Run the PCR product on a 3.5% agarose gel in 1× TBE with Hyperladder V as molecular weight marker (Fig. 3b). There will be only one DNA band at ~65 bp position containing the adapter-linked small RNA inserts.
4. Excise the ~65 bp band, electroelute the DNA, and perform one phenol:chloroform:isoamyl alcohol extraction.
5. Ethanol-precipitate the DNA, and dissolve the DNA in 87 μL of distilled water.

3.4 Concatamerization of the PCR Products

3.4.1 *Ban*I Digestion

1. Set up the digestion reaction: 87 μL of the ~65 bp PCR product, 10 μL of NEB buffer and 4 and 3 μL of *Ban*I endonuclease.
2. Perform the reaction at 37 °C for 18 h in a temperature-controlled water bath.
3. Run the digested product on a 3.5% agarose gel in 1× TBE with Hyperladder V as molecular weight marker (Fig. 4a). There will be only one DNA band at ~45 bp position containing the digested DNA, excise the ~45 bp band, electro-elute the DNA, and perform one phenol:chloroform:isoamyl alcohol extraction.
4. Ethanol-precipitate the DNA, and dissolve the DNA in 86 μL of distilled water.

3.4.2 Concatamerization of the *Ban*I Digested Product

1. Set up the reaction as follows: 86 μL of the *Ban*I digested product, 10 μL of the T4 DNA ligase buffer, and 4 μL of T4 DNA ligase.
2. Perform the reaction at RT overnight.
3. Run the DNA on a 2.5% agarose gel in 1× TBE (Fig. 4b). Concatamers will be observed as smears from 45–1,000 bp region. Excise the DNA smear between 400–1,000 bp and extract the DNA using Qiagen Gel Extraction kit.
4. Dissolve the DNA in 100 μL of distilled water.

3.4.3 End-Filling the Concatamerized Product for pGEM-T Cloning

1. Set up the end-filling reaction: 100 μL of concatamerized DNA, 4 μL of dNTP mix, 50 μL of 10× REDTaq PCR reaction buffer, 333.5 μL of distilled water, and 12.5 μL of REDTaq DNA polymerase.
2. Gently mix and incubate the reaction mixture at 72 °C for 30 min, purify the DNA using Qiagen PCR Purification Kit.
3. Dissolve the DNA in 30 μL of distilled water.

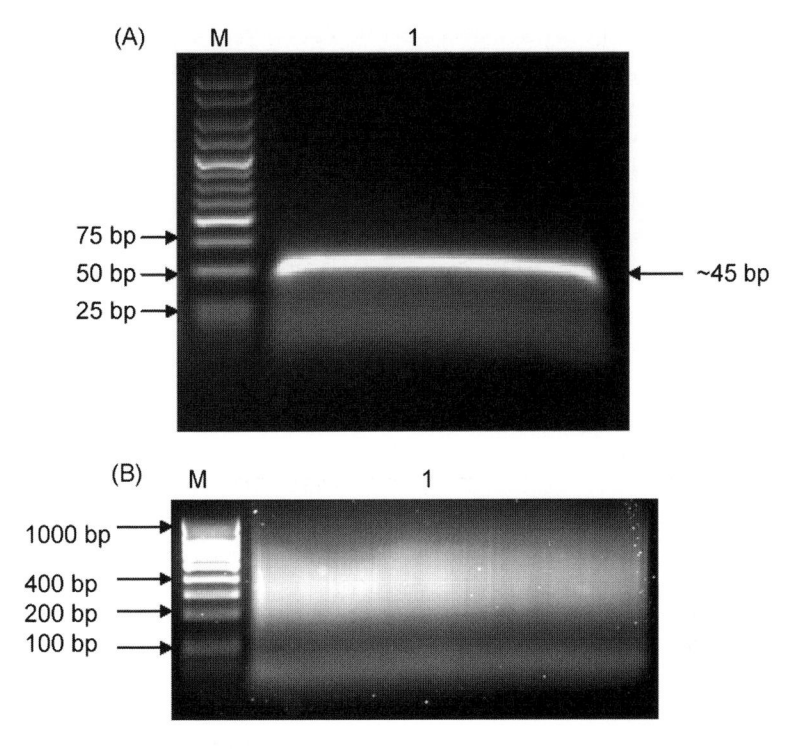

Fig. 4 Concatamerization of the PCR product. (**a**) *Ban*I digested PCR product to reveal the sticky ends. *Lane M*: Hyperladder V molecular weight marker; *lane 1*: *Ban*I digested DNA (~40–45 bp). (**b**) Concatamerization of the PCR products. There is a smear between 45–1,000 bp region indicating successful ligation reaction. DNA product of 400–1,000 bp is excised for cloning and sequencing. *Lane M*: Hyperladder IV molecular weight marker; and *lane 1*: DNA Concatamer smear

3.5 Cloning the DNA Concatamers to pGEM-T Vector and Sequencing Analysis

1. Clone the end-filled DNA into the pGEM-T plasmid using the pGEM-T Easy system I following the manufacturer's protocol except for incubating the ligation reation at RT overnight to maximize ligation efficiency.
2. Transform the plasmid to JM109 competent cells following the manufacturer's protocol and screen the white bacterial colonies using PCR with T7 and SP6 primers (Fig. 5).
3. Grow the positive bacterial colonies in 5 mL of Luria Broth in the presence of ampicillin. Extract and sequence the plasmid using T7 (or SP6) primer following standard protocols.
4. Use BioEdit program (15) to do BLAST-search for the newly obtained small RNA library against the viral genome sequences to identify siRNAs with 100% sequence homology, and to map the siRNAs to the virus genome (14).

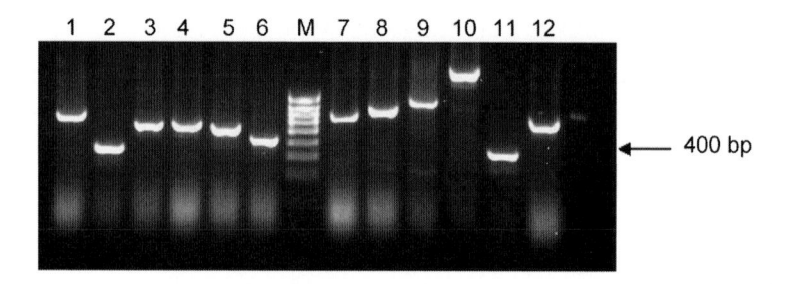

Fig. 5 PCR screening of the white bacteria colonies having small RNA inserts. Only plasmids having inserts of 400–1,000 bp are sequenced. *Lanes 1–12*: PCR products of different bacterial clones; *lane M*: Hyperladder IV molecular weight marker

4   Notes

1. All equipments and chemicals used before RT-PCR (inclusive) should be RNase-free.
2. Before the total RNA isolation, switch on the Beckman centrifuge, install the JA-20 rotor and set the temperature at 4 °C.
3. During RNA isolation, frozen samples should not be thawed before adding Tri Reagent.
4. The Tri Reagent contains phenol so that the total RNA extraction should be done in a fume hood.
5. Extract the small RNAs after visualizing the small RNA band localized with the 19–24 nt RNA size markers (Fig. 2). Absence of this band indicates low concentration of small RNA. More starting materials may be needed in this case.
6. If there is a smear instead of a distinct band in the range of 19–24 nt RNA size markers, do not continue as the RNA is degraded.
7. The 5′ and 3′ adapters should be stored in 10 μL aliquots to avoid repeated freezing and thawing.
8. Always try to store DNA samples after every step, especially the DNA product of the RT-PCR so that you can always reamplify the DNA if cloning is not successful in the first attempt.

References

1. van der Krol, A. R., Mur, L. A., Beld, M., Mol, J. N., and Stuitje, A. R. (1990) Flavonoid genes in petunia: addition of a limited number of gene copies may lead to a suppression of gene expression. *The Plant Cell* **2**, 291–299.
2. Napoli, C., Lemieux, C., and Jorgensen, R. (1990) Introduction of a chimeric chalcone synthase gene into petunia results in reversible co-suppression of homologous genes *in trans*. The *Plant Cell* **2**, 279–289.
3. Bernstein, E., Caudy, A. A., Hammond, S. M., and Hannon, G. J. (2001) Role for a bidentate ribonuclease in the initiation step of RNA interference. *Nature* **409**, 363–366.
4. Hamilton, A. J. and Baulcombe, D. C. (1999) A species of small antisense RNA in posttranscriptional gene silencing in plants. *Science* **286**, 950–952.
5. Zamore, P. D., Tuschl, T., Sharp, P. A., and Bartel, D. P. (2000) RNAi: double-stranded RNA directs the ATP-dependent cleavage of mRNA at 21 to 23 nucleotide intervals. *Cell* **101**, 25–33.

6. Elbashir, S. M., Lendeckel, W., and Tuschl, T. (2001) RNA interference is mediated by 21- and 22-nucleotide RNAs. *Genes and Development* **15,** 188–200.

7. Elbashir, S. M., Martinez, J., Patkaniowska, A., Lendeckel, W., and Tuschl, T. (2001) Functional anatomy of siRNAs for mediating efficient RNAi in *Drosophila melanogaster* embryo lysate. *The EMBO Journal* **20,** 6877–6888.

8. Hammond, S. M., Bernstein, E., Beach, D., and Hannon, G. J. (2000) An RNA-directed nuclease mediates post-transcriptional gene silencing in *Drosophila* cells. *Nature* **404,** 293–296.

9. Doench, J. G., Petersen, C. P., and Sharp, P. A. (2003) siRNAs can function as miRNAs. *Genes and Development* **17,** 438–442.

10. Hammond, S. M., Caudy, A. A., and Hannon, G. J. (2001) Post-transcriptional gene silencing by double-stranded RNA. *Nature Reviews Genetics* **2,** 110–119.

11. Aukerman, M. J. and Sakai, H. (2003) Regulation of flowering time and floral organ identity by a MicroRNA and its APETALA2-like target genes. *The Plant Cell* **15,** 2730–2741.

12. Chen, X. (2004) A MicroRNA as a translational repressor of APETALA2 in arabidopsis flower development. *Science* **303,** 2022–2025.

13. Hutvagner, G. and Zamore, P. D. (2002) RNAi: nature abhors a double-strand. *Current Opinion in Genetics and Development* **12,** 225–232.

14. Molnar, A., Csorba, T., Lakatos, L., Varallyay, E., Lacomme, C., and Burgyan, J. (2005) Plant virus-derived small interfering RNAs originate predominantly from highly structured single-stranded viral RNAs. *Journal of Virology* **79,** 7812–7818.

15. Hall, T. A. (1999) BioEdit: a user-friendly biological sequence alignment editor and analysis program for Windows 95/98/NT. *Nucleic Acids Symposium Series* **41,** 95–98.

Chapter 17
Solution Structure Probing of RNA Structures

Marc R. Fabian and K. Andrew White

Abstract Single-stranded RNA plant viruses not only code for viral proteins within their RNA genomes, they often maintain elaborate RNA secondary structures. These structures can be integral to a variety of viral processes, such as viral translation, genome replication, subgenomic mRNA transcription, and genome encapsidation. RNA secondary structures may function to recruit and bind *trans*-acting protein factors, or may become part of higher order tertiary RNA structures, which themselves may be functionally relevant. To fully understand such viral RNA elements and their mechanisms of action, it is necessary to first determine their secondary structures.

Computer. modeling based on free energy minimizing principles is generally used as an initial approach to predict potential RNA secondary structures in a sequence. The most popular program is *mfold*, which is available for free to the public at http://www.bioinfo.rpi.edu/applications/mfold/old/rna/form1.cgi (1). Though useful starting points, the *mfold*-predicted structures must be confirmed by more direct experimental approaches. Solution structure probing of RNA is a commonly utilized technique that provides information regarding the secondary structure of an RNA primary sequence in solution. This process involves treating the RNA of interest with enzymes or chemicals that modify RNA differentially based on its secondary structure. The modified sites, at either single- or double-stranded regions, can be subsequently identified by primer extension and gel electrophoresis. Data from solution structure probing experiments can be superimposed onto a computer-predicted structure to further help confirm or refine the predicted RNA secondary structure model.

Keywords Viral RNA; RNA genome; RNA elements; RNA secondary structure; stem-loop

1 Introduction

Solution structure probing is a valuable tool for mapping functionally important RNA secondary structures in single-stranded RNA virus genomes. This biochemical technique takes advantage of a barrage of ribonucleases and chemicals that

From: *Methods in Molecular Biology, Vol. 451, Plant Virology Protocols:*
From Viral Sequence to Protein Function
Edited by G.D. Foster, I.E. Johansen, Y. Hong, and P.D. Nagy © Humana Press, Totowa, NJ

cleave or modify an RNA sequence, respectively. Specifically, these enzymes and chemicals recognize different nucleotides within single-stranded or helical regions. Once cleaved or modified, RNA sequences can be analyzed by performing primer-extension reactions and separating the products by gel electrophoresis along with a sequencing ladder. Although many different ribonucleases and chemicals can be utilized, this chapter will outline only four in particular: the ribonucleases RNase T_1 and V_1, and the chemicals CMCT [1-cyclohexyl-3-(2-morpholinoethyl)carbodii mide metho-*p*-toluene sulfonate] and DEPC (diethylpyrocarbonate).

RNase T_1 cleaves phosphodiester bonds after the 3' phosphate of a single-stranded guanidine (and to a lesser extent single-stranded uridine). RNase V_1 generally cleaves nucleotides that exist within double-stranded or stacked regions of RNA structure, and produces a fragment with a 5' phosphate. CMCT reacts with single-stranded uridine at N-3 (and to a lesser extent guanidine at N-1). DEPC reacts with single-stranded adenine at N-7. Once cleaved or modified, treated RNAs can be analyzed via radioactive primer-extension reactions. The reverse transcriptase will extend on treated RNAs up to the point of cleavage or modification. In this chapter, we outline procedures that can be used to analyze RNA secondary structures present within the plus-sensed single-stranded RNA plant viruses.

2 Materials

2.1 RNA Solution Structure Mapping

1. 5× RNase T_1/V_1 Buffer: 125 mM Tris-HCl, pH 7.5, 25 mM $MgCl_2$, 0.5 mM EDTA, 1 M NaCl. Store in aliquots at−20 °C
2. Torula Yeast RNA (5 µg µl⁻¹) (Sigma)
3. RNase T_1 (0.05 U µl⁻¹) (GE Healthcare Life Sciences)
4. RNase V_1 (0.07 U µl⁻¹) (GE Healthcare Life Sciences)
5. 1-cyclohexyl-3-(2morpholinoethyl)carbodiimide metho-*p*-toluene (CMCT) (50 mg ml⁻¹). (Sigma Aldrich)
6. 10% (v/v) diethylpyrocarbonate (DEPC) (Sigma Aldrich)
7. 3 M NaOAc, pH 5.5
8. 0.5 M EDTA (ethylenediaminetetraacetic acid)
9. 10% (w/v) SDS (Sodium dodecyl sulphate)
10. Phenol:cholorform:isoamyl alcohol (PCI) (25:24:1) (v:v:v). Store at 4 °C. This solution is corrosive so use with caution.
11. 95% ethanol
12. 70% ethanol

2.2 Primer Extension Analysis

1. Superscript II 5x reverse transcriptase buffer (Invitrogen): 250 mM Tris-HCl, pH 8.3, 375 mM KCl, 15 mM $MgCl_2$

2. 1 mM dNTP mix (dCTP, dTTP, dGTP only)
3. 10 mM dNTP mix (dATP, dTTP, dCTP, dGTP)
4. 100 mM DTT (dithiothreitol)
5. [α-^{35}S]dATP > 1000 Ci mmol^{-1} (GE Healthcare Life Sciences). Use appropriate safety precautions when using radioactive materials
6. Superscript II reverse transcriptase (200 U μl^{-1}) (InVitrogen)
7. stop solution: 95% fromamide, 20 mM EDTA, 0.05% bromophenol blue, 0.05% xylene cyanol FF
8. 10 pmol μl^{-1} complementary primer oligonucleotide

2.3 DNA Sequencing Ladder

1. Sequenase Version 2.0 DNA Sequencing Kit (GE Healthcare Life Sciences)
2. [α-^{35}S]dATP > 1,000 Ci mmol^{-1} (GE Healthcare Life Sciences)
3. Stop solution (see Sect. 2.2)
4. 10 pmol μl^{-1} complementary primer oligonucleotide

2.4 Denaturing Polyacrylamide Gel Electrophoresis

1. 40% acrylamide/bis solution (29:1). Store at 4 °C. Caution, unpolymerized acrylalmide is a neurotoxin that can be absorbed through the skin (avoid contact and wear gloves)
2. 10X TBE (Tris, borate, EDTA) running buffer: 108 g Tris base, 55 g boric acid, 40 ml of 0.5 M EDTA, pH 8.0, in 1 l. Store at room temperature
3. Urea (electrophoresis grade)
4. 10% (w/v) Ammonium persulfate (APS): prepare 10% solution in water; store at 4 °C. Make fresh every 2 weeks
5. *N,N,N,N'*-Tetramethyl-ethylenediamine (TEMED, Sigma Aldrich). Store at 4 °C
6. SA sequencing gel system (BRL)

3 Methods

3.1 Preparation of RNA Samples

1. Mix 14 pmol of in vitro transcribed RNA with 3 μg yeast RNA, 5 μl of 5x RNase T_1/V_1 buffer and DEPC-treated ddH$_2$O in a total volume of 20 μl in a microfuge tube. Prepare one tube for each of the different treatments planned with the RNases or chemicals. Note, the chemical reactions are also carried out in the RNaseT$_1$/V$_1$ buffer.
2. To promote proper folding of the RNA, incubate the RNA mixtures at 65 °C for 2 min, followed by 37 °C for 10 min and then leave at 25 °C for 10 min (*see* note 1).

3. Prepare an extraction solution in microfuge tube for each RNA transcript tube: 10 μl of 3 M NaOAc, pH 5.5, 1 μl of 0.5 M EDTA, 1 μl of 10% SDS, 2 μl yeast RNA (5 μg μl^{-1}), 66 μl DEPC-treated ddH$_2$O, and 100 μl PCI (*see* note 2).

3.2   Treatment of RNA Samples

RNA sample treatments are outlined in the subheadings below. These treatments listed below have been designed to favor one-hit kinetics, where one molecule of RNA is modified or cleaved only once during the reaction. In general, one-hit kinetics is favored by lower concentrations of the enzymes and chemicals or shorter reaction times. Depending on the source and activity of the enzymes, it may be necessary to reoptimize reaction conditions.

3.2.1   Treatment of RNA with RNase T1

1. Add 1 μl of 0.025 U μl^{-1} RNase T$_1$ to RNA mixture following the annealing step. Incubate at 25 °C for 1 min.
2. For methods in Sects. 3.2 and 3.3, create a negative control where RNA is incubated for 1 min with 1 μl of ddH2O in place of RNase T$_1$ or V$_1$.

3.2.2   Treatment of RNA with RNase V1

1. Add 1 μl of 0.035 U μl^{-1} RNase V$_1$ to RNA mixture following the annealing step. Incubate at 25 °C for 1 min.

3.2.3   Treatment of RNA with DEPC

1. Add 2 μl of 10% (v/v) DEPC to RNA mixture following the annealing step. Allow it at room temperature for 15 min. Incubate at 25 °C for 15 min.

3.2.4   Treatment of RNA with CMCT

1. Add 2 μl of CMCT (50 mg ml^{-1}) and 2 μl of fresh 0.2 M NaOH to RNA mixture following the annealing step. Incubate at 25 °C for 15 min.
2. As a control to this reaction, create a parallel reaction containing 2 μl of 0.2 M NaOH and 2 μl of ddH$_2$O instead of CMCT. Incubate the control at 25 °C for 15 min.

3.3   Isolation of Treated RNA Transcripts

1. Following treatments, immediately add RNA mixture to microfuge tubes containing the prepared extraction solution.

2. Vortex for 30 s, and centrifuge at ~20,000 g for 2 min.
3. Remove the aqueous layer from each tube and transfer to a new microfuge tube containing 250 µl of 95% ethanol. Incubate at −80 °C for 60 min.
4. Centrifuge at ~20,000 g for 15 min, and wash the pellet (that is barely visible) with 150 µl of 70% ethanol.
5. Resuspend dried pellets in 8 µl of DEPC-treated ddH$_2$O.

3.4 Primer Extension of Treated RNA Transcripts

1. Assemble in a microfuge tube 1 µl of RNase/chemically-treated RNA with 0.5 µl primer (10 pmol µl) that is complementary to the RNA transcript ~50 bases downstream of the region that is to be analyzed.
2. Anneal the primer by heating the mixture for 1 min at 90 °C and allow it to remain at room temperature for 5 min. Then centrifuge it in a microfuge.
3. Add the following mixture to the tube: 2 µl of 5x reverse transcriptase buffer, 0.5 µl of 1 mM dNTP mixture (dCTP, dGTP, dTTP), 0.5 µl of 100 mM DTT, 1 µl (10 µCi) [α-^{35}S]dATP, 3 µl DEPC-treated ddH$_2$O, and 1 µl Superscript II reverse transcriptase (200 U µl^{-1}). Incubate the mixture at 42 °C for 5 min.
4. Add 0.5 µl of 10 mM dNTP mixture (dATP, dGTP, dCTP, dTTP), and incubate for an additional 20 min at 42 °C.
5. Stop the reaction by adding 7 µl of stop solution.

3.5 Manual DNA Sequencing Ladder Reaction

This procedure is conducted using a DNA sequencing kit made by GE Healthcare Life Sciences; however, sequencing kits from other companies can also be used. Follow the instructions provided with the specific kit to generate the sequence ladder using the same primer used above for primer extension. For the template, use a plasmid containing a cDNA insert of the viral sequence that spans the RNA sequence of interest (and also contains the primer binding site).

3.6 Polyacrylamide Gel Preparation

1. These instructions are for a BRL Life Technologies Model SA vertical polyacylamide apparatus.
2. Clean a set of glass sequencing plates with soap, then acetone, then isopropanol.
3. Prepare a 0.25-mm thick, 6% polyacrylamide gel by mixing in a 250 ml flask 22.5 g Urea, 4.5 ml 10x TBE, 9 ml 40% acrylamide (19:1 acrylamide:bis), and 15 ml ddH$_2$O. Heat flask at 37 °C for ~5 min to dissolve urea.

4. Before pouring the gel, add 180 μl of 10% (w/v) APS and 20 μl of TEMED to the acrylamide solution and swirl to mix.

5. Using a 25 ml pipette, dispense the acrylamide solution in between the glass plates. Tap the glass plates over the region of acrylamide to prevent bubble formation within the gel.

6. When pouring is complete, insert the smooth edge of a sharks-tooth comb into the top of the gel to a depth of ~5 mm. The acrylamide gel should polymerize roughly after 60 min. After polymerization, the comb can be removed, washed, and inserted in the reverse orentiation (i.e., tooth side down). The teeth should just barely enter the gel so as to guarantee contact with it.

7. Next, remove the clamps, place the gel into the gel apparatus and fill the top and bottom reservoirs with 1x TBE buffer.

8. Pre-run the gel for 60 min at 1,600 V, 30 mA, 50 W to allow the gel to heat up to its proper running temperature.

9. After the pre-run, flush wells with running buffer to remove diffused urea and load sequencing and primer extension samples. Run the samples at the pre-run settings until the xylene cyanol dye in the samples runs to one inch above the bottom of the gel.

10. When the electrophoresis run is completed, remove the glass plates from the apparatus and place horizontally in a large plastic tube. Gently, pry apart the glass plates using a plastic spatula.

11. Place the glass plate with the attached gel facing glass-down in the plastic container. Take a sheet of 3 MM paper slightly larger than the size of the gel and dampen it completely with ddH$_2$O.

12. Place the damp 3 MM paper on top of the gel. Take a 25 ml pipette and gently roll it over the 3 MM paper to increase its contact with the gel. Starting from one corner of the gel, gently peel back the 3 MM paper horizontally to that corner and see if the gel is attached to the 3 MM paper. If so, continue pulling until the gel is completely removed from the glass. If not, place back down, reroll gently with the pipette or choose another corner to begin.

13. Once removed from the glass plate, place a sheet of saran wrap over the gel and place it in a gel drier for 60 min at 80 °C.

14. Once dry, expose the gel to X-ray film or phosphoimage.

3.7   Data Analysis

Analysis of the autoradiogram should allow for determination of residues that were reactive to the enzymes or chemicals. The presence of bands in the various lanes must be compared with the appropriate negative control lanes to ensure that the bands produced are not nonspecific. For assignment of bands to corresponding residues in the sequencing ladder, remember that the primer-extension reaction stalls at the residue 3′ to the modified base for the single-stranded probes. Bands that are visible and more intense than background and control lanes should be considered as positive modifications. In some cases, the modified bands and sequencing ladder do not align perfectly. However, the correct assignment can normally be made, based on the relative position and the predicted identity of the residue that is preferentially modified by the enzyme or chemical. The modified positions can then be plotted onto an *mfold*-predicted secondary structure to assess the validity of the proposed model.

3.8 Example of Results from Solution Structure Probing of a Viral RNA Element

Figure 1 depicts an autoradiogram showing solution structure analysis of a *Tomato bushy stunt virus* RNA element located at the 5′ terminus of its RNA genome (2). In

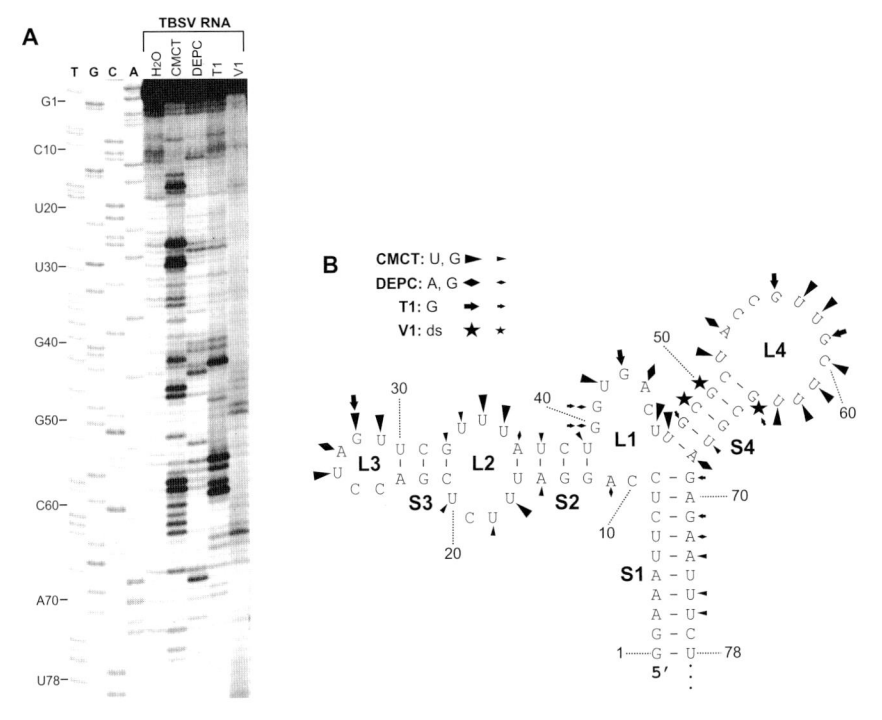

Fig. 1 An example of solution structure probing of the 5′ terminal region of a viral RNA (2). (**a**) RNA transcripts were subjected to chemical (CMCT and DEPC) and enzymatic (RNase T1 and V1) modifications, and the products generated were analyzed by primer extension. The products were separated in a 6% polyacrylamide gel in the presence of 8 M urea along with a sequence ladder generated with the same primer used for reverse transcription. The positions of selected nucleotides in the (+)-strand sequence are indicated at the *left*. (**b**) The results of the solution structure probing are mapped onto the predicted RNA secondary structure. Probes that generated primer extension products at greater than approximately twofold the level of the H_2O control were scored as positive hits. Strong and weak reactivities (as determined by the relative differences in darker and lighter band intensities within a given treatment, respectively) are indicated by large and small symbols (see key in *panel B*), respectively. The 78 nt long RNA domain is subdivided into a series of stems (S1 through S4) and loops (L1 through L4)

general, there is good correlation between the *mfold* computer-predicted structure shown and the reactivity of the enzymes and chemicals used to analyze the RNA. Single-stranded regions were most reactive with the single-stranded specific reagents RNase T1, CMCT, and DEPC. In contrast, RNase V_1 did not react efficiently with stem-1 (possibly due to inaccessibility of the enzyme), which is predicted to contain a long double-stranded region. Despite this negative result, stem-1 does indeed form in this RNA (as demonstrated by compensatory mutational analysis) (2). This example illustrates an important possible limitation of this method – where probing results do not correlate with the actual structure. Therefore, although this approach can be useful for assessing potential RNA structures, it should be used in conjunction with other types of analyses such as compensatory mutational analysis where RNA secondary structures are disrupted and restored and their corresponding functional activities are assessed.

4   Notes

1. Isolation and purification of an in vitro transcribed RNA often leads to a homogenous population of RNA molecules forming a heterologous population of secondary structures, some that may in fact be biologically inactive. Heating, followed by a slow cooling, allows the RNA to denature and facilitates renaturing into a more homogenous population of thermodynamically favorable (and hopefully biologically active) structures.
2. It is important to prepare the RNA extraction mixture in advance of the treatments, as reaction times for RNase treatments incubation times are very short.
3. When designing a primer to create both a sequencing ladder and to use for primer extension reactions, its position relative to the RNA segment of interest is important. In general, primers will start to read sequence clearly about 50 nucleotides from the primer. Therefore, the primer should be designed to pair to a region at least 50 nucleotides downstream of the region of interest.

Acknowledgments   M.R.F. was supported by an NSERC postgraduate scholarship. Our research is supported by NSERC, PREA, and CRC.

References

1. Zuker, M., Mathews, D.H., and D.H. Turner. 1999. Algorithms and Thermodynamics for RNA Secondary Structure Prediction: A Practical Guide. In RNA Biochemistry and Biotechnology, 11–43, J. Barciszewski and B.F.C. Clark, eds., NATO ASI Series, Kluwer Academic Publishers, Boston.
2. Wu, B., Vanti, W.B., and White, K.A. (2001). An RNA domain within the 5′ untranslated region of the *Tomato bushy stunt virus* genome modulates viral RNA replication. *J. Mol. Biol.* 305:741–756.

Chapter 18
RNA Encapsidation Assay

Padmanaban Annamalai and A.L.N. Rao

Abstract Analysis of viral RNA encapsidation assay provides a rapid means of assaying which of the progeny RNA are competent for packaging into stable mature virions. Generally, a parallel analysis of total RNA and RNA obtained from purified virions is advisable for accurate interpretation of the results. In this, we describe a series of in vivo assays in which viral RNA encapsidation can be verified. These include whole plants inoculated either mechanically or by Agroinfiltration and protoplasts. The encapsidation assay described here is for an extensively studied plant RNA virus, brome mosaic virus, and can be reliably applied to other viral systems as well as with appropriate buffers. In principle, the encapsidation assay requires purification of virions from either symptomatic leaves or transfected plant protoplasts followed by RNA isolation. The procedure involves grinding the infected tissue in an appropriate buffer followed by a low speed centrifugation step to remove the cell debris. The supernatant is then emulsified with an organic solvent such as chloroform to remove chlorophyll and cellular material. After a low seed centrifugation, the supernatant is subjected to high speed centrifugation to concentrate the virus as a pellet. Depending on the purity required, the partially purified virus preparation is further subjected to sucrose density gradient centrifugation.

Following purification of virions, encapsidated RNA is isolated using standard phenol-chloroform extraction procedure. An important step in the encapsidation assay is the comparative analysis of total and virion RNA preparations by Northern hybridization. This would allow the investigator to compare the number of progeny RNA components synthesized during replication vs. encapsidation. Northern blots are normally hybridized with radioactively labeled RNA probes (riboprobes) for specific and sensitive detection of desired RNA species.

Keywords Encapsidation; RNA virus; RNA packaging

From: *Methods in Molecular Biology, Vol. 451, Plant Virology Protocols:*
From Viral Sequence to Protein Function
Edited by G.D. Foster, I.E. Johansen, Y. Hong, and P.D. Nagy © Humana Press, Totowa, NJ

1   Introduction

Encapsidation or assembly of plant viral genome into stable particles by virus-encoded coat protein is considered to be a crucial event in the life cycle of a given virus (17). Although this process was initially envisioned to protect viral nucleic acids from extra cellular environment, advances made in handling viral genomes through reverse genetics revealed that encapsidation is obligatory to fulfill two important roles: cell-to-cell and long distance spread within a susceptible host plant and transmission by insect vectors to healthy hosts (10, 12, 18). Consequently, study of encapsidation provides fundamental knowledge concerning how viruses assemble into infectious particles and eventually help to develop strategies to curb spread of viral diseases.

Plant viruses exhibit a variety of genome organizations (mono, bi, or tripartite) and particle morphologies (icosahedral, flexious, and rod-shaped) (15). Literature is replete with review articles concerning the organization of plant viral genomes and their replication (1, 3, 9, 13). Following initial replication of a given viral genome, genes required to perpetuate the infection process are expressed either by proteolytic processing of the polyprotein translated from the genomic RNA (eg., Monopartite potyviruses; 11) or via subgenomic RNA synthesis (eg., Monopartite tobacco mosaic virus or tripartite cucumber mosaic virus; 16). Mature virions of viruses that express their genes via polyprotein processing contain only the genomic RNA (eg., Potyviruses; 5). By contrast, encapsidation in viruses that express their genes via subgenomic RNA synthesis are highly selective. For example, although TMV and CMV express their coat protein genes via subgenomic RNA, only in the later case the subgenomic genomic RNA is efficiently encapsidated (15, 17). By definition, subgenomic RNAs are genetically redundant and are always generated from the replicated genomic RNAs. Thus, reasons for their selective encapsidation in one case but not in the other are not well understood. Therefore, a comparative analysis of total and virion RNA profiles obtained from infected hosts will be informative. Although the procedures described later are optimized for RNA encapsidation analysis of brome mosaic virus (BMV), they are applicable to other plant viruses as well.

2   Construction of Biologically Active Clones (4)

2.1   Materials

T7/T3 vector, PCR machine, PCR reaction kit (Ambion, or Strategene or other kit), restriction enzymes, Rapid ligation kit (Promega, or Stratagene), competent cells, LB medium, and appropriate antibiotics for selection.

2.2   Preparation RNA Transcripts (In Vitro Transcription)

Construction of a cDNA clone from which biologically active RNA transcripts can be synthesized in vitro using either T7 or T3 polymerase promoter is almost a prerequisite for many molecular biology-related experimentation. Because cDNA clones are not directly infectious (8), placement of the viral sequence downstream of a promoter of a RNA polymerase is essential for in vitro synthesis of infectious RNA transcripts (2, 14). Alternatively, desired gene of interest can also be amplified by PCR with specific primers that contain polymerase promoter site and these templates can be used for in vitro transcription. The methods for cloning the gene(s) of interest are available in most of the molecular biology books such as Sambrook and Russel (20). Later we describe a procedure that is routinely used in our laboratory for in vitro synthesis of RNA transcripts (19) that can be used for in vitro encapsidation assays. Note that these reactions can be set up without incorporating a 5′ cap-like structure, which is not required for in vitro assembly. Furthermore, absence of 5′ cap-like structure will give high yield of RNA transcripts.

1. Lineraize the plasmid DNA with an appropriate restriction enzyme downstream of the insert to be transcribed.
2. Once the plasmid DNA was completely digested, extract the reaction mixture with phenol/chloroform and precipitate the DNA with ethanol followed by a 70% ethanol wash and drying the sample in speed vacuum.
3. Resuspend the digested plasmid DNA in TE buffer at a concentration of $1\,\mu g\,\mu l^{-1}$.
4. Using a commercially available kit (for eg., MEGAscript, Ambion), set up the transcription reaction at 37 °C (approximately for 1–2 h). Follow the instructions according to the manufacturer of the kit.
5. After incubation, DNA template is removed either with DNase (20) or by LiCl precipitation (19).
6. Centrifuge the contents at 4 °C for 15 min at maximum speed (12,000–15,000 rpm) to pellet RNA.
7. Remove the supernatant and wash the pellet with 70% ethanol.
8. Dry RNA pellet by speed-vac and dissolve with known amount of RNase free water.
9. Determine RNA concentration by spectrophotometer. It is imperative to verify the integrity of RNA transcripts by agarose gel electrophoresis prior to using them for in vitro assembly assays. Store RNA transcripts at −20 °C or −80 °C.

3   Virus Purification

3.1   Purification of BMV Virus Particles

Materials

BMV extraction buffer: 0.5 M NaAc; 0.08 M MgAc pH 4.5 and add 1/100 volume of B mercaptoethanol just before use. Store the buffer at 4 °C.

BMV suspension buffer: Dilute BMV extraction buffer to 1/10 with sterile distilled water.

Sterile mortars and pestles, sterile centrifuge tubes, chloroform and acid-washed sand (Sigma).

1. Collect BMV-infected leaves either barley, *Chenopiodium quinoa*, or *Nicotiana benthamiana*.
2. Grind leaves thoroughly in extraction buffer (1.0 ml g^{-1} leaf) and add 0.5 g acid-washed sand to facilitate easy grinding and breaking of cells.
3. Filter the extract through muslin cloth and collect flurry. Again ground the portion retained on cheese cloth with extraction buffer. Repeat filtration through cheese cloth.
4. Transfer the solution to centrifuge tubes and add equal volume of prechilled chloroform and vortex for 5 min at room temperature.
5. Centrifuge the emulsified solution at 10,000 g for 15 min at 4 °C.
6. Transfer supernatant to clean sterile beaker and make sure that no chloroform is left in the supernatant.
7. Transfer supernatant to sterile ultracentrifuge tubes.
8. Centrifuge at 30,000 rpm for 2.5–3 h in a high speed Beckman centrifuge.
9. Remove the supernatant completely and suspend pellet in desired volume of (200–500 μl) BMV suspension buffer.
10. Subject the above partially purified virus to 5–25% sucrose density gradient centrifugation. Finally, measure the concentration of the virus using spectrophotometer at OD at 260 nm.

3.2   Preparation of Coat Protein Subunits

Materials

Required stock solutions: 1 M Tris-HCl, pH 7.5, 50 mM EDTA, 100 mM DTT, and 100 mM PMSF in isoprophyl alcohol.

1x Dialysis buffer: 0.5 M CaCl$_2$, 50 mM Tris HCl, pH 7.5, 1.0 mM EDTA, 1.0 mM DTT, and 0.5 mM PMSF.

Dialysis membrane (20) or dialysis cassettes.

3.3   Dissociation of Coat Protein from Purified Virus Particles

1. Prepare 1,000 ml of 1x dialysis buffer.
2. Prepare dialysis membranes according to Sambrrok et al. (20).
3. Dispense required concentration of purified virus into a dialysis bag. Test for any holes in the dialysis bag, prevent leakage during dialysis.
4. Place the virus containing dialysis bag or dialysis cassette in a beaker containing dialysis buffer.
5. Dialyze 24 h at 4 °C while stirring. A cloudy precipitate (representing viral RNA) will appear.
6. After 24 h, a cloudy precipitate (viral RNA) must appear in the dialysis bag. Collect the solution from the dialysis bag/cassette and centrifuge at 12,000 g for

15 min at 4 °C to pellet viral RNA. Cloudy precipitate will form a pellet and this pellet can be used to recover viral RNA by reextracting with phenol/chloroform followed by ethanol precipitation.

7. Collect supernatant and centrifuge at 220,000 g in a Beckman TL 100 centrifuge for 2 h at 4 °C to pellet any undissociated virus particles.
8. Collect the supernatant.
9. Determine the concentration of the coat protein subunits by measuring at OD 254 and 280 nm or by other methods such as Bradford assay.
10. Use coat protein subunits immediately or can be stored at 4 °C for 1–2 weeks.
11. Verify the authenticity and integrity of coat protein by 12–16% SDS-PAGE followed by Western blot analysis.

3.4   In Vitro Assembly of Empty and RNA Containing Virions

Materials

1. RNA transcripts
2. RNA assembly buffer: 50 mM NaCl; 50 mM Tris-HCl, pH 7.2; 10 mM KCl; 5.0 mM MgCl$_2$; 1.0 mM DTT
3. Re-assembly buffer: 1.0 M Nacl; 50 mM NaAc, pH 4.8; 1.0 mM EDTA; and 1.0 mM DTT
4. Dialysis buffer B: 1.0 M NaCl; 20 mM Tris-HCl, pH 7.5, 1.0 mM EDTA; 1.0 mM DTT and 1.0 mM PMSF

3.5   In vitro assembly of empty virions

1. Empty virions of BMV can be assembled in vitro by dialyzing only coat protein subunits of about 200–500 μg against reassembly buffer at 4 °C for 24 h.
2. Dispense the reaction mixture to Centricon-100 microconcentrators (Amicon, Beverley, MA; Follow instructions supplied by the manufacturer for using microconcentrators)
3. Centrifuge at 2,000 g at 4 °C for 30 min
4. Wash the reaction mixture by adding 1.5 ml of reassembly buffer
5. Repeat the above step two more times
6. Finally elute the virions by centrifugation at 4 °C for 5 min
7. Assess virion assembly under an electron microscope (see later)

3.6   In Vitro Assembly of RNA Containing Virions

1. Prepare RNA transcripts to be assembled and calculate the concentration
2. Mix the coat protein subunits and RNA transcript at a ratio of 1:5 (wt/wt)

3. Dispense above mixture to a dialysis bag/cassette and properly secure to avoid any leaks
4. Prepare 1,000 ml of RNA 1x assembly buffer
5. Place the dialysis bag into beaker and stir
6. Dialyze the assembly reaction at 4 °C for 24 h
7. After 24 h, collect the mixture containing assembled virions and add 1.5 ml of RNA assembly buffer
8. Pass this mixture through Centricon-100 column and centrifuge at low speed (2,000 g) for 30 min
9. Wash the column once with 1.5 ml of RNA assembly buffer at 4 °C for 30 min
10. Repeat the above washing step
11. Elute the virions by centrifugation at 4 °C for 5 min
12. Verify the concentration at OD 260 nm

3.7 Electron Microscopy

1. Adjust the concentration of virus particles to 10–25 µg ml^{-1}
2. Place a drop of solution containing virus particles in a glow discharged carbon coated copper grid
3. Add 10 µl of 1% uranyl acetate and leave it for a minute
4. Wash once with sterile distilled water and allow drying for 1–5 min
5. Examine grids under an electron microscope

3.8 Extraction Viral RNA from Encapsidated Virions

Materials

Eppendorf tubes (1.5 ml), bentonite (2.5 mg ml^{-1}), 20% SDS, phenol/chloroform/isoamyl alcohol-PCI (25:24:1, v/v) and ethanol.

1. Collect in vitro assembled virions and transfer them to 1.5 ml eppendorf tube
2. Add 1/10 vol bentonite and 20% SDS; Mix well
3. Add equal volume of phenol/chloroform/isoamyl alcohol and vortex for 5 min
4. Centrifuge at 12,000 g for 20 min at 4 °C
5. Collect supernatant and precipitate RNA by adding 1/10th of 3 M NaOAc (pH 5.2) and 2.5 volume of cold 100% ethanol
6. Keep at −20 °C for overnight or at −80 °C for 3 h
7. Centrifuge at 12,000 g for 20 min at 4 °C
8. Wash pellet with 70% ethanol and dry
9. Dissolve pellet in RNase-free water and estimate concentration of the RNA by spectrophotometer

4 Assaying RNA Encapsidation in Plant Protoplasts

Materials

Barley or Chenopodium quinoa or Nicotiana benthamiana leaves; 0.55 M mannitol in water pH 5.9; 0.55 M sucrose in water; Sterilize the solution by filtration. Polyethylene glycol-40%, (weigh 40 g PEG-2000, add 50 ml sterile distilled water, 10 ml of 5% MES (morpholineethanesulfonic acid) buffer (pH 5.9), and 1 ml of 0.3 M $CaCl_2$. Briefly warm the mixture in microwave oven and stir the mixture until it dissolved. Adjust the pH 5.9 with 0.1 M KOH. Finally, filter the solution using 0.45-μm pore size filter. Store at room temperature. Use immediately or store upto 3 weeks).

Enzymes: Cellulose, Macerozyme, Driselase, and Macerozyme, BSA

4.1 Protoplast Culture Medium

Solution A (1,000×): Dissolve 0.00249 g copper sulphate, 0.0166 g potassium iodide (KI), and 26.648 g magnesium sulphate in 100 ml of 0.055 M mannitol, pH 5.9. Check the pH (5.8) of the solution and sterilize by autoclaving.

Solution B (1,000×): Dissolve 2.712 g of potassium phospahte (KH_2PO4) and 10.111 g potassium nitrate (KNO_3) in 100 ml sterile distilled water and adjust pH to 6.5 with 1N KOH. Sterilize by autoclaving.

Solution C (1,000×): Dissolve 14.70 g of calcium chloride in 100 ml of 0.55 M mannitol pH 5.9. Adjust the pH to about 6.2 and sterilize the solution.

Gentamycin: Prepare 10 mg ml^{-1} stock solution in sterile water. Store at −20 °C.

Cephaloridine: Prepare 30 mg ml^{-1} stock solution in sterile water. Keep at −20 °C.

4.2 Preparation of Protoplasts

1. Prepare the enzyme solution: For 0.5 g leaf tissue, use 0.25 g cellulose, 12.5 mg BSA, and 12.5 mg macerozyme. Add 12.5 ml 0.55 M mannitol, pH 5.9 and stir at room temperature. Filter the solution through 0.45-μm pore size filter.
2. Collect 0.5 g leaves of barley plants (5–6 days old). Slice the leaves lengthwise and then crosswise with a razor blade. Incubate the sliced (1 mm^2) leaf materials in the enzyme solution for 3–4 h at 28 °C.
3. After 3–4 h, decant the enzyme solution containing the protoplasts into beaker. Filter through gauze (300–350 μm) into another beaker. Transfer the solution to a sterile 50 ml polypropylene tube underlaid with 10 ml of 0.55 M sucrose. Use 12.5 ml protoplast solution per tube.
4. Centrifuge in a Beckmann centrifuge at 400 rpm for 5 min at 20 °C.

5. Remove the protoplasts from the interface of mannitol and sucrose. (They will constitute a dark green band at the interface.) Transfer them into another 50 ml tube containing 10–15 ml 0.55 M mannitol, pH 5.9. Mix gently and centrifuge at 600 rpm for 4 min at 20 °C. Remove most of the supernatant and gently resuspend the cells in the remaining liquid. Add 10 ml 0.55 M mannitol, pH 5.9 and repeat the wash as described earlier.

6. Resuspend the protoplasts in a known volume of 0.55 M mannitol, pH 5.9 and determine the number of viable protoplasts with a hemacytometer. [Stain the protoplasts with fluorescein diacetate by combining 1–2 drops of protoplasts suspension solution with 1–2 µl of fluorescein diacetate (5 mg ml⁻¹ in acetone). Count the bright fluorescent cells].

7. Centrifuge at 600 rpm for 3 min at 20 °C. Remove most of the supernatant and keep the volume as quite small and add either virus or RNA transcript.

4.3 Transfect the Protoplasts with Viral RNA or Transcripts

1. Gently shake the pellet to resuspend the protoplasts
2. Add RNA transcript or/viral RNA and then add 150 µl of 40% PEG
3. Gently shake for 10 min, and add two drops of 0.55 M mannitol, pH 5.9. Gently mix the suspension
4. Continue to add mannitol dropwise over the next 5–10 min until the volume has reached 1.5 ml. Incubate on ice for 15 min
5. Pellet the protoplasts at 600 rpm for 3 min at 20 °C
6. Wash once with 1 ml 0.55 M mannitol, pH 5.9
7. Resuspend 1×10^5 protoplasts in 1 ml culture medium containing 0.55 M mannitol (pH 5.9), a 1x concentration each of solutions A, B, and C, 10 µg of Gentamycin and 0.3 mg Cephaloridine.
8. Place the transfected protoplasts in a culture plate and keep under fluorescent lamp for 20–24 h.

4.4 Isolation of Assembled Virions and Extraction of Encapsidated RNA

Materials

Protoplast lysis buffer: 100 mM glycine, 10 mM EDTA, 100 mM NaCl (pH 9.5), 2% SDS, bentonite 2.5 mg ml⁻¹; phenol/chloroform (25:24, v/v); 100% ethanol.

1. Collect the transfected protoplasts by centrifugation at 600 rpm for 3 min at 20 °C.
2. Discard the supernatant and add 250 µl protoplasts lysis buffer and add 250 µl phenol/chloroform.

3. Vortex for 5 min at room temperature
4. Centrifuge at 12,000 rpm for 10 min at 4 °C.
5. Collect the supernatant and repeat the phenol/chloroform extraction.
6. Collect the supernatant and add 1/10th of 3 M NaoAc (pH 5.6) and add 2.5 volumes of 100% cold ethanol.
7. Mix the contents and keep at −80 °C for 15–30 min.
8. Pellet the RNA by centrifugation (12,000 rpm) for 20 min at 4 °C. Wash the pellet with 70% ethanol, dry pellet, and dissolve the RNA in 25 µl of water.

5   RNA Encapsidation Assay in Planta (Agroinfiltration)

First, to facilitate agroinfiltration, the genome of RNA virus was placed in the expression vector (binary vector) under the control of the CaMV 35S promoter and transformed into suitable agrobacterium strains. Transformed agrobacterial cultures are grown to log phase, collected by low speed centrifugation, and resuspended in an infiltration solution (10 mM $MgCl_2$, 10 mM MES). The suspension is used further for leaf infiltration. After 2–6 days, the infiltrated leaves can be used for the analysis of gene of interest and its encapsidation. The agroinfiltration system has several advantageous over the conventional methods. Using agroinfiltration, the expression and encapsidation of two or more genes can be analyzed by different combinations of agrobacteria containing genes of interest (6, 7). For example, in the case of bipartite or tripartite viruses, coexpression of individual viral RNAs can be achieved by simply mixing agrobacterium strains each containing one of the RNAs prior to inoculation.

Materials: Binary expression vector, agrobacterium strain, LB medium, antibiotics (50 µg ml^{-1} Kanamycin, 10 µg ml^{-1} Rifampicin), 10 mM $MgCl_2$, 10 mM MES pH 5.6; 100 mM Acetosyringone in methanol.

5.1   *Construction of T-DNA Based Plasmids*

1. Select the appropriate T-DNA vector (for example pCASS vector).
2. Amplify the desired gene of interest by PCR having unique restriction sites at both ends. This will facilitate to clone the gene of interest in to T-DNA vector.
3. Digest the T-DNA vector and the PCR product.
4. Purify the digested product by Gel elution
5. Check the concentration of eluted producs for ligation.
6. Ligate the PCR produt with linearized T-DNA vector, transform it to *E. coli*, and plate on LB agar plates amended with Kanamycin or appropriate selection marker.

7. Incubate the plates for overnight at 37 °C.
8. Pick up few colonies and grow on LB liquid medium with appropriate antibiotics for overnight at 37 °C.
9. Extract the plasmid DNA, and screen by restriction analysis for the presence of gene of interest.
10. Once the positive clone has been identified (the plasmid DNA can be used to transform into agrobacterium).

5.2   Transformation into Agrobacterium

1. Take 4–6 μg plasmid DNA (T-DNA construct having the gene of interest) and add to the agrobacterium competent cells (EHA105 or).
2. Keep on ice for 45 min.
3. Freeze the competent cells containing plasmid DNA on liquid nitrogen for 1 min.
4. Thaw the cells at 37 °C for 3 min.
5. Add growth medium (LB medium) and incubate in a shaker for 3–4 h at 28 °C.
6. Centrifuge the mixture (at low speed 2,000 rpm for 5 min).
7. Discard most of the supernatant and suspend the cells with little less than 0.1 ml medium.
8. Plate the mixture on LB agar plate supplemented with appropriate antibiotics.
9. Incubate the plates at 28 °C for 2–3 days.
10. Several colonies will appear on plates after 2–3 days.
11. Screen the colonies for presence of T-DNA plasmid (by mini preparation and restriction analysis).
12. Select the positive transformant for further studies.
13. Alternatively, store the positive transformant of agrobacterium by making glycerol stock and store at −80 °C.

5.3   Agroinfiltration

1 Steak appropriate transformed A. tumefaciens on LB agar plates supplemented with appropriate antibiotics and incubate at 28 °C for 2 days.
2. A single colony to be inoculated into 5 ml LB broth with 50 μg ml^{-1} of kanamycin and 10 μg ml^{-1} of rifampicin (or use an antibiotic of the helper Ti plasmid encoded resistance) and keep the tubes at 28 °C in an orbital shaker at 250–300 rpm for 2 days.
3. Inoculate 1 ml of fresh culture into 50 ml LB broth supplemented with kanamycin 50 μg ml^{-1} as well as rifampicin 10 μg ml^{-1}, 10 mM MES pH 5.6, and 100 μM acetosyringone.
4. Keep the flask at 28 °C for 16 h in an orbital shaker, which can rotate at 250–300 rpm.

5. Check the absorbance of the fresh culture at OD_{600}. The OD_{600} of the culture must have reached to 1.0.
6. Transfer the fresh culture to screw cap oak ridge tube or sterile Falcon tube and centrifuge at 5,000 rpm for 10 min at room temperature in a Beckman or Table top centrifuge.
7. Decant the supernatant and resuspend the bacterial pellet with 50 ml of 10 mM $MgCl_2$.
8. Centrifuge at 5,000 rpm for 10 min at room temperature and decant the supernatant.
9. Resuspend the cells with 50 ml of 10 mM $MgCl_2$ and add 100 μM acetosyringone (from 100 mM acetosyringone stock) and gently mix well.
10. Check the final optical density at 600 nm (OD_{600}). The final OD_{600} must be of 1.0. When two or more A. tumifaciens strains needed to infiltrate together, grow the each strain independently as mentioned above and mix equal amount prior to infiltration.
11. The bacterial culture should be kept at room temperature for at least 3 h without shaking.
12. After 3 h, the culture is ready for infiltration. Infiltration should be performed with a 1 ml syringe without needle.
13. Two or three expanded leaves of young seedlings of N. benthamiana (5 leaves stage, 2- to 3-week-old plants) should be ideal for infiltration.
14. Perform the infiltration by gently punching the tip of the syringe on the backside of the leaf with blocking by finger from the other side. Gently and slowly push the syringe barrel. Note the bacterial suspension spreads into the intracellular space of the leaf and further spreads up to tip of leaf. Once this has been done, infiltrate into the other leaf. After agroinfiltration, transfer plants to greenhouse.
15. Collect the leaf samples from 2 days and analyze the expression and encapsidation of gene of interest by extracting total RNA and virion RNA.
16. Analyze RNAs by Northern blot hybridization using specific probes.

6 Analysis of Encapsidated RNA by Northern Hybridization

6.1 Materials

The materials include agarose, gel casting tray, gel tank, 10× MOPS buffer (0.2 M MOPS, 40 mM NaOAc, 5 mM EDTA, adjust pH 7.0 with solid NaoH, filter the solution), 37% formaldehyde; 20× SSC (3 M sodium chloride, 0.3 M trisodium citrate), nylon membrane, blotting unit, sample buffer: 10× MOPS buffer/formaldehyde/formamide/H_2O (0.5 ml Formamide; 0.18 ml formaldehyde, 0.1 ml 10x MOPS buffer, 0.22 ml H_2O).

Hybridization solution: For 10 ml: 4.0 ml 5 M NaCl; 1.0 ml 20% SDS; 2.0 ml Denhardt's solution; 0.3 ml salmon sperm DNA; 0.2 ml yeast tRNA and 2.5 ml water.

6.2   Preparation of Formaldehyde-Denatured RNA Gel

1. Prepare 1.2% agarose gel by melting 2.4 g agarose in 174 ml RNAse free distilled water.
2. Use microwave until it completely dissolved.
3. Allow the mixture cool for 5–10 min.
4. Add 20 ml, 10× MOPS, and 6 ml of 37% formaldehyde and throughly mix and pour into the gel tray.

6.3   Sample Preparation and Electrophoresis

1. Take known amount of viral RNA (0.5–1.0 μg), add 10–15 μl sample buffer, and mix thoroughly.
2. Heat the reaction mixture containing viral RNA at 65 °C for 10 min and cool the reaction mixture for 5 min on ice.
3. Add 2 μl of 6× loading dye.
4. Load the sample into the well.
5. Electrophorese the sample for 2–3 h at 100 V or until the dye front reached 2/3 of the way to the bottom of gel.

6.4   RNA Transfer from Gel to Nylon Membrane

1. Rinse the gel in 7× SSC for 10 min. RNA transfer from gel to a nylon membrane to be performed with transVac vaccum blotting unit (Amersham). Alternatively, transfer can be done by conventional methods.
2. After transfer, wash the membrane briefly with 7× SSC and dry at room temperature.
3. Place the membrane in UV cross linker (Strategene) for optimal cross link.
4. To certify equal loading of RNA samples, stain the membrane with methylene blue solution (0.04% methylene blue in 0.5 M sodium acetate, pH 5.2).
5. Estimate the equal amount of RNA by visualization.

6.5   Preparation of Radiolabeled RNA Probes

1. Use sterile Eppendorf tube to mix the following components:

   Transcription buffer (5×) 4.0 μl
   100 mM Tris-HCl (pH 8.3), 250 mM KCl, 25 mM MgCl2, BSA (1 mg ml^{-1})
   DTT (0.1 M) 2.0 μl

RNA guard (40 U ml^{-1}) 1.0 µl
ATP, CTP, GTP (2.5 mM each) 4.0 µl
UTP (100 µM) 2.4 µl
[32P]UTP (~3,000 Ci mmol^{-1}) 4.0 µl
DNA template linearized (1 µg µl^{-1}) 1.0 µl
T7 or T3 RNA polymerase (25 U µl^{-1}) 1.0 µl
Water 0.6 µl

2. Mix the contents by vortexing and briefly centrifuge.
3. Incubate the reaction mixture at 37 °C for 1 h.
4. Terminate the reaction by adding 20 µl TE buffer.
5. Extract once with an equal volume of phenol/chloroform.
6. Collect the supernatant, add 1.0 µl carrier Yeast RNA (1 µg ml^{-1}) ½ volume of 7.5 ammonium acetate and 2.0 volume of cold ethanol.
7. Incubate at −80 °C for 1 h.
8. Centrifuge the tubes at 4 °C for 20 min.
9. Wash the pellet with 70% ethanol, dry, and suspend in 50 µl water.
10. Use the probes immediately or store at −80 °C for latter use.

6.6 Prehybridization and Hybridization

Materials: Prehybridization solution, Hybridization oven, 2× SSC, SDS.

Prehybridization Solution: Deionized Formamide 10.0 ml, SDS 10% 2.0 ml, Denhardt's solution (50×) 2.0 ml, Denatured Salmon Sperm DNA (10 mg ml^{-1}) 0.3 ml, NaCl (5.0 M) 4.0 ml, Sterile distilled water 1.7 ml

1. Place the membrane into hybridization bottle.
2. Add 5–10 ml prehybridization solution and keep the tubes at 55 °C or 65 °C for 3–6 h.
3. Label the RNA probe using p^{32}-UTP.
4. Add the probe directly to prehybridization mixture and hybridize for overnight or at least 12–16 h at 65 °C.
5. After hybridization, wash the membrane once with 2× SSC; 02% SDS at room temperature for 30 min.
6. Wash the membrane twice with 0.2× SSC and 0.2% SDS at 65 °C for 30 min.
7. Remove the membrane from the hybridization bottle and air dry for 5–10 min.
8. Wrap with saran wrap and expose to PhosphorImager Casette or expose to X-ray film.
9. Analyze the results.

Acknowledgments This study was supported by a grant GM 064465–01A2 from the National Institutes of Health to A.L.N. Rao.

References

1. Ahlquist, P. 1992. Bromovirus RNA replication and transcription. Curr Opin Genet Dev 2:71–6.
2. Ahlquist, P. and M. Janda. 1984. cDNA cloning and in vitro transcription of the complete brome mosaic virus genome. Mol Cell Biol 4:2876–82.
3. Ahlquist, P., A. O. Noueiry, W. M. Lee, D. B. Kushner, and B. T. Dye. 2003. Host factors in positive-strand RNA virus genome replication. J Virol 77:8181–6.
4. Allison, R. F., M. Janda, and P. Ahlquist. 1988. Infectious in vitro transcripts from cowpea chlorotic mottle virus cDNA clones and exchange of individual RNA components with brome mosaic virus. J Virol 62:3581–8.
5. Allison, R. F., J. C. Sorenson, M. E. Kelly, F. B. Armstrong, and W. G. Dougherty. 1985. Sequence determination of the capsid protein gene and flanking regions of tobacco etch virus: Evidence for synthesis and processing of a polyprotein in potyvirus genome expression. Proc Natl Acad Sci USA 82:3969–72.
6. Annamalai, P., and A. L. Rao. 2006. Delivery and expression of functional viral RNA genomes in planta by agroinfiltration, pp. 16B.2.1–2.15. In T. Downey (ed.), Current Protocols in Microbiology, vol. 1. Wiley.
7. Annamalai, P., and A. L. Rao. 2005. Replication-independent expression of genome components and capsid protein of brome mosaic virus in planta: A functional role for viral replicase in RNA packaging. Virology 338:96–111.
8. Boyer, J. C., and A. L. Haenni. 1994. Infectious transcripts and cDNA clones of RNA viruses. Virology 198:415–26.
9. Buck, K. W. 1999. Replication of tobacco mosaic virus RNA. Philos Trans R Soc Lond B Biol Sci 354:613–27.
10. Canto, T., D. A. Prior, K. H. Hellwald, K. J. Oparka, and P. Palukaitis. 1997. Characterization of cucumber mosaic virus. IV. Movement protein and coat protein are both essential for cell-to-cell movement of cucumber mosaic virus. Virology 237:237–48.
11. Carrington, J. C., S. M. Cary, and W. G. Dougherty. 1988. Mutational analysis of tobacco etch virus polyprotein processing: Cis and trans proteolytic activities of polyproteins containing the 49-kilodalton proteinase. J Virol 62:2313–20.
12. Carrington, J. C., K. D. Kasschau, S. K. Mahajan, and M. C. Schaad. 1996. Cell-to-cell and long-distance transport of viruses in plants. Plant Cell 8:1669–81.
13. Dreher, T. W. 1999. Functions of the 3′-untranslated regions of positive strand RNA viral genomes. Annu Rev Phytopathol 37:151–74.
14. Dreher, T. W., A. L. Rao, and T. C. Hall. 1989. Replication in vivo of mutant brome mosaic virus RNAs defective in aminoacylation. J Mol Biol 206:425–38.
15. Hull, R. 2002. Plant Virology. Academic Press.
16. Kao, C. C., and K. Sivakumaran. 2000. Brome mosaic virus, good for an RNA virologist's basic needs. Mol Plant Pathol 1:91–98.
17. Rao, A. L. 2006. Genome packaging by spherical plant RNA viruses. Annu Rev Phytopathol 44:61–87.
18. Rao, A. L. and G. L. Grantham. 1995. Biological significance of the seven amino-terminal basic residues of brome mosaic virus coat protein. Virology 211:42–52.
19. Rao, A. L. N., R. Duggal, F. Lahser, and T. C. Hall. 1994. Analysis of RNA replication in plant viruses, pp. 216–236. In K. W. Adolph (ed.), Methods in Molecular Genetics: Molecular Virology Techniques, vol. 4. Academic Press, Orlando, Florida.
20. Sambrrok, J. and D. L. Russel. 2001. Molecular Cloning: A Laboratory manual.. Cold Spring Harbor Laboratory Press, Cold Spring Harbor, NY.

Section 3
Protein Analysis and Investigation of Protein Function

Chapter 19
Surface Plasmon Resonance Analysis of Interactions Between Replicase Proteins of Tomato Bushy Stunt Virus

K.S. Rajendran and Peter D. Nagy

Abstract Replication of the viral RNA genome performed by the viral replicase is the central process during the viral infection cycle (Nagy and Pogany, see earlier chapter four). Most RNA viruses assign one or more proteins translated from their own genomes for assembling the viral replicase complex, which consists of the viral RNA, viral proteins, and several subverted host proteins embedded in cellular membranes. Understanding the various biochemical activities of the replication proteins can lead to target identification for human intervention to control viral infections or the damage to the host cells. The replicase proteins of tomato bushy stunt virus (TBSV) are selected as model system to study the dynamics of interactions between viral replicase proteins using surface plasmon resonance (SPR) analysis. The SPR assay provides real-time protein interaction data by measuring the change in refractive index at the surface of the sensor chip due to the change in mass resulting from the interaction between the immobilized protein and the protein that is being passed over the immobilized chip surface. SPR-based biosensor BIAcore X was used to carry out TBSV replicase protein interaction studies.

Keywords Tomato bushy stunt virus; Replicase proteins; Protein–protein interaction; BIAcore X; Surface plasmon resonance; Protein interaction kinetics

1 Introduction

RNA viruses code for one or more replicase proteins, which are necessary for the replication of their genetic material. The viral replicase is a multisubunit enzyme consisting of virus-coded proteins, including the RNA-dependent RNA polymerase (RdRp), and host proteins (1–3). The viral subunits of the replicase complex usually contain multiple functional domains distributed among them. The interactions between the viral replication proteins are one of the important forces in the assembly and function of the viral replicase complex (4–6). Therefore, identification of domains involved in such interactions and the kinetics of interactions assume greater practical significance for devising viral control strategies.

From: *Methods in Molecular Biology, Vol. 451, Plant Virology Protocols: From Viral Sequence to Protein Function*
Edited by G.D. Foster, I.E. Johansen, Y. Hong, and P.D. Nagy © Humana Press, Totowa, NJ

Surface Plasmon Resonance (SPR)-based optical biosensors are now widely recognized and are well established tools to obtain quantitative as well as qualitative data on interactions between biological macromolecules (7–9). The kinetics and energetics of interaction data is collected in real time with nanomolar quantities of proteins and it obviates the need for labeling the interacting partners. Although there are several biosensors available in the market, BIAcore instruments are the most widely used.

The principle behind this method is SPR, which measures changes in the refractive index on the sensor chip surface when interactions between the two molecules (one fixed on to the chip surface and the other passed over the immobilized chip surface in a buffer) occur. The change in the refractive index is then plotted in a graph, called sensogram, with time on the X-axis and resonance units on the Y-axis. The data from the sensogram can be used to extract the association and dissociation constants of protein complexes.

In this chapter, we describe the use of BIAcore X instrument to study the interactions between replicase proteins of tomato bushy stunt virus (TBSV). TBSV is a small, messenger-sense, single-stranded RNA virus (10). The recombinant TBSV replicase proteins p33 and p92 were expressed in *E. coli* as C-terminal fusion to maltose binding protein (MBP). Affinity purification of MBP-p33 and MBP-p92 was done using amylose-resin chromatography. The data collected from the SPR assay were analyzed using BIAevaluation provided by the manufacturer.

2 Materials

2.1 pH Scouting

1. Purified recombinant MBP-p33C protein. We used the C-terminal half of p33, which is soluble when expressed in *E. coli* (6, 11). The protein solution was centrifuged for 15 min at 14,000 g to remove any particulate matter before using in BIAcore X instrument to avoid clogging of the microfluidic flow circuit.
2. CM-5 biosensor chip (BIAcore, Piscataway, NJ).
3. Immobilization buffer: sodium acetate, 10 mM, pH 4.0, 4.5, 5.0, 6.0. All buffers and solutions used in BIAcore X instrument are degassed and filtered using 0.22 μm filter. Alternatively, a set of sodium acetate buffers can be purchased from BIAcore Inc, NJ.
4. Running buffer: 10 mM HEPES pH 7.4, 150 mM NaCl, 3 mM EDTA, and 0.005% Tween-20. This buffer is also available from BIAcore Inc, NJ, as HBS-EP.
5. Wash solution: 50 mM NaOH.

2.2 Protein Immobilization on Sensor Chip

1. Affinity purified recombinant MBP-p33C and MBP proteins.
2. Snake skin dialysis membrane (Pierce).

3. Sodium acetate, 10 mM, pH 5.0.
4. CM-5 sensor chip (BIAcore, Piscataway, NJ).
5. Amine coupling kit (BIAcore Inc, NJ). It includes 115 mg *N*-hydroxy-succinimide (NHS), 750 mg 1-ethyl-(3-dimethylaminopropyl)-carbodiimide hydrochloride (EDC), and 10.5 ml 1 M ethanolamine–HCl pH 8.5. Stock solutions of 0.1 M and 0.4 M NHS and EDC, respectively, are prepared by resuspending the salt in sterile water and stored in single-use aliquots at −20 °C.
6. BIA normalizing solution from BIAcore Inc, NJ. This solution contains 70% (w/w) glycerol.
7. Running buffer: 10 mM HEPES pH 7.4, 150 mM NaCl, 3 mM EDTA, and 0.05% Tween-20.
8. Wash solution: 50 mM NaOH.

2.3 Interaction Analysis

1. Affinity purified recombinant MBP-p33C and MBP proteins.
2. Running buffer: 10 mM HEPES pH 7.4, 150 mM NaCl, 3 mM EDTA, and 0.05% Tween-20. This buffer can also be purchased from BIAcore Inc, NJ.
3. Regeneration buffer: 50 mM NaOH.

2.4 Data Analysis

1. BIAevaluation 4.1 software from BIAcore Inc, NJ.

3 Methods

There are four major steps in BIAcore-based SPR experiments suitable for analysis of protein–protein interactions: (1) obtaining pure protein samples, (2) immobilizing one of the binding partners on the chip surface, (3) running the other protein (or the same protein if self-interaction is the goal of the study) over the immobilized surface of the chip to obtain the original data, and (4) data analysis. The second and third steps involve the core BIAcore X instrument and a computer (PC), which operates the BIAcore X control software. Users are required to inject samples and buffers to the injection port on the instrument and run the software. The data analysis is done using BIAevaluation software 4.1 provided with the instrument. The analyzed data can be exported as spreadsheet and nicely presentable graphs can be made using Microsoft Excel software.

3.1   Familiarize with BIAcore X Instrument and the Software

Before starting the actual SPR assay, it is very important to first get familiarized with the basic operations of the BIAcore system, especially docking/undocking of sensor chip, sample loading, BIAcore X control software, and BIAevaluation software. The "BIAcore X getting started" kit from BIAcore Inc, NJ, is also very helpful for the first time user for training in the basic procedures involved in using the instrument. The kit includes all the buffers, reagents, and proteins.

3.2   Preparation of Protein Samples for the SPR Assay

1. TBSV replicase proteins, namely p33, p92, and their truncated versions, were expressed in *E. coli* as MBP fusion and purified using amylose resin affinity chromatography (6, 11).
2. The concentration of proteins is quantified using Bio-Rad protein assay reagent.
3. The purity of the protein samples were tested in SDS-polyacrylamide gel electrophoresis (SDS-PAGE). If there are multiple proteins in the sample, the accuracy of the kinetic data obtained with such protein samples may not be correct.
4. To keep the buffer composition of the protein solution as close to buffers used in SPR assay as possible, the proteins are dialyzed against BIAcore assay buffers. The proteins to be fixed on sensor chip are dialyzed against 10 mM sodium acetate buffer pH 5.0 and the proteins to be passed over the chip surface are dialyzed against running buffer in a snake skin dialysis membrane – 10 kDa molecular weight cut off (Pierce). Alternatively, the protein samples can be diluted in appropriate SPR assay buffers.

3.3   pH Scouting to Determine Appropriate Immobilization pH

pH scouting is referred to the experimental procedure of finding the appropriate immobilization buffer pH and is performed on new CM-5 chip without modifying its surface. The level of protein immobilization is affected by the electrostatic attraction of proteins to the chip surface and this attraction is referred to as preconcentration. The carboxymethylated dextran matrix of CM-5 chip is negatively charged at pH values above approximately 3.5. Therefore, the pH of the immobilization buffer should be above 3.5 and below the isoelectric point of the protein – in our case 5.5 for MBP-p33C to achieve efficient immobilization.

1. The CM-5 sensor chip should be equilibrated at room temperature 30 min before use.
2. Switch on the BIAcore X instrument and the PC and run the BIAcore X control software.
3. The buffer inlet tube is inserted into priming buffer and an empty beaker is placed under waste outlet to collect the liquid waste.

4. A "Dock" dialogue box will appear when the software is started and there is no chip in the instrument. To dock the sensor chip, open the chip port cover, pull the slider, place the chip in the slider and push the slider into the instrument, close the port cover, and click "Dock" in the software.

5. Prime the instrument using running buffer three times. A "Prime" dialogue box from the software guides users to complete this step.

6. When new chip is docked, it is important to normalize the signal response using BIAnormalizing solution. Choose Tools > Working Tools > Normalize and inject 100 µl BIAnormalizing solution.

7. Dilute the MBP-p33C protein to 1 µM in immobilization buffers at different pH levels, such as pH 4.0, 4.5, 5.0, and 6.0.

8. Start a sensogram by choosing Run > Run Sensogram. In the "Flowcell" dialogue box, choose single "Detection mode" and FC1 (Flow cell 1) "Flow Path" and set the flow rate at 10 µl min⁻¹.

9. Inject 10 µl of 1 µM MBP-p33 protein diluted in immobilization buffer pH 4.0, following sample injection procedures recommended by the manufacturer.

10. After the baseline of the sensogram is stabilized, repeat the injections with MBP-p33 diluted in immobilization buffer pH 4.5. Follow this step for MBP-p33 diluted in buffers at pH 5.0 and 6.0

11. After all the sample applications, inject 10 µl wash solution (50 mM NaOH) to clean the sample loop off the protein samples, stop data collection by selecting Run > Stop Sensogram, and save the data.

12. Compare the difference in electrostatic attraction between different buffers as shown in the sensogram in Fig. 1. Buffers at pH 4.5 and 5.0 facilitate highest preconcentration of MBP-p33C protein on chip surface. Since amine coupling is more efficient with uncharged amino groups on the protein at higher pH values and the theoretical pI of MBP-p33C is 5.5, we chose pH 5.0 for our immobilization procedure.

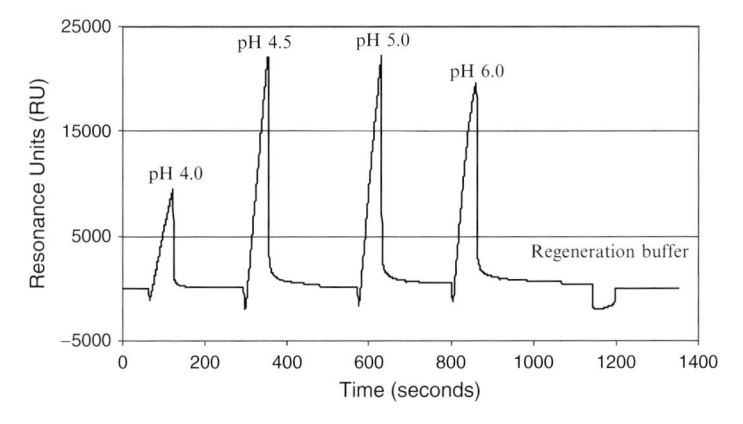

Fig. 1 The sensogram shows the effect of the pH of the immobilization buffer on electrostatic attraction of MBP-p33C to the chip surface

3.4 Immobilization of MBP-p33C on CM-5 Sensor Chip

Immobilization is performed using amine coupling kit from BIAcore following the manufacturer's protocols. An "Amine Coupling" surface preparation procedures wizard guides the user through the immobilization procedure with instructions for every step. BIAcore X divides the chip surface into two flow cells, each of which can be used to fix different proteins. We immobilized 8,000 RU MBP-p33C in Flow cell 1 (Fc1) and 8,000 RU MBP in Fc2 for qualitative (yes/no) binding analysis.

1. Thaw EDC and NHS stock solutions on ice and place ready-to-use ethanolamine–HCl on ice.
2. Choose Surface preparation procedures from "Tools" menu and select "Amine Coupling".
3. In the "Flow Cell" box, select "single" detection mode and "Fc1" flow path. Set the flow rate to 5 µl min^{-1}.
4. Mix NHS and EDC to 1:1 and inject 35 µl (7 min). This step activates the dextran matrix to give reactive esters.
5. Inject 35 µl of 1 µM MBP-p33C diluted in immobilization buffer pH 5.0.
6. Inject 35 µl ethanolamine–HCl to deactivate any remaining active esters.
7. Inject 5 µl of regeneration buffer (50 mM NaOH) to condition the immobilized chip surface. Stop the sensogram and save the data.
8. Using View > Baseline and View > Reference Line commands in BIAcore X control software determine the level of MBP-p33C bound to the chip (Fig. 2). This level is used to calculate the theoretical maximum binding capacity (R_{max}) of immobilized chip surface.
9. Repeat step 2–8 to prepare control surface in Flow cell 2 with MBP. This control surface is used to filter out noises in the experiment that result from nonspecific binding of proteins to chip surface or bulk effect due to differences in running buffer and protein solution.
10. For kinetic studies, we immobilized 250 RU of MBP-p33C in FC 1 and MBP in FC2 by following the steps 2–9 (12).

3.5 Binding Analysis

The chip prepared with one of the binding partner immobilized on its surface can be used for up to 100 times in binding assays. We prepared p33C protein chip and tested several mutant proteins of p33 and p92 proteins for their interaction with p33C. Here we describe the self-interaction between p33C molecules.

3.5.1 Qualitative Binding Analysis

1. Dock the CM-5 chip immobilized with 8,000 RU MBP-p33C in Flow cell 1 and 8,000 RU MBP in Flow cell 2 (6).

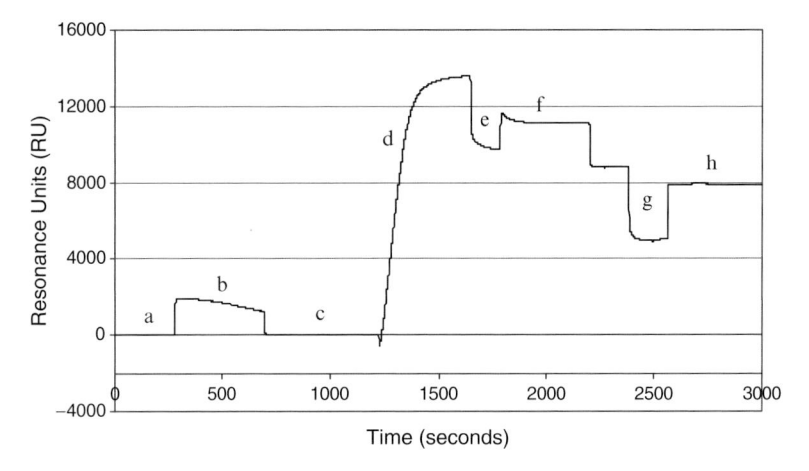

Fig. 2 Sensogram showing MBP-p33C immobilization. (a) Starting baseline, (b) EDC/NHS activation of dextran matrix, (c) baseline after surface activation, (d) covalent coupling of MBP-p33C to dextran matrix, (e) loosely bound MBP-p33C is washed away by immobilization buffer, (f) deactivation by ethanolamine, (g) surface conditioning by regeneration buffer (50 mM NaOH), (h) actual immobilization level of MBP-p33C, which is 8,000 RU in this case

2. Prime the system with running buffer as guided by the "Prime" wizard in the software.
3. Start a sensogram from BIAcore X control software and in the "Flowcell" dialogue box choose multichannel detection mode, flow path Fc1–2 and Fc2 (MBP) as reference cell.
4. Set the flow rate at $20 \mu l \, min^{-1}$.
5. Dilute MBP-p33C to $1 \mu M$ in the running buffer.
6. Inject $60 \mu l$ (3 min) of $1 \mu M$ MBP-p33C in the running buffer over the p33C chip surface. The protein flows through both Fc1 (coupled with p33C) and Fc2 (coupled with MBP) and a sensogram for each flow cell is generated in real time.
7. After 3 min of protein injection, allow the running buffer to pass for additional 3 min to record the dissociation phase of protein interactions.
8. Inject $10 \mu l$ of regeneration buffer to remove any bound proteins and bring the chip surface back to base level for the next binding assay.
9. Stop the sensogram and save the data. Undock the sensor chip from the instrument and store it in refrigerator.
10. Open the data file in BIAevaluation software, select both sensogram Fc1 and Fc2 to display graph, remove any air spikes and the regeneration phase data set to highlight only the association and dissociation phases of the interactions in the graph as shown in Fig. 3a,b.
11. Zero baseline on Y-axis using "Y-transform" menu.
12. Subtract the experimental data (MBP-p33C) from reference surface (MBP) using Y-transform menu. Any nonspecific binding of p33C to MBP or dextran matrix chip surface or bulk effect resulting from differences in running buffer and protein solution is canceled out by this subtraction based on data from the reference surface.

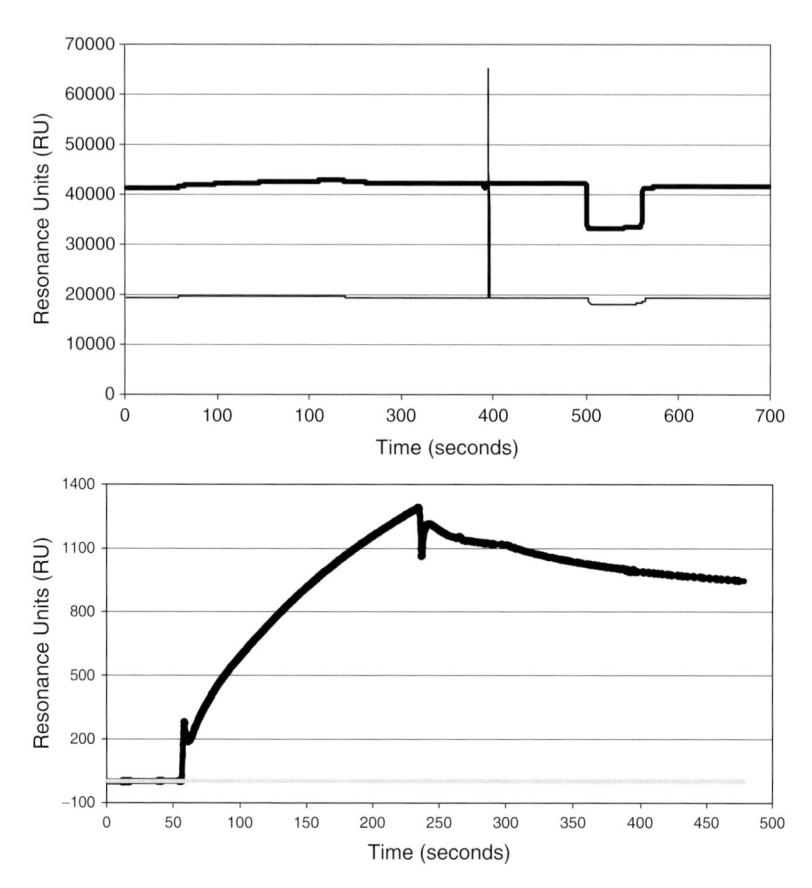

Fig. 3 (**a**) Sensogram as generated in real time by BIAcore X control software. The darker line represents data from p33C-coupled surface (Fc1) and lighter line represents data from MBP-coupled reference surface (Fc2). Note that the air spike in Fc2 and also the regeneration phase burry the most important data showing association and dissociation phases of the binding curve. (**b**) Sensogram after processing by using BIAevaluation 4.1 software. The experimental sensogram (Fc1) is subtracted from reference sensogram (Fc2) and the air spikes and regeneration phases are removed

13. Export the resultant data file after the above modifications as text document and use Microsoft Excel program to make the necessary changes to prepare publication quality graphs.

3.5.2 Kinetic Analysis

1. Dock the CM-5 chip immobilized with 250 RU MBP-p33C in Flow cell 1 and 250 RU MBP in Flow cell 2 (12).
2. Prime the system with running buffer as guided by the "Prime" wizard.
3. Start a sensogram from BIAcore X control software and in the "Flowcell" dialogue box choose multichannel detection mode, flow path Fc1–2 and Fc2 (MBP) as reference cell.
4. Set the flow rate at $40 \, \mu l \, min^{-1}$.

5. Dilute MBP-p33C to 25, 50, 75, 100, 250, and 500 nM in the running buffer.
6. Inject 100 μl (2.5 min) of 25 nM MBP-p33C in the running buffer over the p33C chip surface. The protein flows through both Fc1 (coupled with p33C) and Fc2 (coupled with MBP) and a sensogram for each flow cell is generated in real time.
7. After protein injection, allow the running buffer to pass for additional 3 min to record the dissociation phase of protein interactions.
8. Inject 10 μl of regeneration buffer to remove any bound proteins and bring the chip surface back to base level for the next binding assay.
9. Stop the sensogram and save the data.
10. Repeat steps 6–9 for 50, 75, 100, 250 and 500 nM p33C proteins.
11. Open the data file in BIAevaluation software, select both sensogram Fc1 and Fc2 to display graph, remove any air spikes and the regeneration phase data set to highlight the association and dissociation phases of the interactions in the graph.
12. Zero baseline on Y-axis using "Y-transform" menu.
13. Subtract the experimental data (MBP-p33C) from data obtained with the reference surface (MBP) using Y-transform menu and save.
14. Repeat steps 12 and 13 for sensograms generated for each protein concentration.
15. Add all the reference-subtracted sensograms from step 13 in one data file and overlay using "Overlay Plot" menu.
16. Align the curves on X-axis to the injection-start time point using X-transform menu.
17. Fit the data using "Simultaneous ka/kd" menu and follow the instructions as guided by this application wizard.
18. Evaluate the curve-fitting first by visual assessment of deviations of experimental data from mathematically fitted data. The data shown in Fig. 4 represent a well-fitted experimental data.

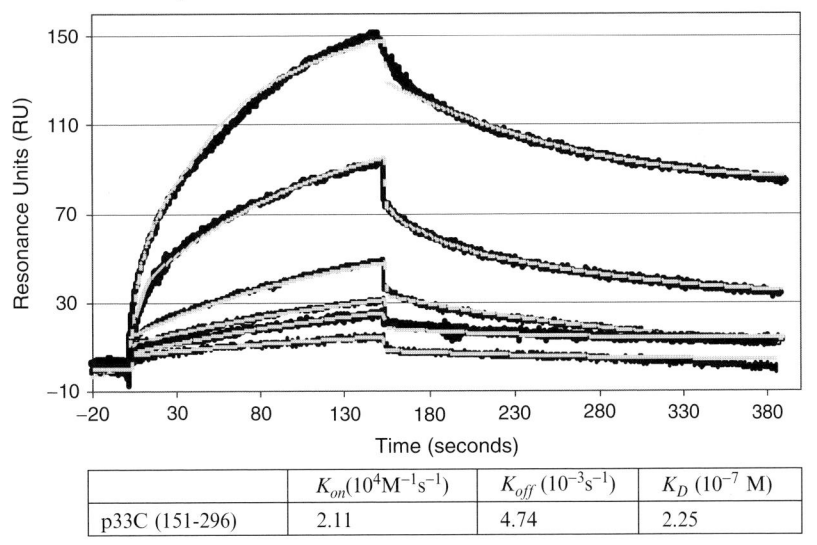

	$K_{on} (10^4 M^{-1} s^{-1})$	$K_{off} (10^{-3} s^{-1})$	$K_D (10^{-7} M)$
p33C (151-296)	2.11	4.74	2.25

Fig. 4 Fitted Kinetic data showing interaction between p33C–p33C proteins. Varying concentrations of highly purified p33C were used (25, 50, 75, 100, 250, and 500 nM) to obtain the above data. The experimental data is fitted to 1:1 binding model using BIAevaluation software. The dark line represents experimental data and the grey line represents mathematical fitting. The table below shows the kinetic parameters derived from experimental data

19. Export the resultant kinetic data file after fitting as text document and use Microsoft Excel program to make the necessary changes to prepare publication quality graphs.

4   Notes

1. Before starting the BIAcore X instrument, go through the maintenance logbook to be sure that the instrument is in good condition and the signal quality in the flow cells is good. Desorbing and sanitization need to be done to clear the flow path off protein or other residues.
2. The purity of the protein is a very important factor for kinetic analysis. The active concentration of impure sample is very difficult to quantitate. Impure samples also contribute to nonspecific binding and bulk effect. Therefore, it is essential to use highly pure protein samples for kinetic analysis. Purity is not a major problem in qualitative (yes/no) type of binding assay provided a reference surface is prepared with a known nonbinding mutant protein.
3. The response signal of the protein being passed on the chip surface is dependent on protein molecular weight. Therefore, BIAcore recommends immobilizing the lower molecular weight proteins onto the chip.
4. The nonspecific binding of active proteins or impurities to chip surface or to immobilized protein on the chip and bulk refractive change due to differences in buffer composition during association phase can be normalized by subtracting the reference cell response from experimental data. Therefore, reference cell preparation assumes critical importance in SPR experiments. The good reference surfaces are the ones immobilized with mutant test protein that lost binding ability. In our case, we found that MBP in reference cell works well for this purpose because all our assay proteins are MBP fusions.
5. All the buffers used for SPR experiments should be filtered and degassed daily to avoid clogging the micro flow circuit. If the buffers are not degassed, there will be frequent air spikes in the sensogram making it difficult to get reproducible information from the experimental data.
6. Biochemical characteristics such as aggregation state of proteins and stoichiometry of the protein interaction influence the outcome of the kinetic analysis. We checked the heterogeneity of our protein samples in gel filtration column and found majority of the proteins eluted in a single peak.
7. Regeneration conditions in our experiments are different for different binding partners. In general, 50 mM NaOH worked well. But for some truncated p33 mutant proteins, we used 10 mM glycine–HCl pH 2.5, which provides milder regeneration condition than 50 mM NaOH. BIAcore sells a set of regeneration buffers and the BIAcore handbooks also list many other regeneration solutions to try.

Acknowledgments   The authors thank Dr. J. Pogany for discussion. This work was supported by NIH-NIAID.

References

1. Ahlquist, P. (2002) *Science* **296,** 1270–3.
2. Buck, K. W. (1996) *Adv Virus Res* **47,** 159–251.
3. Nagy, P. D., and Pogany, J. (2006) *Virology* **344,** 211–20.
4. Goregaoker, S. P., Lewandowski, D. J., and Culver, J. N. (2001) *Virology* **282,** 320–8.
5. O'Reilly, E. K., Paul, J. D., and Kao, C. C. (1997) *J Virol* **71,** 7526–32.
6. Rajendran, K. S., and Nagy, P. D. (2004) *Virology* **326,** 250–61.

7. Crouch, R. J., Wakasa, M., and Haruki, M. (1999) *Methods Mol Biol* **118,** 143–60.
8. Goodrich, T. T., Wark, A. W., Corn, R. M., and Lee, H. J. (2006) *Methods Mol Biol* **328,** 113–30.
9. Pattnaik, P. (2005) *Appl Biochem Biotechnol* **126,** 79–92.
10. White, K. A., and Nagy, P. D. (2004) *Prog Nucleic Acid Res Mol Biol* **78,** 187–226.
11. Rajendran, K. S., and Nagy, P. D. (2003) *J Virol* **77,** 9244–58.
12. Rajendran, K. S., and Nagy, P. D. (2006) *Virology* **345,** 270–9.

Chapter 20
Biochemical Approaches for Characterizing RNA–Protein Complexes in Preparation for High Resolution Structure Analysis

Raúl C. Gomila and Lee Gehrke

Abstract RNA–protein interactions control viral RNA replication, transcription, translation, and particle assembly. Progress toward understanding the functional significance of RNA–protein complexes in the viral life cycle is hindered by the lack of high resolution structural information. Challenges to acquiring structural data include RNA's inherent instability and conformational plasticity, coupled with the comparatively high cost of generating large quantities of RNA for biophysical experiments. The potential for successful structure determination is increased by conducting biochemical experiments that outline interacting domains and identify key residues. These approaches are aimed at defining and characterizing RNA and protein substrates that are suitable for high resolution structural analysis.

Keywords Structure; RNA–protein interaction; Virus; RNA; Peptide; RNA structure; Crystallography

1 Introduction

RNA–protein interactions have important structural and functional roles at every stage in the life cycle of positive strand RNA viruses. As illustrated by recent work on reovirus (1–4), the coupling of biochemistry and structural biology is a powerful experimental approach that provides significant mechanistic insight into understanding viral entry, RNA translation, RNA replication, and particle assembly. After identifying an interesting RNA–protein interaction, investigators often set a goal of solving a high-resolution structure. However, the potential for fulfilling that goal is uncertain because of the technical challenges of X-ray crystallography (5) and nuclear magnetic resonance spectroscopy.

The argument presented in this chapter is that the potential for success in determining structure is increased substantially by first mapping the interacting RNA and protein domains and identifying key nucleotide/amino acid residues. While biochemical experiments do not guarantee a structure, they provide a wealth of relevant structural and functional information while defining substrates that are suitable for

From: *Methods in Molecular Biology, Vol. 451, Plant Virology Protocols: From Viral Sequence to Protein Function*
Edited by G.D. Foster, I.E. Johansen, Y. Hong, and P.D. Nagy © Humana Press, Totowa, NJ

initial structural analysis. Here we provide detailed descriptions of the some of the methods that have been used to characterize dengue virus RNA–protein interactions and alfalfa mosaic virus RNA–coat protein interactions (6–15), the latter of which led to the determination of a high-resolution cocrystal structure (16).

2   Materials

1. Buffer A: 10 mM Na_2HPO4 pH 7.2, 100 mM NaCl, 0.1 mM EDTA, 5% glycerol, 7 mM β-mercaptoethanol.
2. Poly-Prep Column (Bio-Rad) (10 ml)
3. Affinity column elution buffers: 15 mM Tris–HCl pH 7.5 containing 250 mM, 500 mM, 1 M or 2 M NaCl.
4. Desalting and concentration column: Nanosep 3 K centrifugal ultrafiltration device (Pall).
5. PBST: phosphate-buffered saline (PBS) containing 0.1% Tween-20.
6. 10× HBB buffer: 240 mM HEPES-KOH pH 7.5, 250 mM NaCl, 50 mM $MgCl_2$, 70 mM β-mercaptoethanol.
7. 8 M guanidine chloride (Mallinckrodt).
8. HBB-guanidine buffers: 1× HBB containing various concentrations of guanidine chloride for denaturing and renaturing proteins in the Northwestern blotting analysis (see text).
9. HYB100: 20 mM HEPES-KOH pH 7.5, 200 mM KCl, 2.5 mM $MgCl_2$, 0.1 mM EDTA, 0.05% NP40, 7 mM β-mercaptoethanol.
10. Sepharose gel swelling buffer: 1 mM HCl (ice-cold).
11. 30 ml Buchner funnel with fritted disc.
12. In vitro transcription: done with kits available from a number of suppliers.
13. 4× RNA renaturation buffer: 40 mM Tris–HCl pH 7.5, 200 mM NaCl, 12 mM $MgCl_2$, 0.4 mM EDTA.

3   Methods

The methods described here are aimed at characterizing an RNA–protein or RNA–peptide complex using biochemical methods, with a further goal of defining substrates appropriate for high resolution structure analysis using X-ray crystallography. The steps include (1) in vitro RNA transcription and purification, (2) Northwestern blotting analysis to identify proteins that bind to an RNA fragment, (3) RNA affinity chromatography to purify and identify RNA-binding proteins, (4) electrophoretic mobility bandshift analysis to estimate protein–RNA- or peptide–RNA-binding affinity, (5) hydroxyl radical footprinting with peptides to define the protein-binding domain on the RNA, and (6) enzymatic structure mapping with peptides to identify features of RNA secondary structure that may contribute to protein binding.

3.1 RNA Transcription, Gel Purification, and Renaturation (see Note 1)

1. RNA is prepared by in vitro transcription. To prepare radioactive RNA probe, a standard 20 µl transcription reaction is supplemented with 20 µCi of α-^{32}P-nucleotide. Following the transcription reaction, the labeled RNA is purified by gel electrophoresis.
2. RNA purification by gel electrophoresis. Full-length RNA is separated from premature termination products and unincorporated label (if probe is being prepared) by electrophoresis into a polyacrylamide/urea gel. The size of the gel is 20 cm by 20 cm by 0.8 mm thick. For RNAs 50–200 nucleotides in length, a 10% polyacrylamide-urea (National Diagnostics) gel is used. For sample loading, add an equal volume of formamide-dye loading buffer (supplied in the transcription reaction kit) to the transcription reaction. Heat the mixture for 5 min at 95° and load immediately onto the gel. Run gel at 10 W (constant power) in 1× TBE buffer until the bromophenol blue dye runs off the gel with the free nucleotide (see note 2).
3. RNA is visualized in the gel using ultraviolet light shadowing. A thin-layer chromatography plate with fluorescent indicator (AnalTech) is placed behind the wrapped gel, and ultraviolet light from a hand-held lamp is directed at the gel. The region of the gel containing the RNA will absorb the UV light, giving the appearance of a shadow on the thin-layer plate. A clean razor blade is used to make a rectangle of cuts around the band – quickly to minimize exposure and possible UV damage to the RNA. The RNA band (without plastic wrap) is transferred to a clean eppendorf tube.
4. Elute the labeled RNA. Crush the acrylamide gel slice in the eppendorf tube using a pipette tip as a pestle. Next, add 200 µl of gel elution buffer, and place the tube on a rocking table in the cold room overnight. To collect the RNA, the gel tube is transferred to a 37 °C water bath for 10 min, and then centrifuged at top speed in a microcentrifuge for 5 min. The supernatant is transferred to a clean tube. A Geiger counter can be used to monitor the efficiency of the elution. If needed, the pelleted gel pieces can be washed with an additional 50–100 µl of gel elution buffer.
5. Perform phenol extraction and precipitate the RNA. The eluate is extracted twice with phenol–chloroform. The RNA is precipitated from the aqueous phase by adding 1/10 volume of 3 M sodium acetate (pH 5.2) and 3 volumes of ethanol. Just before experimental use, the precipitated RNA is sedimented, washed twice with 70% ethanol, dried briefly, and solubilized in 20 µl RNase-free water.
6. Determine the RNA specific activity. One-half microliter of the solubilized RNA is analyzed by liquid scintillation counting, and 1 µl is used for determination of RNA concentration by ultraviolet light spectrophotometry. Approximately 500,000 cpm µl^{-1} is expected for RNA probes.
7. RNA renaturation. Prior to experimental use, the RNA is diluted to a final concentration in 1× REN buffer, heated for 2 min at 90 °C, and slow-cooled (see note 3).

3.2 Northwestern Analysis to Identify RNA-Binding Proteins

For some applications, the interacting domains of the RNA and protein may be
known prior to use of methods described here. In other cases, the characterization
may be at an earlier stage wherein there is evidence of RNA–protein interactions
without specific information on the interacting regions or without prior identification of the protein(s) that bind a particular RNA. A rapid screen for proteins that
bind to an RNA fragment can be done using Northwestern blotting methods (17).
Proteins are separated by SDS-polyacrylamide gel electrophoresis, transferred to a
membrane, renatured, and then probed using a radiolabeled RNA fragment.

1. SDS-PAGE and transfer: Proteins are separated by electrophoresis into a 10%
   gel (BioRad) and are transfered to nitrocellulose membrane using standard western blotting methods and equipment.
2. Nonspecific interaction sites are blocked by soaking the membrane in 5% nonfat
   milk (Blotto) in PBST for 1 h at room temperature.
3. Wash the membrane in 1× HBB for 10 min.
4. Wash the membrane twice for 10 min at room temperature using HBB containing 6 M guanidine chloride.
5. Gradually renature the bound proteins by washing the membrane sequentially in
   1× HBB buffers containing 3 M, 1.5 M, 0.75 M, 0.375 M, and 0.187 M guanidine
   chloride. Each of these washes is done for 10 min at room temperature. The volume needed for each wash solution is approximately 10 ml for a 9 cm × 7 cm
   membrane. Do not reduce the time allocated to these washes.
6. Wash the membrane once in 1× HBB for 10 min at room temperature.
7. Wash the membrane twice in HYB100 at room temperature, 10 min each.
8. Probe solution is prepared by adding 2 µl of the labeled RNA (see Sect. 3.1.4;
   about 1×10^6 cpm) to 2 ml of HYB100. The membrane is transferred to a heat-
   sealable pouch and the 2 ml of probe solution is added. The membrane is incubated with probe for 4 h at room temperature on a rocking table. The membrane
   is then washed three times for 10 min each in 10 ml HYB100 buffer, wrapped in
   plastic film, and exposed to X-ray film overnight.

An example of a representative Northwestern blot analysis is presented in Fig. 1.

3.3 Identification and Purification of RNA-Binding Proteins Using Affinity Chromatography

RNA affinity chromatography is a gentle and relatively rapid approach for enriching
a population of RNA-binding proteins. The methods for RNA affinity chromatography described here are based on published work (18). Starting with an RNA fragment
of potential regulatory interest, such as a 5′ or 3′ untranslated region, interacting proteins can be enriched from a cell lysate, followed by further characterization. We have
found that affinity chromatography on Sepharose gave the best signal with lowest

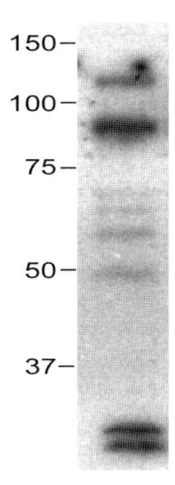

Fig. 1 Northwestern blot. Proteins present in approximately 10 µg of HeLa cell cytoplasmic extract were separated by electrophoresis into a 10% SDS-polyacrylamide gel and then transferred to a nitrocellulose membrane. Bound proteins were gradually denatured and renatured as described in the Methods section. The RNA probe for the analysis is the 3′ untranslated region of dengue fever virus RNA that was radiolabeled during in vitro transcription in the presence of α-32P UTP. The analysis revealed a number of protein bands that interacted with the labeled dengue virus RNA

background of nonspecific RNA-binding proteins; however, alternate methods have been described (19).

1. The matrix for affinity chromatography is cyanogen bromide-activated Sepharose 4B (Amersham). Swell sufficient matrix for both a preclear (no RNA) column and for an RNA affinity column. Weigh approximately 0.6 g of dry Sepharose, which will make about 2 ml of swelled matrix. Add the 0.6 g Sepharose to 20 ml of 1 mM HCl in a 50 ml plastic tube. Place the slurry on a rocking table at room temperature for 1 h.

2. Pour the slurry onto the Buchner funnel with fritted disc, and wash the Sepharose with 100 ml of ice-cold 1 mM HCl. Once the matrix is swollen, we have to proceed immediately with the coupling step.

3. RNA coupling. After washing the Sepharose with 1 mM HCl, the matrix is suspended and washed off the Buchner funnel using 20 ml of 10 mM Tris, pH 6.8 (see note 4). Equal volumes of the slurry are transferred to each of two 15 ml plastic tubes. For the RNA affinity column, approximately 100 µg of RNA is added to one tube of Sepharose matrix (see note 5). Both tubes containing Sepharose slurry are placed on a rocking table in the cold room overnight.

4. Following the overnight coupling incubation, sediment the two Sepharose preparations by brief centrifugation in a clinical centrifuge. Remove the supernatant 10 mM Tris solution. Replace the Tris solution with 10 ml Buffer A and then resuspend and decant each of the resuspended Sepharose matrix slurries into individual 10 ml polyprep columns.

5. Affinity chromatography. As a preclearing step to minimize nonspecific interactions, cell extracts are first passed through the control (no-RNA) column four

times. After the fourth pass, the collected eluate is then passed through the RNA affinity column four times.

6. Both columns are washed with ten column volumes (10 ml) of Buffer A.

7. Proteins are eluted from the two columns with buffers containing increasing NaCl concentrations. Each elution is done with two column volumes (about 2 ml), using a buffer that contains 15 mM Tris–HCl pH 7.5 plus 250 mM, 500 mM, 1 M, and 2 M NaCl (see note 6).

8. Desalting and sample concentration: From the 2 ml eluates, 500 µl aliquots are loaded into separate Nanosep columns, which are then centrifuged at 14,000 rpm for 10–20 min to reduce the volume. The loading and centrifugation cycle is repeated until the total 2 ml eluate volumes are reduced to about 25 µl. The concentrated protein solution is then diluted to 500 µl in Buffer A in the Nanosep device to normalize the buffer concentrations. The volume is then reduced to less than 100 µl by further centrifugation in the Nanosep column.

9. Analysis of eluted proteins by SDS-polyacrylamide gel electrophoresis. Eluted and concentrated protein samples are analyzed by electrophoresis into a 10% SDS-polyacrylamide gel, followed by staining with silver or Coomassie dye. Samples from the RNA-affinity column and the negative control column are run side-by-side for comparison. Bands that are specific for the RNA affinity column are potential specific RNA-binding proteins. A representative protein gel is presented in Fig. 2. The arrows indicate proteins specific to the RNA affinity column preparation.

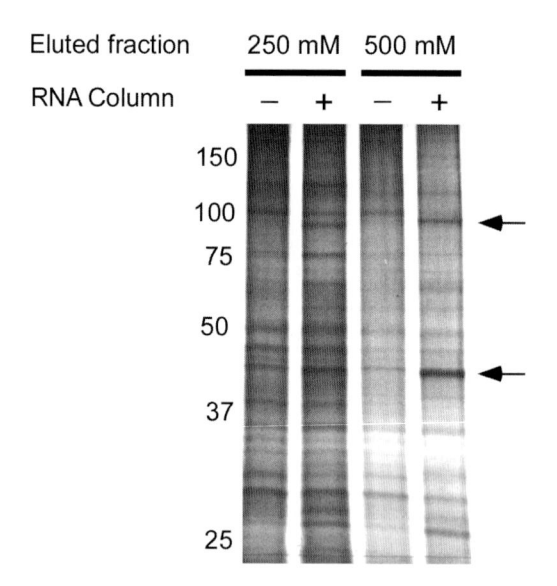

Fig. 2 Proteins eluted from the RNA affinity chromatography are separated using SDS-PAGE, followed by silver staining to visualize the individual protein bands. At least two bands (arrows) are enriched in the samples eluted from the 500 mM fraction of the RNA column(+) as compared to the matrix-only preclearing column (−)

10. Identification of bound proteins by mass spectroscopy. Gel slices containing protein bands that are specific to the RNA affinity column are excised and processed for mass spectrometry.
11. Protein expression. Following the identification of the RNA-binding protein, detailed experiments will require the availability of purified expressed protein. The methods for protein expression are beyond the scope of this review; however, clones for many proteins can be obtained as cDNAs from I.M.A.G.E. Consortium (http://image.llnl.gov/), which are available for purchase from ATCC. As an alternate approach, chemically synthesized peptides representing RNA-binding domains can be used (20) (see note 7).

3.4 Electrophoretic Mobility Bandshift Analysis to Estimate Protein–RNA- or Peptide–RNA-Binding Affinity

The affinity of the RNA–protein interaction is relevant to understanding the function of the complex and also for evaluating the potential for structure analysis. Methods for estimating RNA–protein-binding affinity include nitrocellulose filter retention (21), electrophoretic mobility bandshift analysis (EMSA) (22, 23), and fluorescence quenching (24). Detailed methods for native gel electrophoretic analysis of nucleic acid–protein complexes have been published elsewhere (22, 25, 26).

Fig. 3 shows an example (27) of EMSA data where the binding of a 26-amino acid peptide from the alfalfa mosaic virus coat protein to 39-nucleotide fragments of alfalfa mosaic virus RNA (11) is analyzed. The apparent dissociation constant

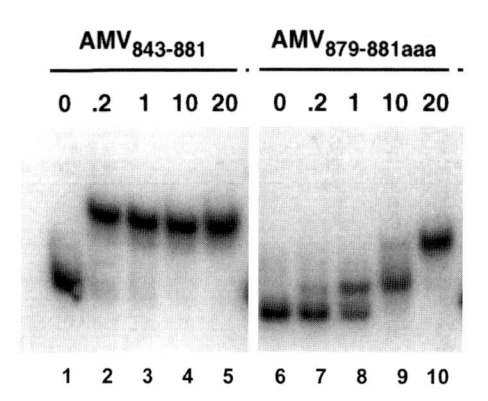

Fig. 3 Electrophoretic mobility bandshift analysis of a 26 amino acid peptide from the alfalfa mosaic virus coat protein bound to a 39-nucleotide 3′-terminal viral RNA fragment. Lanes 1–5: the wild-type RNA sequence presented in Fig. 4A. Lanes 6–10: a variant RNA wherein nucleotides 879–881 have been changed to AAA, which diminishes peptide-binding affinity. The numbers above the lanes refer to the concentration (micromolar) of alfalfa mosaic virus N-terminal peptide CP26 in the binding reactions

for the RNA–peptide interaction is defined as the peptide concentration at which 50% of the RNA is shifted into RNA–peptide complex (23).

Compare lanes 1–5 with lanes 6–10. Note that for the wild-type RNA fragment (lanes 1–5), the Kd cannot be determined accurately from these data because more than half of the RNA is shifted into complex at every concentration. It can be concluded, however, that the Kd is less than 0.2 μM. The three nucleotide changes in the variant RNA (lanes 6–10) diminish the affinity for the peptide. The apparent Kd of the interaction is approximately 1 μM (lane 8). A phosphorimager or film scanning instrument can be used for quantitative analysis of the binding kinetics (see note 8).

3.5 Mapping the Protein-Binding Domain on RNA Using Peptides and Hydroxyl Radical Footprinting

Defining the RNA nucleotides that are in contact or in close proximity to the bound protein is useful for learning how the protein recognizes the RNA for binding and also for defining a minimal binding site for structure analysis. Structure determination often requires insertion of heavy metal nucleotide derivatives to facilitate phase determination. In the absence of footprinting information, random insertions may disrupt the RNA–protein complex, thereby wasting time and resources. Hydroxyl radical footprinting provides high resolution mapping data to define potential contact areas between RNA and protein.

The methods for performing hydroxyl radical footprinting have been described in detail previously (28), and an example (27) of hydroxyl radical footprinting data is presented in Fig. 4b. Several points can be noted. First, the untreated RNA (lane 1) is intact, showing no detectable RNA hydrolysis. Second, the T1 footprint under denaturing conditions (lane 2), when coupled with the hydroxyl radical ladder (lane 4), permits unequivocal correlation of the band pattern with the nucleotide sequence (Fig. 4a). Third, the protection patterns, indicated by the brackets, demonstrate regions of the RNA that are in contact or in close proximity to the peptide. The use of a modified peptide is shown in this figure, where a single amino acid change R17K disrupts peptide binding (lane 7).

3.6 Enzymatic Structure Mapping with Peptides to Identify Features of RNA Secondary Structure that May Contribute to Protein Binding

Understanding the relationships between the structure of an RNA–protein complex and the functional significance is aided by defining elements of RNA secondary structure. RNA secondary structure analysis requires a unique end label, which can be either 5′ or 3′. The preparation of intact end-labeled RNA is critical for structure mapping analysis. Detailed protocols for conducting enzymatic structure mapping

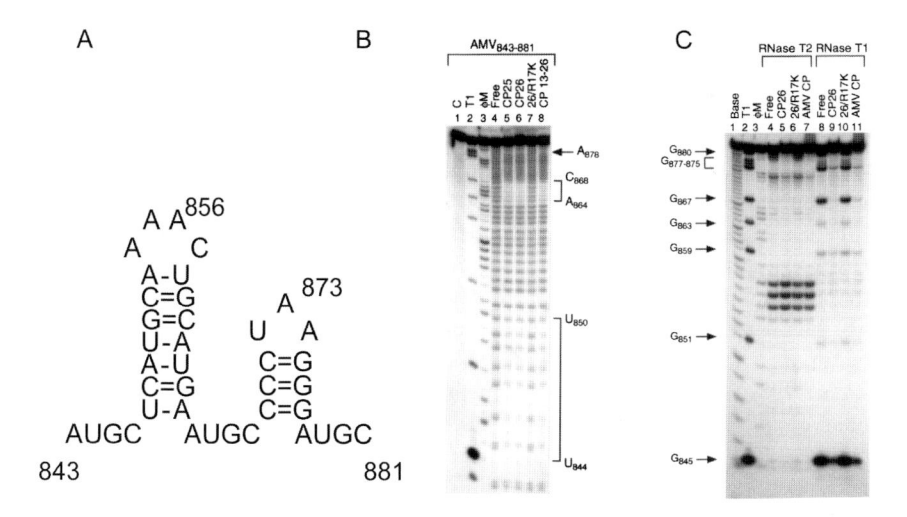

Fig. 4 Alfalfa mosaic virus coat protein–RNA interactions analyzed by hydroxyl radical footprinting and enzymatic structure mapping. (a) Nucleotide sequence and secondary structure map of the 3′ terminal 39 nucleotides of alfalfa mosaic virus RNA 4. (b) Hydroxyl radical footprinting analysis of peptide binding to the 39-nucleotide RNA fragment. Lane 1: RNA only; Lane 2: ribonuclease T1 map, denaturing conditions; Lane 3: RNA digested with ribonuclease PhyM under denaturing conditions in urea to cleave primarily at adenosine and uridine bases; Lane 4: RNA (without protein or peptide) reacted with hydroxyl radicals to generate a ladder; Lane 5: hydroxyl radical footprinting of RNA bound to a peptide representing amino acids 1–25 of the viral coat protein; Lane 6: hydroxyl radical footprinting of RNA bound to a peptide representing amino acids 1–26 of the viral coat protein; Lane 7: 26/R17K, hydroxyl radical footprinting of RNA bound to a nonbinding peptide (13) representing amino acids 1–26 of the viral coat protein but containing a point mutation converting arginine 17 to lysine; Lane 8: hydroxyl radical footprinting of RNA bound to a peptide representing amino acids 1–13 of the viral coat protein. (c) Secondary structure mapping using single strand-specific ribonucleases T2 and T1. Lane 1: alkaline hydrolysis ladder; AMV CP: full-length alfalfa mosaic virus coat protein. Other labels are as shown in panel B. Lanes 4–7: RNA or RNA–peptide complexes digested with the T2 ribonuclease (nondenaturing conditions); lanes 8–11: RNA or RNA–peptide complexes digested with ribonuclease T1 (nondenaturing conditions)

have been published (29). An example of secondary structure mapping data is presented in Fig. 4c, where single-stranded nucleases T1 and T2 have been used (27).

4 Concluding Remarks

This chapter describes methods that can be used to define and biochemically characterize RNA–protein complexes. The emphasis is on understanding the functional significance of the complex while providing the groundwork for potential high resolution structure analysis. These biochemical methods define the interacting domains on the RNA and protein and can be used to minimize the sizes of RNA and protein molecules to be used for structure determination.

A summary of the data generated by applying these methods for characterizing the alfalfa mosaic virus RNA–coat protein interaction (6–16, 27, 28, 30) is presented in Fig. 5. Hydroxyl radical footprinting identified the shaded areas as potential coat protein-binding regions, and these interacting domains were later confirmed by the crystal structure (16). The footprinting data allowed us to limit the size of the RNA used for crystallography to 39-nucleotides, while peptide experiments defined arginine-17 as a key interacting residue (13). The base pairing pattern and secondary structure identified in the structure was, in large part, predicted from experimental RNA structure mapping (27, 31). In vitro genetic selection was also extremely valuable for identifying elements of the RNA sequence and structure that were critical for protein binding (6, 9, 12). The combined biochemical data strongly suggested that, although the two base-paired stems were required for binding, the loop nucleotides capping the two hairpins were not critical determinants. We used this information in the structure analysis to predict where to insert bromouridine nucleotides (Fig. 5, inserted at the circled positions) to permit phase determination without disrupting the complex.

Technology for obtaining high resolution structures of RNA and RNA protein complexes has improved significantly and now includes high throughput instrumentation for faster screening of conditions that yield crystals. Nonetheless, it is difficult to predict with any certainty if a complex will be suitably well-behaved to permit structure determination. Similarly, it is challenging to state exactly how much biochemical information is needed prior to initiating attempts at growing crystals. Among the relatively small number of RNA–protein complexes that have been examined by either NMR or crystallographic methods, most have been subjected to significant biochemical characterization prior to structure determination (23, 32–41). The implication is that the probability of successful structure analysis correlates positively with characterizing the complex using biochemical methods.

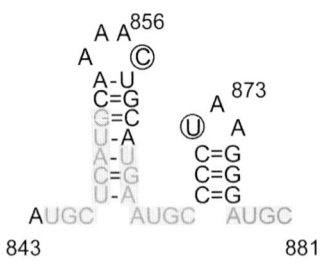

Fig. 5 Partial summary of experimental data. The shaded nucleotides were protected by peptide or coat protein from enzymatic or hydroxyl radical cleavage. The circled nucleotides at positions 857 and 871 are sites where bromouridine was inserted for phase determination during the X-ray crystallographic analysis

Notes

1. Minimizing potential RNase contamination at all stages of the experiments is critical. In addition to using ribonuclease-free water and reagents, all equipment such as gel electrophoresis boxes, combs, and glass plates should be treated carefully to remove or avoid RNase contamination. We use reverse-osmosis water, which is then passed through polishing deionizing tanks and a charcoal filter before being irradiated with ultraviolet light (Hydro Systems). Autoclaved water is sterile, but not necessarily RNase-free. Treating water with 0.1% diethylpyrocarbonate (DEP) and followed by autoclaving is effective, but adds an additional step. Tris buffers cannot be treated with DEP. Many RNA transcription kits and other biochemical kits include RNase-free water, and that water should be used and restricted to use with RNA-related applications. Mixed use of buffers, loading solutions, etc. for DNA and RNA use can introduce RNase contamination and should be avoided. All glassware should be baked overnight at 200–250 °C. Gel electrophoresis boxes can be wiped with commercially available towels (RNase-Zap, Ambion), followed by rinsing with RNase-free water. Gel combs should also be cleaned and rinsed before use, and the electrode buffer should be made with RNase-free water. We use commercially available 10× TBE buffers without further treatment. Plasticware in our laboratory is rinsed with 5% acetic acid, then with RNase-free water before use. RNA should be stored either as a precipitate at −20 °C, or in aqueous solution at −80 °C.

2. Prerun the gel for 30 min before loading. Following the electrophoresis run, the lower buffer chamber will contain the unincorporated radioactivity, and care must be taken to avoid contaminating other surfaces. To convert the radioactive gel buffer to solid waste for easier disposal, carefully pour the buffer into a plastic bin packed with paper towels and then left to air-dry in an appropriate shielding.

3. RNA precipitation in ethanol or RNA storage in aqueous solution at −80° can cause RNA to aggregate. The renaturation process is useful for disrupting the aggregates and slow cooling in a solution containing magnesium will facilitate RNA folding. We have found, however, that more stringent conditions are sometimes needed. For example, the dengue virus RNA fragment used for the RNA affinity column aggregates unless the RNA is quick-cooled rather than slow-cooled. The presence or absence of RNA aggregates can be monitored by native gel electrophoresis in agarose or polyacrylamide.

4. The activated Sepharose reacts with primary amine groups to form covalent bonds that link the RNA to the matrix. The conditions for the coupling are aimed at maintaining RNA stability and preventing formation of an excessive number of covalent bonds between the RNA and matrix, which could impede protein binding by steric hindrance. Coupling is done at pH 6.8 because RNA stability decreases at alkaline pH. To control the number of covalent bonds formed, we combined the coupling and blocking of the reactive sites on the matrix by using 10 mM Tris buffer. The amine groups in the dilute Tris buffer compete with the RNA for coupling sites. The result is that the RNA is covalently linked while remaining accessible for protein binding.

5. Approximately five transcription reactions, 20 μl in volume each, are generally sufficient to generate 100 μg of RNA for the affinity column coupling.

6. The affinity columns do not store well and should be used immediately after the coupling and washing reactions are completed. Reuse of the columns is not recommended.

7. The use of peptides instead of the full-length protein facilitates the characterization of the complex because (1) it avoids the need for protein expression and purification, and (2) it simplifies testing mutants to evaluate the importance of individual amino acids. Characterizing the alfalfa mosaic virus RNA–coat protein interaction (11, 13, 16, 28) was simplified significantly by using peptides representing the RNA-binding domain of the viral coat protein instead of the full-length coat protein. Like many plant viral coat proteins, the alfalfa mosaic virus coat protein has a highly basic N-terminal "arm," which was suspected to be the RNA-binding domain. Conveniently, a simple trypsin cleavage removed the N-terminus and also disrupted RNA–protein binding, thus identifying it as a 25-amino acid RNA-binding domain (42). Depending on the

size of the RNA-binding protein, it may be useful to synthesize peptides (30 amino acids in length) that cover suspected RNA-binding domains. Though not inexpensive, the synthesis can be cost-effective when considering the costs of labor-intensive and time-consuming protein expression and purification.

8. A common error in determining the affinity of RNA–protein complexes is to use RNA concentrations that are in excess of the dissociation constant. The analysis should be done using extremely low RNA concentrations, and in protein excess. Carey and Uhlenbeck (21, 43) discuss the thermodynamic basis.

Acknowledgments Patricia Ansel-McKinney contributed in generating the data presented in Figs. 3 and 4. This work was supported by the National Institutes of Health (GM42504) and The Ellison Medical Foundation (ID-SS-0147–01). NIH predoctoral fellowship was given to RCG (GM64985).

References

1. Tao, Y., Farsetta, D. L., Nibert, M. L., and Harrison, S. C. (2002) *Cell* **111,** 733–45.
2. Liemann, S., Chandran, K., Baker, T. S., Nibert, M. L., and Harrison, S. C. (2002) *Cell* **108,** 283–95.
3. Olland, A. M., Jane-Valbuena, J., Schiff, L. A., Nibert, M. L., and Harrison, S. C. (2001) *Embo J* **20,** 979–89.
4. Reinisch, K. M., Nibert, M. L., and Harrison, S. C. (2000) *Nature* **404,** 960–7.
5. Ke, A., and Doudna, J. A. (2004) *Methods* **34,** 408–14.
6. Boyce, M., Scott, F., Guogas, L. M., and Gehrke, L. (2006) *J Mol Recognit* **19,** 68–78.
7. Petrillo, J. E., Rocheleau, G., Kelley-Clarke, B., and Gehrke, L. (2005) *J Virol* **79,** 5743–51.
8. Guogas, L., Laforest, S., and Gehrke, L. (2005) *J Virol* **79,** 5752–61.
9. Rocheleau, G., Petrillo, J., Guogas, L., and Gehrke, L. (2004) *J Virol* **78,** 8036–46.
10. Laforest, S. M., and Gehrke, L. (2004) *Rna* **10,** 48–58.
11. Ansel-McKinney, P., and Gehrke, L. (1998) *J Mol Biol* **278,** 767–85.
12. Houser-Scott, F., Ansel-McKinney, P. A., Cai, J. M., and Gehrke, L. (1997) *J Virol* **71,** 2310–19.
13. Ansel-McKinney, P., Scott, S. W., Swanson, M., Ge, X., and Gehrke, L. (1996) *EMBO J* **15,** 5077–84.
14. Houser-Scott, F., Baer, M. L., Liem, K. F., Jr., Cai, J. M., and Gehrke, L. (1994) *J Virol* **68,** 2194–205.
15. Baer, M., Houser, F., Loesch-Fries, L. S., and Gehrke, L. (1994) *EMBO J.* **13,** 727–35.
16. Guogas, L. M., Filman, D. J., Hogle, J. M., and Gehrke, L. (2004) *Science* **306,** 2108–11.
17. Blackwell, J. L., and Brinton, M. A. (1995) *J Virol* **69,** 5650–8.
18. Copeland, P. R., and Driscoll, D. M. (1999) *J Biol Chem* **274,** 25447–54.
19. Rouault, T. A., Hentze, M. W., Haile, D. J., Harford, J. B., and Klausner, R. D. (1989) *Proc Natl Acad Sci USA* **86,** 5768–72.
20. Frankel, A. D. (1994) *in* "RNA–protein interactions" (Nagai, K., Mattaj, I. W., and Glover, D. M., Eds.), RL Press, New York, pp. 221–47.
21. Carey, J., and Uhlenbeck, O. C. (1983) *Biochemistry* **22,** 2610–5.
22. Carey, J. (1991) *Methods Enzymol* **208,** 103–17.
23. Calnan, B. J., Tidor, B., Biancalana, S., Hudson, D., and Frankel, A. D. (1991) *Science* **252,** 1167–71.
24. Paoletti, A. C., Shubsda, M. F., Hudson, B. S., and Borer, P. N. (2002) *Biochemistry* **41,** 15423–8.
25. Silver, S. C., and Hunt, S. W., 3rd (1993) *Mol Biol Rep* **17,** 155–65.
26. Lane, D., Prentki, P., and Chandler, M. (1992) *Microbiol Rev* **56,** 509–28.
27. Ansel-McKinney, P. (1996), Ph.D. Thesis, Harvard University, Cambridge, MA, pp. 249.

28. Ansel-McKinney, P., and Gehrke, L. (1997) *in* "Analysis of mRNA Formation and Function" (Richter, J. D., Ed.), Academic Press, New York, pp. 285–303.
29. Knapp, G. (1989) *Methods Enzymol* **180,** 192–212.
30. Swanson, M., Ansel-McKinney, P., Houser-Scott, F., Yusibov, V., Loesch-Fries, L. S., and Gehrke, L. (1998) *J Virol* **72,** 3227–34.
31. Quigley, G. J., Gehrke, L., Roth, D. A., and Auron, P. E. (1984) *Nucl Acids Res* **12,** 347–66.
32. Batey, R. T., and Williamson, J. R. (1996) *J Mol Biol* **261,** 536–49.
33. Batey, R. T., and Williamson, J. R. (1996) *J Mol Biol* **261,** 550–67.
34. Puglisi, J. D., Chen, L., Blanchard, S., and Frankel, A. D. (1995) *Science* **270,** 1200–03.
35. Price, S. R., Ito, N., Oubridge, C., Avis, J. M., and Nagai, K. (1995) *J Mol Biol* **249,** 398–408.
36. Oubridge, C., Ito, N., Teo, C. H., Fearnley, I., and Nagai, K. (1995) *J Mol Biol* **249,** 409–23.
37. Chen, L., and Frankel, A. D. (1995) *Proc Natl Acad Sci USA* **92,** 5077–81.
38. Tan, R. Y., and Frankel, A. D. (1994) *Biochemistry* **33,** 14579–85.
39. Battiste, J. L., Tan, R. Y., Frankel, A. D., and Williamson, J. R. (1994) *Biochemistry* **33,** 2741–47.
40. Tan, R., Chen, L., Buettner, J. A., Hudson, D., and Frankel, A. D. (1993) *Cell* **73,** 1031–40.
41. Puglisi, J. D., Tan, R., Calnan, B. J., Frankel, A. D., and Williamson, J. R. (1992) *Science* **257,** 76–80.
42. Zuidema, D., Bierhuizen, M. F. A., and Jaspars, E. M. J. (1983) *Virology* **129,** 255–60.
43. Carey, J., Cameron, V., de Haseth, P. L., and Uhlenbeck, O. C. (1983) *Biochemistry* **22,** 2601–10.

Chapter 21
Probing Interactions Between Plant Virus Movement Proteins and Nucleic Acids

Tzvi Tzfira and Vitaly Citovsky

Abstract Most plant viruses move between plant cells with the help of their movement proteins (MPs). MPs are multifunctional proteins, and one of their functions is almost invariably binding to nucleic acids. Presumably, the MP–nucleic acid interaction is directly involved in formation of nucleoprotein complexes that function as intermediates in the cell-to-cell transport of many plant viruses. Thus, when studying a viral MP, it is important to determine whether or not it binds nucleic acids, and to characterize the hallmark parameters of such binding, i.e., preference for single- or double-stranded nucleic acids and binding cooperativity and sequence specificity. Here, we present two major experimental approaches, native gel mobility shift assay and ultra violet (UV) light cross-linking, for detection and characterization of MP binding to DNA and RNA molecules. We also describe protocols for purification of recombinant viral MPs over-expressed in bacteria and production of different DNA and RNA probes for these binding assays.

Keywords Binding cooperativity; DNA; gel mobility shift; movement-protein–nucleic acid complexes; RNA; UV light cross-linking

1 Introduction

One of the general biochemical properties of many (and perhaps most) cell-to-cell movement proteins (MPs) of plant viruses is their ability to interact with nucleic acids. This protein activity makes biological sense because the main function of MPs is to transport the viral genome from the infected cell to the surrounding healthy cells, and the most direct way for MP to achieve this goal is simply to associate with the nucleic acid molecule and transport it through plasmodesmata. Since the ability to bind single-stranded (ss) RNA and DNA was originally demonstrated for MP of *tobacco mosaic virus* (TMV) (1), MPs from a large number of very diverse plant viruses, such as tobamoviruses, caulimoviruses, dianthoviruses, alfamoviruses, tospoviruses, umbraviruses, bromoviruses, cucumoviruses, fabaviruses, sobemoviruses, carmoviruses, necroviruses, tombusviruses, geminiviruses, hordeiviruses,

Edited by G.D. Foster, I.E. Johansen, Y. Hong, and P.D. Nagy © Humana Press, Totowa, NJ

potexviruses, pomoviruses, and luteoviruses, have been shown to exhibit nucleic acid-binding activities (reviewed in 2). Table 1 illustrates specific examples of viral MPs that bind nucleic acids and shows that most MPS exhibit the following four common characteristics of this binding: preference for single-stranded nucleic acids, comparable affinity toward ssRNA and ssDNA, cooperativity, and lack of sequence specificity. On the other hand, some viral MPs can also bind dsDNA and show preference for certain topological forms of the DNA molecules (Table 1). Furthermore, MPs of the viruses, such as *cowpea mosaic virus* (CPMV) (reviewed by 3), thought to move between cells exclusively as viral particles rather than as MP-viral genome complexes, have not been shown to possess nucleic acid-binding activities; instead, they may interact with the whole virions via MP-CP binding (4, 5).

Thus, when initiating a study of a plant virus MP, it is important to determine whether or not it interacts with nucleic acids and characterize the general parameters of this interaction. Here, we present protocols for biochemical assays that detect MP-nucleic acid-binding and define its hallmark features. It is important to note that, once the MP-nucleic acid-binding is demonstrated and initially characterized, additional studies can be performed that focus on detailed kinetic and structural aspects of this process. Although the methodology of such advanced studies is beyond the scope of this chapter, their experimental and conceptual approaches have been described and discussed in numerous papers, reviews, and monographs (e.g. 6, 7–11).

2 Materials

2.1 Equipment and Consumables

1. Environmentally controlled shaker (37 °C) for culturing *Escherichia coli*
2. Spectrophotometer for measuring optical density of bacterial cultures
3. Polymerase chain reaction (PCR) thermocycler for production of DNA probes
4. French press with a small (3 ml) chamber for breaking bacterial cells
5. 70 °C water bath
6. 56 °C water bath
7. Hot plate
8. Microfuge
9. 4 °C cold room or large refrigerator (cold box)
10. High speed centrifuge (e.g., Sorvall or Beckman)
11. Disposable 1–3 ml syringes with G26 needles
12. Dialysis tubing with 10 kDa molecular mass cut-off
13. Stir plates with stir bars
14. Vertical gel electrophoresis box, glass plates, and comb with 5-mm-wide teeth suitable for polyacrylamide gel electrophoresis (PAGE) (*see* note 1)
15. Horizontal gel electrophoresis box and comb with 5-mm wide teeth suitable for agarose gel electrophoresis
16. Power supply with leads

Table 1 Nucleic acid-binding properties of some plant viral MPs

Genus	Virus	MP	Binding to				Cooperativity	Sequence specificity	Reference
			ssRNA	ssDNA	dsRNA	dsDNA			
Alfamovirus	AMV	P3	++	++	−	−	+	No	(32–34)
Bromovirus	BMV	3a	++	++	−	−	++	No	(35, 36)
Caulimovirus	CaMV	P1	++	+	N.R.	−	++	No	(12, 37)
Carmovirus	TCV	p8	++	N.R.	N.R.	N.R.	++	No	(38, 39)
	CarMV	p7	++	N.R.	N.R.	N.R.	++	No	(40)
Cucumovirus	CMV	3a	+	+	−	−	+	No	(41–43)
Dianthovirus	RCNMV	35 kDa	++	++	N.R.	−	++	No	(44–46)
Fabavirus	BBWV-2	VP37	+	+	−	−	++	No	(47)
Geminivirus (bipartite)	SLCV	BV1/BC1	+/±	++/±	N.R.	−	N.R.	No	(48)
	BDMV	BV1/BC1	N.R.	++/−	N.R.	++/++	N.R.	2–9 kb open circles	(49)
Hordeivirus	PSLV	63 kDa TGBp1	++	N.R.	N.R.	N.R.	+	No	(50)
	BSMV	58 kDa TGBp1	++	−	++	−	N.R.	No	(51)
	BNYVV	42 kDa TGBp1	++	++	++	++	N.R.	No	(52)
Ilarvirus	PNRSV	32 kDa	++	−	±	−	++	No	(53)
Luteovirus	PLRV	17 kDa	++	++	N.R.	−	N.R.	No	(54)
Necrovirus	TNV	p7a	++	++.	N.R.	−	N.R.	N.R.	(55)
Pomovirus	PMTV	51 kDa TGBp1	++	N.R.	N.R.	N.R.	N.R.	No	(56)
		13 kDa TGBp2	++	N.R.	N.R.	N.R.	N.R.	No	(56)

(continued)

Table 1 (continued)

Genus	Virus	MP	Binding to				Cooperativity	Sequence specificity	Reference
			ssRNA	ssDNA	dsRNA	dsDNA			
Potexvirus	PVX	25 kDa TGBp1	+	N.R.	N.R.	N.R.	+	No	(50, 57)
	BaMV		+	N.R.	N.R.	N.R.	N.R.	No	(58)
	FoMV	28 kDa TGBp1	++	N.R.	N.R.	N.R.	N.R.	No	(59)
	WClMV	26 kDa TGBp1 26 kDa TGBp1	++	N.R.	N.R.	N.R.	−	No	(60)
Sobemovirus	CoMV	P1	+	N.R.	N.R.	−	N.R.	No	(61)
Tobamovirus	TMV	30 kDa	++	++	N.R.	−	++	No	(1, 10)
	TVCV	29 kDa	++	N.R.	N.R.	N.R.	N.R.	No	(62)
Tombusvirus	TBSV	P22	++	N.R.	N.R.	N.R.	+	N.R.	(63)
Tospovirus	TSWV	NSm	++	N.R.	−	N.R.	N.R.	No	(64)
Umbravirus	GRV	ORF4	++	++	N.R.	−	−	No	(43)

++, binding comparable to that of TMV MP; +, binding weaker than that of TMV MP; ±, very weak binding; −, no binding; N.R., not reported. Plant virus genera are according to (65). *AMV*, Alfalfa mosaic virus; *BMV*, Brome mosaic virus; *CaMV*, Cauliflower mosaic virus; *TCV*, Turnip crinkle virus; *CarMV*, Carnation mottle virus; *CMV*, Cucumber mosaic virus; *RCNMV*, Red clover necrotic mosaic virus; *BBWV-2*, Broad bean wilt virus 2; *SLCV*, Squash leaf curl virus; *BDMV*, Bean dwarf mosaic virus; *PSLV*, Poa semilatent virus; *BSMV*, Barley stripe mosaic virus; *BNYVV*, Beet necrotic yellow vein virus; *PLRV*, Potato leaf roll virus; *PNRSV*, Prunus necrotic ringspot virus; *TNV*, Tobacco necrosis virus; *PMTV*, Potato mop-top virus; *PVX*, Potato virus X; *BaMV*, Bamboo mosaic virus; *FoMV*, Foxtail mosaic virus; *WClMV*, White clover mosaic virus; *CoMV*, Cocksfoot mottle virus; *TMV*, Tobacco mosaic virus; *TVCV*, Turnip vein clearing virus; *TBSV*, Tomato bushy stunt virus; *TSWV*, Tomato spotted wilt virus; *GRV*, Groundnut rosette virus. Modified from ref. (2); Copyright 2004 from *The Ins and Outs of Nondestructive Cell-to-Cell and Systemic Movement of Plant Viruses* by Waigmann, Ueki, Trutnyeva, and Citovsky. Reproduced by permission of Taylor & Francis Group, LLC., http://www.taylorandfrancis.com

17. Vacuum gel dryer
18. Blotting paper (Whatman)
19. UV light cross-linker (e.g., Stratalinker 1,800 from Stratagene, Inc.) or a germicidal UV light lamp
20. X-ray film, film cassette, and intensifying screen for autoradiography. Alternatively, a PhosphorImager with its cassette and intensifying screen can be used to reduce exposure time and facilitate digital acquirement of the image
21. CCD gel documentation system or Polaroid camera with a UV light table
22. NucTrap probe purification columns (Stratagene)
23. Ice buckets
24. Work space for handling radioactive isotopes

2.2 Media, Antibiotics, Buffers, Enzymes, and Other Chemicals

1. Double-distilled water (ddH2O), autoclaved
2. Stock solution of 0.5 M isopropyl-beta-d-thiogalactopyranoside (IPTG) in ddH2O (see note 2). Aliquot and store at −20 °C for up to 30 days
3. 1,000× stock solutions of antibiotics: 20 mg ml^{-1} kanamycin or 100 mg ml^{-1} ampicillin in ddH2O (see note 2) and 25 mg ml^{-1} solution of chloramphenicol in ethanol. Aliquot and store at −20 °C for 30 days
4. Stock solution of 0.5 M ethylenediaminetetraacetic acid (EDTA), pH 8.0. Autoclave and store at room temperature
5. Stock solution of 1 M tris (hydroxymethyl) aminomethane hydrochloride (Tris/HCl), pH 8.0. Autoclave and store at room temperature
6. Stock solution of 1 M dithiothreitol (DTT) stock solution in ddH2O. Aliquot and store at −20 °C
7. Stock solution of 1 M phenylmethanesulphonyl fluoride (PMSF) stock solution in methanol or dimethylsulfoxide (DMSO). Prepare fresh before use
8. Buffer L: 10 mM Tris/HCl pH 8.0, 200 mM NaCl, 1 mM EDTA, 10% glycerol. Prepare freshly before use. Add DTT and PMSF to 1 mM each immediately before use
9. Buffer L with 1 M NaCl. Prepare fresh before use
10. Buffer L with 1 M NaCl and 4 M urea (see note 3). Prepare fresh before use
11. Luria broth (LB) liquid medium: 5 g yeast extract, 10 g tryptone, and 10 g NaCl in one liter of ddH2O. Autoclave and store at room temperature
12. LB solid medium: same as LB liquid medium, only add 15 g l^{-1} agar before autoclaving. Store at room temperature
13. 2× yeast/tryptone (YT) liquid medium: 16 g tryptone, 10 g yeast extract, and 5 g NaCl in 1 l of ddH$_2$O. Autoclave and store at room temperature
14. Agarose-molecular biology grade (Fisher Scientific or any other brand).
15. Ethidium bromide 10 mg ml^{-1} in ddH$_2$O. Filter through a Whatman paper and store at room temperature in the dark (can wrap the container in aluminum foil). It is carcinogenic, so exercise caution

16. Tris/Borate/EDTA (TBE) buffer 10× stock solution: mix 108 g Tris base, 55 g boric acid, and 40 ml 0.5 M EDTA; check pH which should be around 8.0. Autoclave and store at room temperature.

17. Loading buffer 5× stock solution for native PAGE and agarose gel electrophoresis: 50% glycerol, 0.04% bromophenol blue (tracking dye) in TBE. Store at 4 °C.

18. 20% sodium dodecyl (lauryl) sulfate (SDS) in ddH$_2$O. Autoclave and store at room temperature.

19. SDS PAGE 5× sample buffer: 10% SDS, 25% beta-mercaptoethanol, 50% glycerol, 300 mM Tris pH 6.8, 0.04% bromophenol blue. Store at 4 °C.

20. SDS PAGE 4× stacking gel buffer: 0.5 M Tris-HCl pH 6.8, 0.4% SDS. Store at 4 °C.

21. SDS PAGE 4× resolving gel buffer: 1.5 M Tris-HCl pH 8.8, 0.4% SDS. Store at 4 °C.

22. SDS PAGE 10× running buffer: 1.92 M glycine, 250 mM Tris base, 1% SDS. Store at 4 °C.

23. Acrylamide, 30% stock solution (acrylamide 28.2%/bis acrylamide 0.8%) in ddH$_2$O. Store at 4 °C. It is a neurotoxin, so exercise caution.

24. Acrylamide polymerizing reagent: 10% ammonium persulfate (APS) in ddH2O. Can store at 4 °C up to 1 week.

25. Acrylamide polymerizing reagent: N,N,N',N'-tetramethylethylenediamine (TEMED) (Pierce).

26. Protein staining solution: 0.2% Coomassie brilliant blue R-250 (Fisher) in 35% methanol, 10% acetic acid in ddH$_2$O. Filter through a Whatman paper and store at room temperature. Can be reused 2–3 times.

27. Protein destaining solution: 10% methanol, 10% acetic acid in ddH$_2$O. Store at room temperature.

28. Protein molecular weight markers such as BlueRanger Prestained Protein Molecular Weight Marker Mix (Pierce).

29. Coomassie (Bradford) Protein Assay Kit (Pierce).

30. Forward and reverse PCR primers specific for the selected DNA probe.

31. Deoxyribonucleotide triphosphate (dNTP) stocks for PCR (2 mM each of the four dNTPs as individual stocks).

32. A dNTP radioactively labeled with α-^{32}P; for example [α-^{32}P]dATP or [α-^{32}P]dCTP (400 Ci mmol^{-1}) (Amersham or Perkin-Elmer).

33. ATP radioactively labeled with γ-^{32}P, i.e., [γ-^{32}P]ATP (3,000 Ci mmol^{-1}) (Amersham or Perkin-Elmer)

34. UTP radioactively labeled with α-^{32}P, i.e., [α-^{32}P]UTP (3,000 Ci mmol^{-1}) (Amersham or Perkin-Elmer)

35. *Taq* DNA polymerase (see note 4)

36. GFX PCR purification kit (Amersham) or PCR Purification Kit (Qiagen)

37. Bacteriophage T4 DNA polymerase (Promega or New England Biolabs)

38. Bacteriophage T4 polynucleotide kinase (Promega or New England Biolabs).

39. Bacteriophage T7 RNA polymerase (Promega or New England Biolabs) and 5 mM stocks of each of the four nucleotide triphosphates (NTPs) for in vitro transcription, or an in vitro transcription kit (see note 5).

40. Bacteriophage T4 gene 32 protein (United States Biochemical Corp. or New England Biolabs) and *E. coli* ssDNA-binding protein (United States Biochemical Corp.), two known ssDNA-binding proteins (6) for use as positive controls
41. Proteinase K (New England Biolabs)
42. Bacteriophage M13mp18 ssDNA and M13mp18 dsDNA (replicative form, RF) (New England Biolabs)
43. RNase A (Qiagen)

2.3   Bacterial Strains and Plasmids

1. *Escherichia coli* strains BL21(DE3), BL21(DE3)pLysS, or BL21(DE3)pLysE (see note 6). Can be purchased from Novagen.
2. A bacteriophage T7 RNA polymerase bacterial expression vector for expression of recombinant MP (e.g., pET series vectors, see note 7). Can be purchased from Novagen.
3. A vector for in vitro transcription of RNA probes from a bacteriophage T7 RNA polymerase promoter (e.g., pBluescript series vectors from Stratagene).

3   Methods

Studies of MP-nucleic acid interactions naturally begin with purification of MP, followed by production of DNA and RNA probes. As in many in vitro approaches, the preferred source of MP is a recombinant protein purified from bacteria. The major caveat in MP expression and purification is to develop a protocol that yields a reasonably soluble preparation. Here, we present such protocol developed for TMV MP; however, different viral MPs may require different approaches and specific modifications of the expression protocol. For probes, we use relatively short (100–800 nucleotides) radioactively-labeled ssDNA, dsDNA, and RNA as well as long (2,000–8,000 nucleotides) molecules. Because most MPs known to bind nucleic acids bind them without sequence specificity (see Table 1), the probes do not need to be of viral origin, whereas potential sequence specificity can be examined using competitor DNA and RNA derived from the viral genome.

Initial detection and characterization of MP-nucleic acid binding focuses on four goals: demonstration of the nucleic acid-binding activity, identification of the preferred binding substrate (i.e., DNA or RNA, single-stranded or double-stranded), determination whether the binding is cooperative or random, and investigation of sequence-specificity of the interaction. Technically, these objectives are most reliably addressed by two in vitro binding assays: native gel mobility shift (also called gel retardation and band shift) in which binding is detected by reduced electrophoretic mobility of protein–nucleic acid complexes, and ultra violet (UV) light cross-linking in which binding is detected as covalent attachment of protein to cross-linked oligonucleotides (e.g., 1, 8, 10, 12). In addition, other techniques (not described here), such as nitrocellulose filter binding (13), coimmunoprecipitation (14), or electron microscopy visualization of the MP–nucleic acid complexes (10, 11), can occasionally be used.

3.1 Expression and Purification of Recombinant MP

1. Subclone the open reading frame of the MP to be tested into a bacteriophage T7 RNA polymerase bacterial expression vector (see note 7).
2. Freshly transform MP-expressing construct into BL21(DE3)pLysE cells (see note 8), plate on LB solid medium supplemented with the appropriate antibiotics (see notes 6, 7), and grow overnight at 37 °C.
3. Next day, inoculate a starter culture in 5 ml of LB liquid medium (see note 9) with the appropriate antibiotics. Grow for about 2 h at 37 °C with 250 rpm shaking.
4. Dilute the starter culture to optical density $A_{600} = 0.1$ into 20 ml of 2× YT liquid medium (see note 10) with the appropriate antibiotics. Grow for about 2 h at 37 °C with 250 rpm shaking until A_{600} reaches 0.7–1.0.
5. Add 20 µl 0.5 M IPTG (to final concentration of 0.5 mM) and let grow for another 3 h; centrifuge (10,000 g, 10 min, 4 °C) and save cell pellet. The pelleted cells can be processed immediately or stored at −70 °C for several months.
6. Resuspend cell pellet in 2 ml of buffer L, keep on ice. Break cells using French press in a 3-ml cell at 20,000 psi. Shear DNA by passing the lysate 4–5 times through a syringe with G26 needle. Transfer to 1.5-ml polypropylene microfuge tubes, centrifuge (12,000 g, 5 min, 4 °C), and save pellet (see note 11).
7. Resuspend the pellet in 1 ml of buffer L with 1 M NaCl by passing it 2–3 times through a syringe with G26 needle. Centrifuge as above. Save pellet (see note 11).
8. Resuspend as above in 0.5 ml of buffer L with 1 M NaCl and 4 M urea. Incubate for 10 min at 70 °C. Centrifuge as above. Save pellet (see note 12).
9. Resuspend as above in 1.5 ml buffer L with 1 M NaCl and 4 M urea and incubate for 15 min at 56 °C. Centrifuge as above. Save supernatant (see note 12).
10. Place the supernatant in a dialysis tube and dialyze for 2 h at 4 °C against 2 l of buffer L, then change buffer and dialyze for additional 2 h at 4 °C. Remove from the dialysis tube (see note 13), determine protein concentration (for example, using a Coomassie Protein Assay Kit), and aliquot. The purified MP can be assayed for the nucleic acid-binding activity immediately or stored at −70 °C for several months. This procedure yields 0.2–0.5 mg of purified MP. We also recommend to confirm the purity and the expected electrophoretic mobility of the isolated MP by SDS PAGE (15). Figure 1a illustrates such analysis of different stages of the TMV MP purification protocol, from total cell extract to the purified protein.

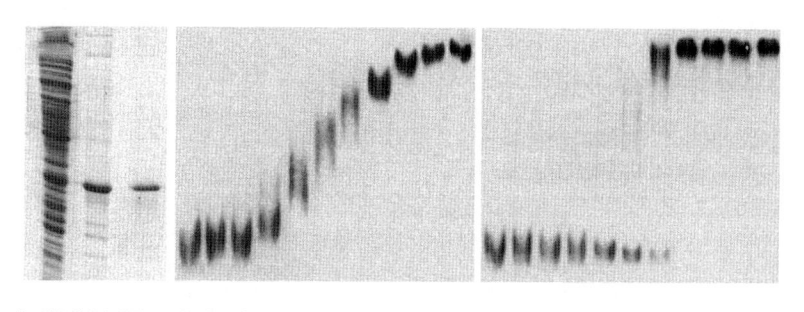

Fig. 1 SDS PAGE analysis of purified TMV MP and native polyacrylamide gel mobility shift analyses of low and high cooperativity of protein binding to ssDNA. (**a**) Purification of recombinant TMV MP. The protein was purified from inclusion bodies following over-expression from pPETP30 plasmid in *E. coli* (1) and analyzed on a 12.5% SDS polyacrylamide gel using Coomassie blue staining of protein bands. *Lane 1*, total bacterial cell lysate; *lane 2*, pellet fraction of the lysate containing TMV MP inclusion bodies; *lane 3*, purified TMV MP obtained by solubilization of protein aggregates. *Arrowhead* indicates the position of the 30 kDa TMV MP. (**b**) Low cooperativity binding of *E. coli* ssDNA binding protein to ssDNA. (**c**) High cooperativity binding of TMV MP to ssDNA. Radioactively labeled ssDNA probe (40 ng) was incubated with increasing amounts of protein followed by gel mobility shift assay on a 4% native polyacrylamide gel. *Lanes 1–11*, protein-to-ssDNA probe weight ratios 0:1, 0.5:1, 0.7:1, 1:1, 2.5:1, 5.0:1, 10:1, 20:1, 50:1, 75:1, and 100:1. *Arrowheads* indicate the positions of complete, fully saturated protein-ssDNA complexes and of the free ssDNA probe. All experimental conditions were as described in this chapter

3.2 Preparation and Labeling of Nucleic Acid Probes for Binding Assays

Interactions of MP with nucleic acids should be assayed using DNA and RNA probes. MP binding to nucleic acids can be assayed using shorter (70–1,000 nucleotides), radioactively end-labeled probes as well as long (2,000–7,000 nucleotides), unlabeled probes. The former are used in mobility shift assays on native polyacrylamide gels, while the latter are suitable for simple agarose gel electrophoresis (see note 14). Generally, DNA probes are prepared by PCR, but, because MP-DNA binding is often sequence nonspecific, it is possible to use commercially available ssDNA and dsDNA, such as genomic ssDNA and replicative dsDNA of bacteriophage M13mp18 (available, for example, from New England Biolabs), as probes for mobility shift assays on agarose gels. In the case of dsDNA preparations, they can be converted to ssDNA probes simply by separating the dsDNA strands following brief boiling and quick chill and storage on ice.

For very small probes, 70–100 nucleotide-long synthetic oligonucleotides can be used (10), completely eliminating the need for PCR-based preparation of the DNA probes. In this case, individual oligonucleotides serve as ssDNA probes, and dsDNA probes are made by annealing complementary oligonucleotides.

Similarly to binding to DNA, MP binding to RNA can be detected by native gel mobility shift assays. In addition, it can be easily analyzed by UV light cross-linking followed by SDS PAGE. Both assays utilize RNA probes generated by in vitro transcription and labeled radioactively along their entire length by including a radioactive nucleotide in the transcription reaction.

3.2.1 Preparation of DNA Probes

1. Design15–17 nucleotide-long forward and reverse PCR primers specific for the selected DNA probe (e.g., viral genome for sequence-specific probes, or an unrelated plasmid for sequence nonspecific probes). Normally, the primers are designed to generate a 300–1,000-bp PCR product, but longer or shorter fragments are also suitable for use as probes.
2. Prepare a PCR cocktail with a total volume of 25 μl volume containing 100 ng template DNA, 0.2 mM of each of the four dNTPs, 0.2 μM of the forward primer, 0.2 μM of the reverse primer, and 2 U of *Taq* DNA polymerase with 2.5 μl of Taq 10× reaction buffer (see note 15).
3. Perform PCR with the following program of the thermocycler: 3-min denaturation at 94 °C, 30 cycles of 30 s at 94 °C and 2 min at 70 °C, and 2 min at 70 °C.
4. The PCR products are purified using the GFX PCR purification kit (Amersham) or PCR Purification Kit (Qiagen) to remove dNTPs, primers, and enzyme (see note 15).
5. Analyze a sample of the PCR products (1–2 μl) on an agarose gel, followed by ethidium bromide staining, to determine the yield of the amplification, verify the expected size of the PCR product, and estimate the amount of amplified DNA based. Calculate the latter based on the known amounts of DNA standards electrophoresed on the same gel.

3.2.2 Radioactive Labeling of DNA Probes

1. DsDNA probes produced by PCR are radioactively end-labeled using T4 DNA polymerase (see note 16) using standard molecular biology protocols (16, 17). Briefly, prepare a mixture with a total volume of 20 μl containing 1–2 μg DNA, 0.1 mM of each of the three dNTPs, 2 μCi of an aqueous solution of the fourth dNTP labeled with α-^{32}P, and 1 U μg^{-1} DNA of T4 DNA polymerase with 2 μl of T4 DNA polymerase 10× reaction buffer. Incubate for 10 min at 37 °C, and stop the reaction by heating for 5 min at 70–75 °C. This labeling reaction should yield probes with a specific activity of approximately 2×10^7 cpm μg^{-1} (see note 17). Purify the probe from unincorporated dNTPs using a NucTrap column.
2. Synthetic oligonucleotide probes are radioactively end-labeled using T4 polynucleotide kinase using standard molecular biology protocols (16, 17). Briefly, prepare a mixture with a total volume of 50 μl containing 1–2 μg oligonucleotide DNA, 150 μCi of an aqueous solution of [γ-^{32}P]ATP, and 20 Richardson units of T4 polynucleotide kinase with 5 μl of T4 polynucleotide kinase 10× reaction buffer. Incubate for 30 min at 37 °C, and stop the reaction by heating for 5–10 min at 70–75 °C. This labeling reaction should yield probes with a specific activity

of approximately 10^6 cpm μg^{-1} (see note 17). Purify the probe from unincorporated dNTPs using a NucTrap column.

3.2.3 Preparation of Radioactively-Labeled RNA Probes

1. Select a template for the RNA probe, which can be either a viral genome-specific fragment or an unrelated, nonspecific sequence, and subclone it under the bacteriophage T7 RNA polymerase promoter of an in vitro transcription vector (see note 18).
2. Linearize the template construct by cleavage with a restriction endonuclease at a site located immediately downstream of the probe template sequence.
3. To make RNA probes for gel mobility shift assay, prepare a mixture with a total volume of 20 μl containing 1–2 μg linearized template DNA, 0.5 mM of each of GTP, CTP, and ATP, 2.5 μM (15 μCi) of an aqueous solution of [α-^{32}P]UTP, 25 μM unlabeled UTP, and 20 U of T7 RNA polymerase with 2 μl of T7 RNA polymerase 10× reaction buffer (see note 19). Incubate for 1 h at 37 °C. This labeling reaction should yield probes with a specific activity of approximately $2–5 \times 10^6$ cpm μg^{-1} (see note 17). Alternatively, a complete in vitro transcription kit can be used according to the manufacturer's instructions (see note 5). Purify the probe from unincorporated NTPs using a NucTrap column per manufacturer's instructions (see note 20).
4. To make RNA probes for UV light cross-linking assay, prepare a mixture with a total volume of 20 μl containing 1–2 μg linearized template DNA, 0.5 mM of each of GTP, CTP, and ATP, 10 μM (60 μCi) of an aqueous solution of [α-^{32}P]UTP, and 20 U of T7 RNA polymerase with 2 μl of T7 RNA polymerase 10× reaction buffer (see note 19). Incubate for 1 h at 37 °C. This labeling reaction should yield probes with a specific activity of approximately $2–5 \times 10^8$ cpm μg^{-1} (see note 17). Alternatively, a complete in vitro transcription kit can be used according to the manufacturer's instructions (see note 5). Purify the probe from unincorporated NTPs using a NucTrap column per manufacturer's instructions (see note 20).

3.3 MP-Nucleic Acid-Binding Assays

3.3.1 Gel Mobility Shift Assays

3.3.1.1 Mobility Shift on Native Polyacrylamide Gels

1. Cast a 4% native polyacrylamide gel. For full-size/mini gels, combine: 2.7/1.7 ml 30% acrylamide and 17.3/10.8 ml 1× TBE, mix, add 20/12.4 μl TEMED and 200/125 μl 10% APS, mix, pour into the gel frame, and immediately insert comb (see note 21). Allow it to remain at room temperature until the acrylamide polymerizes completely (see note 22).

2. In a 1.5-ml polypropylene microfuge tube, combine 40 ng of radioactively-labeled DNA or RNA probe (see notes 23 and 24), 1 μg of purified MP (see note 25) in a total volume of 15 μl buffer L and incubate on ice for 10–30 min. See notes 26 and 27 for a recommended set of initial experimental and control-binding reactions.

3. Add 4 μl of 5× native gel loading buffer, mix, load into the gel well (see note 28), and electrophorese at the electric field strength of 8 V cm^{-1} (see note 29) at 4 °C (i.e., in a cold room or a cold box) until the tracking dye has migrated 1/2–2/3 way down the gel.

4. Remove the gel from the gel box, remove side spacers, and, using a spatula, slowly separate the glass plates. Place a sheet of blotting paper over the gel and carefully peel the gel off the glass plate. Cover the exposed side of the gel with saran wrap and dry in a gel dryer under vacuum (see note 30).

5. Expose the dried gel to X-ray autoradiography film for overnight at −70 °C with an intensifying screen. Alternatively, analyze the gel using PhosphorImager. See note 31 for general guidelines on how to interpret the results of this experiment.

3.3.1.2 Mobility Shift on Agarose Gels

1. Make a 0.3% solution of agarose in 1× TBE, boil until agarose dissolves fully, cool to 55 °C, and pour into a horizontal gel tray, insert comb. Allow it to remain at room temperature until agarose solidifies completely (see note 22).

2. In a 1.5-ml polypropylene microfuge tube, combine 1 μg of unlabeled probe (see notes 23 and 32), 25 μg of purified MP (see note 25) in a total volume of 15 μl buffer L, and incubate on ice for 10–30 min. See note 26 for a recommended set of initial experimental and control-binding reactions.

3. Add 4 μl of 5× native gel loading buffer, mix, load into the gel well, and electrophorese at the electric field strength of 5–7 V cm^{-1} (see note 29) at room temperature until the tracking dye has migrated to the end of the gel.

4. Stain the gel (see note 33) for 10 min at 4 °C in 2 μg ml^{-1} of ethidium bromide in 1× TBE, and destain for 1–3 h at 4 °C in 1× TBE supplemented with 1.5 M NaCl (see note 34).

5. Visualize the stained probe and record the image using a CCD gel documentation system or a Polaroid camera with a UV light table. See note 31 for general guidelines on how to interpret the results of this experiment.

3.3.2 UV Light Cross-Linking Assay

1. Cast a 12.5% resolving SDS polyacrylamide gel. For full-size/mini gels, combine: 6.7/2.1 ml 30% acrylamide, 4.0/1.25 ml 4× resolving gel buffer, and 5.3/1.65 ml ddH$_2$O, mix, add 15/2.5 μl TEMED and 60/25 μl 10% APS, mix, pour into the gel frame, and immediately overlay with ddH$_2$O (see note 35). Leave at room temperature until the acrylamide polymerizes completely (see note 22).

2. Aspirate ddH$_2$O from the polymerized resolving gel and cast a stacking polyacrylamide gel. For full-size/mini gels, combine: 0.75/0.375 ml of 30% acrylamide,

1.25/0.625 ml of 4× stacking gel buffer, and 3/1.5 ml ddH$_2$O, mix, add 5/2.5 µl TEMED and 50/25 µl of 10% APS, mix, pour into the gel frame, and immediately insert comb (see note 21). Leave at room temperature until the acrylamide polymerizes completely (see note 22).

3. In a 1.5-ml polypropylene microfuge tube, combine 10 ng of RNA probe (see note 23), 0.2 µg of purified MP (see note 24) in a total volume of 15 µl buffer L, and incubate on ice for 10–30 min. See notes 25 and 27 for a recommended set of initial experimental and control-binding reactions.

4. Open the reaction tubes, place them in a UV light cross-linker, and irradiate with 1.8 J of UV light. Alternatively, incubate the samples for 30 min on ice at a distance of 6 cm under a germicidal UV light lamp.

5. Add 1 µl of 0.5 mg ml^{-1} RNase A in ddH$_2$O and incubate for 30 min at 37 °C to digest RNA probe unprotected by the bound protein.

6. Add 4 µl of 5× SDS gel sample buffer, mix, load into the gel well (see notes 28 and 36), and electrophorese at the electric field strength of 10 V cm^{-1} (see note 29) until the tracking dye has migrated to approximately 0.5 cm from the bottom of the gel.

7. Remove the gel from the gel box, remove side spacers, and, using a spatula, slowly separate the glass plates. With a razor blade, remove and discard the bottom 0.5–1 cm of the resolving gel, which contains radioactive nucleotides derived from the digested unbound probe. Also, the stacking gel – which is sticky and difficult to handle and which should not contain the protein bands of interest – can be removed and discarded.

8. Place the resolving gel in a glass or plastic box with the protein staining solution, stain for 30 min at room temperature with gentle shaking (see note 37). Remove staining solution, add destaining solution and destain at room temperature with gentle shaking until molecular weight marker bands are visible (see note 38).

9. Remove the gel from the destaining solution, place it on a sheet of blotting paper, cover the other side of the gel with saran wrap, and dry in a gel dryer under vacuum (see note 30).

10. Expose the dried gel to X-ray autoradiography film for overnight at −70 °C with an intensifying screen. Alternatively, analyze the gel using PhosphorImager. See notes 39 and 40 for general guidelines on how to interpret the results of this experiment.

3.4 Characterization of MP-Nucleic Acid-Binding Cooperativity and Sequence Specificity

3.4.1 Binding Cooperativity

For proteins that bind nucleic acids without sequence specificity – such as most viral MPs (*see* Table 1) – nucleic acid molecules present a continuous lattice of potential binding sites, rather than individual discrete and isolated binding sites as in sequence-specific binding. Sequence nonspecific binding of proteins to nucleic acid lattices occurs in two major modes: random and cooperative. In the random-binding mode, every nucleotide initiates attachment of protein with the same probability such that protein molecules that bind independently of each other are randomly distributed on the nucleic acid lattice (18). Thus, at subsaturating concentrations of protein, all protein molecules associate with all probe molecules with comparable affinity. In native gel mobility shift assays, which represent the major and simplest tool for determination of the protein–nucleic acid-binding mode, random protein binding is detected as a ladder of discrete protein-probe complexes, the size of which increases with the increase in protein concentration (13, 19) (illustrated in Fig. 1b).

In the cooperative-binding mode, protein association with the nucleic acid is not random; instead, protein molecules tend to bind in long clusters such that some nucleic acid molecules become fully coated with the protein, while others are still protein-free. Thus, cooperatively binding proteins exhibit an "all-or-none" behavior in native gel mobility shift assays, i.e., at subsaturating concentrations of protein, only two species of probe exist, free probe and maximally shifted complete protein-probe complexes, and no intermediate bands representing partly coated probe are detected (e.g., 1, 13) (illustrated in Fig. 1c). On the basis of this rationale, potential cooperativity (or the lack thereof) of MP binding to nucleic acids can be easily determined from the dose response to protein concentration in gel mobility shift assays.

1. Prepare a 4% native polyacrylamide gel or a 0.3% agarose gel as described above.
2. In 1.5-ml polypropylene microfuge tubes, set up a series of reactions using the constant amount of probe as described earlier for mobility shift assays on native polyacrylamide or agarose gels. In each tube, vary only the amount of added protein. We recommend using the following protein-to-probe weight-to-weight ratios: 0:1, 0.2:1, 0.5:1, 0.7:1, 1:1, 2.5:1, 5.0:1, 10:1, 20:1, 50:1, 75:1, and 100:1 (see notes 41 and 42). For suggested positive controls that generate reference, gel mobility shift patterns characteristic for high and low cooperativity-binding modes, see note 43.
3. Incubate on ice for 10–30 min.
4. Perform the gel mobility shift assay and detect formation of protein-probe complexes as described earlier. See note 44 for general guidelines on how to interpret the results of this experiment, and note 45 on how to use these data to estimate the minimal size of the nucleic acid-binding site of the tested protein.

3.4.2 Sequence Specificity of Binding

Most viral MPs bind nucleic acids irrespective of their sequence (see Table 1). It is possible, however, that, in addition to its general nucleic acid-binding ability, an MP may exhibit preferential binding toward the viral genome. Relative affinity of MP to various RNA and DNA sequences can be accurately assessed from binding competition experiments (e.g., 1) using unlabeled specific and nonspecific competitors.

1. Prepare a 4% native polyacrylamide gel or a 12.5% SDS polyacrylamide gel as described earlier (see note 46).

2. In 1.5-ml polypropylene microfuge tubes, set up a series of reactions as described earlier for mobility shift assays on native polyacrylamide gels or for UV light cross-linking. Each tube should contain constant amounts of protein and probe and increasing amounts of unlabeled nucleic acid competitor (see notes 47 and 48). We recommend using the following competitor-to-probe weight-to-weight ratios: 0:1, 0.5:1, 1:1, 5:1, 10:1, 50:1, 100:1, 500:1, and 1,000:1 (see note 41). At least two different competitors should be used: specific (e.g., viral genomic sequences) and nonspecific (e.g., M13mp18 ssDNA or even the unlabeled probe itself).

3. Incubate on ice for 10–30 min.

4. Perform the gel mobility shift assay or UV light cross-linking assay, dry the gel, and detect formation of protein-probe complexes by autoradiography as described earlier.

5. Align the autoradiogram with the dried gel and mark the location of the bands corresponding to the complete protein-probe complexes. Excise the corresponding gel regions with sharp scissors or a razor blade, and determine their radioactivity by counting Cerenkov radiation (no scintillation fluid is required).

6. Calculate the amount of the bound probe as percent of maximal binding (i.e., in the absence of competitor). Then, plot these data as a function of probe-to-competitor weight ratio. From the resulting competition curves, calculate IC50 (inhibitory concentration 50%) values for each unlabeled competitor using the probe-to-competitor weight ratio required to reduce binding by 50%. See note 49 for general guidelines on how to interpret the results of this experiment.

4   Notes

1. Boxes for either full-size gels ($20 \times 20\,cm^2 \times 1\,mm$) or minigels ($10 \times 7\,cm^2 \times 1\,mm$) can be used. These apparati can be purchased from BioRad, BRL, Hoefer, or any other manufacturer, or made in-house.

2. Filter sterilize IPTG, ampicillin, and kanamycin before aliquoting and storage. The choice of antibiotics depends on the bacterial strain and expression vector used to produce MP (see notes 6 and 7).

3. To facilitate preparation of this solution, add all liquid ingredients from stock solutions, i.e., 1 M Tris/HCl pH 8.0, 5 M NaCl, 0.5 M EDTA, and 100% glycerol, add solid urea, and ddH_2O to about 1/2 of the final desired volume, dissolve urea by heating the mixture at 70 °C, bring up to desired volume with ddH_2O and keep on ice until use.

4. Because PCR is used to produce relatively short DNA fragments for use as binding probes, there is no need for expensive high fidelity DNA polymerases, such as *ExTaq* or *Pfu*.

5. In vitro transcription can be performed using the reaction mixture described here. Note that all reagents, such as NTP stocks, used in preparation and handling of RNA should be RNase-free and prepared using RNase-free water (for standard protocols of preparation of RNase-free reagents, see, for example, 16, 17). Alternatively, one can use a complete kit, such as the Riboprobe system (Promega), which is designed for in vitro preparation of high specific activity single-stranded RNA probes.

6. Normally, the bacterial strain BL21(DE3) is sufficient for IPTG-induced over-expression from the bacteriophage T7 TNA polymerase expression vectors. However, viral MPs, potentially due to their strong single-stranded nucleic acid binding activity, are often toxic to bacterial cells; thus, even background levels of MP expression due to "leaky" production of the T7 RNA polymerase from its IPTG-inducible promoter may significantly impair bacterial growth. To suppress the MP expression until the cell cultures are ready for IPTG induction, the BL21(DE3) cells must carry a pLysS or pLysE plasmid. These plasmids, which carry the T7 lysozyme gene under the tet promoter from the pACYC184 plasmid (20), produce T7 bacteriophage lysozyme, which adsorbs and sequesters T7 RNA polymerase (21–23), preventing background expression of MP. Both pLysS and pLysE carry resistance to chloramphenicol, and they are maintained in the BL21(DE3) cells by culturing in the presence of $20 \mu g \ ml^{-1}$ of chloramphenicol. pLysE produces more lysozyme than pLysS, resulting in a tighter control of background expression, but both of them allow good over-expression after IPTG induction.

7. Our protocols utilize recombinant MP over-expressed in *E. coli* using the T7 RNA polymerase system, but any source of purified MP, such expression in baculovirus (24) or in yeast (25) is suitable for these experiments. For T7 RNA polymerase expression, we use pET-based vectors (23, 26) ranging from the early pET3 series to the more recent pET21 and pET28 series to the latest pET53 vectors, which allow fusion of different epitope tags to the over-expressed protein. We prefer to express either an untagged protein from a pET3 vector or a protein tagged at its C-terminus with hexahistidine from pET21 or pET28 vectors; the latter constructs allow further purification of the expressed protein on Ni columns.

8. We noticed that BL21(DE3)pLysE cells harboring the MP expression construct do not express well if used from a frozen glycerol or DMSO stock; thus, we freshly transform bacteria with the expression construct before each experiment. In addition, because BL21(DE3)pLysE cells produce lysozyme (see note 6) and are relatively fragile, we prefer to use the standard heat shock transformation method (e.g., 16, 17) rather than electroporation.

9. Normally, inoculation with a single colony further ensures the genetic homogeneity of the culture. However, since all colonies on the transformation plate derive from the same clone of the expression construct, we use multiple colonies to increase the inoculum and decrease growth time for the starter culture.

10. LB liquid medium can be used instead of 2× YT, but we consistently observe better expression in a richer 2× YT medium.

11. This pellet contains inclusion bodies of the over-expressed MP. With TMV MP, we observe virtually no over-expressed protein in the soluble fraction (supernatant). The insolubility of the inclusion bodies can be advantageous because it allows easy removal of contaminating bacterial proteins by multiple washes with high salt (see Fig. 1a).

12. The first wash with urea solubilizes about one half of the protein and further removes contaminants that have not been washed away by high salt. Importantly, incubation at 70 °C inactivates ssDNA-specific nucleases, which are present in the bacterial lysate, and, if not inactivated, preclude the use of the MP preparation in ssDNA-binding assays. Although half of the produced protein is lost during the first urea solubilization, we recommend this step for optimal purification and better refolding of the remained MP, which is achieved by the second urea solubilization performed at 56 °C (see Fig. 1a).

13. Normally, the MP solution remains clear after dialysis, but if some of the protein precipitates (looks like a fine white dust in the dialysis tube), it can be returned into solution by 2–3 passes through a G26 needle. It is important to note that this MP preparation is not completely bona fide soluble as the protein will precipitate when centrifuged at high speeds; this reduced solubility may be an intrinsic biological property of MP molecules that tend to interact with each other for cooperative binding to single-stranded nucleic acids (1). However, MP prepared using this protocol binds ssDNA and RNA (1), and it enhances plasmodesmal permeability when microinjected into plant tissues (27).

14. Although somewhat more laborious, native PAGE provides better resolution between different protein–nucleic acid complexes.

15. If thermocycler without a heated lid is used, overlay the reaction mixture with 50 μl of mineral oil to prevent evaporation during amplification; thermocyclers with heated lid do not require mineral oil. If mineral oil is used, it should be removed three extractions with one volume of chloroform each, followed by phenol precipitation, before use.

16. This approach is based on the terminal transferase activity of Taq polymerase, used to produce the probe (see above), which adds a single A overhang to each 3'-end of the PCR product and allows such DNA molecules with protruding 3' ends to be easily end-labeled with T4 DNA polymerase. T4 DNA polymerase has 3'-to-5' exonuclease and 5'-to-3' polymerase activities, which remove the 3' overhang and exchange the base of the dsDNA with the corresponding radioactively-labeled nucleotide, resulting in a blunt-ended probe with radioactive label incorporated at or very near to the termini of the DNA molecule (16).

17. To determine specific activity of the DNA or RNA probes, combine 1 μl of 1:10 dilution of the probe with 0.5 ml ice-cold 10% trichloracetic acid (TCA), place on ice for 10 min, collect the precipitate by vacuum filtration on a Whatmann GF/C glass fiber filter (before adding sample, prewet the filter with a small amount of 10% TCA), and read the Cerenkov radiation (do not use scintillation fluid). Calculate the specific activity of DNA probes as follows: total incorporated counts per minute (cpm)/total DNA input into the labeling reaction, where total incorporated cpm is TCA-precipitated cpm × 10 (i.e., dilution of probe sample) × total volume of probe preparation (i.e., 20 μl for dsDNA labeled using T4 DNA polymerase, and 50 μl for oligonucleotides labeled using T4 polynucleotide kinase). Calculate the specific activity of RNA probes as follows: total incorporated cpm/total RNA synthesized, where total incorporated cpm is TCA-precipitated cpm × 200 (i.e., dilution of probe sample × total volume of probe preparation), and total RNA synthesized is percent incorporation (i.e., TCA-precipitated cpm/cpm in the same sample before TCA precipitation × 100) × maximum theoretical RNA yield (i.e., total nmol of labeled + unlabeled limiting NTP × 4 × average molecular mass of a nucleotide within RNA, or for the reaction described here, total nmol of UTP × 4 × ~320 ng nmol^{-1}).

18. For technical simplicity and because most MPs do not interact efficiently with dsRNA (see Table 1), we focus only on ssRNA probes. If necessary, MP binding to dsRNA can be tested as described elsewhere (see references in Table 1).

19. Factors to be considered for maximal efficiency of the in vitro transcription reaction: Salt concentrations exceeding 50 mM may inhibit the T7 RNA polymerase activity; DTT, which is required for the T7 RNA polymerase activity, is unstable, even when frozen; thus, supplementing the reactions with the final concentration of 10 mM of freshly-made DTT may increase the reaction yield. Higher yields may also be obtained by raising NTP concentration to 4 mM each, adding Mg^{2+} to 4 mM above the total concentration of NTPs. Also, supplementing the reaction with inorganic pyrophosphatase (e.g., from New England Biolabs) to a final concentration of 4 U ml^{-1} will solubilize pyrophosphate precipitate that sequesters Mg^{2+}.

20. Alternatively, RNA probes can be purified on denaturing polyacrylamide gels as described elsewhere (e.g., 28).

21. Completely fill the gel frame with the acrylamide solution. The gel should be cast and comb inserted quickly because, once APS is added, the polymerization process begins. When inserting the comb, slightly tilt the gel to allow release of air bubbles and avoid their trapping under the comb teeth, which should be completely inserted into the gel solution. Do not disturb the gel until it has polymerized completely; polymerization can be easily detected as slight shrinking of the gel in the areas between the comb teeth and appearance of a thin layer of water (released following polymerization) above these areas. If gel does not polymerize, use freshly made 10% APS. Although degassing the acrylamide solution is often recommended, we find that this step can be safely omitted from our protocol.

22. Usually, this step takes about 30 min.

23. Because most viral MPs bind both ssDNA and RNA without sequence specificity (see Table 1), and because DNA probes are technically simpler to produce and handle than RNA probes, we recommend to begin studies of MP-nucleic acid interactions with sequence nonspecific DNA probes.

24. This experimental design is suitable for both DNA and RNA probes (e.g., 1). Probe amounts are calculated based on their specific activities (see note 17). For dsDNA probes, PCR fragments are used directly, and complementary oligonucleotides should be annealed by mixing them at 1:1 molar ratio, incubating for 5 min at 70 °C, 30 min at room temperature, and keeping on ice until use. For ssDNA probes, oligonucleotides are used directly, and PCR fragments should be denatured by boiling for 5 min and immediately placed on ice until use. RNA probes are used directly.

25. It is difficult to predict the optimal probe-to-MP ratio at which the binding is detected; thus, for initial experiments, we suggest to use a clear excess of MP.

26. Suggested reactions for initial testing of MP binding to DNA using gel mobility shift assays. Experimental reactions: MP + dsDNA probe, MP + ssDNA probe. Negative controls: dsDNA probe alone, ssDNA probe alone, MP + dsDNA probe treated for 30 min at 37 °C with 1 mg ml^{-1} of Proteinase K (New England Biolabs), MP + ssDNA probe treated for 30 min at 37 °C with 1 mg ml^{-1} of Proteinase K, bacteriophage T4 gene 32 protein + dsDNA probe. Positive control: bacteriophage T4 gene 32 protein + ssDNA probe.

27. Suggested reactions for initial testing of MP binding to RNA using gel mobility shift assays. Experimental reaction: MP + RNA probe. Negative controls: RNA probe alone, MP + RNA probe treated for 30 min at 37 °C with 1 mg ml^{-1} of Proteinase K. Positive control: bacteriophage T4 gene 32 protein + RNA probe. Note that, although the T4 gene 32 protein exhibits a much lower affinity to ssRNA than to ssDNA (6, 29), it is still suitable as a positive control.

28. Slowly remove the comb, taking care not to disturb the wells in the polymerized gel, place the gel into the gel box, rinse the wells by filling them with running buffer (1× TBE for native gels or 1× SDS gel running buffer) and removing the buffer with a vacuum aspirator or a Hamilton glass syringe, and, using a micropipette, slowly load the sample into the empty well, allowing it to slide along one side of the well to avoid trapping bubbles. Slowly overlay the loaded sample with running buffer to completely fill the wells. Then, fill both chambers of the gel box with running buffer. Fill the top chamber slowly, and never add running buffer to the top chamber before the samples in the wells have been completely covered with running buffer; this avoids mixing the loaded sample with the running buffer and helps to obtain sharper bands.

29. For vertical gel electrophoresis, total applied voltage is calculated based on the height of the gel, which corresponds to the entire native gel height for gel mobility shift assays and to the height of the resolving gel for UV light cross-linking assays. Alternatively, the vertical gels can be electrophoresed at maximal voltage and constant current of 25 mA; the ability to maintain constant current is found in more expensive power supplies, but electrophoresis at constant current usually yields better resolution of protein bands, especially with SDS polyacrylamide gels. For horizontal gel electrophoresis, total applied voltage is calculated based on the distance between the electrodes.

30. Usually, the drying process is finished after 1–1.5 h at 80 °C. Make sure the gel is completely dry before removing it from the dryer; if vacuum is broken while the gel is still wet, the gel will crack.

31. The results of gel mobility shift assays should be interpreted using the following general guidelines. On polyacrylamide gels, free dsDNA migrates faster than ssDNA, and on agarose gels, free dsDNA migrates slower than ssDNA. Protein-probe complexes always migrate much slower than the corresponding free probe, with large protein–nucleic acid complexes (i.e., multiple protein molecules bound to a molecule of probe) remaining very close to the loading well of the gel (see Fig. 1b). Proteinase K treatment should abolish retardation of the probe. The bacteriophage T4 gene 32 protein, known to bind ssDNA and ssRNA (6, 29), should cause reduced mobility of ssDNA and RNA probes, but not of dsDNA probe.

32. Agarose gel mobility shift assays are best suited for DNA probes detected by ethidium bromide staining (e.g., 19, 30) [although detection by Southern blot hybridization is also possible (31)]. Thus, higher amounts of probe are recommended. Also, for nonspecific probes (see note 23), we recommend commercially available preparations of M13mp18 ssDNA and dsDNA.

33. Low percent agarose gels are very fragile. To reduce handling and risk of breakage, we recommend leaving the gel in the gel tray throughout the staining/destaining procedure. Also, if the gel tray is UV transparent, keep the gel in the tray during visualization of DNA bands.

34. High concentration of NaCl in the destaining buffer helps to dissociate protein from the DNA while ethidium is still present in the gel and is able to intercalate (e.g., stain), allowing to visualize the DNA probe, which had been covered by protein and, thus, more difficult to stain.

35. Fill about two thirds of the gel frame so that enough space is left to fit the teeth of the comb and allow 1–2 cm of stacking gel between the comb and the resolving gel. The gel should be cast and overlaid with ddH$_2$O quickly because, once APS is added, the polymerization process begins. Overlay is required to protect the acrylamide solution in the gel frame from molecular oxygen, which inhibits acrylamide polymerization and causes formation of jagged, uneven gel edge. Some protocols overlay resolving gels with isobutanol, which is light and does not easily mix with acrylamide, but isobutanol should be extensively washed out before casting the stacking gel. In contrast, ddH$_2$O slowly and carefully overlaid to about 0.5–1 cm height above the acrylamide solution requires no wash, and it can be simply aspirated before pouring the stacking gel. After overlaying, do not disturb the gel until it has polymerized completely; polymerization can be easily detected as a clear and sharp interface between ddH$_2$O and the gel. If gel does not polymerize, use freshly made 10% APS. Although degassing the acrylamide solution is often recommended, we find that this step can be safely omitted from our protocol.

36. Remember to load molecular weight markers in one of the wells of the gel.

37. We recommend to perform the staining procedure to fix the gel and visualize molecular weight markers and MP. However, this step is not absolutely necessary, especially if prestained molecular weight markers are used. Instead, the unstained gel can be directly dried and autoradiographed.

38. Use small volumes of staining or destaining solutions, which are just sufficient to cover the gel. The gel is destained faster if the destaining solution is frequently changed or if a small piece of sponge, which absorbs Coomassie blue and removes it from the destaining solution, is placed in the box with the gel.

39. The results of the UV light cross-linking assay should be interpreted using the following general guidelines. Protein–RNA complexes are detected as radioactively-labeled protein bands that represent protein covalently cross-linked to the probe. Note that, because RNase A treatment removes all probe sequences that are not in close contact with the bound protein, very short oligonucleotides remain cross-linked to the protein causing only slight reduction in its electrophoretic mobility. In the absence of MP or following Proteinase K treatment, no radioactively-labeled protein bands should be observed. The bacteriophage T4 gene 32 protein, known to bind ssRNA (6, 29), should produce a radioactively-labeled band.

40. One unique and important advantage of the UV light cross-linking assay, when compared with most other assays for protein–nucleic acid binding, is that it directly identifies the protein species that interacts with the probe, allowing to rule out potential artifacts due to contaminants in the tested protein preparations.

41. We recommend using weight, rather than molar, ratios because, as described in this section, sequence nonspecific binding occurs along the entire nucleic acid lattice and, thus, depends on the total amount of polynucleotides in the reaction, which is better reflected by the weight of the probe than by the number of its molecules.

42. For some MPs, it may be impossible to obtain preparations concentrated enough to allow testing of very high protein-to-probe ratios; however, most MP preparations should be suitable for testing 10:1 to 20:1 ratios, which are normally sufficient to detect cooperative binding (e.g., 1, 13).

43. As positive controls and reference gel mobility shift patterns typical for high and low cooperativity binding, we recommend using commercially available preparations of the T4 gene 32 protein and *E. coli* ssDNA-binding protein. The former represents a paradigm for a protein with high cooperativity of binding whereas the latter exhibits a lower-binding cooperativity,

which allows detection of partly coated probe molecules with intermediate degree of mobility shift (6). Note that, low-binding cooperativity of the *E. coli* ssDNA-binding protein is better expressed at higher salt concentrations, i.e., 200–300 mM NaCl, while at lower salt concentrations, i.e., 20–50 mM NaCl, this protein exhibits higher binding cooperativity (13, 19).

44. Most plant viral MPs that bind ssDNA and ssRNA without sequence specificity exhibit various degrees of binding cooperativity (see Table 1). Thus, one would expect to detect at least some degree of cooperative binding that is manifested as a sharp transition from free, non-retarded probe to protein bound, strongly retarded probe upon increasing the protein-to-probe weight ratio. The higher the binding cooperativity is, the more narrow becomes the range of protein concentrations over which the change from zero to essentially complete protein-probe binding occurs (compare Fig. 1b and c).

45. The gel mobility shift experiments described in this section also define the minimum protein-to-probe weight ratio needed for complete binding. Knowing this ratio, the size of the probe, and the molecular mass of the tested protein allows one to calculate the size of the nucleic acid binding site, i.e., a number of nucleotides associated with one protein molecule. Obviously, this calculation is based on the assumption that all molecules in the tested protein preparation are equally active and, thus, represents the minimal value for the size of the binding site. Although falling short of by far more complicated spectroscopy experiments traditionally used to determine the precise size of nucleic acid-binding sites (e.g., 19), this approach still represents a valuable tool for initial characterization of MP–nucleic acid interactions.

46. Because estimating binding competition requires the use of a labeled probe and unlabeled competitor, gel mobility shift on native polyacrylamide gels and UV light cross-linking are the most suitable assays for these experiments.

47. We suggest using the lowest protein-to-probe weight ratio that yields the complete shift of all probes in the reaction mixture. This ratio minimizes the presence of free, excess protein or probe, and it can be determined from the dose response experiments described for studies of binding cooperativity.

48. To determine true binding competition, rather than displacement, both the competitor and the probe must be present in the reaction mixture before the protein is added.

49. Preferential binding to a specific competitor results in a more efficient competition, which manifests as a shift of the corresponding competition curve toward the lower probe-to-competitor weight ratios and the proportionately reduced value of IC50 when compared with those of a nonspecific competitor. Conversely, sequence nonspecific binding results in virtually identical competition curves and IC50 values for specific and nonspecific competitors.

Acknowledgments We apologize to colleagues whose original works have not been cited due to the lack of space. The work in our laboratories is supported by grants from NIH, NSF, USDA, BARD, and BSF to VC, and by grants from BARD and HFSP to TT.

References

1. Citovsky, V., Knorr, D., Schuster, G., and Zambryski, P. C. (1990) The P30 movement protein of tobacco mosaic virus is a single-strand nucleic acid binding protein. *Cell* **60,** 637–647.

2. Waigmann, E., Ueki, S., Trutnyeva, K., and Citovsky, V. (2004) The ins and outs of non-destructive cell-to-cell and systemic movement of plant viruses. *Crit. Rev. Plant Sci.* **23,** 195–250.

3. Pouwels, J., Carette, J. E., Van Lent, J., and Wellink, J. (2002) Cowpea mosaic virus: effects on host processes. *Mol. Plant Pathol.* **3,** 411–418.

4. Lekkerkerker, A., Wellink, J., Yuan, P., van Lent, J., Goldbach, R., and van Kammen, A. B. (1996) Distinct functional domains in the cowpea mosaic virus movement protein. *J. Virol.* **70,** 5658–5661.

5. Carvalho, C. M., Wellink, J., Ribeiro, S. G., Goldbach, R. W., and Van Lent, J. W. (2003) The C-terminal region of the movement protein of Cowpea mosaic virus is involved in binding to the large but not to the small coat protein. *J. Gen. Virol.* **84,** 2271–2277.
6. Chase, J. W., and Williams, K. R. (1986) Single-stranded DNA binding proteins required for DNA replication. *Annu. Rev. Biochem.* **55,** 103–136.
7. Revzin, A. (1990) The Biology of Nonspecific DNA–Protein Interactions, CRC Press, Boca Raton, FL.
8. Citovsky, V., Guralnick, B., Simon, M. N., and Wall, J. S. (1997) The molecular structure of *Agrobacterium* VirE2-single stranded DNA complexes involved in nuclear import. *J. Mol. Biol.* **271,** 718–727.
9. Abu-Arish, A., Frenkiel-Krispin, D., Fricke, T., Tzfira, T., Citovsky, V., Grayer Wolf, S., and Elbaum, M. (2004) Three-dimensional reconstruction of *Agrobacterium* VirE2 protein with single-stranded DNA. *J. Biol. Chem.* **279,** 25359–25363.
10. Citovsky, V., Wong, M. L., Shaw, A., Prasad, B. V. V., and Zambryski, P. C. (1992) Visualization and characterization of tobacco mosaic virus movement protein binding to single-stranded nucleic acids. *Plant Cell* **4,** 397–411.
11. Fujiwara, T., Giesman-Cookmeyer, D., Ding, B., Lommel, S. A., and Lucas, W. J. (1993) Cell-to-cell trafficking of macromolecules through plasmodesmata potentiated by the red clover necrotic mosaic virus movement protein. *Plant Cell* **5,** 1783–1794.
12. Citovsky, V., Knorr, D., and Zambryski, P. C. (1991) Gene I, a potential movement locus of CaMV, encodes an RNA binding protein. *Proc. Natl. Acad. Sci. USA* **88,** 2476–2480.
13. Citovsky, V., Wong, M. L., and Zambryski, P. C. (1989) Cooperative interaction of *Agrobacterium* VirE2 protein with single stranded DNA: implications for the T-DNA transfer process. *Proc. Natl. Acad. Sci. USA* **86,** 1193–1197.
14. Christie, P. J., Ward, J. E., Winans, S. C., and Nester, E. W. (1988) The *Agrobacterium tumefaciens virE2* gene product is a single-stranded-DNA-binding protein that associates with T-DNA. *J. Bacteriol.* **170,** 2659–2667.
15. Laemmli, U. K. (1970) Cleavage of structural proteins during the assembly of the head of bacteriophage T4. *Nature* **277,** 680–685.
16. Sambrook, J., Fritsch, E. F., and Maniatis, T. (1989) Molecular Cloning: A Laboratory Manual, Cold Spring Harbor Laboratory, Cold Spring Harbor, NY.
17. Ausubel, F. M., Brent, R., Kingston, R. E., Moore, D. D., Smith, J. A., Seidman, J. G., and Struhl, K. (1987) Current Protocols in Molecular Biology, Greene Publishing-Wiley Interscience, New York, NY.
18. McGhee, J. D., and von Hippel, P. H. (1974) Theoretical aspects of DNA-protein interactions: cooperative and non-cooperative binding of large ligands to a one-dimensional homogeneous lattice. *J. Mol. Biol.* **86,** 469–489.
19. Lohman, T. M., Overman, L. B., and Datta, S. (1986) Salt-dependent changes in the DNA binding cooperativity of *Escherichia coli* single strand binding protein. *J. Mol. Biol.* **187,** 603–615.
20. Chang, A. C. Y., and Cohen, S. N. (1978) Constuction and characterization of amplifiable multicopy DNA cloning vehicles derived from the P15A cryptic mimiplasmid. *J. Bacteriol.* **134,** 1141–1156.
21. Studier, F. W., and Moffatt, B. A. (1986) Use of bacteriophage T7 RNA polymerase to direct selective high-level expression of cloned genes. *J. Mol. Biol.* **189,** 113–130.
22. Moffatt, B. A., and Studier, F. W. (1987) T7 lysozyme inhibits transcription by T7 polymerase. *Cell* **49,** 221–227.
23. Studier, F. W., Rosenberg, A. H., Dunn, J. J., and Dubendorff, J. W. (1990) Use of T7 RNA polymerase to direct expression of cloned genes. *Methods Enzymol.* **185,** 60–89.
24. Atkins, D., Roberts, K., Hull, R., Prehaud, C., and Bishop, D. H. L. (1991) Expression of the tobacco mosaic virus movement protein using a baculovirus expression vector. *J. Gen. Virol.* **72,** 2831–2835.
25. Berna, A., Gafny, R., Wolf, S., Lucas, W. J., Holt, C. A., and Beachy, R. N. (1991) The TMV movement protein: role of the C-terminal 73 amino acids in subcellular localization and function. *Virology* **182,** 682–689.

26. Rosenberg, A. H., Lade, B. N., Chui, D.-S., Lin, S.-W., Dunn, J., and Studier, F. W. (1987) Vectors for selective expression of cloned DNAs by T7 RNA polymerase. *Gene* **56**, 125–135.

27. Waigmann, E., Lucas, W. J., Citovsky, V., and Zambryski, P. C. (1994) Direct functional assay for tobacco mosaic virus cell-to-cell movement protein and identification of a domain involved in increasing plasmodesmal permeability. *Proc. Natl. Acad. Sci. USA* **91**, 1433–1437.

28. Stern, B. D., and Gruissem, W. (1987) Control of plastid gene expression: 3′ inverted repeats act as mRNA processing and stabilizing elements, but do not terminate transcription. *Cell* **51**, 1145–1157.

29. Delius, H., Mantell, N. J., and Alberts, B. (1972) Characterization by electron microscopy of the complex formed between T4 bacteriophage gene 32-protein and DNA. *J. Mol. Biol.* **67**, 341–350.

30. Zupan, J., Citovsky, V., and Zambryski, P. C. (1996) *Agrobacterium* VirE2 protein mediates nuclear uptake of ssDNA in plant cells. *Proc. Natl. Acad. Sci. USA* **93**, 2392–2397.

31. Tzfira, T., Vaidya, M., and Citovsky, V. (2001) VIP1, an *Arabidopsis* protein that interacts with *Agrobacterium* VirE2, is involved in VirE2 nuclear import and *Agrobacterium* infectivity. *EMBO J.* **20**, 3596–3607.

32. Schoumacher, F., Giovane, C., Maira, M., Poirson, A., Godefroy-Colburn, T., and Berna, A. (1994) Mapping of the RNA-binding domain of the alfalfa mosaic virus movement protein. *J. Gen. Virol.* **75**, 3199–3202.

33. Schoumacher, F., Erny, C., Berna, A., Godefroy-Colburn, T., and Stussi-Garaud, C. (1992) Nucleic acid binding properties of the alfalfa mosaic virus movement protein produced in yeast. *Virology* **188**, 896–899.

34. Schoumacher, F., Gagey, M. J., Maira, M., Stussi-Garaud, C., and Godefroy-Colburn, T. (1992) Binding of RNA by the alfalfa mosaic virus movement protein is biphasic. *FEBS Lett.* **308**, 231–234.

35. Jansen, K. A., Wolfs, C. J., Lohuis, H., Goldbach, R. W., and Verduin, B. J. (1998) Characterization of the brome mosaic virus movement protein expressed in *E. coli. Virology* **242**, 387–394.

36. Fujita, M., Kazuyuki, M., Kajiura, Y., Dohi, K., and Furusawa, I. (1998) Nucleic acid-binding properties and subcellular localization of the 3a protein of brome mosaic bromovirus. *J. Gen. Virol.* **79**, 1273–1280.

37. Thomas, C. L., and Maule, A. J. (1995) Identification of the cauliflower mosaic virus movement protein RNA binding domain. *Virology* **206**, 1145–1149.

38. Wobbe, K. K., Akgoz, M., Dempsey, D. A., and Klessig, D. F. (1998) A single amino acid change in turnip crinkle virus movement protein p8 affects RNA binding and virulence on *Arabidopsis thaliana. J. Virol.* **72**, 6247–6250.

39. Akgoz, M., Nguyen, Q. N., Talmadge, A. E., Drainville, K. E., and Wobbe, K. K. (2001) Mutational analysis of Turnip crinkle virus movement protein p8. *Mol. Plant Pathol.* **2**, 37–48.

40. Marcos, J. F., Vilar, M., Pérez-Payá, E., and Pallás, V. (1999) *In vivo* detection, RNA-binding properties and characterization of the RNA-binding domain of the p7 putative movement protein from carnation mottle carmovirus (CarMV). *Virology* **255**, 354–365.

41. Vaquero, C., Liao, Y. C., Nahring, J., and Fischer, R. (1997) Mapping of the RNA-binding domain of the cucumber mosaic virus movement protein. *J. Gen. Virol.* **78**, 2095–2099.

42. Li, Q., and Palukaitis, P. (1996) Comparison of the nucleic acid- and NTP-binding properties of the movement protein of cucumber mosaic cucumovirus and tobacco mosaic tobamovirus. *Virology* **216**, 71–79.

43. Nurkiyanova, K. M., Ryabov, E. V., Kalinina, N. O., Fan, Y., Andreev, I., Fitzgerald, A. G., Palukaitis, P., and Taliansky, M. (2001) Umbravirus-encoded movement protein induces tubule formation on the surface of protoplasts and binds RNA incompletely and non-cooperatively. *J. Gen. Virol.* **82**, 2579–2588.

44. Giesman-Cookmeyer, D., and Lommel, S. A. (1993) Alanine scanning mutagenesis of a plant virus movement protein identifies three functional domains. *Plant Cell* **5**, 973–982.

45. Osman, T. A., Thommes, P., and Buck, K. W. (1993) Localization of a single-stranded RNA-binding domain in the movement protein of red clover necrotic mosaic dianthovirus. *J. Gen. Virol.* **74**, 2453–2457.

46. Osman, T. A. M., Hayes, R. J., and Buck, K. W. (1992) Cooperative binding of the red clover necrotic mosaic virus movement protein to single-stranded nucleic acids. *J. Gen. Virol.* **73**, 223–227.

47. Qi, Y. J., Zhou, X. P., Huang, X. Z., and Li, G. X. (2002) *In vivo* accumulation of Broad bean wilt virus 2 VP37 protein and its ability to bind single-stranded nucleic acid. *Arch. Virol.* **147**, 917–928.

48. Pascal, E., Sanderfoot, A. A., Ward, B. M., Medville, R., Turgeon, R., and Lazarowitz, S. G. (1994) The geminivirus BR1 movement protein binds single-stranded DNA and localizes to the cell nucleus. *Plant Cell* **6**, 995–1006.

49. Rojas, M. R., Noueiry, A. O., Lucas, W. J., and Gilbertson, R. L. (1998) Bean dwarf mosaic geminivirus movement proteins recognize DNA in a form- and size-specific manner. *Cell* **95**, 105–113.

50. Kalinina, N. O., Rakitina, D. A., Yelina, N. E., Zamyatnin Jr., A. A., Stroganova, T. A., Klinov, D. V., Prokhorov, V. V., Ustinova, S. V., Chernov, B. K., Schiemann, J., Solovyev, A. G., and Morozov, S. Y. (2001) RNA-binding properties of the 63 kDa protein encoded by the triple gene block of poa semilatent hordeivirus. *J. Gen. Virol.* **82**, 2569–2578.

51. Donald, R. G., Lawrence, D. M., and Jackson, A. O. (1997) The barley stripe mosaic virus 58-kilodalton beta(b) protein is a multifunctional RNA binding protein. *J. Virol.* **71**, 1538–1546.

52. Bleykasten, C., Gilmer, D., Guilley, H., Richards, K. E., and Jonard, G. (1996) Beet necrotic yellow vein virus 42 kDa triple gene block protein binds nucleic acid *in vitro*. *J. Gen. Virol.* **77**, 889–897.

53. Herranz, M. C., and Pallás, V. (2004) RNA-binding properties and mapping of the RNA-binding domain from the movement protein of Prunus necrotic ringspot virus. *J. Gen. Virol.* **85**, 761–768.

54. Tacke, E., Prufer, D., Schmitz, J., and Rohde, W. (1991) The potato leafroll luteovirus 17K protein is a single-stranded nucleic acid-binding protein. *J. Gen. Virol.* **72**, 2035–2038.

55. Offei, S. K., Coffin, R. S., and Coutts, R. H. (1995) The tobacco necrosis virus p7a protein is a nucleic acid-binding protein. *J. Gen. Virol.* **76**, 1493–1496.

56. Cowan, G. H., Lioliopoulou, F., Ziegler, A., and Torrance, L. (2002) Subcellular localisation, protein interactions, and RNA binding of Potato mop-top virus triple gene block proteins. *Virology* **298**, 106–115.

57. Kalinina, N. O., Fedorkin, O. N., Samuilova, O. V., Maiss, E., Korpela, T., Morozov, S. Y., and Atabekov, J. G. (1996) Expression and biochemical analyses of the recombinant potato virus X 25K movement protein. *FEBS Lett.* **397**, 75–78.

58. Wung, C. H., Hsu, Y. H., Liou, D. Y., Huang, W. C., Lin, N. S., and Chang, B. Y. (1999) Identification of the RNA-binding sites of the triple gene block protein 1 of bamboo mosaic potexvirus. *J. Gen. Virol.* **80**, 1119–1126.

59. Rouleau, M., Smith, R. J., Bancroft, J. B., and Mackie, G. A. (1994) Purification, properties, and subcellular localization of foxtail mosaic potexvirus 26-kDa protein. *Virology* **204**, 254–265.

60. Lough, T. J., Shash, K., Xoconostle-Cázares, B., Hofstra, K. R., Beck, D. L., Balmori, E., Forster, R. L. S., and Lucas, W. J. (1998) Molecular dissection of the mechanism by which potexvirus triple gene block proteins mediate cell-to-cell transport of infectious RNA. *Mol. Plant Microbe Interact.* **11**, 801–814.

61. Tamm, T., and Truve, E. (2000) RNA-binding activities of cocksfoot mottle sobemovirus proteins. *Virus Res.* **66**, 197–207.

62. Ivanov, K. I., Ivanov, P. A., Timofeeva, E. K., Dorokhov, Y. L., and Atabekov, J. G. (1994) The immobilized movement proteins of two tobamoviruses form stable ribonucleoprotein complexes with full-length viral genomic RNA. *FEBS Lett.* **346**, 217–220.

63. Desvoyes, B., Faure-Rabasse, S., Chen, M. H., Park, J. W., and Scholthof, H. B. (2002) A novel plant homeodomain protein interacts in a functionally relevant manner with a virus movement protein. *Plant Physiol.* **129**, 1521–1532.
64. Soellick, T., Uhrig, J. F., Bucher, G. L., Kellmann, J. W., and Schreier, P. H. (2000) The movement protein NSm of tomato spotted wilt tospovirus (TSWV): RNA binding, interaction with the TSWV N protein, and identification of interacting plant proteins. *Proc. Natl. Acad. Sci. USA* **97**, 2373–2378.
65. Brunt, A. A., Crabtree, K., Dallwitz, M. J., Gibbs, A. J., Watson, L., and Zurcher, E. J. (1996 onwards). http://biology.anu.edu.au/Groups/MES/vide/

Chapter 22
Movement Profiles: A Tool for Quantitative Analysis of Cell-to-Cell Movement of Plant Viral Movement Proteins

Kateryna Trutnyeva, Pia Ruggenthaler, and Elisabeth Waigmann

Abstract Movement proteins (MPs) are virally encoded factors that mediate transport of viral nucleic acid between plant cells. Many MPs are able to move between cells themselves. This feature serves as the basis for evaluation of the transport activity of individual MPs. MPs are transiently expressed as a fusion to autofluorescent proteins such as green fluorescent protein (GFP) in individual epidermal cells of leaves by biolistic delivery. Expressing cells can be directly monitored for subcellular localization and cell-to-cell movement of the MP:GFP fusion protein into neighboring cells by confocal scanning microscopy. During the time frame of transient expression, numerous cells are evaluated at several time points, and the accumulated data are depicted in a graph termed "movement profile." Thus, a movement profile will provide information on the correlation between subcellular localization of the MP in the expressing cell and the efficiency of cell-to-cell transport, the time course and efficiency of targeting of the MP to plasmodesmata, and the translocation efficiency of the MP into neighboring cells.

Keywords Plasmodesmata; Movement protein; TMV; Microtubules; Subcellular localization; Movement profile; Particle bombardment; Handheld gene gun; Cell-to-cell movement

1 Introduction

To infect a plant, viruses are initially introduced into single host cells during mechanical damage or via vectors such as insects, fungi, and nematodes. There, the virus replicates and moves from the initially infected cell to neighboring cells using plasmodesmata (PD), complex cytoplasmic bridges interconnecting plant cells. This process, termed local or cell-to-cell movement, takes place primarily in mesophyll and epidermal tissues of leaves. To invade the whole plant, viruses cross the boundary into the vascular system and exploit the phloem stream for so-called long distance or systemic movement throughout the plant.

From: *Methods in Molecular Biology, Vol. 451, Plant Virology Protocols: From Viral Sequence to Protein Function*
Edited by G.D. Foster, I.E. Johansen, Y. Hong, and P.D. Nagy © Humana Press, Totowa, NJ

For many years, scientific efforts have focused on elucidating the mechanisms of cell-to-cell spread. The virally encoded movement proteins (MPs) are the central agents in cell-to-cell movement. MPs are encoded by all viruses, but their number and their detailed mode of action varies depending on the viral group. During local movement, viruses spread by one of two clearly distinct mechanisms: viral MPs either interact with the viral genome, frequently RNA, to form ribonucleoprotein complexes and to target those complexes to PD. Ribonucleoprotein complexes are considered the transport form of viral RNA that is translocated through plasmodesmata into adjacent cells. Alternatively, MPs are part of virally induced tubuli that extend through the cell wall, presumably replacing plasmodesmata, and serve as a conduit for spread of viral particles. In the first mechanism, plasmodesmata do not appear to be destroyed or even structurally modified by the movement process, whereas in the second mechanism pathogenic structures, the tubuli, are clearly apparent (for reviews see refs. 1–5).

Viral spread as ribonucleoprotein complexes has been extensively studied in the model virus *Tobacco mosaic virus* (TMV), a tobamovirus, and many of the mechanistic principles elucidated within the TMV system have proven true for other viral genera as well. Thus, the movement protein of TMV, TMV-MP, will also be used here as an example to describe a method for quantitative evaluation of the cell-to-cell movement capacity of viral MPs. TMV-MP is a multifunctional protein with several biological activities: it binds single-stranded nucleic acids (6, 7); associates with cellular structures such as the cytoskeleton (8, 9), the endoplasmic reticulum (ER; refs. 10–12), and plasmodesmata (13); and increases plasmodesmal permeability (14, 15). Most importantly in the context of this chapter, TMV-MP moves itself through PD (16–19), which can be directly visualized when the TMV-MP is fused to the green fluorescent protein (GFP; ref. 20).

A common method to study movement protein function relies on particle bombardment which leads to transient expression of the MP:GFP fusion protein in single epidermal cells of plant leaves. Both subcellular localization and cell-to-cell movement of the viral MP can then be monitored in living cells by confocal laser scanning microscopy. Here, we describe this method for the movement protein of TMV, which is characterized by a particularly complex set of subcellular localization patterns. The relationship between subcellular localization patterns and cell-to-cell movement is evaluated throughout the time range of transient expression, usually up to 3 days after bombardment. Data from approximately 100 cells are collected for each time point. This requires several individual bombardment experiments and involves several generations of plants that should be grown as reproducibly as possible in a controlled environment. In a final step, data are statistically evaluated and graphically presented in a scheme termed "movement profile."

2 Materials

2.1 Chemicals and Supplies for DNA Precipitation onto Microcarriers

1. DNA for transient expression, at a concentration of $1\,\mu g\,\mu L^{-1}$ (see note 1)
2. Double-distilled water (ddH$_2$O), autoclaved
3. Absolute ethanol (see note 2)
4. Gold particles with $1\,\mu m$ diameter (Bio-Rad)
5. Polyvinylpyrrolidone (PVP) stock solution: $20\,mg\,mL^{-1}$ PVP in absolute ethanol
6. PVP working solution: $0.05\,mg\,mL^{-1}$ in absolute ethanol (see note 3)
7. $0.05\,M$ spermidine in ddH$_2$O (see note 4)
8. $1\,M$ CaCl$_2$

2.2 Equipment and Consumables for Tubing Preparation and Biolistic Delivery

2.2.1 Tubing Prep Station (Bio-Rad) and Accessory Components Required for Cartridge Production

1. Tubing Prep Station and tubing support cylinder, assembled according to the manufacturer's manual
2. Syringe adaptor tubing and nitrogen hose
3. 10-mL syringe
4. Nitrogen tank with compressed nitrogen of at least grade 4.8. The nitrogen is required to dry the cartridges.
5. 15-mL disposable polypropylene centrifuge tubes

2.2.2 Handheld Helios™ Gene Gun (Bio-Rad) and Components

1. Handheld Helios™ Gene Gun (Bio-Rad)
2. Moveable Helium tank with compressed helium of at least grade 4.5. Helium is used to accelerate microcarriers during bombardment (see note 5).
3. Helium pressure regulator
4. Helium hose assembly
5. Tefzel Gold Coat™ tubing for cartridge production
6. Tubing cutter and blades
7. Vials for cartridge collection and storage
8. Cartridge holders
9. Barrel liner, to provide a defined distance between gene gun and sample

10. O-rings for barrel liner
11. Diffusion screen
12. 1% agar plates (see note 6)
13. Cartridge extractor tool

2.3 Equipment for Sample Preparation for Confocal Microscopy

1. Microscope slides, plain (Sigma), size 75×25 mm
2. 30×22 mm microscope coverslips
3. Watchmaker forceps
4. Double side sticking tape (see note 7)

2.4 Data Collection and Analysis by Laser Scanning Microscopy

1. Confocal laser scanning microscope
2. Software supplied by the manufacturer
3. Image analysis software, for example, Adobe Photoshop software (Adobe Systems, San Jose, CA)
4. Microsoft Excel for graphical data presentations

2.5 Plasmids and Plants

1. Use plasmids or binary vectors where the MP:GFP construct is expressed from a strong promoter. We routinely use plasmid pRTL-TMV-MP:GFP, containing the strong CaMV 35S promoter (21) (see note 8). Binary vector 35S-mgfp4-ER (22) for expression of ER localized GFP has also been successfully used (see note 9).
2. *Nicotiana* plants (*Nicotiana benthamiana, Nicotiana clevelandii, Nicotiana glutinosa, Nicotiana tabacum*) should be at least 1 month old. Plants that are 6–8 weeks old are routinely used, and mature leaves are chosen for bombardment. Plants are grown in a controlled environment at 22°C with 16 h of light alternating with 8 h of darkness.

3 Methods

3.1 Transient Expression of Viral MPs by Helios™ Gene Gun

The method described here makes use of a Handheld Helios™ Gene Gun (Bio-Rad), which allows bombardment of intact plants in the controlled conditions of a greenhouse or growth chamber. The general protocol provided in the Handheld Helios™ Gene Gun manual was optimized for effective transient transformation of leaf cells of *Nicotiana* plants.

3.1.1 Precipitation of DNA onto Gold Microcarriers

1. Purify plasmid DNA using effective purification kits (see note 1). The final concentration of purified DNA should be $1\,\mu g\ \mu L^{-1}$.
2. Prepare $3.5\,mL$ of $0.05\,mg\ mL^{-1}$ PVP working solution. This amount will be sufficient to coat 80-cm length of Tefzel Gold Coat™ tubing. A tubing length of 80-cm is required for one coating reaction (see also Sect. 3.1.2).
3. Weigh $25\,mg$ of 1-μm gold particles (microcarrier) into a 1.5-mL microfuge tube
4. Add $100\,\mu L$ of $0.05\,M$ spermidine to the gold particles
5. Vortex mixture for a few seconds
6. Add $50\,\mu g$ of the appropriate plasmid DNA to the gold and spermidine mixture
7. Mix DNA with the spermidine–gold mixture by vortexing at full speed
8. Next, while vortexing the mixture at moderate rate, add $100\,\mu L$ of $1\,M$ $CaCl_2$ dropwise (see note 10).
9. Incubate mixture at room temperature for 10 min to facilitate precipitation of the DNA onto gold particles.
10. Spin mixture for 5 s at full speed in a microfuge
11. Remove and discard the supernatant
12. Wash the pellet three times with 1 mL of absolute ethanol
13. After the final ethanol wash, resuspend the pellet in $200\,\mu L$ PVP working solution and transfer this microcarrier suspension into a 15 mL disposable polypropylene centrifuge tube.
14. Adjust the volume of the suspension to $3.5\,mL$ by adding the rest of the prepared PVP working solution.
15. This final suspension is then used to coat 80-cm of tubing, using the Tubing Prep Station from Bio-Rad.

3.1.2 Coating of the Tefzel Gold Coat™ Tubing with Loaded Microcarriers According to the Manual

1. Set up the Tubing Prep Station and connect it to a nitrogen tank
2. Insert an 80-cm long piece of Tefzel Gold Coat™ tubing into the tubing support cylinder via the opening on the right side of the Tubing Prep station until it reaches the O-ring at the opposite end of the tubing support cylinder (see note 11).
3. Open the nitrogen tank and adjust flow to 0.3–0.4-L min^{-1} using the knob on the flowmeter.

4. Let the nitrogen flow through the Tefzel Gold Coat™ tubing for at least 15 min to completely dry the tubing.

5. Turn off the nitrogen flow and remove Tefzel Gold Coat™ tubing from the Tubing Prep Station.

6. Connect an 80-cm piece of dried Tefzel Gold Coat™ tubing to a 10-mL syringe by inserting one end of the tubing into the syringe adaptor tubing fitted to the tip of the 10-mL syringe. Insert the other end of the dried Tefzel Gold Coat™ tubing into the microcarrier suspension (prepared as in Sect. 3.1.1). The microcarrier suspension should be briefly vortexed immediately before the Tefzel Gold Coat™ tubing is inserted to ensure even distribution of the gold particles.

7. Quickly draw the microcarrier suspension into the dried Tefzel Gold Coat™ tubing with the help of the syringe (see note 12).

8. Remove the Tefzel Gold Coat™ tubing from the centrifuge tube. Move the microcarrier suspension to create some solution-free space at each end of the Tefzel Gold Coat™ tubing by sucking with the syringe.

9. Slide the suspension filled Tefzel Gold Coat™ tubing into the tubing support cylinder until it penetrates the O-ring.

10. Let microcarriers settle at the bottom side of the Tefzel Gold Coat™ tubing for 3–5 min, then remove the ethanol supernatant slowly from the tubing at a rate of approximately $1–2\,cm\,s^{-1}$ using the 10-mL syringe. Detach syringe together with adaptor tubing from the Tefzel Gold Coat™ tubing.

11. Rotate the tubing support cylinder to position the settled microcarriers on the top side of the tubing. The gold particles then move by gravity along the sides of the tubing thereby coating the inner surface. After 10–15 s, start rotation of the Tefzel Gold Coat™ tubing. Dry the Tefzel Gold Coat™ tubing in a nitrogen steam while it continues to rotate. The flow rate of nitrogen should be set at approximately $0.35–0.4\,L\,min^{-1}$.

12. Dry tubing for 5 min, then turn off motor and nitrogen flow. Remove tubing from the tubing support cylinder and examine the tubing to verify that the microcarriers are evenly distributed over the length of the tubing (see note 13). Cutoff and discard unevenly or sparsely coated tubing. Cut evenly coated tubing into cartridges using a Tubing Cutter (Bio-Rad). An 80-cm length of tubing yield approximately 32 cartridges. Since we use 25 mg of gold to coat 80-cm length of Tefzel Gold Coat™ tubing, this is equivalent to 0.78 mg of gold per cartridge.

13. Store cartridges at 4°C in storage vials in desiccated environment, capped tightly and wrapped with parafilm (see note 14).

3.1.3 Biolistic Delivery

1. Assemble a Handheld Helios™ Gene Gun according to the instructions of the manufacturer and connect it to the Helium tank.

2. Load cartridges into a cartridge holder and insert the holder into the Gene Gun.

3. Screw the barrel liner into the barrel of the Gene Gun and position diffusion screen at the appropriate place in the barrel liner (see note 15).

Fig. 1 Biolistic delivery of TMV-MP:GFP into a *Nicotiana tabacum* leaf. Note the positioning of the agar plate and the Gene Gun

4. Transform plants directly in the greenhouse or growth chamber by bombarding the bottom side of a mature leaf. Support the leaf on its upper side with the agar plate, then, carefully bend the leaf, so that the bottom side of the leaf becomes accessible to the Gene Gun (Fig. 1) (see note 6). We routinely use a Helium pressure of 250 psi to bombard leaves of *Nicotiana* plants.
5. Gently press the leaf with the barrel liner to the agar plate and discharge the cartridge.

3.2 Analysis by Confocal Microscopy

In principle, both GFP and RFP fusion proteins can be monitored by confocal laser scanning microscopy. However, so far we have been able to detect cell-to-cell movement only for MP:GFP fusion proteins, but not for MP:RFP fusion proteins. Here, we describe as an example, the analysis of subcellular localization and movement capacity of TMV-MP fused to GFP in the host plant *Nicotiana glutinosa*. In each individual bombardment experiment, up to five shots are placed on one to two plants. To obtain a time course, data can be collected at days 1, 2, and 3 after bombardment. For each time point, a section of each of the shots is excised and analyzed, and the information derived from all shots is pooled. Depending on expression frequency, five shots will yield information on 20–30 cells per time point. Thus, to obtain data on a sufficient number of cells (approximately 100 cells), three to five individual bombardment experiments are performed.

3.2.1 Sample Preparation

1. From each shot excise a part of the bombarded area at day 1 after bombardment.
2. Place the excised area onto a glass slide, bottom side up, and cut it into smaller pieces with a maximal area of $1\,cm^2$. Remove large veins.
3. To avoid squeezing of the leaf tissues by the cover slip, attach two layers of double-sided tape to the glass slide at each side of the sample (see note 7). Place a drop of water onto each piece of sample and cover the sample with a coverslip. Completely fill the space under the coverslip with water.
4. Mount sample slide onto the microscope stage of a confocal laser scanning microscope. To find expressing cells, use a filter set appropriate for GFP. The emission filter should be a bandpass filter that cuts out chloroplast autofluorescence.
5. Scan all expressing cells and their neighbors in the sample. Do not select for specific cells, for example, those with a high expression level.
6. Repeat the procedure for other time points, for example, days 2 and 3 after bombardment.

3.2.2 Confocal Settings and Scanning Procedure

1. Excite GFP fluorescence with an Argon-Krypton or Argon laser at 477 or 488 nm. Detect GFP emission between 500 and 550 nm.
2. If an RFP fusion protein has been expressed, excite RFP fluorescence with a Helium–Neon laser at 561 nm or an Argon–Krypton laser at 568 nm. Detect emission between 600 and 620 nm.
3. Use a 40X immersion oil objective to scan expressing cells and neighboring cells. For optical sections, use a step size of $0.8–1\,\mu m$.
4. Assemble confocal stacks from serial optical sections into projections using software supplied by the manufacturer of the confocal laser scanning microscope.
5. Projections can be usually saved as individual files with file type "TIFF." These files are recognized by standard image analysis software such as Adobe Photoshop, and can thus be easily prepared for viewing and printing.

3.3 Data Analysis

3.3.1 Classification of Subcellular Patterns and Evaluation of Cell-to-Cell Movement

Depending on the viral movement protein, different sets of subcellular localization patterns may be observed. When TMV-MP:GFP is transiently expressed in epidermal

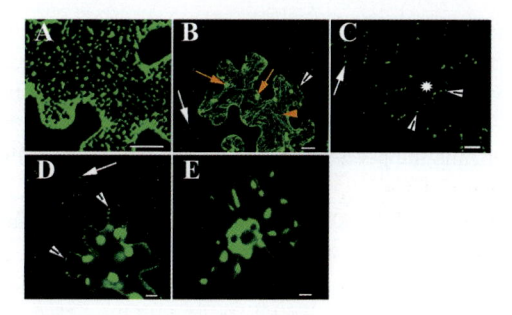

Fig. 2 Intracellular localization patterns of TMV-MP:GFP. (a) ER pattern; (b) MT pattern; (c) PD pattern; (d): PD pattern with large bodies; (e): large irregular spots as a result of overexpression. *Asterisk* designates the expressing cell, *white arrowheads* point to PD localized TMV-MP:GFP in expressing cell, *white arrows* point to TMV-MP:GFP that has moved into neighboring cell, *orange arrowhead* points to microtubules, *orange arrows* point to cortical bodies. Bars = 10 μm

cells of leaves, three main types of subcellular localization patterns became apparent (Fig. 2): (i) ER pattern, characterized by small, regularly spaced cortical TMV-MP: GFP punctae that are associated with the ER (Fig. 2a; ref. 23); (ii) PD pattern, characterized by cell wall associated TMV-MP:GFP punctae, that represent localization to plasmodesmata (Fig. 2c; ref. 24). PD localization may also be observed in conjunction with large bodies (Fig. 2d), cells with this pattern are also grouped into the category PD pattern. (iii) MT pattern, characterized by an extensive filamentous network that has been identified as TMV-MP associated with microtubules (MT; refs. 8, 25), frequently in conjunction with irregular cortical bodies and plasmodesmal localization (Fig. 2b).

Besides those three main patterns, several other rare intracellular localization patterns are observed. For example, some cells express TMV-MP:GFP in large irregular aggregates, which are likely the result of a high level of overexpression (Fig. 2e). These cells or cells with diffuse localization patterns are collectively classified as "other" localization patterns. Typically 5–10% of the analyzed cells fall into this category.

For cell-to-cell movement, presence of viral movement protein in cells surrounding the expressing cells is scored. Cell-to-cell movement manifests as TMV-MP:GFP localized in plasmodesmal punctae in neighboring cells (Fig. 2b–d, white arrows). Using the projection files, each cell is evaluated for subcellular localization and cell-to-cell movement. These data are assembled for each individual bombardment experiment and time point to calculate the percentage of cells with a particular localization pattern and the percentage of movement positive cells for each localization pattern.

1. Use the projection files to evaluate each cell for subcellular localization and cell-to-cell movement.
2. For each individual bombardment experiment and time point, calculate the percentage of cells with a particular expression pattern: $(\%_{pattern}) = 100 \times (N_{pattern}/$

$N_{experiment}$). In this formula, $N_{pattern}$ is the number of cells with a particular pattern; $N_{experiment}$ is the number of expressing cells that were analyzed in the individual bombardment experiment, and $\%_{pattern}$ is the percentage of cells with a particular pattern.

3. In addition, determine the percentage of movement positive cells with a particular localization pattern: $(\%+_{pattern}) = 100 \times (N+_{pattern}/N_{experiment})$. In this formula, $N_{experiment}$ is the number of cells analyzed in the individual bombardment experiment; $N+_{pattern}$ is the number of movement positive cells with a particular pattern, and $(\%+_{pattern})$ is the percentage of movement positive cells with a particular pattern.

3.3.2 Statistical Evaluation

For statistical evaluation, data from three to five individual bombardment experiments are required. Since in each of the three to five individual bombardment experiments a different number of expressing cells has been analyzed, a weighted mean is calculated to determine the overall percentage of cells with a particular pattern or the overall percentage of movement positive cells with a particular pattern for each time point.

1. The weighted mean of the percentage of cells with a particular pattern is defined as $(\%_{pattern})_{mean} = \Sigma_{i=1,...,n}(\%_{pattern})_i(N_{experiment,i}/N_{total})$, where i is an individual bombardment experiment, n is the total number of individual bombardment experiments, $(\%_{pattern})_i$ is the percentage of cells with a particular pattern in the ith individual bombardment experiment, $N_{experiment,i}$ is the number of cells in the ith individual bombardment experiment and N_{total} is the total number of cells analyzed in all bombardment experiments ($N_{total} = \Sigma_{i=1,...,n}N_{experiment,i}$).

2. The mean error m of the weighted mean of the percentage of cells with a particular pattern is calculated as $m = +/-\{[hvv]/N_{total}(n-1)\}^{1/2}$, where v_i is defined by $v_i = (\%_{pattern})_{mean} - (\%_{pattern})_i$ and $[hvv] = \Sigma_{i=1,...,n}N_{experiment,i}v_iv_i$.

3. Similarly, the weighted mean of the percentage of movement positive cells with a particular pattern is defined as $(\%+_{pattern})_{mean} = \Sigma_{i=1,...,n}(\%+_{pattern})_i(N_{experiment,i}/N_{total})$, where i is an individual bombardment experiment, n is the total number of individual bombardment experiments, $(\%+_{pattern})_i$ is the percentage of movement positive cells with a particular pattern in the ith individual bombardment experiment, $N_{experiment,i}$ is the number of cells in the ith individual bombardment experiment and $\mathbf{N_{total}}$ is the total number of cells $N_{total} = \Sigma_{i=1,...,n}N_{experiment,i}$.

4. The mean error m of the weighted mean of the percentage of movement positive cells with a particular pattern is calculated as $m = +/-\{[hvv]/N_{total}(n-1)\}^{1/2}$, where v_i is defined by $v_i = (\%+_{pattern})_{mean} - (\%+_{pattern})_i$ and $[hvv] = \Sigma_{i=1,...,n}N_{experiment,i}v_iv_i$.

5. Use Microsoft Excel to program calculation steps such that only the number of cells $N_{experiment,i}$ and the $(\%_{pattern})_i$ or $(\%+_{pattern})_i$ for each individual bombardment experiment are required as input.

3.3.3 Movement Profile: Graphic Representation of Statistically Evaluated Data

To graphically represent the obtained data as movement profile, Microsoft Excel software is used. For each time point after bombardment and for each of the three main expression patterns, the overall percentage of cells with a particular pattern is represented by bars (Fig. 3).

1. Depict the percentage of cells with an ER pattern by black bars, the percentage of cells with MT pattern by hatched bars, and the percentage of cells with PD pattern by grey bars. Cells that have been grouped into "other" localization patterns are usually not represented.
2. Depict the overall percentage of movement positive cells with a particular pattern by white bars that overlap with the bars for that particular pattern.

With this type of graphic representation, several conclusions about the time course of movement and the relationship between subcellular localization patterns and efficiency of cell-to-cell movement can be drawn. For example, TMV-MP:GFP cell-to-cell movement is primarily connected to a PD pattern, much less to an MT pattern, and not at all to an ER pattern (19). We have defined two parameters that characterize the transport capacity of the MP: (1) targeting frequency and (2) translocation frequency. (1) Targeting frequency is defined by the percentage of cells with a PD pattern and is a measure for how efficiently an MP reaches PD (Fig. 3, grey bars). (2) The translocation frequency is characterized by the percentage of

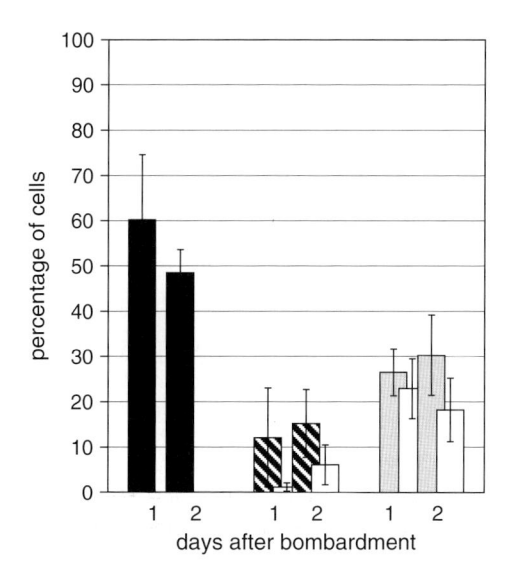

Fig. 3 Movement profile for TMV-MP:GFP expressed in *Nicotiana glutinosa*. *Black bars* show the overall percentage of cells with ER pattern, *hatched bars* show the overall percentage cells with MT pattern, and *grey bars* show the overall percentage of cells with PD pattern. The percentage of movement positive cells within each pattern is represented by overlapping *white bars*

cells with a PD pattern that show cell-to-cell movement. It is a measure for the efficiency of translocation through PD once the TMV-MP:GFP has been correctly targeted to PD (Fig. 3, white bars overlapping grey bars; ref. 19).

In addition, movement profiles are useful to compare the transport activity of an MP in different host plants or in different environmental conditions, or to compare the transport activity of a set of MP mutants.

4 Notes

1. For the preparation of plasmid DNA with a concentration of $1\,\mu g\,\mu L^{-1}$ we use Nucleobond PC 100 Kit (Macherey-Nagel, Düren, Germany) as well as Qiagen Plasmid Midi Kit (100) (Hilden, Germany); both kits yield DNA of sufficient quality for transient expression in epidermal leaf cells.
2. The use of absolute ethanol is important to ensure even loading of the Tefzel Gold Coat™ tubing with microcarriers and fast drying of the coated tube. Store absolute ethanol at −20°C.
3. Prepare PVP working solution in absolute ethanol ($0.05\,mg\,mL^{-1}$) freshly before use.
4. Spermidine (Sigma) is very hygroscopic; therefore, the whole content of a package should be used at once to prepare the solution, which can then be stored in 1-mL aliquots at −20°C.
5. The moveable helium tank allows bombardment of plants directly in the growth chamber or greenhouse. Thus, plants are not stressed by a change in environment.
6. A 1% agar plate is used as support for the leaf during the high-pressure helium shot. To prepare the plate fill a petri dish with 1% agar as full as possible. After the agar has solidified, cover the agar surface with parafilm.
7. We use Scotch double-sided tape (3M Company, St. Paul, Minnesota).
8. In our experience, only fusion proteins with the TMV-MP sequence at the N terminus were functional in cell-to-cell transport.
9. Transient expression from binary vectors is usually less efficient.
10. Select vortex speed such that the solution is only slightly moved.
11. If you plan to do several coatings, you can use an appropriately larger piece of Tefzel Gold Coat tubing™ for drying and cut it afterward into 80-cm long pieces.
12. It is important to work fast to prevent sedimentation of the gold microcarriers at the bottom of the 15-mL polypropylene tube.
13. Even distribution is achieved if the whole tubing shows a slightly brownish color on its inner surface. If distinct brown spots or lines are visible, aggregation of microcarriers has occurred, which reduces the efficiency of transient expression.
14. In our experience, cartridges that are used directly after coating or on the next day yield the highest expression efficiency.
15. It is important to use diffusion screens to reduce wounding of tissues by the high-pressure shot.

Acknowledgments This project was supported by grants from the Austrian Science Foundation (Sfb17, project part 08), the WWTF (LS 123) and the ARC Seibersdorf to E.W.

References

1. Carrington, J. C., Kasschau, K. D., Mahajan, S. K., and Schaad, M. C. (1996) *Plant Cell* **8**, 1669–81.
2. Heinlein, M. (2002a) *Cell. Mol. Life Sci.* **59**, 58–82.

3. Lazarowitz, S. G., and Beachy, R. N. (1999) *Plant Cell* **11**, 535–48.
4. Tzfira, T., Rhee, Y., Chen, M.-H., and Citovsky, V. (2000) *Ann. Rev. Microbiol.* **54**, 187–219.
5. Waigmann, E., Ueki, S., Trutnyeva, K., and Citovsky, V. (2004) *Crit. Rev. Plant Sci.* **23**, 195–250.
6. Citovsky, V., Knorr, D., Schuster, G., and Zambryski, P. C. (1990) *Cell* **60**, 637–47.
7. Citovsky, V., Wong, M. L., Shaw, A., Prasad, B. V. V., and Zambryski, P. C. (1992) *Plant Cell* **4**, 397–411.
8. Heinlein, M., Epel, B. L., Padgett, H. S., and Beachy, R. N. (1995) *Science* **270**, 1983–85.
9. McLean, B. G., and Zambryski, P. (2000) *in* Actin: a dynamic framework for multiple plant cell functions (Staiger, C. J., Baluska, F., Volkmann, D., and Barlow, P. W., Eds.), Kluwer, Dordrecht.
10. Heinlein, M., Padgett, H. S., Gens, J. S., Pickard, B. G., Caspar, S. J., Epel, B. L., and Beachy, R. N. (1998) *Plant Cell* **10**, 1107–20.
11. Mas, P., and Beachy, R. N. (1999) *J. Cell Biol.* **147**, 945–58.
12. Reichel, C., and Beachy, R. N. (1998) *Proc. Natl. Acad. Sci. USA* **95**, 11169–74.
13. Tomenius, K., Clapham, D., and Meshi, T. (1987) *Virology* **160**, 363–71.
14. Wolf, S., Deom, C. M., Beachy, R. N., and Lucas, W. J. (1989) *Science* **246**, 377–79.
15. Waigmann, E., Lucas, W., Citovsky, V., and Zambryski, P. (1994) *Proc. Natl. Acad. Sci. USA* **91**, 1433–37.
16. Crawford, K. M., and Zambryski, P. C. (2001) *Plant Physiol.* **125**, 1802–12.
17. Kotlizky, G., Katz, A., van der Laak, J., Boyko, V., Lapidot, M., Beachy, R. N., Heinlein, M., and Epel, B. L. (2001) *Mol. Plant Microbe Interact.* **14**, 895–904.
18. Waigmann, E., and Zambryski, P. (1995) *Plant Cell* **7**, 2069–79.
19. Trutnyeva, K., Bachmaier, R., and Waigmann, E. (2005) *Virology* **332**, 563–77.
20. Chalfie, M., Tu, Y., Euskirchen, G., Ward, W. W., and Prasher, D. C. (1994) *Science* **263**, 802–05.
21. McLean, B. G., Zupan, J., and Zambryski, P. C. (1995) *Plant Cell* **7**, 2101–14.
22. Haseloff, J., Siemering, K. R., Prasher, D. C., and Hodge, S. (1997) *Proc. Natl. Acad. Sci. USA* **94**, 2122–27.
23. Gillespie, T., Boevink, P., Haupt, S., Roberts, A. G., Toth, R., Valentine, T., Chapman, S., and Oparka, K. J. (2002) *Plant Cell* **14**, 1207–22.
24. Oparka, K. J., Prior, D. A. M., Santa-Cruz, S., Padgett, H. S., and Beachy, R. N. (1997) *Plant J.* **12**, 781–89.
25. McLean, B. G., Zupan, J., and Zambryski, P. C. (1995) *Plant Cell* **7**, 2101–14.

Chapter 23
Analysis of siRNA-Suppressor of Gene Silencing Interactions

Lóránt Lakatos and József Burgyán

Abstract RNA silencing is an evolutionarily conserved system that functions as an antiviral mechanism in higher plants and animals. To counteract RNA silencing, viruses evolved silencing suppressors that interfere with siRNA guided RNA silencing pathway. We used the heterologous *Drosophila* in vitro embryo RNA to analyze the molecular mechanism of suppression of silencing suppressors. We found that different silencing suppressors inhibit the RNA silencing via binding to siRNAs. None of the suppressors affected the activity of preassembled RISC complexes. In contrast, suppressors uniformly inhibited the siRNA-initiated RISC assembly pathway by preventing RNA silencing initiator complex formation. Here, we provide the protocol for the detailed analysis of p19 silencing suppressors of tombusviruses in the heterologous *Drosophila* in vitro system.

Keywords Silencing suppressor; in vitro *Drosophila* RNA silencing system; siRNA; p19; siRNA binding

1 Introduction

RNA silencing is an evolutionary conserved intracellular surveillance system, based on the recognition and targeting of RNAs containing regions that are double-stranded (dsRNA). Natural roles of RNA silencing include genome defense and specification of heterochromatin formation, posttranscriptional inhibition of gene expression by miRNAs, and *trans*-acting siRNAs, and antiviral defense (1, 2).

RNA silencing is induced by dsRNA that is sensed by the RNase III family enzyme DICER. DICER digests dsRNA into 21–26 nt ds silencing interfering RNAs (siRNA), which incorporate into the RNA-induced silencing complex (RISC). Argonaute is the core component of RISC, which first eliminates the passenger strand of the si- or miRNA and then it is able to do the sequence specific degradation and/or recognition of single-stranded target RNA (3, 4).

From: *Methods in Molecular Biology, Vol. 451, Plant Virology Protocols:*
From Viral Sequence to Protein Function
Edited by G.D. Foster, I.E. Johansen, Y. Hong, and P.D. Nagy © Humana Press, Totowa, NJ

Virus-induced silencing leads to the sequence-specific degradation of viral RNA and generation of a mobile silencing signal that activates or potentiates RNA silencing in noninfected cells (5). For successful virus infection, plant and animal RNA viruses are evolved to express proteins to counteract RNA silencing (6, 7).

The well-established *Drosophila* in vitro embryo RNA silencing system recapitulates the ds as well as siRNA-dependent RNA silencing of gene expression (9, 10). Therefore, the *Drosophila* in vitro RNA silencing system is suitable to ask mechanistic questions about the silencing suppressor action. This system allowed us to study the effect of silencing suppressors having RNA-binding activity in the target cleavage assay and electrophoretic mobility shift experiment visualizing RISC and RISC intermediates. Our results showed that ds siRNA-binding suppressors such us p21 of the Beet yellows virus, p19 of Carnation italian ringspot virus, or HcPro of Tobacco etch virus inhibited RISC assembly, but could not interfere with the ss siRNA containing active RISC (10–12).

2   Materials

2.1   siRNA-Binding Assay

1. siRNAs were ordered from Dharmacon (www.dharmacon.com).
2. For the electrophoretic mobility shift experiments, we used the Penguin Electrophoresis System from Owl Technologies ($20 \times 20\,cm^2$ glass plates, 1.5 mm spacers and combs).
3. For native gels 40% acrylamide/bis-acryamide (39:1) stock solution was used. Gels were prepared in 0.5× TBE.

2.2   Target Cleavage Assay

1. Two hours old Drosophila embryos were dechorionated in 50% bleach, washed in water, and dried. Dried embryos were lysed in 1× lysis buffer (30 mM HEPES ph 7.5, 100 mM KoAc, 2 mM $MgCl_2$, 5 mM DTT). Lysate was cleared with centrifugation, then the supernatant was flash frozen in aliquots in liquid nitrogen and stored at −80 °C (8).
2. For native and denaturing gels 40% acrylamide/bis-acrylamide (38:2) stock solution was used.
3. Guanylyl transferase and *S*-adenosyl methionine (SAM) are obtained from Ambion.
4. 2× PK buffer contains 200 mM TRIS-HCl (pH 7.4), 25 mM EDTA (pH 8.0), 300 mM NaCl, and 2% sodium dodecyl sulphate.

5. Energy regenerating system: 10 mM creatine phosphate (cat# 2380) and 10 μg ml^{-1} creatine phosphokinase (cat# 2384) from Calbiochem (8). Creatine kinase stock is prepared by dissolving 20 mg creatine kinase in 1 ml of ice-cold 40 mM TRIS-Acetate (ph 6.8), 200 mM KoAc, 0.2 mM EDTA, 20 mM mercaptoethanol. This solution is then added to an equal volume of ice-cold glycerol.
6. 5′ end cleavage products were quantified with a Genius Image Analyser (Syngene).

2.3 Electrophoretic Mobility Shift Experiment of Extracts with p19

1. For native gels 40% acrylamide/bis-acryamide (39:1) stock solution was used.
2. For the electrophoretic mobility shift experiments, we used the Penguin Electrophoresis System from Owl Technologies (20 × 20 cm^2 glass plates, 1.5 mm spacers and combs).
3. Loading buffer: 1× lysis buffer, 6% ficoll 400, and 0.025% w/v xylane cianol.

3   Methods

Since a plant-derived in vitro RNA silencing system, which can be programmed by exogenous siRNAs, is not available, we adopted the heterologous *Drosophila* in vitro RNAi system to test the silencing suppressor activity. The *Drosophila* in vitro RNA silencing system can be efficiently programmed with siRNAs characterized as 21 nt dsRNAs having 2 nt 3′ overhangs. Thus, this system allows us to test the effect of a siRNA-binding silencing suppressor to RISC assembly and single-stranded RNA containing active RISC. In this system, p19 powerfully inhibited RISC assembly but not active RISC as shown by target cleavage experiments. In the *Drosophila* in vitro system, RISC assembly and RISC intermediate complexes can be visualized by electrophoretic mobility shift experiment that led us to determine that p19 competes with DICER2-R2D2 silencing initiator complex for siRNAs thus hampering the first step of RISC assembly (11).

3.1   siRNA-Binding Assay

1. One strand of the siRNA is phosphorilated with γ-^{32}P ATP in a 20 μl reaction using 20 U of T4 polynucleotide kinase. The 5′-phosphorylated complementary strand in five times molar excess was added and then the reaction was heated to 95 °C for 1 min and cooled slowly to anneal the strands. Duplexes were purified by PAGE on a native 15% polyacrylamide containing 0.5× TBE gel. The labeled

siRNA binding assay

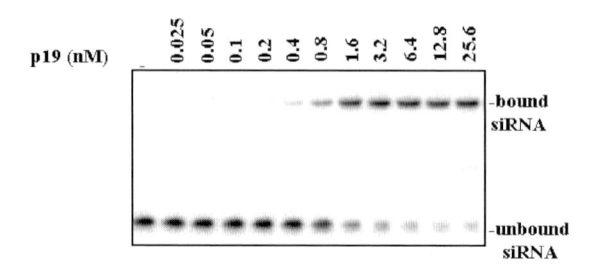

Fig. 1 p19 binds ds siRNA. siRNA-binding reactions were assembled by adding siRNA and series of dilutions of p19 in 1× lysis buffer supplemented with 0.2% Tween

duplex was cut out of the gel and eluted in solution containing 0.3 M NaCl then precipitated with ethanol. Generally, the yield is about 50% of the input siRNA. This siRNA can be used for both siRNA-binding assays and programming *Drosophila* embryo extracts.

2. p19 protein was expressed as a GST fusion protein and isolated on glutathione-Sepharose 4B resins, according to the manufacturer's instructions (Amersham Bioscience). From the beads p19 was liberated by thrombin cleavage, the supernatant was dialysed against 1× lysis supplemented with 10% glycerol then quantified with Bradford assay (12).

3. In vitro binding reactions were assembled on ice. In a 10 μl reaction, siRNA at 10 pM final concentration were used in 1× lysis buffer containing 0.02% of Tween 20 (12). p19 was used as a series of dilutions in 0.025–33.6 nM concentration. Reactions were then incubated at 25 °C for 15 min and then loaded Fig. 1.

3.2 Target Cleavage Assay

1. Target RNA was transcribed in vitro in a 100 μl reaction. The RNeasy kit was used to purify the RNA from the in vitro transcription reaction. Then, the target RNA was cap-labeled in the presence of α-32P GTP and 1 mM SAM with recombinant vaccinia virus guanylyl transferase for 2 h at 37 °C. Gel purification of the cap-labeled target RNA was carried out on a 12% acrylamide gel containing 8 M urea. Labeled target RNA was identified by autoradiography, then cut out of the gel and eluted in 2× PK buffer for overnight. The eluate was phenol extracted and precipitated with 3 vol of ethanol and quantified as described (13).

2. Ten microliters in vitro reactions were assembled on ice in 1× lysis buffer containing 10% glycerol, about 10 μg of embryo extracts, 5 nM siRNA, 0.5 nM cap-labeled target RNA, and 1 mM ATP. To keep the ATP level constant, ATP regenerating system was used. To assay the silencing suppressor activity in this system, purified p19 of CIRV was added to the reactions in series of dilutions

made in 1× lysis. When the effect of p19 was tested to RISC assembly, embryo extract, inducer siRNA, and target RNA were added simultaneously (direct competition experiment). When the effect of silencing suppressors on preassembled RISC activity was analysed, siRNA was preincubated with extracts to allow RISC formation for 10 min, then target RNA and suppressor proteins were added (indirect competition experiment). Reactions were incubated at 25 °C for 1 h.

3. RNA was isolated by adding 200 μl 2× PK buffer containing 80 ng ml^{-1} Proteinase K to one reaction. Tubes are incubated at 65 °C for 15 min, phenol–chloroform extracted and precipitated with 2.5 vol of ethanol.

4. RNA was resolved on a 12% acryamide gel containing 8 M urea.

5. RISC activity was measured by quantification of the 5′-end product of siRNA-directed cleavage of the target RNA (Fig. 2 upper panel, Fig. 3 upper panel).

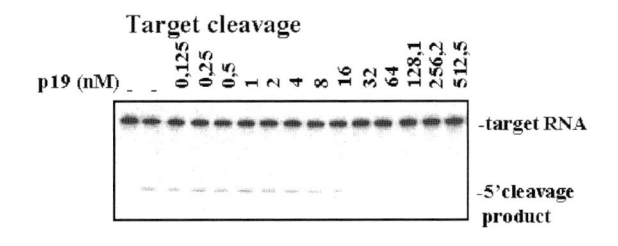

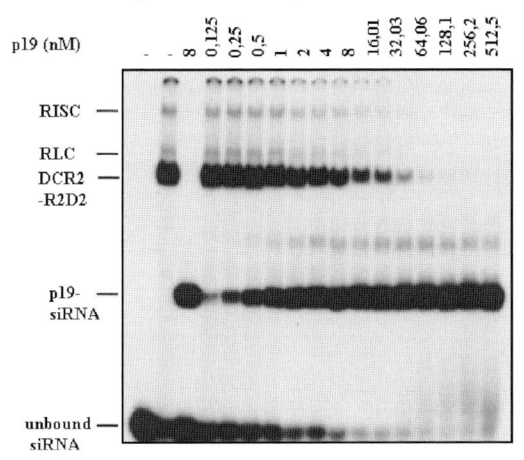

Fig. 2 p19 inhibits RISC activity by affecting RISC formation in the direct competition experiment. Target cleavage experiment reactions were assembled by adding siRNA, series of dilutions of p19, and target RNA simultaneously to the extracts. Reactions were the same for the electrophoretic mobility experiment except target RNA was omitted. p19 sequesters siRNA, thus inhibiting RNA silencing, which is manifested by the disappearance of the 5′ end cleavage product. p19 competes with the silencing initiator DCR2-R2D2 complex for siRNA binding. Diminished siRNA-DCR2-R2D2 complex formation leads to reduced RISC assembly. RISC activity in the target cleavage experiment correlates well with the amount of RISC in the electrophoretic mobility experiment (Fig. 2 lower panel, Fig. 3 lower panel)

Target cleavage

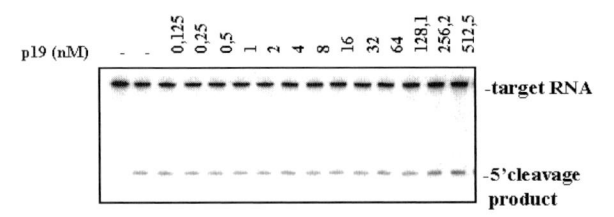

Electrophoretic mobility shift

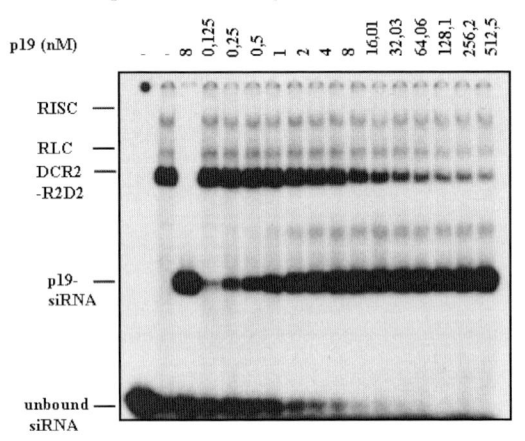

Fig. 3 p19 is not able to inhibit preassembled RISC. For target cleavage experiment, reactions were assembled by adding first siRNA to the extracts and reaction were incubated until RISC assembled. Then series of dilutions of p19 and target RNA were added. Reactions were the same for the electrophoretic mobility experiment except target RNA was omitted. Target cleavage is not affected by p19, since RISC contains single stranded RNA. Consistently, the amount of RISC has not been changed as shown by the electrophoretic mobility experiment. The amount of RISC loading complex (RLC) changed by the administration of p19 but a less extent than the ds siRNA containing DCR2-R2D2. This can be due to the fact that RLC contains ds as well as ss siRNA (14)

3.3 Electrophoretic Mobility Shift Experiment

1. Reactions for the direct and indirect competition electrophoretic mobility shift experiments were assembled and incubated as for the target cleavage experiments, except the cap-labeled target RNA was omitted.
2. Reactions were diluted with equal volume of loading buffer then loaded onto a prechilled (4 °C) 4.0% native acrylamide gel containing 1× TBE. Gels were usually run in the cold room for 5 h.
3. Gels were then dried and exposed to a storage phosphor screen (Molecular Dynamics Typhoon Phosphorimager, Amersham Biosciences).

4 Notes

1. Creatine kinase stock solution can be used for 3–4 weeks.
2. RISC is a multiple turnover enzyme. By limiting the target RNA concentration (0.5 nM), we used single turnover conditions; therefore, in the target cleavage experiments we measure the absolute amount of RISC.
3. Quantification of the band corresponding to RISC in the electrophoretic mobility shift assay let us to define the absolute amount RISC also, which makes target cleavage and electrophoretic mobility shift experiments comparable.
4. To evaluate the effect of silencing suppressors in the target cleavage and the electrophoretic mobility shift experiments comparable, we used the same batch of embryo extract and labeled siRNA preparation for both experiments.

Acknowledgments This work was supported by grants from the Hungarian Scientific Research Fund (OTKA; T046728 and OTKA; T048852), the "RIBOREG" EU project (LSHG-CT-2003503022), and the Scientia Amabilis Foundation. L.L. is a recipient of the Bolyai János Fellowship.

References

1. Matzke, M.A. and Matzke, A.J. (2004) Planting the seeds of a new paradigm. PLoS Biol, 2, E133.
2. Mello, C.C. and Conte, D., Jr. (2004) Revealing the world of RNA interference. Nature, 431, 338–342.
3. Tomari, Y. and Zamore, P.D. (2005) Perspective: machines for RNAi. Genes Dev, 19, 517–529.
4. Matranga, C., Tomari, Y., Shin, C., Bartel, D.P. and Zamore, P.D. (2005) Passenger-strand cleavage facilitates assembly of siRNA into Ago2-containing RNAi enzyme complexes. Cell, 123, 607–620.
5. Baulcombe, D. (2004) RNA silencing in plants. Nature, 431, 356–363.
6. Silhavy, D. and Burgyan, J. (2004) Effects and side-effects of viral RNA silencing suppressors on short RNAs. Trends Plant Sci, 9, 76–83.
7. Voinnet, O. (2005) Induction and suppression of RNA silencing: insights from viral infections. Nat Rev Genet, 6, 206–220.
8. Tuschl, T., Zamore, P.D., Lehmann, R., Bartel, D.P. and Sharp, P.A. (1999) Targeted mRNA degradation by double-stranded RNA in vitro. Genes Dev, 13, 3191–3197.
9. Nykanen, A., Haley, B. and Zamore, P.D. (2001) ATP requirements and small interfering RNA structure in the RNA interference pathway. Cell, 107, 309–321.
10. Lakatos, L., Szittya, G., Silhavy, D. and Burgyan, J. (2004) Molecular mechanism of RNA silencing suppression mediated by p19 protein of tombusviruses. Embo J, 23, 876–884.
11. Lakatos, L., Csorba, T., Pantaleo, V., Chapman, EJ., Carrington, JC., Liu, YP., Dolja, VV., Fernández Calvino, L., López-Moya, JJ., Burgyán, J. (2006) Small RNA binding is a common strategy to suppress RNA silencing by several viral suppressors. EMBO J, 25(12), 2768–80.
12. Vargason, J., Szittya, G., Burgyan, J. and Hall, T.M. (2003) Size selective recognition of siRNA by an RNA silencing suppressor. Cell, 115, 799–811.
13. Haley, B., Tang, G. and Zamore, P.D. (2003) In vitro analysis of RNA interference in *Drosophila melanogaster*. Methods, 30, 330–336.
14. Matranga C, Shin C, Bartel DP, Zamore PD (2005) Passenger-strand cleavage facilitates assembly of siRNA into Ago2-containing RNAi enzyme complexes. Cell, 123(4):607–20.

Chapter 24
Phosphorylation Analysis of Plant Viral Proteins

Kristiina M. Mäkinen and Konstantin I. Ivanov

Abstract Posttranslational modification of proteins is a key regulatory mechanism in a variety of cellular processes. This chapter outlines the concepts and methods used to investigate protein phosphorylation and its physiological relevance during plant virus infection. Rather than providing an exhaustive review of the experimental protocols for protein phosphorylation analysis, we focus on methods that can be used to study phosphorylation of viral proteins. We address the following points: how to determine that a viral protein of interest is phosphorylated; how to map the phosphorylation sites; how to identify the protein kinase(s) involved. Finally, we describe a number of useful strategies to evaluate the biological significance of phosphorylation.

Keywords Protein phosphorylation; Plant virus protein; Phosphorylation site mapping; Kinase identification; Plant virus infection

1 Introduction

A multitude of posttranslational modifications including, but not limited to, formation of disulfide bonds, cleavage by proteinases, phosphorylation, glycosylation, and acylation regulate protein function within living cells. Phosphorylation involves the enzymatic transfer of a phosphoryl group from adenosine triphosphate to a protein, with the aid of an enzyme called protein kinase. The *Arabidopsis thaliana* genome encodes for approximately 1,000 protein kinases (1). During viral infection, the most important virus-specific functions such as virion assembly (2) and dissociation (3) may be regulated by protein phoshporylation. Evidence suggests that phosphorylation of plant virus-encoded movement proteins may play an important role in viral movement (4). For example, phosphorylation of tobacco mosaic virus movement protein regulates cell-to-cell movement of the virus (5). Analysis of viral protein phosphorylation is typically difficult because of low abundance and low stoichiometry of phosphorylation. In addition, this task is frequently complicated by the presence of multiple differentially phosphorylated protein isoforms. Therefore, in most cases, a viral protein phosphorylation study presents a formidable challenge to a researcher.

From: *Methods in Molecular Biology, Vol. 451, Plant Virology Protocols: From Viral Sequence to Protein Function*
Edited by G.D. Foster, I.E. Johansen, Y. Hong, and P.D. Nagy © Humana Press, Totowa, NJ

2   Materials

2.1   Covalent Cross-Linking of Antibodies to Protein A(G)-Sepharose

1. Protein A- or protein G-Sepharose (slurry in 20% ethanol, Amersham Biosciences, Uppsala, Sweden). For other suppliers, verify that the albumin-binding region of protein G has been genetically deleted.
2. Sodium Phosphate buffer: 0.1 M sodium phosphate, pH 8.0
3. Cross-linking buffer: 0.2 M triethanolamine, pH 8.2
4. Dimethyl pimelimidate (DMP, Pierce Biotechnology, Rockford, IL)
5. Stop buffer: 0.1 M ethanolamine, 1 M glycine, pH 8.5
6. PBS (phosphate-buffered saline, 137 mM NaCl, 2.7 mM KCl, 1.4 mM KH_2PO_4, 8.1 mM Na_2HPO_4, pH 7.2–7.4)
7. Antibody elution buffer: 0.1 M glycine–HCl, pH 2.5
8. IP buffer: 20 mM Tris–HCl, pH 7.5, 150 mM NaCl, 0.5% Nonidet P-40

2.2   Metabolic Cell Labeling

1. Man-pp: 0.5% 2-morpholinoethanesulfonic acid (MES)-buffer supplied with 2% sucrose, 200 mM mannitol, and a mixture of salts analogous to B5 salts but without phosphate-containing salts; pH 5.7 adjusted with KOH; autoclave for 15 min in 120 °C.

2.3   Immunoprecipitation Coupled with Enzymatic Dephosphorylation

1. Lambda protein phosphatase (λ-PPase) reaction buffer: 50 mM Tris–HCl, 100 mM NaCl, 2 mM $MnCl_2$, 2 mM dithiothreitol, 0.1 mM EGTA, pH 7.5
2. Lambda protein phosphatase (New England Biolabs, Ipswich, MA)

2.4   Two-Dimensional Phosphopeptide Mapping and Phosphoamino Acid Analysis

1. Digestion buffer: 10% (v/v) acetonitrile in freshly made 50 mM ammonium bicarbonate, pH 8.0
2. pH 1.9 buffer: 0.58 M formic acid, 1.36 M glacial acetic acid
3. pH 3.5 buffer: 0.5% (v/v) pyridine, 0.87 M glacial acetic acid

2.5 Immunocomplex Kinase Assay

1. Kinase buffer: 20 mM Hepes, pH 7.4, 100 mM NaCl, 10 mM $MgCl_2$, 2 mM $MnCl_2$ or 2 mM $CaCl_2$ (depending on the metal preference of the kinase under investigation)

2.6 Double-Label Immunofluorescence Confocal Microscopy

1. Dulbecco's medium (PBS supplied with 0.5 mM $MgCl_2$ and 0.9 mM $CaCl_2$)
2. Fixing solution: 500 mM mannitol in 0.5% 2-morpholinoethanesulfonic acid (MES)-buffer, pH 5.7 (adjusted with KOH) containing paraformaldehyde (4%) and glutaraldehyde (0.2%) (electron microscopy grade, Sigma). If necessary, heat-dissolve carefully in a fume hood on a stirring hot plate.
3. Wash buffer: Dulbecco's medium supplied with 0.2% BSA
4. Quenching solution: 0.1% $NaBH_4$ in PBS
5. Permeabilization solution: 0.1% Triton X-100 in PBS
6. Antibody dilution buffer: 5% BSA and 0.1% bovine gelatine in Dulbecco's medium

2.7 Infection with the Mutant Virus

1. GFP-tagged infectious viral DNA/cDNA
2. A high-fidelity DNA polymerase such as Phusion High Fidelity DNA polymerase (Finnzymes) or Pfu turbo (Stratagen) or, alternatively, a commercial site-directed mutagenesis kit, e.g., Quick Change (Stratagen)

2.8 Silencing of the Protein Kinase

1. Gateway™ recombinational cloning reagents (Invitrogen)
2. pHELLSGATE vectors developed by CSIRO (Australia's National Science Agency)
3. Tobacco rattle virus vector developed at the Sainsbury Laboratory (Norwich, UK)
4. Host plant-specific cDNA-library

3   Methods

3.1   Analysis of the Protein Phosphorylation State

Immunoprecipitation is a widely used method for analysis of protein posttransla-tional modifications, including phosphorylation. Not only does it allow examination of the in vivo phosphorylation status of the protein of interest, but also, in conjunction with microsequencing, allows characterization of coimmunoprecipitating protein kinases or phosphatases. Quality of the antibodies raised against the protein of interest is crucial to the success of immunoprecipitation.

3.1.1   Covalent Cross-Linking of Antibodies to Protein A(G)-Sepharose

Conventional immunoprecipitation method relies on a strong noncovalent interac-tion between antibody and protein A- or protein G-Sepharose (see note 1). This interaction is disrupted during antigen elution at the last stage of the immuno-precipitation protocol, resulting in sample contamination with released antibody. Antibody release from protein A(G)-Sepharose represents a major problem if molecular weight of a precipitated protein is similar to that of antibody heavy or light chains. Sample contamination with antibody is also an important issue when precipitated protein is to be directly analyzed by mass spectrometry. The best solu-tion to the problem of antibody release is covalent cross-linking of the antibody to protein A(G)-Sepharose. The following protocol provides guidelines for how to perform the cross-linking.

1. Vortex 500 μl of protein A(G)-Sepharose slurry. Transfer the slurry to a screw-cap microcentrifuge tube and resuspend in 900 μl of cold sodium phosphate buffer. Pellet the beads by brief (30 s, 10,000 rpm) centrifugation in a micro-fuge, and carefully aspirate supernatant to obtain packed beads. A small vol-ume of buffer should be left over the beads. Repeat the wash three times with 1 ml sodium phosphate buffer and resuspend beads in 800 μl of the same buffer.
2. Add to the beads 100–200 μl serum or 150–250 μg purified IgG in a maximum volume of 200 μl. Mix bead suspension on a rotator at +4 °C for 30 min to 1 h to bind antibody to protein A(G)-Sepharose.
3. Pellet the beads by brief centrifugation; wash three times with 1 ml sodium phos-phate buffer as in step 1 and resuspend the beads in 1 ml cross-linking buffer.
4. Mix the bead suspension in cross-linking buffer for 5 min at room temperature on a rotator, spin down in a microfuge, and wash twice more with 1 ml cross-linking buffer as described earlier.
5. Resuspend the beads in 1 ml cross-linking buffer containing 32 mM cross-linker dimethyl pimelimidate (DMP, 8.3 mg DMP per milliliter of buffer; see note 2). Allow the cross-linking reaction to proceed for 45 min at room temperature on a rotator.

6. Pellet the beads by brief centrifugation and carefully remove supernatant without disturbing the pellet.
7. Resuspend the bead pellet in 1 ml stop buffer, spin down in a microfuge, carefully remove supernatant, and resuspend beads in 1 ml of the same buffer.
8. Mix the bead suspension in stop buffer for 1 h at room temperature on a rotator, spin down in a microfuge, and resuspend in 1 ml PBS.
9. Pellet the beads by brief centrifugation and wash twice with 1 ml PBS.
10. To remove unbound antibody, resuspend the bead pellet in 1 ml antibody elution buffer, spin down in a microfuge, and carefully remove supernatant without disturbing the pellet.
11. Wash the bead pellet three times with 1 ml IP buffer. After the last wash, resuspend beads once again in IP buffer to make ~50% slurry. The resulting immunoaffinity matrix is ready for use in immunoprecipitation and can be stored for several weeks at +4 °C. The matrix may be reused several times for immunoprecipitation of the same antigen.

3.1.2 Metabolic Cell Labeling with Radioactive Orthophosphate

The following section describes metabolic labeling of virus-infected protoplasts with $^{32}P_i$ or $^{33}P_i$. The protoplasts can be subsequently lysed and used for immunoprecipiatation of radioactively-labeled phosphoproteins. Since the labeling involves considerable amounts of radioactivity, it must be performed with proper shielding, protective clothing, and monitoring. It is strongly advisable that the whole procedure is first tested with unlabeled cells to optimize the use of supplies, equipment, and radioactive waste disposal.

1. Electroporate protoplasts with viral RNA, DNA, or infectious cDNA. Incubate virus-infected and control mock-infected protoplasts for 1–2 h in phosphate-free Man-pp or similar medium at a concentration of 10^5 protoplasts per milliliter.
2. Add carrier-free ^{32}P- or ^{33}P-orthophosphate (0.25–1 Ci ml^{-1}) and incubate protoplasts for 8 h to overnight to allow for phosphate incoproration. Lyse the cells as described in Sect. 3.1.3.1.

Alternatively, metabolic labeling of phosphoproteins can be performed in leaf tissue. For this purpose, leaves of virus-infected and mock-infected plants are cut in disks (about 1 cm in diameter) and soaked in 25 mM Hepes buffer, pH 6.8, containing 0.5–1 mCi ml^{-1} of carrier-free ^{32}P- or ^{33}P-orthophosphate. Vacuum is applied until the leaf disks darken and the mixture is incubated overnight at room temperature. Following removal of the incubation solution, leaf disks are lysed as described in Sect. 3.1.3.2. The yield of labeled protein from leaf disks is usually lower than that from protoplasts, since leaf cells cannot be efficiently starved for phosphate.

3.1.3 Cell Lysis

Efficient cell lysis is critical for successful immunoprecipitation. The following paragraph describes two major approaches to cell lysis. The first approach is based on the use of a lysis buffer containing mild detergent Nonidet P-40. This approach works well with virus-infected protoplasts. However, virus-infected leaf tissue is often difficult to solubilize in mild lysis buffer. In this case, it is useful to first lyse cells in SDS-containing buffer and then dilute the lysate to an SDS concentration compatible with immunoprecipitation. The latter approach can also be used when the studied protein is hydrophobic or when immunoprecipitation is carried out with antibodies raised against denatured proteins.

When studying protein phosphorylation, phosphatase inhibitors such as orthovanadate and fluoride must be added to a lysis buffer to prevent target protein dephosphorylation during cell lysis and subsequent protein isolation. Lysis of radio-actively-labeled cells has to be carried out with proper shielding, protective clothing, and monitoring.

3.1.3.1 Cell Lysis Using Mild Detergent

1. Transfer protoplasts to a screw-cap centrifuge tube, pellet them by centrifugation ($1,800\,g$, 5 min), and carefully remove medium without disturbing the pellet. If protoplasts are metabolically labeled with ^{32}P or ^{33}P, do not use vacuum aspirator to remove medium because of the risk of contamination with radioactive aerosols.
2. Resuspend cell pellet in cold IP buffer supplemented with 50 mM NaF, 1 mM Na_3VO_4, 5 mM EDTA, and protease inhibitors (e.g., complete EDTA-free inhibitor cocktail tablets from Roche).
3. Transfer cell suspension to screw-cap microcentrifuge tube(s) and briefly sonicate on ice ($2 \times 10\,s$). In the case of radioactively-labeled protoplasts, use a water bath sonicator rather than a sonicator with a probe.
4. Clear the lysate by centrifugation in a microfuge for 15 min at 14,000 rpm at +4 °C. Carefully transfer supernatant to a new tube. The lysate is now ready to be used for immunoprecipitation.

3.1.3.2 Cell Lysis by Boiling in SDS-Containing Buffer

1a. *For protoplasts*: Pellet protoplasts as described in the previous protocol, step 1. Resuspend pellet in IP buffer supplemented with 50 mM NaF, 1 mM Na_3VO_4, 5 mM EDTA, and 1% SDS. Boil immediately for 10 min.
1b. *For leaf tissue*: Thoroughly rinse leaf discs with water, dry them on filter paper, and cut into as small pieces as possible. Transfer the resulting leaf fragments to screw-cap microcentrifuge tube(s) and vortex in IP buffer supplemented with 50 mM NaF, 1 mM Na_3VO_4, 5 mM EDTA, and 1% SDS. Boil immediately for 10 min.

2. Clear the boiled lysate by centrifugation in a microfuge for 15 min at 14,000 rpm at +4 °C. Carefully transfer supernatant to new tube(s) and dilute (1:10) with cold IP buffer supplemented with 50 mM NaF, 1 mM Na_3VO_4, 5 mM EDTA, protease inhibitors, and 3% bovine serum albumin (BSA). The final concentration of SDS in the lysate should be no more than 0.1%. The lysate is now ready to be used for immunoprecipitation.

3.1.4 Immunoprecipitation Coupled with Enzymatic Dephosphorylation

The following section provides a detailed description of an optimized immunoprecipitation protocol that could serve as a basis for analysis of viral protein phosphorylation and identification of coprecipitated kinase(s). To deterime the in vivo phosphorylation status of a protein of interest, we often use immunoprecipitation coupled with enzymatic dephosphorylation. In this approach, the immunoprecipitated protein is treated with lambda protein phosphatase (λ-PPase) to strip the bound phosphate from phosphoserine, phosphothreonine, or phosphotyrosine (see note 3). The treatment is carried out while the protein is still bound to immunoaffinity beads. By comparing the extent of radioactive label incorporated into the immunoprecipitated protein with that incorporated into the same protein after it has been dephosphorylated with λ-PPase, one can determine the in vivo phosphorylation status of the protein. In the case of unlabeled immunoprecipitated protein, it is possible to check whether the protein is phosphorylated by comparing its electrophoretic mobility before and after phosphatase treatment (see note 4). In an alternative approach, Western blotting with anti-phosphoamino acid antibodies could be used to comparatively analyze the isolated, unlabeled protein and the same protein treated with λ-PPase (see note 5). The disappearance or decrease in intensity of a band corresponding to the phosphatase-treated protein is diagnostic of phosphorylation. If the desired protein can be immunoprecipitated in significant amounts, yet another approach may be taken. The proteins may be resolved by gel electrophoresis and stained with a fluorescent stain that selectively detects phosphoproteins (see note 6). As in the case of Western blotting, the disappearance or decrease in intensity of a stained band corresponding to the phosphatase-treated protein is diagnostic of phosphorylation.

The immunoprecipitation protocol described below is routinely used in our laboratory to isolate different phosphoproteins in nanogram to microgram quantities. We normally immunoprecipitate the desired protein from two identical cell lysates prepared according to the method described earlier. One of the two immunoprecipitates is treated with λ-PPase, while the other one remains untreated. In a pilot experiment, we also perform control immunoprecipitation of mock-infected cell lysate to confirm that the isolated protein is virus-specific.

1. Vortex 500 μl of protein A(G)-Sepharose slurry. Transfer the slurry to a screw-cap microcentrifuge tube and resuspend in 1 ml of cold IP buffer. Gently pellet the beads by brief (30 s, 10,000 rpm) centrifugation in a microfuge, and carefully aspirate supernatant to obtain packed beads. A small volume of buffer should be

left over the beads. Repeat the wash three times with 1 ml IP buffer and resuspend beads in the same buffer to make 50% slurry.

2. Eliminate nonspecific contaminants that can potentially bind to protein A(G)-Sepharose by incubating the cell lysate (obtained as described in Sect. 3.1.3) with 50 µl of 50% protein A(G)-Sepharose slurry in IP buffer for 1 h in a cold room on a rotator.

3. Pellet the beads by brief (30 s, 10,000 rpm) centrifugation in a microfuge and carefully transfer the supernatant to a new tube without touching the beads.

4. Add 50 µl of DMP-crosslinked immunoaffinity matrix (50% slurry, prepared as described in Sect. 3.1.1) and incubate overnight in a cold room on a rotator.

5. Pellet the beads by brief (30 s, 10,000 rpm) centrifugation and wash three times by resuspending the beads in 1 ml of cold IP buffer supplemented with 50 mM NaF, 1 mM Na_3VO_4, 5 mM EDTA, and protease inhibitors.

5a. *For the sample to be treated with λ-PPase*: Pellet the beads by brief (30 s, 10,000 rpm) centrifugation and wash twice by resuspending the beads in 1 ml of cold IP buffer supplemented with protease inhibitors only.

5b. Pellet the beads by brief (30 s, 10,000 rpm) centrifugation and wash twice by resuspending the beads in 1 ml of cold λ-PPase reaction buffer.

5c. Pellet the beads by brief (30 s, 10,000 rpm) centrifugation, remove supernatant and resuspend the beads in 30 µl of λ-PPase reaction buffer.

5d. Add 400 U of λ-PPase, mix by pipetting up and down and incubate for 30 min at +30 °C.

6. Pellet the beads by brief (30 s, 10,000 rpm) centrifugation and wash twice by resuspending the beads in 1 ml of cold IP buffer without Nonidet P-40.

7. Pellet the beads by brief (30 s, 10,000 rpm) centrifugation and wash twice by resuspending the beads in 1 ml of 5 mM Tris–HCl buffer, pH 7.5.

8. Pellet the beads by brief (30 s, 10,000 rpm) centrifugation and carefully remove supernatant. Remove the remaining supernatant using an insulin syringe and immediately add 30 µl of 0.1% (v/v) trifluoroacetic acid (TFA) to elute the protein from the beads. Mix by pipetting up and down, and allow the sample to stay for 3 min at room temperature and mix again.

9. Collect the eluted protein quantitatively using an insulin syringe and transfer the eluate to a new tube. Repeat elution twice more with 30 µl of 0.1% TFA. Pool the eluted fractions or, alternatively, analyze each fraction separately (see note 7).

3.2 Phosphorylation Site Mapping

To elucidate the physiological role of phosphorylation, it is important to identify the specific amino acid residues that are phosphorylated in vivo. The following methods can be used for this purpose.

3.2.1 Two-Dimensional Electrophoresis

The first step in phosphorylation site mapping is to determine whether a protein of interest is phosphorylated on single or multiple residues. This can be achieved by comparative analysis of phosphorylated and dephosphorylated protein by two-dimesional (2D) electrophoresis. This two-step method separates proteins according to two independent properties: isoelectric point and molecular weight. In the first dimension, proteins are separated according to their isolectric point by immobilized pH gradient gel electrophoresis. In the second dimension, they are separated according to their approximate molecular weight using sodium dodecyl sulfate-polyacrylamide gel electrophoresis (SDS-PAGE). Detailed protocols for 2D electrophoresis are available in refs. (6, 7). Every phosphate group added to a protein increases its net negative charge. Conversely, addition of each phosphate group changes the isoelectric point of a protein so that it migrates to a different position in a pH gradient. Given the dynamic nature of protein phosphorylation, various differentially phosphorylated forms of a protein can be resolved in a two-dimensional gel. Presence of several distinct spots in the first dimension usually indicates phosphorylation at multiple sites (see note 8). Phosphatase treatment converts these several spots to a single spot corresponding to dephosphorylated protein. Several methods can be applied to detect phosphoproteins in two-dimensional gels. These include autoradiography of ^{32}P- or ^{33}P-radiolabeled proteins and staining with selective phosphoprotein stains (see note 6). If large quantities of purified phosphoprotein are available, conventional staining methods such as silver staining, Coomassie Blue, and colloidal Coomassie Blue may also be used. Finally, 2D gels can be blotted onto a membrane and probed with anti-phosphoamino acid antibodies or an antibody against the desired protein.

3.2.2 Two-Dimensional Phosphopeptide Mapping, Phosphoamino Acid Analysis, and Radioactive Phosphate-Release Sequencing

This section outlines general procedures for analysis of phosphorylated proteins by two-dimensional (2D) peptide mapping in conjunction with phosphoamino acid analysis and radioactive phosphate-release sequencing. Peptide mapping is a powerful technique for identification of amino acid residues that are modified by phosphorylation. The high sensitivity and reproducibility of this method make it suitable for routine laboratory use. Since the method requires incorporation of radioactive label (^{32}P or ^{33}P) into the phosphorylated protein of interest, proper safety precautions, including shielding and radioactive waste disposal, should be taken during sample preparation and handling. 2D phosphopeptide mapping relies on separation of radioactively-labeled peptides by electrophoresis and chromatography on thin-layer cellulose (TLC) plates. One way to obtain peptides for analysis is to enzymatically digest the radioactively-labeled phosphoprotein recovered from one- or two-dimensional gels (for details, see ref. 8). Alternatively, the radioactively-labeled phosphoprotein may be first transferred to a membrane and then digested with trypsin or other protease. We normally perform tryptic digestion of proteins immobilized on a polyvinylidene fluoride (PVDF) membrane. The following protocol contains instructions on how to perform the digestion:

1. After SDS PAGE or 2D electrophoresis, electrophoretically transfer radiolabeled protein to Immobilon-P membrane (PVDF, Millipore, Billerica, MA).
2. Localize the target protein "band" (for one-dimensional gels) or "spot" (for two-dimensional gels) on the membrane by Ponceau S staining and/or autoradiography.
3. Carefully excise the band using a single-edge razor blade. Wash the membrane piece several times with 10 mM Tris–HCl, pH 8.0, to remove Ponceau S. Rinse the membrane with deionized water, cut into smaller pieces, and soak for 30 min at 37 °C in 0.5% polyvinylpyrrolidone-360 (Sigma, St. Louis, MO) in 100 mM acetic acid to block nonspecific absorption of trypsin.
4. Wash the membrane pieces five times with 200 µl of 10% (v/v) acetonitrile and then once with 200 µl of digestion buffer.
5. Completely immerse the membrane pieces in 50 µl of digestion buffer containing 0.05 µg µl^{-1} trypsin (Sequencing Grade Modified Trypsin; Promega, Madison, WI). Incubate for 4 h at 37 °C and then overnight at the same temperature after a second addition of trypsin (2.5 µg).
6. Thoroughly vortex the tube with membrane pieces for 2 min to help release peptides into solution (see note 9). Transfer the digest to a new tube. Wash the membrane pieces twice with 100 µl of 80% (v/v) acetonitrile (with thorough vortexing or sonication) and pool the washes.
7. Dry the digest in a centrifugal vacuum evaporator and clean by repeated lyophilization and resuspension in 100 µl of water.
8. Resuspend the peptide pellet in 5 µl of 2D electrophoresis buffer (for buffer composition, see ref. 8), and spot onto a 20 × 20 cm^2 cellulose thin layer chromatography (TLC) plate (VWR International, Buffalo Grove, IL).

A detailed protocol on how to perform 2D separation of proteolytic digests by electrophoresis and chromatography on TLC plates can be found in (8). Briefly, peptides are separated in the first dimension using the Hunter thin layer electrophoresis system (e.g., HTLE-7000; C.B.S. Scientific, Del Mar, CA). Electrophoresis in the first dimension is followed by ascending chromatography in the second dimension. After 2D peptide separation is complete, the TLC plate is air-dried and the positions of phosphopeptides are detected by autoradiography. Individual phosphopeptides are then eluted from the plate for further analysis. The efficiency of phosphopeptide elution from cellulose can be checked by liquid scintillation counting. A detailed description of the elution procedure is available in (9). The type of phosphorylated amino acid within the eluted peptide can be identified by 2D phosphoamino acid analysis as described in (8). Briefly, the eluted material is hydrolysed in 6N HCl at 110 °C, lyophilized, resuspended in a buffer containing nonlabeled phosphoamino acid standards, and spotted onto a TLC plate. Thin-layer electrophoresis is performed in pH 1.9 buffer in the first dimension and in pH 3.5 buffer in the second dimension. Phosphoamino acid standards are visualized by ninhydrin spraying and the positions of radioactive phosphoamino acids are detected by autoradiography. The phosphoamino acid composition of the peptide is determined by matching the spots on the autoradiogram with the ninhydrin-stained standards on the TLC plate. As a next step, radioactive phosphate-release sequencing can be used to determine the position of the phosphorylated amino acid within the eluted phosphopeptide. The phosphopeptide is coupled to Sequelon-AA membrane (Millipore) according to the manufacturer's instructions and sequenced on a gas phase sequencer. Released phenylthiohydantoin derivatives from each cycle of Edman degradation are spotted on a TLC plate, and the radioactivity of each spot is quantified using a phosphorimager. This method positionally places the phosphoamino acid within the sequenced phosphopeptide. Thus, 2D phosphoamino acid analysis and radioactive phosphate-release sequencing can be used in parallel to determine the type of phosphorylated amino acid and its position within the phosphopeptide. If the phosphoprotein sequence is known, this information is usually enough to identify the exact location of phosphorylated residue(s) (Fig. 1).

3.2.3 Mass Spectrometry

In recent years, mass spectrometry (MS) has emerged as a powerful method for phosphorylation site mapping. Two ionization techniques, matrix-assisted laser desorbtion-ionization (MALDI) and electrospray ionization (ESI), in combination with different mass analyzers have been used for the identification of phosphopeptides derived from proteolytic digests of phosphoproteins. This paragraph presents a short summary of several MS techniques used for identification of phosphorylation sites in proteins, without trying to cover all aspects.

The key to successful analysis of phosphorylated proteins by MS lies in sample preparation. Low stoichiometry and often low abundance of phosphorylated peptides makes it difficult to detect them in a proteolytic digest containing a large number of nonphosphorylated peptides. The situation is further aggravated by the presence of peptides derived from contaminating proteins. Therefore, special emphasis should be placed on sample purification. Using the above-described immunoprecipitation protocol, we routinely obtain protein for MS analysis that is substantially less contaminated with detergent and released antibody compared with the same protein obtained with a conventional protocol. The amount of protein required for phosphorylation analysis is generally higher than that required for MS-based protein identification. As a general rule, there should be enough protein to be clearly detected in a Coomassie-stained gel. After immunoprecipitation, we lyophilize the eluted protein in a centrifugal vacuum evaporator to remove TFA, dissolve the dried pellet in 100 μl of water, lyophilize again, and proceed with trypsin digestion (see notes 10 and 11) according to the following protocol:

1. Redissolve protein pellet in 3 μl of 50% acetonitrile in freshly made 100 mM ammonium bicarbonate, pH 8.0. Mix by pipetting up and down, let the sample to stay for 10 min at room temperature and mix again.

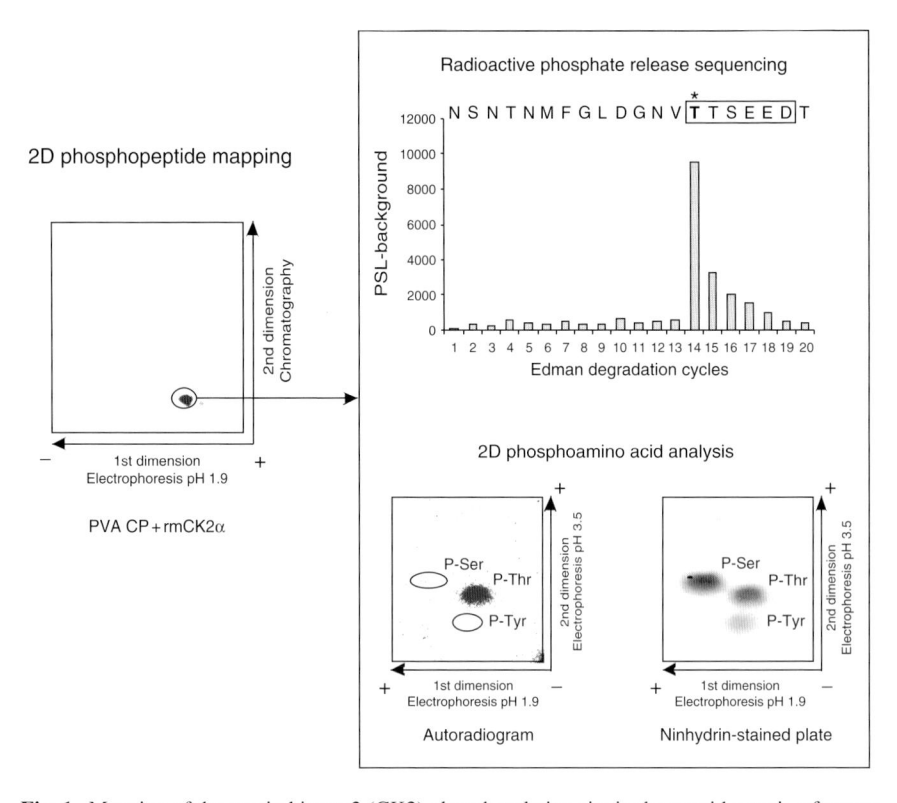

Fig. 1 Mapping of the casein kinase 2 (CK2) phosphorylation site in the capsid protein of potato virus A (PVA CP). Shown on the *left* is a 2D peptide map of PVA CP phosphorylated by the recombinant α-catalytic subunit of CK2 from maize (rmCK2α) in the presence of [γ-^{32}P]ATP. Trypsin-digested peptides were separated by thin layer electrophoresis in the first dimension and chromatography in the second. The circled phosphopeptide was recovered from the plate and subjected to phosphate-release sequencing (inset box, *top*) and 2D phosphoamino acid analysis (inset box, *bottom*). Radioactivity released by each cycle of Edman degradation was quantified by phosphorimaging after subtraction of the background. Amino acids are shown in one-letter code above the graph. The phosphorylated threonine residue is shown in *boldface* and is marked with an *asterisk*. The phosphoamino acid composition of the peptide was determined by thin layer 2D electrophoresis followed by autoradiography. The circled regions on the autoradiogram (inset box, *bottom left*) indicate the positions where phosphoamino acid markers migrated, as determined by ninhydrin staining (inset box, *bottom right*). (Reproduced from ref. 18 with permission from American Society of Plant Biologists)

2. Add 27 µl of sequencing grade modified trypsin solution to a final protease:protein ratio of 1:100 to 1:20 (w/w). Prepare the solution in freshly made 100 mM ammonium bicarbonate, pH 8.0. The final concentration of acetonitrile in the reaction mixture should not exceed 5%.

3. Incubate 16–18 h at +37 °C. After incubation, the digest is ready for analysis by liquid chromatography electrospray ionisation mass spectrometry (LC-ESI-MS) (see note 12).

Efficient chromatographic separation of hydrophilic phosphopeptides critically depends on the type of LC column. Columns exhibiting superior retention of hydrophilic peptides (such as Atlantis dC_{18} (Waters, Milford, MA) or Zorbax SB-C18 (Agilent, Palo Alto, CA) are the columns of choice for phosphopeptide separation. Since phosphopeptides are often present in low amounts and are poorly ionized in comparison to their nonphosphorylated counterparts, unambiguous analysis of minor phosphopeptides is best carried out using the highly sensitive nanoscale electrospray ionization source. After the masses of LC-separated peptides are experimentally determined, they need to be assigned to certain peptide sequences within the protein. Phosphopeptides can be identified by comparing experimentally measured peptide masses with the masses predicted theoretically. A mass difference of 79.966 Da occurs when phosphate is added to serine, threonine, or tyrosine residue and suggests that the peptide is phosphorylated. It is common that some peptides contain more than one phosphorylation site; therefore, peptide masses differing from theoretical masses by 159.932 Da and even 239.898 Da should also be selected for further analysis. Ion signal intensities of selected peptides should be compared with those of the same peptides derived from the phosphatase-treated protein. Substantial decrease in intensity or disappearance of the peptide ion signal after phosphatase treatment is a good indication that the selected peptide indeed contains phosphorylation site(s). At the same time, treatment with phosphatase may increase the intensity of the ion signal corresponding to the dephosphorylated form of the peptide. To identify the phosphorylated amino acid residue(s) within the selected peptides, remaining sample aliquots may be subjected to peptide fragmentation sequencing by LC-ESI-tandem mass spectrometry (MS/MS). During peptide fragmentation in a collision cell, phosphoserine and phosphothreonine undergo neutral loss of phosphoric acid (98 Da) through a process called β-elimination. β-elimination of phosphoserine generates dehydroalanine with a characteristic mass of 69 Da, while β-elimination of phosphothreonine generates dehydroamino-2-butyric acid with a characteristic mass of 83 Da (Fig. 2). In contrast to phosphoserine and phosphothreonine, no β-elimination of phosphotyrosine occurs under the same conditions. Consequently, presence of the fragment ion -H_3PO_4 ($-98 Da^{***}$) and the mass difference of 69 Da between neighboring ion peaks in a peptide fragmentation spectrum indicates phosphoserine, while the mass difference of 83 Da indicates phosphothreonine. Presence of the mass difference of 243 Da indicates phosphotyrosine. Thus, peptide sequencing by LC-ESI-MS/MS provides effective means for phosphorylation-site mapping within the selected candidate peptide.

Despite the relative simplicity of the above-described method, phosphorylation-site mapping by MS is often very challenging. One limitation of the above method is that it is carried out in positive ion mode. Phosphopeptides usually exhibit low response in MS in positive ion mode because of negative charge of phosphate group. Greater relative ion intensities of phosphopeptides can be obtained in negative ion mode. However, in negative ion mode, nonphosphorylated acidic peptides may produce high background. Furthermore, MS/MS peptide fragmentation cannot be carried out in negative ion mode. Therefore, analysis of phosphorylation by MS may require a combination of several experimental methods. These methods are discussed in detail in refs. (10, 11).

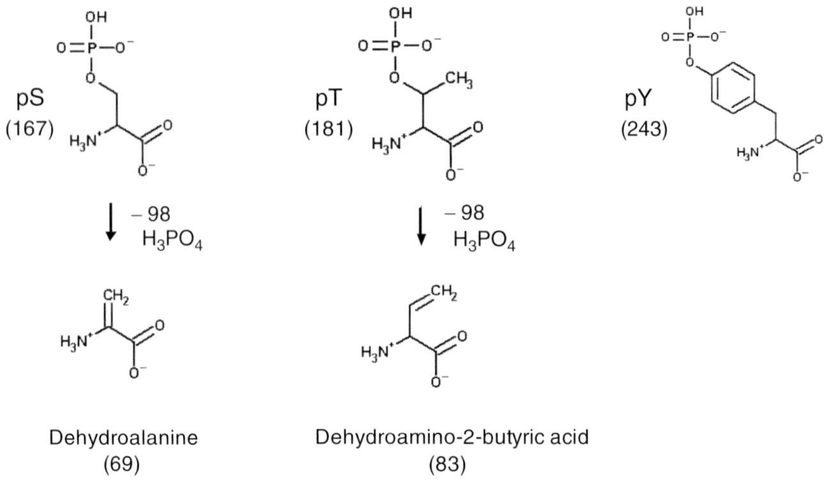

pS (167)

pT (181)

pY (243)

−98 H_3PO_4

−98 H_3PO_4

Dehydroalanine (69)

Dehydroamino-2-butyric acid (83)

Fig. 2 The chemical mechanism for β-elimination of phosphoric acid from phosphoserine and phosphothreonine. Residual masses of respective amino acids are shown in *parentheses*

Characterization of phosphorylated peptides by MS may be also complicated by the low abundance of phosphopeptides in complex peptide mixtures. Therefore, several approaches have been introduced to enrich phosphopeptides and thus increase their signal-to-noise ratio. Immobilized metal-ion affinity chromatography (IMAC) is the most common approach used for selective enrichment of phosphopeptides from peptide mixtures. This approach allows removal of nonphosphorylated peptides from the digest so that only phosphopeptides are analyzed by MS (see note 13). IMAC, however, requires substantially more sample compared with direct LC-ESI-MS analysis. A detailed description of the methodology behind IMAC can be found in ref. (11). An alternative to IMAC is the use of titanium dioxide (TiO_2) columns (12). Yet another approach to phosphopeptide enrichment is based on chemical modification of phosphorylated residues, followed by reversible protein biotinylation, proteolytic digestion, and affinity isolation of biotinylated peptides (13). All of the above-described approaches allow for substantial phosphopeptide enrichment; however, none of them is generally applicable to all analyses. To summarize the above section, successful mapping of phosphorylation sites by MS highly depends on stoichiometry of phosphorylation, sample amount and quality, and may require a combination of several MS techniques.

3.3 Protein Kinase Identification

After the amino acid residue(s) that are phosphorylated in vivo are determined, it is important to identify the protein kinase(s) responsible for the phosphorylation. This task is typically difficult because the kinase(s) in question usually cannot be identified simply by analyzing the phosphorylation consensus sequence. The two most common approaches to kinase identification are, on the one hand, the use of selective cell-permeable kinase inhibitors, and on the other, analysis of kinase activity that coprecipitates with the target protein. A new approach based on the use of kinase siRNA libraries is currently being developed. However, since this approach is not yet fully established, we will not discuss it in this chapter.

3.3.1 Selective Protein Kinase Inhibitors

The choice of an appropriate kinase inhibitor is not a simple matter; it depends on various factors, such as inhibitor selectivity/specificity, type of cells used, etc. As a first step toward kinase identification, one should use selective and cell-permeable inhibitors for each kinase specific for the experimentally determined consensus sequence. When the search is narrowed to a certain kinase family, isozyme-selective inhibitors, such as PKC isozyme inhibitors, should be used. The amount of inhibitor is also very important, since nonspecific off-target effects can occur at higher inhibitor concentrations. We recommend choosing the inhibitor concentration on the basis of published data. Other factors, such as duration of incubation, inhibitor stability, etc. should also be considered. An appropriate control should be included to account for nonspecific effects of solvent used to solubilize the inhibitor.

3.3.2 Immunocomplex Kinase Assay

The purpose of immunocomplex kinase assay is to determine whether the phosphorylated protein can be isolated in a complex with a protein kinase. The assay principle is rather simple. The target antigen precipitated by the antibody coprecipitates the interacting protein kinase, i.e., the kinase is bound to the target antigen, which in turn is captured by the antibody cross-linked to protein A(G)-Sepharose. The resulting immune complex is incubated in a kinase buffer in the presence of $[\gamma\text{-}^{32}\text{P}$ or $^{33}\text{P}]\text{ATP}$ and the target antigen is analyzed for radiolabel incorporation. The first part of the assay is conducted according to the protocol previously described for immunoprecipitation (see protocol in Sect. 3.1.4, steps 1–5). After that, the immunocomplex is subjected to in vitro kinase reaction as follows:

1. Pellet the beads by brief (30 s, 10,000 rpm) centrifugation and wash twice by resuspending the beads in 1 ml of cold kinase buffer.
2. Pellet the beads by brief (30 s, 10,000 rpm) centrifugation, remove supernatant, and resuspend the beads in 30 µl of kinase buffer supplemented with 10 µM ATP and 2.5 µCi $[\gamma\text{-}^{32}\text{P}$ or $^{33}\text{P}]\text{ATP}$ (see note 14). Incubate for 30 min at +30 °C.
3. Pellet the beads by brief (30 s, 10,000 rpm) centrifugation, wash the beads with 1 ml of cold IP buffer, centrifuge again, and discard the radioactive supernatant.
4. Add 30 µl of 2× concentrated SDS PAGE sample buffer, resuspend and boil for an additional 5 min to elute the protein from the beads.
5. Collect the eluted protein quantitatively using an insulin syringe and transfer the eluate to a new tube. Analyze the eluate by gel electrophoresis and autoradiography.

3.3.3 In-Gel Kinase Assay

If an active kinase coimmunoprecipitates with the protein of interest, the enzyme can be further characterized using an in-gel kinase assay. This method, described in detail in ref. (14), involves copolymerization of a kinase substrate in the separating layer of an SDS-PAGE gel. The coimmunoprecipitated kinase is loaded on such a modified gel and electrophoresis is performed under denaturing conditions. After electrophoresis is complete, SDS is removed from the gel and the kinase is allowed to refold by various gel treatments. The gel is incubated in a kinase buffer containing [γ-^{32}P or ^{33}P]ATP, washed to remove unincorporated label, fixed and analyzed by autoradiography to reveal phosphorylation of the gel-incorporated substrate. When the kinase in the gel phosphorylates the substrate, it produces a band on the film. Alignment of the band with molecular weight markers allows one to determine the approximate size of the kinase. Once its size is determined, the kinase can be separated by conventional gel electrophoresis and identified by peptide microsequencing or mass spectrometry.

3.3.4 Double-Label Immunofluorescence Confocal Microscopy

If an antibody against the identified kinase is available, double-label immunofluorescence confocal microscopy can be employed to demonstrate colocalization of the kinase and its substrate in virus-infected cells. The following protocol can be used for this type of analysis:

1. Infect plants with the virus and allow enough time for systemic infection to develop.
2. Isolate protoplasts from the infected leaf material.
3. Suspend protoplasts in the Fixing Solution and fix for 80 min with one change of solution after 30 min.
4. Wash fixed protoplasts first with 30% Dulbecco/70% Man-pp, then 50% Dulbecco/50% Man-pp and finally with Dulbecco supplied with 0.2% BSA.
5. To quench the background fluorescence, incubate the cells with 0.1% $NaBH_4$ in PBS for 10 min. Wash with PBS to remove the quenching reagent.
6. Permeabilize the cells by treating them with 0.1% TritonX-100 in PBS for 20 min.
7. Block nonspecific binding of antibodies with 5% BSA and 0.1% bovine gelatine in Dulbecco's medium for 1.5 h in +4 °C. At this stage, we recommend to check with a microscope whether the protoplasts are in good condition.
8. Incubate the fixed protoplasts with specific primary antibodies overnight in Dulbecco's medium at +4 °C. Note that primary antibodies need to be obtained from two different animals (e.g., mouse and rabbit). Include appropriate controls to check for the specificity of labeling.
9. Wash the cells with three changes of Dulbecco's medium.

10. Incubate the cells for 5 h with secondary antibodies conjugated to different fluorochromes (e.g., Alexa Fluor 594 and Alexa Fluor 488 from Molecular Probes, Eugene, OR).
11. Wash the cells thoroughly (e.g., six times) with Dulbecco's medium.
12. Coat microscope slides with poly-lysine solution (Sigma) by pipetting a drop of solution onto a slide. Remove excess liquid and let the slides dry. Allow the protoplasts (suspended in 200 μl of Dulbecco's medium) to settle on the slide for 30 min. Remove excess medium with filter paper. Mount the cells with 0.1% 1,4-diazabicyclo[2.2.2]octane DABCO (Sigma) in 50% PBS/50% glycerol. Cover the cells with a cover slip and seal the borders between glasses with nail polish.
13. Analyze with a laser scanning confocal microscope equipped with filters for both fluorochromes (note 15).

3.4 Physiological Role of Phosphorylation

The most important objective of any protein phosphorylation study is to determine whether a phosphorylation event detected in vivo has a physiological role. To address this objective, one first needs to identify the phosphorylation sites and the kinase involved. It is then possible to introduce amino acid exchange mutations into each phosphorylation site and study the resulting phenotype. In the case of plant viruses, specific mutations are introduced into viral infectious cDNA (icDNA). Phosphorylation-deficient mutations are produced by changing the affected serine, threonine, or tyrosine residues to alanine. Alternatively, phosphorylated residues may be replaced by aspartate or tyrosine mimicking the electrostatic and steric effects of phosphorylation (15). It is important to note, however, that mutagenesis alone does not conclusively prove that the phenotype change is indeed caused by the absence of phosphorylation. Additional experiments are often required to exclude the possibility that a point mutation exerts its effect through alteration of the target protein conformation. This may be done, for example, by comparing virus infection phenotypes in plants infected with a mutant virus and in plants where expression of a gene encoding the responsible kinase is suppressed by RNA silencing.

3.4.1 Infection with the Mutant Virus

1. If possible, tag viral infectious cDNA (icDNA) with green fluorescent protein (GFP) to follow the progress of infection.
2. Choose unique restriction sites in the icDNA for subcloning of the fragment that encodes the phosphorylated residues. Note that the fragment has to be later back-cloned into the icDNA. Subclone the smallest possible icDNA fragment, because it should be later analyzed by sequencing.

3. Perform PCR-based site-directed mutagenesis to generate the desired phosphor-ylation site mutations. Use a thermostable DNA polymerase with proofreading activity (e.g., Phusion high fidelity DNA polymerase, Finnzymes, Espoo, Finland or Pfu turbo, Stratagene, La Jolla, CA). If you use a commercial site-directed mutagenesis kit (e.g., Quick Change from Stratagene), follow the man-ufacturer's instructions for primer design and temperature cycling conditions.

4. Sequence each mutagenesis product to verify the desired mutation. Transfer the mutated fragment back to the icDNA. Verify the resulting construct by restric-tion analysis or, if needed, by sequencing.

5. Inoculate young plants with 10 μg of the mutated icDNA by particle-mediated bombardment using the Helios Gene Gun or the PDS-1000/He particle delivery system (Bio-Rad, Hercules, CA). Use wild type icDNA and nonreplicating viral cDNA as controls.

6. Quantify the accumulation of viral capsid protein in the inoculated and systemati-cally infected leaves (note 16) using antibody sandwich-enzyme-linked immuno-sorbent assay (DAS-ELISA). Make a standard curve using purified virions.

7. Follow the cell-to-cell and long distance movement of the GFP-tagged mutant virus (see Fig. 3) using a fluorescent microscope (e.g., Leica MZFLIII, Wetzlar, Germany).

8. To investigate a possible role of phosphorylation in virus replication, electropo-rate protoplasts with the mutated icDNA. Quantify virus accumulation in proto-plasts by DAS-ELISA or quantitative PCR.

3.4.2 Silencing of the Protein Kinase

Expression of specific genes can be transiently suppressed using RNA silencing, also called RNA interference (RNAi). This is a powerful tool, which can be used to

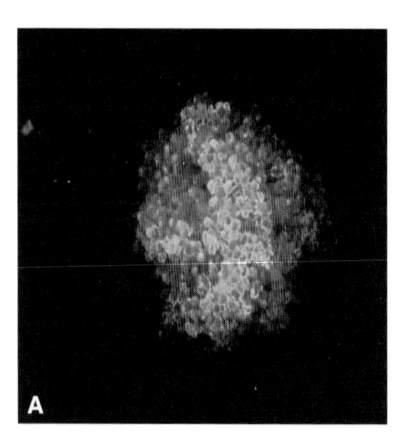
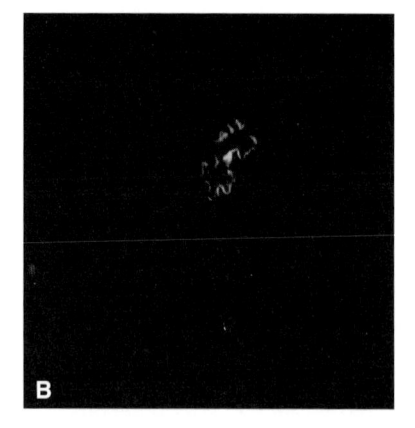

Fig. 3 GFP fluorescence in leaves of *Nicotiana benthamiana* at 4 days after inoculation with wild-type GFP-tagged potyvirus (**a**) and its CK2 phosphorylation-deficient mutant (**b**). Note that the mutant virus is unable to spread from cell to cell

study the biological functions of host protein kinases. Hairpin RNAs with a double-stranded "stem" are very effective for RNA silencing. For example, the pHELLSGATE vectors, developed by CSIRO (Australia's National Science Agency), in conjunction with Gateway cloning technology (Invitrogen) can be used to produce constructs that express double-stranded RNA molecules targeting plant kinases. pHELLSGATE vectors are particularly useful for determining the functions of individual members of a gene family, or genes involved in complex biochemical or developmental pathways (16). Viral vectors provide an alternative way to silence genes. Tobacco rattle virus vector, developed at the Sainsbury Laboratory (Norwich, UK), is a widely used viral silencing vector (17). The following section outlines common procedures used to silence kinase genes.

1. Generate PCR primers for amplification of the kinase gene or its fragment from a host plant-specific cDNA-library. If needed, extend the primers with additional nucleotides required for cloning of a PCR product into the corresponding silencing vector.
2. Clone the correct PCR fragment into the silencing vector. Introduce the resulting construct into a host plant by agroinfiltration or particle-mediated bombardment.
3. Infect upper leaves with the virus a few days after plant inoculation with the silencing vector. The right timing should be determined experimentally and may require optimization.
4. Follow the progress of infection as in Sect. 3.4.1 steps 6 and 7.

Notes

1. Protein A and protein G have different IgG binding specificities, depending on the IgG source. For example, unlike protein A, protein G binds to human IgG_3 and mouse IgG_1. On the other hand, protein A binds more strongly to polyclonal IgG from guinea pig and dog.
2. The vial containing DMP should be warmed to room temperature before opening.
3. The immunoprecipitated protein can be treated in parallel with a phoshatase specific either for Ser(P)/Thr(P) or for Tyr(P) residues (e.g., PPase-2A, *Yersinia* PTP, LAR, TC PTP, etc.). By comparing the effects of treatment with different phosphatases, it may be possible to determine whether the immunoprecipitated protein is phosphorylated on Ser/Thr or Tyr residues.
4. Altered electrophoretic mobility of a phosphatase-treated protein is diagnostic of phosphorylation. The opposite is not true; lack of altered mobility is not evidence of lack of phosphorylation.
5. Western blotting with specific antibodies against phosphoserine, phosphothreonine, or phosphotyrosine will not only determine whether a protein of interest is phosphorylated, but will also identify the type of phosphorylated residue(s).
   Note, however, that some commercial antibodies against phosphoserine may not perform as expected. Furthermore, because phosphoserine and phosphothreonine are structurally rather similar, not all commercial antibodies are able to distinguish well between these two phosphorylated amino acids.
6. For specific phosphoprotein staining, we use Pro-Q Diamond phosphoprotein stain (Molecular Probes, Eugene, OR). Although the most intense bands in a Pro-Q Diamond-stained gel usually correspond to phosphoproteins, nonspecific background staining may also be observed

with high levels of sample. Therefore, we stain the same gel for total protein with a quantitative SYPRO Ruby stain (Molecular Probes, Eugene, OR) to distinguish between a low-abundance phosphoprotein and high-abundance nonphosphorylated protein. We stain the gels according to the manufacturer's protocol and acquire gel images with a Fuji FLA-5100 system equipped with a 532 nm laser and a 575 nm long pass filter (Fujifilm, Tokyo, Japan).

7. Protein elution with 0.1% (v/v) TFA is fully compatible with mass spectrometry analysis, unlike other elution methods (e.g., elution with 2× Laemmli SDS sample buffer or 0.1 M glycine–HCl, pH 2.5).

8. In a two-dimensional gel, protein isoforms can sometimes be mistaken for differentially phosphorylated protein. Therefore, strong denaturants such as 8 M urea should be used for rehydration of immobilized pH gradient strips.

9. Very large hydrophobic peptides may be retained on the membrane. If those peptides contain phosphorylation sites, these sites will not be represented on a phosphopeptide map.

10. Trypsin is usually the first choice of protease for peptide digestion. However, other proteases (V8, Asp-N, Lys-C, etc.) cleaving at different sites may also be used. Note, however, that proteases other than trypsin are much more likely to generate miscleavages.

11. Prior to digestion with trypsin, it may be necessary to eliminate disulfide bonds within peptides and between peptides. Several different chemical reactions can be used to reduce disulfide bonds and prevent their reformation. The latter can be accomplished by alkylation of free sulfhydryl groups. Note, however, that an additional purification step is required if the protein has been reduced and alkylated to remove excess reagents and byproducts.

12. Depending on the type of MS analysis, it may be necessary to remove ammonium bicarbonate from a tryptic digest. This may be achieved by multiple rounds of lyophilization and resuspension in water. Alternatively, tryptic digestion can be carried out in 10 mM ammonium bicarbonate, pH 8.0, and further analyzed without desalting. In the latter case, pH of the buffer should be closely monitored.

13. Phosphopeptide enrichment by IMAC may be hampered by nonspecific binding of acidic peptides to metal affinity matrix. Such nonspecific binding can be decreased through peptide methylation.

14. The immunocomplex kinase assay may also be performed using purified recombinant protein as a kinase substrate. In this case, the protein should be added to the kinase reaction mixture to a final concentration of 0.15–0.5 mg ml^{-1}.

15. Colocalization of the kinase and its substrate must be detected within the same optical section. Otherwise, the seeming "colocalization" may in fact be caused by signal overlap in the absence of physical interaction between the two proteins.

16. The mutated virus may have a movement-deficient phenotype or may be completely unable to replicate. If no phenotype difference is observed in one plant species, try another host plant. The effect of phosphorylation-mimicking mutations on virus infection can be plant-specific (5).

Acknowledgments The authors thank Dr. Pietri Puustinen for help with writing the immunostaining protocol, Dr. Leena Valmu and Mr. Alun Parsons for critical reading of the manuscript. This work was supported by the Academy of Finland (grant 206870).

References

1. Xing, T., Quellet, T., and Miki, B.L. (2002) Towards genomic and proteomic studies of protein phosphorylation in plant-pathogen interactions. *Trends Plant Sci.* **7**, 224–230.
2. Law, L.M.J., Everitt, J.C., Beatch, M.D., Holmes, C.F.B., and Hobman, T.C. (2003) Phosphorylation of rubella virus capsid regulates its RNA binding activity and virus replication. *J. Virol.* **77**, 1764–1771.

3. Morrison, E.E., Wang, Y., and Meredith, D.M. (1998) Phosphorylation of structural components promotes dissociation of the herpes simplex virus type 1 tegument. *J. Virol.* **72**, 7108–7114.

4. Lee, J., and Lucas, W.J. (2001) Phosphorylation of viral movement proteins: regulation of cell-to-cell trafficking. *Trends Microbiol.* **9**, 5–8.

5. Waigmann, E., Chen, M.H., Bachmaier, R., Ghoshroy, R., and Citovsky, V. (2000) Regulation of plasmodesmal transport by phosphorylation of tobacco mosaic virus cell-to-cell movement protein. *EMBO J.* **19**, 4875–4884.

6. Görg A. (2004) *2D Electrophoresis. Principles and Methods*, GE Healthcare, Little Chalfont, UK.

7. Görg, A. and Weiss, W. (1998) High-resolution two dimensional electrophoresis of proteins using immobilized pH gradients, in *Cell biology: a laboratory handbook, Second edition* (Celis, J.E., ed.), Academic press, vol. 4, 386–396.

8. Boyle, W.J., van der Geer, P., and Hunter, T. (1991) Phosphopeptide mapping and phosphoamino acid analysis by two-dimensional separation on thin-layer cellulose plates. *Methods Enzymol.* **201**, 110–149.

9. Sefton, B.M. (1997) Phosphopeptide mapping and identification of phosphorylation sites, in *Current protocols in molecular biology* (Ausubel, F.M., Brent, R., Kingston, R.E., Moore, D.D., Seidman, J.G., Smith J.A., Struhl, K., eds.), Wiley, Hoboken, NJ, p. 18.9.10.

10. Carr, S.A., Annan, R.S., and Huddleston, M.J. (2005) Mapping posttranslational modifications of proteins by MS-based selective detection: application to phosphoproteomics. *Methods Enzymol.* **405**, 82–115.

11. Corthals, G.L., Aebersold, R., and Goodlett, D.R. (2005) Identification of phosphorylation sites using microimmobilized metal affinity chromatography. *Methods Enzymol.* **405**, 66–81.

12. Larsen, M.R., Thingholm, T.E., Jensen, O.N., Roepstorff, P., and Jørgensen T.J.D. (2005) Highly selective enrichment of phosphorylated peptides from peptide mixtures using titanium dioxide microcolumns. *Mol. Cell. Proteomics* **4**, 873–886.

13. Adamczyk, M., Gebler, J.C., and Wu, J. (2001) Selective analysis of phosphopeptides within a protein mixture by chemical modification, reversible biotinylation and mass spectrometry. *Rapid Commun. Mass Spectrom.* **15**, 1481–1488.

14. Wooten M.W. (2002) In-gel kinase assay as a method to identify kinase substrates. *Sci. STKE* 15.

15. Dean, A.M., and Koshland, D.E. (1990) Electrostatic and steric contributions to regulation at the active site of isocitrate dehydrogenase. *Science* **249**, 1044–1046.

16. Helliwell, C. and Waterhouse, P. (2003) Constructs and methods for high-throughput gene silencing in plants. *Methods* **30**, 289–95.

17. Ratcliff, F., Martin-Hernandez, A.M., and Baulcombe, D.C. (2001) Tobacco rattle virus as a vector for analysis of gene function by silencing. *Plant J.* **25**, 237–245.

18. Ivanov, K.I., Puustinen, P., Gabrenaite, R., Vihinen, H., Rönnstrand, L., Valmu, L., Kalkkinen, N., and Mäkinen, K. (2003) Phosphorylation of the potyvirus capsid protein by protein kinase CK2 and its relevance for virus infection. *Plant Cell* **15**, 2124–2139.

Chapter 25
Analysis of Interactions Between Viral Replicase Proteins and Plant Intracellular Membranes

Hélène Sanfaçon and Guangzhi Zhang

Abstract Replication of the genome of positive-strand RNA plant viruses takes place in membrane-bound complexes that contain viral replicase proteins, viral RNA, and host proteins. Many viral replicase proteins play a crucial role in the assembly of replication complexes at intracellular membranes. They are integral membrane proteins that interact directly with the membranes and bring other proteins and the viral RNA to the complex via protein–protein or protein–RNA interactions. In this chapter, we describe subcellular fractionation methods that determine whether viral proteins are integral membrane proteins *in planta*. Differential centrifugation techniques are used to produce membrane-enriched fractions, which can then be analyzed for the presence of viral replicase proteins by immunoblotting. Confirmation of the membrane-association is obtained by membrane flotation assays and treatment of membrane-enriched fractions with high salt or high pH followed by detection of the viral proteins. Because many plant viruses replicate in association with the endoplasmic reticulum (ER), we also discuss two techniques to specifically analyze the interaction of viral proteins with these membranes. These techniques are continuous sucrose-gradient fractionation in the presence or absence of 3 mM Mg^{2+} and glycosylation assays.

Keywords Membrane–protein interaction; Membrane flotation assay; Subcellular fractionation; Glycosylation assay; Sucrose-gradient fractionation; Plant virus; Replication complex; RNA replication

1 Introduction

Positive-strand RNA viruses constitute the vast majority of plant viruses. Viral RNA synthesis occurs in replication complexes, which are associated with intracellular membranes and contain viral and host proteins and the viral RNA (1, 2). Many viruses encode integral membrane proteins that act as membrane anchors for the replication complex. These proteins are targeted to intracellular membranes in the absence of other viral proteins. The membrane anchors interact with other viral

From: *Methods in Molecular Biology, Vol. 451, Plant Virology Protocols:*
From Viral Sequence to Protein Function
Edited by G.D. Foster, I.E. Johansen, Y. Hong, and P.D. Nagy © Humana Press, Totowa, NJ

components (viral RNA and/or other viral replicase proteins) to redirect them toward the replication complex. Known examples of integral membrane replicase proteins include the bromovirus 1a protein, the tombusvirus p33 protein, the potyvirus 6 kDa protein, and the comovirus and nepovirus NTB-VPg protein (3–8). The identification of integral membrane viral replicase proteins is an important step in the elucidation of the assembly and architecture of viral replication complexes.

To identify viral membrane proteins, complementary methods often give the most convincing results. When specific antibodies are available, immunogold-labeling combined with electron microscopy is the method of choice to determine the subcellular localization of the protein. However, in the case of highly hydrophobic membrane proteins, it is often difficult to produce antibodies. A useful alternative is to fuse viral proteins to small epitope tags (e.g., HA epitope) or to larger proteins (e.g., green fluorescent protein, GFP), for which commercial antibodies are available (see Chap. 3–31). Tagged proteins can be introduced into infectious clones or expressed by agroinfiltration (see Sect. 4 of this book). In addition, fusion of viral proteins to fluorescent proteins allows direct visualization of their subcellular localization by confocal microscopy. The above-mentioned methods will identify the subcellular compartments to which each viral protein is targeted but they do not provide information on the nature of the interaction of the protein with the membrane. Integral membrane proteins, which are targeted to the membranes through direct interaction of the protein with the lipid bilayer of the membrane, are not distinguished from peripheral membrane proteins, which are brought to the membranes through protein–protein interaction with a membrane-associated protein. In this chapter, we provide techniques to confirm the association of proteins with membranes and to determine whether they are integral membrane proteins. Since many plant viruses replicate in association with the endoplasmic reticulum (ER) (1), we also provide two techniques to analyze the association of replicase proteins with the ER *in planta*.

Subcellular fractionation of plant extracts produces membrane-enriched fractions, which can be analyzed for protein content by denaturing polyacrylamide gel electrophoresis (SDS-PAGE) (9) followed by immunoblotting. The easiest but most crude method to produce membrane-enriched fractions is through differential centrifugation. A short spin at low gravitational force eliminates the nuclei, cell wall debris, and other large organelles (chloroplast, mitochondria). The resulting supernatant (S3) fraction is then centrifuged at a higher gravitational force to produce a supernatant (S30) enriched in soluble cytoplasmic proteins and a pellet (P30) enriched in intracellular membranes. However, it should be noted that the presence of proteins in the P30 fraction does not prove that they are membrane-associated as protein aggregation can also result in their separation in this fraction. Thus, other methods are necessary to confirm that proteins are membrane-associated. In membrane-flotation assays, S3 or P30 fractions are resuspended in a buffer containing 72% sucrose and overlaid with a two step-sucrose gradient. Because of their low density, membranes float at the interface between the 65% and 10% sucrose layers, while soluble proteins or aggregated proteins remain at the bottom of the gradient (5).

To determine whether membrane proteins are integral or peripheral, membrane-enriched fractions can be extracted under various conditions. Treatment of the fraction with high salt allows the solubilization of peripheral proteins, while integral proteins or luminal proteins (proteins translocated into the lumen of the membranes) remain associated with the membranes (10). Under high-pH conditions, peripheral and luminal proteins but not integral membrane proteins are released from the membranes.

Continuous sucrose gradients can be used to determine whether the protein is associated with ER membranes. The ER is a very large organelle with many specialized compartments (11). The rough ER is associated with ribosomes and this association is lost at low magnesium concentration. This property can be used to identify ER-associated proteins. In the presence of Mg^{2+}, the rough ER sediments toward the bottom of a 20–45% sucrose gradient, while in the absence of Mg^{2+}, the dissociation of ribosomes will result in a shift of ER fractions toward the top of the gradient (12). To confirm the separation of ER in the gradients, plant ER resident proteins (e.g., the endogenous Bip protein, a plant chaperone translocated into the lumen of the ER) can be used as a marker (11).

In the case of ER-targeted viral proteins containing transmembrane domains, glycosylation assays can be used to confirm that the protein is retained in the ER and to study the topology of the protein in the membrane. N-glycosylation sites are characterized by the consensus sequence Nx(S/T) and are recognized by the ER luminal glycosyl-transferase provided that they are translocated inside the ER lumen and that they are physically separated from the transmembrane domain of the protein by at least 12–14 amino acids (13, 14). Glycosylation of the protein at a naturally occurring or introduced N-glycosylation site results in an increase in the apparent molecular mass of approximately 3 kDa. Therefore, N-glycosylation sites can be used as convenient markers to identify segments of the protein that are translocated into the lumen of the membranes. To confirm that glycosylation has occurred, two methods can be used. First, the glycosylation site can be eliminated by site-directed mutagenesis (see Chap. 3–27) and second, the protein can be treated with an endoglycosidase, which removes the sugar moiety. ER resident glycoproteins can be distinguished from glycoproteins that are transported to other organelles of the secretory pathway (e.g., Golgi) by their sensitivity to endoglycosidase H (endoH). EndoH releases high-mannose carbohydrate side chains synthesized in the ER but does not recognizes complex oligosaccharides produced in the Golgi (15).

2 Materials

2.1 Subcellular Fractionation

1. Plant tissues: *Nicotiana benthamiana* leaves expressing individual viral proteins tagged to GFP
2. Homogenization buffer 1: 50 mM Tris-HCl, pH 8, 10 mM KCl, 3 mM $MgCl_2$, 1 mM EDTA, 1 mM DTT, 0.1% BSA, 0.3% dextran, 13% (w/v) sucrose,

Complete protease inhibitor (Roche, one tablet/50 mL). Store at −80 °C in 2–5 mL aliquots

3. Mortar and pestle chilled at 4 °C
4. Miracloth (Rose Scientific)

2.2   SDS-PAGE

1. Separating buffer: 1.5 M Tris-HCl, pH 8.8
2. Stacking buffer: 1 M Tris-HCl, pH 6.8
3. 30% acrylamide stock solution: Dissolve 58.4 g of acrylamide and 1.6 g of bis-acrylamide in 200 mL distilled water (dH$_2$O), filter, store at 4 °C in a dark bottle. (Acrylamide is a neurotoxin when unpolymerized. Care should be taken while handling).
4. 10% Ammonium persulfate (APS) in dH$_2$O. Store at −20 °C in 1 mL aliquots
5. N, N, N, N′-tetramethyl-ethylenediamine (TEMED, Bio-Rad)
6. Tert-amyl alcohol (Sigma), store at room temperature
7. Protein loading buffer (2×): 100 mM Tris-HCl, pH 6.8, 4% SDS, 0.04% bromophenol blue, 2% β-mercaptoethanol, 20% glycerol, 25 mM EDTA. Store at −20 °C in 1 mL aliquots
8. Running buffer (10×): 250 mM Tris, 1.92 M glycine, 1% (w/v) SDS. Dissolve 30.3 g Tris base, 144.1 g glycine, and 10 g SDS in 1 L dH$_2$O (no need to adjust pH). Store at room temperature
9. SDS-PAGE molecular weight standards (Bio-Rad) and/or prestained protein molecular weight marker (Invitrogen)

2.3   Western Blot

1. Transfer buffer: 25 mM Tris base, 190 mM Glycine, 20% methanol. Dissolve 3.03 g Tris base and 14.41 g glycine in 800 mL dH$_2$O (no need to adjust pH). Methanol (200 mL) is added in a subsequent step (see Sect. 3.3).
2. PVDF membranes (Bio-Rad, protein sequencing grade)
3. Phosphate-buffered saline with Tween-20 (PBS-T): Prepare a 10× stock solution of PBS: 1.38 M NaCl, 0.027 M KCl, 0.1 M Na$_2$PO$_4$, 0.018 M KH$_2$PO$_4$, pH 7.4. Dilute 100 mL of 10× PBS with 900 mL dH$_2$O and add 1 mL Tween 20.
4. Blocking buffer: 5% (w/v) powder skim-milk in PBS-T
5. Primary antibody: GFP antibody (BD Bioscience) diluted 1:8,000 in PBS-T
6. Secondary antibody: peroxidase-conjugated Affinipure goat anti-mouse antibody (Jackson ImmunoResearch) diluted 1:15,000 in PBS-T
7. Enhanced chemiluminescent (ECL) detection reagents and Hyperfilm ECL (Amersham)

8. Protein gel destaining buffer: 45:45:10 methanol:water:acetic acid. Store at room temperature
9. Protein gel staining buffer: 0.1% (w/v) Coomassie brilliant blue R250 in destaining buffer. Store at room temperature

2.4   Biochemical Treatments

1. 0.1 M Na_2CO_3, pH 11.3
2. 1 M NaCl

2.5   Membrane Flotation Assay

1. NTE buffer: 100 mM NaCl, 10 mM Tris-HCl, pH 8.0, 1 mM EDTA, Complete protease inhibitor (Roche, 1 tablet per 50 mL). Store at −80 °C in 2–5 mL aliquots
2. Sucrose solutions (85%, 65%, and 10%) (w/v) in NTE buffer. Chill to 4 °C before use

2.6   Sucrose Gradient Fractionation of ER-Associated Protein in the Presence or Absence of Mg^{2+}

1. Homogenization buffer 2: 50 mM Tris-HCl, pH 8; 10 mM KCl, 3 mM $MgCl_2$, 1 mM DTT, 0.1% BSA, 0.3% dextran, 13% sucrose (w/v), Complete protease inhibitor (Roche, one tablet/50 mL). Store at −80 °C in 2–5 mL aliquots
2. Homogenization buffer 3: 50 mM Tris-HCl, pH 8; 10 mM KCl, 1 mM EDTA, 1 mM DTT, 0.1% BSA, 0.3% dextran, 13% sucrose, Complete protease inhibitor (Roche, one tablet/50 mL). Store at −80 °C in 2–5 mL aliquots
3. Sucrose solutions (20% and 45%) (w/v) in homogenization buffer 2 and 3
4. Gradient maker (VWR)

2.7   Deglycosylation of ER-Associated Transmembrane Proteins Containing N-Glycosylation Sites Using Endoglycosidase H

1. Protein loading buffer (4×): 200 mM Tris-HCl, pH 6.8, 8% SDS, 0.08% bromophenol blue, 4% β-mercaptoethanol, 40% glycerol, 50 mM EDTA. Store at −80 °C in 1 mL aliquots
2. Endoglycosidase H (Endo H, Roche)
3. EndoH buffer: 100 mM Na-citrate (pH 5), 100 mM β-mercaptoethanol, 1% Triton X-100

3   Methods

This protocol assumes the availability of virus-infected plant leaves or of agroinfiltrated plant leaves that transiently express individual viral proteins. For the purpose of this protocol, we will describe the detection of viral proteins fused to the GFP protein using commercially available anti-GFP antibodies.

3.1   Subcellular Fractionation

1. Gently grind leaf tissue in homogenization buffer (1 g of fresh weight/4 mL of buffer) in a mortar and pestle.
2. Filter homogenate through miracloth and centrifuge at 3,700 g for 10 min at 4 °C.
3. Discard the pellet, which contains nuclei, cell wall debris, and large organelles. Transfer the supernatant (S3 or postnuclear fraction) to a fresh centrifuge tube. A portion of this fraction may be kept for further analysis. Centrifuge the remainder of the S3 fraction at 30,000 g for 20 min at 4 °C.
4. Collect the supernatant or S30 fraction, which contains the soluble content of the cell and is enriched in soluble proteins.
5. Resuspend the pellet in a volume of homogenization buffer equal to that of the S30 fraction. This resuspended pellet is the P30 fraction, which is enriched in intracellular membranes and membrane-associated proteins.
6. Separate equal volumes of S30 and P30 fractions by SDS-PAGE (Sect. 3.2), transfer the proteins from the gel to a PVDF membrane, and detect the proteins by immunoblotting (Sect. 3.3). Membrane proteins should be present predominantly in the P30 fraction (note 1).

3.2   SDS-PAGE

(These instructions assume the use of Bio-Rad mini-protein 3 apparatus. Users of other gel systems can adjust according to the manufacturer's instructions.)

1. Clean the glass plates with a detergent and rinse thoroughly with distilled water. Rinse the plates again with 95% ethanol and air-dry. Set up the gel plates with the Bio-Rad clamps and place in casting stand.
2. Prepare a 0.75-mm thick separating gel using the mixture corresponding to the desired percentage of acrylamide as outlined in Table 1 (note 2). Add the first four ingredients, swirl the mixture gently, add 10% APS and TEMED, and swirl again. Pour the gel immediately after adding the last two components, leaving space for a stacking gel. Be careful not to trap any air bubbles in the gel. Overlay with *tert*-amyl alcohol. The gel should polymerize in 20–30 min.

Table 1 Recipes for preparing separating gel for SDS-PAGE

	Percentage of gel		
Solution components	8%	12%	15%
Separating buffer	1.3[a]	1.3	1.3
30% acrylamide stock	1.3	2.0	2.5
dH$_2$O	2.3	1.6	1.1
10% SDS	0.05	0.05	0.05
10% APS	0.05	0.05	0.05
TEMED	0.003	0.002	0.002

[a]Numbers correspond to the volume of each component (in milliliter) required to obtain a solution with a final volume of 5 ml which is sufficient for casting one 0.75-mm thick gel

3. After polymerization, pour off the *tert*-amyl alcohol and rinse the top of the gel several times with distilled water. Tilt the gel on its side and use a tissue to drain the water completely.
4. Prepare 2 mL of 5% stacking gel by mixing 0.26 mL stacking buffer, 0.34 mL 30% acrylamide-stock solution, 1.36 mL dH$_2$O, 20L 10% SDS, 20L 10% APS and 2L TEMED. Pour the stacking gel and insert the comb at an angle to avoid trapping bubbles. The gel should polymerize in 20–30 min.
5. Prepare 1 L running buffer by diluting 100 mL of 10× running buffer in 900 mL dH$_2$O.
6. Remove the comb once the stacking gel has polymerized. Wash the wells with running buffer using a syringe.
7. Protein sample preparation: Mix 50 μL of each sample with an equal volume of 2× protein loading buffer. Heat in a boiling water bath for 5 min and centrifuge at 14,000 g for 5 min. Collect the supernatant (note 3).
8. Clamp gel to electrode stand. Place in tank and add running buffer to the top of the inner chamber. Pour running buffer to the outer chamber until the level is above the bottom of the glass plate. Load 5–10 μL of each sample in each well using gel loading tips (Bio/Can). Leave one lane for molecular weight markers on one side (note 4).
9. Place the lid on the tank and connect to a power supply. Turn on the power and run the gel at 200 V until the dye front just runs off (about 40 min).

3.3 Western Blot

(These instructions assume the use of Bio-Rad mini trans-blot transfer cell.)

1. Prepare a sheet of PVDF membrane and four sheets of Whatman filter paper cut just larger than the size of separating gel. Wet the PVDF membrane in 200 mL methanol and add methanol with membrane to 800 mL of transfer buffer. Mix well and pre-cool the mixture at −20 °C for 1 h prior to use.
2. Wet filter papers and sponge pads in transfer buffer.

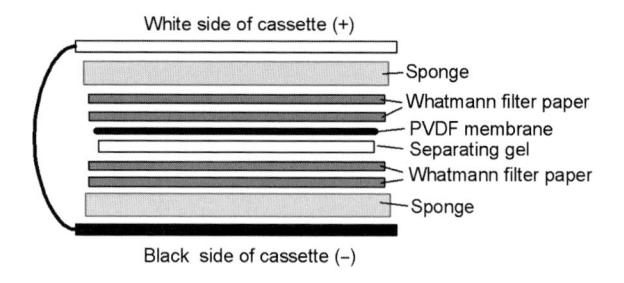

White side of cassette (+)

Sponge
Whatmann filter paper
PVDF membrane
Separating gel
Whatmann filter paper
Sponge

Black side of cassette (–)

Fig. 1 Assembly of the Western blot transfer cassette. The different layers inside the transfer cassette (represented by the *black* and *white boxes* connected by a *black line*) are indicated on the *right*. (+): Anode, (–): Cathode

3. Disconnect the gel unit from the power supply and disassemble. Remove the stacking gel.
4. Assemble the sandwich in the cassette as shown in Fig. 1. This step is accomplished in a tray filled with the transfer buffer to ensure that all components remain wet. This will also allow the gel to be equilibrated in transfer buffer. When closing the cassette, care should be taken to ensure that no air bubbles are trapped in the sandwich. Use a glass tube to scroll out any bubbles.
5. Place the transfer cassette into the tank. Add the cooling unit filled with ice to allow cooling of the system during the transfer. Pour the transfer buffer just over the top wire. Use a stir bar to keep buffer well-mixed during transfer.
6. Put the lid on the tank and connect to a power supply. Transfer at 100 V for 1 h or at 50 V overnight at 4 °C (note 5). Replace the cooling unit during the transfer if the ice has melted.
7. Once the transfer is complete, disassemble the cassette by removing first the sponge pad and filter papers from the black side and then the gel. Collect the membrane and cut a corner to mark its orientation.
8. Block the PVDF membrane in 20 mL of PBST-skim milk (5%) for 1 h at room temperature on a rocking platform or overnight at 4 °C.
9. Pour off the blocking solution and briefly rinse the membrane three times with 20 mL of PBST buffer. Add the GFP antibody diluted in 10 mL of PBST to the membrane. Incubate for 1 h at room temperature on a rocking platform (note 6).
10. Pour off the primary antibody and briefly rinse the membrane two times with 20 mL of PBST followed by 3 × 5 min washes in 50 mL of PBST at room temperature.
11. Dilute the goat anti mouse antibody in 10 mL of PBST and add to the membrane. Incubate for 1 h at room temperature on a rocking platform.
12. Pour off the secondary antibody and repeat the rinsing and washing steps as in step 9.
13. During the final wash, mix 0.35 mL of ECL detection solution 1 with 0.35 mL of detection solution 2. Drain the excess wash buffer from the membrane and place it protein side up on a clean transparent acetate sheet. Do not blot dry the filter. Pipette the mixed detection solution evenly on the membrane. Incubate for 1 min at room temperature (note 7).

14. Drain off excess detection solution from the membrane. Overlay the membrane with a new acetate sheet. Gently remove any trapped air bubbles.
15. Place the wrapped membrane protein side up in an X-ray film cassette and place a sheet of Hyperfilm ECL on top of the membrane. Close the cassette and expose for a suitable length of time, typically a few min. Develop the film. All these steps should be carried out in a dark room.

3.4 Biochemical Treatment of P30 Fraction

1. Resuspend P30 fractions in 0.1 M Na_2CO_3, pH 11, or 1 M NaCl.
2. Incubate on ice for 20–30 min with mixing every 5 min.
3. Centrifuge at 30,000 g for 20 min at 4 °C.
4. Collect the supernatant fraction. Resuspend the pellet in an equal volume of homogenization buffer.
5. Separate equal volumes of each fraction by SDS-PAGE and probe by Western blotting as described earlier. A typical result is shown in Fig. 2 (note 8).

3.5 Membrane Flotation Assay

1. Set up sucrose gradient as shown in Fig. 3a. Mix 300 µL of S3 or P30 fractions with 1.6 mL of 85% sucrose (the final concentration of sucrose will be 71.5%).

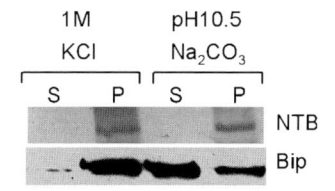

Fig. 2 Typical behavior of an integral and luminal membrane protein after incubation of membrane-enriched fractions with high salt (1 M NaCl) or high pH (pH 11.5) solutions. Membrane-enriched fractions (P30) were obtained from cucumber plants infected with *Tomato ringspot nepovirus* (ToRSV). The fractions were extracted with 1 M NaCl and 0.1 M Na_2CO_3 (pH 11.5) and separated into soluble (S) and membrane-enriched (P) fractions by centrifugation at 30,000 g. Proteins present in the fractions were separated by SDS-PAGE (12% polyacrylamide) and transferred to PVDF membranes. Home-made antibodies (anti-NTB antibodies as indicated on the right) were used to detect the presence of the ToRSV NTB-VPg protein, an ER membrane protein that has been suggested to act as a membrane anchor for the replication complex (8). After both treatments, the protein remains in the pellet fraction, confirming that it is an integral membrane protein. As a control, anti-Bip antibodies (provided by Dr. Chrispeels) were used to detect the Bip protein, an endogenous ER luminal protein. As expected for a luminal protein, the Bip protein is released in the supernatant after treatment at high pH but remains associated with the pellet fraction in conditions of high salt. Reprinted after modification from ref. 8 (with permission from the American Society for Microbiology)

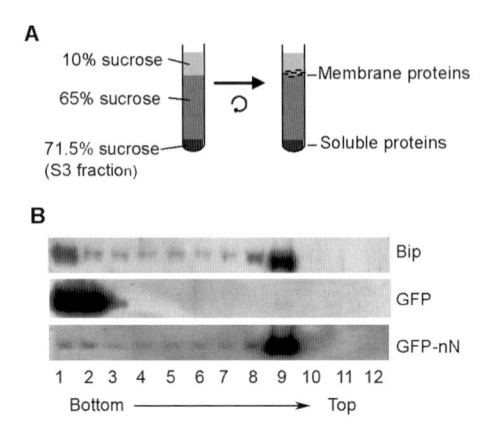

Fig. 3 Membrane flotation assays: (**a**) Diagram showing the layers of sucrose step gradient. After centrifugation, the membrane proteins (represented by *dashed lines*) are expected to float to the interface between 10% and 65% sucrose layers. (**b**) Typical result: *N. benthamiana* plants were agroinfiltrated to express various GFP fusion proteins. Four days after agroinfiltration, S3 fractions were obtained and subjected to membrane flotation analysis. Twelve fractions were collected from the bottom of the gradient and equal amount of each fraction were separated by SDS-PAGE (12% polyacrylamide) and subjected to immunoblotting. The ER-associated Bip protein is detected predominantly in fraction 9 corresponding to the interface between the 10% and 65% sucrose layers. The free GFP remains at the bottom of the gradient (fractions 1 and 2), as expected for a soluble protein. A GFP fusion protein containing the N-terminal membrane association domain from the ToRSV NTB-VPg protein (GFP-nN) cofractionated with the Bip protein in fraction 9. Reprinted after modification from ref. 5 (with permission from the American Society for Microbiology)

Place the mixture at the bottom of a 12 mL swinging bucket centrifuge tube (e.g., SW41, Beckman).

2. Overlay the mixture with 7 mL of 65% sucrose and then 3.1 mL of 10% sucrose using a 10 mL syringe. Care should be taken not to disturb the interface between each layer.

3. Centrifuge at 100,000 g for 18 h at 4 °C.

4. Puncture the bottom of the tube with a 26–28 G needle and collect 12 1-mL fractions from the bottom of the tube. Membrane protein should float to the interface between the 10% and 65% sucrose solution (fractions 8–9). Soluble proteins will remain at the bottom of the gradient (fractions 1–2).

5. Separate equal volumes of each fraction by SDS-PAGE and probe by Western blotting as described earlier. A typical result is shown in Fig. 3b.

3.6 Sucrose Gradient Fractionation of ER-Associated Protein in the Presence or Absence of Mg^{2+}

1. Prepare two 20–45% sucrose gradients in homogenization buffer 2 (containing 3 mM $MgCl_2$) and homogenization buffer 3 (containing 1 mM EDTA but no

MgCl$_2$) using a gradient maker as shown in Fig. 4a. Place 4.5 mL of 20% sucrose solution in the chamber closest to the gradient and 4.5 mL of 45% sucrose solution in the other chamber. A stir bar in the 20% sucrose gradient solution (not shown in the figure) is used to ensure gradual concentration of the sucrose solution in the

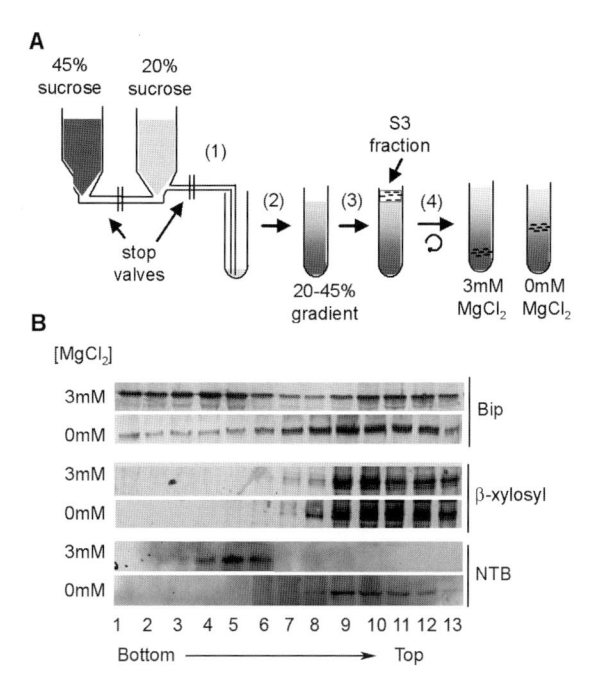

Fig. 4 Sucrose gradient fractionation of ER-associated proteins in the presence or absence of Mg^{2+}. (**a**) Diagram showing the different steps involved in the preparation of continuous 20–45% sucrose gradient. These steps include pouring of the gradient using a gradient maker (1–2), layering of the S3 fraction over the gradient (3), and centrifugation of the gradient (4). In the presence of Mg^{2+}, proteins associated with the rough ER (shown by the *dashed lines*) are found at the bottom of the gradient. In the absence of Mg^{2+}, they shift toward the top of the gradient. (**b**) Typical result: Postnuclear fractions (S3 fractions) were prepared from cucumber plants infected with *Tomato ringspot nepovirus* (ToRSV) in the presence or absence of 3 mM MgCl$_2$ and were fractionated on 20 to 45% sucrose gradients. Fractions were collected from the bottom of the gradient and equal amount of each fraction were loaded on a 12% SDS-polyacrylamide gel. Proteins present in each fraction were detected by Western blotting using antibodies specific for Bip (an ER marker), β-xylosyl-containing proteins (a Golgi marker), and the ToRSV NTB-VPg protein (anti-NTB antibodies). The concentration of MgCl$_2$ used in each sucrose gradient is shown on the *left*. In the presence of MgCl$_2$, the Bip protein separates in two peaks. The *first peak* (fractions 2–5) corresponds to proteins associated with the rough ER. The *second peak* (fractions 9–12) possibly corresponds to proteins associated with other ER compartments. In the absence of MgCl$_2$, the first peak shifts toward the top of the gradient resulting in the separation of the Bip protein in a single peak (fractions 8–12). Golgi-associated proteins remain at the top of both gradients (fractions 9–13, β-xylosyl). The ToRSV NTB-VPg protein separates in fractions 4–6 in the presence of MgCl$_2$, and shifts toward the top of the gradient (fractions 9–12) in the absence of MgCl$_2$. This shift mirrors that observed for the Bip protein and confirms that the protein is associated with the ER. Reprinted after modification from ref. 8 (with permission from the American Society for Microbiology)

second chamber. Fill the gradient from the bottom of a 12 mL swinging bucket centrifuge tube (e.g., SW41, Beckman). This step must be achieved very slowly, making sure that no air bubbles are trapped between the two chambers or between the chambers and the ultracentrifuge tube. Once the gradient has been poured, place at 4 °C overnight. Be careful not to disrupt the gradient during transport.

2. Prepare two S3 fractions as described in Sect. 3.1 using homogenization buffer 2 or 3 (note 9).

3. Carefully overlay 2.5 mL of each S3 fraction on the 20–45% sucrose gradient prepared in the corresponding buffer (note 10).

4. Centrifuge the two gradients at 143,000 g for 4 h at 4 °C. Puncture the tube at the bottom using a 26–28 G needle and collect 13 1-mL fractions from the bottom of the tube.

5. Separate equal volumes of each fraction by SDS-PAGE and probe by Western blotting as mentioned earlier using GFP antibodies to detect the GFP fusion protein and an antibody specific for an ER resident protein (e.g., Bip chaperone protein) to identify fractions containing ER membranes (note 11). A typical result is shown in Fig. 4b.

3.7 Deglycosylation of ER-Associated Transmembrane Proteins Containing N-Glycosylation Sites Using Endoglycosidase H

1. Add 5 L of 4× protein loading buffer to 15 L of a P30 fraction in an Eppendorf tube. Boil for 5 min and centrifuge at 14,000 g for 5 min.

2. Collect the supernatant and transfer 4 L aliquots in two separate Eppendorf tubes.

3. To the first tube, add 45 L of Endo H buffer. To the second tube, add 45 L of Endo H buffer and 1.5 L of Endo H (1 U).

4. Incubate the reactions at 37 °C overnight.

5. Add 16 L of 4× protein loading buffer to each tube and boil 5 min. Centrifuge at 14,000 g for 5 min.

6. Collect the supernatants. Load 20 L of each sample on an SDS-polyacrylamide gel and probe by Western blotting as described earlier (note 12). A typical result is shown in Fig. 5.

4   Notes

1. Many integral membrane proteins exist as oligomers within the membrane. These oligomers are often at least partially resistant to denaturing conditions, thus dimers and other oligomeric forms of the protein are often detected by SDS-PAGE in addition to the monomers (16) (e.g., Fig. 5).

2. When the protein concentration in plant extracts is low, a larger volume of samples may be needed to detect the protein by Western blotting. In that case, a 1.5-mm thick gel instead of a 0.75-mm thick gel should be prepared. For proteins in the 20–60 kDa range, a 12% acrylamide gel is appropriate. If the expected size of a protein is larger than 60 kDa, the concentration of

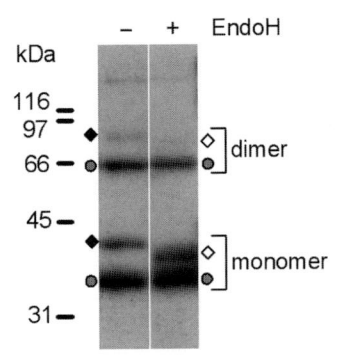

Fig. 5 Deglycosylation assay of an ER-associated protein containing an *N*-glycosylation site. The C-terminal region of the ToRSV NTB-VPg protein contains a transmembrane domain followed by a section translocated in the lumen of the membranes. This region of the protein was fused in frame with the GFP protein to produce the GFP-cNV3 protein. A naturally occurring *N*-glycosylation site is present in the luminal section of the protein (5, 8, 17). *N. benthamiana* plants expressing the GFP-cNV3 protein by agroinfiltration were extracted to produce P30 fractions. The fractions were treated with Endoglycosydase H. After the treatment, proteins were separated by SDS-PAGE (12% polyacrylamide), and detected by Western blotting using anti-GFP antibodies. Migration of molecular mass standards is shown on the *left*. Multiple forms of the proteins are detected including monomeric and dimeric forms (as indicated on the *right*). *Black* and *white diamonds* indicate a glycosylated form of the protein, which decreases in size after treatment with Endo H. The *grey circles* indicate a truncated form of the protein, which is not glycosylated. The sensitivity of the glycosylation to Endo H confirms that the protein is associated with the ER. Reprinted after modification from ref. 5 (with permission from the American Society for Microbiology)

acrylamide in the gel can be reduced to 8% to improve the separation and subsequent transfer to PVDF membranes of the protein. On the other hand, if a protein is smaller than 20 kDa, the concentration of acrylamide in the gel should be increased to 15% to prevent proteins from running off the gel.

3. Because of their hydrophobicity, some membrane proteins tend to aggregate during boiling. These large aggregates are unable to enter the gel and remain at the well or at the junction between the stacking and the separating gel. When this happens, one solution may be to incubate the samples with loading buffer at 37 °C for 30 min instead of boiling.

4. During Western blotting, prestained markers are particularly useful. They can be left on the membrane during the various washes and incubations with antibodies and will remain clearly visible at the end of the procedure. However, their migration is not as precise as that of other molecular weight standards (such as the broad range markers from Bio-Rad). When accurate evaluation of protein size is required, it is recommended to use these markers instead of or in addition to the prestained markers. In that case, the portion of the PVDF membrane corresponding to the lane containing these markers should be cut prior to incubation in blocking buffer. Stain this piece of the membrane with Coomassie Blue staining solution for 2–3 min, destain for 5–10 min in destaining buffer and let it dry.

5. When transferring proteins from a 1.5-mm thick gel and/or when transferring larger proteins (>60 kDa), longer transfer may be needed. In these cases, we routinely use the following conditions: 30–60 min at 100 V, reduce voltage to 50 V, and continue transfer overnight.

6. This monoclonal antibody has very high affinity for GFP and gives very little background with endogenous plant proteins. Because of the absence of background, it may be difficult to align the film on the membrane after exposure. We routinely use a permanent marker to out-

line the film on the acetate sheet before developing. Alternatively, it may also be useful to stick a small piece of fluorescent tape next to the membrane before exposing. When using fusions to the HA epitope, the rat HA antibody (Roche) used at a dilution of 1:1,000 is also an excellent commercially available antibody for use with plants. When using home-made antibodies, it is often necessary to conduct pilot experiments with different dilutions of the antibody to obtain optimal detection with minimal background. It is also necessary to use a healthy plant extract as a control to determine whether endogenous plant proteins react with the antibody.

7. If the membrane is dry during the addition of the detection reagents, blotching or white areas will be observed. It is better to leave some washing buffer on the membrane rather then to be overzealous in removing excess liquid from the membrane.

8. After treatment with high salt or high pH solutions, it is also possible to use the membrane flotation assay to demonstrate that the proteins are still associated with the membrane.

9. It is important to be gentle at this step. Too much grinding in the mortar and pestle may result in altering the integrity of the membranes.

10. Do not load too much material on the gradient. When the proteins are in low concentration, it may be tempting to concentrate the sample before loading on the sucrose gradient. However, we have found that this often results in poor separation of the different types of membranes on the gradient.

11. Separation of ER membranes will vary slightly from one gradient to another. It is important to use an ER resident protein as a marker to identify fractions containing ER membranes in each gradient. A viral protein associated with ER membranes should be detected in the same fractions as the ER marker in both gradients and should shift toward the top of the gradient in the absence of Mg^{2+} in parallel with the ER marker.

12. The presence of the deglycosylation buffer in the sample affects the migration of the proteins on SDS-PAGE, resulting in less sharp banding patterns than in the original sample. For this reason, it is important to have a control tube incubated in the presence of the buffer but in the absence of enzyme for comparison to the digested sample. It is also not recommended to evaluate the size of proteins based on their migration after the deglycosylation assay.

Acknowledgments   The authors thank Mrs. Joan Chisholm for critical reading of the manuscript, Dr. M. Chrispeels (University of California, San Diego) for the gift of anti-Bip antibodies, and Dr. A. Sturm (Friedrich Miescher Institute) for the gift of anti-β-xylosyl antibodies. This work was supported in part by an NSERC research grant awarded to HS.

References

1. Sanfacon, H. (2005) Replication of positive-strand RNA viruses in plants: contact points between plant and virus components. *Can J Bot* **83**, 1529–1549.

2. Salonen, A., Ahola, T., and Kaariainen, L. (2005) Viral RNA replication in association with cellular membranes. *Curr Top Microbiol Immunol* **285**, 139–173.

3. Noueiry, A. O., and Ahlquist, P. (2003) Brome Mosaic Virus RNA Replication: Revealing the Role of the Host in RNA Virus Replication. *Annu Rev Phytopathol* **41**, 77–98.

4. Carette, J. E., van Lent, J., MacFarlane, S. A., Wellink, J., and van Kammen, A. (2002) Cowpea mosaic virus 32- and 60-kilodalton replication proteins target and change the morphology of endoplasmic reticulum membranes. *J Virol* **76**, 6293–6301.

5. Zhang, S. C., Zhang, G., Yang, L., Chisholm, J., and Sanfacon, H. (2005) Evidence that insertion of Tomato ringspot nepovirus NTB-VPg protein in endoplasmic reticulum membranes is directed by two domains: a C-terminal transmembrane helix and an N-terminal amphipathic helix. *J Virol* **79**, 11752–11765.

6. Schaad, M. C., Jensen, P. E., and Carrington, J. C. (1997) Formation of plant RNA virus replication complexes on membranes: role of an endoplasmic reticulum-targeted viral protein. *Embo J* **16**, 4049–4059.

7. White, K. A., and Nagy, P. D. (2004) Advances in the molecular biology of tombusviruses: gene expression, genome replication, and recombination. *Prog Nucleic Acid Res Mol Biol* **78**, 187–226.

8. Han, S., and Sanfacon, H. (2003) Tomato ringspot virus proteins containing the nucleoside triphosphate binding domain are transmembrane proteins that associate with the endoplasmic reticulum and cofractionate with replication complexes. *J Virol* **77**, 523–534.

9. Laemmli, U. K. (1970) Cleavage of structural proteins during the assembly of the head of bacteriophage T4. *Nature* **227**, 680–685.

10. Sankaram, M. B., and Marsh, D. 1993. Protein-lipid interactions with peripheral membrane proteins, p. 127–162. *In* A. Watts (ed.), Protein-lipid interactions. Elsevier Science Publishers B. V., Amsterdam.

11. Staehelin, L. (1997) The plant ER: a dynamic organelle composed of a large number of discrete functional domains. *Plant J* **11**, 1151–1165.

12. Wienecke, K., Glass, R., and Robinson, D. G. (1982) Organelles involved in the synthesis and transport of hydroxyproline-containing glycoproteins in carrot root disks. *Planta* **155**, 58–63.

13. Nilsson, I. M., and von Heijne, G. (1993) Determination of the distance between the oligosaccharyltransferase active site and the endoplasmic reticulum membrane. *J Biol Chem* **268**, 5798–5801.

14. Kasturi, L., Eshleman, J. R., Wunner, W. H., and Shakin-Eshleman, S. H. (1995) The hydroxy amino acid in an Asn-X-Ser/Thr sequon can influence N-linked core glycosylation efficiency and the level of expression of a cell surface glycoprotein. *J Biol Chem* **270**, 14756–14761.

15. Maley, F., Trimble, R. B., Tarentino, A. L., and Plummer, T. H., Jr. (1989) Characterization of glycoproteins and their associated oligosaccharides through the use of endoglycosidases. *Anal Biochem* **180**, 195–204.

16. DeGrado, W. F., Gratkowski, H., and Lear, J. D. (2003) How do helix-helix interactions help determine the folds of membrane proteins? Perspectives from the study of homo-oligomeric helical bundles. *Protein Sci* **12**, 647–65.

17. Wang, A., Han, S., and Sanfacon, H. (2004) Topogenesis in membranes of the NTB-VPg protein of Tomato ringspot nepovirus: definition of the C-terminal transmembrane domain. *J Gen Virol* **85**, 535–545.

Chapter 26
Membrane and Protein Dynamics in Virus-Infected Plant Cells

Michael Goodin, Romit Chakrabarty, and Sharon Yelton

Abstract In terms of functional genomics research, *Nicotiana benthamiana*, more so than other model plants, is highly amenable to high-throughput methods, especially those employing virus-induced gene silencing and agroinfiltration. Furthermore, through recent and ongoing sequencing projects, there are now upward of 18,000 unique *N. benthamiana* ESTs to support functional genomics research. Despite these advances, the cell biology of *N. benthamiana* itself, and in the context of virus infection, lags behind that of other model systems. Therefore, to meet the challenges of diverse cell biology studies that will be derived from ongoing functional genomics projects, a series of methods relevant to the characterization of membrane and protein dynamics in virus-infected cells are provided here. The data presented here were derived from our studies with plant rhabdoviruses. However, the employed techniques should be broadly applicable within the field of plant virology. We report here on the use of a novel series of binary vectors for the transient or stable expression of autofluorescent protein fusions in plants. Use of these vectors in conjunction with advanced microscopy techniques such as fluorescent recovery after photobleaching and total internal fluorescence microscopy, has revealed novel insight into the membrane and protein dynamics of virus-infected cells.

Keywords *Nicotiana benthamiana*, agroinfiltration, autofluorescent protein, FRAP, TIRFM, laser scanning confocal microscopy

1 Introduction

Nicotiana benthamiana Domin ($n = 19$; $1C = 3.20\,\text{pg}$), a plant native to Australia, has emerged as a powerful model for molecular and cell biology studies, particularly in the elucidation of host–pathogen interactions (1–3). We are developing novel vectors and microscopy methods to support *N. benthamiana* cell biology research, with particular emphasis on the development of binary vectors for the transient or stable expression of expression autofluorescent protein fusions in plants (4) as well as methods for studying protein and membrane dynamics in virus-infected cells (5).

From: *Methods in Molecular Biology, Vol. 451, Plant Virology Protocols: From Viral Sequence to Protein Function*
Edited by G.D. Foster, I.E. Johansen, Y. Hong, and P.D. Nagy © Humana Press, Totowa, NJ

Our particular justification for developing these resources is largely borne out of tests of our working hypothesis that the mechanism by which pathogens trigger changes in host gene expression in a compatible interaction is related to both the cellular localization of pathogen-encoded proteins as well that of host-encoded gene products in response to infection. Tests of this hypothesis are dependent upon the availability of tools that permit facile and unambiguous determination of the cellular loci at which proteins accumulate. These experiments are being conducted with the plant rhabdoviruses, *Sonchus yellow net virus* (SYNV), and *Potato yellow dwarf virus* (PYDV). Both SYNV and PYDV are in the genus *Nucleorhabdovirus* in the family *Rhabdoviridae*. These enveloped viruses with monopartite, minus-sense, single-stranded RNA genomes include some of the greatest threats to human, animal, and plant health (6–8). We have previously reported that infection of *N. benthamiana* plants, which express GFP targeted to the endoplasmic reticulum, with SYNV or PYDV results in accumulation of GFP in nuclei of virus-infected cells (5). To probe the relationship between relocalized membranes and sites of viral protein accumulation, we have employed a novel set of plant expression vectors and microscopy techniques. The protocols provided below should be applicable to a wide variety of investigations into virus-induced membrane and protein dynamics in plant cells.

The binary vectors described in this protocol were derived from the pSAT vectors reported by Tzfira et al. (9). However, unlike the original pSAT vectors, which required cloning genes of interest into an intermediate plasmid followed by sub-cloning into a binary vector, our modified vectors permit recombination-cloning directly into a binary vector that carry a marker that can be selected in transformed plants. Therefore, these new vectors, presently called pAFPs (Fig. 1) can be used in high-throughput transient expression studies. Any constructs of interest can then be used to generate transgenic plants without the need for further subcloning.

In addition to protein localization and in vivo interaction studies, the pAFP vectors can be used in conjunction with advanced microscopy techniques such as fluorescence recovery after photobleaching (FRAP) and total internal reflectance fluorescence microscopy (TIRFM). Briefly, FRAP involves the photobleaching of a fluorescent molecule (e.g., GFP) in a region of interest followed by measuring the rate of fluorescence recovery (10). Commonly, FRAP results, like those presented here, are analyzed qualitatively to determine if protein mobility is "rapid" or "slow" or some other characteristic such as the presence of binding interactions, existence of an immobile fraction, or if particular chemical treatment affects fluorescence recovery (10). In contrast to FRAP, which can be used to study protein and membrane dynamics at any point in the z-plane through a sample, TIRFM is suitable for experiments conducted at the surface of membranes (11). The principle of TIRFM derives from the properties of a light beam passing through a medium with a refractive index n_1 (e.g., glass) to an interface with a second medium of refractive index $n_2 < n_1$ (e.g., the cytoplasm). Total internal reflection occurs at all angles of incidence Θ that are greater than a critical angle $\Theta_c = \arcsin(n_2/n_1)$. However, despite being totally reflected, the incident beam establishes an "evanescent wave," which is an electromagnetic field that penetrates into the second medium and decays

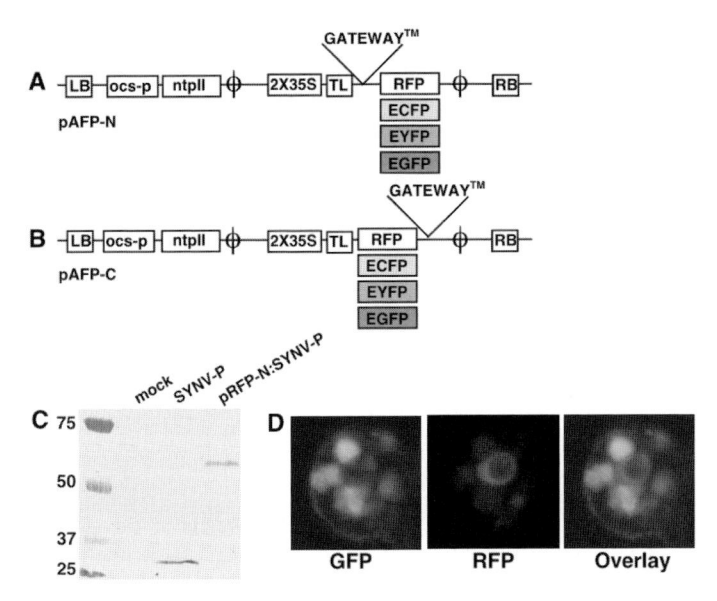

Fig. 1 Partial maps of novel binary vectors for transient or stable (kanamycin selection) expression of autofluorescent protein fusions (AFPs) in plant cells. Vectors for expression of amino- (**a**; pAFP-N) or carboxy-terminal fusions (**b**; pAFP-C) to monomeric DsRed (RFP), or the enhanced versions of the cyan, yellow, and green variants of GFP are now available. All of these vectors contain expression cassettes flanked by the *Agrobacterium tumefaciens* Ti plasmid left (LB) and right borders (RB). Expression of AFP fusions is promoted by a duplicated cauliflower mosaic virus 35S promoter (2 × 35S) and mRNA transcripts contain a 5′ translational enhancer from Tobacco etch virus (TL). Transcripts are terminated by the CaMV terminator. Resistance of transformed plant cells to kanamycin is conferred by the *ntpII* gene flanked by the octopine synthase promoter and terminator. The construction of these new binary vectors will be published elsewhere. (**c**) Polyclonal antibodies raised in chickens were used to probe nitrocellulose membranes onto which proteins from mock-inoculated, SYNV-infected, and pRFP-N-SYNV-agroinfiltrated leaves were transferred after electrophoretic separation. Immunoblots were developed with rabbit anti-chicken antibodies linked to alkaline phosphatase. (**d**) Demonstration of the new vectors in plant cells. Green channel (left panel; GFP), red channel (middle panel; RFP), and overlay (right panel; Overlay) of confocal micrographs of an SYNV-infected nucleus in 16c leaf epidermal cell in which RFP-P was transiently expressed from pRFP-N. Fluorescence from RFP is detected in a large intranuclear body, perhaps corresponding to viroplasm

exponentially with the distance z from the interface, according to the relationship $d = (\lambda/4\pi) \, (n_1^2 \sin_2 \Theta - n_2^2)^{1/2}$ (where λ corresponds to the wavelength of light). The depth of the evanescent wave (d) can be adjusted between 70 nm and 300 nm depending on the excitation wavelength employed (11). Although currently not widely adapted in plant virology, our recent experience with these techniques suggests that both FRAP and TIRFM will become increasingly important in this field.

2 Materials

2.1 Plant Material, Growth Conditions, and Virus Inoculation Procedures

Transgenic "16c" *N. benthamiana* plants (12) that constitutively express the mgfp5-ER variant of the green fluorescent protein targeted to the endoplasmic reticulum (ER; 13), under the control of the *Cauliflower mosaic virus* 35S promoter, were grown in a greenhouse under ambient conditions. Plants were mechanically inoculated with virus when they had 4–6 fully expanded leaves and typically used at 10–14 days post inoculation.

2.2 PCR Amplification for Gateway® Cloning

We typically conduct in planta protein expression using a novel set of binary vectors, presently designated pAFP, that utilize recombination-mediated cloning using Gateway technology (Invitrogen, LaJolla, CA; Fig. 1). When referring to specific derivatives of these vectors the "A" in pAFP is replaced with the first letter of a relevant autofluorescent protein. For example, pAFP vectors expressing red fluorescent protein (RFP) fusions are reported as pRFP constructs (Fig. 1). To take advantage of this system, *att* sites must be introduced into the 5′ and 3′ ends of the open reading frames of interest. This is most readily achieved by conducting two rounds of PCR using a high-fidelity DNA polymerase. In the first PCR, gene-specific primer pairs are used, with each primer containing approximately half of the 5′ or 3′ *att* sites. A second round of PCR introduces the remainder of the *att* sites. We prefer this two-step approach as it drastically reduces the costs of primers, particularly in high-throughput expression studies where hundreds to thousands of genes need to be cloned.

1. PCR reaction was performed using an iCycler (BioRad, USA).
2. This cloning procedure will be demonstrated for the SYNV-P protein gene. In theory, the P-specific sequences in the forward and reverse primers (in bold) can be replaced to amplify any gene of interest. First-round PCR was conducted with primers, P-*att*B1-Forward 5′-AAAAAGCAGGCTTA**ATGGAAATCGAT CCAAATTACGTTAACC**-3′ and P-*att*B2-Reverse 5′-AGAAAGCTGGGTAC **GCCTTCTTTGGGTCAATAAGAACTA**-3′. The second-round of PCR utilizes universal primers for adding the complete *att*B site (2-*att*B1-Forward 5′-G GGGACAAGTTTGTACAAAAAAGCAGGCT-3 ′ and 2-*att*B2-Reverse 5′-GGGGACCACTTTGTACAAGAAAGC TGGGTA-3′).
3. PCR reactions were performed using Phusion™ High Fidelity DNA polymerase (Finnzymes, Finland) and Phusion High Fidelity buffer.
4. PCR products were verified by electrophoresis on 0.8% (w/v) agarose gels with TAE (40 mM Tris–acetate and 2 mM EDTA) buffer.

2.3 BP Clonase™ Reaction

Following amplification of genes of interest, the PCR products must be introduced into a "donor" plasmid. This is done by BP Clonase™-mediated in vitro recombination.

1. The donor vector pDONR 221 (Invitrogen Corp., USA) was used for cloning of PCR product using BP Clonase™ II enzyme mix (Invitrogen Corp., USA).
2. Enzyme dilution buffer: 10 mM Tris–HCl buffer, pH 8.5.
3. Reactions were terminated by the addition of proteinase K (Invitrogen Corp., USA).
4. One Shot OmniMAX2 T1 phage-resistant cells (Invitrogen Corp., USA) were used for transformation.
5. S.O.C. medium (2% tryptone, 0.5% yeast extract, 10 mM NaCl, 2.5 mM KCl, 10 mM MsSO$_4$, and 20 mM glucose).

2.4 Screening of Entry Clones

1. PCR reactions were conducted using an iCycler (BioRad, USA).
2. Forward and reverse primers (P-*att*B1-Forward 5′-AAAAAGCAGGCTTA**ATG GAAATCGATCCAAATTACGTTAACC**-3′ and P-*att*B2-Reverse 5′-AGAAAGCTGGGTA**CG CCTTCTTTGGGTCAATAAGAACTA**-3′).
3. The other components in the PCR reaction included 10 mM dNTPs (Invitrogen Corp., USA), DyNAzyme™ EXT DNA Polymerase (Finnzymes, Finland) and DyNAzyme buffer (10×).
4. The sequence verification of the Entry clones was done using M13 Forward (5′-GTAAAACGACGGCCAG-3′) and M13 Reverse (5′-CAGGAAACAGCTATGAC-3′) primers and Big Dye[R] technology (Applied Biosystems, USA).
5. Antibiotic stock: 50 mg mL^{-1} Kanamycin monosulfate (Fisher Scientific, USA)

2.5 LR Clonase™ Reaction

Validated entry clones can be recombined into a wide variety of vectors for protein expression in bacterial, yeast, insect, or plant cells. Similar to the cloning of PCR products, transfer of genes from entry clones into "destination" vectors (Fig. 1) is recombination-mediated using LR Clonase™.

1. LR Clonase™ II enzyme (Invitrogen Corp., USA).
2. To adjust the volume of the LR reaction, 10 mM Tris–HCl buffer, pH 8.5 was used.
3. Reaction was terminated by proteinase K (Invitrogen Corp., USA).
4. One Shot OmniMAX™2 T1 phage-resistant cells (Invitrogen Corp., USA) were used for transformation.

5. S.O.C. medium (2% tryptone, 0.5% yeast extract, 10 mM NaCl, 2.5 mM KCl, 10 mM MgSO$_4$, and 20 mM glucose.

2.6 Screening of Expression Clones and Transformation in Agrobacterium LBA4404

1. Plasmid DNA mini-prep from 3 mL of a 5 mL overnight culture was purified using QIAprep® Spin Miniprep Kit (QIAGEN Sciences, USA).
2. *Agrobacterium tumefaciens* strain LBA4404 made chemically competent (20 mM CaCl$_2$).
3. LB medium: 1% (w/v) yeast extract, 1% (w/v) bacto-tryptone, and 0.5% (w/v) NaCl.
4. Antibiotic stocks: 25 mg µL^{-1} Rifampin (Fisher Bioreagents, USA), 100 mg µL^{-1} spectinomycin sulfate (MP Biomedicals Inc., USA), and 300 mg mL^{-1} streptomycin sulfate (MP Biomedicals Inc., USA).

2.7 SDS-Polyacrylamide Gel Electrophoresis (SDS-PAGE)

1. Separating buffer: 1.5 M Tris–HCl, pH 8.8. Store at room temperature.
2. Stacking buffer: 0.5 M Tris–HCl, pH 6.8. Store at room temperature.
3. Acrylamide/bis-acrylamide stock solution (30% T, 2.6% C). To reduce risk of exposure to this neurotoxin we purchase acrylamide as a "ready-to-use" solution (Bio-Rad, Hercules, CA).
4. Catalyst: *N,N,N,N'*-Tetramethyl-ethylenediamine (TEMED)
5. Ammonium persulfate: 10% (w/v) aqueous solution preferably made fresh or stored at 4 °C for up to one week.
6. Water-saturated isobutanol. Mix equal volumes of water and isobutanol, shake vigorously, and allow phases to separate. Use the top layer. Store at room temperature.
7. 4× sample preparation buffer: Mix 2 mL water, 2 mL stacking buffer, 0.8 mL glycerol, 3.2 mL 10% (w/v) SDS, 0.8 mL β-mercaptoethanol, and 0.2 mL of 0.05% (w/v) bromophenol blue.
8. Running buffer: 5× stock can be prepared by dissolving 15 g Tris base, 72 g glycine, and 5 g SDS in water to a final volume of 1 L.

2.8 Western Blotting for Detecting AFP Fusions

1. Transfer buffer: 25 mM Tris, 192 mM glycine, 20% (v/v) methanol, 0.05% (w/v) SDS.
2. Nitrocellulose membrane (0.45 µm pore size; AmershamBioscience, Sweden).

3. Tris-buffered saline with Tween (TBS-T). Prepared as a 10× stock [1.37 M NaCl, 27 mM KCl, 250 mM Tris–HCl, pH 7.4, 1% (v/v) Tween-20].
4. Blocking buffer: 1× TBS-T containing 5% (w/v) non-fat dry milk powder.
5. Primary antibody. Chicken anti-SYNV-P diluted 1:2000 in TBS-T.
6. Secondary antibody: Rabbit anti-chicken-alkaline phosphatase secondary antibody diluted 1:30,000 in TBS-T (Sigma).
7. Alkaline phosphatase buffer: 100 mM Tris–HCl, 100 mM NaCl, 5 mM MgCl2, pH 9.5. Stored at 4 °C.
8. Dye solution: NBT (Nitro Blue Tetrazolium); prepared as a 50 mg mL^{-1} stock in 70% (v/v) N,N-dimethyl-formamide (DMF). BCIP (5-bromo-4-chloro-3-indolyl phosphate) prepared as a 25 mg mL^{-1} stock in 100% DMF. Store both stocks at 4 °C.

2.9   Isolation of Protoplasts from Nicotiana benthamiana

1. Protoplast isolation solution: 0.7 M mannitol.
2. Cellulase (Onozuka R-10) and Macerozyme R-10 (Research Products International).
3. Sieve set, 35, 45, 60, and 120 mesh (Bel-Art Products).
4. Nylon cloth (50 μm; Small Parts, Inc.).
5. Protoplast culture medium (1 L): 1 mL 1,000× vitamin stock (added after autoclaving), 0.5 mL 2,000× hormone stock, 4.4 g Murashige-Skoog salts, 34.2 g sucrose, 0.58 g MES, 72.8 g mannitol. Adjust pH of final solution to 5.8 using 1 N KOH. Autoclave, cool to room temperature and store at 4 °C.
6. Vitamin stock (1,000×, 20 mL): 20 mg thiamine-HCl, 10 mg pyridoxine-HCl, 10 mg nicotinic acid, 2.0 g myo-inositol. Stir 1 h at room temperature to dissolve myo-inositol. Make 1 mL aliquots and store at −20 °C.
7. Hormone stock (2,000×, 50 mL): 20 mg 2,4-dichlorophenoxyacetic acid, 20 mg kinetin, 2.5 mL 1 M KOH, and 50 mL sterile water. Vortex to dissolve. Store at −20 °C.

2.10   Fluorescence Microscopy

1. No. 1 or No. 1½ coverslips and slides from Corning (Corning, NY).
2. Samples for confocal microscopy and FRAP experiments conducted using approximately 25 mm^2 sections of leaf tissue mounted in water.
3. Imaging should be conducted in the epidermal layer on the abaxial surface of the leaf.
4. TIRFM was conducted using protoplasts prepared from virus-infected or mock-inoculated leaves.

3   Methods

3.1   PCR Amplification for Gateway® Cloning

Autofluorescent protein fusions were expressed from modified pSAT vectors (Tzfira et al., 2005) in order to produce a novel set of binary vectors for transient or stable expression of autofluorescent proteins (AFPs) in plant cells. The details of this construction and validation of these vectors, which permit recombination-mediated cloning of genes of interest using Gateway technology, will be published elsewhere. Briefly, the Gateway destination module from vectors pSAT6-DEST-EGFP-N1 (Genbank accession AY818370) and pSAT6-DEST-EGFP-C1 (Genbank accession AY818372) were subcloned into all N1 and C1 pSAT6 variants (Tzfira et al., 2005). Binary vectors were then generated by subcloning the PspP1 fragments of the Gateway compatible derivatives into pRCS-2-ntpII (9). For convenience here, we will refer to derivatives used in this study that express cyan, yellow, or red fluorescent protein fusions as pCFP-C1, pRFP-C1 or pRFP-N1, and pYFP-C1, respectively.

1. Prior to cloning into binary vectors, genes of interest were amplified by a two-step PCR procedure to introduce *att* sites at their 3′ and 5′ termini.
2. The first PCR reaction was conducted with primers (P-*att*B1-forward and P-*att*B2-reverse 5′) that include the P gene-specific sequence and part of the *att*B sequence. Subsequently, the second PCR reaction is done with universal primers (2-*att*B1-forward 5′ and 2-*att*B2-reverse) that include the entire *att*B sequence.
3. A 50 µL PCR reaction was conducted using Phusion™ High Fidelity DNA Polymerase (Finnzymes, Finland) that includes 10 µL of Phusion HF buffer (5×), 1 µL of 10 mM dNTPs, 50 pmol of forward and reverse gene-specific primers (first set of primers), 1 unit of Phusion™ DNA polymerase, and template DNA (20–30 ng). PCR cycling conditions included 2 min hot start at 98 °C, 1 cycle of initial denaturation for 2 min at 94 °C, 25 cycles of amplification at 94 °C for 30 s, 45 °C for 60 s, 72 °C for 90 s, and 1 cycle of final extension at 72 °C for 5 min. The PCR products were gel purified following electrophoresis through 0.8% (w/v) agarose gel. The second PCR reaction (same conditions mentioned above) is conducted with universal primers and the final PCR product (that included the SYNV-P gene flanked by *att*B sequence) is again gel purified after running in 0.8% (w/v) agarose gel.

3.2   BP Clonase™ Reaction

1. PCR products produced as described above were cloned into pDONR using recombination mediated by BP Clonase™. In a 1.5 mL microcentrifuge tube the reaction components included the *att*B-PCR product (15–150 ng) and 1 µL of

pDONR 221 vector (150 ng). The volume was adjusted to 8 μL by addition of 10 mM Tris–HCl buffer, pH 8.5. Finally, 2 μL of BP Clonase™ II enzyme was added to the above components and mixed well. The reaction was incubated at 25 °C overnight. Furthermore, 1 μL of proteinase K (2 μg μL^{-1}) solution was added to terminate the reaction and the sample was incubated at 37 °C for 10 min.

2. Mix 2 μL of the BP reaction with 50 μL of competent One Shot® OmniMAX™ 2 T1 phage resistant cells (Invitrogen Corp, USA). Incubate on ice for 10–15 min and then heat shock cells at 42 °C for 30 s in a water bath.

3. Return the cells back to ice for 2 min and then add 250 μL of S.O.C medium. Incubate the cells at 37 °C for 1 h with constant shaking. The cells are then plated onto selection plates (Kanamycin, 50 mg L^{-1}).

3.3 Screening the Entry Clones

1. Tranformation reactions were incubated overnight on agar plants amended with the appropriate antibiotics. Transformed colonies were then analysed for the presence of the SYNV-P gene. The colonies were screened by colony PCR. Template DNA is prepared by resuspending a part of a colony in a 1.5 mL microfuge tube containing 100 μL water. After brief vortex the cells are heat lysed in a microwave for 3 min. The PCR reaction (50 μL) included 5 μL of DyNAzyme™ (Finnzymes, Finland) buffer (10×), 1 μL of 10 mM dNTPs, 50 pmol of forward and reverse primers (P-*att*B1-forward and P-*att*B2-reverse), 2 μL of lysed cell as a source for DNA template, and 0.5 μL of DyNAzyme™ EXT DNA polymerase (1 U μL^{-1}). PCR cycling conditions included 1 cycle of initial denaturation for 2 min at 94 °C, 25 cycles of amplification at 94 °C for 30 s, 45 °C for 60 s, 72 °C for 90 s, and 1 cycle of final extension at 72 °C for 5 min.

2. The PCR product was loaded onto 0.8% (w/v) agarose gel to check the presence of the SYNV-P amplified fragment. Next, the positive entry clones for SYNV-P were further analysed by enzyme digestion.

3. Finally, the clones were sequenced by Big Dye® Terminator v3.1 Cycle Sequencing Kit (Applied Biosystems, USA) and ABI PRISM 310 Genetic Analyzer AME Bioscience (AME Bioscience, Norway).

3.4 LR Clonase™ Reaction

1. Once the entry clones have been validated by sequencing, the inserts must be mobilized from the entry vector into the destination vector using LR Clonase™-mediated recombination.

2. In a 1.5 mL microcentrifuge tube the reaction components included the entry clone (50–150 ng) and 1 μL of binary pRFP-N destination vector (150 ng). The

volume was adjusted to 8 µL by addition of 10 mM Tris–HCl buffer, pH 8.5. Finally, 2 µL of LR Clonase™ II enzyme was added to the above components and mixed well.

3. Incubate reaction tubes at 25 °C overnight.
4. Add 1 µL of proteinase K solution to terminate the reaction and incubate at 37 °C for an additional 10 min.
5. Mix 2 µL of the LR reaction with 50 µL of competent One Shot OmniMAX™2 T1 phage-resistant cells (Invitrogen Corp, USA). Keep the cells on ice for 10–15 min, heat shock by incubating at 42 °C for 30 s. Next, place the cells back to ice for 2 min and then add 250 µL of S.O.C medium. Incubate the cells at 37 °C for 1 h with constant shaking. The cells are then plated onto selection plates (50 mg L^{-1} spectinomycin and 20 mg L^{-1} streptomycin).

3.5 Screening the Expression Clones and Transformation in A. tumefaciens Strain LBA4404

1. The same PCR-based colony screen was followed here as in the case for screening of entry clones in order to confirm the presence of recombinant vectors.
2. The expression clone is then transformed into *Agrobacterium tumefaciens* strain LBA4404. The procedure involves addition of 10 µL of mini-prep DNA of recombinant pAFP clones to chemically competent *A. tumefaciens* LBA4404 strain (80 µL). The cells are first frozen in liquid Nitrogen and then thawed at 37 °C for 5 min. Next, 500 µL of LB medium is added and the culture is incubated at 28 °C for 3 h. The cells are plate on selection plates (25 mg L^{-1} rifampin, 100 mg L^{-1} spectinomycin, and 300 mg L^{-1} streptomycin).
3. The transformed *A. tumefaciens* LBA4404 colonies start appearing after 2–3 days. The colonies are streaked on fresh selection plates and are ready for plant transformation after further incubation for 1–2 days.

3.6 Transient Expression of Proteins by Agroinfiltration

1. Streak *Agrobacterium* transformed with expression vector of choice onto LB plates amended with the appropriate antibiotics and incubate at 28 °C.
2. Following incubation, use an inoculating loop to harvest cells and resuspend in agroinfilatration buffer (10 mM MES, 10 mM MgCl$_2$, pH 5.9)
3. Adjust cell suspension to an O.D.$_{600}$ of 0.6–1.0.
4. Add acetosyringone to 150 uM (note 1).
5. Incubate at room temperature (18–22 °C) for 2–3 h.
6. Fill a 1 mL (tuberculin) syringe barrel with the cell suspension to infiltrate, and gently appress the tip to the abaxial surface of the leaf. Infiltrate leaf by gently

depressing the plunger while maintaining a good seal between the syringe tip and leaf.

7. Mark infiltrated leaf area as appropriate (note 2).
8. Incubate plants with illumination at 18–28 °C for 48 h.
9. Agroinfiltrated tissues are suitable for microscopy or biochemical analyses for at least 96 h post infiltration.
10. Increased levels of protein expression and/or extension the length of time tissues are suitable for experiments can be achieved if a suppressor of RNA silencing is coinfiltrated with the constructs of interest (note 3).
11. If there is a need to coexpress two or more proteins by agroinfiltration simply mix equal parts of suspensions of cells transformed with different constructs.

3.7 Counterstaining Live Cells with Cell-Permeant Dyes

1. Adjust an appropriate volume (0.3 mL/leaf) of MES buffer to 2.25 ug 4′,6-diamidino-2-phenylindole (DAPI) or other cell-permeant dye as needed (note 4).
2. Infiltrate leaf tissue as for agroinfiltration (i.e., appress the syringe to the abaxial surface of the leaf with sufficient pressure to make a good seal). Tissue that has been previously infiltrated with *Agrobacterium* is suitable for counter-staining with DAPI after 48 h.
3. Incubate plants in the dark for 20 min prior to examination. Leaves infiltrated with DAPI are suitable for microscopy for at least 24 h.
4. Excise an approximately 25 cm^2 section of leaf with a one-sided razor blade or scalpel and mount on a standard microscope slide. Mount the tissue in 2–3 drops of water and cover with a coverslip.
5. Examine tissue by standard epifluorecence or confocal microscopy (Fig. 2).

3.8 SDS-PAGE

1. This protocol assumes the use of a Bio-Rad Mini-ProteanII gel system and the use of glass plates with 0.75 mm spacers.
2. Prepare two 0.75 mm thick, 10% separating gels by mixing 3.34 mL of water, 2.0 mL of separating buffer, 80 ul of 10% SDS, and 2.6 mL of acrylamide/bis stock. Add 80 ul of 10% APS and 4 uL of TEMED. Pour the gel to a level 1 cm below the bottom of the level taken by the lane spacers. Overlay the gel with water-saturated butanol. The gel should polymerize within 20–30 min.
3. Pour off the isobutanol and rinse the gel with water.
4. Prepare a 4% stacking gel by mixing 1.8 mL of water, 0.75 mL of stacking buffer, 40 uL of 10% SDS buffer, and 0.39 mL acrylamide/bis stock. Add 60 uL of 10% (w/v) APS and 3 uL of TEMED. Fill atop stacking gel and insert spacers. The gel should polymerize within 20–30 min.

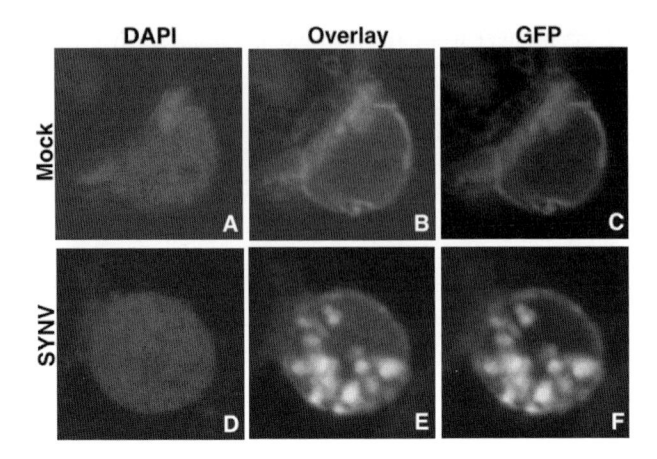

Fig. 2 Confocal micrographs of rhabdovirus-induced nuclear inclusions in 16c *N. benthamiana* plants inoculated with SYNV (**a–c**) or mock-inoculated (**d–f**). (**a–c**) and (**d–f**): Detection of DAPI and GFP fluorescence, and the overlay of these images, in single nuclei of epidermal cells of SYNV- and mock-inoculated leaves, respectively. Scale bar = 5 µm. Colored versions of these micrographs have been published (5)

5. Remove the combs, rinse the wells with water and assemble the gels into the holder.
6. Add 1× running buffer to the upper and lower chambers of the gel unit and load 5–20 uL of sample to each well. Include one lane of prestained molecular weight markers.
7. Complete assembly of the gel unit and connect to a power supply. Run at 200 V for 75–90 min or until the tracking dye just starts to elute from the bottom of the gel.

3.9 Western Blotting

1. As for SDS-PAGE, this protocol assumes the use of the Bio-rad mini-Protean system.
2. Assemble in the following order on the black side of the gel cassette: (a) fiber pad, (b) 1 sheet 3 mm filter paper, (c) separating gel (stacking gel is removed), (d) nitrocellulose, cut to the same dimensions as the gel, (e) 1 sheet 3 mm filter paper, and (f) fiber pad.
3. Close the gel cassette and insert into cassette holder, keeping the black side of the cassette against the black side of the transfer apparatus.
4. Insert ice pack and then fill the apparatus with transfer buffer.
5. Connect assembled apparatus to a power supply and transfer at 0.25A for 60 min.

6. Disassemble gel and incubate nitrocellulose in 10–15 mL blocking buffer for 15 min on a rocker. Make sure prestained markers are visible on the nitrocellulose before proceeding as this indicates how well the proteins transferred from the gel to the nitrocellulose.

7. Add primary antibody to blocking buffer and continue to rock slowly at room temperature for 2–3 h or incubate at 4 °C overnight.

8. Dispose of antibody solution and rinse nitrocellulose with three 5 min washes of 20 mL TBS-T with rocking.

9. Incubate nitrocellulose for 2 h to overnight in secondary antibody diluted 1:30,000 in 10 mL blocking buffer.

10. Dispose of antibody solution and rinse nitrocellulose with three 5 min washes of 20 mL TBS (no Tween) with rocking.

11. Drain buffer and add dye solution (12 mL alkaline phosphatase buffer and 50 uL each of the NBT and BCIP stock solutions). Cover the gels to reduce exposure to light (aluminum foil works well) and incubate at room temperature with rocking.

12. After sufficient color development, rinse nitrocellulose in water to stop the reaction. Dry the blot on paper towels and scan when dry to obtain a permanent digital record.

3.10 Confocal Microscopy

All confocal microscopy and FRAP experiments were performed using an Olympus FV1000 laser scanning confocal microscope. This microscope is equipped with lasers that permit wavelengths ranging from 405–633 nm to be used. Additionally, the microscope is equipped with a simultaneous (SIM) scanner with a dedicated 405 nm laser for photobleaching/photoconversion experiments.

1. Fluorescence from DAPI, GFP, and RFP was acquired sequentially using the 405 nm (diode laser), 488 nm (multiline argon laser), and 543 nm (red He-Ne laser) laser lines, respectively.

2. Imaging at high-magnification is conducted exclusively with water immersion lenses (note 5).

3. Images are typically acquired at a pixel resolution of 512 × 512 or 1,024 × 1,024 at a scan rate of 10 μs/pixel. The speed at which images are acquired depends upon the rate at which nuclei in cells move. Most often the movement is in the z-plane, so multiple images can be acquired after refocusing. However, nuclei may occasionally exhibit rapid movement in x–y or x–y–z planes.

4. Confocal images in TIFF format using Olympus Fluoview software (Olympus).

5. Image analysis was conducted using Photoshop version 7.0 and Canvas 8.0.

3.11 Fluorescence Recovery After Photobleaching (FRAP)

This protocol assumes the use of an Olympus FV1000 laser scanning confocal microscope equipped with a simultaneous (SIM)-scanner with a dedicated 405 nm laser.

1. About 25 mm square sections of leaf tissue were mounted on glass slides in water and covered with a glass coverslip.
2. Imaging for FRAP experiments was conducted using a PLAPO60XWLSM/1.0 objective and 488 nm laser line from a multiline argon laser set at 0.3% power.
3. Select regions-of-interest (ROIs; typically 1 µm diameter circles) and photobleach for 50 ms using a 405 nm diode laser, set at full power, delivered via the FV1000 SIM scanner.
4. In a typical experiment (Fig. 3), two images were acquired prior to photobleaching followed by an additional 7 images to monitor fluorescence recovery. Images were acquired at a pixel resolution of 256 × 256 and a scan rate of 2 µs/pixel. Complete experiments were acquired in 3,852 ms (Fig. 3).

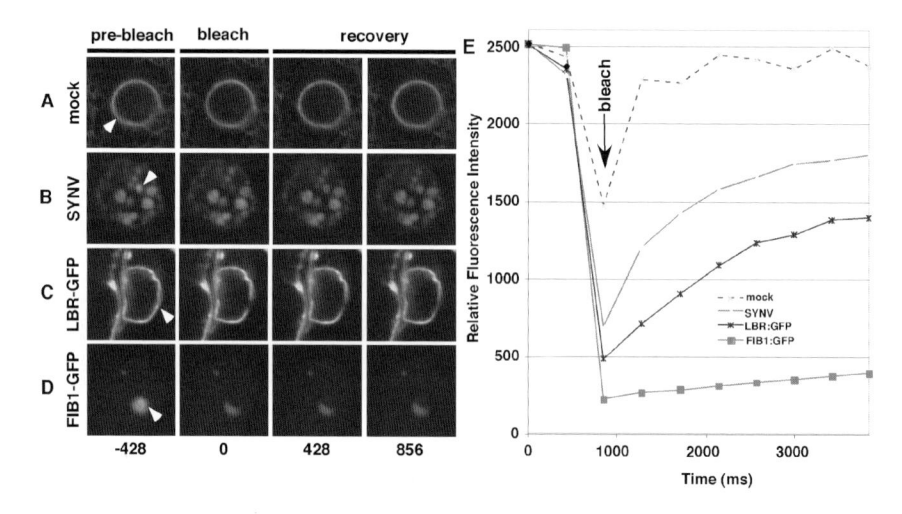

Fig. 3 Fluorescence recovery after photobleaching of *N. benthamiana* nuclear membranes and proteins. (**a–d**): Confocal micrographs taken prior to (prebleach) and after (recovery) a 50 ms pulse of 405 nm laser illumination (bleach). The white arrowhead points to the region to be bleached. (**a**) Nuclear envelope of nuclei in cells of mock inoculated 16c plants expressing endomembrane-targeted GFP. (**b**) SYNV-infected nucleus in a leaf cell of 16c plant. (**c**) Nuclear envelope of wild-type *N. benthamiana* in which inner nuclear membrane is marked by transient expression of a human lamin B receptor (14). (**d**) Nucleus of wild-type *N. benthamiana* cell in which the *Arabidopsis thaliana* nucleolar marker Fibrillarin1 was expressed as a GFP fusion (5). (**e**) Quantitative presentation of normalized fluorescence data corresponding to the micrographs shown in (**a–d**). Fluorescence recovery was monitored for 2,800 ms following bleach of the regions of interest shown in (**a–d**)

5. FRAP experiments were repeated at least twice for each ROI, with approximately 2 min between bleaching events in order to allow full recovery of fluorescence in bleached ROIs (note 6).
6. Quantitative fluorescence data can be exported in Excel format and confocal images in TIFF format using Olympus Fluoview software (Olympus).
7. Image analysis was conducted using Photoshop version 7.0 and Canvas 8.0.
8. Replicated fluorescence intensity data was averaged and these data were normalized across experiments (Fig. 3).

3.12 Isolation of Protoplasts from Nicotiana benthamiana Leaves for TIRFM

1. Select SYNV-infected plants that are 10–14 dpi and use young leaves showing symptoms. Choose healthy control plants that have 10–12 fully expanded leaves.
2. Weigh out 1 g of leaf tissue per plant and put in large glass petri dish.
3. Cut leaf tissue into 5 mm squares using a razor blade in a 50 mL solution containing 1% (w/v) cellulase and 0.1% (w/v) macerozyme.
4. Cover petri dish and incubate protoplasts at 25 °C in the dark for 12 h with gentle shaking.
5. Pour the solution through the sieve set (35–120 mesh), prewetted with ultrapure water (resistivity of 18.2 MΩ-cm), into a 30 mL centrifuge tube using a funnel (note 7).
6. Pour solution through a 50 μm nylon cloth into a new 30 mL tube.
7. Centrifuge for 5 min at 900 rpm. Remove and discard supernatant by pipetting.
8. Gently disperse the protoplasts in 25 mL of 0.7 M mannitol and centrifuge at 800 rpm for 5 min. Remove and discard supernatant.
9. Gently disperse the protoplasts in 10 mL of 0.7 M mannitol and centrifuge at 650 rpm for 3 min. Remove all but 0.5–1 mL of supernatant.
10. Protoplasts can immediately be used for TIRFM or kept in protoplast culture medium at room temperature for no longer than 24 h.

3.13 TIRFM Examination of ER Membranes in N. benthamiana

1. Protoplasts were mounted on glass slides and gently overlaid with a coverslip.
2. TIRFM was conducted using a Nikon Inverted Microscope TE2000E inverted microscope equipped with CFI Plan Apo TIRF 60×/1.45 oil and CFI Plan Apo TIRF 100×/1.45 oil objectives (Fig. 4).

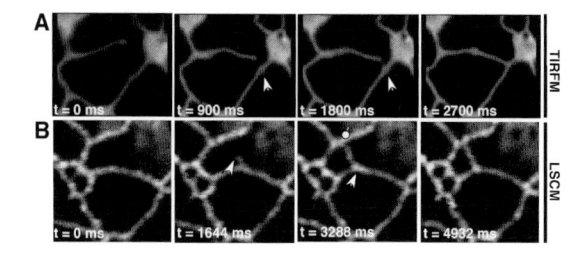

Fig. 4 Comparison of TIRFM and LSCM for examination of ER tubules in *N. benthamiana* protoplasts. (**a**) Time course analysis of ER tubule fusion determined by TIRFM. (**b**) Time course analysis of ER tubule fusion determined by LSCM. TIRFM provides superior resolution and faster image acquisition compared to LSCM, which permits novel ER membrane dynamics to be viewed with ease

3. Excitation of GFP was accomplished using the 488 nm line of a multiline argon laser.
4. Controlling software for image acquisition was Metamorph ver 6.2 (Molecular Devices Corporation, Sunnyvale, CA).

4 Notes

1. A convenient 0.1 M stock of acetosyringone can be prepared in 50% (v/v) ethanol and stored indefinitely at −20 °C.
2. Markers with indelible ink "Sharpies" or small leaf punches, made with a disposable pipette tip, can be used to mark sites of infiltration.
3. Coexpression of the tomato bushy stunt virus p19 protein, or another strong suppressor of RNA silencing can enhance protein expression levels in agroinfiltrated cells (15).
4. The effective concentrations of cell-permeant dyes used in this manner must be determined empirically. In addition to DAPI we have used CellTrace™ BODIPY® TR methyl ester for staining endomembranes (5).
5. The use of water immersion lenses is critical to successful high-magnification live-cell confocal microscopy. This is because the use of conventional oil immersion objectives when imaging cellular details and activities at a micrometer distances from the specimen cover glass often results in artifacts, including severe spherical aberration.
6. We have conducted up to seven FRAP experiments in the same ROI without any obvious photodamage effects on cells or statistically significant changes in FRAP kinetics.
7. When filtering the protoplasts through the 50 μm cloth into the 30 mL centrifuge tube, it is a great advantage to have the tip of the funnel touching the side of the tube so that the cells "slide" down into the tube. This reduces sheer forces and thus improves recovery of intact protoplasts.

Acknowledgments This work was made possible by USDA/CSREES grant 2005-01612 awarded to M.G. and NSF grant 0500710 awarded to Peter Nagy, Mark Farman, Michael Goodin, and Sharyn Perry. We also wish to thank the Nagy lab for protocols and assistance related to protoplast isolation.

References

1. Escobar NM, Haupt S, Thow G, Boevink P, Chapman S, Oparka K. (2003) High-throughput viral expression of cDNA-green fluorescent protein fusions reveals novel subcellular addresses and identifies unique proteins that interact with plasmodesmata. Plant Cell. 15:1507–1523.
2. Liu Y, Schiff M, Dinesh-Kumar SP. (2004) Involvement of MEK1 MAPKK, NTF6 MAPK, WRKY/MYB transcription factors, COI1 and CTR1 in N-mediated resistance to tobacco mosaic virus. Plant J. 38:800–809.
3. Mueller LA, Solow TH, Taylor N, Skwarecki B, Buels R, Binns J, Lin C, Wright MH, Ahrens R, Wang Y, Herbst EV, Keyder ER, Menda N, Zamir D, Tanksley SD. (2005) The SOL genomics network. A comparative resource for Solanaceae biology and beyond. Plant Physiol. 138:1310–1317.
4. Goodin MM, Dietzgen RG, Schichnes D, Ruzin S, Jackson AO. (2002) pGD vectors: versatile tools for the expression of green and red fluorescent protein fusions in agroinfiltrated plant leaves. Plant J. 31:375–383.
5. Goodin M, Yelton S, Ghosh D, Mathews S, Lesnaw J. (2005) Live-cell imaging of rhabdovirus-induced morphological changes in plant nuclear membranes. Mol Plant-Microbe Interact. 18:703–709.
6. Jackson AO, Francki RIB, Zuidema D. (1987) Biology, structure and replication of plant rhabdoviruses. In: Wagner RR (ed.). The Rhabdoviruses. Plenum Press: New York, NY.
7. Jackson AO, Dietzgen RG, Goodin MM, Bragg JN, Deng M. (2005) Biology of plant rhabdoviruses. Annu Rev Phytopathol. 43:623–660.
8. Hogenhout SA, Redinbaugh MG, Ammar el-D. (2003) Plant and animal rhabdovirus host range: a bug's view. Trends Microbiol. 11:264–271.
9. Tzfira T, Tian GW, Lacroix B, Vyas S, Li J, Leitner-Dagan Y, Krichevsky A, Taylor T, Vainstein A, Citovsky V. (2005) pSAT vectors: a modular series of plasmids for autofluorescent protein tagging and expression of multiple genes in plants. Plant Mol Biol. 57:503–516.
10. Sprague BL, McNally JG. (2005) FRAP analysis of binding: proper and fitting. Trends Cell Biol. 15:84–91.
11. Schneckenburger H. (2005) Total internal reflection fluorescence microscopy: technical innovations and novel applications. Curr Opin Biotechnol. 16:13–18.
12. Ruiz MT, Voinnet O, Baulcombe DC. (1998) Initiation and maintenance of virus-induced gene silencing. Plant Cell. 10:937–946.
13. Siemering KR, Golbik R, Sever R, Haseloff J. (1996) Mutations that suppress the thermosensitivity of green fluorescent protein. Curr Biol. 6:1653–1663.
14. Irons SL, Evans DE, Brandizzi F. (2003) The first 238 amino acids of the human lamin B receptor are targeted to the nuclear envelope in plants. J Exp Bot. 54:943–950.
15. Voinnet O, Rivas S, Mestre P, Baulcombe D. (2003) An enhanced transient expression system in plants based on suppression of gene silencing by the p19 protein of tomato bushy stunt virus. Plant J. 33:949–956.

Chapter 27
Site-Directed Mutagenesis of Whole Viral Genomes

Li Liu and George P. Lomonossoff

Abstract This chapter introduces an efficient and accurate site-directed mutagenesis protocol, which allows the color selection of mutants through the simultaneous activation or deactivation of the α-peptide of β-galactosidase. It uses double-stranded plasmid DNA as the mutational template. This protocol can efficiently create mutations of large inserts at multiple sites simultaneously and can be used to perform multiple rounds of mutation on the same construct. Thus, constructs containing whole open-reading frames and whole viral genomes can be subjected to site-directed mutagenesis and used for subsequent functional studies.

Keywords Site-directed mutagenesis; dsDNA templates; T4 DNA polymerase; T7 DNA polymerase; *Dpn*I; *Escherichia coli mut*S strain

1   Introduction

Site-directed mutagenesis has become an essential tool in the study of the structure and functions of nucleic acids and proteins. Techniques using thermostable polymerases (PCR-based methods) or T7 and T4 DNA polymerase (high fidelity methods) have undergone extensive development. PCR methods provide a quick way of obtaining mutant DNA, though it is difficult to perform simultaneous mutagenesis at multiple sites, and the approach often introduces unwanted additional mutations.

With the high fidelity methods, one of the main challenges is the selection of the mutated sequence from a background of the original template molecules. An early efficient method for achieving this involved using dU-containing single-stranded DNA as a template (1, 2). However, the requirement for dU-containing single-stranded DNA is a significant drawback as production of the template is often time-consuming and sometimes difficult. As a result, a considerable amount of effort has been put into developing methods that can use normal plasmid double-stranded DNA as a template. Among the methods that have been developed are those of Deng and Nickoloff, which involves the elimination of a unique restriction site (unique site elimination, USE) to protect the newly synthesized DNA from the

From: *Methods in Molecular Biology, Vol. 451, Plant Virology Protocols:*
From Viral Sequence to Protein Function
Edited by G.D. Foster, I.E. Johansen, Y. Hong, and P.D. Nagy © Humana Press, Totowa, NJ

restriction enzyme digestion (3, 4), Ohmori's method to repair an RNA-binding site mutation at the ColE1 replication origin to recover DNA replication in normal bacteria strains (5), Li's method, which uses a uracil-containing double-stranded DNA template and *Dpn*I treatment to reduce the background of wild-type DNA in vitro and in vivo, respectively (6), and Xin's method, which uses two rounds of denaturing, annealing, and synthesis in vitro followed by treatment with *Dpn*I to eliminate wild-type methylated and mixed hemimethylated DNA, allowing only the newly synthesized double-stranded DNA to survive the digestion (7).

Although each of the protocols has its advantages, USE requires the presence of a unique restriction site, which is not always available, Ohmori's method needs a replication-defective vector prepared in a specific bacterial strain, and the use of *Dpn*I, although quite efficient in PCR-related site-directed mutagenesis (8), has variable results with methylated and hemimethylated DNA (9). Furthermore, the two rounds of mutagenesis often produce a very low rate of newly synthesized double-stranded DNA.

One situation that often occurs is the requirement to conduct several rounds of sequential mutagenesis on the same DNA fragment. This can be done to produce a series of related mutants or to produce revertants to confirm that a phenotype is really caused by the intended mutation. To carry out further mutagenesis, recloning is frequently needed for most of the mutagenesis protocols. To avoid this, Lesley and Bohnsack developed a method, which can perform multiple rounds of mutagenesis without recloning the target DNA by using two selectable markers (10). However, as a result the vector size is substantially increased, two separate selection steps are required to select the two markers, and the simultaneous mutation efficiency is decreased.

We have recently developed a new and simple protocol to perform site-directed mutagenesis of whole viral genomes (11). The method uses α-complementation for color selection as a screenable marker, high temperature treatment for quick template denaturation, T4 or T7 DNA polymerase for high fidelity DNA extension, and *Dpn*I treatment and a mismatch-repairing defective bacterial strain to reduce background. Application of this method to a plant virus (11–13), a fungal mitochondria virus and several animal viruses (L. Liu, unpublished data) confirmed that it is an efficient way of introducing multiple mutations simultaneously and carrying out multiple rounds of mutagenesis including the production of revertants. The method should be generally suitable for the manipulation of large DNA fragments and whole viral genomes and will enable multiple rounds of mutagenesis to be performed without the need for subcloning. A comparison of the steps of this new method with some existing procedures is shown in Fig. 1.

2   Materials

2.1   *Enzymes*

All enzymes are from England Biolabs, Inc., Beverly, MA:

1. Polynucleotide kinase
2. T4 DNA polymerase

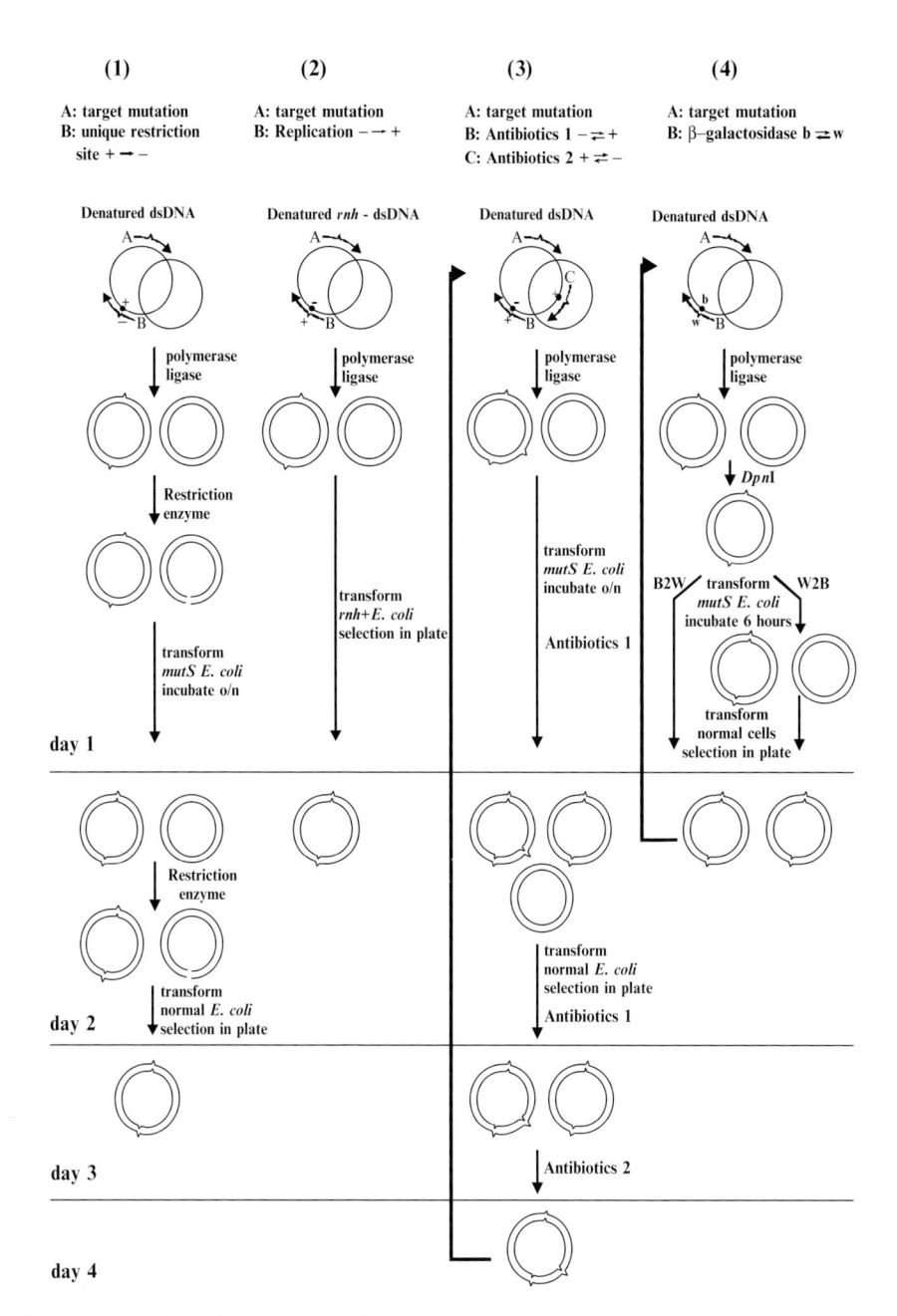

Fig. 1 Comparison of time-courses of the various mutagenesis methods that use double-stranded DNA as template. Denatured DNA templates are illustrated as two single *circles*, with primers for target mutations (**a**) and for selections (**b** and **c**) indicated by *curved arrows*. The protocols that can perform multiple rounds of mutations are indicated by a *thick arrow* showing that the cycle can be repeated. *Column 1*: Protocol based on restriction site elimination (3). The selection primer eliminates a unique restriction site. *Column 2*: Replication-defect protocol (5). The selection primer repairs a defect that allows the DNA replicate in *rnh- E. coli* cells but fails to replicate in normal cells. *Column 3*: Antibiotic resistance gene repairing protocol (10). The system uses two selection primers for antibiotics resistance, with one primer repairing defective resistance gene 1 and the other inactivating resistance gene 2. *Column 4*: α-complement-based protocol (11) described in detail here

3. T7 DNA polymerase
4. T4 DNA ligase
5. *Dpn*I, diluted to 1 U µL^{-1} with enzyme dilution buffer

2.2 Bacterial Strains

1. DH5α (Promega): φ80d*lac*ZΔM15, *rec*A1, *end*A1, *gyr*A96, *thi*-1, *hsd*R17 (r$_k^-$, m$_k^+$), *sup*E44, *rel*A1, *deo*R, Δ(*lac*ZYA-*arg*F) U169, *pho*A
2. XLblue mutS Kan^s (Stratagene): Δ(*mcr*A)183, Δ(*mcr*CB-*hsd*SMR-*mrr*)173, *end*A1, *sup*E44, *thi*-1, *gyr*A96, *rel*A1, *lac* *mut*S::Tn10 (Tet^r), [F′ *pro*AB *lac*I^{q}ZΔM15]

2.3 Solutions and Buffers

1. Suspension buffer: 50 m*M* Tris-HCl, pH 8.0; 10 m*M* EDTA
2. Alkaline buffer: 0.2N NaOH, 1% (w/v) SDS, 0.2% (v/v) Triton X-100
3. Neutralization buffer: 60% (v/v) 5 M KCl, 11.5% (v/v) glacial acetic acid
4. TE buffer: 10 m*M* Tris-HCl, pH 8.0, 1 mM EDTA
5. ATP (Sigma, St. Louis, MO) 10 mM stock, store at −20 °C
6. dNTP (Sigma, St. Louis, MO) 5 mM stock of a mixture of dATP, dCTP, dGTP, and dTTP; Store at −20 °C
7. Annealing buffer (10×): 200 mM Tris-HCl, pH 7.6, 100 mM MgCl, 500 mM NaCl; Store at −20 °C
8. Basic synthesis buffer (10×): 100 mM Tris-HCl, pH 7.6, 20 mM DTT; Store at −20 °C
9. Enzyme dilution buffer: 50 mM KCl, 10 mM Tris-HCl, pH 7.6, 0.1 mM EDTA, 1 mM DTT, 200 µg mL^{-1} BSA, and 50% glycerol; Store at −20 °C
10. X-gal (Phamacia) 2% (w/v) solution in dimethylformamide (Sigma, St. Louis, MO). Store at −20 °C
11. IPTG (Phamacia) 2% (w/v) solution in H$_2$O. Store at −20 °C

2.4 Plasmids

Two plasmids, which contain either an intact (pM81B) or a disrupted (pM81W) *lac*Z open reading frame and multicloning sites (Fig. 2), are used to subclone DNA fragments and conduct site-directed mutagenesis. The multicloning sites are flanked by restriction sites *Pac*I and *Asc*I, which are compatible with insertion into the plant transformation and agroinfection vector pBINPLUS (14). Thus the effects of target mutagenesis can readily be functionally assessed in plants.

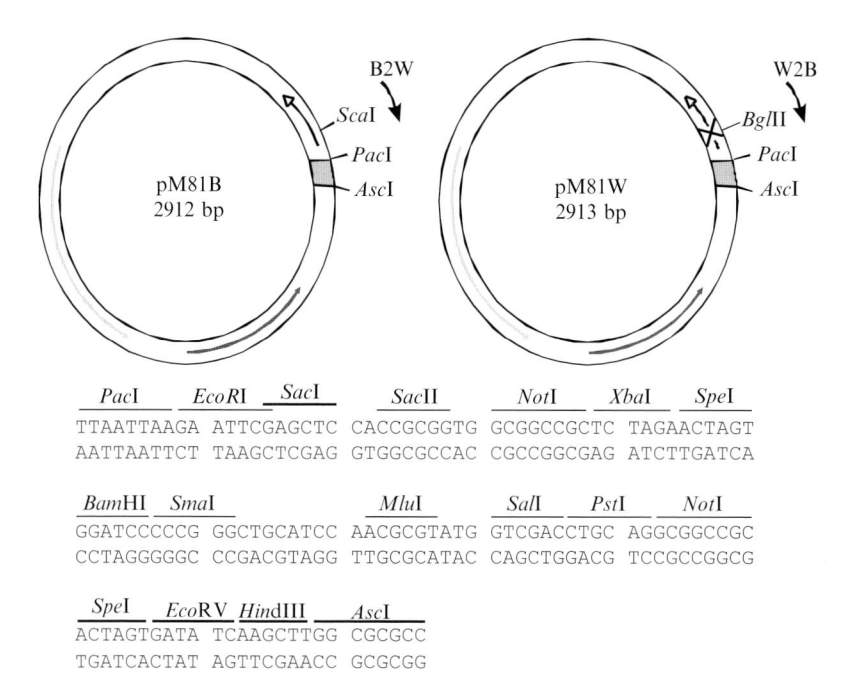

PacI	EcoRI	SacI	SacII	NotI	XbaI	SpeI
TTAATTAAGA	ATTCGAGCTC	CACCGCGGTG	GCGGCCGCTC	TAGAACTAGT
AATTAATTCT	TAAGCTCGAG	GTGGCGCCAC	CGCCGGCGAG	ATCTTGATCA

BamHI	SmaI		MluI	SalI	PstI	NotI
GGATCCCCCG	GGCTGCATCC	AACGCGTATG	GTCGACCTGC	AGGCGGCCGC
CCTAGGGGGC	CCGACGTAGG	TTGCGCATAC	CAGCTGGACG	TCCGCCGGCG

SpeI	EcoRV	HindIII	AscI
ACTAGTGATA	TCAAGCTTGG	CGCGCC
TGATCACTAT	AGTTCGAACC	GCGCGG

Fig. 2 Plasmids for site-directed mutagenesis. The plasmid is depicted by *double circles* and the *lacZ* gene is represented by an *arrow* with an open head. The screenable primers used in each plasmid are indicated outside the circles with orientations indicated. The mutation that disrupts the open reading frame of the *lacZ* gene creates a *Bgl*II site is indicated by an X, while the mutation that restores it creates a *Sca*I site. The *Col*E1 replication origin (*black arrow*) and ampicillin resistant gene (*grey arrow*) are also shown. The multiple cloning sites are indicated by a *grey box*, with two outmost restriction sites *Pac*I and *Asc*I indicated. The sequences of the multiple cloning sites are listed below the plasmid map, and the restriction enzyme names are included above their recognition sequences. Please note there are two sites of *Not*I and *Spe*I sites

2.5 Oligonucleotides

There are two selection oligonucleotide primers (forward orientation from *Pac*I to *Asc*I) designed to alternately disrupt or restore the β-galactosidase ORF. Oligonucleotide B2W (blue to white; 5′ GAC GGC CAG TGA GAT CTA TTG GCG TAA TCA TG 3′) is used with pM81B-derived plasmids to disrupt the *lacZ* gene and introduce a *Bgl*II site, and W2B (white to blue; 5′ GAC GGC CAG TGA GTA CTT TGG CGT AAT CAT G 3′) is applied with pM81W-derived plasmids to restore *lacZ* gene expression and introduce a *Sca*I site.

Target primers are designed with the orientation equivalent to that of B2W and W2B.

3 Methods

The present method makes use of α-complementation as a selecting marker. The two plasmids, pM81W and pM81B, contain the *lac*Z expression cassette upstream of the multiple cloning sites (Fig. 2). The plasmids differ in that the *Lac*Z gene in pM81B is intact, while that in pM81W is disrupted by a frame-shift mutation. Thus when plated on X-Gal-containing plates, pM81B and pM81W give blue and white colonies, respectively. Conversion between the active and inactive forms of the *Lac*Z gene can be effected in either direction by using primer W2B and B2W, as appropriate, in the mutagenesis procedure, with *Sca*I or *Bgl*II restriction sites being created alternately. If mutagenesis is carried out with a combination of the appropriate *Lac*Z-specific primer (the selection primer) and a primer specific for the target region (the target primer), blue/white selection will be linked to the introduction of the desired mutation. The color selection can be used for multiple rounds of mutagenesis by using the two *Lac*Z primers alternately.

3.1 Oligonucleotide Phosphorylation

1. Mix 1 μL (100 pmol) of primer in 2.5 μL of 10× polynucleotide kinase buffer, 2 μL of 10 mM ATP, 19.5 μL of H_2O, and 1 μL (10 U μL^{-1}) of T4 polynucleotide kinase (GIBCO BRL). The final concentration of phosphorylated oligonucle-otides is 4 pmol μL^{-1}.
2. Incubate at 37 °C for 30 min
3. Stop the reaction by incubating at 70 °C for 10 min
4. Store the phosphorylated primers at −20 °C

3.2 Template DNA Preparation

Template DNA is prepared by the mini-prep method modified from the protocol described by Sambrook et al. (15). If desired, the DNA can be further purified by using mini-prep columns (Qiagen GmbH, Germany) and its concentration determined by spectrophotometry at 260 nm (see note 1).

1. Culture bacteria in 10 mL of LB medium containing carbenicillin (100 μg mL) at 37 °C overnight
2. Collect bacteria by centrifugation for 5 min at 4,000 rpm
3. Resuspend the pellets in 200 μL of suspension buffer and transfer to Eppendorf tubes
4. Add 200 μL alkaline buffer and mix gently until the solution is clear
5. Add 450 μL neutralization buffer and gently mix
6. Centrifuge for 10 min at full speed in a micro-centrifuge

7. Transfer the supernatants to Eppendorf tubes
8. Add 600 μL of isopropanol
9. Centrifuge for 5 min
10. Wash the pellets with 70% (v/v) ethanol
11. Discard the supernatant after centrifugation for 2 min
12. Air-dry and dissolve in 100 μL of TE buffer, pH 8.0, containing 20 μg mL^{-1} RNase A

3.3 Template Denaturing and Annealing

1. Mix template plasmid DNA (0.05 pmol) (see note 1) with 2 μL of annealing buffer, 1 μL of W2B or B2W oligonucleotide, 1 μL of (each) target oligonucleotide. The final volume is 20 μL
2. Heat the mixture at 95 °C for 3 min
3. Immediately put into ice-bath for 1 min to anneal the primers and template (see note 2)

3.4 Synthesis

1. Mix 20 μL of the above mixture (see Sect. 3.3.) and 10 μL of synthesis solution, which contains 3 μL of basic synthesis buffer, 3 μL of ATP, 1.5 μL of dNTP, 2.5 U T7 DNA polymerase (see note 3), 2.5 U T4 DNA ligase
2. Incubate at 37 °C for 90 min

3.5 DpnI Treatment

1. Mix the whole reaction (30 μL see Sect. 3.4) with 1–2 U DpnI
2. Incubate at 37 °C water bath for 15–20 min (see note 4)
3. Transfer to an ice-bath
4. Transform 200 μL of XLblue mutS competent cells (transformation efficiency is more than 1 × 10^6 colonies per microgram plasmid).

3.6 Tranformation

1. Add the DpnI-treated DNA to a 1.5-mL Eppendorf tube containing 200 μL of competent XMLmutS cells (see note 5)
2. Incubate on ice for 10–30 min

3. Heat shock at 42 °C water bath for 1.5 min
4. Immediately transfer to ice
5. Keep on ice for 1.5 min

3.7 DNA Preparation

1. Transfer the transformed cells to 10 mL prewarmed LB broth containing 100 μg mL^{-1} carbenicillin
2. Incubate with shaking for 6 h (see note 6)
3. Harvest the cells by centrifugation
4. Transfer bacteria to a 1.5-mL Eppendorf tube
5. Extract plasmid DNA on a mini-prep scale (see Sect. 3.2) and dissolve in 30 μL TE buffer, pH 8.0, containing 20 μg mL^{-1} RNAse A

3.8 Selection

To obtain a homogenous population of mutant DNA, the extracted DNA is used to transform 100 μL of *E. coli* strain DH5α for color selection. After transformation (see Sect. 3.4), cells are transferred to a Petri dish containing 20 mL of LB-agar with carbenicillin (100 μg mL^{-1}), X-gal (40 μL stock), and IPTG (5 μL stock; see note 7). After overnight incubation at 37 °C, blue or white colonies, according to which screenable oligonucleotide was used, are selected for confirmation of the incorporation of the target mutations.

4 Notes

1. DNA dissolved in TE buffer containing RNase A (20 μg mL^{-1}) can be used without further purification without affecting the mutagenesis efficiency.
2. Better results can be obtained for large constructs (>6 kp) when the DNA is first denatured by NaOH using the following procedure: DNA sample containing 4 μL of 2N NaOH in a total 20 μL is denatured at room temperature for 5 min, and precipitated by adding 10 μL of 3 M NaOAc and 100 μL of 100% ethanol. The sample is centrifuged at 13,000 rpm for 15 min, washed once with 70% ethanol, and dissolved in up to 16 μL H$_2$0. Continue with the standard protocol by adding primer and annealing buffer followed by the heat treatment.
3. T4 DNA polymerase (2.5 U per reaction) can be used for the site-directed mutagenesis with the same high efficiency.
4. As the concentration of commercial *Dpn*I is too high (normally 10–20 U μL^{-1}), dilutions using enzyme dilution buffer are needed to produce a 1 U μL^{-1} *Dpn*I stock, e.g., 1 μL of *Dpn*I (20 U μL^{-1}) is added with 19 μL of enzyme dilution buffer. The stock can be stored at −20 °C for several months.
5. Competent cells are prepared by a modified one-step DMSO method developed by Chung et al. (16). Fifty microliters of cell stock (in 50% glycerol) stored at −20 °C are incubated in 10 mL

of LB medium at 37 °C overnight. Then 1 mL of freshly incubated cells is transferred to a flask containing 100 mL of LB medium and incubated at 37 °C for 3–4 h (or until the OD600 reaches 0.5–0.6). The cells are harvested by centrifugation at 2,000 rpm for 10 min and resuspended in 2 mL of LB, which is then added to a mixture of 6 mL LB, 0.5 mL 1 M $MgSO_4$, 0.5 mL DMSO (taken from a 1 mL aliquot of a fresh batch of DMSO stored at −80 °C), 2 mL 50% PEG3500, and 1 mL glycerol. The cells are divided into 220 µL aliquots in 1.5-mL Eppendorf tubes and stored at −80 °C.

6. When the primer B2W is used in the mutagenesis reaction, homogeneous mutants can be selected on a Petri dish containing X-gal and IPTG without the need to extract the DNA. After the incubation with shaking for 3–4 h at 37 °C, 1–5 µL (depending on the transformation efficiency) of the cells are mixed with 200 µL H_2O and transferred to a Petri dish, and carry on the remaining procedures of Sect. 3.8.

7. X-gal and IPTG can be freshly added to the cells using the following steps: the cells are mixed with 100 µL H_2O, then with 40 µL of X-gal and 5 µL of IPTG and transferred to a Petri dish containing 20 mL of LB-agar with carbenicillin (100 µg mL^{-1}).

Acknowledgments This work was funded under the BBSRC LINK scheme, the EC Framework 5 Quality of Life Programme (contract No. QLK2-CT-2002-01050) and the EC Framework 6 "Pharma-planta" project.

References

1. Kunkel, T. A., Bebenek, K., and McClary, J. (1991) Efficient site-directed mutagenesis using uracil-containing DNA, *in* "Methods in Enzymology", Academic Press, Inc., pp. 125–139.
2. Kunkel, T. A. (1985) Rapid and efficient site-specific mutagenesis without phenotypic selection. *Proc. Natl. Acad. Sci. USA* **82**, 488–492.
3. Deng, W. P., and Nickoloff, J. A. (1992) Site-directed mutagenesis of virtually any plasmid by eliminating a unique site. *Anal. Biochem.* **200**, 81–88.
4. Wong, F., and Komaromy, M. (1995) Site-directed mutagenesis using thermostable enzymes. *BioTechniques* **18**, 1034–1038.
5. Ohmori, H. (1994) A new method for strand discrimination in sequence-directed mutagenesis. *Nucleic Acids Res.* **22**, 884–885.
6. Li, F., Liu, S. L., and Mullins, J. I. (1999) Site-directed mutagenesis using uracil-containing double-stranded DNA templates and DpnI digestion. *Biotechniques* **27**, 734–738.
7. Xin, W., Huang, D. W., Zhang, Y. M., and Geng, L. (2004) DNA mutagenesis using T4 DNA polymerase and DpnI restriction endonuclease. *Anal. Biochem.* **329**, 151–153.
8. Weiner, M. P., Costa, G. L., Schoettlin, W., Cline, J., Mathur, E., and Bauer, J. C. (1994) Site-directed mutagenesis of double-stranded DNA by the polymerase chain reaction. *Gene* **151**, 119–123.
9. Lu, L., Patel, H., and Bissler, J. J. (2002) Optimizing *Dpn*I digestion conditions to detect replicated DNA. *Biotechniques* **33**, 316–318.
10. Lesley, S. A., and Bohnsack, R. N. (1994) Site-directed mutagenesis using the altered sites II systems, *in* "Promega Notes", Vol. 46, Madison, WI, p.06.
11. Liu, L., and Lomonossoff, G. P. (2006) A site-directed mutagenesis method utilising large double-stranded DNA templates for the simultaneous introduction of multiple changes and sequential multiple rounds of mutation: application to the study of whole viral genomes. *J. Virol. Met.* **137**, 63–71.
12. Liu, L., Canizares, M. C., Monger, W., Perrin, Y., Tsakiris, E., Porta, C., Shariat, N., Nicholson, L., and Lomonossoff, G. P. (2005) Cowpea mosaic virus-based systems for the production of antigens and antibodies in plants. *Vaccine* **23**, 1788–1792.

13. Liu, L., Grainger, J., Canizares, M. C., Angell, S. M., and Lomonossoff, G. P. (2004) Cowpea mosaic virus RNA-1 acts as an amplicon whose effects can be counteracted by a RNA-2-encoded suppressor of silencing. *Virology* **323**, 37–48.
14. van Engelen, F. A., Moltoff, J. W., Conner, A. J., Nap, J.-P., Pereira, A., and Stiekema, W. J. (1995) pBINPLUS: an improved plant transformation vector based on pBIN19. *Transgenic Res.* **4**, 288–290.
15. Sambrook, J., Fritsch, E. F., and Maniatis, T. (Eds.) (1989) Molecular cloning: A laboratory manual. Second edition. Cold Spring Harbor Laboratory Press, Cold Spring Harbor Laboratory Press.
16. Chung, C. T., Niemela, S. L., and Miller, R. H. (1989) One-step preparation of competent *Escherichia coli*: Transformation and storage of bacterial cells in the same solution. *Proc. Natl. Acad. Sci. USA* **86**, 2172–2175.

Chapter 28
Viral Protein–Nucleic Acid Interaction: South (North)-Western Blot

Huanting Liu

Abstract Maize streak virus (MSV) genome has four open reading frames. C1 and C2 encoded by the complementary sense are required for virus replication, while V1 and V2 encoded by virion sense are required for infectivity. V1 encodes movement protein (MP), while V2 encodes coat protein (CP). Deletion or mutation of MSV CP does not prevent virus replication in single cells or protoplasts but leads to a loss of infectivity in the inoculated plant suggesting that MSV CP is required for virus movement. Towards understanding the role of MSV CP and MP in virus movement, the interaction of MSV CP and MP with viral DNA was investigated using the South-western assay. Wild type and truncated MSV CPs and MP were expressed in *E. coli* and the expressed CPs and MP were used to investigate interaction with single-stranded (ss) and double-stranded (ds) DNA. The results showed MSV MP does not bind DNA in the assay while MSV CP bound ss and ds viral and *uid*A DNA in a sequence non-specific manner.

Keywords Maize streak virus; Coat protein; Movement protein; South-western blotting; North-western blotting; CP-DNA interaction

1 Introduction

Interactions of viral proteins with nucleotides and nucleic acids are important in a number of respects. They are involved in encapsidation (1–3), viral gene expression and regulation (4), and also provide a means for the virus to subvert the cellular environment in the host thus ensuring optimal conditions for replication (5–9). Replication of the virus is absolutely dependent on such proteins and they are often smaller and less complicated in terms of their subunit composition than cellular analogues, providing easy targets to analyze the replication mechanism (10–12). Unlike animal viruses, plant viruses move from infected cells into adjoining healthy cells through the plasmodesmata. This process is mediated by virus-encoded movement protein (13). It is well documented that the MPs interact with viral DNA or RNA and move viral DNA or RNA from cell to cell. Moreover, for

From: *Methods in Molecular Biology, Vol. 451, Plant Virology Protocols:*
From Viral Sequence to Protein Function
Edited by G.D. Foster, I.E. Johansen, Y. Hong, and P.D. Nagy © Humana Press, Totowa, NJ

some plant DNA viruses, movement protein also moves the viral DNA into the nucleus for replication. Biological and biochemical analysis of these interactions led to a clear understanding how plant virus spreads within its hosts (13–18).

Methods have been developed to analyze protein–nucleic acid interactions. Band shift (gel retardation) or electrophoretic mobility shift assay (EMSA) is the most widely used method. In this method, a labelled nucleic acid fragment containing a recognition site for a DNA-binding protein is incubated with a protein extract and then subjected to non-denaturing polyacrylamide gel electrophoresis (PAGE). The binding of a protein to nucleic acid is identified by an altered electrophoretic mobility of the nucleic acid (19, 20). However, the protein, which is analyzed using this method, must electrophorese as a single band under non-denaturing conditions. A protein that has multimeric forms under non-denaturing conditions or has a poor solubility is difficult to analyze using EMSA. To analyze these multimeric or insoluble proteins, an alternative method called the 'South-western' assay (21) is developed. Proteins are first fractionated by SDS-PAGE and then transferred onto a membrane. After renaturation, a labelled nucleic acid probe is incubated with the membrane and the protein–nucleic acid interaction is identified by detection of the labelled probe on the protein bands. In South-western assay, a labelled DNA probe is used, while in north-western assay a RNA probe is used. A number of viral DNA or RNA-binding proteins have been identified and analyzed using South(north)-western analysis (1, 2, 6, 8, 14, 18, 22–24).

The protocols presented in this chapter describe the analysis of MSV CP- and MP-DNA interactions using South-western assay and the identification of the CP-DNA binding domain using truncated CP derivatives.

2 Materials

2.1 Virus Isolate

A cloned Nigerian isolate of MSV, pMSV-Ns (25), was the source of all MSV DNA used throughout the work described.

2.2 Protein Expression and E. coli Strains

1. E. coli BL21(DE3) chemical competent cells.
2. Expression constructs pETCPwt, pETCP201 (MSV CP with 20 aa truncated from the N-terminus), pETCP801 (MSV CP with 80 aa truncated from the N-terminus), pETCP214 (MSV CP with 140 aa truncated from the C-terminus) are as described (23) and pETMP, a construct expressing His-tagged MSV MP is as described (26).

3. L-broth medium: Dissolve 10 g of tryptone, 5 g of yeast extract, 5 g of NaCl in 900 mL of distilled water and adjust to pH 7.4, add water upto 1 L and autoclave.

4. Kanamycin stock (50 mg mL^{-1}): Dissolve 50 mg of kanamycin (Melford Laboratories Ltd, UK) in 0.5 mL of sterilized water and then top up to 1 mL, stored in aliquots at −20°C, added to medium or agar dishes as required.

5. L-Agar-Kan dishes: L-agar dishes containing 50 μg mL^{-1} of kanamycin.

6. 1 M IPTG solution: Dissolve 0.238 g of IPTG (Melford Laboratories Ltd, UK) in 0.5 mL sterilized water, adjust the volume to 1 mL and stored in single use aliquots at −20°C.

7. Shaking incubator.

2.3   SDS-PAGE

1. Thirty percent acrylamide/bis solution (29:1, 3.3% C, this is a neurotoxin when unpolymerized, handle with care and not to receive exposure).

2. 1.5 M Tris-HCl, pH 8.8: Dissolve 18.16 g of Trizma base (Sigma) in 80 mL distilled water, adjust pH to 8.8 using 5 M HCl and top up to 100 mL with distilled water.

3. 1.0 M Tris-HCl, pH 6.8: Dissolve 12.11 g of Trizma base (Sigma) into 80 mL distilled water, adjust pH to 6.8 using 5 M HCl and top up to 100 mL with distilled water.

4. 10% SDS: Dissolve 10 g of sodium dodecyl sulphate (Sigma) into 100 mL of distilled water (see note 1).

5. Ammonium persulfate: 10% solution in water and immediately freeze in single use aliquots (200 μL) at −20°C.

6. N,N,N,N'-Tetramethylethylenediamine (TEMED, Sigma, see note 2).

7. Water-saturated isobutanol: Mix equal volumes of water and isobutanol in a glass bottle and allow separation. Use the top layer.

8. Running buffer (5×): 125 mM Tris, 960 mM glycine, 0.5% (w/v) SDS. Store at room temperature.

9. Prestained protein markers (Invitrogen).

10. 2× Laemmli loading buffer: 100 mM Tris-HCl pH 6.8, 4% SDS, 20% glycerol, 200 mM α-mercaptoethanol and 0.03% (w/v) bromophenol blue (27). Store in aliquots at −20°C.

11. Coomassie stain solution (1 L): Dissolve 1.0 g of Coomassie blue R-250 (Sigma) in 450 mL of methanol, add 450 mL of water and 100 mL of glacial acetic acid, filter through 3MM Whatman filter disc. Store at room temperature.

12. Destain solution (1 L): Mix 200 mL of methanol with 100 mL of glacial acetic acid and 800 mL of water. Store at room temperature.

13. Mini-PROTEAN II gel system (Bio-Rad).

2.4   Synthesis of Virus-Specific Radioactive-Labelled Probes

1. DNA labelling kit (Ready-To-Go™, Amersham).
2. MSV sequence-specific probe DNA fragments (2.7 Kbp *BamH* I fragment from pMSV-Ns).
3. Non-MSV sequence *uid*A (*gus*) gene (1.8 Kbp *gus* gene fragment excised from pJIT166) (28).
4. α-[^{32}P]-dCTP (111 TBq mmol^{-1}, 370 kBq µL^{-1}, Dupont).
5. MicroSpin Columns (S-200, Amersham).
6. Heating block.

2.5   South-Western Blot

1. MSV CPs and MP samples, *E. coli* singles-stranded DNA binding protein (SSB, 1 mg mL^{-1}), fraction V bovine serum albumin (BSA, 1 mg mL^{-1}), lysozome (1 mg mL^{-1}) and cytochrome C (1 mg mL^{-1}).
2. Transfer buffer: 39 mM glycine, 48 mM Tris, 20% methanol and 0.037% SDS. Store at room temperature.
3. Nitrocellulose membrane (Millipore), 3MM Chr chromatography paper (Whatman).
4. Renaturation buffer: 50 mM Tris-HCl pH 7.5, 100 mM KCl, 1% Triton X-100, 10% glycerol and 1 mM ZnCl$_2$.
5. Reaction buffer: 50 mM Tris-HCl pH 7.5, 0.1% Triton X-100, 10% glycerol, 0.1 mM ZnCl$_2$ and 250 mM KCl.
6. Platform shaker.
7. Simi-Dry Transfer Cells (Bio-Rad).
8. X-ray film (Fuji).
9. Film developer.

2.6   Radioactive Probe Stripping

1. Radioactive probe stripping buffer: 500 mM KCl, 0.1% Triton X-100.
2. 1× PBS.

2.7   Western Blotting

1. Tris-buffered saline (TBS): Prepare 10× stock with 1.37 M NaCl, 27 mM KCl, 250 mM Tris-HCl, pH 7.4. Dilute 100 mL of stock with 900 mL of water for use.
2. Tris-buffered saline with Tween (TBS-T): Prepare 10× stock with 10× TBS plus 1% Tween-20. Dilute 100 mL of stock with 900 mL of water for use.

3. Blocking buffer: 5% (w/v) non-fat dry milk in 1× TBS-T.
4. Antibody dilution buffer: 1× TBS-T supplemented with 2% BSA (w/v) and 0.02% sodium azide.
5. Anti-MSV and NSV MP sera (26) (see note 3).
6. Secondary antibody: Anti-rabbit IgG-alikline phosphatase (Sigma, see note 4)
7. Alkaline phosphatase buffer: 100 mM NaCl, 5 mM $MgCl_2$, 100 mM Tris-Cl, pH 9.5.
8. NBT stock solution: Dissolve 0.5 g of nitro blue tetrazolium in 10 mL of 70% dimethylformamide. Stock is stable in room temperature.
9. BCIP stock solution: Dissolve 0.5 g of 5-bromo-4-chloro-3-indolyl phosphate in 10 mL of 100% dimethylformamide. Stock is stable in room temperature.
10. Platform shaker.

3 Methods

The full process of South(north)-western blotting includes three segments, protein sample preparation, protein–nucleic acid interaction and protein identification. Protein samples used for South-western or north-western analysis can be obtained from different sources. For virus structure proteins, such as virus coat protein, these samples can be obtained from the disrupted virus particles or even obtained from the infected tissue. For most non-structural proteins, recombinant protein technique is a commonly used method to produce those samples. Many expression systems have been developed to express the recombinant protein in *E. coli*, insect cells, yeast, mammalian and plant cells. The system used to produce a protein is solely dependent on the research purpose. For example, if post-translational modification is important, eukaryotic expression system should be used. In South-western analysis, *E. coli* expression system is a commonly used method to produce the target proteins, especially the pET system (29), which is one of the most popular system used because of its efficiency and flexibility. Protein–nucleic acid interaction is usually investigated using radioactive-labelled probes. Recently, non-radioactive probes are applicable, although the sensitivity sometime is not as high as the radioactive probes. Identification of the target proteins is important if a protein is overexpressed using recombinant techniques. The most commonly used method is immunological identification using a specific antibody. If a specific antibody is unavailable, expression of a protein with a tag could be an advantage by which the target protein can be identified by using commercially available antibody raised against the attached tag (such as Hexahistidines). If equipments are available, the expressed protein can be identified by mass spectrometric analysis (30).

In South(north)-western analysis, it is important that control proteins are applied to the assay. Single-stranded DNA-binding protein (SSB) and bovine serum albumin (BSA) are commonly used as controls. However, positive-charged lysozome and cytochrome C possess advantage to indicate if the protein–nucleic acid interaction is simply caused by basic charges under the experimental conditions.

3.1   Preparation of the MSV CP and CP Derivatives and MP (see note 5)

3.1.1   Transformation of *E. coli* Strains

1. Pipette 50 μL of *E. coli* BL21 (DE3) competent cells into six 1.5 mL pre-chilled Eppendorfs. Add 10 ng of constructs pETCPwt, pETCP201, pETCP801, pETCP214, pETMP and pET3a vector (mock) DNA, respectively.
2. Mix gently, leave on ice for approximately 30 min.
3. Set water bath at 42°C.
4. Heat-shock the cells by incubating the tubes in the water bath (42°C) for 1.5 min and then quickly moving back on ice for 15 min.
5. Add 0.3 mL of LB into the tubes and incubate in a shaking incubator for 1 h at 37°C.
6. Spread 0.1 mL of transformed cells onto an L-agar-Kan dish and incubate overnight at 37°C.
7. Grow overnight cultures by inoculating a single colony into 10 mL of LB containing 50 μg mL^{-1} of kanamycin in a universal and incubating overnight at 37°C.

3.1.2   Expression of MSV CP and MP

1. Prepare 12 universals with 10 mL of L-broth containing 50 μg mL^{-1} of kanamycin.
2. Inoculate two universals for each construct, respectively, with 100 μL of overnight culture (Sect. 3.1.1, step 5).
3. Grow the culture at 37°C with a shaking speed of 180 rpm for ~3 h (monitor the cell density to reach 0.6 of OD$_{600}$).
4. Induce the protein expression by adding 4 μL of 1 M IPTG into one of the universal and incubate for another 4 h at 37°C, leave another for un-induced control (see note 6).
5. Check the protein expression by SDS-PAGE (Sect. 3.2).
6. Aliquot 200 μL of the cell culture into 0.5 mL Eppendorfs, harvest the cells by centrifugation at 13,000 rpm for 2 min at 4°C, remove and dispose the supernatant after being treated with disinfectant.
7. Keep the pellets at −70°C.

3.2   SDS-PAGE Analysis

These instructions assume the use of a Bio-Rad Mini-PROTEAN II gel system. They are easily adaptable to other formats. The glass plates for the gels are required to scrub clean with a rinsable detergent after use and rinsed extensively with distilled water. They can be kept clean until use in a plastic rack in 70% ethanol and air-dry before use.

1. Mount the glass plates with 1.5 mm spacers onto the clamps and assemble onto a casting stand according to the instruction manual.

2. Prepare two 1.5-mm thick, 12% gel (10 mL) by mixing 4 mL of acrylamide/bis solution, 3.3 mL of water, 2.5 mL of 1.5 M Tris-HCl pH 8.8, 100 μL of 10% SDS, 100 μL of ammonium persulfate solution and 10 μL of TEMED (Table 1). Mix gently and pour the gel (leave space for a stacking gel) and over-lay with water-saturated isobutanol. The gel should polymerise in about 20 min (see note 7).

3. Pour off the isobutanol and rinse the top of the gel twice with water and remove the water carefully.

4. Prepare the stacking gel (3 mL) by mixing 0.5 mL of acrylamide/bis solution, 2.1 mL of water, 380 μL of Tris-HCl pH 6.8, 30 μL of 10% SDS, 30 μL of 10% ammonium persulfate solution and 5 μL of TEMED (Table 2). Pour the stack gel and insert the comb. The stacking gel should polymerise within 15 min.

5. Prepare the running buffer by diluting 100 mL of the 5× running buffer with 400 mL of water in a measuring cylinder. Seal with parafilm and invert to mix.

6. Remove the protein samples from the freezer, resuspend each sample in 50 μL 2× Laemmli loading buffer and boil for 10 min. For BSA, lysozome and cyto-chrome C, 4 μg of protein sample was mixed with 20 μL of loading buffer and boiled for 3 min (see note 8).

7. Once the stacking gels have set, carefully remove the comb and use a 5 mL syringe fitted with a 22-gauge needle to wash and clean the wells with running buffer.

8. Remove the gel clamps from the casting stand and assemble onto the clamp frame, put the frame into the running tank.

9. Add the running buffer to the clamp frame and the buffer tank (wells should be submerged). Load the wells with 10 μL of each sample and one well with pre-stained protein markers.

10. Connect the lid to clamp frame and connect to a power supply. Run the gel at 180 V for about 40 min or stop when the dye fronts (blue) reach the gel end. Stain the gels (Step 11, 12) or transfer the proteins onto the nitrocellulose membrane (see Sect. 3.4).

11. Disassemble the running unit and remove the glass plates from the clamps and carefully remove the gels into 20 mL of Coomassie stain solution and stain for 15 min (mark each gel).

12. Destain the gel in 20 mL of destain solution and change the solution twice until the protein bands are clear. Expression of MSV MP and CPs are shown in Fig. 1a.

3.3 Synthesis Radioactive Labelled DNA Probes (see Note 9)

1. Remove 2 Reaction Mixes of Ready-to-Go labelling kit, and check whether the bead is visible in the bottom of the tube. If not, tap against a hard surface to bring the bead to the bottom of the tube.

Table 1 Solutions for preparing resolving gels for Tris-glycine SDS-polyacrylamide gel electrophoresis

Solution components	Component volumes (mL) per gel mold volume of							
	5 mL	10 mL	15 mL	20 mL	25 mL	30 mL	40 mL	50 mL
6%								
H_2O	2.6	5.3	7.9	10.6	13.2	15.9	21.2	26.5
30% acrylamide mix	1.0	2.0	3.0	4.0	5.0	6.0	8.0	10.0
1.5 M Tris (pH 8.8)	1.3	2.5	3.8	5.0	6.3	7.5	10.0	12.5
10% SDS	0.05	0.1	0.15	0.2	0.25	0.3	0.4	0.5
10% ammonium persulfate	0.05	0.1	0.15	0.2	0.25	0.3	0.4	0.5
TEMED	0.006	0.012	0.018	0.024	0.028	0.036	0.046	0.056
8%								
H_2O	2.3	4.6	6.9	9.3	11.5	13.9	18.5	23.2
30% acrylamide mix	1.3	2.7	4.0	5.3	6.7	8.0	10.7	13.3
1.5 M Tris (pH 8.8)	1.3	2.5	3.8	5.0	6.3	7.5	10.0	12.5
10% SDS	0.05	0.1	0.15	0.2	0.25	0.3	0.4	0.5
10% ammonium persulfate	0.05	0.1	0.15	0.2	0.25	0.3	0.4	0.5
TEMED	0.005	0.010	0.015	0.020	0.025	0.030	0.038	0.046
10%								
H_2O	1.9	4.0	5.9	7.9	9.9	11.9	15.9	19.8
30% acrylamide mix	1.7	3.3	5.0	6.7	8.3	10.0	13.3	16.7
1.5 M Tris (pH8.8)	1.3	2.5	3.8	5.0	6.3	7.5	10.0	12.5
10% SDS	0.05	0.1	0.15	0.2	0.25	0.3	0.4	0.5
10% ammonium persulfate	0.05	0.1	0.15	0.2	0.25	0.3	0.4	0.5
TEMED	0.004	0.008	0.012	0.016	0.02	0.024	0.030	0.036
12%								
H_2O	1.6	3.3	4.9	6.6	8.2	9.9	13.2	16.5
30% acrylamide mix	2.0	4.0	6.0	8.0	10.0	12.0	16.0	20.0
1.5 M Tris (pH 8.8)	1.3	2.5	3.8	5.0	6.3	7.5	10.0	12.5
10% SDS	0.05	0.1	0.15	0.2	0.25	0.3	0.4	0.5
10% ammonium persulfate	0.05	0.1	0.15	0.2	0.25	0.3	0.4	0.5
TEMED	0.004	0.008	0.012	0.016	0.02	0.024	0.030	0.036
15%								
H_2O	1.1	2.3	3.4	4.6	5.7	6.9	9.2	11.5
30% acrylamide mix	2.5	5.0	7.5	10.0	12.5	15.0	20.0	25.0
1.5 M Tris (pH 8.8)	1.3	2.5	3.8	5.0	6.3	7.5	10.0	12.5
10% SDS	0.05	0.1	0.15	0.2	0.25	0.3	0.4	0.5
10% ammonium persulfate	0.05	0.1	0.15	0.2	0.25	0.3	0.4	0.5
TEMED	0.004	0.008	0.012	0.016	0.02	0.024	0.030	0.036

Table 1 and 2 are modified from (31)

2. Denature 45 µL (50 ng) of MSV DNA and *uid*A (gus) DNA in an Eppendorf, respectively, by heating for 2–3 min at 95–100°C.
3. Immediately place on ice for 2 min, then centrifuge briefly.
4. Add the following to the tubes containing the Reaction Mix bead: 45 µL of denatured MSV or *uid*A DNA (25–50 ng); 5 µL of [α-^{32}P] dCTP (3,000 Ci

Table 2 Solutions for preparing 5% stacking gels for Tris-glycine SDS-polyacrylamide gel electrophoresis

Solution components	Component volumes (mL) per gel mold volume of							
	1 mL	2 mL	3 mL	4 mL	5 mL	6 mL	8 mL	10 mL
H2O	0.68	1.4	2.1	2.7	3.4	4.1	5.5	6.8
30% acrylamide mix	0.17	0.33	0.5	0.67	0.83	1.0	1.3	1.7
1.0 M Tris (pH 6.8)	0.13	0.25	0.38	0.5	0.63	0.75	1.0	1.25
10% SDS	0.01	0.02	0.03	0.04	0.05	0.06	0.08	0.1
10% ammonium persulfate	0.01	0.02	0.03	0.04	0.05	0.06	0.08	0.1
TEMED	0.002	0.003	0.004	0.005	0.006	0.007	0.008	0.012

Tables 1 and 2 are modified from (31)

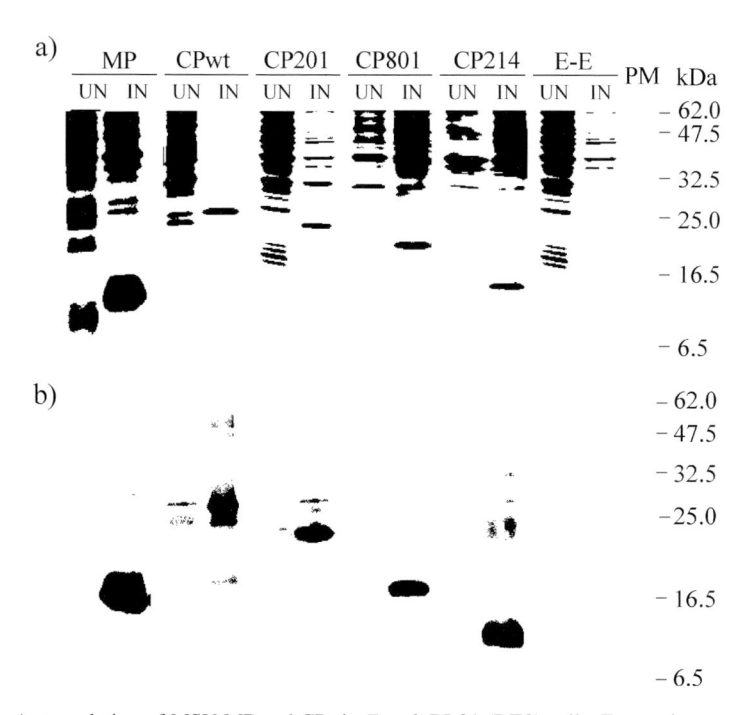

Fig. 1 Accumulation of MSV MP and CPs in *E. coli* BL21 (DE3) cells. Expression was carried out as described in Sect. 3.1.2 with 0.4 mM IPTG induction (IN) or without induction (UN). (**a**) Total cellular proteins were fractionated by SDS-PAGE using a 12% SDS-polyacrylamide gel and stained with 0.15% Coomassie blue. (**b**) Western blotting of the expressed MSV MP and CPs using anti-MSV MP or CP serum. MP: *E. coli* cells expressing MSV; MP: CPwt: *E. coli* cells expressing wild type CP; CP201, CP801 and CP214: *E. coli* cells expressing MSV CP truncation derivative CP201, 801 and 214, respectively; MP: *E. coli* cells expressing MSV MP; E-E: *E. coli* cells transformed with pET3a; PM: protein markers.

mmol^{-1}, 50 μCi), mix by gently pipetting up and down several times, removed by a pulse centrifugation.

5. Incubate at 37°C for 15 min (see note 10).

6. To remove the unlabelled radioactive nucleotide using MicroSpin column, resuspend the resin in the column by vortexing.
7. Loosen the cap one-fourth turn and snap off the bottom closure.
8. Place the column in a 1.5 ml screw-cap microcentrifuge tube for support. Alternatively, cut the cap from a flip-top tube and use this tube as a support.
9. Pre-spin the column at 735g (e.g., 3,000 rpm in an Eppendorf model 5415C variable-speed centrifuge with an 18-position fixed-angle rotor) for 1 min. Start the timer at the same time as you start the microcentrifuge.
10. Place the column in a new 1.5 mL tube, remove and discard the cap and slowly apply the sample to the top-centre of the resin, being careful not to disturb the bed. Spin the column at 735g for 2 min. Collect the double-stranded (ds) probe from the bottom of the support tube.
11. Transfer 20 μL of ds probe into a new tube, heat for 3 min at 100°C and quickly cooled on ice, diluted twice with SDW and used as single-stranded (ss) probe.

3.4 South-Western Analysis

1. Run 4 SDS-PAGE as described in Sect. 3.2 with lanes loaded with pre-stained protein markers, 2 μg of BSA/lysozome and 2 μg of cytochrome C.
2. Soak the gel for 30 min in transfer buffer.
3. Transfer the target proteins to four nitrocellulose filters, respectively, using Semi-Dry Transfer Cell (can transfer four gels simultaneously).

   a. Wet three layers of Whatman paper with transfer buffer and put on the transfer base without any bubble formation.
   b. Put a pre-wetted filter on the top of the papers and then place the gel, another three layers of wet Whatman paper with transfer buffer.
   c. Align the papers with the gel and make sure that there is no bubble within each layer.
   d. Put on the cathode and safety lid.
   e. Set running current about 5 mA cm^{-2} of the gel and voltage not more than 25 V.
   f. Transfer for about 40 min (see note 11).

4. Open the Transfer Cell and check the transfer efficiency with the pre-stained protein markers on the membrane before removing the membrane.
5. Renature the proteins in a small container by incubating the membrane for 10 h at 4°C in 20 mL of renaturation buffer.
6. Incubate the membrane in 20 mL of reaction buffer for another 2 h.
7. Dispose the reaction buffer, add 10 mL of new reaction buffer containing 20 μL of MSV-specific or non-MSV ds probe and 40 μL of diluted ss probe, respectively, into each labelled container and incubate for further 3 h at 4°C. Cover the containers with lead box to prevent the radiation (see note 12).
8. Dispose the reaction buffer into radioactive waste container and wash the membrane five times with reaction buffer at 4°C, each for 10 min.
9. Dry the membrane, and expose to X-ray film at −70°C for 2 h.

10. Develop the film and check the autoradiographs. MSV CP bound ss and ds viral and *uid*A DNA in a sequence nonspecific manner (Fig. 2),while MSV MP does not bind DNA in the assay (Fig. 3a, c). The DNA-binding domain of MSV CP is mapped at the N-terminus of the protein (Fig. 4a, b).

3.5 Stripping the Radioactive Probes

1. Wash the membrane in 20 mL of 500 mM KCl, 0.1% Triton X-100 at room temperature, change the buffer until no radioactivity was detected.
2. Incubate in membrane in 1× TBS for 10 min and kept for Western blot analysis.

3.6 Western Hybridization Analysis

1. Incubate the membrane in 1× TBS for 10 min.
2. Block the filter by putting it into a sealed plastic bag (or a small container) containing 10 mL of blocking buffer and gently agitate on a platform shaker for 2 h.
3. Discard the blocking buffer and immediately put 10 mL of antibody dilution buffer containing 1:1,000 diluted anti-MSV CP or MP serum into the bag, seal the bag and gently agitate on a platform shaker over night at 4°C or 2–4 h at room temperature.
4. Wash the filter with blocking buffer for four times at room temperature, each for 10 min and agitated on platform shaker.
5. Put the filter into another plastic bag containing 10 mL of antibody dilution buffer plus 1:1,000 diluted enzyme-coupled goat anti-rabbit IgG. Seal the bag and agitate on the platform shaker for 2 h at room temperature.
6. Wash the filter four times with 1× TBS-T in a container, each for 10 min on the platform shaker at room temperature (see note 13).
7. Incubate the filter with alkaline phosphatase buffer on platform shaker for 10 min at room temperature.
8. Add 66 µL of NBT stock in 10 µL of alkaline phosphatase buffer, mix well and then add 33 µL of BCIP stock (This chromogenic substrate mixture should be used within 30 min.), put the filter into the substrates agitated on platform shaker at room temperature.
9. Monitor the progress of the reaction carefully. When the bands are of the desired intensity, transfer the filter to a tray containing 50 mL of TBS containing 200 µL of 0.5 M EDTA, pH 8.0.
10. Image the filter and analyze the results. To detect the expressed MSV CPs and MP by Western blotting, the SDS-PAGE fractionated proteins were blotted onto the nitrocellular membrane and Western blotting was carried out as described (Fig. 1b). The CP derivatives and MP was analyzed by Western blotting after South-western assays to confirm the presence of the target proteins (Figs. 3c,d and 4c).

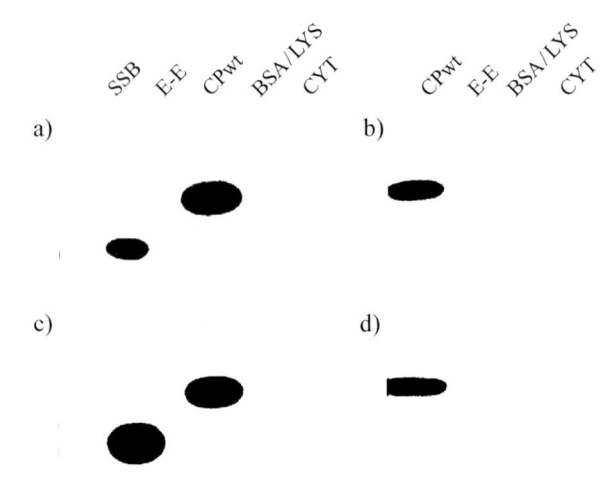

Fig. 2 South-western blotting analysis of MSV CP-DNA interaction. Autoradiographs were obtained using (**a**) MSV dsDNA probe, (**b**) *uid*A gene dsDNA probe, (**c**) MSV ssDNA probe or (**d**) *uid*A gene ssDNA probe. SSB: *E. coli* single-stranded DNA binding protein; E-E: *E. coli* cells transformed with pET3a vector; CPwt: *E. coli* cells expressing wild type of MSV CP; BSA/LYS: mixture of bovine serum albumin/lysozyme; CYT: cytochrome C.

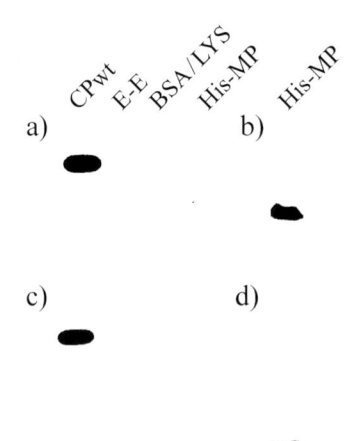

Fig. 3 South-western analysis of MSV MP–DNA interaction. Autoradiographs were obtained using either a MSV dsDNA probe (**a**) or ssDNA probe (**c**). Western blotting of the MP after autoradiography is shown in panels (**b**) and (**d**) indicating the presence of MSV MP in the assay. E-E: *E. coli* cells transformed with pET3a vector; CP wt: *E. coli* cells expressing MSV wild type CP; HIS-MP: *E. coli* cells expressing His-tagged MSV MP and BSA/LYS = Bovine serum albumin and lysozyme.

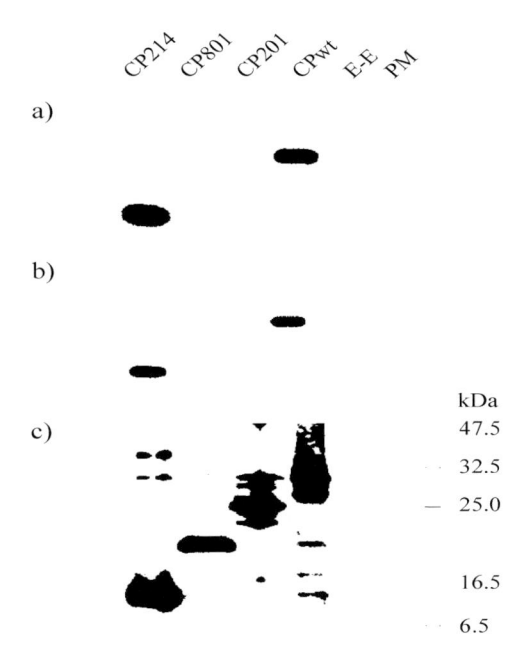

Fig. 4 Location of DNA binding domain analyzed using South-western blotting. South-western analysis using MSV CP and its truncation derivatives revealed that the DNA-binding domain of MSV CP is located at the N-terminus. Autoradiographs were obtained using (**a**) MSV ssDNA probe or (**b**) MSV dsDNA probe. Western blotting of CPs after autoradiography is shown in panel (**c**) indicating the presenc of these proteins. E-E: *E. coli* cells tranformed with pET3a vector; CPwt, CP201, CP801 and CP214: *E. coli* cells expressing MSV CP and its truncation derivatives; PM: protein markers.

4 Notes

1. SDS powder is anionic detergent, it is strongly advised to ware musk and weight this producut in flow hoods.
2. TEMED is usually stored at room temperature. Buy small bottles as it may decline in quality (gels will take longer to polymerise) after opening.
3. Anti-MSV CP and MP sera were produced against purified MSV CP and MP in rabbits. If antibody against the target is not available, express the target protein with a tag and use the anti-tag serum to identify the target proteins.
4. An anti-mouse conjugate should be used if the primary antibody is generated in mice. If chosen to use ECL detection system, a second antibody conjugated to horseradish peroxidase should be used.
5. Disrupted virus particles or MSV-infected tissue can be used for CP samples.
6. If the expression level of the target proteins is low, extend the induction for a few more hours or overnight.
7. Precasted SDS-gels are more convenient to use and available from most suppliers. If manual casting is still in use, Tables 1 and 2 show solution components for different gels. If the protein is smaller, the gel concentration should be higher.
8. In most cases, cells from 200 μL of culture were disrupted by 50 μL of loading buffer, if the loading mix is still viscous after boiling, add more loading buffer and boil for another 3 min.

9. Safety procedure should be followed for handling isotope. For North-western analysis, RNA probe is usually synthesized using T3, T7 or SP6 RNA polymerase. The gene needs to be cloned into a vector containing a T3, T7 or SP6 promoter. The vector was then digested by restriction enzymes and the linearized DNA was purified using a QIAGEN spin column (Qiagen). To syntheses the probe, $1\,\mu L$ of linearized plasmid template, $2\,\mu L$ of $10\times$ transcription buffer, $1\,\mu L$ of ATP, CTP and GTP (each $10\,mM$), $5\,\mu L$ of α-$[^{35}S]$ rUTP ($20\,mCi\ mL^{-1}$, Amersham), $2\,\mu L$ of T3, T7 or SP6 RNA polymerase II and RNase-free H_2O were mixed in a total volume of $20\,\mu L$. The mixture was incubated for 30 min at 37°C and incubated for another 15 min after adding $1\,\mu L$ of RNase-free DNase I. Reaction was stopped by adding $1\,\mu L$ of 0.5 M EDTA and RNA probe was purified using MicroSpin Columns (S-200, Amersham). Instead using radioactive labelling system to produce DNA or RNA probes, non-radioactive labelling system, such as dioxigenin conjugated to UTP (DIG) system (Roche), can be used to syntheses the probes.
10. For some difficult templates, reaction may require up to 30 min, the probes in the reaction may be used directly for hybridization without any purification.
11. If used wet transfer tank, pour 1,000 mL transfer buffer into a glass tray and put the gel holder cassette into the buffer (black side up), put one matrix in the cassette and then three layers of 3MM Whatman paper, the wetted membrane (methanol soak for Millipore membrane), the gel, another three layers of Whatman paper and then another matrix, close the cassette and put into the blotting module and then into buffer tank. Pour the transfer buffer into the tank and insert a cool box containing ice into the tank. Put the lid on and transfer at a constant current of 300 mA for 1 h or 30 mA for overnight.
12. For North-western blotting, RNA probe should be added and all the buffers used should be prepared using RNase-free water. The salt concentration of the reaction buffer affects the binding, reduce the salt concentration if the binding strength is too weak or increase the concentration if the background is high.
13. If using enhanced chemiluminescent (ECL) Western blotting system, transfer the membrane onto a clean square Petri dish, protein side up. Mix 1.5 mL of solution A and 1.5 mL of solution B and then pour onto the membrane, spread the solution mix all over the surface. After 1 min, remove the solution by lifting the membrane with a forceps and putting one side on a piece of paper towel. Wrap the membrane with cling film and fix it into an exposure cassette with sealer tape. Put an ECL film (or an X-ray film) into the cassette in the dark room (Mark the film position on the cassette.). Develop the film after 1 min (put another film and expose longer if the signal is too weak or expose short if the signal is too strong). Check the result by putting the developed film back to the cassette.

Acknowledgements The author thanks Professor Jeffery Davies, Dr. Margaret Boulton for providing the cloned MSV. This work was carried out under MAFF licence PHF 1419a/951/24 and financial supported by John Innes Foundation.

References

1. Stockley, P. (1992). Viral protein–nucleic acid interactions. *Curr. Opin. Struct. Biol.* **2**, 143–149.
2. Thompson, S. R. and Melcher, U. (1993). Coat protein of cauliflower mosaic virus binds to ssDNA. *J. Gen. Virol.* **74**, 1141–1148.
3. Suluja, E., Strokowskaja, L., Zagorski-Ostoja, W., and Palucha, A. (2005). Virus-like particles of potato leafroll virus as potential carrier system for nucleic acids. *Acta Biochim Pol.* **52**, 699–702.
4. Peabody, D. S. (1997). Role of the coat protein–RNA interaction in the life cycle of bacteriophage MS2. *Mol. Gen. Genet.* **254**, 358–364.

5. Munger, K. and Phelps, W. C. (1993). The human papillomavirus E7 protein as a transforming and transactivating factor. *Biochim. Biophys. Acta* **1155**, 111–123.

6. Fontes, E. P., Luckow, V. A., and Hanley-Bowdoin, L. (1992). A geminivirus replication protein is a sequence-specific DNA binding protein. *Plant Cell* **4**, 597–608.

7. Monaghan, A., Webster, A., and Hay, R. T. (1994). Adenovirus DNA binding protein: helix destabilising properties. *Nucleic Acids Res.* **22**, 742–748.

8. Noris, E., Jupin, I., Accotto, G. P., and Gronenborn, B. (1996). DNA-binding activity of the C2 protein of tomato yellow leaf curl geminivirus. *Virology* **217**, 607–612.

9. Bedinger, P., Hochstrasser, M., Jongeneel, C. V., and Alberts, B. M. (1983). Properties of the T4 bacteriophage DNA replication apparatus: the T4 dda DNA helicase is required to pass a bound RNA polymerase molecule. *Cell* **34**, 115–123.

10. Castillo, A. G., Collinet, D., Deret, S., Kashoggi, A., and Bejarano, E. R. (2003). Dual interaction of plant PCNA with geminivirus replication accessory protein (Ren) and viral replication protein (Rep). *Virology* **312**, 381–394.

11. Gutierrez, C., Ramirez-Parra, E., Mar Castellano, M., Sanz-Burgos, A. P., Luque, A., and Missich, R. (2004). Geminivirus DNA replication and cell cycle interactions. *Vet. Microbiol.* **98**, 111–119.

12. Liu, J. Z., Blancaflor, E. B., and Nelson, R. S. (2005). The tobacco mosaic virus 126-kilodalton protein, a constituent of the virus replication complex, alone or within the complex aligns with and traffics along microfilaments. *Plant Physiol.* **138**, 1853–1865.

13. Citovsky, V., Knorr, D., Schuster, G., and Zambryski, P. (1990). The P30 movement protein of tobacco mosaic virus is a single-strand nucleic acid binding protein. *Cell* **60**, 637–647.

14. Deom, C. M., Lapidot, M., and Beachy, R. N. (1992). Plant virus movement proteins. *Cell* **69**, 221–224.

15. Citovsky, V. and Zambryski, P. (1991). How do plant virus nucleic acids move through intercellular connections? *Bioessays* **13**, 373–379.

16. Gilbertson, R. L., Sudarshana, M., Jiang, H., Rojas, M. R., and Lucas, W. J. (2003). Limitations on geminivirus genome size imposed by plasmodesmata and virus-encoded movement protein: insights into DNA trafficking. *Plant Cell* **15**, 2578–2591.

17. Noueiry, A. O., Lucas, W. J., and Gilbertson, R. L. (1994). Two proteins of a plant DNA virus coordinate nuclear and plasmodesmal transport. *Cell* **76**, 925–932.

18. Pascal, E., Sanderfoot, A. A., Ward, B. M., Medville, R., Turgeon, R., and Lazarowitz, S. G. (1994). The geminivirus BR1 movement protein binds single-stranded DNA and localizes to the cell nucleus. *Plant Cell* **6**, 995–1006.

19. Carey, J. (1991). Gel retardation. *Meth. Enzymol.* **208**, 103–117.

20. Crothers, D. M., Gartenberg, M. R., and Shrader, T. E. (1991). DNA bending in protein–DNA complexes. *Meth. Enzymol.* **208**, 118–146.

21. Sukegawa, J. and Blobel, G. (1993). A nuclear pore complex protein that contains zinc finger motifs, binds DNA, and faces the nucleoplasm. *Cell* **72**, 29–38.

22. Citovsky, V. and Zambryski, P. (1993). Transport of nucleic acids through membrane channels: snaking through small holes. *Annu. Rev. Microbiol.* **47**, 167–197.

23. Liu, H., Boulton, M. I., and Davies, J. W. (1997). Maize streak virus coat protein binds single- and double-stranded DNA in vitro. *J. Gen. Virol.* **78**, 1265–1270.

24. Van Wezel, R., Liu, H., Wu, Z., Stanley, J., and Hong, Y. (2003). Contribution of the zinc finger to zinc and DNA binding by a suppressor of posttranscriptional gene silencing. *J. Virol.* **77**, 696–700.

25. Boulton, M. I., King, D. I., Markham, P. G., Pinner, M. S., and Davies, J. W. (1991). Host range and symptoms are determined by specific domains of the maize streak virus genome. *Virology* **181**, 312–318.

26. Liu, H. (1997). Molecular Biology of Maize Streak Virus Movement in Maize. *Ph. D. thesis*, University of East Anglia, Norwich.

27. Laemmli, U. K. (1970). Cleavage of structural proteins during the assembly of the head of bacteriophage T4. *Nature* **227**, 680–685.

28. Guerineau, F., Lucy, A., and Mullineaux, P. (1992). Effect of two consensus sequences preceding the translation initiator codon on gene expression in plant protoplasts. *Plant Mol. Biol.* **18**, 815–818.

29. Studier, F. W., and Moffatt, B. A. (1986). Use of bacteriophage T7 RNA polymerase to direct selective high-level expression of cloned genes. *J. Mol. Biol.* **189**, 113–130.

30. Clauser, K. R., Hall, S. C., Smith, D. M., Webb, J. W., Andrews, L. E., Tran, H. M., Epstein, L. B., and Burlingame, A. L. (1995). Rapid mass spectrometric peptide sequencing and mass matching for characterization of human melanoma proteins isolated by two-dimensional PAGE. *Proc. Natl. Acad. Sci. USA* **92**, 5072–5076.

31. Sambrook, J., Fritsch, E. F., and Maniatis, T. (1989). Molecular Cloning: a Laboratory Manual, 2nd Edition (Cold Spring Harbor, New York: Cold Spring Harbor Laboratory Press).

Chapter 29
Protein–Protein Interactions: The Yeast Two-Hybrid System

Deyin Guo, Minna-Liisa Rajamäki, and Jari Valkonen

Abstract Yeast two-hybrid systems are powerful tools to identify novel protein–protein interactions and have been extensively used to study viral protein interactions. The most commonly used systems are GAL4-based and LexA-based systems. Over the last decade, a range of modifications and improvements have been made to the original yeast two-hybrid system to expand the scope of molecular interaction assays and to eliminate false positives. Detailed protocols are provided for yeast strain storage, yeast transformation, yeast mating, preparation of growth and selection medium, quantitative reporter gene assays (α- and β-galactosidase liquid assays) and detection of fusion protein by Western blot.

Keywords Yeast two-hybrid system; protein–protein interactions; yeast transformation; galactosidase assay; Western blot

1 Introduction

Protein–protein interactions are critical for the vast majority of biological processes and play pivotal roles during the virus infection cycle; for example, in the formation of virus replication complexes, assembly of virions, virus movement between cells and virus transmission by vectors. Analysis of protein–protein interactions is crucial for understanding protein functions and the molecular mechanisms underlying biological processes. For this reason, numerous techniques have been developed to identify and characterize protein–protein interactions, from biochemical approaches such as coimmunoprecipitation and affinity chromatography to molecular genetic approaches such as the yeast two-hybrid system (1).

The yeast two-hybrid system (YTHS), first described by Fields and Song in 1989 (2), has proven itself to be a powerful tool for identifying novel protein–protein interactions and has been used extensively to detect interactions between proteins from many viruses and different cellular organisms. Using YTHS, protein interaction maps have been established for several viruses, including *Escherichia coli* bacteriophage T7 (3), *Vaccinia virus* (4), *Wheat streak mosaic virus* (5); *Hepatitis*

C virus (6), *Potato virus A* and *Pea seed-borne mosaic virus* (7), *Soybean mosaic virus* (8), *Porcine teschovirus* (9), *Human herpesvirus 8* (Kaposi's sarcoma-associated herpesvirus), and *Human herpesvirus 3* (Varicella-zoster virus) (10). Over past decades, YTHS has evolved from low-throughput manual screens to systematic interrogations of entire proteomes in human cells and has been used to explore the global protein–protein interaction network within cells of yeast (11, 12), bacterium *Helicobacter pylori* (13), *Drosophila melanogaster* (14), nematode *Caenorhabditis elegans* (15), the malaria parasite *Plasmodium falciparum* (16), and human (17, 18).

The YTHS is based on the observation that many eukaryotic transcription factors possess two separable domains, the DNA-binding domain (BD), which binds to the upstream activating sequence (UAS), and the activation domain (AD), which recruits the RNA polymerase. Both domains are required for the induction of gene expression, but they do not need to be present within the same protein. Therefore, when two test proteins are fused with BD and AD, respectively, the two domains will be brought into proximity providing that the test proteins can associate with each other, thus reconstituting a functional transcription factor and driving expression of downstream reporter gene (Fig. 1). The most commonly used systems (Table 1) are the GAL4 system, where the BD and AD domains of yeast GAL4 protein are used (2, 19) and the LexA system, where the bacterial repressor protein LexA is used as the BD in combination with the *E. coli* B42 activation domain (20) or human herpes simplex virus VP16 protein (21, 22).

The yeast strains used for YTHS carry auxotrophic mutations in genes required for biosynthesis of amino acids or nucleotides, such as *LEU2*, *TRP1*, *HIS3*, *URA3*, and *ADE2*. Transformation of two-hybrid plasmids or expression of some reporter genes can complement these mutations and allow for positive selection of the yeast transformants and interactions. The reporters in YTHS include two types of genes, one enabling colorimetric readout like *lacZ*, *MEL1*, and *gfp* (green fluorescent protein), and the other being prototrophic markers, such as *HIS3*, *LEU2*, *URA3*, and *ADE2*.

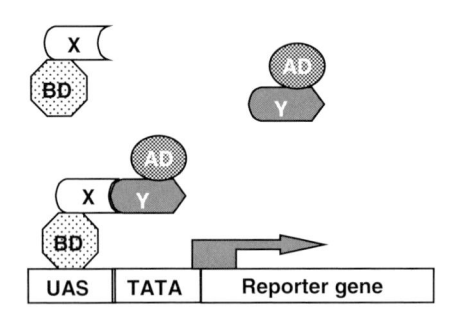

Fig. 1 Principle of the yeast two-hybrid system. The protein X is fused to the DNA-binding domain (BD) of a yeast transcription factor, which binds to the upstream activating sequence (UAS). The protein Y is fused to the transcriptional activation domain (AD), which recruits the RNA polymerase II. If the protein X and Y interact, a functional transcription factor will be reconstituted and thus drive the expression of reporter genes. TATA indicates the yeast minimal promoter

Table 1 Different versions of the yeast two-hybrid systems in common use

System	BD vector/ selection maker	AD vector/ selection maker	Yeast strain (reporter genes)	Provider or reference
GAL4-based systems				
Matchmaker	pGBKT7, pAS2-1, pGBT9/(*TRP1*)	pGADT7, pACT2, pGAD424/ (*LEU2*)	AH109 (*HIS3, ADE2, lacZ, MEL1*) GC-1945 (*HIS3, lacZ*) Y187 (*lacZ*)	Clontech; (19)
ProQuest	pDBLeu/(*LEU2*)	pPC86/(*TRP1*)	MaV203 (*HIS3, URA3, lacZ*)	Invitrogen
HybriZap	pBD-GAL4 Cam/ (*TRP1*)	pAD-GAL4-2.1/(*LEU2*)	YRG-2 (*HIS3, lacZ*)	Stratagene
LexA-based systems				
Hybrid Hunter	pHybLex/Zeo/ (*Zeocin^R*))	pYESTrp/ (*TRP1*)	L40 (*HIS3, lacZ*) EGY48 (*LEU2 + lacZ with pSH18-34*)	Invitrogen
Interaction Trap or DupLEX-A	pEG202, pGilda pNLexA/ (*HIS3*)	pB42AD (pJG4-5)/(*TRP1*)	EGY48 (*LEU2 + lacZ with pSH18-34*)	OriGene, Clontech; (20)
LexA	pBTM116, pLexA, pLexNa/ (*TRP1*)	pVP16/(*LEU2*)	L40 (*HIS3, lacZ*) AMR69 (*lacZ*)	(21, 22)

The widely used versions or variants of the two-hybrid systems are listed in Table 1. YTHS kits including appropriate yeast strains, hybrid vectors, and controls are commercially available from a variety of sources, such as Clontech (Matchmaker systems, http://www.clontech.com), Invitrogen (Hybrid Hunter and ProQuest systems, http://www.invitrogen.com), Stratagene (HybriZAP systems, http://www.stratagene.com), OriGene (DupLEX-A system, http://www.origene.com), and MoBiTec (Grow'n'Glow GFP System, http://www.mobitec.com).

Over the past decade, a range of modifications to the basic principle of original YTHS have been made to expand the scope of molecular interactions that can be identified. One-hybrid system was developed for detecting DNA–protein interactions (23), and three-hybrid systems for RNA–protein interactions (24, 25), and ternary protein complex analysis (26). One-and-a-half hybrid system was designed for analyzing proteins that conditionally bind to DNA (27, 28). As the classical YTHS utilizes a nuclear localization signal (NLS) to guide the fusion proteins to the nucleus, new systems have been generated to detect protein interaction in cytoplasm, such as hSos/Ras recruitment system (29, 30) and the split-ubiquitin system (31, 32). For further characterization of known protein interactions and screening for inhibitors, reverse two-hybrid systems have been developed (33, 34). Two-hybrid systems based on other hosts than yeast are generated, such as bacterial two-hybrid system (35) and mammalian two-hybrid system (36).

This chapter is not intended to give extensive review and description of various versions of the yeast two-hybrid systems and their derivatives. Here, we focus on the key components of the classical GAL4-based yeast two-hybrid system and its main experimental procedures for detecting interactions of limited number of viral and host proteins.

2 Materials

Yeast strains, vectors, and control plasmids can be obtained from commercial sources or research laboratories that originally generated the system or components (Table 1). You can also build up your own system for use in your research with the components from different sources as long as the components are compatible with each other as discussed later (see notes 1–7). In this chapter, we use the most recent version of the Matchmaker system (Clontech) as an example to describe the experimental procedures, which are essentially the same for different variants of the two-hybrid systems.

2.1 Yeast Strains and Plasmids

2.1.1 Yeast Strains

1. AH109 with genotype (37): *MATa, trp1-901, leu2-3,112, ura3-52, his3-200, gal4Δ, gal80Δ, MEL1, LYS2::GAL1*$_{UAS}$*-GAL1*$_{TATA}$*-HIS3, GAL2*$_{UAS}$*-GAL2*$_{TATA}$*-ADE2, URA3::MEL1*$_{UAS}$*-MEL1*$_{TATA}$*-lacZ.* AH109 contains four reporter genes – *HIS3* and *ADE2* for nutritional selection and *MEL1* and *lacZ* for color screening.
2. Y187 with genotype (38): *MATα, leu2-3,112, ura3-52, trp1-901, his3-200, ade2-101, gal4Δ, gal80Δ, met⁻, URA3::GAL1*$_{UAS}$*-GAL1*$_{TATA}$*-lacZ.* Y187 exhibits a higher level of induced β-galactosidase activity than AH109, so use liquid cultures of Y187 for quantitative β-galactosidase assays. Y187 is also used as a mating partner with AH109.

2.1.2 Plasmids

Cloning and control vectors used in the Matchmaker Gal4 system (Clontech) are listed in Table 2. Other plasmids can be used as long as they are compatible. pGBKT7 and pGADT7 are used as cloning vectors and their sequence information is available at Clontech websites. pCL1 encodes the full-length, wild-type GAL4 protein and can be used as a positive control for α- and β-galactosidase assays. pGADT7-T and pGBKT7-53 fusion proteins interact in yeast two-hybrid assay and provide positive control for the interaction. pGADT7-T and pGBKT7-Lam fusion

Table 2 Yeast cloning vectors and control plasmids

Plasmid	GAL4 domain/epitope tag[a]	Yeast selection[b]	Bacterial selection[c]	Usage
pGBKT7	BD, c-Myc	*TRP1*	kan[r]	Cloning
pGADT7	AD, HA	*LEU2*	amp[r]	Cloning
pCL1	GAL4 (BD + AD)	*LEU2*	amp[r]	Positive control
pGADT7-T	AD-HA-T antigen	*LEU2*	amp[r]	Control
pGBKT7-53	BD- c-Myc-p53	*TRP1*	kan[r]	Control
pGBKT7-Lam	BD- c-Myc-lamin C	*TRP1*	kan[r]	Control

[a] AD, GAL4 activation domain; BD, GAL4 DNA-binding domain; HA, hemagglutinin
[b] *LEU2*, leucine prototrophy; *TRP1*, tryptophan prototrophy
[c] amp[r], ampicillin resistance; kan[r], kanamycin resistance

proteins do not interact in yeast two-hybrid assay and can be used as a negative control. pGBKT7-Lam encodes human lamin C that does not interact with most other proteins and can thus be used also as a negative control for interaction with test proteins. As a negative control, remember also to check interaction with your test protein and empty cloning vector.

Detailed information of plasmids and yeast strains can be obtained from different reference books, provider's instruction manuals, or from original authors.

2.2 Buffers and Solutions

1. 10× TE buffer: 0.1 M Tris-HCl, pH 7.5, 10 mM EDTA. Adjust pH to 7.5 and autoclave.
2. 10× LiAc: 1 M lithium acetate. Adjust pH to 7.5 with dilute acetic acid and autoclave.
3. 50% PEG 3350 (polyethylene glycol; Sigma #P-3640): 100 g in 200 mL of sterile deionized H_2O.
4. 100% DMSO (dimethyl sulfoxide; Sigma D-4540).
5. Herring testes carrier DNA (10 mg mL^{-1}): Sonicated (Clontech K1606-A630440) and denatured by boiling for 20 min and immediately placed on ice. Store at −20°C.
6. Glass beads (425–600 μm; Sigma G-8772).
7. 2× Laemmli loading buffer: 125 mM Tris-HCl, pH 6.8, 4% (w/v) SDS, 20% (v/v) glycerol, 2% (v/v) β-mercaptoethanol, 0.2% (w/v) bromophenol blue.
8. 10× TBS: 12.1 g L^{-1} Tris-base, 87.8 g L^{-1} NaCl. Adjust pH to 7.4 with HCl and autoclave.
9. 1× TBS + 0.1% Tween: 1,000 mL 1× TBS, 1 mL Tween-20.

10. Z-buffer: 60 mM Na_2HPO_4, 40 mM NaH_2PO_4, 10 mM KCl, 1 mM $MgSO_4$. Adjust pH to 7.0 and autoclave.
11. X-gal stock solution (20 mg mL^{-1}): Dissolve 20 mg of X-gal (5-bromo-4-chloro-3-indolyl-ß-D-galactopyranoside; Sigma B-4252) per milliliter DMF (*N,N*-dimethylformamide). Divide to aliquots and store in the dark at −20°C.
12. X-α-gal stock solution (20 mg mL^{-1}): Dissolve 20 mg X-α-gal (5-bromo-4-chloro-3-indolyl-α-D-galactopyranoside; Clontech, 630407) per mL DMF (*N,N*-dimethylformamide). Store in glass or polypropylene bottles in the dark at −20°C.
13. Z-buffer/X-gal solution: 10 mL of Z-buffer, 27 µL of ß-mercaptoethanol, 167 µL of X-gal stock solution (20 mg mL^{-1}). Prepare immediately prior to use.
14. NET-buffer: 50 mM Tris-HCl pH 8.0, 5 mM EDTA, 0.4 M NaCl, 100 U mL^{-1} Trasylol (Aprotimin), 1% Nonidet P-40 (NP-40). Store at 4°C.
15. 10% Triton X-100: 1 mL of Triton X-100 added to 9 mL of deionized H_2O.
16. PMSF stock solution: Dissolve 0.1742 g PMSF (phenylmethyl-sulfonyl fluoride; Sigma P7626) in 10 mL isopropanol. Keep in the dark at −20°C (see note 8).
17. ONPG solution (4 mg mL^{-1}): Dissolve 4 mg of ONPG (*o*-nitrophenyl ß-D-galactopyranoside; Sigma N-1127) per milliliter of Z-buffer. Adjust pH to 7.0 and dissolve. Divide to aliquots and store at −20°C.
18. PNP-α-Gal solution (100 mM): 100 mM PNP-α-Gal (*p*-nitrophenyl α-D-galactopyranoside; Sigma N-0877) in deionized H_2O. Dissolve 30.1 g per 10 mL of deionized H_2O and filter-sterilize.
19. 1 M Na_2CO_3 solution.

2.3 Bacterial and Yeast Media

1. LB broth: 10 g L^{-1} bacto-tryptone, 5 g L^{-1} bacto-yeast extract, 10 g L^{-1} NaCl, 15 g L^{-1} agar (for plates only). Adjust pH to 7.0 with 5 N NaOH and autoclave. Add appropriate antibiotic after cooling down to 55°C.
2. YPD medium: 20 g L^{-1} bacto-peptone, 10 g L^{-1} yeast-extract, 20 g L^{-1} agar (for plates only). Add H_2O to make volume to 900 mL; adjust pH to 5.8 and autoclave. Add 100 mL of 20% filter-sterilized glucose.
3. YPDA medium: 20 g L^{-1} bacto-peptone, 10 g L^{-1} yeast-extract, 0.1 g L^{-1} adenine hemisulphate, 20 g L^{-1} agar (for plates only). Add H_2O to make volume to 900 mL; adjust pH to 5.8 and autoclave. Add 100 mL of 20% filter-sterilized glucose. You can also purchase commercially YPD medium or YPD agar medium in ready powder form.
4. SD medium (see note 9): 6.7 g yeast nitrogen base without amino acids. 100 mL of appropriate 10× dropout stock solution (see later). Add H_2O to 960 mL; adjust pH to 5.8 and autoclave. Add 40 mL of 50% glucose (filter-sterilized).

5. 10× Dropout (DO) solution (see note 10): Dissolve the following chemicals in 1,000 mL of deionized H_2O, autoclave and store at 4°C for up to 1 year.

200 mg L^{-1}	L-Adenine hemisulfate salt
200 mg L^{-1}	L-Arginine HCl
200 mg L^{-1}	L-Histidine HCl monohydrate
300 mg L^{-1}	L-Isoleucine
1,000 mg L^{-1}	L-Leucine
300 mg L^{-1}	L-Lysine HCl
200 mg L^{-1}	L-Methionine
500 mg L^{-1}	L-Phenylalanine
2,000 mg L^{-1}	L-Threonine
200 mg L^{-1}	L-Tryptophan
300 mg L^{-1}	L-Tyrosine
200 mg L^{-1}	L-Uracil
1,500 mg L^{-1}	L-Valine

6. SD plates: Mix 20 g agar and 100 mL of appropriate 10× dropout solution, and add H_2O to 860 mL; adjust pH to 5.8 and autoclave. Cool down to 65°C and add 40 mL of 50% glucose (filter-sterilized), 100 mL of yeast nitrogen base without amino acids (6.7 g per 100 mL, filter-sterilized) (see note 11).

7. SD/X-α-gal plates: To include X-α-gal on plate, add 1 mL of X-α-gal (20 mg mL^{-1}) stock to 1 L of appropriate SD medium after autoclaving and cooling the medium to 55°C. Alternatively, you can spread 100 µL of X-α-gal (2 mg mL^{-1}) onto premade 10-cm plate.

8. SD/X-gal plates: Mix 20 g agar in 700 mL of deionized H_2O and 100 mL of appropriate 10× dropout solution. Autoclave and immediately add 100 mL of 20% glucose (filter-sterilized) and 100 mL of yeast nitrogen base without amino acids (6.7 g per 100 mL, filter-sterilized). Allow to cool to 55°C, and add 100 mL of 10× BU salts (see below) and 4 mL of 20 mg mL^{-1} X-Gal dissolved in DMF (*N,N*-dimethyl formamide). For SD/X-gal plates, BU salts must be included in the medium to adjust the pH to ~7, which is closer to the optimal pH for β-galactosidase activity, and to provide the phosphate necessary for the β-gal assay to work.

9. SD/Gal/Raf/X-gal plates: If a yeast strain like EGY48 is used in your system, you must use 2% galactose and 1% raffinose as the carbon sources instead of glucose because this strain uses the inducible *GAL1* promoter controlled by wild-type *GAL4* and *GAL80*. The procedure to make this kind of plates is the same as SD plates with glucose. Allow the medium in the plates to harden at room temperature. Store plates inverted, in a plastic sleeve, in the dark, at 4°C for up to 2 months.

10. 10× BU salts: 70 g L^{-1} $Na_2HPO_4 \cdot 7H_2O$, 30 g L^{-1} NaH_2PO_4. Adjust pH to 7.0 and autoclave; store at room temperature.

2.4   Equipments

The facility for a standard molecular laboratory is sufficient for the yeast two-hybrid assays. However, you need an incubator with and without shaker at 30°C for growing yeast cells and a vortexer with 1.5 mL-tube holder at cold for breaking yeast cells.

3   Methods

3.1   Generation of Yeast Two-Hybrid Constructs

You can use the whole open reading frame (ORF) of the desired gene, fragments or domains of the ORF, or a random library of viral genome for screening protein–protein interactions. Construct your fusion genes in yeast vectors by cloning the test gene fragments in frame with the DNA-binding domain (BD) and activation domain (AD), respectively. Standard molecular biological methods can be applied when making the desired plasmids (39). Sequencing of the whole ORF including the cloning sites is recommended.

If you have your test gene cloned in another vector, it is sometimes possible to use compatible restriction sites present in both vectors to clone the gene into yeast two-hybrid vector. Note, however, that the test gene must be in-frame with DNA-binding or activation domain. Alternatively, gene fragment can be PCR-amplified with specific primers including suitable restriction sites for cloning. In the multiple cloning sites of yeast vectors, you can find several possible restriction sites and pick two suitable ones. The most commonly used and often best are sites for *Eco*RI, *Bam*HI, and *Sal*I presuming that your test gene does not contain the sites. Typically, you can use the same sites for in-frame cloning in both yeast vectors and note also that e.g., *Sal*I and *Xho*I give compatible ends and can be ligated together. After cloning, it is necessary to verify the correct fusion of open reading frames at the junctions. If PCR was adopted at the cloning stage, the whole ORF of the insert sequence including cloning sites should be sequenced to make sure that no extra nucleotide or stop codon or missense mutation is introduced in the constructs.

A random library can be constructed e.g., by making virus cDNA primed with an oligo(dT) or random primers containing a suitable restriction site for yeast vector cloning. An adaptor sequence containing the other useful restriction site is then ligated to other end of the cDNA fragments and the fragments are cloned to yeast vector. Note that now only one third of the clones result in fusion gene with correct reading frame.

For cDNA library screening, the bait gene fragment is cloned into DNA-binding domain vectors like pGBKT7.

Before starting to check the interactions, verify that none of the fusion proteins alone activates the reporter genes by transforming each construct independently into yeast strain. Assay the transformants for auto-activation of reporter genes.

3.2   Growth and Storage of Yeast Strains

Yeast cells are normally grown at 30°C. For long-term storage, yeast strains can be kept in media with 25% glycerol at −70°C. To recover a strain from a frozen stock, streak small amount of frozen cells onto a YPDA or appropriate SD-X agar plate and incubate at 30°C for a couple of days (until diameter of yeast colonies are 1–3 mm). The colonies can be stored for up to 2 months by sealing the plates with Parafilm, then you need to streak a fresh working stock plate. For liquid cultures, use fresh colonies (1- to 3-weeks old) for best results. Pick a single colony (2–3 mm in diameter) and inoculate into medium, vortex vigorously, and incubate at 30°C with shaking (230–250 rpm). Be sure that there is no cell clumps left in inoculum as otherwise the growth may be slow. Before starting, verify your yeast nutritional requirement phenotype by streaking the strain onto SD agar plate without appropriate nutrients. Yeast strain should not grow on plates where nutrients used for plasmid or interaction selection are dropped out.

3.3   Transformation of Yeast Cells

3.3.1   Preparation of Competent Cells (LiAc Method)

1. Inoculate cells from a single colony (2–3 mm in diameter) into 20 mL of YPDA medium and incubate at 30°C with shaking (230–250 rpm) until the OD_{600} = 1–2 (14–18 h).
2. Transfer 5–20 mL of overnight culture to 300 mL of YPDA in a 1-L flask to produce an OD_{600} = 0.2. Incubate at 30°C with shaking (230–250 rpm) for 2–3 h until the OD_{600} of the yeast culture is 0.4–0.6.
3. Prepare and filter-sterilize 1× TE/LiAc and PEG/LiAc solutions.

   10× TE, 10× LiAc, 50% PEG, H2O;
   TE/LiAc (50ml), 5 mL, 5 mL, no, 40 mL
   PEG/LiAc (10mL), 1 ml, 1 mL, 8 mL
   (Each transformation sample needs 0.6 mL)

4. Pellet the cells by centrifugation at 1,000g (GSA rotor, 2,500 rpm) for 5 min at room temperature (20–21°C). (Carefully discard the supernatant as the pellet is easily dissolved.)
5. Wash the pellet with 40 mL of freshly prepared TE/LiAc and recentrifuge as in Step 4.
6. Discard the supernatant and resuspend the pellet in 1.5 mL of 1× TE/LiAc.

To maximize transformation efficiency, use competent cells immediately or within 1 h.

3.3.2   Transformation of Competent Cells

1. For testing interaction between two proteins, set up DNA mix in a sample tube by adding AD fusion plasmid A and BD fusion plasmid B.
2. Check also interaction between A (or B) and an empty plasmid as well as A (or B) and a negative control plasmid.
3. Remember to include to each transformation test at least one positive and negative interaction control.
4. For each test pair, set up the DNA mix in sample tube by adding 0.1–1 µg of each type of plasmid DNA (ca. 3 µL of each plasmid miniprep), together with 100 µg of herring testes carrier DNA.
5. Add 100 µL of the competent cells to each sample tube and mix well.
6. Add 0.6 mL of PEG/LiAc to each tube and vortex to mix.
7. Incubate at 30°C for 30 min with shaking at 200 rpm.
8. Add 70 µL of DMSO (to final concentration of 10%) and mix gently.
9. Heat shock for 15 min in a 42°C water bath (7 min for L40).
10. Chill the cells on ice and centrifuge for 5 s at 14,000 rpm.
11. Remove the supernatant and resuspend the cells in 0.2–0.5 mL of TE or water.
12. Spread 0.1–0.2 mL of the transformation mixture onto each plate (100-mm) containing appropriate synthetic selection medium (for 150-mm plate, use 0.3 mL).

    For example, if one of the yeast plasmids contains *LEU2* and the other *TRP1* nutritional marker, plate the transformation mixture on SD-Leu-Trp plate as this will select colonies that have acquired both plasmids. If you have only one plasmid, use medium where one appropriate nutrient has been left out. To select for strong interactions, plate transformations on selection medium where nutrients (e.g., His, Ade) selecting for interacting partners have been left out in addition to nutrients for plasmid selection.
13. Incubate the plates at 30°C (face down) until colonies grow to 1–2 mm (usually 2–5 days depending on yeast strains and selection pressure).
14. Pick colonies and restreak them on selection medium for testing interacting partners (e.g., for SD-Leu-Trp-His-Ade if *LEU2* and *TRP1* select for different plasmids and *HIS3* and *ADE1* for interaction between two proteins) and grow at 30°C. Seal the master plates with Parafilm and store at 4°C (They can be stored at 4°C for 3–4 weeks.) (see note 12).

3.4   Pairwise Interactions by Yeast Mating

Yeast mating is a convenient alternative to yeast cotranformations and can also be used to screen a pretransformed cDNA library. Plant viruses encode less than 20 proteins. Interactions between any pair of the viral proteins can be tested in a matrix assay by yeast mating. The following procedure is designed for studying interactions between eight different proteins and can be scaled up when necessary.

1. Transform the 8 AD fusion plasmids of interest into yeast strain AH109 (*MATa*) separately and select on SD/-Leu. The resulting yeast strains are named A1 to A8, respectively.
2. Transform the 8 BD fusion plasmids into yeast strain Y187 (*MATα*) separately and select on SD/-Trp. The resulting yeast strains are named B1 to B8, respectively.
3. Pick one colony of each yeast strain and place into one 1.5-mL microcentrifuge tube containing 0.5 mL of YPDA medium. Vortex the tubes and completely resuspend the cells. The yeast colonies used for mating should be larger than 2 mm in diameter and less than 2-months old.
4. Take a sterile and flat-bottom 96-well microtiter plate, and aliquot 100 μL of YPDA medium to each well in columns 1–8 and rows A-H.
5. Pipete 50 μL of A1 cells into each well in column 1, A2 cells into column 2, and so on. Accordingly, aliquot 50 μL of B1 cells into each well in row A, B2 cell in row B, and so on. Then cover the plate with a sterile lid. In this way, the 64 pairwise matings (8 × 8 matrices) are set up.
6. Place the plate on a rotating platform shaker and incubate at 30°C for 6–18 h at 200 rpm.
7. Spread 100 μL of each mating culture on 100-mm SD-Leu-Trp + X-α-gal plates. For selection of strong interactions, use SD-Leu-Trp-His-Ade + X-α-gal.
8. Incubate the plates at 30°C for 3–7 days to allow diploid cells to form visible colonies.
9. Evaluation of the nutritional and reporter phenotypes of diploid cells is the same as the transformed cells.

3.5 Detection of Fusion Proteins by Western Blot Analysis

It is always necessary to confirm the expression of fusion proteins in yeast by performing Western analysis of soluble protein extracts of yeast transformants. Antibodies to the actual proteins tested would provide the most accurate data, but will not allow testing the expression of different proteins in the same blot. Therefore, the antibodies available for detecting the DNA-binding and activation domains in the fusion proteins or antibodies specific for different tags commonly used in yeast two-hybrid fusions are usually employed. Remember to calculate the size of these domains to the estimated protein size when analyzing the bands.

3.5.1 Extraction of Yeast Soluble Proteins for Western Analysis

1. Inoculate cells from a single transformed yeast colony to 5 mL of an appropriate synthetic selection medium and incubate at 30°C with shaking (230–250 rpm) over night or until good culture. As a negative control, inoculate cells from a nontransformed yeast colony (see note 13).

2. Pellet cells from 1.2 mL culture by centrifugation.
3. Add 1:1 ratio (v/v) glass beads and 50 μL of 2× Laemmli loading buffer and 50 μL of H_2O.
4. Vortex vigorously for 6 min at cold.
5. Centrifuge at 14,000 rpm for 10 min.
6. Transfer supernatant to a fresh tube and boil (100°C) for 5 min. Keep tubes on ice.
7. Use 10–20 μL of the sample to Western blot. If you are not using them immediately, store them at −20°C.

3.5.2 Western Analysis

1. Prepare sodium dodecyl sulfate polyacrylamide (SDS-PAGE) gel, load the samples and run the gel. Instructions for preparing and running SDS-PAGE gels are available in standard laboratory manuals, e.g., Sambrook and Russel (39).
2. Transfer the proteins to appropriate membrane by performing Western blot.
3. Block the membrane in 1× TBS with 0.1% Tween-20 and 5% skim milk powder at room temperature with shaking for 1 h (or 4°C for overnight).
4. Dilute the first antibody in 1× TBS with 2.5% milk powder (without Tween 20) and incubate at 4°C with shaking overnight. As a first antibody, you can use antibody specific to the test protein or alternatively, commercially available antibodies recognizing the GAL4 BD and AD or epitope tags present in two-hybrid vectors (e.g., HA epitope in pGADT7 and c-Myc in pGBKT7).
5. Wash the membrane three times 10 min in 1× TBS with 0.1% Tween-20 with shaking.
6. Dilute the second antibody in 1× TBS with 2.5% milk powder (without Tween-20) and incubate at room temperature with shaking for 1 h. Use secondary antibody that is able to recognize the first antibody (e.g., if the first antibody is monoclonal, use anti-mouse antibody) and is linked to suitable enzyme (e.g., horse-radish peroxidase or alkaline phosphatase).
7. Wash the membrane four times, 10 min in 1× TBS with 0.1% Tween-20 with shaking.
8. Develop and detect bands on the membrane by adding substrate solution. The appropriate substrate and method used depend on the kind of secondary antibody used. If there is positive interaction in yeast, but are unable to detect the protein in Western analysis, the interaction might still be true. The protein might, e.g., be toxic to yeast cells and is easily degraded, which make it difficult to detect the protein. In the case of negative interaction, it is always important to detect the protein to draw any conclusions.

3.6 Filter Assay of β-Galactosidase Activity

1. Use fresh yeast colonies (1–3 mm in diameter).
2. Prepare Z buffer/X-gal solution (calculate 2.5 mL for each 100-mm plate).

10 mL Z buffer
27 µL β-mercaptoethanol
167 µL X-gal stock solution (20 mg mL^{-1}).

3. Presoak sterile Whatman #1 or VWR grade 413 filters with Z-buffer: 2 mL for 10-cm plates and 5 mL for 15-cm plates.

   a) Place a sterile dry filter paper over the surface of the agar plate containing the transformant colonies (at least 1–2 mm in diameter) for some minutes and carefully lift the filter from plate when the filter is evenly wetted. Remember to mark the filter and its orientation (e.g., by poking holes with a sterile needle).

   b) Alternatively, you can streak several yeast colonies from one transformation onto one filter. Use a pencil to make and mark grids on the filter so that you can recognize each transferred transformant. This method does not require sterile filter paper and can spare reagents.

4. Submerge the filter paper (colonies facing up) in liquid nitrogen with forceps and wait for 10 s.

5. Transfer the filter (colony side up) on the lid of the Petri dish and allow it to thaw at room temperature.

6. Place the filter on the presoaked filter in Z-buffer/X-gal solution from step 3. Avoid trapping air bubbles under the filter. If the filter looks dry, add a little more Z-buffer/X-gal solution.

7. Incubate the filters at 30°C and check periodically for the appearance of blue colonies/color (30 min–12 h). Colonies showing strong interactions should turn blue in 1–2 h, whereas very weak signals take overnight incubation. Prolonged incubation may result also false positives.

3.7 α- and β-Galactosidase Activity by Liquid Assay

3.7.1 β-Galactosidase Activity by Liquid Assay

1. Inoculate 5 mL of an appropriate liquid selection medium and incubate at 30°C with shaking (230–250 rpm) overnight (14–18 h). Use triplicate of each sample and remember to include positive and negative controls.

2. Vortex the culture tube vigorously to disperse cell clumps and measure OD_{600}. (see note 14).

3. Pellet cells of 1.0 mL of culture by centrifugation at 14,000 rpm for 30 s.

4. Resuspend cells in NET-buffer: 50 µL NET-buffer, 5 µL 10% Triton X-100, 1 µL 100 mM PMSF.

5. Add 0.2 g (ca. 100 µL) glass beads (0.5 mm in diameter).

6. Vortex for 6 min at cold.

7. Add 650 µL of 1× Z buffer with 0.27% (v/v) β-mercaptoethanol and mix well.

8. Add 160 μL of ONPG solution (4 mg mL^{-1}) and mix well.
9. Incubate the mixture at 30°C for 10–60 min.
10. Quench the reaction with 400 μL of 1 M Na_2CO_3.
11. Centrifuge 5 min at 14,000 rpm to remove cell debris.
12. Read absorbance at 420 nm.
13. Calculate the relative activity using the equation (40):

ß-galactosidase units = $1{,}000 \times (OD_{420}/t \times V \times OD_{600})$
where t = time (min) of incubation of the reaction mixture
V = volume (mL) of cell culture used
OD_{600} = A_{600} of 1 mL of culture.

3.7.2 α-Galactosidase Activity by Liquid Assay

1. Inoculate 5 mL of an appropriate liquid selection medium and incubate at 30°C with shaking (230–250 rpm) overnight (14–18 h).
2. Vortex the culture tube vigorously to disperse cell clumps and measure OD_{600} (see note 14).
3. Pellet cells of 1.0 mL of culture by centrifugation at 14,000 rpm for 2 min and carefully transfer the supernatant to a fresh tube.
4. Prepare PNP-α-Gal solution (100 mM) assay buffer: 24 μL assay buffer for each 1-mL assay and 48 μL assay buffer for each 200-μL assay (96-well microtiter plate format). Use triplicate of each sample and remember to include positive and negative controls.
5. Transfer 8 μL of supernatant into a 1.5-mL microcentrifuge tube or alternatively 16 μL into a well of microtiter plate.
6. Add assay buffer: 24 μL into a 1-mL tube/48 μL into a well of microtiter plate
7. Incubate at 30°C for 60 min. Remember to cover the microtiter plate
8. Add stop solution to terminate reaction: 960 μL (1× stop solution) into 1-mL tube or 136 μL (10× stop solution) into a well of microtiter plate
9. Read absorbance at 410 nm
10. Calculate the activity using equation:

α-galactosidase activity = $1{,}000 \times V_f \times OD_{410}/[(\varepsilon \times b) \times t \times V_i \times OD_{600}]$
where t = time (min) of incubation
V_f = volume of assay (200 μL or 992 μL)
V_i = volume of culture medium supernatant added (16 μL or 8 μL)
OD_{410} = A_{410} of the reaction mix
OD_{600} = A_{600} of 1 mL of culture
$\varepsilon \times b$ = p-nitrophenol molar absorbtivity at 410 nm × the light path (cm)
= 10.5 (mL/μmol) for 200-μL format
= 16.9 (mL/μmol) for 1-mL format where b = 1 cm

3.8 Verification of the Protein Interactions by Other Methods

The results generated by YTHS should be verified with independent methods. In YTHS, the proteins are tested in fusion with BD and AD rather than in their natural form, and interactions occur in the nucleus, which may not be the correct subcellular location for the respective protein interaction. There are several possible methods available and they are described in details in other chapters.

Verifying the results of YTHS is not straight-forward because, e.g., in the commonly used *in vitro* tests, protein interactions occur under fundamentally different conditions. Despite these uncertainties, it is more assuring to find two proteins to interact or lack interaction in independent systems. The pull-down assay where the protein of interest is expressed as a fusion with e.g., glutathione-S-transferase (GST) that can be bound to glutathione-beads is often used. If two proteins interact, they should precipitate together with the beads. Alternatively, in coimmunoprecipitation, antibodies are used to precipitate and detect the proteins of interest. Epitope tags produced from yeast vectors and antibodies to these tags are especially useful for this purpose. Immunoprecipitation is carried out with the antibodies to the tag of one of the test proteins. If this protein interacts with the other protein, the latter coprecipitates, which can be detected in Western blot analysis using the anti-tag antibodies.

The crucial validation of two-hybrid results is to show that the two proteins exist in the same subcellular compartment and the interaction is biologically relevant. Colocalization of proteins in plants cells can be analyzed by fluorescence resonance energy transfer (FRET) or immunofluorescence of the two proteins with confocal microscope.

4 Notes

1. Compatibility of different systems: The compatibility depends on the selection markers of individual plasmids, the type of DNA-binding domain used, and the genotype of yeast strains. The general rule is that the DNA-binding domain can bind to the corresponding upstream sequence in the reporter genes, and all the plasmids can be maintained by nutritional or antibiotic resistance selection. For example, the pGADT7 in the Matchmaker system (Clontech) can be replaced with the pVP16 vector of the LexA system (22), and *vice versa*. The same is true for pJG4-5 (OriGene, Clontech) and pPC86 (Invitrogen), pGBKT7 (Clontech) and pBD-GAL4 (Stratagene), and pHybLex/Zeo (Invitrogen) and pEG202 (Origene, Clontech).

2. Evaluation of yeast two-hybrid assay results: Although YTHS is a powerful method for analyzing many protein–protein interactions in a relatively time- and cost-efficient manner, great care should be taken when interpreting the results. Similar to any methods used for detecting protein–protein interactions, YTHS cannot reveal all of the protein interactions taking place in biological processes, thus resulting in "false negatives." The reasons for obtaining false negative results are not clear, but may include the sequestration of one or two proteins by interaction with endogenous yeast proteins, or instability and improper folding of fusion proteins at an unnatural subcellular compartment, or absence of additional cellular factors required for the interaction, or steric constraints on interaction properties and occlusion of

interaction sites by the fusion domains. The latter situation may be avoided by fusing the proteins of interest to different AD or BD domains (e.g., from Gal4 to LexA, or *vice versa*) or changing a usual C-terminal fusion vector to an unusually used N-terminal fusion vector (e.g., pNlexA; see Table 1).

On the other hand, YTHS can also generate false positive interactions that are not of biological relevance. False positive interactions may happen with inherently "sticky" proteins, which interact with many partners in a partially specific manner. This type of false positives can be eliminated by including negative controls in the experiments. Other false positives may result from the changes in the metabolic functioning of yeast induced by fusion protein expression, leading to bias of the reporter assays. False positives (self-activation) can also take place when the AD fusion proteins interact directly with the sequences flanking the GAL4 binding site or transcription factors bound to the specific TATA boxes. This type of false positives can be eliminated by using different reporter genes with different promoter structures, e.g., in the Clontech GAL4 system 3. It is noteworthy to mention that there is one class of false positives that cannot be eliminated by YTHS itself, i.e., the two proteins do interact in yeast cells, but they are localized in different subcellular compartments or expressed in different time periods *in vivo*, so that they do not have chance to meet *in vivo* and such interactions are physiologically irrelevant. Therefore, the ultimate verification of the interactions should be carried out *in vivo*.

3. Self-activation of reporter genes: If one fusion protein with either BD or AD activates the reporter gene activities on its own, it cannot be used directly in the yeast two-hybrid assays. However, such self-activation activity may be eliminated by switching the gene fragment from the AD vector to the BD vector and *vice versa*. Deletion of a part of the gene while keeping the interaction domains of interest may also eliminate or alleviate the self-activation.
4. Background growth: The reporter gene *HIS3* in some yeast strains like L40, Y153, and Y190 carries the original TATA box and thus has constitutive, leaky expression of HIS3, leading to background growth of yeast cells in medium lacking histidine. Such background growth can be suppressed by 2–50 mM of 3-aminotriazole (3-AT).
5. Toxicity of fusion protein expression: The test protein may be toxic to yeast cells. This problem could be avoided by using vectors with weak promoters such as pGBT9 and pGAD424 or inducible promoters such as pJG4-5 in the LexA system. Deletions introduced to the protein may also reduce toxicity.
6. Failures to detect protein interactions: If you fail to detect interactions between two proteins, there could be a variety of reasons. One possibility is that the two test proteins are true noninteractors; however, in many cases, failure to detect interactions are due to false negatives as discussed in Sect. 4.2 or the fusion proteins cannot be localized to the yeast nucleus. If the full-length proteins show negative results, truncated proteins can be tested. Switch from GAL4-based system to LexA-based system or *vice versa* may help to detect more interactions.
7. Strength of protein–protein interactions: In general, the interactions between viral proteins in YTHS tend to be much weaker than those found between yeast proteins. Therefore, one should be prepared to grow the yeast for relatively long time and include appropriate controls. For example, the AD and BD fusions of the viral coat protein that has the inherent ability to interact and form particles, or viral proteins known to form dimers, can be included for comparison. The relative strength of the interactions can be measured by using a quantitative α- or β-galactosidase assay, which shows correlation with the results from *in vitro* affinity measurements (41). These assays are useful for comparison of the variants of the same protein or the different interacting partners of a protein. However, quantitative data from the liquid assays cannot be compared between different host strains carrying different reporter constructs.
8. PMSF is hazardous.
9. SD is the minimal synthetic dropout medium where appropriate nutrient(s) is dropout. Untransformed yeast strains that are auxotrophic cannot grow in SD medium missing one or more of the nutrients, so it can be used for selecting transformants (colonies that have acquired plasmids carrying the nutritional markers or contain the protein interactions that activate the reporter genes).

10. For nutrient selection, leave out appropriate amino acids and/or nucleic bases. For example, if you plan to make SD-Leu selection medium, leave out leucine from the dropout solution; and if you make SD-Leu-Trp-His, leave out leucine, tryptophan, and histidine HCl monohydrate from your dropout solution. Alternatively, you can make 10× dropout-stock without adenine hemisulfate, histidine HCl monohydrate, leucine, and tryptophan. The missing nutrients can be prepared as separate concentrated solutions that are added to SD selection medium as needed (e.g., to make SD-Leu-Trp-His medium, add adenine hemisulfate salt to the dropout stock of SD-Leu-Trp-His-Ade).

11. If you use the yeast nitrogen base without ammonium sulfate, you need to add 5 g ammonium sulfate and 1.7 g of yeast nitrogen base. You can also autoclave yeast nitrogen base, but in that case plates will be softer. Alternatively, you can purchase minimal SD base and minimal SD agar base as well as different dropout (DO) solutions in powder form from commercial sources.

12. Two or more plasmids can be transformed simultaneously, but the efficiency of cotransformation decreases with each added plasmid.

13. Failures to detect fusion protein in Western analysis: Sometimes, it might be difficult to detect the fusion protein in Western blot. You may try to optimize incubation time and temperature for culturing the yeast cells in liquid media as some proteins are degraded by prolonged incubation. Alternatively, collect yeast colonies directly from agar plate followed by suspending in liquid for extraction of yeast proteins. Note: some two-hybrid vectors (e.g., pGAD424 and pGBT9) use truncated *ADH1* promoter that is very weak and leads to low level expression of fusion proteins undetectable in Western analysis.

14. To be at the linear range, it should be between 0.5 and 1.0. If necessary, dilute the cell suspension.

Acknowledgments We thank Dr Qingzhen Liu for critically reading the manuscript and Xiaoyun Wu for compiling the reference list. Deyin Guo's laboratory is supported by the Luojia Professorship Program of Wuhan University and China National Science Foundation (NSFC-30570394 and 30270313). The YTHS studies on plant virus proteins of M.L. Rajamäki and J.P.T. Valkonen are funded by the Academy of Finland (grant 204104, 1102003, 1102134 and 1110797).

References

1. Golemis, E. A., Tew, K. D., and Dadke, D. (2002) Protein interaction-targeted drug discovery: evaluating critical issues. *Biotechniques* **32**, 636–642.
2. Fields, S. and Song, O. (1989) A novel genetic system to detect protein-protein interactions. *Nature* **340**, 245–246.
3. Bartel, P. L., Roecklein, J. A., SenGupta, D., and Fields, S. (1996) A protein linkage map of *Escherichia coli* bacteriophage T7. *Nat. Genet.* **12**, 72–77.
4. McCraith, S., Holtzman, T., Moss, B., and Fields, S. (2000) Genome-wide analysis of vaccinia virus protein-protein interactions. *Proc. Natl. Acad. Sci. USA* **97**, 4879–4884.
5. Choi, I. R., Stenger, D. C., and French, R. (2000) Multiple interactions among proteins encoded by the mite-transmitted wheat streak mosaic tritimovirus. *Virology* **267**, 185–198.
6. Flajolet, M., Rotondo, G., Daviet, L., Bergametti, F., Inchauspe, G., Tiollais, P., Transy, C., and Legrain, P. (2000) A genomic approach of the hepatitis C virus generates a protein interaction map. *Gene* **242**, 369–379.
7. Guo, D., Rajamaki, M. L., Saarma, M., and Valkonen, J. P. T. (2001) Towards a protein interaction map of potyviruses: protein interaction matrixes of two potyviruses based on the yeast two-hybrid system. *J. Gen. Virol.* **82**, 935–939.

8. Kang, S. H., Lim, W. S., and Kim, K. H. (2004) A protein interaction map of soybean mosaic virus strain G7H based on the yeast two-hybrid system. *Mol. Cells* **18**, 122–126.
9. Zell, R., Seitz, S., Henke, A., Munder, T., and Wutzler, P. (2005) Linkage map of protein-protein interactions of Porcine teschovirus. *J. Gen. Virol.* **86**, 2763–2768.
10. Uetz, P., Dong, Y. A., Zeretzke, C., Atzler, C., Baiker, A., Berger, B., et al. (2006) Herpesviral protein networks and their interaction with the human proteome. *Science* **311**, 239–242.
11. Ito, T., Tashiro, K., Muta, S., Ozawa, R., Chiba, T., Nishizawa, M., et al. (2000) Toward a protein–protein interaction map of the budding yeast: a comprehensive system to examine two-hybrid interactions in all possible combinations between the yeast proteins. *Proc. Natl. Acad. Sci. USA* **97**, 1143–1147.
12. Uetz, P., Giot, L., Cagney, G., Mansfield, T. A., Judson, R. S., Knight, J. R., et al. (2000) A comprehensive analysis of protein–protein interactions in *Saccharomyces cerevisiae*. *Nature* **403**, 623–627.
13. Rain, J. C., Selig, L., De, R. H., Battaglia, V., Reverdy, C., Simon, S., et al. (2001) The protein–protein interaction map of *Helicobacter pylori*. *Nature* **409**, 211–215.
14. Giot, L., Bader, J. S., Brouwer, C., Chaudhuri, A., Kuang, B., Li, Y., et al., (2003) A protein interaction map of *Drosophila melanogaster*. *Science* **302**, 1727–1736.
15. Li, S., Armstrong, C. M., Bertin, N., Ge, H., Milstein, S., Boxem, M., et al. (2004) A map of the interactome network of the metazoan *C. elegans*. *Science* **303**, 540–543.
16. LaCount, D. J., Vignali, M., Chettier, R., Phansalkar, A., Bell, R., Hesselberth, J. R., et al. (2005) A protein interaction network of the malaria parasite *Plasmodium falciparum*. *Nature* **438**, 103–107.
17. Rual, J. F., Venkatesan, K., Hao, T., Hirozane-Kishikawa, T., Dricot, A., Li, N., et al. (2005) Towards a proteome-scale map of the human protein–protein interaction network. *Nature* **437**, 1173–1178.
18. Stelzl, U., Worm, U., Lalowski, M., Haenig, C., Brembeck, F. H., Goehler, H., et al. (2005) A human protein–protein interaction network: a resource for annotating the proteome. *Cell* **122**, 957–968.
19. Chien, C. T., Bartel, P. L., Sternglanz, R., and Fields, S. (1991) The two-hybrid system: a method to identify and clone genes for proteins that interact with a protein of interest. *Proc. Natl. Acad. Sci. USA* **88**, 9578–9582.
20. Gyuris, J., Golemis, E., Chertkov, H., and Brent, R. (1993) Cdi1, a human G1 and S phase protein phosphatase that associates with Cdk2. *Cell* **75**, 791–803.
21. Vojtek, A. B., Hollenberg, S. M., and Cooper, J. A. (1993) Mammalian Ras interacts directly with the serine/threonine kinase Raf. *Cell* **74**, 205–214.
22. Hollenberg, S. M., Sternglanz, R., Cheng, P. F., and Weintraub, H. (1995) Identification of a new family of tissue-specific basic helix-loop-helix proteins with a two-hybrid system. *Mol. Cell Biol.* **15**, 3813–3822.
23. Li, J. J. and Herskowitz, I. (1993) Isolation of ORC6, a component of the yeast origin recognition complex by a one-hybrid system. *Science* **262**, 1870–1874.
24. SenGupta, D. J., Zhang, B., Kraemer, B., Pochart, P., Fields, S., and Wickens, M. (1996) A three-hybrid system to detect RNA-protein interactions in vivo. *Proc. Natl. Acad. Sci. USA* **93**, 8496–8501.
25. Putz, U., Skehel, P., and Kuhl, D. (1996) A tri-hybrid system for the analysis and detection of RNA–protein interactions. *Nucleic Acids Res.* **24**, 4838–4840.
26. Zhang, J. and Lautar, S. (1996) A yeast three-hybrid method to clone ternary protein complex components. *Anal. Biochem.* **242**, 68–72.
27. Dalton, S., and Treisman, R. (1992) Characterization of SAP-1, a protein recruited by serum response factor to the c-fos serum response element. *Cell* **68**, 597–612.
28. Naya, F. J., Stellrecht, C. M., and Tsai, M. J. (1995) Tissue-specific regulation of the insulin gene by a novel basic helix-loop-helix transcription factor. *Genes Dev.* **9**, 1009–1019.
29. Aronheim, A., Zandi, E., Hennemann, H., Elledge, S. J., and Karin, M. (1997) Isolation of an AP-1 repressor by a novel method for detecting protein-protein interactions. *Mol. Cell Biol.* **17**, 3094–3102.

30. Broder, Y. C., Katz, S., and Aronheim, A. (1998) The ras recruitment system, a novel approach to the study of protein–protein interactions. *Curr. Biol.* **8**:1121–1124.
31. Johnsson, N., and Varshavsky, A. (1994) Split ubiquitin as a sensor of protein interactions in vivo. *Proc. Natl. Acad. Sci. USA* **91**, 10340–10344.
32. Stagljar, I., Korostensky, C., Johnsson, N., and Te, H. S. (1998) A genetic system based on split-ubiquitin for the analysis of interactions between membrane proteins *in vivo*. *Proc. Natl. Acad. Sci. USA* **95**, 5187–5192.
33. Leanna, C. A., and Hannink, M. (1996) The reverse two-hybrid system: a genetic scheme for selection against specific protein/protein interactions. *Nucleic Acids Res.* **24**, 3341–3347.
34. Vidal, M., Braun, P., Chen, E., Boeke, J. D., and Harlow, E. (1996) Genetic characterization of a mammalian protein–protein interaction domain by using a yeast reverse two-hybrid system. *Proc. Natl. Acad. Sci. USA* **93**, 10321–10326.
35. Joung, J. K., Ramm, E. I., and Pabo, C. O. (2000) A bacterial two-hybrid selection system for studying protein–DNA and protein–protein interactions. *Proc. Natl. Acad. Sci. USA* **97**, 7382–7387.
36. Luo, Y., Batalao, A., Zhou, H., and Zhu, L. (1997) Mammalian two-hybrid system: a complementary approach to the yeast two-hybrid system. *Biotechniques* **22**, 350–352.
37. James, P., Halladay, J., and Craig, E. A. (1996) Genomic libraries and a host strain designed for highly efficient two-hybrid selection in yeast. *Genetics* **144**, 1425–1436.
38. Harper, J. W., Adami, G. R., Wei, N., Keyomarsi, K., and Elledge, S. J. (1993) The p21 Cdk-interacting protein Cip1 is a potent inhibitor of G1 cyclin-dependent kinases. *Cell* **75**, 805–816.
39. Sambrook, J., and Russell, D.W. (2001) Molecular Cloning: A laboratory manual, 3rd ed., vol. 1. Cold Spring Harbor Laboratory Press, Cold Spring Harbor, New York.
40. Miller, J.H. (1972) Experiments in molecular genetics. Cold Spring Harbor Laboratory Press. Cold Spring Harbor, New York.
41. Estojak, J., Brent, R., and Golemis, E. A. (1995) Correlation of two-hybrid affinity data with *in vitro* measurements. *Mol. Cell. Biol.* **10**, 5820–5829.

Chapter 30
NMR Analysis of Viral Protein Structures

Andrew J. Dingley, Inken Lorenzen, and Joachim Grötzinger

Abstract Nuclear magnetic resonance (NMR) spectroscopy is a powerful tool to study the three-dimensional structure of proteins and nucleic acids at atomic resolution. Since the NMR data can be recorded in solution, conditions such as pH, salt concentration, and temperature can be adjusted so as to closely mimic the biomacromolecules natural milieu. In addition to structure determination, NMR applications can investigate time-dependent phenomena, such as dynamic features of the biomacromolecules, reaction kinetics, molecular recognition, or protein folding. The advent of higher magnetic field strengths, new technical developments, and the use of either uniform or selective isotopic labeling techniques, currently allows NMR users the opportunity to investigate the tertiary structure of biomacromolecules of ~50 kDa. This chapter will outline the basic protocol for structure determination of proteins by NMR spectroscopy. In general, there are four main stages: (i) preparation of a homogeneous protein sample, (ii) the recording of the NMR data sets, (iii) assignment of the spectra to each NMR observable atom in the protein, and (iv) generation of structures using computer software and the correctly assigned NMR data.

Keywords Chemical shift; NMR; Protein structure; Structural biology; Viruses

1 Introduction

Innovations in structural biology have profoundly transformed our understanding of the function of biological molecules, and clearly the use of structural biology tools to characterize proteins encoded by plant virus genomes will provide detailed insights into processes such as suppression of posttranscriptional gene silencing. Although X-ray crystallography is considered the most prominent technique for three-dimensional (3D) structure determination at atomic resolution, not all biological molecules are conducive to crystallization or behave the same in the crystalline state as they do in the cellular milieu. Nuclear magnetic resonance (NMR) spectroscopy has therefore emerged as an important complement to X-ray crystallography

From: *Methods in Molecular Biology, Vol. 451, Plant Virology Protocols: From Viral Sequence to Protein Function*
Edited by G.D. Foster, I.E. Johansen, Y. Hong, and P.D. Nagy © Humana Press, Totowa, NJ

as it allows biomacromolecular structure to be determined in conditions that closely mimic the physiological state. Many of the structural genomics consortia rely equally on both methods to solve protein structures (1, 2). In addition, NMR spectroscopy can be used to characterize events that cannot readily be quantified or observed by X-ray crystallography, such as dynamic features of biomacromolecules, reaction kinetics, molecular interactions (e.g., drug binding), or protein folding (3–6).

The first solution-state NMR protein structure was determined in 1984 using methods developed by Wüthrich and co-workers (7). In these early years, 3D protein structure determination using NMR spectroscopy took several months and in some situations years to complete and was restricted to small (10 kDa) well-behaved proteins. However, in the last decade the advance of instrument technology coupled with the development of techniques enables researchers to characterize larger and more complex biological systems much more quickly and with far greater precision. Protein structures of 30–40 kDa are routinely solved, protein complexes up to 900 kDa have been analyzed (8), and membrane proteins in detergent micelle systems are providing structural information that was previously inaccessible (9–11). The primary intent of this chapter is to provide a generic guide for solving protein structures by NMR spectroscopy.

2 Materials

2.1 NMR Sample Preparation

1. Deuterium oxide (D_2O), uniformly ^{13}C-labeled glucose and ^{15}N-labeled salts (i.e., NH_4Cl or $(NH_4)_2SO_4$ (Cambridge Isotope Laboratories, Andover, MA)).
2. The most widely used bacteria host is the B strain *Escherichia coli* BL21 (DE3) and derivates thereof.
3. Luria–Bertani (LB) agar (per liter): Add 10 g tryptone (Difco, Detroit, MI), 5 g yeast extract, 10 g NaCl to 800 mL of H_2O. Adjust pH to 7.5 with NaOH and then add 15 g agar and melt agar into the solution using a microwave. Adjust volume to 1 L with H_2O and sterilize by autoclaving. Store at 4°C.
4. LB broth: The same procedure as LB agar, except no agar is added. Store at 4°C.
5. To produce uniformly $^{13}C/^{15}N$-labeled proteins, the overexpression of the target protein in a bacterial expression system is performed in a medium which contains uniformly ^{13}C-labeled glucose and ^{15}N-labeled NH_4Cl or $(NH_4)_2SO_4$ as the sole carbon and nitrogen sources, respectively. Modifications to the basic M9 minimal medium strategy is used (Table 1) (12, 13). When ^2H enrichment is required, the overexpression of the target protein is performed in D_2O rather than H_2O, although alternative ^2H-labeling methods are available (14).
6. Standard buffers for maintaining the protein sample stability and pH (see note 1).
7. A protein sample between 0.5 and 3 mM in 250–550 µL is required.

Table 1 Recipe for 1 L of minimal growth medium

Basic solution (anhydrous salts, autoclave after preparation)[a]	g/970 mL
KH_2PO_4	13.0
Na_2HPO_4	9.0
K_2HPO_4	10.0
K_2SO_4	2.4
NH_4Cl or $(NH_4)_2SO_4$	1.0
10 mL trace element solution (sterilize by filtration)[b]	g/100 mL
EDTA-Na_2	0.50
$CaCl_2{\cdot}2H_2O$	0.60
$FeSO_4{\cdot}7H_2O$	0.60
$MnCl_2{\cdot}4H_2O$	0.12
$CoCl_2{\cdot}6H_2O$	0.08
$ZnSO_4{\cdot}7H_2O$	0.07
$CuCl_2{\cdot}2H_2O$	0.03
H_3BO_3	0.002
$(NH_4)_6Mo_7O_{24}{\cdot}4H_2O$	0.025
5 mL Aqueous Vitamins (sterilize by filtration)	mg/100 mL
Pantothenic acid	60
Folic acid	40
Nicotinate	500
Pyridoxine HCl	40
Riboflavin	20
Thiamine	100
Cyanobolamin	2
1 M $MgCl_2$ (sterilize by filtration)	1 mL
Uniformly ^{13}C-labeled D-glucose[c,d]	2–5 g L^{-1}

[a] Adjust to pH 7.2–7.4 with KOH. pH should be ~6.7 before adjustment.
[b] Add the EDTA to a fraction of the water. Add each ingredient separately, stir and wait for several minutes. Add the rest of the water, wrap the container in foil and stir overnight. The color should turn from green to gold.
[c] Dissolve in 5–10 mL of water and sterilize by filtration.
[d] Use 8 g L^{-1} for proteins labeled only with ^{15}N.

8. Chemical shift referencing products: 3-trimethylsilyl(2,2,3,3,-$^2H_4$)propionate (d_4-TSP) or tetramethylsilane (TMS, Sigma-Aldrich, St. Louis, MO).
9. Standard high quality (528-PP or better) 5 mm outer diameter NMR tubes (Wilmad, Buena, NJ) or 5 mm symmetrical matched microcells (Shigemi Inc., Allison Park, PA).

3 Methods

3.1 NMR Sample Preparation

The preparation of a sample is frequently the bottleneck for structural studies by NMR spectroscopy. Sample preparation is the process of preparing sufficient quantities of pure (>95%), soluble protein to begin the process of structure determination. Since NMR spectroscopy suffers from intrinsically poor sensitivity compared to other spectroscopic techniques, a protein concentration in the low mM range (0.5–3 mM) is required, and this sample should preferably be stable for at least a few weeks. NMR spectroscopy measures the absorbance of radio frequency radiation that occurs when particular magnetically active nuclei (e.g., ^1H, ^{13}C, ^{15}N, ^{31}P) are placed under the influence of strong magnetic fields. During the early 1980s, ^1H homonuclear experiments were performed with unlabeled samples (i.e., ^1H only) and structure determination was restricted to proteins <10 kDa in size. With the advent of heteronuclear methods coupled with better hardware, the determination of protein structures beyond 10 kDa exploited the benefits of detecting ^1H, ^{13}C, and ^{15}N nuclei in a single heteronuclear multidimensional experiment. More recent advances in NMR methods using ^2H labeling permit proteins beyond 50 kDa to be viable structural targets by NMR spectroscopy (15).

This section will focus on the bacterial overexpression approach, which is currently the standard approach used for ^{15}N-, or $^{15}N/^{13}C$-, or $^{15}N/^{13}C/^2H$ enrichment of proteins (see note 2). There are various protocols available, and no single protocol works effectively for every protein overexpression. For example, changes in the amount of aeration, temperature, cell density, concentration of nutrients, and maintenance of pH during cell growth can sometimes prove vital in maximizing soluble protein expression levels. The protocol below is a starting point from which modifications can be made to improve protein yields (12, 13, 15). Protein overexpression should preferably produce soluble correctly folded protein. Expression of insoluble protein should be avoided as this indicates that the protein may be poorly structured or incorrectly folded. Additionally, purification of insoluble protein material will require a refolding step. Solubility enhancement tags (SET) have been designed to improve the production of the target protein in the soluble form (16, 17).

3.1.1 Protein Overexpression in Minimal Medium Using a Bacteria Host System

1. The cDNA fragment coding for the target protein is cloned into a T7 overexpression system (e.g., pET vector system, Novagen, San Diego, CA) using standard molecular biology tools. This plasmid is transformed into BL21 (DE3) pLysS cells and plated out on to LB-agar plates containing an appropriate antibiotic and incubated overnight (~16 h) at 37°C (see note 3).
2. When growing bacterial cells in M9 minimal medium, it is important to acclimatize the cells to this medium, otherwise protein overexpression can fail. Single colonies are used to inoculate 10 mL LB-broth cultures containing antibiotic. Cultures are placed in a shaking air incubator at 37°C and 200–300 rpm and growth monitored by optical density at 595 nm (OD_{595}). Once cells have entered mid-exponential growth phase, the cells are harvested, resuspended in unlabeled (i.e., $^{14}N/^{12}C$) minimal medium and transferred to a 200 mL unlabeled

minimal medium culture with antibiotic and cell growth continued overnight at 37°C and 200–300 rpm.

3. The 200 mL culture is harvested by centrifugation and the cell pellet resuspended in a small volume (~20 mL) of labeled minimal medium. This cell suspension is used to inoculate a large (e.g., 500–2,000 mL) labeled minimal medium culture which is used for overexpression of the target protein. To improve aeration, large cultures should be performed using baffled shaker flasks or fermentors. Protein expression is under the control of the *lacUV5* promoter, and thus, protein expression is triggered by the addition of the chemical inducer isopropyl β-D-1-thiogalactopyranoside (IPTG) (18) (see note 4). IPTG is added when the cell density has reached 0.6–1.0. Cells are harvested between 2 and 6 h after induction. The harvesting time point should be determined empirically as it will depend on the density of the culture at the time of induction, the length of induction for optimal protein yield, and the response of the host to the expressed protein (see note 5).

4. The cells are resuspended in ~40 mL of Tris buffer, pH 8.0 and frozen at − 20°C.

3.1.2 Protein Purification

The use of affinity tags has dramatically reduced the time required to purify proteins. The most popular affinity tag is the polyhistidine tag which may be fused to either the N or C terminus of the target protein (see note 6). A protease cleavage site (e.g., tobacco etch virus (TEV) protease, thrombin, Factor X) is positioned between the affinity tag and the target protein. Most purification protocols that we have used for recovery of soluble protein involve these generic steps with a polyhistidine tagged protein.

1. The cells are thawed and completely homogenized by passage through a French pressure cell at 20,000 psi. Other homogenization processes that can be equally effective include tip sonicators, microfluidiser apparatus, and mechanical homogenizers. The sample should be kept cool during this process. The cell debris is removed by centrifugation and discarded.

2. The supernatant is loaded on to an immobilized metal affinity column (e.g., nickel charged-NTA resin). Purification is performed at 4°C with a solution flow rate of 0.5–2.0 mL min^{-1}. The impurities are removed by washing the column thoroughly with buffer. Polyhistidine tagged target protein is eluted from the column by washing with increasing concentrations of imidazole (up to 500 mM) or by reducing the pH of the buffer. Purification should be >80%.

3. The polyhistidine affinity tag is removed by digestion with the appropriate protease. Recombinant TEV protease (Invitrogen, Carlsbad, CA) is the preferred choice and is used at ratios between 1:10 and 1:100 TEV:protein (w/w).

4. Preparative size exclusion chromatography is used to further purify the sample (see note 7). Purification is performed at 4°C with a solution flow rate of 0.5–2.0 mL min^{-1}. This step can be used to exchange the protein into the buffer used for NMR analysis.

5. Purified protein is concentrated using centrifugal ultrafiltration vivaspin concentrators (Vivascience Limited, Gloucestershire, UK). The sample volume is reduced to the required NMR concentrations (see note 8).
6. During the protein purification process, small aliquots of samples should be taken for SDS-polyacrylamide gel electrophoresis (SDS-PAGE) analysis.

3.1.3 NMR Sample Preparation

Subsequent to finding suitable buffer conditions (see note 8) and concentrating the protein sample to the required millimolar concentration, the following reagents should be added to the sample:

1. Sodium azide (0.1–1.0 mM) to inhibit growth of bacteria.
2. Protease inhibitors: phenylmethylsulfonyl fluoride (0.01–0.1 mM), EDTA (0.1 mM), and aprotinin (5 μg mL^{-1}). Alternatively, protease inhibitor kits from Roche (Mannheim, Germany) are effective.
3. To prevent oxidation of any cysteine residues, 1–10 mM of either dithiothreitol (DTT) or tris(2-carboxyethyl)phosphine (TCEP) should be freshly added to the sample from a high-concentration stock solution. Deuterated DTT can be purchased from Cambridge Isotope Laboratories (Andover, MA).
4. D$_2$O to a final concentration of 5–7% (v/v). The spectrometer uses the deuterium signal (^{2}H) as a fixed reference frequency.
5. Chemical shift referencing compound: 1 mM TMS or d$_4$-TSP. This compound is added from a concentrated stock solution.
6. Oxygen has a dual deleterious effect on protein samples for NMR analysis because it is a paramagnetic species (which causes the signals in the NMR spectra to broaden) and an oxidizing agent. To prevent these problems, samples should be completely evacuated with nitrogen or other inert gases, and the sample sealed.
7. The final volume should be 250–280 μL for samples in 5 mm symmetrical matched microcells and 500–550 μL for samples in standard 5 mm outer diameter NMR tubes.

3.2 Collection of NMR Data

This section provides a guide for (i) assessing whether the prepared protein sample is suitable for structural analysis by NMR spectroscopy and (ii) the basic approach for collecting data for solving the structure of medium-sized (i.e., ~25 kDa) proteins by NMR spectroscopy. For proteins >30 kDa, where substantial broadening of the peaks occurs due to the slower tumbling rates of the protein, ^{2}H-labeling is required. In contrast, an unlabeled protein sample combined with homonuclear two-dimensional (2D) NMR experiments may be sufficient to determine the structure of small proteins (<10 kDa) (19).

1. NMR experiments are performed in the temperature range between 15 and 37°C.
2. The initial NMR experiments to record are one-dimensional (1D) ¹H spectra using an unlabeled protein sample. These 1D ¹H NMR experiments provide insight into whether the protein is folded and monomeric (or forms higher multimers) and thus tractable for NMR structure determination. Two 1D spectra were recorded in ~5 min on a 600 MHz NMR instrument with sample concentrations of ~0.5 mM (Fig. 1). The top spectrum represents a well-behaved folded protein, whereas the bottom spectrum represents a primarily unstructured protein. The important features of a 1D spectrum that indicate that the protein is folded are good dispersion of the amide proton resonances (7.5–10 ppm) and the upfield shift of resonances (~0 ppm) arising from methyl protons buried within the protein core. Poorly folded proteins give rise to inadequate peak dispersion and resonance broadening due to chemical exchange phenomena (Fig. 1b). Consequently, these proteins are very difficult to study by NMR spectroscopy.

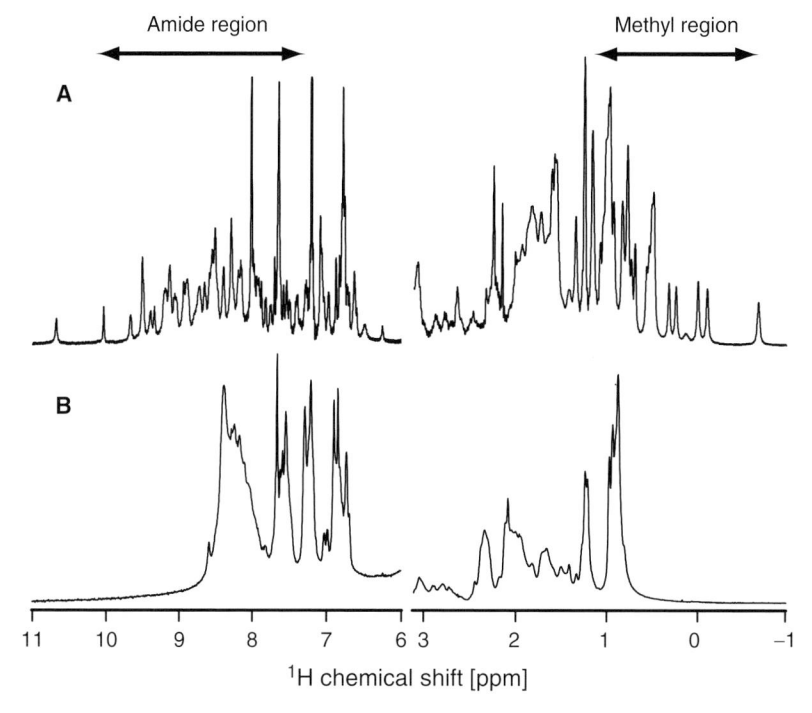

Fig. 1 Regions of 1D spectra of a folded protein and a primarily unstructured protein. (**a**) The 1D is characteristic of a globular protein (sushi domain from the IL-15 alpha receptor, IL-15Rα sushi domain) showing good dispersion of the amide proton resonances (7.5–10 ppm) and upfield shift of resonances (~0 ppm) arising from methyl protons buried within the protein core. (**b**) The 1D of the RNA silencing suppressor C2 protein from tomato yellow leaf curl geminivirus (TYLCV) is indicative of a primarily unstructured protein with poor peak dispersion in the amide region and resonance broadening due to chemical exchange. The amide region (6–11 ppm) of the spectra have been scaled by a factor of three compared to the upfield region of the spectra (−1–3 ppm)

3. The second piece of information that can be obtained from 1D spectra is the T_2 or transverse relaxation time of the amide protons (20). Two 1D ^{1}H spin-echo spectra are recorded with different transverse relaxation delay periods (or spin-echo times). By measuring the change in the peak intensities between the two 1D spectra, an approximate T_2 value is determined using (1):

$$T_2(\text{ms})=2(\Delta_A-\Delta_B)/\ln(I_A/I_B).\qquad(1)$$

Δ_A and Δ_B are the transverse relaxation delay periods used in the 1D spectra (e.g., $100\,\mu s$ and $2.9\,ms$), whereas I_A and I_B are the peak intensities recorded in the 1D spectra (see note 9).

The T_2 value is primarily dependent on the molecular tumbling rate, that is, the size of the protein – a larger protein will have a shorter T_2 value. Since viscosity influences the tumbling rate, raising the sample temperature will decrease the viscosity and therefore increase the T_2 value. This increase in T_2 can be beneficial as the sensitivity of the NMR experiments should improve. As a general rule, structure determination of proteins with $T_2 \geq 12\,ms$ can be solved without the requirement of ^{2}H enrichment.

4. All of the remaining NMR experiments are heteronuclear and require either ^{15}N or ^{15}N/^{13}C-enriched proteins. The 2D ^{1}H–^{15}N heteronuclear single quantum correlation (HSQC) experiment correlates the nitrogen atom of an N–H group with the directly attached proton. As such, a peak is observed for each backbone amide group in the protein and for particular side-chain groups (e.g., tryptophan indole NH$^{\epsilon1}$, asparagine and glutamine NH$_2$ groups) (see note 10). By inspection of the 2D ^{1}H–^{15}N HSQC spectrum and counting the number of peaks arising from the backbone amide groups indicates whether (i) the protein is folded, (ii) multiple conformers exist, (iii) peaks are broadened due to chemical/conformational exchange, or (iv) parts of the protein sequence are not visible at all. For example, the well-dispersed resonances in Fig. 2a are indicative of a stable folded protein. In contrast, Fig. 2b illustrates the spectrum of an essentially unstructured protein, in which, resonances are primarily clustered at ^{1}H random coil chemical shift values (7.8–8.4 ppm) and nonuniform peak intensities are due to chemical exchange.

5. The sequential assignment process involves recording a series of through-bond scalar correlation 3D NMR experiments to facilitate the identification or assignment of resonances to atoms in the protein. The 3D NMR experiments are named in the order that the magnetization is transferred between nuclei. The names of nuclei which are used only for transfer and whose frequencies are not detected are given in parentheses. For example, the HNCA experiment has three orthogonal chemical shift coordinates: the ^{1}H amide chemical shift on one axis (H), the ^{15}N amide chemical shift on another axis (N), and the ^{13}C$_\alpha$ chemical shift on the third axis (CA) (21). The HNCA gives an intraresidue peak for each amide group with its corresponding ^{13}C$_\alpha$ nuclei, and an interresidue peak with the ^{13}C$_\alpha$ of the proceeding residue (Fig. 3). By introducing a third dimension, the

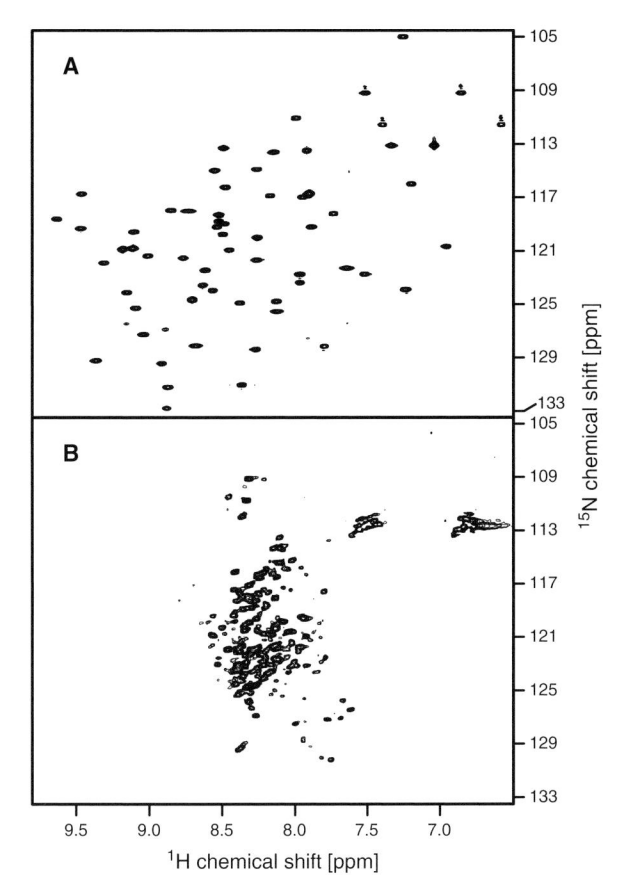

Fig. 2 2D ^{1}H–^{15}N HSQC spectra of (**a**) the IL-15Rα sushi domain and (**b**) the RNA silencing suppressor C2 protein of TYLCV

spectral overlap observed in 2D spectra is eliminated and the unambiguous assignment of resonances is possible. Figure 3 illustrates some common 3D NMR experiments used to obtain both backbone and side-chain assignment information. For sequence specific backbone assignments (i.e., ^{1}H$_N$, ^{1}H$_\alpha$, ^{15}N, ^{13}C$_\alpha$, ^{13}CO) combinations of the HNCA (22), HNCO (22), HN(CO)CA (22), HN(COCA)HA (23), HN(CA)HA (24), CBCA(CO)NH (25), HBHA(CBCACO)NH (26), HCACO (27), and CBCANH are routinely used, whereas for side-chain assignments the HCCH-TOCSY (28), (H)C(CO)NH/H(CCO)NH (29) and ^{15}N-edited TOCSY experiments are required. Although using many of these experiments introduces multiple redundancies, this improves the reliability of correctly assigning the resonances. Depending on sample concentration and hardware, each of these experiments takes between 1 and 3 days to record.

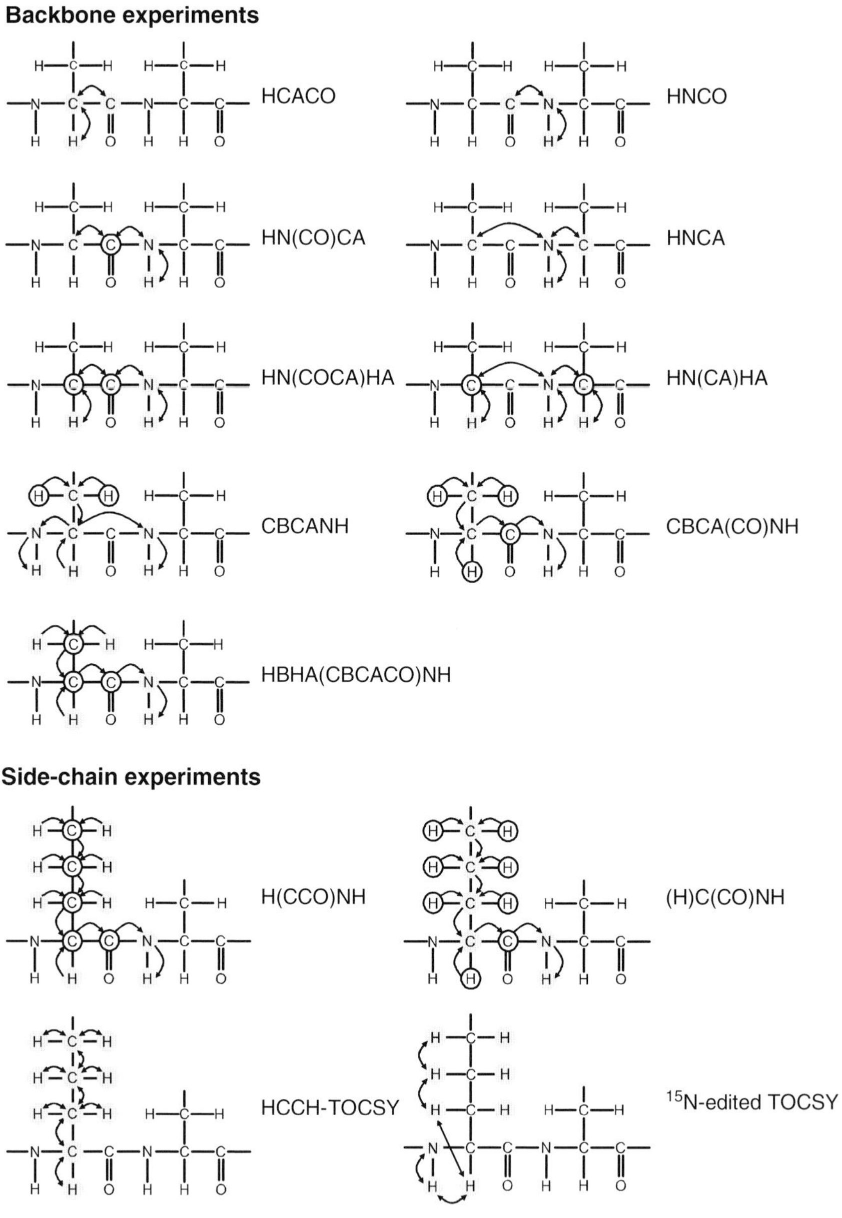

Backbone experiments

Side-chain experiments

Fig. 3 Schematic overview of heteronuclear 3D through-bond scalar correlation experiments used for obtaining backbone and side-chain assignment information. *Shaded* nuclei are frequency labeled, whereas *open circled* nuclei are only involved in the transfer of magnetization between nuclei. *Double-headed arrows* indicate that the experiment is an "out and back" experiment (i.e., the initially excited proton nucleus and the detected proton nucleus are the same) whereas *single-headed arrows* indicate that the experiment starts at a particular proton spin and detects on another proton

6. Assignment of proline residues is achieved using the HA(CA)N and/or HA(CACO)N experiments. These experiments correlate the ^{1}H$_\alpha$ with the intra-residue ^{15}N and with the ^{15}N resonance of the following residue (30).

7. Residues that are difficult to assign but provide many important long-range structural constraints in the structure calculations are the aromatic residues. 3D ^{13}C(aromatic)-edited H(C)CH-TOCSY and ^{13}C(aromatic)-edited NOESY HSQC experiments are used to obtain aromatic side-chain assignments. Alternatively, experiments that correlate the C$_\beta$ with the aromatic carbons or protons are employed (31, 32).

8. 3D ^{15}N-edited NOESY and ^{13}C-edited NOESY spectra are recorded (see note 11). These experiments provide "through-space" connectivities between pairs of protons that are in close spatial proximity (≤ 5 Å) and are the main source of geometric information used in protein structure determination by NMR spectroscopy. Several 3D NOESY spectra should be recorded with different mixing times (typically in the range of 50–250 ms) to determine more precisely the strength of the NOE interaction and to calibrate the distances between proximate pairs of protons. Each 3D NOESY experiment takes 2–4 days to record.

9. Dihedral angle constraints are derived from scalar or J-coupling constant data because simple geometric relationships exist between three-bond J-couplings and dihedral angles (33). Backbone phi (ϕ) angles are determined by recording quantitative J correlation 3D HNHA ($^3J_{HNH\alpha}$) (34), or (HN)CO(CO)NH ($^3J_{C'C'}$) (35) experiments, whereas the psi (Ψ) angles are determined by the HN(CO)CA experiment (36). The chi-one (χ_1) angles are derived from HNHB ($^3J_{NH\beta}$) (37) and ^{15}N–{^{13}C$_\gamma$}/^{13}C–{^{13}C$_\gamma$} spin-echo difference experiments ($^3J_{C'C\gamma}$ and $^3J_{NC\gamma}$) (38). Other methods to extract dihedral angle information include exclusive correlation spectroscopy type measurements and cross-correlated relaxation methods (39). More recently, backbone ϕ and ψ dihedral angle constraints are simply generated from secondary chemical shift data (i.e., ^{1}H$_\alpha$, ^{13}C$_\alpha$, ^{13}C$_\beta$, ^{13}CO, and ^{15}N) using the program TALOS (40).

10. The presence of hydrogen bonds (H-bonds) provides important secondary structure information and therefore valuable constraints in structure calculations. The detection of H-bond couplings (HBCs) is achieved using the long-ranges HNCO experiment (41, 42). This experiment directly detects the presence of an H-bond by the observation of a correlation between the donor amide group and the acceptor carbonyl carbon in backbone N–H...O=C H-bonds (see note 12). The long-range HNCO experiment requires 2–3 days acquisition time.

11. Residual dipolar couplings (RDCs) provide long-range structural information that is not accessible from NOE data (43). The inclusion of RDC constraints in the structure calculations typically improves the precision and accuracy of the structure. RDCs are different to J-couplings in that peaks are split due to interactions between two nearby nuclear dipoles that are differentially oriented relative to the external magnetic field. In solution, dipolar interactions normally average to zero, so RDCs are essentially "invisible." Partial or residual orientation is achieved by introducing an alignment agent to the sample and thereby

these dipolar couplings can be observed. The most commonly used alignment mediums are oriented bicelles, filamentous phages, rod-shaped cellulose particles, n-alkyl-poly(ethylene glycol)/n-alkyl alcohol mixtures, and compressed acrylamide-gel matrices (44). The simplest RDC measured is the one-bond amide group dipolar couplings ($^1D_{NH}$). To measure the $^1D_{NH}$, the 2D IPAP ^1H–^{15}N HSQC experiment is recorded (45) on protein samples in the presence and absence of alignment (0.5 days to record). By taking the difference in the magnitude of the splitting of the peaks observed in the isotropic (nonaligned, observe only the one-bond J-coupling ($^1J_{NH}$)) and anisotropic (aligned, observe) (both $^1J_{NH}$ and $^1D_{NH}$) conditions, the magnitude and sign of the $^1D_{NH}$ is derived for each amide group. Besides the $^1D_{NH}$, four other dipolar couplings associated with the protein backbone are frequently measured. These are the one-bond $C_\alpha C'$ ($^1D_{C\alpha C'}$), $C_\alpha H_\alpha$ ($^1D_{C\alpha H\alpha}$), C'N ($^1D_{C'N}$) and the two-bond $H_N C'$ ($^2D_{HNC'}$) RDCs. A 3D version of the HNCO experiment permits the measurement of the $^1D_{C\alpha C'}$, $^1D_{C'N}$, and $^2D_{HNC'}$ (46). 2D J-modulated ^1H–^{13}C HSQC (47) or 3D HN(CO)CA (46) experiments are recorded for determining the $^1D_{C\alpha H\alpha}$. More recently 3D experiments that use perdeuterated proteins provide a highly accurate approach to measure many one and two-bond RDCs between amide protons and nearby carbon and nitrogen atoms (48, 49).

3.3 Assignment of NMR Spectra

Once all the required NMR experiments are recorded, the information they contain has to be analyzed. Assignment is the process of identifying the resonance frequencies of all ^1H, ^{13}C, and ^{15}N nuclei. The assignment of peaks in the spectra recorded can be the most manually intensive step in solving a protein structure by NMR spectroscopy (see note 13).

1. Raw NMR data are processed using the script-driven software package NMRPipe (50) and analyzed (i.e., spectral display, interactive analysis, assignment) using either the CcpNmr (51) software suite, NMRView (52), or XEASY (53).
2. The first step in the structure determination process is to assign the backbone resonances. In principle, a combination of only two 3D experiments is sufficient for obtaining the majority of the sequential backbone assignment information – one experiment provides both intra- and interresidue correlations between backbone resonance frequencies and a second experiment that provides only interresidue peaks with the same set of backbone frequencies. For the actual assignment process, it is more efficient to reduce the three-dimensional space created by a 3D experiment to a series of 2D planes or "strips." A common assignment approach is to start with the 3D HNCA and HN(CO)CA experiments, since both have ^1H_N, ^{15}N, and $^{13}C_\alpha$ frequency axes. The HNCA spectrum shows at one ^1H_N and ^{15}N position correlations to two $^{13}C_\alpha$ nuclei (Fig. 3). One correlation is to the intraresidue $^{13}C_\alpha$ (i) and the other is to the preceding $^{13}C_\alpha$

(i − 1). The corresponding amide ^{1}H and ^{15}N frequency of the ^{13}C$_\alpha$ (i − 1) frequency is then found. To achieve this, the HNCA spectrum is scanned for a peak representing an intraresidue correlation with a ^{13}C$_\alpha$ frequency that matches the ^{13}C$_\alpha$ (i − 1) frequency of the interresidue peak. By finding this peak, these two residues are linked together and a sequential assignment has been made (although it will not be possible to define which two residues are linked until further peak assignments are completed). In addition to this intraresidue peak, this ^{1}H, ^{15}N position also shows a second correlation to the ^{13}C$_\alpha$ of the preceding residue. The identification of this residue's sequential neighbor is achieved using the same frequency matching procedure. Figure 4 illustrates the sequential assignment process using the HNCA experiment. One problem with this method is that intra- and interresidue peaks need to be distinguished. Typically the intraresidue peak is stronger than the interresidue correlation. However, this is not always the case and sometimes one of the two expected correlations is missing. To discriminate between the intra- versus interresidue correlations, the 3D HN(CO)CA experiment is used in conjunction with the HNCA experiment. The HN(CO)CA at the same 2D ^{1}H, ^{15}N position as in the HNCA identifies one of the two correlations in the HNCA experiment as the i − 1 correlation. The assignment of the ^{1}H, ^{15}N, and ^{13}C$_\alpha$ frequencies along the protein backbone can thus be achieved by combining these two experiments (Fig. 5). Any ambiguities in the assignment process due to peak overlap or the absence of a correlation can be resolved using alternative 3D NMR experiments (see note 14).

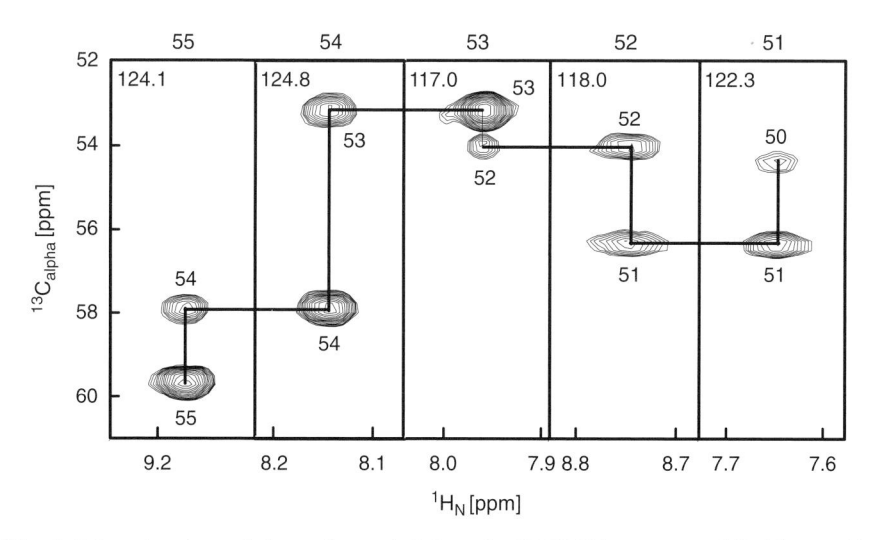

Fig. 4 Selected regions of planes (i.e., strips) from the 3D HNCA spectrum of IL-15Rα sushi domain along the ^{13}C$_\alpha$ axis taken at the ^{1}H, ^{15}N resonance frequencies of Leu[51] to Glu[55]. For each strip, at one ^{1}H$_N$ and ^{15}N position correlations to two ^{13}C$_\alpha$ nuclei are observed. One correlation is to the intraresidue ^{13}C$_\alpha$ (i) and the other is to the preceding ^{13}C$_\alpha$ (i − 1). Resonances are labeled with assignment information. The ^{15}N frequency is given in the top left-hand corner of each strip. The line traces the sequential assignments between each strip

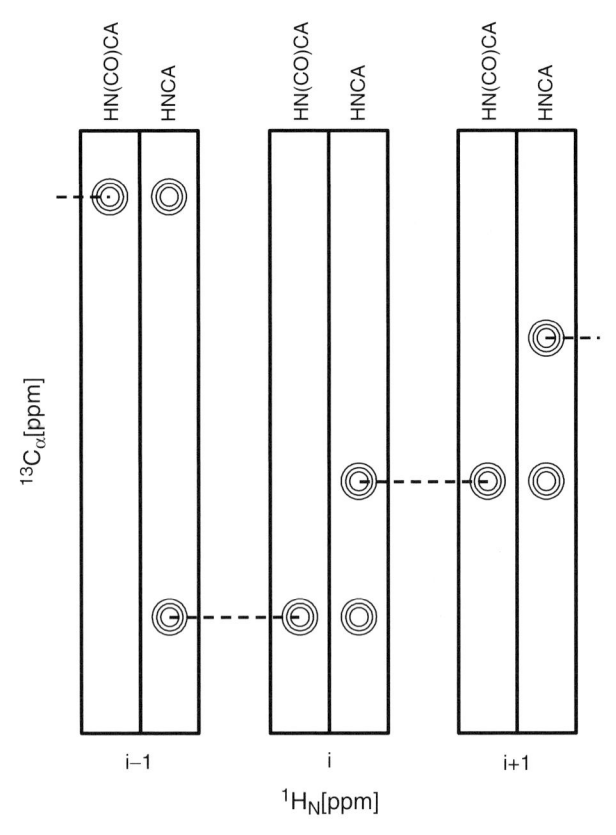

Fig. 5 Schematic of 2D strips taken from the 3D HN(CO)CA and HNCA spectra at ^{1}H, ^{15}N resonance frequencies of residues $i - 1$, i, and $i + 1$. The HNCA gives intra- and interresidue $^{13}C_\alpha$ correlations, whereas the HN(CO)CA provides only interresidue $i - 1$ correlations

3. There are a number of approaches that are used in combination to obtain side-chain assignment information. As for the backbone assignment process, it is easier to reduce the three-dimensional space to a series of 2D "strips" which correspond to peaks in a 2D spectrum (e.g., ^{1}H–^{15}N HSQC) with correlations along the length of the strip due to the transfer of magnetization between the side chain nuclei. As amino acid side-chains have different spin systems due to differences in their chemical composition, they give rise to particular peak connectivity patterns (19). In conjunction with the backbone assignment information, these different connectivity patterns assist in the side-chain assignment process. The 3D (H)C(CO)NH, H(CCO)NH, and ^{15}N-edited TOCSY data is used to assign aliphatic side-chain ^{13}C and ^{1}H resonances which are linked to the previously assigned backbone amide ^{1}H and ^{15}N frequencies (Fig. 6). Another approach is to correlate peaks arising from the side-chain ^{1}H and ^{13}C nuclei in the 3D HCCH-TOCSY and ^{13}C(aromatic)-edited H(C)CH-TOCSY experiments with previously assigned $^1H_\alpha$, $^{13}C_\alpha$, and $^{13}C_\beta$ resonances which are also observed in these experiments.

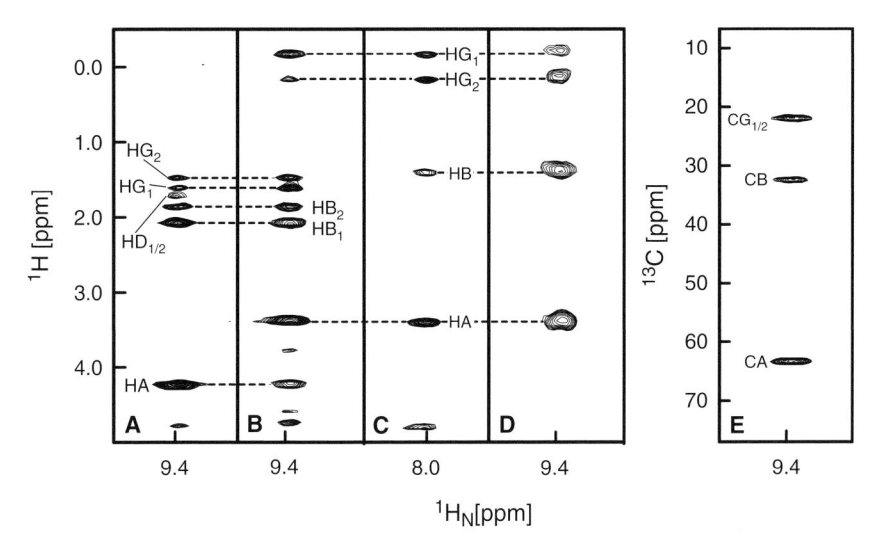

Fig. 6 2D strip plots of the correlations observed for residues Lys[47] (^{15}N frequency = 129.5 ppm) and Val[46] (^{15}N frequency = 123.6 ppm) of the IL-15Rα sushi domain from three-dimensional (**a**) ^{15}N-edited TOCSY (K47), (**b**) ^{15}N-edited NOESY (K47), (**c**) ^{15}N-edited TOCSY (V46), (**d**) H(CCO)NH, and (**e**) (H)C(CO)NH spectra. In the ^{15}N-edited TOCSY data, the amide groups of K47 and V46 show correlations to their side-chain protons, whereas for the 3D H(CCO)NH and (H)C(CO)NH experiments, the K47 amide group correlates with the aliphatic protons (**d**) and carbons (**e**) of the preceding residue (i.e., V46). The assignment data from the ^{15}N-edited TOCSY and H(CCO)NH can be used to assign interresidue and intraresidue NOEs between the amide proton of K47 and nearby protons (*dotted lines*). Resonances are labeled with assignment information

4. The BioMagResBank (BMRB web site) contains characteristic chemical shift ranges, averages, and standard deviations for ^{1}H, ^{15}N, and ^{13}C nuclei for all amino acids. This data is used to check for any incorrect peak assignments.

5. The assignment and quantification of the peaks in 3D NOESY spectra is achieved by using the assignment data that has been obtained from the previous steps. By overlaying 2D strips from the 3D ^{15}N/^{13}C-edited TOCSY and NOESY spectra, the assignment data from the TOCSY spectra can be translated on to the NOESY spectra and intraresidue and interresidue NOEs can be distinguished (Fig. 6). A peak that exists in the NOESY spectra that is absent in the TOCSY spectra indicates an interresidue NOE. The identification of an interresidue NOE involves searching through the assignment list to find the frequency that corresponds to the peak position (see note 15). The objective is to assign and quantitate these NOEs as they provide important structural constraints.

6. The assignment information is mapped (see note 16) on to the NMR spectra recorded for determining RDCs, [3]J-couplings (i.e., dihedral angles) and HBCs. For determining HBCs and [3]J-coupling values, quantitation of peak intensity or volume coupled with the appropriate formulae is required (e.g., [3]J-coupling magnitudes are related to dihedral angles by the Karplus equation), whereas the

difference in the distance (Hz) between the splitting of a peak in the spectra recorded when the protein is aligned versus nonaligned is used to determine RDC values. The backbone chemical shift data can be used by the program TALOS to calculate backbone ϕ and ψ dihedral angle values.

3.4 Structure Generation

Structure generation is driven by computer-intensive processes. NOE-based distances are the main source of structural information. These constraints are supplemented by the dihedral constraints from ^3J-coupling data, H-bond constraints from HBCs or exchange experiments, and RDC constraint information. Ideally, 10–15 constraints per residue give rise to a precise and accurate structure.

1. The chemical shift information is used to predict the secondary structure of the protein. The chemical shift index (CSI) is a statistical method to predict secondary structure elements from recorded chemical shifts by comparison with their corresponding random coil values (54, 55). The CSI values are derived from the $^{13}C_\alpha$, $^{13}C_\beta$, ^{13}CO, and $^1H_\alpha$ chemical shift data and provide an easy approach to determining secondary structure regions within the protein.
2. The intensity of each NOE peak is converted into an approximate distance. The NOE-based distances are divided into three groups:

    (i) Strong (1.8–2.7 Å): intense NOE peaks recorded in a NOESY spectrum with a short mixing time.
    (ii) Medium (1.8–4.0 Å): weak NOE signals that appear in the short mixing time NOESY spectrum.
    (iii) Weak (1.8–5 Å): additional NOE peaks that appear in NOESY spectra recorded with longer mixing times.

3. The NOE distance constraints, along with the other NMR-derived constraints are tabulated. This constraint file is supplemented with covalent geometry constraints (e.g., bond lengths, bond angles, amide-bond planarity, and chirality) about each amino acid.
4. The constraint file (without the H-bond and RDC constraints) is submitted to one or more distance geometry (e.g., DYANA) and/or simulated annealing programs (e.g., CNS/XPLOR) for structure calculation. These programs generate a family of protein structures that are consistent with the experimental constraints. The structure generation process typically involves iterative cycles of the geometry optimization process (see note 17). In the final stages of structure refinement, RDCs and H-bond constraints are added to the calculation.
5. PROCHECK-NMR or WHAT IF are used to assess the quality of the structure. These programs check various structural parameters including bond lengths and angles, dihedral angles, and measure atomic contacts. Validation of the structure is an important step and should not be ignored.

6. The following criteria indicate that the protein structure has been solved to good precision: (i) no persistent NOE violations >0.5 Å, (ii) a root mean square deviation (RMSD) of <0.5 Å for backbone atoms (calculated over the family of structures) and an RMSD of <1.0 Å for all atoms, (iii) ≤10% dihedral angle violations (PROCHECK-NMR), (iv) limited number of deviations from idealized covalent geometry (i.e., bonds, angles), (v) final number of bad contacts should be small.

4 Notes

1. Strong peaks in the NMR spectra appear when using buffers that have chemical compositions that include nonlabile hydrogen atoms (e.g., MES, glycine). These solute signals can sometimes complicate the acquisition and analysis of the NMR spectra. To circumvent this potential problem, there are a number of choices: (i) use solutes, if possible, with chemical structures that contain no hydrogen atoms (e.g., sodium phosphate), (ii) use deuterium-labeled solutes, and (iii) acquire isotope-editing NMR experiments which "filter" the signal through another nucleus such as ^{13}C or ^{15}N.

2. Although bacterial expression remains the most convenient, economical method for preparing isotopically labeled proteins, it is not always possible to obtain samples in this manner, particularly if there are problems with protein folding (e.g., complex disulphide bond pattern), bacterial toxicity, or the protein is packaged into insoluble aggregates (inclusion bodies). In addition, many eukaryotic proteins undergo posttranslational modifications that are absent in bacterial expression systems. Alternative expression systems for obtaining correctly folded soluble protein include methylotrophic yeast (e.g., *Pichia pastoris*), mammalian, and baculovirus hosts (15, 56, 57). More recent advances include using in vitro or cell-free protein synthesis platforms that are based on using either *E. coli* or wheat germ transcription and translational machinery (15, 58–61). Not only are these cell-free systems potentially cost-effective and more efficient than cell-based expression systems, they offer the opportunity to selectively label particular amino acids and introduce noncanonical amino acids into the protein sequence (56).

3. The plasmid confers a particular antibiotic resistance and this is used as the selection marker. The antibiotic should be freshly prepared and stored at −20°C. The concentration of the antibiotic used in the culture is between 50 and 100 µg mL^{-1}.

4. The optimal concentration of IPTG used for induction of soluble protein expression should be assayed. This can be performed by growing cultures in medium and inducing protein expression using different concentrations of IPTG (e.g., 0–10 mM). The cells are harvested, ruptured, and the soluble and insoluble material used to detect protein expression using SDS-PAGE. Very high levels of expression can lead to the production of insoluble protein aggregates or packaging of the target protein into inclusion bodies. This can be a problem, as a refolding step will be required and may not successfully lead to the purification of active correctly folded protein. An increase in the level of soluble protein expression may be achieved by reducing the concentration of IPTG used and decreasing the temperature from 37°C to as low as 15°C. Conversely, it may be advantageous to express the protein in inclusion bodies as this protects a loosely structured protease-sensitive protein from degradation.

5. Similar approaches are required when growing cells in D_2O for ^2H enrichment of the protein. Cells should progressively be trained to grow in higher ratios of D_2O/H_2O. Protein production in deuterated media usually results in lower yields than observed with nondeuterated media and expression times are considerably longer due to the isotope effect. As such, careful optimization of growth conditions is extremely important to ensure maximum protein yields when using deuterated minimal media.

6. Other affinity tags available include cellulose binding protein, glutathione-s-transferase, maltose binding protein, and strep-tag (62). The choice of affinity tag can affect the yield and solubility/stability of the expressed protein. Consequently, a number of different affinity tags should be tested, including positioning the tags at either terminus.

7. To ensure sample homogeneity, additional purification phases may be required prior to the final size exclusion chromatography step.

8. Although the protein may appear stable at lower concentrations in the buffer selected for the NMR studies, during the concentration process protein may precipitate or form large micro-aggregates. This may not be visible till the sample has been transferred to the NMR tube and some initial NMR experiments performed. Aggregation of the protein prevents the recording of good quality NMR spectra. As a rule, some effort must be made in finding the optimal conditions for protein stability that precludes precipitation. This may include sampling numerous temperatures, pH values (i.e., different buffers), and salt conditions. As the pH of the solution affects the rate of exchange of backbone amide protons with the solvent H_2O (63), and thus their NMR "visibility" (the amide group is key in many 2D/3D heteronuclear NMR experiments), the lowest pH value must be chosen which allows native conformation, tolerable solubility, and negligible aggregation. The addition of cosolutes such as glycerol, osmolytes (e.g., glycine), mild detergents (e.g., CHAPS), and reducing agents (e.g., DTT, TCEP) may also increase protein stability. The process of sourcing optimal conditions is called sample conditioning (64) and involves small amounts of protein using the microdialysis button test (65) or microdrop screen (66).

9. An alternative and more informative measure of a protein's macromolecular state is to measure the translational diffusion coefficient (67, 68). By using standard proteins to generate a calibration curve, information on the molecular weight and shape of the target protein can be derived (68).

10. Since proline residues do not have amide groups, these residues will not give rise to a correlation in the 2D ^1H–^{15}N HSQC.

11. The NOESY experiments are also used to assist and confirm the assignments made during the sequential assignment process.

12. The long-range HNCO experiment is inherently insensitive and applicable to proteins below ~20 kDa. Slowly exchanging amide protons, deduced from amide proton exchange experiments (69), are also used to determine the presence of H-bonds.

13. There are a number of semiautomated or fully automated approaches available (70). These approaches work best with good-quality NMR data using triple-resonance experiments.

14. Although the combination of the HNCA, and HN(CO)CA can be used to provide sequence specific backbone assignments, the dispersion of the peaks in the $^{13}C_\alpha$ spectrum is rather poor, and there is usually more than one $^{13}C_\alpha$ resonating at one particular frequency. This is especially the situation for the most crowded region of the spectrum (i.e., ~55 ppm) and with large proteins (i.e., more signals) where resonance overlap is an issue. Consequently, the backbone assignment process becomes ambiguous. The solution to this problem is to exploit other chemical shift frequencies such as the $^1H_\alpha$ by combining the HN(CA)HA/HN(COCA)HA or HNCA/HCACO experiments using the same procedure as outlined for the HNCA/HN(CO)CA combination. Alternatively, the chemical shift of the $^{13}C_\beta$ nucleus using the CBCANH/CBCA(CO)NH or HNCA/CBCA(CO)NH experiments can be exploited. The $^{13}C_\beta$ is a particularly good nucleus to target since the $^{13}C_\beta$ chemical shifts are spread over a much wider frequency range compared to the $^{13}C_\alpha$. For proteins >30 kDa in size, ^2H-labeling combined with TROSY-based (71) triple-resonance experiments are used for obtaining sequence specific backbone assignments.

15. The assignment of NOESY spectra can be challenging, as there are often ambiguous NOE correlations that arise from ^1H nuclei having the same or nearly identical chemical shifts. Programs, such as ARIA (72) have been developed to assist with ambiguous NOE data and the generation of protein structures using various molecular dynamics engines. Besides ARIA, there are a number of programs such as NOAH/DIAMOD (73) and CANDID/ATNOS (74)

that combine automated assignment of NOESY spectra and structure generation. These programs require a resonance assignment list which is generated manually from the sequential assignment of the through-bond scalar correlation 3D NMR experiments acquired.

16. Small variations in (i) data set sizes (i.e., resolution), (ii) sample conditions (e.g., pH, salt), and (iii) temperature, can lead to small changes in chemical shifts between spectra. As such, care should be taken when mapping assignment data from one spectrum to another.

17. It is possible to use ambiguous NOE data without using automated programs such as ARIA (see note 15). In the initial structure generation step, a low-resolution structure is calculated from a subset of NOE data that is interpreted unambiguously. Using this structure, it is possible to employ iterative methods to resolve many of the NOE ambiguities. For example, take a NOE peak that could be ascribed to a through-space interaction between protons A and B or protons A and C. Once a low-resolution structure is available, it is usually possible to distinguish between these two possibilities. If protons A and C are >5 Å apart while protons A and B are <5 Å apart, it is obvious that the peak arises from a NOE between protons A and B.

Acknowledgments This work was supported by a grant from the Deutsche Forschungsgemeinschaft (SFB 415).

References

1. Snyder, D. A., Chen, Y., Denissova, N. G., Acton, T., Aramini, J. M., Ciano, M., et al. (2005) Comparisons of NMR spectral quality and success in crystallization demonstrate that NMR and X-ray crystallography are complementary methods for small protein structure determination. *J. Am. Chem. Soc.* **127**, 16505–16511.

2. Yee, A. A., Savchenko, A., Ignachenko, A., Lukin, J., Xu, X., Skarina, T., et al. (2005) NMR and X-ray crystallography, complementary tools in structural proteomics of small proteins. *J. Am. Chem. Soc.* **127**, 16512–16517.

3. Wuthrich, K. (1995) NMR – this other method for protein and nucleic acid structure determination. *Acta Crystallogr. D Biol. Crystallogr.* **51**, 249–270.

4. Redfield, C. (2004) NMR studies of partially folded molten-globule states. *Methods Mol. Biol.* **278**, 233–254.

5. Mittermaier, A. and Kay, L. E. (2006) New tools provide new insights in NMR studies of protein dynamics. *Science* **312**, 224–228.

6. Carlomagno, T. (2005) Ligand-target interactions: what can we learn from NMR? *Annu. Rev. Biophys. Biomol. Struct.* **34**, 245–266.

7. Williamson, M. P., Havel, T. F., and Wuthrich, K. (1985) Solution conformation of proteinase inhibitor IIA from bull seminal plasma by 1H nuclear magnetic resonance and distance geometry. *J. Mol. Biol.* **182**, 295–315.

8. Fiaux, J., Bertelsen, E. B., Horwich, A. L., and Wuthrich, K. (2002) NMR analysis of a 900K GroEL GroES complex. *Nature* **418**, 207–211.

9. Fernandez, C., Hilty, C., Wider, G., and Wuthrich, K. (2002) Lipid-protein interactions in DHPC micelles containing the integral membrane protein OmpX investigated by NMR spectroscopy. *Proc. Natl. Acad. Sci. U S A* **99**, 13533–13537.

10. Ahn, H. C., Juranic, N., Macura, S., and Markley, J. L. (2006) Three-dimensional structure of the water-insoluble protein crambin in dodecylphosphocholine micelles and its minimal solvent-exposed surface. *J. Am. Chem. Soc.* **128**, 4398–4404.

11. Chill, J. H., Louis, J. M., Miller, C., and Bax, A. (2006) NMR study of the tetrameric KcsA potassium channel in detergent micelles. *Protein Sci.* **15**, 684–698.

12. Reilly, D., and Fairbrother, W. J. (1994) A novel isotope labeling protocol for bacterially expressed proteins. *J. Biomol. NMR* **4**, 459–462.

13. Cai, M., Huang, Y., Sakaguchi, K., Clore, G. M., Gronenborn, A. M., and Craigie, R. (1998) An efficient and cost-effective isotope labeling protocol for proteins expressed in Escherichia coli. *J. Biomol. NMR* **11**, 97–102.

14. Kay, L. E. and Gardner, K. H. (1997) Solution NMR spectroscopy beyond 25 kDa. *Curr. Opin. Struct. Biol.* **7**, 722–731.

15. Goto, N. K. and Kay, L. E. (2000) New developments in isotope labeling strategies for protein solution NMR spectroscopy. *Curr. Opin. Struct. Biol.* **10**, 585–592.

16. Zhou, P., Lugovskoy, A. A., and Wagner, G. (2001) A solubility-enhancement tag (SET) for NMR studies of poorly behaving proteins. *J. Biomol. NMR* **20**, 11–14.

17. Huth, J. R., Bewley, C. A., Jackson, B. M., Hinnebusch, A. G., Clore, G. M., and Gronenborn, A. M. (1997) Design of an expression system for detecting folded protein domains and mapping macromolecular interactions by NMR. *Protein Sci.* **6**, 2359–2364.

18. Studier, F. W. and Moffatt, B. A. (1986) Use of bacteriophage T7 RNA polymerase to direct selective high-level expression of cloned genes. *J. Mol. Biol.* **189**, 113–130.

19. Wuthrich, K. (1986) NMR of proteins and Nucleic Acids, Wiley, New York.

20. Sklenar, V. and Bax, A. (1987) Spin-echo water suppression for the generation of pure-phase two-dimensional NMR Spectra. *J. Magn. Reson.* **74**, 469–479.

21. Ikura, M., Kay, L. E., and Bax, A. (1990) A novel approach for sequential assignment of ^{1}H, ^{13}C, and ^{15}N spectra of proteins: heteronuclear triple-resonance three-dimensional NMR spectroscopy. Application to calmodulin. *Biochemistry* **29**, 4659–4667.

22. Grzesiek, S. and Bax, A. (1992) Improved 3D triple-resonance NMR techniques applied to a 31 kDa protein. *J. Magn. Reson.* **96**, 432–440.

23. Clubb, R. T. and Wagner, G. (1992) A triple-resonance pulse scheme for selectively correlating amide ^{1}HN and ^{15}N nuclei with the ^{1}H alpha proton of the preceding residue. *J. Biomol. NMR* **2**, 389–394.

24. Clubb, R. T., Thanabal, V., and Wagner, G. (1992) A new 3D HN(CA)HA experiment for obtaining fingerprint HN-Halpha peaks in ^{15}N- and ^{13}C-labeled proteins. *J. Biomol. NMR* **2**, 203–210.

25. Grzesiek, S. and Bax, A. (1992) Correlating backbone amide and side chain resonances in larger proteins by multiple relayed triple resonance NMR. *J. Am. Chem. Soc.* **114**, 6291–6293.

26. Grzesiek, S. and Bax, A. (1993) Amino acid type determination in the sequential assignment procedure of uniformly ^{13}C/^{15}N-enriched proteins. *J. Biomol. NMR* **3**, 185–204.

27. Kay, L. E., Ikura, M., Tschudin, R., and Bax, A. (1990) Three-dimensional triple-resonance NMR spectroscopy of isotopically enriched proteins. *J. Magn. Reson.* **89**, 496–514.

28. Bax, A., Clore, G. M., and Gronenborn, A. M. (1990) ^{1}H–^{1}H correlation via isotropic mixing of ^{13}C magnetization, a new three-dimensional approach for assigning ^{1}H and ^{13}C spectra of ^{13}C-enriched proteins. *J. Magn. Reson.* **88**, 425–431.

29. Grzesiek, S., Anglister, J., and Bax, A. (1993) Correlation of backbone amide and aliphatic side-chain resonances in ^{13}C/^{15}N-enriched proteins by isotropic mixing of ^{13}C magnetization. *J. Magn. Reson. Ser. B* **101**, 114–119.

30. Wang, A. C., Grzesiek, S., Tschudin, R., Lodi, P. J., and Bax, A. (1995) Sequential backbone assignment of isotopically enriched proteins in D_2O by deuterium-decoupled HA(CA)N and HA(CACO)N. *J. Biomol. NMR* **5**, 376–382.

31. Yamazaki, T., Forman-Kay, J. D., and Kay, L. E. (1993) Two-dimensional NMR experiments for correlating carbon-13-beta and proton-delta/epsilon chemical shifts of aromatic residues in ^{13}C-labeled proteins via scalar couplings. *J. Am. Chem. Soc.* **115**, 11054–11055.

32. Grzesiek, S. and Bax, A. (1995) Audio-frequency NMR in a nutating frame. Application to the assignment of phenylalanine residues in isotopically enriched proteins. *J. Am. Chem. Soc.* **117**, 6527–6531.

33. Karplus, M. (1959) Contact electron-spin coupling of nuclear magnetic moments. *J. Chem. Phys.* **30**, 11–15.

34. Vuister, G. W. and Bax, A. (1993) Quantitative J correlation: a new approach for measuring homonuclear three-bond J(HNHα) coupling constants in ^{15}N-enriched proteins. *J. Am. Chem. Soc.* **115**, 7772–7777.

35. Grzesiek, S. and Bax, A. (1997) A three-dimensional NMR experiment with improved sensitivity for carbonyl-carbonyl J correlation in proteins. *J. Biomol. NMR* **9**, 207–211.
36. Permi, P., Kilpeläinen, I., and Annila, A. (2000) Determination of backbone angle ψ in proteins using a TROSY-based α/β-HN(CO)CA-J experiment. *J. Magn. Reson.* **146**, 255–259.
37. Archer, S. J., Ikura, M., Torchia, D. A., and Bax, A. (1991) An alternative 3D NMR technique for correlating backbone ^{15}N with side chain Hß resonances in larger proteins. *J. Magn. Reson.* **95**, 636–641.
38. Hu, J.-S., Grzesiek, S., and Bax, A. (1997) Two-dimensional NMR methods for determining 1 angles of aromatic residues in proteins from three-bond J_{CC} and J_{NC} couplings. *J. Am. Chem. Soc.* **119**, 1803–1804.
39. Schwalbe, H., Carlomagno, T., Hennig, M., Junker, J., Reif, B., Richter, C., and Griesinger, C. (2001) Cross-correlated relaxation for measurement of angles between tensorial interactions. *Methods Enzymol.* **338**, 35–81.
40. Cornilescu, G., Delaglio, F., and Bax, A. (1999) Protein backbone angle restraints from searching a database for chemical shift and sequence homology. *J. Biomol. NMR* **13**, 289–302.
41. Cordier, F. and Grzesiek, S. (1999) Direct observation of hydrogen bonds in proteins by inter-residue $^{3h}J_{NC}$ scalar couplings. *J. Am. Chem. Soc.* **121**, 1601–1602.
42. Cornilescu, G., Hu, J.-S., and Bax, A. (1999) Identification of the hydrogen bonding network in a protein by scalar couplings. *J. Am. Chem. Soc.* **121**, 2949–2950.
43. Tjandra, N. and Bax, A. (1997) Direct measurement of distances and angles in biomolecules by NMR in a dilute liquid crystalline medium. *Science* **278**, 1111–1114.
44. Bax, A. and Grishaev, A. (2005) Weak alignment NMR: a hawk-eyed view of biomolecular structure. *Curr. Opin. Struct. Biol.* **15**, 563–570.
45. Ottiger, M., Delaglio, F., and Bax, A. (1998) Measurement of J and dipolar couplings from simplified two-dimensional NMR spectra. *J. Magn. Reson.* **131**, 373–378.
46. de Alba, E., Suzuki, M., and Tjandra, N. (2001) Simple multidimensional NMR experiments to obtain different types of one-bond dipolar couplings simultaneously. *J. Biomol. NMR* **19**, 63–67.
47. Tjandra, N. and Bax, A. (1997) Measurement of dipolar contributions to $^{1}J_{CH}$ splittings from magnetic-field dependence of J modulation in two-dimensional NMR spectra. *J. Magn. Reson.* **124**, 512–515.
48. Wu, Z. and Bax, A. (2002) Measurement of long-range ^{1}H-^{1}H dipolar couplings in weakly aligned proteins. *J. Am. Chem. Soc.* **124**, 9672–9673.
49. Meier, S., Haussinger, D., Jensen, P., Rogowski, M., and Grzesiek, S. (2003) High-accuracy residual ^{1}HN-^{13}C and ^{1}HN-^{1}HN dipolar couplings in perdeuterated proteins. *J. Am. Chem. Soc.* **125**, 44–45.
50. Delaglio, F., Grzesiek, S., Vuister, G. W., Zhu, G., Pfeifer, J., and Bax, A. (1995) NMRPipe: a multidimensional spectral processing system based on UNIX pipes. *J. Biomol. NMR* **6**, 277–293.
51. Vranken, W. F., Boucher, W., Stevens, T. J., Fogh, R. H., Pajon, A., Llinas, M., Ulrich, E. L., Markley, J. L., Ionides, J., and Laue, E. D. (2005) The CCPN data model for NMR spectroscopy: development of a software pipeline. *Proteins* **59**, 687–696.
52. Johnson, B. A. (2004) Using NMRView to visualize and analyze the NMR spectra of macromolecules. *Methods Mol. Biol.* **278**, 313–352.
53. Bartels, C., Xia, T.-H., Billeter, M., Güntert, P., and Wüthrich, K. (1995) The program XEASY for computer-supported NMR spectral analysis of biological macromolecules. *J. Biomol. NMR* **5**, 1–10.
54. Wishart, D. S., Sykes, B. D., and Richards, F. M. (1992) The chemical shift index: a fast and simple method for the assignment of protein secondary structure through NMR spectroscopy. *Biochemistry* **31**, 1647–1651.
55. Wishart, D. S. and Sykes, B. D. (1994) The ^{13}C chemical-shift index: a simple method for the identification of protein secondary structure using ^{13}C chemical-shift data. *J. Biomol. NMR* **4**, 171–180.

56. Bruggert, M., Rehm, T., Shanker, S., Georgescu, J., and Holak, T. A. (2003) A novel medium for expression of proteins selectively labeled with ^{15}N-amino acids in Spodoptera frugiperda (Sf9) insect cells. *J. Biomol. NMR* **25**, 335–348.

57. Wood, M. J. and Komives, E. A. (1999) Production of large quantities of isotopically labeled protein in Pichia pastoris by fermentation. *J. Biomol. NMR* **13**, 149–159.

58. Ozawa, K., Headlam, M. J., Schaeffer, P. M., Henderson, B. R., Dixon, N. E., and Otting, G. (2004) Optimization of an Escherichia coli system for cell-free synthesis of selectively N-labeled proteins for rapid analysis by NMR spectroscopy. *Eur. J. Biochem.* **271**, 4084–4093.

59. Vinarov, D. A., Lytle, B. L., Peterson, F. C., Tyler, E. M., Volkman, B. F., and Markley, J. L. (2004) Cell-free protein production and labeling protocol for NMR-based structural proteomics. *Nat. Methods* **1**, 149–153.

60. Vinarov, D. A. and Markley, J. L. (2005) High-throughput automated platform for nuclear magnetic resonance-based structural proteomics. *Expert Rev. Proteomics* **2**, 49–55.

61. Guignard, L., Ozawa, K., Pursglove, S. E., Otting, G., and Dixon, N. E. (2002) NMR analysis of in vitro-synthesized proteins without purification: a high-throughput approach. *FEBS Lett.* **524**, 159 162.

62. Waugh, D. S. (2005) Making the most of affinity tags. *Trends Biotechnol.* **23**, 316–320.

63. Wuthrich, K. and Wagner, G. (1979) Nuclear magnetic resonance of labile protons in the basic pancreatic trypsin inhibitor. *J. Mol. Biol.* **130**, 1–18.

64. Bagby, S., Tong, K. I., and Ikura, M. (2001) Optimization of protein solubility and stability for protein nuclear magnetic resonance. *Methods Enzymol.* **339**, 20–41.

65. Bagby, S., Tong, K. I., Liu, D., Alattia, J. R., and Ikura, M. (1997) The button test: a small scale method using microdialysis cells for assessing protein solubility at concentrations suitable for NMR. *J. Biomol. NMR* **10**, 279–282.

66. Lepre, C. A. and Moore, J. M. (1998) Microdrop screening: a rapid method to optimize solvent conditions for NMR spectroscopy of proteins. *J. Biomol. NMR* **12**, 493–499.

67. Altieri, A. S., Hinton, D. P., and Byrd, R. A. (1995) Association of biomolecular systems via pulsed field gradient NMR self-diffusion measurements. *J. Am. Chem. Soc.* **117**, 7566–7567.

68. Dingley, A. J., Mackay, J. P., Chapman, B. E., Morris, M. B., Kuchel, P. W., Hambly, B. D., et al. (1995) Measuring protein self-association using pulsed-field-gradient NMR spectroscopy: application to myosin light chain 2. *J. Biomol. NMR* **6**, 321–328.

69. Wagner, G. (1983) Characterization of the distribution of internal motions in the basic pancreatic trypsin inhibitor using a large number of internal NMR probes. *Q. Rev. Biophys.* **16**, 1–57.

70. Altieri, A. S. and Byrd, R. A. (2004) Automation of NMR structure determination of proteins. *Curr. Opin. Struct. Biol.* **14**, 547–553.

71. Pervushin, K., Riek, R., Wider, G., and Wuthrich, K. (1997) Attenuated T2 relaxation by mutual cancellation of dipole-dipole coupling and chemical shift anisotropy indicates an avenue to NMR structures of very large biological macromolecules in solution. *Proc. Natl. Acad. Sci. USA* **94**, 12366–12371.

72. Linge, J. P., O'Donoghue, S. I., and Nilges, M. (2001) Automated assignment of ambiguous nuclear overhauser effects with ARIA. *Methods Enzymol.* **339**, 71–90.

73. Oezguen, N., Adamian, L., Xu, Y., Rajarathnam, K., and Braun, W. (2002) Automated assignment and 3D structure calculations using combinations of 2D homonuclear and 3D heteronuclear NOESY spectra. *J. Biomol. NMR* **22**, 249–263.

74. Herrmann, T., Guntert, P., and Wuthrich, K. (2002) Protein NMR structure determination with automated NOE-identification in the NOESY spectra using the new software ATNOS. *J. Biomol. NMR* **24**, 171–189.

Chapter 31
Localization of Viral Proteins in Plant Cells: Protein Tagging

Sophie Haupt, Angelika Ziegler, and Lesley Torrance

Summary This chapter describes techniques for in vivo imaging of fluorescent fusion proteins in living cells by confocal laser scanning microscopy (CLSM). Methods are provided for (i) producing the constructs for transient expression from plasmids or virus-based vectors, (ii) introduction of constructs to plant epidermal cells; (iii) imaging of the expressed proteins by CLSM and image processing, and (iv) studying the expression in the presence of agents that affect the integrity or function of cytoskeletal elements. Notes are provided to aid comprehension and indicate problems.

Keywords In vivo imaging; Green fluorescent protein; Potato mop-top virus; Binary vectors; Run-off transcripts; Microprojectile bombardment; Confocal laser scanning microscopy; Virus vectors

1 Introduction

In vivo imaging of plant virus-encoded proteins fused to fluorophores such as green fluorescent protein (GFP) provides information on the roles and functions of the proteins in virus infection processes. GFP-tagged proteins can be expressed in living cells from virus-based vectors, transiently in the absence of virus or as transgenes on plant transformation. Such studies have generated new information and stimulated new lines of investigation. For example, the *Tobacco mosaic virus* (TMV) 30K movement protein was shown to localize to plasmodesmata (PD), seen as punctate spots of green fluorescence at the cell wall and to move to neighboring cells (1). It was also shown that increase in the size exclusion limit (SEL) of *Nicotiana tabacum* plasmodesmata was restricted to the leading edge of TMV infection (2). GFP-labeled triple gene block (TGB) proteins (Fig. 1) of Potex- and Hordeiviruses have revealed a wealth of information regarding membrane association and trafficking (3–6). Coexpression of individual TGB with GFP also showed that Potato virus X TGB2 increased the SEL of PD (7). The ORF3 protein encoded by Groundnut rosette umbravirus was shown to localize to nucleoli when fusion

From: *Methods in Molecular Biology, Vol. 451, Plant Virology Protocols:*
From Viral Sequence to Protein Function
Edited by G.D. Foster, I.E. Johansen, Y. Hong, and P.D. Nagy © Humana Press, Totowa, NJ

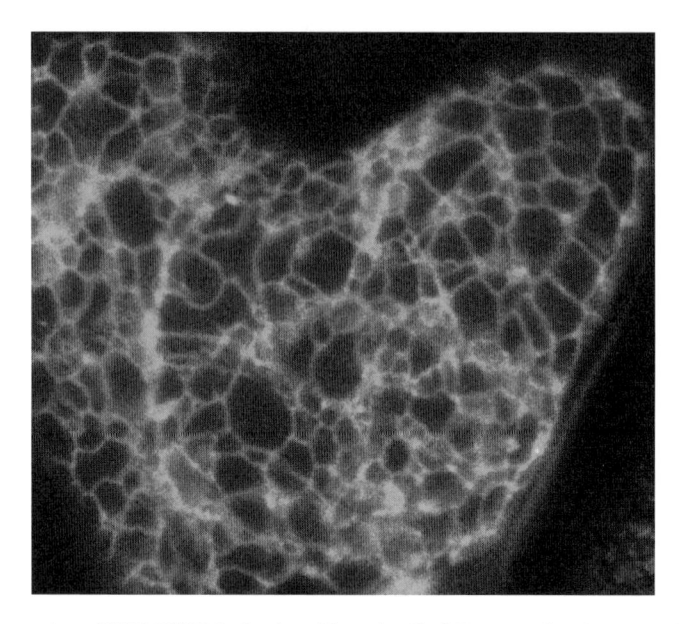

Fig. 1 Expression of GFP-TGB3 fusion in epidermal cell of *Nicotiana benthamiana*. GFP-TGB3 labels the cortical ER at 1–2 days post bombardment

proteins were expressed from virus-based vectors (8, 9). High-throughput methods have been developed where fusions of green fluorescent protein with random, partial cDNAs from a *Nicotiana benthamiana* library were cloned into a TMV-based vector, the proteins were expressed in plants and the lesions rapidly screened using confocal laser scanning microscopy (CLSM). This method revealed many novel localizations and unique proteins, some of which are putative novel plasmodesmatal proteins (10). Additional information, for example, on the role of microtubules or the actin cytoskeleton, can be obtained by expression of fusion proteins in transgenic plants where the subcellular organelles are also labeled; or by the application of various chemical inhibitors that affect the integrity or function of these subcellular structures, or dyes that label membranes and organelles (6, 11).

In this chapter, we describe methods to visualize plant virus proteins in living cells using CLSM. Methods for construct preparation, introduction into plants and localization of expressed proteins in epidermal cells by CLSM are given. The constructs were designed so that the proteins were expressed as fusions to either GFP or monomeric red fluorescent protein (mRFP) under the control of the cauliflower mosaic virus 35S promoter or from a TMV-based vector. Delivery methods include particle bombardment or infiltration of *Agrobacterium* transformed with plasmid of interest. Techniques are described to examine the effects of chemical inhibitors or dyes on fluorescent protein localization and to investigate association with subcellular organelles and cytoskeletal structures. Finally, notes on problems and pitfalls of CLSM examination are provided including dual fluorophore imaging and the avoidance of cross-talk between emission spectra and the acquisition and processing of images.

It should be noted that transient expression of fluorescent fusion proteins represents an artificial situation and care should be taken when interpreting the results (12). For example, the fluorescent fusion proteins are large and a GFP fusion at one extremity of a protein may produce inactive molecules whereas fusion at the other extremity may not. In addition, artifacts such as protein aggregation may result from high levels of overexpression of heterologous proteins in epidermal cells. Therefore, it is useful to employ several different experimental approaches to advance a hypothesis.

2 Materials

2.1 Cloning of Fusion Constructs

2.1.1 Splicing by Overlap Extension

The method involves three separate PCR reactions (13). The two fragments produced in the first-stage reactions will have overlapping sequence and they form the template for the second stage. Four primers are required for each construct, two flanking primers and two hybrid primers (Fig. 2). When the two first-stage products are mixed, they can partially anneal and contribute to the amplification of the hybrid gene.

1. Primers: Provided lyophilized by supplier. Make stock solutions in distilled sterile water and keep frozen.
2. dNTP: Keep concentrated frozen stocks in the freezer. Keep working stock solutions 10 mM dNTP in small aliquots to avoid repeated cycles of thawing and freezing.
3. PCR reaction buffer: The buffers are supplied with the polymerase enzyme.
4. $MgCl_2$: The amount of $MgCl_2$ required must be optimized for each experiment, so it is advisable to use supplier that provides the PCR reaction buffer with separate $MgCl_2$ solution.
5. Polymerase: There are many good quality heat-stable DNA polymerases available from different suppliers, including proofreading enzymes (e.g., *Pfu* (Stratagene) or Phusion High Fidelity (New England Biolabs)).
6. TBE buffer (composition); TBE agarose gel
7. Vectors. We routinely use the plasmid vector pRTL2 (14) for transient expression from a CaMV 35S promoter. Alternatively, virus-based vectors such as TMV30B (15) and binary vectors such as pGREEN (16) can be used.
8. Medium for bacterial growth (e.g., LB)
9. Antibiotics (e.g., ampicillin). Keep frozen stocks.
10. Competent *Escherichia coli* cells
11. Ligase and ligase buffer

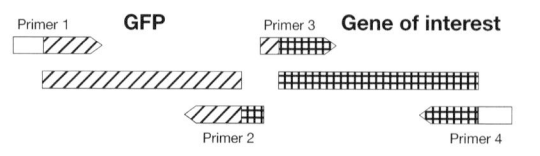

Fig. 2 Overlap extension PCR

Fig. 3 Purpose-built device for biolistic bombardments

2.2 Bombardment/Biolistic Inoculation

1. Plasmid DNA or transcripts from viral constructs
2. 100% Ethanol
3. Micro-gold carrier (Bio-Rad Laboratories, Hercules, CA). A stock solution is made by washing ~50 mg of the microcarrier in 1 mL of 100% ethanol for 1 h, followed by four washes in sterile water. After spinning down, the cleaned microcarrier was resuspended in 1 mL of sterile water.
4. Discharge assembly (13-mm Plastic Swinney Filter Holder, PALL Gelman Laboratory, Ann Arbor, Mi)
5. 6–8 week old plants
6. Handgun (17), (Fig. 3)

2.3 Imaging

1. Double-sided adhesive tape (Sellotape GB, Dunstable, UK)
2. Microscope slides
3. Confocal laser scanning microscope Leica SP2 with Leica water dipping lenses for live cell work (CLSM, Leica Microsystems, Heidelberg, Germany)

2.4 Infiltration with Drugs/Dyes

1. 10-μL pipette tips (Eppendorf)
2. Disposable 1-mL syringes (BD Plastipak™, Becton Dickinson, UK)
3. DAPI (D1360, Molecular Probes, OR, USA) stock solution 1 mg mL^{-1} in H$_2$O at −20°C, working solution 1 μg mL^{-1} H$_2$O
4. FM4–64 (T3166, Molecular Probes, OR, USA) stock solution in water at −20°C, working solution 1–5 μL in distilled water.
5. Latrunculin B (L5288, SIGMA, UK) disrupts actin-mediated processes, stock solution 1 mg mL^{-1} DMSO at −20°C, working solution 10 μg mL^{-1} in H$_2$O
6. Brefeldin A (B7651, SIGMA, UK) induces blocks in the secretory pathway and reabsorbs the Golgi back into the ER, stock solution 10 mg mL^{-1} in Methanol at −20°C, working solution 10 μg mL^{-1} in H$_2$O

2.5 Image Processing

1. Leica SP2 confocal software (Leica Microsystems, Heidelberg, Germany)
2. Adobe Photoshop 8.0 (Mountain View, CA)

3 Methods

3.1 PCR Amplification

1. Perform two separate 50 μL PCR reactions to amplify the GFP sequence (primers 1 and 2) and the target gene (primers 3 and 4). Each reaction contains 5 μL of 10X PCR reaction buffer (without MgCl$_2$), 1 μL of 50 mM MgCl$_2$, 2 μL of 10 mM dNTP, 2 μL of the relevant primers (10 mM), 1 μL of template, 0.5 μL of Taq polymerase (Roche), and sterile distilled water to a final volume of 50 μL in a 0.2-mL PCR tube.
2. Use a thermocycler (e.g., Eppendorf Mastercycler personal) and 25 cycles of amplification (94°C for 30 s, 60°C for 1 min, 72°C for 2 min), followed by an extension at 72°C for 10 min (see note 6).

3. Load a 5-μL aliquot onto a 1.2% TBE agarose gel. Stain gel in ethidium bromide solution. The no-template control reaction should not produce a PCR product. The GFP amplification should give a 700 bp band.
4. Set up a PCR reaction containing 2 μL of each primary PCR product, 10X buffer, primers 1 and 4, dNTP, $MgCl_2$, water, and polymerase. Set up a negative control reaction without primary PCR products.
5. Use a thermocycler and 25 cycles of amplification (as step 2)
6. Load a 5-μL aliquot onto a 1.2% TBE agarose gel. The expected PCR product should be the combined size of the target gene and the GFP sequence. For the negative control reaction, no product should be visible.

3.2 Cloning of PCR Product to Vector

1. Digest PCR product with the relevant restriction enzymes. Load reaction mix onto 1.2% TBE-agarose gel. After visualization with ethidium bromide, gel-purify the DNA (note 8). The Qiagen Min Elute gel Extraction Kit and the Zymogen Zymoclean Gel DNA Recovery Kit both work well.
2. Ligate digested PCR product and vector DNA of choice (pRTL-2 for transient expression, or a virus-based vector such as TMV 30B, or a binary vector such as pGREEN).
3. Transform competent cells (e.g., XL10 Gold, Stratagene) and plate on selective medium. Incubate plates overnight at 30°C (note 9)
4. Check colonies for presence of the insert.
5. Make a stock of plasmid DNA from 100 mL cultures and prepare plasmid DNA using a Qiagen Plasmid Hispeed Midiprep Kit.

3.3 Transformation of Agrobacterium by Electroporation

1. Grow an overnight culture of *Agrobacterium* at 28°C
2. Dilute the overnight culture 10 times in fresh medium and grow until OD_{600nm} of 0.5
3. Centrifuge at 4,000 rpm for 10 min at 4°C
4. Wash cells in cold sterile distilled water and centrifuge as above
5. Wash cells in 10% glycerol (cold) and centrifuge as above
6. Resuspend cells in 10% glycerol (1/100 original volume). Aliquots of competent cells can be stored frozen at −80°C for at least 5 months
7. Mix 1 μL of DNA with a 100-μL aliquot of competent cells
8. Electroporate at 2.5 kV, 200 W, and 25 uF
9. Add SOC medium and let the cells recover at 28°C for 1 h with shaking
10. Plate on selective medium

3.4 Agrobacterium Cultures

1. Grow *Agrobacterium* culture overnight at 28°C in 5 mL of LB containing the relevant antibiotic.
2. Use this culture to inoculate 50 mL of LB containing antibiotics, 10 mM of MES, and 20 μM of acetosyringone.
3. Pellet cells by centrifugation and resuspend pellet in a solution containing 10 mM of $MgCl_2$, 10 mM of MES, and 150 μM of acetosyringone. The concentration of Agrobacterium should be 0.5 OD_{600}.
4. Leave solution at room temperature for 2–3 h and load into a 2-mL syringe. For infiltration see Sect. 3.7.

3.5 Mutagenesis

The introduction of mutations can help with elucidating role and function of a protein of interest. The Stratagene QuikChange Site-Directed Mutagenesis Kit (Cat #200523) (18) is based on synthetic oligonucleotide primers and Dpn *I* digestion of methylated plasmid DNA. This results in a very low background of wild type plasmid. The system allows point mutations and deletions or insertions of single or multiple adjacent amino acids. Due to a high fidelity DNA polymerase (*Pfu* Turbo) unwanted second-site errors are virtually eliminated.

3.6 Bombardment/Biolistic Inoculation

1. Mix 0.5–1.0 μL of DNA (40–50 ng μL^{-1}) with 5–10 μL of 100% ethanol (add 1 μL at a time).
2. Vortex 11 μL of Tungsten Gold into the DNA mixture
3. Load between 2 and 4 μL onto the nozzle and let the ethanol evaporate
4. Put nozzle into the holder and screw it tight into the gun
5. Raise plant on stage until target leaf is 2–3 cm under the nozzle
6. Use up to four shots per loading inoculating areas at the tip of the leaf while avoiding the major veins (intercostal fields, see Fig. 4).
7. Support the leaf with your hand when discharging the gun to avoid tearing (note 13)

3.7 Infiltration of Agrobacteria, Dyes, and Drugs

1. Choose a fully mature "source" leaf on a 6–8 week old plant
2. Pierce a small hole into the leaf at the tip avoiding major veins (intercostal fields, see Fig. 2), using a small pipette tip

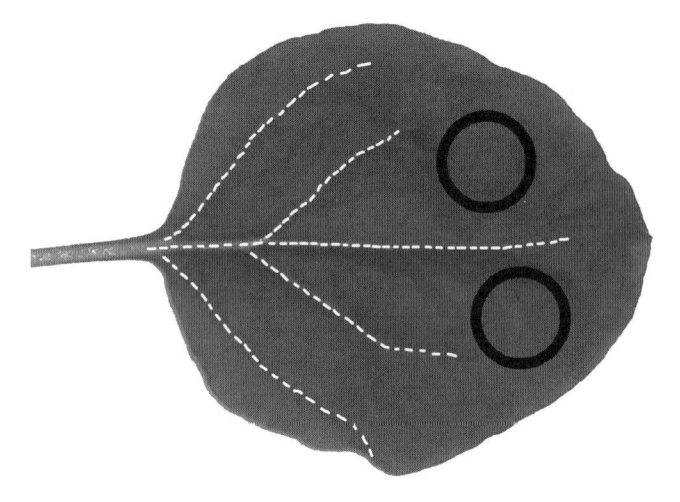

Fig. 4 Fully mature source leaf from *N. benthamiana*. Dotted lines mark the major veins, circles show the preferred areas for infiltration and biolistic inoculation

3. Infiltrate by pressing a syringe without a needle firmly against the leaf surface while supporting the leaf with your hand during the process to avoid tearing.
4. Infiltrated area will appear shiny
5. Leave plant under light source and well watered for

DAPI-20	60 min
FM4-64	0.5–2 h
BFA	1–2 h
Latrunculin	1–2 h
Agrobacterium	1–2 days

6. Detach leaf and examine under the confocal microscope.

3.8 Imaging

1. Cover microscope slide with adhesive tape
2. Cut out leaf area of interest and press gently but firmly onto microscope slide until it is completely flattened using your thumb or palm of your hand.
3. View under a confocal microscope (Leica SP2) using the following excitation and emission wavelengths

	PMT position	Excitation (nm)	Emission (nm)
GFP	1	488	500–530
mRFP	2	561	570–635
Chlorophyll	3	488	670–700
DAPI	1	405	450–460
FM4–64	1	561	620–650

4. After microscopy, images are processed for size and resolution using Adobe Photoshop 8.0.
5. See notes 14–20 for microscopy.

4 Notes

1. Primers have to be designed from known sequence of the target genes (see Fig. 4). Primers will usually contain restriction sites to allow cloning of the final PCR product. Most restriction enzymes will not cut efficiently when the recognition site is at the end of a sequence. Several base pairs can be added to alleviate the problem. Alternatively, the PCR product can be cloned into a holding vector such as pGEM-T (Promega) via the A-overhang generated by the DNA polymerase (non-proofreading polymerases). The insert can then be excised from the vector plasmid DNA and ligated to pRTL-2 or any other vector of choice. If a linker of a few amino acids is desired between the GFP and the gene of interest, the sequence can be incorporated in the primers.

2. Usually, the suppliers of primers will provide useful information, such as primer molecular weight and Tm. This is helpful for making the working dilutions and deciding on annealing temperatures.

3. Use thin-wall PCR tubes, the heat will be transferred faster to the reaction mix. Allow 1 min extension time per kilobase of sequence.

4. Use aerosol resistant pipette tips to avoid contamination and include a negative control reaction.

5. Keep the number of cycles as low as possible, this will result in fewer misincorporated nucleotides.

6. An extension step at 72°C for 30 min will result in a greater number of white colonies, when the PCR product is ligated to an A–T cloning vector such as pGEM-T that allows for blue–white selection.

7. Sequence the final product and compare with the sequence of the original template to check for changes introduced by the PCR process.

8. The Qiagen Min Elute gel Extraction Kit and the Zymogen Zymoclean Gel DNA Recovery Kit both work well. Both have the advantage that the purified DNA can be recovered in a very small amount of buffer (8 µL), which results in a higher concentration of DNA. Ethidium bromide is toxic and a potent mutagen. Avoid any direct contact.

9. In our hands, incubation at 30°C as opposed to 37°C increases the plasmid stability.

10. Binary plasmids: A plasmid containing the left and right borders of T-DNA and replication origins for both *E. coli* and *Agrobacterium*, carries the gene to be transferred and a selectable marker. The plasmid is introduced by conjugation to *Agrobacterium* containing a helper plasmid that provides the *vir* functions.

11. Acetosyringone actively induces the transfer of T-DNA from *Agrobacterium* to the plant by inducing the *vir* functions.

12. The position of the GFP (e.g., N or C terminal) can have an effect on the function of the gene studied.

13. Keep plant to bombardment in a growth chamber with high humidity 3 days prior to bombardment to increase infectivity.

14. When using combinations of GFP and RFP-labeled fusion genes, sequential scanning must be used to avoid cross-talking because of overlapping emission spectra.

15. Lambda scanning can be used to identify the emission spectrum of a specific fluorochrome (e.g., autofluorescence or GFP) and help to optimize the detection. This also helps in discriminating autofluorescence from fluorescence originated by fluorophores like GFP or mRFP.

16. When imaging moving objects like ER or Golgi, it is not advisable to use the "Average" function to improve the signal-to-noise ratio. However, decreasing the laser speed from 400 to 200 Hz will considerably improve image quality.

17. Using the 488 nm excitation wavelength for GFP results in autofluorescence from chloroplasts. By imaging the chloroplasts with a separate photo multiplier tube (PMT), the autofluorescence can be subtracted from the images.

18. To improve image quality, always keep the gain as low as possible. This will also keep the background noise low. For optimization of gain and threshold use the "Qlut" function.

19. When scanning sequentially, separate PMT have to be used for emission spectra.

20. The following Web sites have good tutorials for the Leica SP2: http://www.hi.helsinki.fi/amu/AMU%20Cf_tut/cf_tut_part2–6c.htm http://www.confocal-microscopy.com/WebSite/SC_LLT.nsf

References

1. Crawford, K.M. and Zambryski, P.C. (2001) Non-targeted and targeted protein movement through plasmodesmata in leaves in different developmental and physiological states. *Plant Phys.* **125**, 1802–1812.

2. Oparka, K.J., Prior, D.A., Santa Cruz, S., Padgett, H.S. and Beachy, R.N. (1997) Gating of epidermal plasmodesmata is restricted to the leading edge of expanding infection sites of *tobacco mosaic virus* (TMV). *Plant J.* **12**, 781–789.

3. Solovyev, A.G., Stroganova, T.A., Zamyatnin, A.A., Fedorkin, O.N., Schiemann, J. and Morozov, S.Y. (2000) Subcellular sorting of small membrane-associated triple gene block proteins: TGBp3-assisted targeting of TGBp2. *Virology.* **269**, 113–127.

4. Zamyatnin, A.A., Solovyev, A.G., Sablina, A.A., Agranovsky, A.A., Katul, L., Vetten, H.J., Schiemann, J., Hinkkanen, A.E., Lehto, K. and Morozov, S.Y. (2002). Dual-colour imaging of membrane protein targeting directed by poa semilatent virus movement protein TGBp3 in plant and mammalian cells. *J. Gen. Virol.* **83**, 651–662.

5. Zamyatnin, A.A. Jr., Solovyev, A.G., Savenkov, E.I., Germundsson, A., Sandgren, M., Valkonen, J.P.T. and Morozov, S.Y. (2004). Transient coexpression of individual genes encoded by the triple gene block of *potato mop-top virus* reveals requirements for TGBp1 trafficking. *Mol. Plant Microbe Interact.* **17**, 921–930.

6. Haupt, S., Cowan, G.H., Ziegler, A., Roberts, A.G., Oparka, K.J. and Torrance, L. (2004) Two plant–viral movement proteins traffic in the endocytic recycling pathway. *Plant Cell.* **17**: 164–181.

7. Tamai, A. and Meshi, T. (2001). Cell-to-cell movement of potato virus X: The role of p12 and p8 encoded by the second and third open reading frames of the triple gene block. *Mol. Plant Microbe Interact.* **14**, 1158–1167.

8. Ryabov, E.V., Oparka, K.J., Santa Cruz, S., Robinson, D.J. and Taliansky, M.E. (1998) Intracellular location of two groundnut rosette umbravirus proteins delivered by PVX and TMV vectors. *Virology.* **242**, 303–313.

9. Ryabov, E.V., Kim, S.H., Taliansky, M. (2004) Identification of a nuclear localization signal and nuclear export signal of the umbraviral long-distance RNA movement protein. *J. Gen. Virol.* **85**, 1329–1333.

10. Escobar N.M., Haupt, S., Thow, G., Boevink, P., Chapman, S. and Oparka, K. (2003) High-throughput viral expression of cDNA–green fluorescent protein fusions reveals novel subcellular addresses and identifies unique proteins that interact with plasmodesmata. *Plant Cell.* **15**: 1507–1523.

11. Gillespie, T., Boevink, P., Haupt, S., Roberts, A.G., Toth, R., Valentine, T., Chapman, S. and Oparka, K.J. (2002) Functional analysis of a DNA-shuffled movement protein reveals that microtubules are dispensable for the cell-to-cell movement of *tobacco mosaic virus. Plant Cell.* **14**: 1207–1222.

12. Waigmann, E., Ueki, S., Trutuyeva, K. and Citovsky, V. (2004) The ins and outs of non-destructive cell-to-cell and systemic movement of plant viruses. *Crit. Rev. Plants.* **23**, 195–250.

13. Horton, R.M., Hunt, H.D., Ho, S.N., Pullen, J.K. and Pease, L.R. (1989) Engineering hybrid genes without the use of restriction enzymes: gene splicing by overlap extension. *Gene.* **77**, 61–68.

14. Restrepo, M.A., Freed, D.D. and Carrington, J.C. (1990) Nuclear transport of plant potyviral proteins. *Plant Cell.* 2, 987–998.

15. Shivprasad, S., Pogue, G.P. and Lewandowski, D.J. (1999) Heterologous sequences greatly affect foreign gene expression in tobacco mosaic virus-based vectors. *Virology.* **255**, 312–323.

16. Hellens R.P., Edward, A., Leyland, N.R., Bean, S. and Mullineaux, P.M. (2000) pGREEN: a versatile and flexible binary Ti vector for Agrobacterium – mediated plant transformation. *Plant Mol. Biol.* 42, 819–832.

17. Gal-On, A., Meiri, E., Elman, C., Gray, D.J. and Gaba, V. (1997) Simple hand-held devices for the efficient infection of plants with viral-encoding constructs by particle bombardment. *J. Virol. Methods.* **64**, 103–110.

18. QuikChange XL Site Directed Mutagenesis Kit. Instruction manual. Stratagene

Section 4
Microscopy/GFP/Protein Tagging/ Infections Clones and Other Such Tools

Chapter 32
Construction of Infectious Clones for RNA Viruses: TMV

Abstract The generation of infectious clones is routinely the first step for reverse genetic studies of RNA plant virus gene and sequence function. The procedure given here, details the creation of cDNA clones of tobacco mosaic virus, from which infectious transcripts can be generated in vitro with T7 RNA polymerase. The procedure describes methods for virion purification, viral RNA extraction, reverse transcription, PCR amplification of genomic cDNA fragments, generation of a full-length cDNA clone under the con'rol of a T7 promoter, in vitro transcription, and infectivity testing.

Keywords TMV; Tobamovirus; RT-PCR; T7 RNA polymerase; In vitro transcription; Infectious cDNA clone

1 Introduction

Tobamoviruses have a single-stranded positive-sense, RNA genome of about 6.5 kb. The generation of a full-length cDNA clone from which infectious transcripts can be generated in vitro or in vivo is the most frequent starting point for reverse genetic studies of these plant RNA viruses. The first RNA plant virus for which such clones were produced was brome mosaic virus (1). Soon after this, two groups produced full-length cDNA clones of tobacco mosaic virus (TMV) and tomato mosaic virus (ToMV), from which infectious RNA transcripts could be produced in vitro (2, 3). Both groups pieced together cDNA fragments to produce full-length cDNAs, under the control of a modified lambda P_R promoter, allowing in vitro transcription of linearized templates with *Escherichia coli* RNA polymerase. The presence of nonviral nucleotides, especially at the 5′ end and to a lesser extent at 3′ end of the transcripts, was found to have a deleterious effect on transcript infectivity. It was also found that incorporation of a cap structure at the 5′ ends of transcripts was required for transcript infectivity. Constructs were subsequently created from which higher yields of infectious, full-length transcripts could be produced in vitro by replacement of the lambda promoter sequence with minimal promoter sequences for the more efficient SP6 or T7 bacteriophage RNA polymerases (4, 5).

From: *Methods in Molecular Biology, Vol. 451, Plant Virology Protocols: From Viral Sequence to Protein Function*
Edited by G.D. Foster, I.E. Johansen, Y. Hong, and P.D. Nagy © Humana Press, Totowa, NJ

Efforts have been made to produce a cheaper and more robust method of infection than in vitro transcription and manual inoculation. Successes with other plant viruses led to attempts to infect plants through manual inoculation of DNA comprising full-length, cDNA clones under the control of the cauliflower mosaic virus (CaMV) 35S promoter to achieve in vivo transcription. Although some host species could be infected by this method, such ToMV or TMV constructs were not infectious in their natural hosts, tomato and tobacco (6, 7). Tobacco plants were infected with a TMV cDNA under the control of the 35S promoter through microprojectile bombardment (8), which transfers the DNA into the nucleus for transcription with greater efficiency than manual inoculation. Another approach that does not require expensive biolistic equipment and can give more pervasive infections on inoculated leaves has also been tested: in this method *Agrobacterium tumefaciens* was used to deliver T-DNAs harboring a TMV cDNA under the control of the CaMV 35S promoter and gene 7 polyadenylation signal (9). As in earlier studies, infectivity was dependent on transcripts initiating at the first viral nucleotide. Although the infectivity of the *Agrobacterium*-delivered construct was improved through the inclusion of a self-cleaving ribozyme sequence to remove nonviral sequences from the 3′ ends of the RNA transcripts, and optimization of bacterial growth conditions to enhance T-DNA transfer, huge numbers of bacteria ($>10^8$) were required to initiate individual lesions. Additionally, lesion development was retarded with respect to an RNA inoculum control.

More recently, *Agrobacterium* delivery of T-DNAs containing cloned cDNAs of the crucifer-infecting tobamovirus turnip vein clearing virus (TVCV) has been used to initiate infections more effectively (10). Further work by Marillonnet et al. has highlighted why previous attempts to transcribe tobamoviruses in vivo may not have been effective. On the hypothesis that tobamoviruses do not normally undergo a nuclear phase, Marillonnet et al. modified possible targets within the viral genome of the nuclear RNA processing machinery (11). The introduction of 43 silent nucleotide substitutions in a region at the 3′ end of the RNA-dependent RNA polymerase gene gave a 33-fold enhancement in infectivity. The inclusion of introns, up to 19, in the TVCV cDNA sequence had an even more dramatic effect, increasing infectivity about 1000-fold in *Nicotiana benthamiana* and by an even greater factor in *Nicotiana tabacum*. Vectors based on these modified constructs should provide a valuable production tool, but their optimization represents a considerable investment of time and effort.

For a faster and easier development process it is recommended that full-length cDNA clones are generated under the control of the T7 RNA bacteriophage promoter for in vitro transcription of infectious transcripts. Although working with RNA requires some precautions to prevent RNase degradation, this can be easily accomplished. Kits are also now available that allow the production of large amounts of full-length RNA with a high percentage of capped transcripts. In addition, it is possible to stabilize tobamovirus in vitro transcripts and enhance their infectivity by up to two orders of magnitude through their in vitro reassembly with purified TMV coat protein (12). Infectious clones were previously generated by the onerous procedure of piecing together cDNA fragments under the control of a chosen promoter. However, with improved reverse transcriptases and the availability of proofreading, thermostable DNA polymerases, it is now feasible to generate infectious, full-length cDNA clones through reverse transcription and PCR amplification of the complete viral genome (13). Although infectious clones may be obtained by this route, the large numbers of PCR cycles that may be required to amplify adequate DNA can result in a high proportion of clones containing disabling errors. Therefore, it is recommended that overlapping 5′ and 3′ genomic cDNA fragments are produced to minimize the number of PCR cycles required and that these are then ligated into a suitable cloning vector in a three-way ligation.

The procedure given below (Fig. 1) describes methods for virus purification; extraction of viral RNA from virions; reverse transcription (RT) of viral RNA; PCR amplification of overlapping 5′ and 3′ cDNA fragments, in the process introducing a T7 promoter sequence adjacent to the 5′ end of the viral sequence and terminal restriction enzyme sites for cloning purposes; cloning of amplified viral genomes and testing the infectivity of transcripts derived from the clones obtained. In the procedure, it is assumed that the viral 5′ and 3′ end sequences, and some internal sequence data, are known. For this procedure, it is necessary to know two restriction enzymes that do not have sites within the viral cDNA sequence and one restriction enzyme that has a unique site located near the middle of the viral genome within the region of overlap of the two PCR products. If these are not known, some preliminary mapping experiments will need to be carried out. In the below procedure, which takes TMV strain U1 as an example, the internal *Bam*HI (position 3332 in the viral genome) site, and *Pst*I and *Kpn*I as absent sites are used.

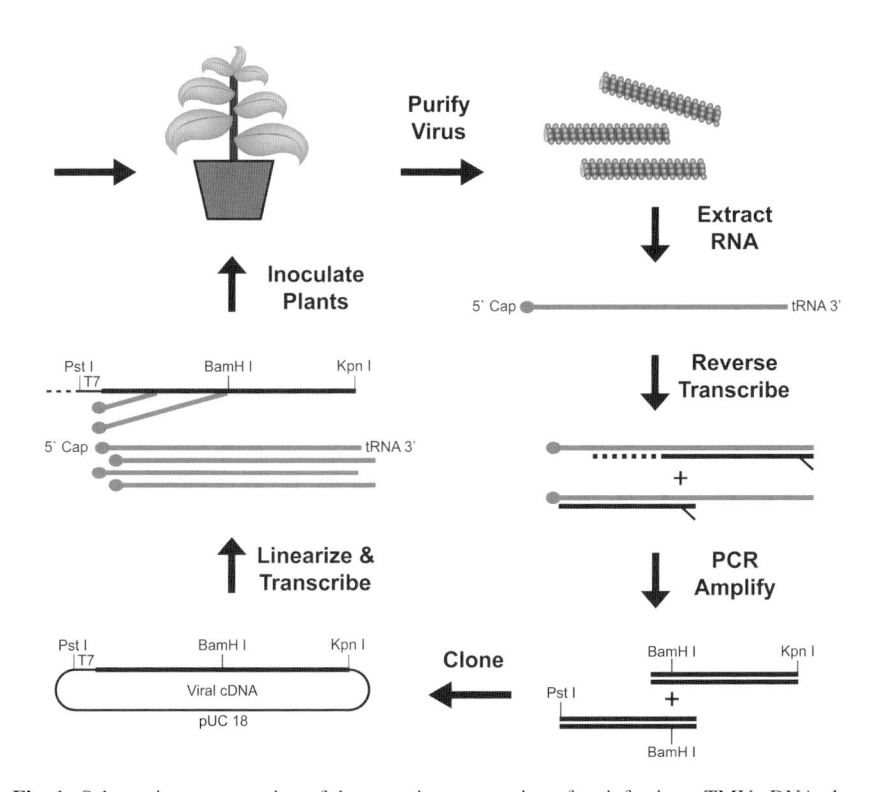

Fig. 1 Schematic representation of the steps in construction of an infectious, TMV cDNA clone

2   Materials

2.1   *Virus Purification*

1. 0.5 M Phosphate buffer: Prepare a 0.5 M solution of disodium hydrogen ortho-phosphate and adjust the pH to 7.2 with 0.5 M potassium dihydrogen orthophosphate. Autoclave and store at room temperature.
2. Extraction buffer: Add 1% (v/v) 2-mercaptoethanol to 0.5 M phosphate buffer just before use (see note 1)
3. Acid washed sand
4. Miracloth (Calbiochem, San Diego, CA)
5. Butan-1-ol
6. 20% (w/v) polyethylene glycol (PEG) (average mol. wt. 8,000). Autoclave and store at room temperature
7. 10 mM Phosphate buffer: prepared by 50-fold dilution of 0.5 M phosphate buffer
8. 5 M sodium chloride. Autoclave and store at room temperature.

2.2   *Viral RNA Extraction*

1. TLES buffer: 50 mM Trizma®-HCl pH 9.0 (Sigma, St. Louis, MO), 150 mM LiCl, 5 mM EDTA, 5% SDS (see note 2).
2. Phenol–chloroform–isoamyl alcohol mixture (25:24:1) saturated with 100 mM Tris pH 8.0 (Sigma)
3. 3 M Sodium acetate pH 5.2
4. Isopropanol
5. Nuclease-free water
6. Absolute ethanol
7. Type I gel loading solution, 6X concentrate (Sigma)
8. UltraPure™ Agarose (Invitrogen, Carlsbad, CA)
9. Tris–Borate–EDTA (TBE) buffer, 10X concentrate (Sigma)
10. SYBR Safe™ DNA gel stain, 10,000X concentrate, and Safe Imager™ blue-light transilluminator (Invitrogen) (see note 3).
11. DNA molecular weight marker ladder

2.3   *Reverse Transcription*

1. Nuclease-free water
2. 10 µM Primer complementary to the 3′ end of the virus (see note 4)
3. 10 µM Primer complementary to internal viral sequence (see note 5)
4. 10 mM dNTP mix (10 mM each of dATP, dCTP, dGTP, and dTTP at neutral pH)
5. SuperScript™ III reverse transcriptase (Invitrogen)

6. RNaseOUT™ recombinant RNase inhibitor (Invitrogen)
7. Acid washed 150–212 µm glass beads (Sigma): add two volumes of water and autoclave
8. Sepharose™ CL-6B (GE Healthcare Life Sciences, Little Chalfont, UK). Wash matrix six times with water: each time shake to thoroughly resuspend matrix, allow to settle under gravity and decant the supernatant. Finally, resuspend in a volume of water equal to the settled matrix volume, dispense 100 mL aliquots and autoclave at 120°C for 20 min.

2.4 Amplification of Overlapping 5′ and 3′ Terminal Fragments

1. 10 µM Sense primer equivalent to the 5′ end of the virus (see note 6)
2. 10 µM Sense primer equivalent to internal viral sequence (see note 5)
3. Expand High Fidelity PCR System (Roche Diagnostics GmbH, Mannheim, Germany)

2.5 Cloning of Amplified Fragments

1. *Bam*HI, *Kpn*I, and *Pst*I restriction endonucleases and buffers (New England BioLabs, Beverly, MA)
2. pUC18 plasmid (Fermentas, Hanover, MD) (see note 7)
3. Calf intestinal alkaline phosphatase (New England BioLabs)
4. 0.2 M Ethylene glycol-bis(2-aminoethylether)-*N,N,N′,N′*-tetraacetic acid (EGTA): dissolve by adjusting pH to 8.0 with sodium hydroxide and filter sterilize
5. QIAEX® II gel extraction kit (QIAGEN GmbH, Hilden, Germany)
6. 10 mM Trizma®-HCl pH 8.0
7. T4 DNA ligase (Invitrogen)
8. XL1-Blue electroporation-competent cells (Stratagene, La Jolla, CA)
9. 100 mg mL⁻¹ ampicillin: dissolve in water, filter sterilize, and store at 4°C.
10. 90-mm Petri dishes containing Luria Agar (Sigma) supplemented with 100 µg mL⁻¹ ampicillin
11. LBG Medium: LB broth (Sigma) supplemented with 1 g L⁻¹ glucose. Dispense 5 mL aliquots to small glass bottles and autoclave.
12. QIAprep® spin miniprep kit (QIAGEN)

2.6 Testing Infectivity of Full-length, cDNA Clones

1. T7 mMessage mMachine™ transcription kit (Ambion, Austin, TX)
2. Gel loading buffer II (Ambion)
3. Aluminum oxide powder, DURALUM® (FEPA F400) microgrits (Washington Mills Electro Minerals Company, Niagara Falls, NY)

3 Methods

3.1 Virus Purification

1. Inoculate virus to expanded leaves of *N. tabacum* cv. Xanthi nn at the five-to-six leaf stage (see note 8).
2. After 1–2 weeks harvest inoculated and systemically infected leaves showing pervasive symptoms of virus infection.
3. Homogenize 20 g of infected leaf tissue in 30 mL of extraction buffer using a pestle and mortar with a little acid washed sand to facilitate grinding (see note 9).
4. Filter the homogenate through Miracloth to remove particulates and collect the filtrate in polypropylene centrifuge tubes.
5. Add butan-1-ol (0.8 mL per 10 mL of filtrate). Cap tubes and mix by inversion every 2 min for 15 min.
6. Separate organic and aqueous phases by centrifugation at 10,000*g* for 30 min at 12°C. Collect the aqueous phase, avoiding any pelleted material or the highly pigmented organic phase.
7. Add 20% PEG solution to give a final concentration of 4%, mix tube contents by inversion and incubate on ice for 15 min.
8. Pellet virus by centrifugation at 10,000*g* for 15 min at 4°C. Discard the supernatant and, after pulse centrifugation, pipette off residual liquid from above the whitish, pelleted virus.
9. Resuspend the pelleted virus in 8 mL of 10 mM phosphate buffer through the gentle use of a small, Dounce tissue grinder with a Teflon pestle.
10. Transfer the resuspension to a fresh polypropylene centrifuge tube and add 1.7 mL of 5 M NaCl and 2.42 mL of 20% PEG. Cap the tube and mix the contents by inversion. Incubate on ice for 15 min prior to pelleting the virus again by centrifugation at 10,000*g* for 15 min at 4°C. Decant the supernatant and, after pulse centrifugation, pipette off residual liquid.
11. Resuspend the white, viral pellet in 1 mL of 10 mM phosphate buffer. Dilute an aliquot of the final suspension, ca. 10 μL, to 1 mL and measure the absorbance at 260 and 280 nm. An A_{260}/A_{280} ratio of about 1.19 is expected and the yield of virus can be estimated using an extinction coefficient ($E_{1cm}^{0.1\%}$) value of three. Dilute the final resuspension to 10 mg mL^{-1} and store at 4°C.

3.2 Viral RNA Extraction

1. Pipette 0.25 mL aliquots of a fresh 10 mg mL^{-1} TMV preparation (see note 10) to four 2-mL microcentrifuge tubes. To each tube add 0.75 mL of TLES buffer and mix contents by inversion.
2. Add 0.9 mL of phenol/chloroform to each tube, vortex, and incubate at 37°C for 15 min with occasional inversion.

3. Separate phases by centrifugation at 13,000g for 5 min and collect the upper aqueous phase, avoiding any denatured protein at the interface. Repeat phenol/chloroform extraction twice more using a volume equal to the aqueous phase.

4. Transfer the aqueous phases to 1.5-mL microcentrifuge tubes and extract with an equal volume of chloroform. Vortex and separate phases as above.

5. Collect the aqueous phases of ca. 0.4 mL to fresh 1.5-mL tubes. To each of the four tubes, add one-tenth volume of 3 M sodium acetate pH 5.2 and one volume of isopropanol. Mix the tube contents and incubate on ice for 5 min.

6. Pellet the viral RNA by centrifugation at 13,000g for 30 min at 4°C. Pipette off the supernatant, pulse spin, and pipette off any residual liquid.

7. Dissolve each pellet in 0.1 mL of nuclease-free water. Rapid dissolution of the pellet is facilitated by pipetting a stream of water at the side of the pellet to dislodge it from the bottom of the tube.

8. After vortexing to complete solvation, pool the four solutions. Reprecipitate the viral RNA by adding 40 µL of 3 M sodium acetate and 1.1 mL of ethanol. Mix the tube contents by inversion and incubate at −20°C for 1 h.

9. Pellet the viral RNA by centrifugation at 13,000g for 30 min at 4°C. Pipette off liquid from above the pellet and wash with 0.5 mL of cold 70% ethanol. Centrifuge for a further 5 min, pipette off liquid from above the pellet, pulse spin, and pipette off any residual liquid. Dry the pellet under vacuum for 3 min and finally dissolve the pellet in 0.1 mL of nuclease-free water.

10. Dilute 5 µL of the final solution to 1 mL with water and measure the absorbance at 260 nm. Calculate the yield of RNA assuming that a 40 µg mL^{-1} solution of RNA gives an A$_{260}$ of 1.

11. Dilute final solution with water to give a 0.5 mg mL^{-1} solution. To confirm the integrity of the recovered RNA aliquots, 0.5–2 µg can be electrophoresed on a nondenaturing, horizontal gel. Using a microwave oven, melt 0.8% agarose in 1X TBE, and after cooling add SYBR Safe$^{™}$ stain to a final concentration of 1X prior to gel casting. Load samples in 1X Type I gel loading solution with a DNA molecular weight marker ladder and after electrophoresis visualize on a blue-light transilluminator. Viral RNA should run as a single major band with little or no downward smearing, indicative of RNA degradation or fragmentation (Fig. 2). Unless the RNA solution is to be used immediately, it should be stored frozen.

3.3 Reverse Transcription

1. Set up two annealing reactions to prime cDNA synthesis of the viral RNA. One reaction should be primed with an oligonucleotide complementary to the 3′ end of the viral sequence and which incorporates a restriction enzyme site absent from the viral sequence. The other oligonucleotide should be completely complementary to a portion of the viral sequence that is just 3′ of a unique restriction enzyme site located near the middle of the viral sequence. For each reaction

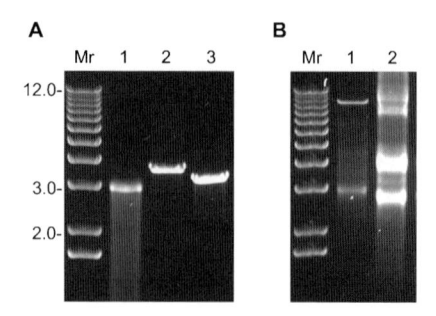

Fig. 2 Nondenaturing gel analysis of TMV-related nucleic acids. (a) RNA extracted from TMV particles (lane 1) and RT-PCR products comprising the 5′ half (lane 2) and 3′ half (lane 3) of the TMV genome. (b) T7 RNA polymerase in vitro transcription products from a linearized, full-length, TMV cDNA clone (lane 1) and from a circular plasmid containing a ribozyme sequence after the full-length, TMV cDNA clone (lane 2). DNA size markers (M_r) are shown (1 Kb Plus DNA ladder, Invitrogen)

    pipette to a nuclease-free 0.5-mL microcentrifuge tubes, 2 μL of 10 μM primer, 2 μL of 0.5 mg mL⁻¹ viral RNA, and 8 μL of nuclease-free water. Mix and incubate at 70°C for 10 min prior to snap-cooling on ice.

2. After 1 min on ice, pulse spin the tubes and add the following components to the reactions: 1 μL 10 mM dNTP mix, 4 μL 5X First-Strand Buffer, 1 μL 0.1 M DTT, 1 μL RNaseOUT™, and 1 μL SuperScript™ III. Mix components by gently pipetting up and down, and incubate reaction at 50°C for 2 h.

3. At the end of the reaction period, add 30 μL of water. To purify the reaction products add 35 μL of phenol/chloroform, vortex, centrifuge at 13,000*g* for 5 min and collect approximately 45 μL of the aqueous phase. Remove salts and primers by spin dialysis (see note 11).

3.4   Amplification of Overlapping 5′ and 3′ Terminal Fragments

1. Set up two PCR reactions with Expand High Fidelity PCR System basically according to the manufacturer's instructions. For amplification of the 5′ fragment of the viral genome, pipette the following to a thin-walled PCR tube: 10 μL of purified products from first-strand reaction primed with the internal primer, 2 μL of 10 mM dNTPs, 3 μL of 10 μM 5′ end primer, 3 μL of 10 μM internal complementary primer, and 32 μL of water. Likewise for amplification of the 3′fragment, pipette the following to a thin-walled PCR tube: 10 μL of purified products from first-strand reaction primed with the 3′ end primer, 2 μL of 10 mM dNTPs, 3 μL of 10 μM internal sense primer, 3 μL of 10 μM 3′ end complementary primer and 32 μL of water. Further, prepare a dilution of the enzyme mix (77 μL water, 20 μL 10X Expand High Fidelity buffer with 15 mM MgCl₂, 3 μL Expand High Fidelity enzyme mix) and, after gentle mixing by pipetting up and down, place this on ice.

2. Set up the following profile on a thermal block cycler: an initial 2 min denaturation at 94°C; 10 cycles comprising a 94°C denaturation of 15 s, an annealing period of 30 s commencing at 45°C, and rising by 1°C per cycle, and a 2.5 min elongation at 68°C; two further cycles comprising a 94°C denaturation of 15 s, an annealing period of 30 s at 55°C, and an elongation period at 68°C that commences at 2.5 min and rising by 5 s per cycle; and a final elongation at 68°C of 7 min.

3. Start the thermal cycling profile and pause the machine when the block reaches 94°C. Place the two thin-walled PCR tubes in the block and leave for about 15 s for the contents to come to temperature. Then add 50 μL of the diluted enzyme mix to each of the two tubes, mixing by gently pipetting up and down, and continue the thermal cycling.

4. At the end of the thermal cycling, pulse spin the PCR tubes and transfer their contents to two 0.5-mL microcentrifuge tubes. Purify the reaction products by extraction with 75 μL of phenol/chloroform. Collect approximately 90 μL of the aqueous phase and desalt by spin dialysis.

5. Take 5 μL aliquots of the products and electrophoresce with a ladder of DNA size standards on a 1% agarose/1X TBE/1X SYBR Safe™ gel. Check that the PCRs have produced the expected products of 3.5 and 3.2 kbp (Fig. 2) and estimate the yields. Roughly 1 μg of each of the DNAs will be required for cloning purposes: if inadequate product has been produced, the number of amplification cycles used should be increased, whereas if excess product has been produced, the number of cycles used can be reduced to minimize the risk of any PCR-induced mutations.

3.5   Cloning of Amplified Fragments

1. Set up digests of the two purified PCR products in 0.5-mL microcentrifuge tubes: increase the volume of the purified products to 88 μL with water, add 10 μL of *Bam*HI NEBuffer and 2 μL of *Bam*HI (20 U μL⁻¹). Mix and incubate at 37°C for 3 h. Purify the digestion products by phenol/chloroform extraction and spin dialysis. Set up secondary digests in 0.5-mL microcentrifuge tubes: to 88 μL of the *Bam*HI digested 5′ end PCR product, add 10 μL of NEBuffer 3, and 2 μL of *Pst*I (20 U μL⁻¹); to 88 μL of *Bam*HI digested 3′ end PCR product add 10 μL of NEBuffer 1, and 2 μL of *Kpn*I (10 U μL⁻¹). Mix the tube contents and incubate at 37°C for 3 h. At the end of this period, purify the products of the secondary digests by phenol/chloroform extraction and spin dialysis.

2. Digest 5 μg of pUC18 plasmid in a 0.5-mL microcentrifuge tube: increase the volume of the purified plasmid DNA to 88 μL with water, add 10 μl of NEBuffer 3, and 2 μL of *Pst*I (20 U μL⁻¹). Mix, incubate at 37°C for 3 h, and purify the digested products by phenol/chloroform extraction and spin dialysis. Set aside 10 μL of the *Pst*I digested plasmid for use as a quantification control. Set up a secondary digest: 78 μL of *Pst*I digested pUC18 dialysate, 10 μL water, 10 μL of NEBuffer 1, and 2 μL of *Kpn*I. Mix and incubate at 37°C. After 2 h, add 2 μL

of calf intestinal alkaline phosphatase ($10\,U\,\mu L^{-1}$), mix and incubate for a further 1 h at 37°C. At the end of this period, inactivate the phosphatase by adding $10\,\mu L$ of 0.2 M EGTA and incubating at 70°C for 10 min. Purify dephosphorylated products by two phenol/chloroform extractions and spin dialysis.

3. Load the purified vector and inserts in 1X Type I gel loading buffer on to a preparative 1% agarose/1X TBE/1X SYBR Safe™ gel with a DNA molecular weight marker ladder and electrophoresce at $8\,V\,cm^{-1}$. At the end of the electrophoresis period, visualize the nucleic acids on a blue-light transilluminator. Excise the three ca. 3 kbp bands with a razor blade and purify the DNA fragments with a QIAEX® II gel extraction kit according to the manufacturer's instructions, except finally elute the DNA fragments in $25\,\mu L$ of 10 mM Trizma®-HCl pH 8.0. To 5-μL aliquots of the purified vector and inserts add $3\,\mu L$ of water and $2\,\mu L$ of Type I gel loading solution 6X concentrate. Run 5.0, 2.5, and $1.25\,\mu L$ aliquots of these dilutions with 4.0, 2.0, 1.0, and $0.5\,\mu L$ aliquots of the previously set aside PstI digested pUC18 and a DNA molecular weight marker ladder on a 1% agarose/1X TBE/1X SYBR Safe™ DNA gel. After electrophoresis, visualize on a blue-light transilluminator and estimate yields.

4. Set up a 20-μL reaction in a 0.5-mL microcentrifuge tube to ligate the inserts to the vector: mix 100 ng of each of the digested insert fragments and 50 ng of the digested and dephosphorylated vector; add water to give a volume of $15\,\mu L$; add $4\,\mu L$ of 5X DNA Ligase Reaction Buffer and mix by pipetting up and down; add $1\,\mu L$ of T4 DNA ligase ($1\,U\,\mu L^{-1}$) and mix again by pipetting up and down. At the same time, set up a control ligation reaction without either of the inserts. Incubate the ligations at room temperature for 2 h and then overnight at 12°C.

5. Purify the products of the ligations by spin dialysis and transform $4\,\mu L$ aliquots into XL1-Blue Electroporation-Competent Cells according to the manufacturer's instructions. Plate aliquots of the transformations on to Luria Agar plates containing $100\,\mu g\,mL^{-1}$ of ampicillin and incubate overnight at 37°C (see note 12).

6. It is to be hoped that there will be many more colonies on the plates derived from the ligation with inserts than on the plates derived from the control ligation. Pick 10 colonies from the former plates into ten 5-mL aliquots of LBG containing $100\,\mu g\,mL^{-1}$ ampicillin and grow the bacteria overnight at 37°C with agitation.

7. Prepare DNA from the 10 overnight cultures using a QIAprep® Spin Miniprep kit according to the manufacturer's instructions. To confirm that the clones contain the desired inserts and that the restriction enzyme sites used for cloning are intact, diagnostic digests should be carried out on each of the miniprep DNAs using the three restriction enzymes used in the cloning procedure (see note 13).

3.6 Testing Infectivity of full-length, cDNA Clones

1. Set up separate reactions to linearize the template DNAs (see note 14): mix in a 0.5-mL microcentrifuge tube $10\,\mu L$ of miniprep DNA, $33\,\mu L$ of water, $5\,\mu L$ of NEBuffer® 1, and $2\,\mu L$ of KpnI. Incubate reactions overnight at 37°C (see note 15).

2. Purify the products of the linearization reactions by phenol/chloroform extraction and spin dialysis. Run 1 μL aliquots of the purified products on a 0.8% agarose/1X TBE/1X SYBR Safe™ gel to confirm the efficacy of the linearization reactions.

3. Assemble 20 μL transcription reactions of the linearized templates in 0.5-mL microcentrifuge tubes at room temperature (see note 16) using a T7 mMessage mMachine kit basically according to the manufacturer's instructions: in order, add 6 μL of *Kpn*I linearized template, 10 μL of 2X NTP/Cap mix, 2 μL of 10X Reaction Buffer, and 2 μL of Enzyme Mix. Mix by gently pipetting up and down and incubate at 37°C. After 1 h, pipette 0.5 μL aliquots from the transcription reactions to tubes containing 8 μL of nuclease-free water and 2 μL of Gel Loading Buffer II. Mix and electrophoresce at 10 V cm^{-1} on a 0.7% agarose/1X TBE/1X SYBR Safe™ gel with a DNA molecular weight marker ladder. After electrophoresis, visualize the transcription reaction products to check the yield and integrity of the transcripts (Fig. 2).

4. After incubating transcription reactions at 37°C for 2 h, dilute the products two-fold with water and proceed to inoculate plants immediately. Dust the leaves of *N. benthamiana*, *N. tabacum* cv. Xanthi nn and *N. tabacum* cv. Xanthi nc (NN) plants at the five-to-six leaf stage with aluminum oxide powder (see note 17). Mark expanded leaves for inoculation by puncturing with a pipette tip. Pipette 5 (*N. benthamiana*) or 10-μL aliquots of the diluted transcription products to interveinal, basal regions of the marked leaves. Using a gloved finger, gently stroke the whole leaf surface to spread the inoculum (see note 18). After 5 min, gently water the leaf surfaces. Propagate plants at 28°C or below with a 16-h light period and 8-h dark period.

5. Inspect plants at daily intervals. *N. benthamiana* and *N. tabacum* cv. Xanthi nn act as the primary indicators of transcript infectivity: *N. benthamiana* should develop systemic infection symptoms after 3 days and may undergo top-necrosis; systemic mosaic symptoms should appear at a later time point on *N. tabacum* cv. Xanthi nn. Infectious transcripts should produce necrotic local lesions on *N. tabacum* cv. Xanthi nc, which have *N* gene resistance against TMV, provided the plants are propagated at below 28°C. The number of lesions produced is indicative of the relative infectivity of the transcripts, so a clone producing high numbers of necrotic lesions on this host should be chosen for further studies (see note 19).

4   Notes

1. Anyone using this protocol should familiarize themselves with the health and safety hazards related to 2-mercaptoethanol, butan-1-ol, phenol, chloroform, isoamyl alcohol, and diethylpyrocarbonate. All manipulations involving these chemicals should be carried out in a fume hood.

2. For steps involving RNA, solutions certified as nuclease-free should be purchased from a reputable supplier. Alternatively, solutions can be treated with diethylpyrocarbonate to inactivate RNase, although this chemical is incompatible with some others, such as Trizma®. Additionally, for steps involving RNA, pipette tips, and microcentrifuge tubes that are certified nuclease-free should be used and gloves should be worn at all times. Gel equipment used for analysis of RNAs should be thoroughly cleaned with detergent, prior to brief treatment with 0.5 M sodium hydroxide and washing with water.

488   S.N. Chapman

3. SYBR Safe™ is suggested as an alternative, less hazardous, nucleic acid stain to ethidium bromide. In addition blue-light illumination of SYBR Safe™-stained DNA is less mutagenic than UV illumination of ethidium bromide stained DNA.

4. Designing a primer complementary to the 3′ end of tobamoviruses is complicated by the presence of tRNA-like structures and GC richness (14). Care should thus be taken to avoid primers that form hairpins or have the capacity to form strong terminal dimers, which can significantly reduce PCR yields. However, the GC-rich nature of the 3′ end means that the primer can be relatively short, but still have an initial T_m of about 45°C, calculated according to nearest-neighbor method and taking in to account salt concentration (15). In the sequence of the primer used here (5′ TTT-T*GG-TAC-C*TG-GGC-CCC-TAC-CG 3′) bases complementary to TMV sequence are underlined and the introduced KpnI site italicized. An extra four bases are added to the 5′ end of the primer to allow efficient cutting of the amplification product by *Kpn*I.

5. Designing internal primers is facile as there is some freedom of choice in their position. The terminal primers are initially partially mismatched, therefore, for effective amplification, the internal primers should be designed so that their T_ms lie between the initial, mismatched T_m and the final, matched T_m of their paired terminal primer. In this example a sense primer (5′ TTA-ACC-CCT-ACA-CCA-GTC-TCC-ATC-A 3′) at position 3,231 and an antisense primer (5′ CTG-TTG-CCT-GGG-AGA-CAC-TTA-TCA-T 3′) at position 3,508, either side of the internal *Bam*HI site, are chosen.

6. Unlike the 3′ end, the 5′ end sequence of TMV is relatively unstructured and devoid of G residues, but does contain octanucleotide and CAA repeats (16). In the sequence of the primer used here (5′ TTT-T*CT-GCA-GTA-ATA-CGA-CTC-ACT-ATA*-GTA-TTT-TTA-CAA-CAA-TTA-CCA-ACA-ACA-A 3′) bases equivalent to TMV sequence are underlined, the introduced *Pst*I site italicized and the T7 promoter sequence underlined and italicized. As before, an extra four bases are added to the 5′ end of the primer to allow efficient cutting of the amplification product by *Pst*I. Due to the length of this primer, it should be more rigorously purified (e.g., through PAGE) to remove truncated and deleted oligonucleotide forms.

7. As expression of plant viral RNA-dependent RNA polymerases may be toxic to *E. coli*, resulting in poor plasmid yields or plasmid instability, a vector should be chosen that does not place the 5′ end of the virus close to a bacterial promoter, such as the *lac*Z promoter found in many cloning vectors. Hence, the two original T7-driven TMV cDNA clones produced in the laboratory of Prof. W.O. Dawson, pTMV004 and pTMV007 (17), which are alternatively orientated in the pUC polylinker, were cloned in to pUC18 and pUC19, respectively.

8. Higher yields of TMV can be obtained by propagating plants at a high temperature, that is, 33°C.

9. The extraction buffer should not be placed on ice as this may cause the salts to precipitate.

10. Use of a fresh virus preparation is recommended for RNA extraction as the encapsidated RNA can become degraded over time.

11. Spin dialysis (18), based on the principle of gel exclusion chromatography, is recommended as a fast and reliable alternative to ethanol precipitation for removing salts and low molecular weight compounds from small volumes of nucleic acid solutions. The bottom of a 0.5-mL or 1.5-mL microcentrifuge tube, depending on the sample volume, is semipierced with a 23-gauge hypodermic needle; approximately 25 µL of glass beads in water are pipetted to the bottom of the tube using a 1-mL pipette; 10 sample volumes of resuspended Sepharose™ CL-6B are pipetted on top of the glass beads; the column is placed inside a 1.5-mL microcentrifuge tube with a hole in the bottom and this is in turn placed inside a 15-mL Falcon™, conical-bottomed centrifuge tube for support; the assembly is centrifuged at 250*g* (at the radius of the column) in a swing-out, bench-top centrifuge for 2 min to remove excess water and to give a packed column volume of five times the sample volume. A volume of water equivalent to the sample volume is pipetted to the centre of the upper, matrix surface and the assembly centrifuged again. The lower 1.5-mL microcentrifuge tube, with a hole in the bottom, is exchanged for an intact 1.5-mL tube for sample collection; the sample is applied to the

centre of the matrix and the assembly centrifuged again; the dialyzed sample is recovered from the 1.5-mL collection tube.

12. Colour screening with IPTG and X-gal may be used to identify clones with inserts; however, this will result in increased *lacZ* promoter activity, which may increase bacterial toxicity effects.

13. In some instances, yields of plasmids containing TMV cDNAs from liquid culture can be very poor. In such cases, the author has found that dispersing bacterial colonies in liquid, plating them to fresh solid media, incubating the plates overnight at 37°C, and preparing DNA from bacteria collected from the plate can significantly improve yields.

14. Although in this example linearization of template DNA is indicated, it is recommended that as soon as a suitable clone is identified that a ribozyme sequence (19) is introduced at the 3′ end of the viral cDNA sequence to obviate this step in future experiments. This can be accomplished simply by digesting an infectious clone with *Kpn*I and ligating in a pair of annealed oligonucleotides, with *Kpn*I compatible termini, to introduce the following sequence at the 3′ end of the virus: 5′ <u>GTA-x-C</u>CC-GGA-TGT-GTT-TTC-CGG-GCT-GAT-GAG-TCC-GTG-AGG-ACG-AAA-CCT-GGA-<u>GTA-C</u> 3′. The *Kpn*I compatible termini are underlined and the ribozyme cleavage site marked with an "x."

15. Linearization overnight is recommended so that no uncut plasmid is left, as circular DNA serves as a better template than linear DNA.

16. Transcription reactions should be assembled at room temperature and in the order stated as the transcription buffer contains spermidine that can coprecipitate DNA.

17. Perhaps, obviously, plants should not have been recently watered from above, as residual water on the leaves will significantly dilute the inoculum.

18. For manual inoculations, leaves should be supported with a gloved hand below and the inoculum spread over the leaf by gently drawing a gloved finger from the base of the leaf to the apex. Ten wipes should be sufficient to wet the whole surface and more than this may result in excessive tissue damage.

19. If an infectious clone is not found in the first screen, larger numbers can be screened by transcribing pools of five templates.

Acknowledgments This work was supported by the Scottish Executive for the Environment and Rural Affairs Development.

References

1. Ahlquist, P., French, R., Janda, M., and Loesch-Fries, L. S. (1984) Multicomponent RNA plant virus infection derived from cloned viral cDNA. *Proc. Natl. Acad. Sci. USA* **81**, 7066–7070.

2. Dawson, W. O., Beck, D. L., Knorr, D. A., and Grantham, G. L. (1986) cDNA cloning of the complete genome of tobacco mosaic virus and production of infectious transcripts. *Proc. Natl. Acad. Sci. USA* **83**, 1832–1836.

3. Meshi, T., Ishikawa, M., Motoyoshi, F., Semba, K., and Okada, Y. (1986) *In vitro* transcription of infectious RNAs from full-length cDNAs of tobacco mosaic virus. *Proc. Natl. Acad. Sci. USA* **83**, 5043–5047.

4. Kumagai, M. H., Turpen, T. H., Weinzettl, N., Della-Cioppa, G., Turpen, A. M., Donson, J., et al. (1993) Rapid, high-level expression of biologically active α-trichosanthin in transfected plants by an RNA viral vector. *Proc. Natl. Acad. Sci. USA* **90**, 427–430.

5. Holt, C. A., and Beachy, R. N. (1991) *In vivo* complementation of infectious transcripts from mutant tobacco mosaic virus cDNAs in transgenic plants. *Virology* **181**, 109–117.

6. Weber, H., Haeckel, P., and Pfitzner, A. J. P. (1992) A cDNA clone of tomato mosaic virus is infectious in plants. *J. Virol.* **66**, 3909–3912.

7. Shintaku, M. H., Carter, S. A., Bao, Y., and Nelson, R. S. (1996) Mapping nucleotides in the 126-kDa protein gene that control differential symptoms induced by two strains of tobacco mosaic virus. *Virology* **221**, 218–225.
8. Dagless, E. M., Shintaku, M. H., Nelson, R. S., and Foster, G. D. (1997) A CaMV 35S promoter driven cDNA clone of tobacco mosaic virus can infect host plant tissue despite being uninfectious when manually inoculated onto leaves. *Arch. Virol.* **142**, 183–191.
9. Turpen, T. H., Turpen, A. M., Weinzettl, N., Kumagai, M. H., and Dawson, W. O. (1993) Transfection of whole plants from wounds inoculated with *Agrobacterium tumefasciens* containing cDNA of tobacco mosaic virus. *J. Virol. Methods* **42**, 227–240.
10. Marillonnet, S., Giritch, A., Gils, M., Kandzia, R., Klimyuk, V., and Gleba, Y. (2004) *In planta* engineering of viral RNA replicons: efficient assembly by recombination of DNA modules delivered by *Agrobacterium*. *Proc. Natl. Acad. Sci. USA* **101**, 6852–6857.
11. Marillonnet, S., Thoeringer, C., Kandzia, R., Klimyuk, V., and Gleba, Y. (2005) Systemic *Agrobacterium tumefasciens*-mediated transfection of viral replicons for efficient transient expression in plants. *Nat. Biotechnol.* **23**, 718–723.
12. Toth, R. L., Pogue, G. P., and Chapman, S. (2002) Improvement of the movement and host range properties of a plant virus vector through DNA shuffling. *Plant J.* **30**, 593–600.
13. Rabindran, S., Robertson, C., Achor, D., German-Retana, S., Holt, C. A,. and Dawson, W. O. (2005) Odontoglossum ringspot virus host range restriction in *Nicotiana sylvestris* maps to the replicase gene. *Mol. Plant Pathol.* **6**, 439–447.
14. Rietveld, K., Linschooten, K., Pleij, C. W. A., and Bosch, L. (1984) The three-dimensional folding of the tRNA-like structure of tobacco mosaic virus RNA. A new building principle applied twice. *EMBO J.* **3**, 2613–2619.
15. Wetmur, J. G. (1991) DNA probes: applications of the principles of nucleic acid hybridization. *Crit. Rev. Biochem. Mol. Biol.* **26**, 227–259.
16. Gallie, D. R., and Walbot, V. (1992) Identification of the motifs within the tobacco mosaic virus 5′-leader responsible for enhancing translation. *Nucleic Acid Res.* **20**, 4631–4638.
17. Lewandowski, D. J., and Dawson, W.O. (1998) Deletion of internal sequences results in tobacco mosaic virus defective RNAs that accumulate to high levels without interfering with replication of the helper virus. *Virology* **251**, 427–437.
18. Murphy, G., and Kavanagh, T. (1988) Speeding-up the sequencing of double-stranded DNA. *Nucleic Acid Res.* **16**, 5198–5198.
19. Haseloff, J., and Gerlach, W. L. (1988) Simple RNA enzymes with new and highly specific endonuclease activities. *Nature* **334**, 585–591.

Chapter 33
Construction of Infectious cDNA Clones for RNA Viruses: Turnip Crinkle Virus

Eugene V. Ryabov

Abstract Reverse genetic approach is widely used in virology as it makes possible direct identification of viral gene function and uses RNA genomes as vectors. Production of infectious cDNA clones is an essential step in developing a reverse genetic system for an RNA virus. Here, we present rapid method for generation of infectious cDNA clone for Turnip crinkle virus (TCV). The infectious cDNA clone could be used for production of in vitro transcripts with the T7 RNA polymerase which could be used for infection of plants or plant cell protoplasts. The procedure described here includes purification of TCV, viral RNA extraction, reverse transcription, PCR amplification of the full-length cDNA copy of TCV linked to a T7 RNA polymerase promoter, cloning into a plasmid vector, in vitro transcription, and selection of infectious clones.

Keywords Plant RNA virus; Turnip crinkle virus; Carmovirus; Infectious cDNA clone; Genomic RNA; Reverse transcription; Polymerase chain reaction; T7 RNA polymerase; In vitro transcription; Reverse genetic system

1 Introduction

Reverse genetic approach has revolutionized study of plant viruses since the mid-1980s (1), making direct identification of viral gene function possible. Production of infectious cDNA clones of RNA genomes of plant viruses is an essential step in developing a reverse genetic system for an RNA virus. *Turnip crinkle virus* (TCV), a member of Carmovirus family, has an icosahedral virus particle about 30 nm in diameter, containing approximately 4 kb positive-sense, single-stranded genomic RNA. The genomic RNA of TCV contains five open reading frames. Two 5′ proximal ORFs encode an RNA replicase (p28 and p88 readthrough product), the ORF3 and ORF4 encode cell-to-cell movement proteins, and the 3′ proximal ORF encodes coat protein, which is also involved in protection against host RNA response, acting as a suppressor of RNA silencing (2).

From: *Methods in Molecular Biology, Vol. 451, Plant Virology Protocols: From Viral Sequence to Protein Function*
Edited by G.D. Foster, I.E. Johansen, Y. Hong, and P.D. Nagy © Humana Press, Totowa, NJ

TCV is one of the best-studied members of Carmovirus group. Historically, interest in research of TCV was promoted by its economic importance. Later, TCV was used as a model to study fundamental processes of plant–virus interactions such as viral cell-to-cell movement (3, 4, 6) and development of RNA silencing (7, 8). Progress in study of molecular biology of TCV, as well as a number of other RNA viruses, became possible due to development of full-length cDNA clone of its RNA genome, which allowed genetic manipulation and precise dissection of gene functions. Although at least two full-length infectious clones of TCV have been designed (5, 8), further study of TCV isolates which exhibit different biological properties may require generation of TCV full-length cDNA clones of those isolates.

Here, we describe a method for production of a full-length infectious cDNA clone of TCV which we had used for cloning of the UK strain of this virus (8). This method was based on the whole genome amplification of TCV genome, the approach which allowed rapid generation of infectious clones of various strains and isolation of the individual components of virus population, which may simultaneously occur in the same plant. In brief, our method includes the following steps:

1. Small scale isolation of TCV particles from individual plants and subsequent RNA extraction
2. Reverse transcription to produce a full-length first strand cDNA copy of TCV genomic RNA
3. Amplification of the full-length cDNA of TCV with the T7 RNA polymerase promoter sequence fused to the 5′ end of TCV cDNA
4. Assessment of the infectivity of the in vitro transcript generated by using the T7-TCV cDNA PCR fragment as a template
5. Cloning of the TCV cDNA fragment into a plasmid vector if the TCV cDNA PCR fragment proved to be a template of infectious TCV in vitro RNA transcripts
6. Inoculation of plants with the in vitro RNA transcripts produced from the individual TCV cDNA clones, with the amplified TCV cDNA fragment and with the wild type TCV RNA
7. Assessment of the infectivity and symptoms induced by individual clones and selection of infectious clones of interest and subsequent sequencing of the selected infectious TCV cDNA clones

2 Materials

All solutions are prepared with MilliQ autoclave water. Microcentrifuge tubes and disposable pipette tips used should be certified as RNAse and DNAse free and should be autoclaved prior to use. Essential laboratory equipment include ultracentrifuge; microcentrifuge, preferably refrigerated; PCR block; spectrophotometer or Nanodrop Spectrophotometer (NanoDrop Technologies, Wilmington, DE); agarose gel equipment with power supply; UV transilluminator; 4°C fridge or cold room;

−20°C and −85°C freezers; 37°C incubator; orbital shaker with temperature control; 42°C water bath; autoclave; and pestle and mortar.

2.1 Virus Purification and RNA Extraction

1. 50 mM sodium phosphate buffer, pH 7.6. Autoclave and store at room temperature
2. Sodium thioglycolate
3. Butan-1-ol
4. Miracloth (Calbiochem, San Diego, CA)
5. RNA extraction buffer: Tris–HCl, pH 8.5, 100 mM EDTA, 2% SDS
6. Phenol–chloroform–isoamyl alcohol mixture (25:24:1) saturated with 100 mM Tris–HCl pH 8.0
7. Chloroform
8. 3 M sodium acetate, pH 5.2
9. Absolute ethanol
10. Agarose
11. Ethidium bromide stock (10 mg mL^{-1}), store at 4°C, protect from light
12. Tris–Borate–EDTA (TBE) Buffer, 10X concentration

2.2 Reverse Transcription

1. Nuclease-free water
2. Reverse transcription primer: 2 µM primer "TCV-PacRT-R" with PacI restriction site preceding the sequence complementary to the 3′ end of the TCV genomic RNA, 5′- GGTGTTAATTAAGGGCAGGCCCCCCCCCCGCGCGAG-3′
3. 5 mM dNTP mix (5 mM of each dNTP, dATP, dCTP, and dTTP; pH 7.0)
4. M-MuLV Reverse Transcriptase (New England Biolabs, Inc., MA)
5. RNase OUT recombinant RNAse inhibitor (Invitrogen)

2.3 Amplification of the TCV cDNA

1. Forward PCR primer: 10 µM of oligonucleotide primer "TCV-T7F" containing the T7 RNA polymerase promoter sequence preceding the sequence identical to eighteen 5′-proximal nucleotide of TCV RNA 5′-GCTATAATACGACTCACTATAGGTAATCTGCAAATCCCTGC-3′
2. Reverse PCR primer: 10 µM of oligonucleotide primer "TCV-PacPCR-R" identical to eighteen 5′-proximal nucleotides of the primer "TCV-PacRT-R", 5′-GGTGTTAATTAAGGG-3′
3. Platinum® *Pfx* DNA Polymerase (Invitrogen)

2.4 Plant Inoculation

1. PacI restriction endonuclease (New England Biolabs)
2. T7 mMessage mMachine transcription kit (Ambion, Austin, TX)
3. Nicotiana benthamiana seeds
4. Arabidopsis thaliana seeds
5. Levingtons M2 potting compost.
6. Aluminum oxide powder, DURALUM (FEPA F400) 400 microgrits (Washington Mills Electro Minerals Company, NY)

2.5 Cloning of Amplified cDNA Fragments

1. QIAEX II Gel Extraction Kit (QIAGEN GmbH, Hilden, Germany)
2. Zero blunt TOPO PCT cloning Kit (Invitrogen): contains vector pCRII TOPO
3. Chemically competent *Escherichia coli* TOP10 cells (Invitrogen)
4. 50 mg mL^{-1} kanamycin, dissolved with sterile deionized water, filter sterilized, stored at 20°C
5. 90-mm Petri dishes with Luria Broth (LB) agar (Sigma) supplemented with kanamycin, 30 mg mL^{-1}
6. LB media
7. QIAprep Spin Miniprep kit (Qiagen)

3 Methods

3.1 Virus Purification and Isolation of Genomic RNA

1. Grow *N. benthamiana* plants at 16-h light period, at 20–27°C in Levingtons M2 potting compost.
2. Inoculate 2–3-week-old *N. benthamiana* with TCV inoculum. At this stage, *N. benthamiana* plants normally have three to four true leaves and are most susceptible to TCV.
3. Grow inoculated N. benthamiana plants for 10 days.
4. Harvest systemically infected leaves showing symptoms of TCV infection.
5. Homogenize 6g of freshly harvested TCV-infected leaf tissue with 24 mL of 50 mM sodium phosphate buffer, pH 7.6, containing 0.15 w/v sodium thioglycolate with a pestle and mortar.
6. Express the homogenate through Miracloth to remove debris and collect the filtrate in 50 mL polypropelene centrifuge tube.

7. Add butan-1-ol (0.9 mL per 10 mL of filtrate) and vortex for 5 min, and incubate for 1 h at 4°C.

8. Separate organic and aqueous phases by centrifugation at 10,000g at 4°C for 15 min. Carefully collect upper aqueous phase.

9. Pellet virus by centrifugation at 35,000 rpm for 3 h 30 min at 4°C in a Beckman 50.2 Ti rotor.

10. Carefully discard the supernatant by pouring, place the tube on filter paper to remove residual liquid.

11. Resuspend the virus pellet with 200 μL of 50 mM sodium phosphate buffer, pH 7.6 and transfer to 1.5-mL microcentrifuge tube.

12. Add 100 μL of RNA extraction buffer, 300 μL of phenol–chloroform–isoamyl alcohol mixture (25:24:1), cap the tube and vortex vigorously for 5 min.

13. Centrifuge at 12,000g for 5 min to separate the phases. Collect the upper aqueous phase, transfer to fresh 1.5-mL microcentrifuge tube and repeat the phenol–chloroform extraction until no visible interphase remains (normally two more times).

14. Transfer the aqueous phase to fresh 1.5-mL microcentrifuge tube, add equal volume of chloroform (about 250 μL), vortex, and separate phases by centrifugation as above.

15. Collect the aqueous phase (about 250 μL) to fresh 1.5-mL microcentrifuge tube, add one-tenth volume of 3 M sodium acetate pH 5.2 and three volumes of cold absolute ethanol.

16. Mix and incubate at −70°C for 30 min.

17. Precipitate the viral RNA by centrifugation at 13,000g for 30 min at 4°C. Remove the supernatant by pipetting.

18. Add 1 mL of cold 70% ethanol, centrifuge for 5 min at 13,000g, remove supernatant by pipetting, avoid picking up the pellet.

19. Pulse-spin the tube and remove the rest of supernatant.

20. Air-dry the viral RNA pellet for about 30 min.

21. Resuspend the white RNA pellet with 50 μL of nuclease-free sterile water, prepare 10-μL aliquots, and store at −70°C before use.

22. In order to measure concentration of the RNA preparation, dilute 10-μL aliquot of the RNA solution to 1 mL with sterile nuclease-free water and measure the absorbance at 260 nm. Calculate RNA concentration, 40 μg mL^{-1} of single-stranded RNA gives one unit of absorbance at 260 nm. If using Nanodrop Spectrophotometer, apply 1 μL of undiluted RNA preparation for measurement.

23. In order to assess integrity of RNA run aliquot of the RNA preparation containing approximately 1–5 μg of RNA in 1% nondenaturing agarose gel in 1X TBE buffer containing 0.5 μg mL^{-1} of ethidium bromide. Visualize on UV transilluminator. TCV RNA should migrate as a single band with some smearing (Fig. 1a).

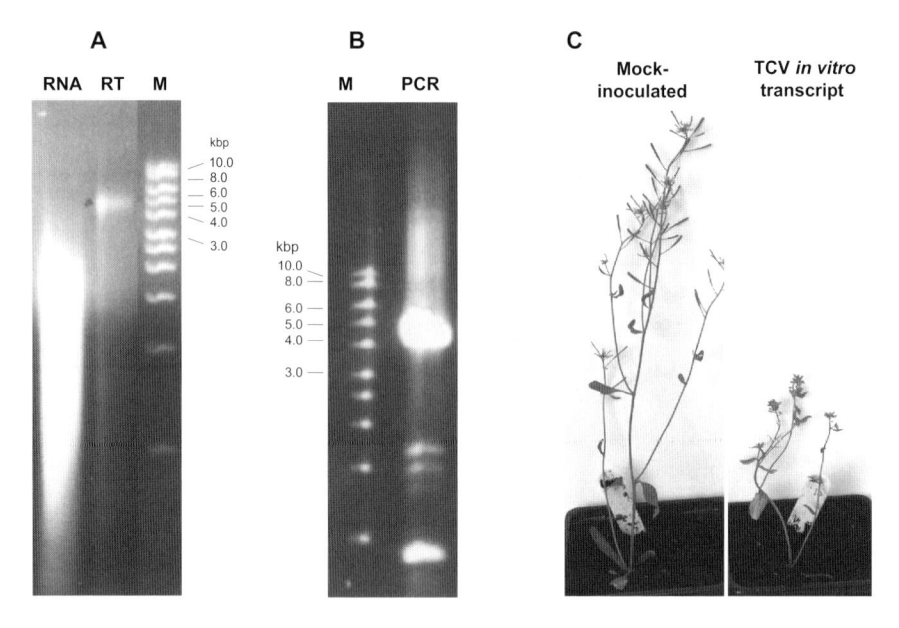

Fig. 1 (**a**) Agarose gel electrophoresis of TCV RNA (lane RNA, 5 μg) and the reverse transcription reaction (lane RT, 0.2 μg), M – DNA ladder; (**b**) agarose gel electrophoresis of the product of the TCV cDNA amplification by PCR (lane PCR 5 μL), M – DNA ladder; (**c**) *Arabidopsis thaliana* (Columbia) plants 12 days postinoculation, mock inoculated (right), and inoculated with the in vitro T7 RNA polymerase generated transcript derived from the cloned TCV cDNA (right)

3.2 *Reverse Transcription (see notes 1 and 2)*

1. Combine in a 100-μL thin-wall nuclease-free PCR tube 5 μL of 1 μg μL⁻¹ viral RNA, 1 μL of 2 μM oligonucleotide primer "TCV-PacRT-R", add 14 μL of RNAse-free water. Mix well by pipetting and place to the PCR heating block preset at 65°C for 3 min.
2. Transfer the tube from the heating block to the room temperature and let it cool down to room temperature, approximately 2 min.
3. Keep the reaction at room temperature and add to the reaction the following: 5 μL mix of dNTPs (5 μM each), 5 μL 10X First Strand Buffer supplied with the M-MuLV Reverse Transcriptase (New England Biolabs, Inc., MA), 17 μL of sterile nuclease-free water, 1 μL RNase OUT recombinant RNAse inhibitor (Invitrogen).
4. Mix well by pipetting, transfer immediately to the heating block set at 37°C.
5. After 30-s incubation, add 2 μL of M-MuLV Reverse Transcriptase (New England Biolabs, Inc., MA) which should be mixed well with the rest of the reaction by pipetting. Incubate at 37°C for 30 min, then increase temperature to 42°C and incubate additional 1 h.
6. To assess quality of the reverse transcription reaction, run 5 μL in 1% nondenaturing agarose gel in 1X TBE buffer containing 0.5 μg mL−1 of ethidium bromide along with 1–5 μg of the TCV RNA preparation (see Sect. 3.1). Visualize on UV transilluminator (Fig. 1a) (see note 3).

3.3 Amplification of TCV cDNA (see note 4)

1. Set up PCR reaction with Platinum® *Pfx* DNA Polymerase (Invitrogen) according to the manufacturer's instruction. For amplification the of the full-length cDNA copy of TCV genome combine following in the 200 µL PCR tube placed on ice: 2 µL of the First Strand cDNA reaction from the previous step, 37 µL sterile nuclease free water, 5 µL 10X Pfx Amplification buffer, 1.5 µL 10 mM dNTP mixture, 1 µL of 50 mM MgSO4, 1.5 µL mixture of amplification primers "TCV-T7F" and "TCV-PacPCR-R" (see note 5).

2. Mix well by pipetting on ice.

3. Add 2 µL (5 U) of *Pfx* DNA polymerase and mix by pipetting. Cap the tube, if necessary, centrifuge briefly to collect the content.

4. Set up the following programme on a thermal cycler: an initial denaturation at 94°C for 2 min; 10 cycles comprising of denaturation at 94°C for 30 s, annealing at 55°C for 3 min, and extension at 68°C for 7 min; then 15 cycles including denaturation at 94°C for 30 s, annealing at 55°C for 1 min, and extension at 68°C for 7 min; a final extension at 68°C for 10 min.

5. To make sure of the hot start of amplification, place the reaction tube in the cycler after it is heated up to 90°C.

6. Run the programme, after the completion keep the reaction at 4°C or on ice.

7. Run a 5 µL aliquot of the PCR reaction in a 1% agarose gel in 1X TBE (containing 0.5 µg mL^{-1} of ethidium bromide) in parallel with a DNA molecular weight standard. Visualize DNA in UV transilluminator. The major product should be approximately 4.0 kbp (Fig. 1b).

8. Load the whole PCR reaction on to a preparative 1% agarose gel (1X TBE, 0.5 µg mL^{-1} of ethidium bromide) along with a DNA molecular weight standard, run at 5 V cm^{-1}. At the end of the run, visualize the DNA on UV transilluminator and excise the approximately 4.5 kbp product with the scalpel blade.

9. Isolate the DNA fragment with QIAEX® II Gel Extraction Kit (QIAGEN) according to the manufacturer's instruction. Elute with 20 µL of sterile nuclease-free water.

10. Assess concentration of the PCR fragment by agarose gel electrophoresis, using DNA molecular weight ladder of known concentration as a standard or by Nanodrop spectrophotometer (see note 6).

3.4 Assessment of Infectivity of the RNA Transcript Derived from the Amplified TCV cDNA

The infectivity of the RNA transcripts derived from the T7 RNA polymerase promoter-TCV cDNA fragments produced by whole genome amplification could be assessed prior to their cloning into plasmid vector to ensure that the amplified product contains infectious cDNA of TCV genomic RNA.

1. Set up 20-μL transcription reactions with the T7 RNA polymerase promoter-TCV cDNA fragment as a template in 500-μL nuclease-free microcentrifuge tubes using a T7 mMessage mMachine kit (Ambion) according to the manufacturer's instructions. Combine at room temperature 0.5–1 μg of the template in 6 μL of nuclease-free water, 10 μL of 2X NTP/CAP mix, 2 μL of 10X Reaction Buffer and 2 μL of Enzyme mix.
2. Mix by pipetting and spin in the microcentrifuge to collect the reaction in the tube bottom.
3. After 1 h to analyze the integrity and yield of RNA transcript, run 0.5 μL aliquot of the reaction on 0.8% agarose gel (1X TBE, 0.5 μg mL−1 of ethidium bromide) and visualize in UV transilluminator.
4. Continue incubation of the rest of reaction at 37°C up to 2 h.
5. Dilute the transcript two folds with nuclease-free water and proceed to inoculation of plants immediately (see note 7).
6. Lightly dust the leaves of *N. benthamiana* and *A. thaliana* plants with aluminum oxide powder.
7. Pipette 5 μL of diluted reaction to the middle of dusted leaves.
8. Immediately gently stroke the whole leaf surface with the inoculum with gloved finger, do not apply excessive force. As a control, inoculate plants with the TCV RNA preparation.
9. After 5–10 min gently apply water on the inoculate leaf surface, shield inoculated plants from direct light by covering with paper for 24 h.
10. Remove the shield and keep plants in the glasshouse at 16-h light period, at 20°C/27°C.
11. Assess infection of plants after 7–10 days (see note 8).

3.5 Cloning of the T7–TCV cDNA Fusion in Plasmid Vector and Assessment of Infectivity of Individual Clones

The system based on use of *Vaccinia* virus topoisomerase I (9), Zero Blunt® TOPO PCR Cloning kit (Invitrogen), proved to be very efficient for cloning of long TCV cDNA fragments with blunt ends into a plasmid vector.

1. To perform TOPO ® Cloning reaction combine in the microcentrifuge tube at room temperature, mix 50 ng of the purified PCR fragment in 4 μL of nuclease-free water, 1 μL Salt Solution (supplied with the kit).
2. Add 1 μL (10 ng) of pCR® II-BluntTOPO® vector.
3. Mix reaction gently with a pipette and incubate at room temperature for 15 min.
4. After the completion of reaction, place the mixture on ice and proceed to transformation using One Shot ® (Invitrogen) competent cells.
5. Set up 42°C water bath, thaw a vial with SOC medium (supplied with the kit) at room temperature, and transfer one tube with One Shot chemically competent TOP10 (Invitrogen) E. coli cells from −85°C freezer to ice to thaw.
6. Add 2 μL of the TOPO cloning reaction to the competent cell suspension (50 μL).

7. Mix gently and incubate on ice for 15 min.
8. Heat-shock the cells for 30 s at 42°C water bath, immediately transfer to ice, incubate for 2 min and add 250 μL of SOC medium.
9. Shake tube horizontally at 37°C (200 rpm) for 1 h in an orbital shaker.
10. Spread the transformation mixture on LB plates with 30 μg mL^{-1} kanamycin (see note 9) and incubate overnight at 37°C.
11. Pick 24 colonies and inoculate 5 mL LB medium containing 50 μg mL^{-1} kanamycin, incubate the bacterial cultures at 37°C in an orbital shaker (200 rpm) overnight.
12. Isolate plasmid DNA from the overnight cultures using a QIAprep Spin Miniprep kit (Qiagen) according to manufacturer's instruction.
13. Digest the plasmid DNA preparations with *Eco*RI (*Eco*RI restriction sites flank insertion site in pCR®II-BluntTOPO® vector). Select clones which contain the insert with the total length about 4,000 bp (the vector size is 3,519 bp). Note that due to the variation between TCV isolates EcoRI site(s) are present in different positions of the TCV cDNA.

3.6   *Linearization of Recombinant Plasmid DNA*

1. Linearize the plasmid preparations selected according to the restriction map. For example, mix in a 1.5-mL microcentrifuge tube 20 μL of plasmid DNA (about 5 μg), 20 μL of NEBufer I (New England Biolabs), 2 μL of BSA (New England Biolabs), 156 μL of water and 2 μL (20 U) of PacI restriction endonuclease (New England Biolabs).
2. Incubate the reaction for 4 h at 37°C.
3. Run an aliquot of the restriction reactions in 1% agarose gel to ensure that digestion is complete. If uncut plasmid remains, add additional 2 μL of PacI and continue reaction for another 4 h.
4. Add 200 μL of phenol–chloroform–isoamyl alcohol mixture (25:24:1), cap the tube, and vortex for 5 min.
5. Centrifuge at 12,000g for 5 min to separate the phases.
6. Collect and transfer the upper aqueous phase to a fresh 1.5-mL microcentrifuge tube.
7. Add 200 μL of chloroform, vortex and separate phases by centrifugation as above.
8. Collect the aqueous phase (about 200 μL) in a fresh 1.5-mL microcentrifuge tube, add one-tenth volume of 3 M sodium acetate (pH 5.2) and three volumes of cold absolute ethanol.
9. Mix well, and incubate at −70°C for 30 min.
10. Precipitate the linearized plasmids by centrifugation at 13,000g for 30 min at 4°C.
11. Remove the supernatant with a pipette.
12. Add 1 mL of cold 70% ethanol and centrifuge for 5 min at 13,000g.
13. Remove supernatant with a pipette avoiding picking up the pellet, pulse-spin the tube and remove the rest of supernatant.

14. Air-dry for about 30 min.
15. Resuspend the transparent DNA pellet with 10 μL of nuclease-free sterile water store at −20°C before use.

3.7 Testing Infectivity of Full-length TCV cDNA Clones

It is expected that not all full-length cDNA clones of TCV will be infectious, therefore, a number of the TCV clones (five to ten) should be tested to select infectious TCV clones which induce symptoms identical to a wild type TCV of a certain strain.

1. For each selected TCV cDNA clone set up a 20-μL transcription reaction using *PacI*-lincarized plasmid as a template and a T7 mMessage mMachine kit according to manufacturer's instructions. Combine the following at room temperature: 1 μg of the linearized plasmid in 6 μL of nuclease-free water, 10 μL of 2X NTP/CAP mix, 2 μL of 10X Reaction Buffer, and 2 μL of Enzyme mix.
2. Mix by pipetting and spin briefly in the microcentrifuge.
3. After 1 h, to analyze the integrity and yield of RNA transcript, run 0.5 μL aliquot of the reaction on 0.8% agarose gel (1X TBE, 0.5 μg mL^{-1} of ethidium bromide) and visualize on UV transilluminator.
4. Continue incubation of the rest of reaction at 37°C up to 2 h.
5. Dilute the transcript twofold with nuclease-free water and proceed to inoculation of plants immediately.
6. Lightly dust the leaves of *N. benthamiana* and/or *A. thaliana* plants with aluminum oxide powder.
7. Pipette 5 μL of diluted reaction to the middle of dusted leaves. Immediately gently stroke the whole leaf surface with the inoculum with gloved finger, do not apply excessive force.
8. After 5–10 min, gently apply water on the inoculate leaf surface, shield inoculated plants from direct light by covering with paper for 24 h.
9. Remove the shield and incubate plants in the glasshouse at 16-h light period, at 20°C/27°C. Use either wild type virus inoculum or RNA extract from TCV virus preparation as a control.
10. Assess inoculated plants daily, systemic symptoms would first appear in Arabidopsis after 5–7 days in the plants onto which infectious transcripts were applied. Later, 10–12 days postinoculation, TCV infection causes severe stunting of the bolts (Fig. 1c).
11. Select clones which induce symptoms identical to those of wild type virus. TCV infection should be confirmed by either RT-PCR of by electron microscopy of the sap extracts.
12. Sequence the selected clones containing the infectious TCV cDNA by using the set of primers designed according to the complete genome sequences of TCV isolates available in the GeneBank (Accession numbers M22445, AY312063).

4   Notes

1. Integrity of input RNA is absolutely crucial for the production of full-length cDNA of TCV genomic RNA which is more than 4 kb long. Therefore, fresh RNA preparation should be used if possible and RNAse inhibitors should be included into a reaction mix. To decrease a proportion of nonspecific annealing and initiation of the first cDNA strands from in the positions other than TCV RNA 3′ terminus, use low concentration of a specific reverse transcription primer and add reverse transcriptase to the reaction which is preheated at least to 37°C.

2. Synthesis of the cDNA of TCV RNA should be primed with an oligonucleotide primer complementary to the 3′ part of the TCV genome with the restriction enzyme site which is not present in TCV genome, *PacI*.

3. The main product of the First strand cDNA reaction, an RNA–DNA heteroduplex, should run as a single band, approximately 4,500 nucleotide, according to the double-stranded DNA molecular weight standards. The mobility of this heteroduplex is significantly lower than that of TCV RNA, there is also much less smearing than in the case of RNA (Fig. 1a).

4. Amplification by PCR allows production of sufficient amount of the full-length cDNA copy for cloning and for initial assessment of infectivity. To minimize a number of errors introduced into the product following amplification, a proofreading thermostable DNA polymerase should be used, for example, Platinum® *Pfx* DNA Polymerase (Invitrogen) or Phusion (New England Biolabs, and the number of amplification cycles should be minimized. To ensure production of sufficient amount of the fragment, a relatively high amount of the first strand reaction, ca. 0.2 μg per 50 μL reaction should be used. Production of nonspecific, short products should be minimized by hot start of amplification, low concentration of primers, and increased annealing temperature.

5. Primer "TCV-T7F" contains the T7 RNA polymerase promoter sequence preceding the sequence identical to 18 5′ proximal nucleotides of TCV genomic RNA as a forward primer. Primer "TCV-PacPCR-R" is identical to 18 nucleotide 5′-proximal sequence to the Primer "TCV-PacRT-R" used in the reverse transcription reaction, 10 μM each.

6. The purified blunt-end DNA product comprised of the T7 RNA polymerase promoter sequence fused to the 5′ terminus of the full-length TCV cDNA could be cloned into appropriate vector.

7. For inoculation of *N. benthamiana* plants at the stage as described in Sect. 3.1 and *A. thaliana* (Columbia) plants grown in the similar conditions at the mature rosette stage before bolting.

8. If plants inoculated with the in vitro transcript from the PCR product show the same symptoms as control plants infected with RNA isolated from TCV preparation, proceed with cloning of the T7 RNA polymerase promoter-TCV cDNA fragment into a plasmid vector.

9. To get at least one plate with well-separated colonies, use two plates and apply 30 μL and 270 μL of the transformation reaction.

Acknowledgments   This work was supported by the Biotechnology and Biological Sciences Research Council.

References

1. Ahlquist, P., French, R., Janda, M., and Loesch-Fries, L. S. (1984) Multicomponent RNA plant virus infection derived from clonal viral cDNA. *Proc. Natl. Acad. Sci. USA* **80**, 7066–7070.

2. Carrington, J. C., Heaton, L. A., Zudema, D. B., Hillman, I., and Morris, T. J. (1989) The genome structure of turnip crinkle virus. *Virology* **281**, 219–226.

3. Cohen, Y., Gisel, A., and Zambrysky, P. C. (2000) Cell-to-cell and systemic movement of recombinant green fluorescent protein-tagged Turnip crinkle virus. *Virology* **273**, 258–266.

4. Hacker, D. L., Petty, I. T., Wei, N., and Morris, T. J. (1992) Turnip crinkle virus genes required for RNA replication and virus movement. *Virology* **186**, 1–8.

5. Heaton, L. A., Carrington, J. C., and Morris, T. J. (1989) Turnip crinkle virus infection from RNA synthesized in vitro. *Virology* **170**, 214–218.

6. Li, W. -Z., Qu, F., and Morris, T. J. (1998) Cell-to-cell movement of turnip crinkle virus is controlled by two small open reading frames that function in trans. *Virology* **244**, 405–416.

7. Qu, F., Ren, T., and Morris, T. J. (2003) The coat protein of turnip crinkle virus suppresses posttranscriptional gene silencing at an early initiation step. J. *Virol.* **77**, 511–522.

8. Ryabov, E. V., Van Wezel, R., Walsh, J., and Hong, Y. (2004) Cell-to-cell, but not long-distance spread of RNA silencing that is induced in individual epidermal cells. *J. Virol.* **78**, 3149–3154.

9. Shuman, S. (1994) Novel approach to molecular cloning and polynucleotide synthesis using vaccinia DNA topoisomerase. *J. Biol. Chem.* **269**, 31731–31734.

Chapter 34
Construction of Infectious Clones for DNA Viruses: Mastreviruses

Margaret I. Boulton

Abstract To characterize a virus at the molecular and biological levels, it is necessary to produce an infectious clone. For most of the *Geminiviridae*, cloning of the genome is relatively easy because of their small genomes and the presence of the virus double-stranded (replicative) DNA form in infected plants. Indeed, the presence of conserved sequences between species in the genera *Begomovirus*, *Curtovirus*, and *Topocuvirus* allows the PCR amplification of most genomes using degenerate "universal" primers. Unlike the other genera, no universal primers are reported that are suitable for all mastreviruses and alternative, more time-consuming methods must be used.

This chapter describes a method that has proven successful for the preparation and testing of infectious clones for a wide range of mastreviruses. It has been designed to ensure its applicability for laboratories throughout the world. Methods are presented for the isolation of total plant DNA and the purification of the replicative (cccDNA) form of the virus using a commercially available plasmid purification kit. Restriction enzyme digestion of the purified DNA using a restriction enzyme with a unique site in the viral genome allows the cloning of a full-length copy of the genome into a high copy number vector, thereby providing a template for sequence analysis and further cloning. The only efficient method for confirming infectivity of mastrevirus clones is using agroinoculation (also termed agroinfection). This requires the production of a multimeric copy of the genome in a T-DNA binary vector, transformation of specific *Agrobacterium* strains with the binary vector clone, and inoculation of specific regions of seedlings, or seeds, of the appropriate host species. These specific requirements are described and discussed.

Keywords *Mastrevirus*; Geminivirus; Dimer clones; Multimeric insert; Binary vector; Agroinoculation; Agroinfection; Infectious clone; *Agrobacterium* transformation; Infectivity; Dicots; Monocots; Cereals; Grasses

1   Introduction

In order to study the biology of a virus at the molecular level, it is necessary to produce an infectious clone of the viral genome and then to sequence it and determine its coding potential. This will allow mutagenesis of specific regions of the genome and the subsequent introduction of the mutated or wild type genomes into host plants for studies on host–virus interactions (reviewed for mastreviruses in ref. 1). In this chapter, methods for the isolation, cloning and inoculation of the genomes of species within the genus *Mastrevirus* (family *Geminiviridae*) will be described. Table 1 shows the species of the genus (*Maize streak virus*, MSV, is the type species) and indicates their host range (dicotyledonous plants or members of the *Poaceae*). Infectious clones are available for the majority of these species, the purification methods used to isolate the viral DNA for cloning, and their genome sequences, are detailed in the table. Detailed information on the virion structure,

Table 1 Species in the genus *Mastrevirus* and the DNA isolation methods used to produce genome-length clones[a]

Species	Accession number[b]	Abbreviation	Method for ds DNA isolation[c]
Bean yellow dwarf virus	Y11023	BeYDV	"2 step" PCR of total plant DNA
Chloris striate mosaic virus[d]	M20021	CSMV	cccDNA preparation
Digitaria streak virus	M23022	DSV	Second strand synthesis of ssDNA
Maize streak virus	X01633	MSV	Second strand synthesis of ssDNA and cccDNA preparation
Miscanthus streak virus	D01030	MiSV	cccDNA preparation
Panicum streak virus	X60168	PanSV	cccDNA preparation
Sugarcane streak virus[e]	S64567	SSV	cccDNA preparation
Sugarcane streak Egypt virus	AF037752	SSEV	RE digestion, gel purification
Sugarcane streak Reunion virus	AF072672	SSREV	RE digestion, gel purification
Tobacco yellow dwarf virus	M81103	TYDV	cccDNA preparation
Wheat dwarf virus	X82104	WDV	Guanidinium thiocyanate and alkaline lysis

[a] For many of the species several different strains have been characterized. For the purposes of this manuscript, only one example has been provided for each species and isolates in the genus have not been included.

[b] In general, the nucleotide sequence accession number of the best characterized strain has been provided.

[c] cccDNA preparation: Total infected plant DNA was subjected to cesium chloride–ethidium bromide density gradient ultracentrifugation to purify viral covalently closed circular (ccc) dsDNA. RE digestion, gel purification: Total infected plant DNA was subjected to restriction enzyme digestion. After agarose gel electrophoresis and ethidium bromide staining, the approximately 2.7 kb (linearized genome) band was eluted.

[d] The infectivity of this clone has not been reported.

[e] This clone was not infectious.

genome organization, and host ranges of the mastreviruses and other geminivirus genera (*Begomovirus*, *Curtovirus*, and *Topucovirus*) is available (2) and at http://www.danforthcenter.org/iltab/geminiviridae. The functions of the mastrevirus products are reviewed in (1).

The circular DNA genomes of the geminiviruses facilitate their cloning and analysis at the molecular level. This is particularly true for members of the genus *Begomovirus* as there are "universal primers" (3) that allow amplification of the entire genome, even from total plant DNA. Unfortunately, such primers are not available for amplification of all mastreviral genomes. However, once part of the viral DNA sequence is available, for example, by PCR amplification using degenerate primers based on the conserved sequence of a closely related virus (4), partially overlapping primers can be used to amplify a full-length copy of the genome. This "two-step" method may be particularly useful if the amount of infected tissue is limited and/or the viral double-stranded (ds) DNA accumulates at low levels in infected plants. If closely related viruses have not yet been characterized, alternative techniques of producing dsDNA for cloning purposes are available.

The monopartite genomes of mastreviruses comprise circular single-stranded (ss) DNA of between 2.5 and 2.8 kb, which can be isolated from purified virions and used in vitro as a template for the production of dsDNA and subsequent cloning of the DNA. This second-strand synthesis is particularly useful for uncharacterized mastrevirus genomes as all mastrevirus species encapsidate a short DNA primer-like molecule (5, 6), thereby obviating the need for sequence information for the purchase of primers. This technique was used for the production of an infectious clone of *Digitaria streak virus* (7) and can be useful if virions are the only available source of mastreviral DNA.

Second-strand synthesis has been generally superseded by the use of the viral dsDNA, which is produced as a replication intermediate and is present at relatively high levels in geminivirus-infected tissues. The viral dsDNA can often be detected by agarose gel electrophoresis and ethidium bromide staining of total DNA extracted from maize leaves infected with strains of MSV that cause severe symptoms. If this is the case, the circular dsDNA forms can be linearized by restriction enzyme (RE) digestion prior to electrophoresis, thereby allowing direct isolation from the gel of a digested genome suitable for cloning. This method is routinely used in my laboratory for many MSV isolates which accumulate in high levels in the plant but is less suitable for mastreviruses which are present at lower titres. In order that the methods described in this chapter are universally applicable to the cloning of all mastrevirus genomes, a method for enrichment of the covalently closed circular (ccc) or supercoiled dsDNA will be described. The technique is based upon the alkaline lysis method used for bacterial plasmid DNA (8) and modification of methods used in my laboratory and that of Palmer et al. (9). Total DNA is extracted from mastrevirus-infected plants using standard phenol–chloroform extraction followed by precipitation with ethanol or isopropanol. After alkaline lysis, the cccDNA is purified using an anion exchange column. This is achieved most conveniently using a commercially available plasmid preparation kit. If such kits are not available, alternative methods of enrichment for cccDNA can be employed, such as that described for the cloning of *Wheat dwarf virus* (WDV) using guanidine thiocyanate and alkaline lysis (10) or by cesium chloride–ethidium bromide density gradient ultracentrifugation (11), although the latter employs highly toxic chemicals, requires an ultracentrifuge, and larger amounts of infected tissue. If the sequence of the mastrevirus is not known, the dsDNA is subjected to RE digestion to identify those enzymes that cut only once in the genome and then is used for cloning. In this chapter, a single full-length copy of the viral genomic DNA will first be ligated to a high-copy vector such as pUC to provide a source of the full-length genomic DNA at high concentration. The insert will then be purified and ligated into a binary cloning vector. The insert will be provided in excess compared to the vector to allow selection of a tandemly repeated (dimer) copy of the viral genome. This method ensures that there is only a single (cloned) version of the genome present in the binary vector and that a high-copy template is available to facilitate nucleotide sequence analysis.

Once cloned, sequencing of the small genome is straightforward. However, it is necessary to prove the infectivity of the cloned genome. In general, the genomic DNA of the mastreviruses is not mechanically transmissible, although a "vascular puncture" technique (12) has been used to introduce the genomic and cloned DNA of MSV into maize seeds. The most efficient method of inoculation requires the insertion of a greater-than-unit-length copy of the viral genome between the T-DNA borders of a binary vector, allowing the genome to be transferred to the plant by *Agrobacterium*-mediated T-DNA transfer. This technique is referred to as "agroinoculation" or "agroinfection" (13, 14). The repeated copy allows the escape of a complete progeny genome either by homologous recombination, or by replicational release; in the latter case, the conserved hairpin structure containing the geminivirus origin of replication must be duplicated in the construct (15). For viruses that may have low virulence (often, but not always, indicated by the presence of mild symptoms), it is best to ensure that both release mechanisms are possible and because of this, a method involving the cloning of a tandem dimeric (head to tail) copy of the genome will be described. We have used a wide variety of binary vectors and have detected little, or no, difference in their relative efficiency. Thus, the binary vector should be chosen based upon the availability of appropriate restriction enzyme sites or the antibiotic selection required.

The *Agrobacterium* strain used is important for those mastreviruses that infect members of the *Poaceae* as it has been shown that octopine strains are unsuccessful for agroinoculation of maize, wheat, and rice, and probably most cereals and grasses (14, 16–18). Although early studies used wild type strains for cereal agroinoculation (these "oncogenic" strains do not cause tumors on the *Poaceae*), more recently disarmed derivatives have been used and this is preferable for reasons of biosafety.

It is best to use immature plants for agroinoculation, because plants increasingly become more resistant to virus infection with age, and young plants may be more susceptible to *Agrobacterium*. Successful agroinoculation of maize has been obtained from 2 to 30 days after germination, but inoculation of 10-day-old seedlings is the simplest method. The site of inoculation depends on whether dicot or monocot plants are used, and both methods will be described in the text.

The entire procedure is summarized below:

Isolation of Total Plant DNA
↓
Purification of Viral cccDNA
↓
Cloning of a Full-length Mastrevirus Genome
↓
Cloning of a Multimer of the Mastrevirus Genome into a Binary Vector
↓
Transformation of *Agrobacterium tumefaciens*
↓
Agroinoculation of Host Plants

2 Materials

2.1 Isolation of Total Plant DNA

1. Infected leaf material (see note 1)
2. Mortar (approximately 15 cm) and pestle, liquid nitrogen (see note 2)
3. DNA extraction buffer: 50 mM Tris–HCl, 50 mM NaCl, 25 mM EDTA, pH 7. Autoclave the buffer and then add SDS to 1% (w/v).
4. Tris buffered phenol–chloroform pH 8 (Phenol:Chloroform:Isoamyl alcohol 25:24:1, saturated with 10 mM Tris, 1 mM EDTA, Sigma-Aldrich) (see note 3)
5. Sterilized tubes: polypropylene tubes (SS34, Sorval, or similar), 1.5-mL microcentrifuge tubes
6. Chloroform (see note 4)
7. Isopropanol (propan-2-ol)
8. Ethanol (absolute and 70%)
9. Sorval centrifuge RC5B or similar (suitable for tubes containing greater than 25 mL and a speed of 12,000g) (see note 5)

2.2 Purification of Viral cccDNA by Alkaline Denaturation and Anion Exchange Column Chromatography

1. Qiagen plasmid mini kit (Qiagen) or an equivalent commercially available kit containing solutions, microcentrifuge tubes, and anion exchange columns
2. Sterilized 1.5-mL microcentrifuge tubes
3. Isopropanol (propan-2-ol)
4. 70% ethanol
5. Microcentrifuge

6. Equipment and reagents for agarose gel electrophoresis: Low electroendosmosis agarose, 1X Tris–borate–EDTA (TBE) buffer, ethidium bromide (see note 6) and commercially available molecular weight markers containing a band of known concentration (e.g., 1 Kb DNA Ladder, Invitrogen), "Minigel" apparatus, UV transilluminator

2.3 Cloning of a Full-length Mastrevirus Genome Using Unique Restriction Enzyme Sites

1. Sterilized microcentrifuge tubes
2. Electrophoresis equipment and reagents, see step (6) of Sect. 2.2
3. Restriction enzymes and buffers and appropriate water baths or heating blocks
4. Commercially available DNA isolation kit (e.g., QIAEX II gel extraction kit or equivalent)
5. Commercially available vector DNA (high-copy number plasmid, e.g., pUC (19)).
6. Alkaline phosphatase (preferably shrimp alkaline phosphatase, see note 28).
7. T4 DNA ligase and appropriate buffer
8. Transformation-competent *Escherichia coli* cells
9. Reagents for selection of recombinants: Petri plates, Luria broth (LB) agar (see note 10). Antibiotic stock solutions (1000X concentration, for pUC this is 100 mg mL^{-1} carbenicillin). Stock solutions of 1 M isopropyl β-D-thiogalactopyranoside (IPTG) and X-gal (40 mg mL^{-1}) (see note 7). Sterilized toothpicks
10. For colony PCR (optional): Commercially available M13 reverse and forward primers and PCR kit (see note 8)
11. Sterilized glycerol

2.4 Cloning of a Tandem Dimeric Copy of the Genome into a T-DNA Binary Vector

1. Qiagen plasmid mini kit as described in step (1) of Sect. 2.2
2. Appropriate REs (the unique cutter within the viral genome, e.g., *Bam*HI, and other enzymes that cut within the vector, but not the viral genome) as determined in step (1) of Sect. 3.3 (also see note 32)
3. Electrophoresis reagents and equipment as in step (6) of Sect. 2.2
4. Commercially available DNA isolation kit (e.g., QIAEX II gel extraction kit or equivalent)
5. Commercially available binary vector DNA (e.g., pBIN19 (20, 21))
6. Alkaline phosphatase (preferably shrimp alkaline phosphatase (SAP), see note 28)
7. T4 DNA ligase (high concentration, 5–20 U μL^{-1}) and appropriate buffer
8. Transformation-competent *E. coli* cells

9. Reagents for selection of recombinants as in step (9) of Sect. 2.3 except that the antibiotic selection for pBIN19 is kanamycin (stock solution 50 mg mL^{-1}) (see note 9)
10. For colony PCR (optional): Purchased reverse and forward primers that recognize the viral genome sequence (see note 37), and a commercially available PCR kit
11. Sterilized glycerol

2.5 Transformation of Agrobacterium tumefaciens

1. MG/L broth (22) (see note 10)
2. *Agrobacterium* strain C58C1(pGV3850) (23)
3. Purified recombinant binary plasmid DNA (at least 1 μg isolated from the clone selected in Sect. 3.4, and control (pBIN19) DNA)
4. Appropriate antibiotic stock solutions, 50 mg mL^{-1} rifampicin, 100 mg mL^{-1} carbenicillin, and 50 mg mL^{-1} kanamycin (see notes 7 and 9)
5. Sterilized culture tube
6. 250 mL Erlenmeyer flask
7. Spectrophotometer (not essential, see note 11)
8. SS34 tubes, or similar, and centrifuge (e.g., Sorvall RC5B)
9. Sterilized 20 mM CaCl$_2$
10. Eppendorf tubes
11. Liquid nitrogen
12. Long-handled forceps
13. Sterilized glycerol
14. Purchased reverse and forward primers that recognize the viral genome sequence, and a commercially available PCR kit

2.6 Agroinoculation of Host Plants

1. *Agrobacterium* containing the multimeric copy of the viral genome (prepared in Sect. 3.5)
2. Prepared petri plates containing LB agar supplemented with 50 μg mL^{-1} kanamycin and 50 μg mL^{-1} rifampicin (see note 12)
3. Seedlings of the species to be inoculated, in appropriately sized plant pots (see note 13)
4. Spatula and toothpicks
5. Microcentrifuge tubes
6. Hamilton syringe (50 μL) with beveled needle (Sect. 3.6.1 only) (see note 14)
7. Needles, either entomological needles glued to an appropriate support (see note 15) or fine sewing needles (Sect. 3.6.2 only)
8. Appropriate growth facilities (see note 16)

3   Methods

3.1   Isolation of Total Plant DNA

1. Precool a mortar and pestle by the addition of approximately 20 mL of liquid nitrogen. Add more liquid nitrogen and approximately 10 g of fresh, or frozen, infected leaf material and grind to a fine powder (see notes 1 and 2).
2. Once the liquid nitrogen has evaporated, add 1 volume (approximately 10 mL) of DNA extraction buffer and mix by grinding intermittently until the mixture thaws (see note 17).
3. Immediately after the mixture is thawed, add 10 mL (or a volume equal to the DNA extraction mix added) of phenol–chloroform reagent and mix well by grinding (see note 3).
4. Transfer the extract to a 30-mL polypropylene tube and centrifuge at 12,000*g* for 10 min at 4°C in a Sorvall SS34 rotor to pellet unwanted plant material (see note 5).
5. Transfer the aqueous layer (see note 18) to a clean polypropylene tube, add an equal volume of phenol–chloroform reagent and mix well. Centrifuge as before. Repeat this extraction and centrifugation step once more so that the interphase between the aqueous phase and the phenol–chloroform phase contains little or no white (proteinaceous) layer.
6. Transfer the aqueous phase to a clean polypropylene tube and add an equal volume of chloroform. Mix and centrifuge as before.
7. Transfer the aqueous phase to a clean tube, add 0.7 volume isopropanol and mix gently to precipitate the nucleic acid which will be seen as a whitish stringy material.
8. Pellet the nucleic acid at 17,000*g* for 20 min. Remove all of the supernatant.
9. Rinse the pellet by mixing it with 5 mL cold 70% ethanol and centrifuging for 5 min at 17,000*g*.
10. Carefully remove all of the ethanol with a pipette and allow the pellet to dry for approximately 15 min at room temperature. This nucleic acid extract is then subjected to alkaline lysis and anion exchange chromatography using the Qiagen plasmid mini kit as described in Sect. 3.2 (see note 19).

3.2   Purification of Viral cccDNA by Alkaline Denaturation and Anion Exchange Column Chromatography

This step is most conveniently done using a commercially available plasmid preparation kit.

1. Gently resuspend the pellet in 500 µL Buffer P1. Divide the mixture and transfer half to each of two microcentrifuge tubes. To each tube add 250 µL Buffer P2. Mix gently, by inversion. Incubate at room temperature for 5 min (see note 20).

2. Add 350 µL Buffer N3. Mix immediately by inversion.
3. Centrifuge for 10 min in a microcentrifuge at 17,000g (~13,000 rpm) to pellet the denatured chromosomal DNA (see note 21).
4. Transfer the supernatants to a clean microcentrifuge tube.
5. Precipitate the supernatant by adding 0.8 volume isopropanol, pellet by centrifugation at 17,000g in a microcentrifuge for 15 min.
6. Repeat steps 1–4, except that in step 1 each pellet should be resuspended in 250 µL Buffer P1 (see note 22).
7. Pipette each supernatant onto a QIAprep spin column.
8. Centrifuge (17,000g) for 30–60 s. Discard the flowthrough.
9. Add 500 µL Buffer PB to each column and centrifuge for 30–60 s. Discard the flowthrough (see note 23).
10. Add 750 µL Buffer PE to each column. Centrifuge for 30–60 s.
11. Discard the flowthrough, centrifuge for an additional 1 min to remove residual PE (wash) buffer.
12. Place the columns in clean microcentrifuge tubes. Elute the cccDNA by adding 50 µL Buffer EB (10 mM Tris–Cl, pH 8.5) to each column. Leave for 1 min, and then centrifuge 1 min. Store these tubes (A1 and A2) containing cccDNA on ice.
13. Elute any residual DNA from the columns into a clean microcentrifuge tube by adding 250 µL Buffer EB to each column, and repeat the incubation and centrifugation described in step 12 (see note 24).
14. Add 0.7 volume isopropanol to these two tubes, mix gently, and pellet the DNA by centrifugation at 17,000g for 15 min.
15. Discard the supernatant and wash the pellet by adding 250 µL cold 70% ethanol and centrifuging for 10 min.
16. Remove all of the supernatant and air dry the pellet for approximately 15 min. Resuspend the pellet in 25 µL Buffer EB and store these tubes (B1 and B2) on ice.
17. Confirm the presence of cccDNA by agarose gel electrophoresis of 1–2 µL of the DNA from each of tubes A1, A2, B1, B2) and ethidium bromide staining (see note 25).

3.3 Cloning of a Full-length Mastrevirus Genome Using Unique Restriction Enzyme Sites

1. Digest aliquots of the cccDNA (prepared in Sect. 3.2) each with a different restriction enzyme. Subject the samples to gel electrophoresis in 1% agarose and stain with ethidium bromide to determine which enzymes linearized the viral DNA and therefore have a single restriction site within the viral genome (see note 26).
2. Insert preparation: Select an enzyme that has a unique restriction site in the viral genome and is present in the multiple cloning site of the high copy number

vector (for the purposes of this Chapter, *Bam*HI and the vector pUC (20), database accession numbers L09136, L09137 will be chosen). Digest up to 40 μL of the cccDNA with the unique cutter (e.g., *Bam*HI). Subject the digest to electrophoresis and recover the linearized genome using a commercial kit (e.g., the QIAEX II gel extraction kit, see note 27).

3. Vector preparation: Digest up to 5 μg of vector (pUC) DNA with the same enzyme (e.g., *Bam*HI) as that selected for linearizing the viral genome. Dephosphorylate the vector using shrimp alkaline phosphatase (SAP) to reduce religation of the vector. Heat inactivate the SAP at 65°C for 15 min and confirm the success of the digestion using agarose gel electrophoresis. Recover the linearized DNA using a commercial kit as in step 2 (see note 28).

4. Assess the approximate concentration of the insert and vector DNA by gel electrophoresis of an aliquot of each preparation along with a marker of known concentration.

5. Ligate the insert and vector at a molar ratio of 3:1, respectively (see note 29).

6. Transformation of *E. coli*: Prepare, or purchase, competent *E. coli* cells (see note 30). Transform using the ligation mix from step 5 and plate one-tenth and nine-tenth of the cells each on L agar containing the appropriate antibiotic (for pUC, 100 μg mL⁻¹ carbenicillin), IPTG (1 mM), and X-Gal (40 μg mL⁻¹) to select for (the white colored) recombinant clones.

7. Confirm the presence of the insert, either by restriction enzyme digestion and electrophoresis of miniprepped DNA prepared using a commercial plasmid mini prep kit (e.g., with *Bam*HI as this will also confirm that the insert can be excised, but see note 23) or, more rapidly, by "colony PCR" (see note 31).

The selected recombinant provides the source of a cloned full-length genome suitable for further cloning and analysis (see note 32).

3.4   Cloning of a Tandem Dimeric Copy of the Genome into a T-DNA Binary Vector

1. Purify recombinant plasmid DNA from the clone selected in Sect. 3.3, using a commercial kit.

2. Insert preparation: Digest approximately 10 μg recombinant plasmid DNA using the restriction enzyme (e.g., *Bam*HI) selected for the insert preparation in Sect. 3.3 and an additional enzyme that does not cut the viral genome but has a recognition site(s) inside the vector backbone. Recover the linearized genome fragment from agarose as described in step (2) of Sect. 3.3 (see note 33).

3. Binary vector preparation: Digest 5 μg binary vector DNA (here, pBIN19, (20, 21) see note 34) with the enzyme used to prepare the viral genome fragment. Dephosphorylate and recover the linearized vector from agarose as described in step (3) of Sect. 3.3, except that 0.7% agarose should be used to allow the separation of any uncut and linearized forms of the relatively large (approximately 12 kb) vector.

4. Assess the approximate concentration of the insert and vector fragments as described in step (4) of Sect. 3.3.

5. Ligate the insert and vector at a molar ratio of at least 10:1, respectively. Use 5 U ligase and 100 ng of pBIN19 (see note 35).
6. Transformation of *E. coli*: Use high-efficiency competent cells produced from a *RecA⁻* strain and the ligation mix from step 5 (see note 36). Plate the cells as described in step (6) of Sect. 3.3 except that the carbenicillin should be replaced by 50 μg mL⁻¹ kanamycin. Recombinant colonies will be white.
7. Select clones containing a multimeric insert by restriction enzyme digestion using enzymes flanking the insert, or by colony PCR (see notes 37 and 32). The restriction enzymes chosen should not have recognition sites within the viral genome (refer to data obtained in steps (1) and (2) of Sect. 3.3 and see note 32).

A recombinant clone containing a multimeric copy of the viral genome should be selected and this will be used for DNA purification and transformation of *Agrobacterium* (see note 38). The clone should be stored in glycerol at −70°C (see note 31).

3.5 Transformation of A. tumefaciens by the Freeze–Thaw Method

3.5.1 Preparation of Competent *Agrobacterium* Cells

1. Inoculate 5 mL of MG/L broth containing 50 μg mL⁻¹ rifampicin and 100 μg mL⁻¹ carbenicillin with *Agrobacterium* strain C58C1(pGV3850) (23). Incubate at 25°C with shaking for approximately 16 h.
2. Transfer 2 mL of the *Agrobacterium* culture to 50 mL MG/L broth in a 250-mL Erlenmeyer flask. Incubate at 25°C with shaking, until the cells reach an OD_{600} between 0.5 and 1.0 (see note 11).
3. Transfer the culture to sterile centrifuge tubes (e.g., Sorval SS34 tubes). Chill on ice for up to 20 min.
4. Pellet the cells by centrifuging the culture at 10,000g for 10 min in a SS34 rotor (or similar) at 4°C. Discard the supernatant.
5. Gently resuspend the cells in 5 mL ice cold 20 mM $CaCl_2$ and repeat the centrifugation. Discard the supernatant.
6. Gently resuspend the cells in 1 mL ice cold 20 mM $CaCl_2$. Aliquot 150 μL of the cells into each of three chilled Eppendorf tubes for transformation (see note 39).

3.5.2 Transformation of Competent *Agrobacterium* Cells

The freeze–thaw method (modified from ref. 24 will be described (see note 40))

1. Prepare purified plasmid DNA (from Sect. 3.4, and also the vector DNA, see notes 38 and 39) at a concentration of approximately 100 ng⁻¹ μg μL⁻¹ in sterile distilled water (SDW) or 1X TE.

2. Add 1.2 mL MG/L medium to a culture tube.
3. Add up to 1 μg plasmid DNA to the transformation-competent *Agrobacterium* cells, add an equal volume of SDW to a third tube (see note 39), mix gently, and freeze by lowering the Eppendorf tube into liquid nitrogen for approximately 5 min. Use long-handled forceps to hold the tube during the freezing process.
4. Thaw the cells at room temperature, leave at room temperature for 5–10 min.
5. Transfer the cells to the culture tube containing 1.2 mL MG/L medium and incubate, shaking, at 25°C for 2–4 h.
6. Pellet the cells in a microcentrifuge for 2 min at 10,000*g*. Discard the supernatant.
7. Resuspend the cells in 1 mL MG/L medium and plate 100 μL, and the concentrated remainder, onto plates containing MG/L agar and the antibiotics appropriate for selection of the binary plasmid construct within the selected *Agrobacterium* strain (for pBIN19 and C58C1(pGV3850), 50 μg mL⁻¹ each of rifampicin and kanamycin and 100 μg mL⁻¹ carbenicillin (see note 41)). Incubate plates at 25°C for 2–3 days until the colonies are sufficiently large for sampling.
8. Confirm the presence of a viral insert by colony PCR using primers able to amplify only from a multimeric genome (see note 37).
9. Store the recombinant in glycerol at −70°C.

3.6   Agroinoculation of Host Plants

This section is divided into two parts because the method for inoculation of monocot seedlings differs from that used for dicot plants.

3.6.1   Agroinoculation of Poaceae

1. Preparation of plant material: Sow four seeds into peat-based compost in approximately 12 cm pots. Sow just under the surface of the compost and grow under appropriate conditions until 2–3 leaves have emerged (see note 42).
2. Preparation of inoculum: 2–3 days prior to plant inoculation sample the glycerol stock of *Agrobacterium* using the wide end of a sterile toothpick. Plate the bacteria thickly onto L agar containing 50 μg mL⁻¹ kanamycin and 50 μg mL⁻¹ rifampicin and incubate at 25°C for 2–3 days (see note 43).
3. Inoculation of plants: Collect the *Agrobacterium* by scraping carefully across the agar with the flat edge of a spatula. Transfer the bacteria to a microfuge tube and add approximately 1 mL SDW and mix, by inversion, to produce a uniform "milky" suspension. Use a Hamilton syringe with beveled needle (see note 14) to inject approximately 30 μL of bacterial suspension into the base of the seedling. Three inoculations (approximately 10 μL each) should be done, one into the centre of the stem just above the base of the plant, a second

approximately 3 mm above this, and a third injection which is done vertically down the stem toward the base of the plant (see note 44 and Fig 1b–d). Inoculate some plants of each species with *Agrobacterium* containing pBIN19 without an insert.

4. Grow the plants under conditions appropriate to the species under test. Examine plants for symptoms from 7 days after infection. Continue to examine plants for up to 2 months. The symptoms will be apparent on leaves that develop subsequent to inoculation (see note 45).

3.6.2 Agroinoculation of Dicot Species

1. Preparation of plant material: Sow seeds of the chosen species, and prick out the seedlings, as soon as they can survive being handled, into peat-based compost in 7 cm pots. The plants can be inoculated when the first three leaves have emerged, but are not fully expanded (see note 46 and Fig. 1f, g).
2. Preparation of inoculum: This should be done as described in step (2) of Sect. 3.6.1.
3. Inoculation of plants: Prepare the inoculum by plating and culturing the *Agrobacterium* as described in step (3) of Sect. 3.6.1. The bacteria will be used directly from the petri plate. Inoculate the plants by collecting a small amount (sufficient to cover a circle approximately 1–2 mm in diameter) of bacteria on a toothpick and smear the bacteria onto the plant where a petiole joins the stem. Prick approximately five times through the bacteria into the stem using a fine needle (see notes 47 and 48 and Fig. 1f, g).
4. Grow the plants and examine them for symptoms as described in step (4) of Sect. 3.6.1 (see note 48).

4 Notes

1. To obtain the best results (the "cleanest" DNA and the highest yield of cccDNA), the youngest leaves that show virus symptoms should be sampled and either used immediately or stored frozen at −70°C. However, dsDNA can be extracted from older tissues and samples stored at −20°C, or even at 4°C for 2 days, can yield suitable dsDNA. Grinding of the tissue is facilitated by cutting the leaves into 1 cm strips immediately prior to freezing, or grinding.
2. If liquid nitrogen is not available, dry ice (solid carbon dioxide) can be used. In this case, precool the mortar and pestle and the tissue in dry ice and keep the mortar and pestle on dry ice throughout the grinding of the plant material (see Sect. 3.1).
3. The phenol–chloroform mixture is toxic and therefore it is recommended that this is purchased, rather than prepared in the laboratory. Wear safety spectacles and two pairs of gloves and dispose of the outer pair if they become contaminated. All handling of this reagent should be done in a fume hood. Dispose of liquids and contaminated materials correctly.
4. Chloroform is carcinogenic and should be handled only in a fume hood. Dispose of waste correctly.
5. If an appropriate centrifuge is not available, it is possible to carry out the procedure by decanting the plant extract into multiple microcentrifuge tubes and using a microcentrifuge. It is important that the tubes are resistant to phenol and chloroform.

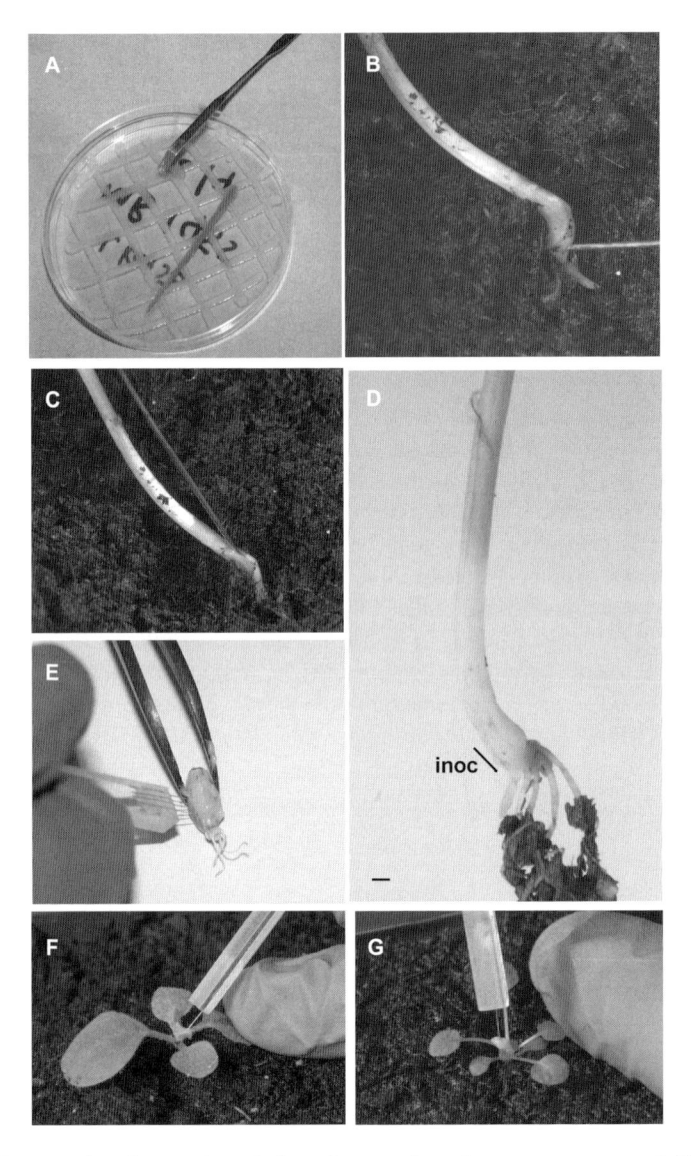

Fig. 1 The procedure for agroinoculation of mastrevirus clones to monocot and dicot hosts. (a) A thickly streaked plate of *Agrobacterium* prepared for inoculation. Bacteria are collected either by scraping the cells into a tube for inoculation of monocots, or by collecting bacteria with a toothpick (dicots and cereal seeds). (b) "Horizontal" (c) "vertical" inoculation of a wheat seedling in the area just above the root emergence zone. (d) Injection of inoculum should be within the area encompassed by the line (inoc). Inoculation of (e) imbibed barley seeds, (f) *Nicotiana benthamiana* and (g) *Arabidopsis thaliana* using an entomological needle, multiple stabs should be made through the *Agrobacterium*. For barley seeds, inoculation is done along the embryonic axis. Bar in panel D = 1 mm

6. Ethidium bromide is a powerful mutagen and a dilute stock solution should be purchased (e.g., $500 \mu g$ mL^{-1}, Sigma). Wear gloves and dispose of contaminated materials correctly.

7. Antibiotic stock solutions should be prepared at a final concentration of 1000X. The antibiotic solutions described here are suitable for pUC. If other vectors are used, the appropriate antibiotics should be substituted (see note 34). Antibiotics are harmful, gloves should be worn and the powder should be weighed out in a fume hood. It is rarely necessary to filter sterilize the stocks for molecular biological use, but SDW (where appropriate) and sterilized containers should be used to prepare the stocks. Carbenicillin is soluble in water. Store at $-20°C$. A 1 M stock of IPTG can be prepared in SDW, X-gal (40 mg mL^{-1}) should be dissolved in dimethylformamide (DMF). Store IPTG and X-Gal at $-20°C$. DMF is toxic, wear gloves and handle in a fume hood. Alternatives to X-gal are now available which do not require DMF solubilization.

8. The PCR kit and primers are necessary only if colony PCR is used for screening the recombinants. In this case, ensure that the primers are appropriate for the vector used. The web sites of most retailers of molecular biology reagents provide information on (or database accession numbers for) the sequences of their vectors. Universal M13 primers are suitable for use with pUC; an approximately 200 bp product is obtained when no insert is present in the clone.

9. Kanamycin is toxic, take appropriate precautions (see note 7). Stock (1000X) is 50 mg mL^{-1} kanamycin in SDW. Store at $-20°C$.

10. MG/L medium is particularly good for culture of a wide variety of *Agrobacterium* strains. The composition (1 L) is 500 mL LB, 10 g mannitol, 2.32 g sodium glutamate, 0.5 g KH$_2$PO$_4$, 0.2 g NaCl, 0.2 g MgSO$_4$.7H$_2$O, 2 μg biotin, pH 7.

LB medium (1 L) comprises 10 g tryptone, 5 g yeast extract, 10 g NaCl, pH 5. All media should be sterilized by autoclaving.

11. A spectrophotometer is not essential, although it is recommended. It is used to determine the OD of the *Agrobacterium* cultures used for the production of competent cells. In the absence of a spectrophotometer, grow the cultures until they are slightly cloudy, producing a "silky" cloud of cells when held up to the light and swirled.

12. Rifampicin is toxic, take appropriate precautions (see note 7). Make a 500X stock solution (25 mg mL^{-1}) in methanol and store at $-20°C$. Rifampicin is light sensitive and should be stored in a dark bottle or a tube covered by aluminum foil.

13. To determine the infectivity of the clone, the plant species chosen should be that from which the viral DNA was extracted. Where possible, the same cultivar or ecotype should be used (but see note 44).

14. Hamilton syringes are available from many suppliers. It is important to purchase one with a beveled end so that the needle will pierce the plant stem. If a Hamilton syringe is not available, use a disposable 1 mL syringe fitted with the smallest gauge needle. The Hamilton syringe will cause less damage to the plants and will facilitate inoculation of grasses with fine stems.

15. Headless entomological needles can be purchased in a wide variety of sizes via the internet from Watkins and Doncaster, UK (http://www.watdon.com/study.htm) and other suppliers. These can be glued to small perspex or plastic rods using cyanoacrylate glue (be careful, this is toxic) in order to facilitate their use. Entomological pins are also available, but often in limited sizes that are too thick for this procedure. When available, the pins are easier to handle and do not require additional support. Entomological needles/pins are preferable to sewing needles because they are finer and cause less damage to the plant.

16. In many countries, the insertion of viral genomes into *Agrobacterium* is subject to biosafety guidelines. For example, in the UK agroinoculation experiments must be carried out under containment level B (level 2) conditions.

17. The plant extract should be thawed, but only allowed to approach room temperature immediately prior to the addition of the phenol–chloroform reagent. Extended incubation at room temperature will lead to polyphenol oxidase activity and a brown DNA pellet which affects subsequent purification steps. The mix should form a viscous liquid at room temperature, if the mix is not liquid, additional DNA extraction buffer should be added at this point.

18. The upper layer is the aqueous phase. Take care not to collect any of the white interphase as this will result in the need for more phenol–chloroform extractions. The extract should be mixed well (conveniently by repeatedly inverting the tube for approximately 20 s), but not sufficiently vigorously to shear the chromosomal DNA or to produce bubbles in the interphase.

19. For some mastreviruses, particularly the MSV strains that produce severe symptoms, further purification may be unnecessary because the viral DNA is visible following electrophoresis of total DNA samples and ethidium bromide staining. If this is the case, restriction digestion can be done using total DNA. If a commercial kit is not available, the alkaline lysis method described in ref. 25 can be used to purify cccDNA. If the pellet is very brown or yield is low, it is possible that an alternative extraction procedure will be required. There are several alternative methods available for extraction of DNA from plants with high phenolic content (26).

Section 3.1 can be completed in approximately 4 h.

20. Throughout this procedure, solutions should be mixed gently, but thoroughly, by inverting the tube. Shaking can decrease the efficiency of cccDNA purification. If the total nucleic acid pellet cannot be completely resuspended in this volume of P1 buffer, more buffer can be added, but the amounts of buffers P2 and N3 must be increased proportionately.

21. If a microcentrifuge capable of 17,000g is not available, the procedure should be carried out using the highest speed attainable.

22. The repetition of steps 1–4 is generally necessary for DNA extracted from plant samples to remove contaminating plant DNA and carbohydrates which can decrease the efficiency of column purification.

23. This step is specific to plant material and helps to remove any carbohydrate and impurities remaining in the plant extract.

24. This step may increase the yield of cccDNA. If no additional DNA is detected in these tubes after step 17, they may be discarded.

25. The amount of cccDNA will be determined by the amount of replicative DNA present in the infected tissue, the tissue used for extraction (the age of the leaves, their structure, and composition affect the efficiency of DNA extraction) and the efficiency of the DNA extraction and purification procedure. It is likely that some contaminating host DNA will still be present but this will not usually affect the ensuing procedures. If, after modifying the DNA extraction technique (see note 19), viral cccDNA is not visible, alternative procedures for the cloning of the viral genome can be used, for example, a two-step PCR method. Degenerate primers to a conserved region of the mastreviral genome (ref. (4) and refer to accession numbers in Table 1) may amplify a region of the viral DNA from the preparation produced in Sect. 3.1 or 3.2. Sequencing of the product will allow the design of overlapping or abutting primers that can be used to amplify the entire genome (27). If this approach is taken, overlapping primers that contain a restriction site (present in the viral genome at this position) will facilitate downstream cloning procedures. The primers should not be designed within the region encompassed by the degenerate primer.

Section 3.2 can be completed in approximately 4 h.

26. Tubes A1 and A2 can be pooled. The choice of restriction enzymes should be based on the sites present in the multiple cloning site (MCS) of the vector used for cloning the monomeric construct (and also those present in the binary vector). In this chapter, *Bam*HI will be chosen as the enzyme having a unique restriction site in the viral genome. The absence of a unique cutter with a site in the MCS of pUC will require a search of the literature for vectors with appropriate sites. Do not clone the viral insert into compatible sites which do not allow excision of an intact viral genome sequence from the recombinant.

27. Any commercial kit recommended for the isolation of DNA fragments from agarose gel may be used. Note that some kits require special electrophoresis conditions (the type of buffers and agarose), this will be mentioned in the manufacturer's protocol. There are many published techniques for the recovery of DNA from agarose without the use of a kit (http://www.protocol-online.org/prot/Molecular_Biology/DNA/DNA_Extraction__Purification/DNA_

Extraction_from_Agarose_Gel). Limit exposure of the stained DNA to UV irradiation as this can cause nicking of the DNA.

28. Steps 2 and 3 can be done in parallel. The vector should be digested completely and dephosphorylated with SAP to prevent its re-ligation. SAP is the preferred enzyme as it is functional in most RE buffers (confirm suitability of the buffers by reference to the manufacturer's data sheet) and can be heat inactivated at 65°C for 15 min. Agarose gel purification of the linearized vector ensures that any uncut vector is removed from the preparation.

29. The amount of vector may be limited by the amount of insert obtained in steps (2) and (3) of Sect. 3.3. It is preferable to use 50 ng vector (requiring 150 ng insert), but successful cloning can be accomplished with as little as 10 ng. Adhere to the manufacturer's protocol for ligation conditions.

30. If *E. coli* cells are to be purchased, cells appropriate to this step and to step (6) of Sect. 3.4 should be selected. Thus, they should be recombination deficient (to encourage retention of a multimeric copy of the genome) and of high competence (to improve the efficiency of transformation with larger binary vectors). Electrocompetent cells are more efficient than chemically prepared cells, but require an electroporator and the purification, or dilution, of the ligation mix to remove salts that cause arcing. When plating the remaining nine-tenth of the transformation mixture, it may be necessary first to reduce the volume by microcentrifugation for 30 s at 10,000*g* to pellet the cells. Plating two concentrations helps to prevent bacterial overgrowth if the ligation and transformation are efficient.

31. The method for colony PCR is as follows: Pick approximately half of the colony with a sterile toothpick and transfer to 50 μL SDW in a microcentrifuge tube. Boil (or heat at 95°C) for 10 min to lyse the bacteria. Centrifuge at 17,000*g* for 5 min to pellet cell debris. Use 5 μL of the supernatant in a 50 μL amplification reaction. See note 8 for details of appropriate primers.

Once the recombinant has been selected it should be stored at −70°C in glycerol. To produce a glycerol stock, part of the colony should be inoculated into 5 mL L broth containing the appropriate antibiotics and incubated, with shaking, at 37°C for approximately 16 h. Add 500 μL of culture to 500 μL sterilized glycerol in a microcentrifuge tube, mix well and immediately store frozen, preferably at −70°C. This mixture can be sampled with a toothpick without thawing the culture.

32. At this point, it is advisable to obtain some sequence data for the cloned viral genome. This will facilitate downstream analyses such as identification of dimer clones. It will also confirm that the clone is a mastrevirus genome and, if sequencing is commenced from each end of the insert (using the universal primers described in note 8), it can be compared with other mastrevirus sequences to confirm that the two ends should join to give a complete genome. If this analysis suggests that a segment of the genome is missing, it is likely that there were two RE recognition sites sufficiently close to each other to produce an apparently full-size genome following the eletrophoretic analysis described in steps (1) and (2) of Sect. 3.3. To identify whether sequence is missing, amplify the cccDNA or total DNA by PCR using primers that flank the "missing" genomic region and sequence the product. These two procedures together will provide the full sequence of the mastrevirus genome. If a full-length genome has not been cloned in Sect. 3.3, the data will facilitate the recloning of the full genome from the total or cccDNA using PCR or RE digestion.

33. Cutting the vector within its backbone is only necessary when the vector is close in size to the expected size of the insert. For mastreviruses, the expected genome (insert) size is 2.5–2.8 kb and pUC is approximately 2.7 kb. The similar sizes make electrophoretic separation of the insert and vector impossible. Larger vectors are available, but it is important that if these are selected the efficiency of transformation is high in order to ensure that the recombinant clone is obtained. For pUC, when possible, choose an enzyme site present in the ampicillin resistance gene as this will yield fragments that separate well from the insert fragment.

34. The binary vector pBIN19 and its derivatives have been widely used for cloning geminiviral genomes. However, many binary vectors are available and many have been used successfully in the author's laboratory. If a different vector is chosen, take care that the antibiotic selection present on the vector will be appropriate for selection of *Agrobacterium* recombinant clones;

several *Agrobacterium* strains have natural antibiotic resistance, or have resistance engineered into the modified Ti plasmid. For example, a binary vector selectable by carbenicillin resistance is not suitable for use with *Agrobacterium* strain C58C1(pGV3850).

35. The aim is to produce concatameric (multiple) inserts into the binary vector. The proportion of multimeric inserts is increased by the use of high concentrations of insert and ligase, a minimum ligation reaction volume, and overnight ligations at temperatures ≤16°C. The amount of insert required can be calculated using the following formula:

$$[(\text{ng vector} \times \text{size of insert in kb}) \div \text{size of vector in kb}]$$
$$\times \text{molar amount of (insert} \div \text{vector)} = \text{ng insert.}$$

For the current example where the approximate sizes of the insert and vector are 2.7 and 12 kb, respectively:

$$[(100\,\text{ng vector} \times 2.7\,\text{kb insert}) \div 12\,\text{kb vector}] \times (10 \div 1) = 225\,\text{ng insert}$$

36. The use of high concentrations of DNA and ligase can inhibit the transformation reaction. Half of the ligation reaction should be used in the first instance. If transformation is not successful, try inactivating the ligase, using less ligation reaction and/or cells of higher competency, and electroporation. If electroporation is used, it will be necessary to either dilute the ligation mix with SDW or purify it and either resuspend or elute it in SDW to prevent arcing.

37. A multimeric insert will normally be found within the first 10 colonies tested. Occasionally, digestion will reveal the presence of three copies of the genome, if no dimers are obtained, the trimer-containing clone can be used. Colony PCR is not efficient for large inserts and therefore viral genome-specific primers that will produce a fragment only if a multimeric genome copy is present should be selected, that is "upstream" and "downstream" of the RE site used for cloning. This will require knowledge of the mastreviral genome (insert) sequence (see note 32). The dimers are likely to be oriented "head-to-tail" (the required orientation) as "head-to-head" constructs are generally unstable. The orientation can be confirmed by digestion or by PCR as described above. In the unlikely event that a dimer is not obtained, alternative, two-step cloning procedures can be used to clone a partial multimeric copy of the genome (14, 28). Although we have monomeric constructs that are infectious (if the cloning site lies within nonessential regions such as specific parts of the short intergenic region), this approach is not recommended.

38. The copy number of the vector will determine the yield of DNA obtained in a miniprep. For low copy number plasmids (e.g., pBIN19 and many other binary vectors), modify the procedure as outlined in the manufacturer's protocol and consider using either a midiprep protocol or perform more than one miniprep per construct. For transformation of *Agrobacterium*, at least 1 μg DNA should be prepared.

Long-term storage of the recombinant clone should be in glycerol at −70°C (see note 31 but note that *Agrobacterium* should be cultured at 25°C for approximately 48 h).

39. The cells are best used on the day that they are prepared. Although it is possible to store them at 4°C for a day, or frozen, this will decrease their competence. In total, three tubes will be required for (i) the dimeric construct DNA, (ii) pBIN19 DNA (to be used as a control for agroinoculation), and (iii) a tube to which water, rather than DNA, will be added, to act as a control for the selection of transformants in step (2) of Sect. 3.5.

40. The freeze–thaw method of transformation is easy and quick and does not require specialized equipment, although it is considerably less efficient than the electroporation and triparental mating methods. Electroporation is very efficient but requires clean DNA, triparental mating is time consuming. Preparation of electroporation-competent cells is described in (29) and the triparental mating method in (30).

41. The pBIN19 control should be selected on the same plates. To hasten growth of the *Agrobacterium*, or if transformation efficiency is low, the concentrations of carbenicillin and rifampicin can be reduced to 50 and 25 μg mL^{-1}, respectively. To test the viability of the competent cells, plate a small aliquot of the untransformed cells onto agar plates without kanamycin.

42. The seeds should be sown approximately 10–21 days before the expected date of inoculation, depending on the species to be tested and the growth conditions. The inoculation of monocots is done at the base of the plant and to facilitate this the seeds should be sown near the soil surface. Soil can be removed from plants sown deeply, but extra care will be needed because contaminating soil may block the fine syringe needle used for inoculation.

43. The growth of bacteria on plates has the advantage that the bacteria can be collected in the absence of antibiotics which can cause chlorosis of inoculated plants. An example of thickly plated *Agrobacterium* is shown in Fig. 1a. If shaking cultures are used, the bacteria should be pelleted by centrifugation, washed once in culture medium, and resuspended in SDW before use. It is not necessary to determine the OD of the bacterial suspension used for inoculation as in our experience a 100-fold variation in concentration has little effect on infection efficiency. Induction of *Agrobacterium vir* expression by acetosyringone is not necessary, as we have not detected increased infection efficiencies for any plants inoculated with preinduced bacteria. If necessary, cultures grown on plates may be stored at 4°C for up to 1 week.

44. Inoculation of wheat is shown in Fig. 1b–d, inoculations should be successful within the area delineated by the line (inoc). "Mock inoculation" using *Agrobacterium* containing only pBIN19 is useful to differentiate between inoculation damage and virus infection symptoms such as dwarfing. If the plant has more than one tiller, repeat the inoculation to all, or most, tillers. The "vertical" inoculation (Fig 1d) is especially useful for larger plants. The needle should be inserted until a slight resistance is felt, at which point the suspension should be expelled. Do not push the needle further down toward the root. If the stem of the plant is very thin, or the plant is small, only perform the first two, horizontal inoculations. For vegetatively propagated plants (e.g., sugarcane) very young plants should be inoculated and more robust needles and more injections may be needed. Agroinoculation of maize is very efficient, and if maize is likely to be a host of the mastrevirus, it should be included. Agroinoculation of maize is described in detail in Boulton (1995). Agroinoculation of barley seedlings is inefficient, even when using *Agrobacterium* strains (e.g., AGL1) ideal for barley transformation; we have obtained higher infection efficiencies by agroinoculating seed (Fig. 1e) using a modification of ref. 12. Some of the suspension may leak out of the plant, but this will not affect infection efficiency as long as the leakage is not caused by the needle being pushed completely through the stem. The inoculation procedure causes significant damage to plants with small stems, but once the procedure is optimized, most will survive.

Agroinoculation efficiency is temperature dependent during the inoculation and for approximately 36 h afterward. The plants should be maintained at 25°C, or less, throughout this period.

45. Agroinoculation causes some dwarfing and distortion of plants and may increase tillering which may resemble some virus symptoms (e.g., the main symptom of WDV is stunting). It is important to compare mock inoculated and inoculated plants.

If no symptoms are apparent, the plants could still be infected and should be tested for the presence of virus (if PCR is used, take care to allow for the potential presence of contaminating virus sequences in any residual inoculum). If infection is not obtained in the original host, alternative *Agrobacterium* strains may be tried as *Agrobacterium*-mediated T-DNA transfer to cereals is strain specific (14, 16, 17). Alternatively agroinoculation of seed may be tried. The possibility of having isolated a noninfectious clone cannot be ignored, but at this point the sequence of the monomer clone should be known and any abnormalities in genome sequence or coding capacity have been identified by comparison with other sequenced mastreviruses (Table 1 and see note 32).

46. An example of the inoculation procedure for *Nicotiana benthamiana* is shown in Fig. 1f. A general rule for agroinoculation of dicot plants is to inoculate very young plants but a compromise between damage (resulting in plant death) and age must be reached. For initial studies, plants at various ages should be tested.

47. A proportion of plants should be mock inoculated (see note 44) and inoculations should be done at 25°C (see note 44).

48. For very small plants such as *Arabidopsis*, inoculation can be done around the plant apex (Fig. 1g) although this will cause some lethality, especially before the methodology is optimized. The plants will need to be supported firmly to prevent them from being pulled out of the compost. The phenotype of mock-inoculated plants should be compared with plants inoculated with the construct containing the viral genome (see note 45). If symptoms are not produced in inoculated plants, testing of asymptomatic inoculated plants can be carried out as described in note 45 and the sequence of the clone compared as outlined in note 45. Although *Agrobacterium* C58C1(pGV3850) has a wide host range, it may still show some host specificity. If tumors are produced after stabbing the original host with wild type C58, it is likely that the disarmed strain is competent for T-DNA transfer. If tumors are not obtained, super-virulent strains should be tested for oncogenicity and the related disarmed strain used for agroinoculation.

References

1. Boulton, M. I. (2002) Functions and interactions of mastrevirus gene products. *Physiol. Mol. Plant Pathol.* **60**, 243–255.
2. Fauquet, C. M., Mayo, M. A., Maniloff, J., Desselberger, U. and Ball, L. A. (2005) *Virus Taxonomy: Eighth Report of the International Committee on Taxonomy of Viruses.* Elsevier.
3. Briddon, R. W. and Markham, P. G. (1994) Universal primers for the amplification of dicot-infecting geminiviruses. *Mol. Biotechnol.* **1**, 202–205.
4. Rybicki, E. P. and Hughes, F. L. (1990) Detection and typing of maize streak virus and other distantly related geminiviruses of grasses by polymerase chain reaction amplification of a conserved viral sequence. *J. Gen. Virol.* **71**, 2519–2526.
5. Donson, J., Morris-Krsinich, B. A. M., Mullineaux, P. M., Boulton, M. I. and Davies, J. W. (1984) A putative primer for second-strand synthesis of maize streak virus is virion-associated. *EMBO J.* **3**, 3069–3073.
6. Morris, B. A. M., Richardson, K. A., Haley, A., Zhan, X. and Thomas, J. E. (1992) The nucleotide sequence of the infectious cloned DNA component of Tobacco yellow dwarf virus reveals features of geminiviruses infecting monocotyledonous plants. *Virology* **187**, 633–642.
7. Donson, J., Gunn, H. V., Woolston, C. J., Pinner, M. S., Boulton, M. I., Mullineaux, P. M., and Davies, J. W. (1988) *Agrobacterium*-mediated infectivity of cloned Digitaria streak virus DNA. *Virology* **162**, 248–250.
8. Ish-Horowitz, D. and Burke, J. F. (1982) Rapid and efficient cosmid vector cloning. *Nucleic Acids Res.* **9**, 2989–2998.
9. Palmer, K. E., Schnippenkoetter, W. H., and Rybicki, E. P. (1998) Geminivirus isolation and DNA extraction, in *Methods in Molecular Biology* **81**, *Plant Virology Protocols: From Virus Isolation to Transgenic Resistance* (Foster, G. D. and Taylor, S. C., eds), Humana Press Inc, Totowa, NJ, pp. 41–52.
10. Bendahmane, M., Schalk, H-J. and Gronenborn, B. (1995) Identification and characterization of Wheat dwarf virus from France using a rapid method for geminivirus DNA preparation. *Mol. Plant Pathol.* **85**, 1449–1455.
11. Sunter, G., Coutts, R. H. A. and Buck, K. W. (1984) Negatively supercoiled DNA from plants infected with a single-stranded DNA virus. *Biochem. Biophys. Res. Commun.* **118**, 747–752.
12. Redinbaugh, M. G. (2003) Transmission of *Maize streak virus* by vascular puncture inoculation with unit-length genomic DNA. *J. Virol. Methods* **109**, 95–98.
13. Grimsley, N., Hohn, T., Davies, J. W. and Hohn, B. (1987) *Agrobacterium*-mediated delivery of infectious maize streak virus into maize plants. *Nature* **325**, 177–179.
14. Boulton, M. I., Buchholz, W. G., Marks, M. S., Markham, P. G. and Davies, J. W. (1989) Specificity of *Agrobacterium*-mediated delivery of maize streak virus DNA to members of the Gramineae. *Plant Mol. Biol.* **12**, 31–40.

15. Stenger, D. C., Revington, G. N., Stevenson, M. C. and Bisaro, D. M. (1991) Replicational release of geminivirus genomes from tandemly repeated copies: Evidence for rolling circle replication of a plant viral DNA. *Proc. Natl. Acad. Sci. USA* **88**, 8029–8033.
16. Boulton, M. I. and Davies, J. W. (1990) Monopartite geminiviruses: markers for gene transfer to cereals, in *Aspects of Applied Biology* **24**, *The Exploitation of Micro-organisms in Applied Biology* (Ball, S. F. L., Knott, C. A., Pink, D., Sopp, P., Spaull, A. F., and Tatchell, G. M., (eds), Association of Applied Biologists, Wellesbourne, Warwickshire, pp. 79–86.
17. Dasgupta, I., Hull, R., Eastop, S., Poggi-Pollini, C., Blakebrough, M., Boulton, M. I. and Davies, J. W. (1991) Rice tungro bacilliform virus DNA independently infects rice after *Agrobacterium*-mediated transfer. *J. Gen. Virol.* **72**, 1215–1221.
18. Heath, J. D., Boulton, M. I., Raineri, D. M., Doty, S. L., Mushegian, A. R., Charles, T. C., Davies, J. W. and Nester, E. W. (1997) Discrete regions of the sensor protein VirA determine the strain-specific ability of *Agrobacterium* to agroinfect maize. *Mol. Plant Microbe Interact.* **10**, 221–227.
19. Yanisch-Perron, C., Vieira, J. and Messing, J. (1985) Improved M13 phage cloning vectors and host strains: nucleotide sequences of the M13mp18 and pUC19 vectors. *Gene* **33**, 103–119.
20. Bevan, M. (1984). Binary *Agrobacterium* vectors for plant transformation. *Nucleic Acids Res.* **12**, 8711–8721.
21. Frisch, D. A., Harris-Haller, L. W., Yokubaitis, N. T., Thomas, T. L., Hardin, S. H. and Hall, T.C. (1995) Complete sequence of the binary vector Bin19. *Plant Mol. Biol.* **27**, 405–409.
22. Lichtenstein, C. P. and Draper, J. (1985) in *DNA cloning: A Practical Approach*, Vol. 2 (Glover, D. M., ed), IRL, Washington, DC, pp. 67–119.
23. Zambryski, P., Joost, H., Genetello, C., Leemans, J., Van Montague, M. and Schell, J (1983) Ti plasmid vector for the introduction of DNA into plant cells without alteration of their normal regeneration capacity. *EMBO J.* **12**, 2143–2150.
24. Chen, H., Nelson, R. S. and Sherwood, J. L. (1994) Enhanced recovery of transformants of *Agrobacterium tumefaciens* after freeze-thaw transformation and drug selection. *Biotechniques* **16**, 664–669.
25. Sambrook, J., Fritsch, E. F. and Maniatis, T. (1989) *Molecular Cloning: A Laboratory Manual*, 2nd edition, Cold Spring Harbor Laboratory Press, Cold Spring Harbor, NY.
26. Hills, P. N. and Van Staden, J. (2002). An improved DNA extraction procedure for plant tissues with a high phenolic content. *S. Afr. J. Bot.* **68**, 549–550.
27. Briddon, R. W., Prescott, A. G., Luness, P., Chamberlin, L. C. L., and Markham, P. G. (1993). Rapid production of full-length, infectious geminivirus clones by abutting primer PCR (AbP-PCR). *J. Virol. Methods* **43**, 7–20.
28. Boulton, M. I. (1995). *Agrobacterium*-mediated transfer of geminiviruses to plant tissues, in *Methods in Molecular Biology* **49**, *Plant Gene Transfer and Expression Protocols* (Jones, H., ed), Humana Press, Totowa, NJ, pp. 77–93.
29. Mattanovich, D., Ruker, F., Machado, A. C., Laimer, M., Regner, F., Steinkellner, H., Himmler, G. and Katinger, H. (1989) Efficient transformation of *Agrobacterium* spp by electroporation. *Nucleic Acids Res.* **17**, 6747.
30. Ditta, G. Stanfield, S., Corbin, D., Helinski, D. R. Broad host range DNA cloning system for gram-negative bacteria: construction of gene bank of *Rhizobium meliloti*. *Proc. Natl. Acad. Sci. USA* **77**, 7347–7351.

Chapter 35

Construction of Infectious Clones of Double-Stranded DNA Viruses of Plants Using Citrus Yellow Mosaic Virus as an Example

Qi Huang and John S. Hartung

Abstract Double-stranded DNA (dsDNA) viruses of plants are believed to be plant pararetroviruses. Their genome is replicated by reverse transcription of a larger than unit-length terminally redundant RNA transcript of the viral genomic DNA using the virus-encoded replicase. In order to produce a cloned, infectious viral genome, the clone must be constructed in a binary vector and be longer than the full, unit-length viral genome. The clone can then be transferred by *Agrobacterium*-assisted inoculation into a suitable host plant to induce virus infection.

Keywords *Agrobacterium*; Badnaviruses; Caulimoviruses; Plant pararetroviruses

1 Introduction

Plant viruses with a dsDNA genome belong to the family *Caulimoviridae*, in the genera *Caulimovirus*, *Badnavirus*, "Rice tungro bacilliform-like viruses," "Legume-infecting viruses," and "Cassava vein mottle-like viruses" (1). Caulimoviruses have isometric particles with a genome size of approximately 8 kb. They are transmitted in nature by aphids, but can also be easily transmitted mechanically by rubbing onto the leaves of host plants. Important caulimoviruses include *Cauliflower mosaic virus* (CaMV), among the best characterized of plant viruses and type species of the genus, *Dahlia mosaic virus* and *Carnation etched ring virus*. All members of *Caulimovirus* have a fairly limited host range, causing mostly mottle or mosaic symptoms on certain vegetables, ornamentals, or weeds (2).

Badnaviruses affect a wide range of tropical plant species including such economically important crops as banana, citrus, cacao, sugarcane, and yam. They are characterized by nonenveloped bacilliform particles with a genome size of 7.1–7.6 kb. Most of the badnaviruses are transmitted by mealybugs. *Citrus yellow mosaic virus* (CYMV), *Sugarcane bacilliform virus*, *Cacao swollen shoot virus*, and *Banana streak virus* are important members of *Badnavirus* with *Commelina yellow mottle virus* (CoYMV) as the type species (3).

From: *Methods in Molecular Biology, Vol. 451, Plant Virology Protocols: From Viral Sequence to Protein Function*
Edited by G.D. Foster, I.E. Johansen, Y. Hong, and P.D. Nagy © Humana Press, Totowa, NJ

All plant dsDNA viruses are considered to be plant pararetroviruses since they replicate their genome by reverse transcription of a larger than unit-length terminally redundant transcript of the viral genomic DNA using the virus-encoded replicase. The host cytosolic initiator methionine tRNA (tRNA^{met}) is presumed to serve as the primer for the reverse transcriptase in the synthesis of minus-strand DNA, since the genomes of all viruses in the *Caulimoviridae* sequenced to date contain a putative tRNA^{met}-binding site. This is a 13–16 nucleotide region complementary to the consensus sequence of plant tRNA^{met} located at the origin of viral replication (3).

Infectious clones delivered efficiently and successfully into plants by *Agrobacterium* are important for studies of gene function in these viruses and their potential as plant gene vectors. While the native or cloned CaMV DNA is infectious by mechanical inoculation of leaves, DNA preparations of other members of the *Caulimoviridae* are not. In the case of CaMV, purified viral DNA, whether intact or linearized, is as infectious as the purified virus (4). Cloned CaMV DNA, either full-length or containing 1.4 or 2 copies of the CaMV genome, is infectious only after it is excised from the vector plasmid. Inoculations are made by mechanical abrasion of leaves of turnip (*Brassica rapa*) with a solution containing more than 1 μg of purified plasmid-derived CaMV DNA per plant (4–8). In contrast to CaMV, neither the cloned *Rice tungro bacilliform virus* genome excised from the vector, the purified badnavirus CoYMV, nor a clone that contained 1.3 copies of CoYMV genome, were infectious by mechanical inoculation of their respective host plants (9, 10).

Agrobacterium-mediated inoculation was developed to offer an alternative means to transfer cloned DNA copies of viral genomes into plants to induce virus infection. This method has been used successfully for a wide range of virus groups including the caulimo- and badnaviruses (3, 6, 9, 10). For caulimoviruses, the method is much more efficient and effective than mechanical inoculation with either native or cloned CaMV DNA (6). For noncaulimoviruses, the method is especially useful since the viruses and their DNAs can not be mechanically inoculated in their host plants. For plant pararetroviruses, clones used for *Agrobacterium*-mediated inoculation require a greater than unit-genome length insert (1.1–2 copies of viral genome) between the T-DNA borders of an *Agrobacterium* binary vector. Such a construct will give rise to a larger than unit-length terminally redundant transcript (genome length plus 180 and 120 nucleotide transcripts in the cases of CaMV and CoYMV, respectively (10, 11)). This requirement is essential in order for the clone to be infectious.

The procedures required to produce infectious clones for plant pararetroviruses can be divided into the following four steps:

1. Cloning of the full length viral genome
2. Cloning of a larger than full-length viral genome into a binary vector and introducing it into Agrobacterium by transformation
3. Infection of host plants by Agrobacterium-assisted inoculation
4. Confirmation of virus infection of host plants

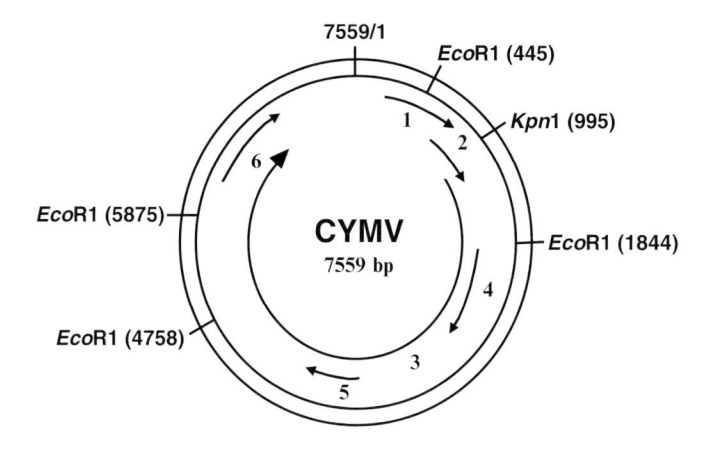

Fig. 1 Organization of the circular CYMV genome. The outer double circle represents the dsDNA genome. *Arrows* indicate the deduced open reading frames 1–6. The *Kpn*I and *Eco*RI restriction sites used to construct genomic clones are shown. Numbers refer to their positions within the genome

The *Badnavirus* CYMV (Fig. 1) will be used to illustrate steps 2 and 3, based mainly on the methods described by Huang and Hartung (3). The freeze–thaw method to transform the binary plasmid into *Agrobacterum* is based on An et al. (12). The described procedures will be particularly useful for noncaulimoviruses since they can not be transmitted mechanically. They will also be useful for caulimoviruses for more efficient transfer of their cloned DNA.

2 Materials

1. Purified viral DNA in sterile TE buffer (10 mM Tris–HCl and 1 mM EDTA, pH 8.0)
2. Restriction endonucleases and buffers, as supplied by manufacturer
3. 0.8% agarose gel in 1X Tris–acetate–EDTA (TAE) buffer (50X TAE: 242 g Tris base, 57.1 mL glacial acetic acid, 100 mL 0.5 M EDTA, pH 8.0)
4. Ethidium bromide solution: purchased as a 10-mL solution at 10 mg mL^{-1} solution from Qbiogene Inc. (Carlsbad, CA). Store at room temperature. Wear appropriate gloves when working with solutions that contain this dye since it is mutagenic and toxic.
5. Cloning vectors pBluescript KS$^+$ (Stratagene, La Jolla, CA) and pUC18 (Invitrogen, Carlsbad, CA)
6. T4 DNA ligase (1 U μL^{-1}) and 5X reaction buffer (Invitrogen, Carlsbad, CA)
7. Chemically competent *Escherichia coli* cells such as the One Shot INVαF′ cells (Invitrogen, Carlsbad, CA)
8. SOC medium: 2% bactotryptone, 0.5% yeast extract, 10 mM NaCl, 2.5 mM KCl, autoclaved, 10 mM MgCl$_2$, 20 mM glucose, filter sterilized

9. X-gal: 5-bromo-4-chloro-3-indolyl-β-D-galactoside, 20 mg mL^{-1} in dimethyl-formamide, stored at −20°C. Light sensitive
10. IPTG: isopropylthio-β-D-galactoside, 100 mM in water, filter sterilized. Store as 500-μL aliquots at −20°C
11. Luria–Bertani (LB) broth: 1% bactotryptone, 0.5% yeast extract, 1% NaCl, autoclaved
12. LB agar plates: Add 15 g agar to 1 L of LB broth before autoclaving
13. LB + Amp medium: LB medium containing 50 μg mL^{-1} of ampicillin, sodium salt
14. LB + Kan plates: LB agar plates containing 50 μg mL^{-1} of kanamycin sulfate
15. Sterile 80% glycerol
16. Binary vector pBI101.2 (see note 1)
17. *Agrobacterium tumefaciens* strain C58C1
18. Sterile 0.1 M CaCl2
19. Liquid nitrogen

3   Methods

3.1   *Cloning of the Full-length Viral Genome*

1. Digest 0.5 μg of purified viral DNA to completion in a 20-μL volume at 37°C for 1 h with a restriction enzyme that cuts only once in the viral genome. This will linearize the DNA. For CYMV, the enzyme could be *Kpn*I (see note 2).
2. Run the digest on a 0.8% agarose gel, visualize the fragment using ethidium bromide and UV irradiation, and excise it from the gel (add 0.5 μg of ethidium bromide to an 80-mL gel before pouring the gel) (see note 3).
3. Purify the fragment using your preferred technique (see note 4).
4. Similarly digest 0.5 μg of a cloning vector such as the pBluescript KS^{+} (pBS) with the same restriction enzyme, and purify the vector DNA as described above.
5. Add to a microcentrifuge tube 10 μL of linearized viral DNA, 1 μL of a 1:10 dilution of the prepared cloning vector, 3 μL of 5X ligation buffer, and 1 μL of T4 DNA ligase. Incubate overnight at 16°C.
6. Stop the reaction by incubating the ligation mixture at 65°C for 15 min, and then cool it on ice.
7. Transform 2.5 μL of the ligated DNA into 50 μL of a suitable competent strain of *E. coli* (see note 5). Incubate on ice for 30 min. Heat shock the cells at 42°C for 40 s, then incubate on ice for 2 min. Add 250 μL of SOC to the cells and shake at 225 rpm for 1 h at 37°C.
8. Plate 50 μL of cells with 10 μL of 20 mg mL^{-1} X-gal and 2.5 μL of 100 mM IPTG onto LB + Amp plates. Incubate at 37°C, inverted, overnight.
9. Pick white-colored colonies and grow overnight in 2 mL of LB + Amp medium.
10. Use 1.5 mL of the overnight culture for miniprep plasmid isolation using your preferred technique (see note 4), and confirm that the clone is as expected by restriction mapping. Store the purified DNA of the positive clone at −20°C.

11. Mix the remaining 0.5 mL of the overnight culture of the positive clone with 0.5 mL of sterile 80% glycerol, and store the mixture at −70°C for future use.

3.2 Cloning of a Larger than Full-length Viral Genome into a Binary Vector and Transformation of Agrobacterium

3.2.1 Construction of a Larger than Full Genome Length Clone in a Binary Vector

A flow diagram of the cloning procedure is shown in Fig. 2.

1. Digest the CYMV full-length clone, pCYMV, with *Bam*HI and *Kpn*I. Separate the *Bam*HI/*Kpn*I fragment by agarose gel electrophoresis, excise and purify the fragment from the gel, and clone the fragment into pUC18 to create pBK using the methods described above.
2. Digest pBK with *Sal*I and *Kpn*I, and purify the *Sal*I/*Kpn*I fragment of pBK (indicated as fragment A in Fig. 2).
3. Digest pCYMV with *Kpn*I and *Xba*I, and purify the *Kpn*I/*Xba*I fragment of pCYMV (fragment B in Fig. 2).
4. Ligate fragments A and B together with the binary vector pBI101.2 that was digested with *Sal*I and *Xba*I to create pBICYMV, a clone that contains 1.4 copies of the CYMV genome, using the methods described above.

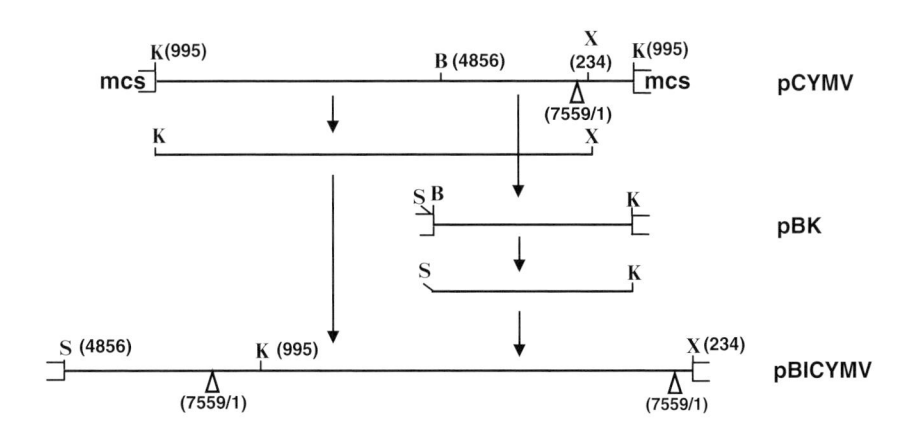

Fig. 2 Construction of pBICYMV containing a 1.4 CYMV genome insert. Restriction endonuclease sites *Bam*HI (B), *Kpn*I (K), *Sal*I (S), and *Xba*I (X) are indicated. The multiple cloning site (MCS) of the vector is also indicated. The *open arrowhead* represents the putative plant tRNA^{met}-binding site. Numbers refer to the position of the restriction endonuclease and the tRNA-binding sites within the CYMV genome

5. Transform 2.5 µL of the ligated DNA into 50 µL of a suitable competent strain of *E. coli* as described above, and plate 50 µL of the cells onto LB + Kan plates; X-gal and IPTG need not be added to the plates as all the colonies will be white. Incubate at 37°C, inverted, overnight.

6. Purify DNA from the colony grown on the LB + Kan plates and confirm the presence of the 1.4 CYMV genome insert in pBICYMV by restriction digestion, as described above.

7. Use the purified pBICYMV DNA for transformation of *Agrobacterium*. Store the DNA at −20°C for future use.

3.2.2 Transformation of the Greater Than Unit-Length Viral Genome Clone into *Agrobacterium* by the Freeze–Thaw Method (see note 6)

All of the following procedures should be completed under aseptic conditions.

1. Grow *Agrobacterium tumefaciens* C58C1 in 2 mL of LB medium in a 15-mL culture tube overnight at 28°C with shaking at 225 rpm.

2. Add 200 µL of the overnight culture to 2 mL of LB medium, and shake at 250 rpm at 28°C until the value of OD_{600} reaches 0.5–1.0.

3. Transfer the culture to a prechilled microcentrifuge tube, and centrifuge at 13,000 rpm at 4°C for 5 min.

4. Discard the supernatant. Resuspend the cells in 1.5 mL of ice-cold 0.1 M $CaCl_2$.

5. Centrifuge again at 13,000 rpm at 4°C for 5 min.

6. Discard the supernatant. Resuspend the pellet in 400 µL of ice-cold 0.1 M $CaCl_2$. Dispense 100-µL aliquots into three prechilled microcentrifuge tubes.

7. Add about 1 µg of either pBICYMV or pBI101.2 (as a vector only control) plasmid DNA, or 5 µL of sterile water (as a negative control) to *Agrobacterium* cells in each of the three labeled tubes.

8. Freeze the cells by placing the tubes in liquid nitrogen in an ice bucket, and then incubate the tubes in a 37°C water bath for 1 min. Mix the contents of the tubes with a quick tap, then continue to incubate for another 5 min at 37°C.

9. Add 900 µL of LB medium to each of the tubes and incubate at 28°C, with shaking, for 2–4 h (see note 7).

10. Centrifuge the tubes at 13,000 rpm for 2 min. Discard the supernatant and resuspend the cells in 100 µL of LB medium.

11. Plate the cells on a LB + Kan plate. Incubate the plate, inverted, at 28°C overnight.

12. Look for transformed colonies in 2–3 days.

13. Confirm that the transformants are positive by restriction digestion and comparison with the vector only transformants.

14. Grow single colonies of positive transformants in 2 mL LB + Kan broth at 28°C with shaking for 2 days. Mix 0.5 mL of the 2-day culture of the transformants with 0.5 mL of sterile 80% glycerol in a 1.7-mL microcentrifuge tube, and store the tube at −70°C for future use.

3.3 Agrobacterium-Assisted Inoculation of Host Plant

3.3.1 Preparation of the *Agrobacterium* Inoculum

1. Dispense 25 mL of LB + Kan (used to maintain the binary plasmid) medium into two sterile 250-mL Erlenmeyer flasks (see note 8).
2. Inoculate one flask with a single colony of *Agrobacterium tumefaciens* C58C1 containing the larger than full-length viral genome clone pBICYMV and the other with C58C1 containing the binary vector only pBI101.2.
3. Incubate the flasks at 28°C with shaking for 16–48 h until a turbid culture is obtained.

3.3.2 Inoculation Procedure

1. Wound stems of 2-year-old sweet orange seedlings (see notes 9 and 10) with three sets of 20 stem slashes using a disposable scalpel that has been dipped into the saturated *Agrobacteium* inoculum. Also wound 10 leaves of each plant with a needle press inoculation tool that has also been dipped into the *Agrobacterium inoculum* (see note 11).
2. Keep the inoculated plants under appropriate growth conditions.

3.4 Confirmation of Virus Infection of Host Plants

1. Symptoms: Observe the inoculated plants at regular intervals for development of disease symptoms (see note 12). Compare symptoms with those occurring on plants infected by graft-inoculation in the same greenhouse as positive controls, as well as with healthy plants and plants inoculated with the binary vector only as negative controls.
2. Detection of virus particles: Confirm the presence of virus particles by electron microscopy.

4 Notes

1. The binary vector is no longer available from Clontech, but can be purchased from The *Arabidopsis* Biological Resource Center at the Ohio State University. This vector is based on pBIN19 but contains the promoterless GUS gene and *nos* termination sequence. In addition to pBIN19 based vectors, other binary vectors may be used as long as the vector contains desired restriction sites for cloning.
2. If the viral genome contains more than one unique restriction site, choose one with protruding ends for cloning (like the *Kpn*I site in CYMV). It will be much easier to clone a linearized viral

fragment with protruding rather than blunt ends. To exclude the possibility that two closely spaced "unique" restriction sites exist within the genome, a restriction fragment containing the "unique" restriction site (like the *Eco*RI fragment from nucleotide 445–1844 in CYMV in Fig. 1) should be cloned and sequenced to verify that the restriction site is truly unique within the genome.

3. Minimize the length of time the gel is exposed to UV light to avoid damage to DNA.

4. In addition to the methods described in ref. 13, there are many commercial kits available, all of which should give a satisfactory result.

5 *E. coli* competent cells can be prepared either by the method of Hanahan (14) or purchased directly from commercial suppliers like the One Shot INVαF′ cells from Invitrogen (Carlsbad, CA).

6. The greater than unit-length clone can also be transformed into *Agrobacterium* by electroporation (15) or triparental mating (16). We chose to use the freeze–thaw method because it is quick, easy, reliable, and does not require specialized equipment, although its transformation frequency is low (approximately 10^3 transformants per mg DNA).

7. This incubation period allows the *Agrobacterium* to express the antibiotic resistance genes. The contents in the microcentrifuge tubes can also be transferred to sterile 15-mL medium tubes to allow more adequate aeration.

8. Always include antibiotics appropriate to the binary plasmid in the culture medium in order to maintain it successfully in *Agrobacterium*.

9. Some antibiotics such as spectinomycin and streptomycin, may cause severe chlorosis of leaves even when only small amounts remain in the culture medium used for inoculation. In that case, the *Agrobacterium* inoculum needs to be washed with LB or other appropriate broth without antibiotics before it is used for inoculation.

10. The plant used should be a host for the virus on which the larger than genome size clone is based, as well as for the *Agrobacterium* strain used for inoculation. Suitable strains of *Agrobacterium* for such plants can be found by searching the literature.

11. This inoculation procedure works well with woody plant species. Depending on the plant species used for inoculation, abrasion of leaves, toothpick stabs on stems, or stem injection of the inoculum can also be used to deliver *Agrobacterium*.

12. The time required for symptom development will vary depending on the virus and host used. For CYMV, it could take 5 months, while for CoYMV, only 16 days may be needed.

References

1. Pringle, C. R. (1998) The universal system of virus taxonomy of the International Committee on Virus Taxonomy (ICTV), including new proposals ratified since publication of the Sixth ICTV Report in 1995. *Arch. Virol.* **143**, 203–210.

2. Gronenborn, B. (1987) The biology of cauliflower mosaic virus and its application as plant gene vector. In Plant DNA Infectious Agents (Hohn, T. and Schell, J., ed.), Vienna, Springer, pp. 1–29.

3. Huang, Q. and Hartung, J. S. (2001) Cloning and sequence analysis of an infectious clone of Citrus yellow mosaic virus that can infect sweet orange via *Agrobacterium*-mediated inoculation. *J. Gen. Virol.* **82**, 2549–2558.

4. Lebeurier, G., Hirth, L., Hohn, T., and Hohn, B. (1980) Infectivities of native and cloned DNA of cauliflower mosaic virus. *Gene* **12**, 139–146.

5. Howell, S., Walker, L. L., and Dudley, R. K. (1980) Cloned cauliflower mosaic virus DNA infects turnips (Brassica rapa). *Science* **208**, 1265–1267.

6. Grimsley, N., Hohn, B., Hohn, T., and Walden, R. (1986) "Agroinfection", an alternative route for viral infection of plants by using the Ti plasmid. *Proc. Natl. Acad. Sci. USA* **83**, 3282–3286.

7. Bouhida, M., Lockhart, B. E. L., and Olszewski, N. E. (1993) An analysis of the complete sequence of a sugarcane bacilliform virus genome infectious to banana and rice. *J. Gen. Virol.* **74**, 15–22.
8. Jacquot, E., Hagen, L. S., Jacquemond, M., and Yot, P. (1996) The open reading frame 2 product of cacao swollen shoot badnavirus is a nucleic acid-binding protein. *Virology* **225**, 191–195.
9. Dasgupta, I., Hull, R., Eastop, S., Poggi-Pollini, C., Blakebrough, M., Boulton, M. I., and Davis, J. W. (1991) Rice tungro bacilliform virus DNA independently infects rice after *Agrobacterium*-mediated transfer. *J. Gen. Virol.* **72**: 1215–1221.
10. Medberry, S. L., Lockhart, B. E. L., and Olszewski, O. E. (1990) Properties of *Commelina yellow mottle virus's* complete DNA sequence, genomic discontinuities and transcript suggest that it is a pararetrovirus. *Nucleic Acids Res.* **18**, 5505–5513.
11. Corey, S. N., Lomonossoff, G. P., and Hull, R. (1981) Characterization of cauliflower mosaic virus DNA sequences which encode major polyadenylated transcripts. *Nucleic Acids Res.* **9**, 6735–6747.
12. An, G., Ebert, P. R., Mitra, A., and Ha, S. B. (1988) Binary vectors. In *Plant Molecular Biology Manual*, vol. A3 (Gelvin, S. B. and Schilperoot, R. A., ed.), Dordrecht, Kluwer, pp. 1–19.
13. Sambrook, J., and Russell, D. W. (2001) *Molecular Cloning: A laboratory Manual*. Cold Spring Harbor Laboratory, Cold Spring Harbor, NY.
14. Hanahan, D. (1985) Techniques for transformation of *E. coli*. In *DNA Cloning*: A Practical *Approach*, vol. 1 (Glover, G. M., ed.), IRL, Oxford, pp. 109–135.
15. Mersereau, M., Pazour, G. J., and Das, A. (1990) Efficient transformation of *Agrobacterium tumefaciens* by electroporation. *Gene* **90**, 149–151.
16. Ditta, G., Stansfield, S., Cobbin, D., and Helinski, D. R. (1980) Broad host range DNA cloning system for gram negative bacteria: construction of a gene bank of *Rhizobium meliloti*. *Proc. Natl. Acad. Sci. USA* **77**, 7347–7351.

Chapter 36
Insertion of Introns: A Strategy to Facilitate Assembly of Infectious Full Length Clones

I. Elisabeth Johansen and Ole Søgaard Lund

Abstract Some DNA fragments are difficult to clone in *Escherichia coli* by standard methods. It has been speculated that unintended transcription and translation result in expression of proteins that are toxic to the bacteria. This problem is frequently observed during assembly of infectious full-length virus clones. If the clone is constructed for transcription in vivo, interrupting the virus sequence with an intron can solve the toxicity problem. The AU-rich introns generally contain many stop codons, which interrupt translation in *E. coli*, while the intron sequence is precisely eliminated from the virus sequence in the plant nucleus. The resulting RNA, which enters the cytoplasm, is identical to the virus sequence and can initiate infection.

Keywords Infectious clone; Intron; Potyvirus; SOEing

1 Introduction

Infectious cDNA clones are useful for analysis of various properties of plant viruses. The cDNA clone allows introduction of specific mutations and exchange, deletion, or insertion of genes or gene fragments. Unfortunately, full-length cDNA clones of many RNA viruses have proved difficult or even impossible to clone and amplify in *Escherichia coli*. Full-length clones have frequently turned out to be noninfectious, possibly due to spontaneous mutations or deletions, which reduces toxicity to *E. coli*. The cause for the toxicity to *E. coli* is not known, but it has been observed that artificially introduced deletions; insertions or frame shifts can alleviate the cloning problems (1, 2).

Insertion of an intron into the virus sequence has proven to be an efficient way to facilitate maintenance of the cDNA in *E. coli* (1, 3, 4, 5). Upon inoculation of the cDNA to the host cell, the intron is precisely spliced from the precursor nuclear messenger RNA (pre-mRNA) in the nucleus, and a true copy of the viral RNA is transported to the cytoplasm where infection can start. The function of clones with introns requires cloning of the viral cDNA between promoter and terminator

From: *Methods in Molecular Biology, Vol. 451, Plant Virology Protocols:
From Viral Sequence to Protein Function*
Edited by G.D. Foster, I.E. Johansen, Y. Hong, and P.D. Nagy © Humana Press, Totowa, NJ

sequences, which are active in the host cell. The 35S promoter and NOS terminator are frequently used in plants (see Chaps. 32 and 33 for a description of construction of infectious clones of RNA viruses). Cloning of the viral cDNA in this context also allows inoculation by agroinfiltration, if the transcription cassette is moved to a binary vector (see Chap. 38 for a description of agroinoculation).

Construction of infectious clones with introns has been reported for viruses belonging to the genera *Potyvirus* (1, 4, 6) *Tobravirus* (7, 8), *Coronavirus* (3), and *Flavivirus* (5). In theory, intron insertion should be applicable for clones of viruses, which are replicated from an RNA template in the cytoplasm.

The sequences required for pre-mRNA intron splicing lie mainly within the intron. These include consensus sequences at the 5′ (GU) and 3′ (AG) ends of the intron; the branchpoint located 18–40 nucleotides upstream of the 3′ splice site and AU-rich sequence elements (9). Furthermore, introns are characterized by high U or AU content compared to more GC-rich exons. In the sequence next to the intron, there is a high representation of A and G at the two positions next to the 5′ end of the intron and G at the first positions 3′ to the intron (10). This is shown as a consensus sequence in Fig. 1. Therefore, it is probably safest to insert introns at positions in the virus sequence matching the AG/G consensus.

A   Monocotyledons

	exon			intron				exon	
G	19	7	**78**	**100**			**100**	**57**	20
A	40	**65**	7			**100**		21	22
U	7	14	7	**99**				11	44
C	34	14	8	1				11	14
Consensus	A	G		**G**	**U**	**A**	**G**	**G**	

B   Dicotyledons

	exon			intron				exon	
G	17	8	**79**	**100**			**100**	**60**	23
A	36	**62**	9			**100**		15	17
U	15	20	9	**98**				11	40
C	31	10	2	1				13	20
Consensus	A	G		**G**	**U**	**A**	**G**	**G**	

Fig. 1 Consensus sequence of plant intron splice sites based on data from (Simpson et al., 1993). (**a**) Consensus in monocotelydons based on the percentage occurrence of nucleotides at exon–intron borders, (**b**) consensus in dicotelydons based on the percentage occurrence of nucleotides at exon–intron borders

There are minor differences between monocot and dicot introns (10), and therefore, it is recommended to select introns from the class in which the infectious clone is going to be analyzed. Furthermore, it should be noted that the monocot sequences included in the analysis by (10) almost exclusively represent introns from the order *Poalis*. Some differences have been observed between species within the dicots (9). However, at present there is no evidence that differences in dicot introns are important for their use in virus clones. For example, *Arabidopsis* introns were used in *Tobacco mosaic virus*, which efficiently infected two species of *Nicotiana* (11), and an intron from *Solanum tuberosum* is spliced in *Pisum sativum* and *Capsicum annuum* (1, 6).

Introns can also be inserted into native *Pst*I restrictions sites as described in detail by (1). Here, we describe a more flexible method based on overlap extension PCR (12).

2 Materials

2.1 Isolation of Genomic DNA as Template for Intron Amplification

1. Plant material of a monocot or a dicot, depending on the host of the virus
2. Liquid nitrogen and safety glasses
3. 2X CTAB buffer: 1.4 M NaCl, 2% cetyl trimethylammonium bromide (CTAB), 100 mM Tris pH 8.0, 20 mM EDTA, 2% polyvinylpyrrolidone (PVP), 0.2% β-mercaptoethanol (added just before use)
4. Waterbath at 65°C
5. Chloroform/isoamylalcohol (24:1)
6. Cold isopropanol, 3 M sodium acetate, and 70% cold EtOH
7. Or use a commercial kit for plant DNA purification

2.2 Identification of Suitable Introns

1. Computer access to NCBI and related resources
2. Sequence data of virus region flanking the intron insertion site

2.3 In silico Test of the Intron Insertion

1. Program to assemble virus sequence with the intron. Either just a word processing program or a designated program for management of sequences

2. Computer access to 'NetPlantGene Server' http://www.cbs.dtu.dk/services/NetPGene/(13) or another splice site prediction program

2.4 Primers and Design

1. Commercial supplier of oligonucleotide primers

2.5 Amplification of Introns and Virus Sequences Flanking the Introns

1. Overlapping cDNA clones covering the complete genome of the virus. Make a dilution of 10–100 ng µL^{-1} plasmid DNA as template for PCR
2. A complete nucleotide sequence of the virus
3. PCR primers (3^{FW} and 4^{RV}) annealing to 5′ and 3′ ends of the intron (Fig. 2)
4. PCR primers to amplify the virus sequence 5′ (1^{FW} and 2^{RV}) and 3′ (5^{FW} and 6^{RV}) to the intron insertion site (Fig. 2)
5. Heat stable DNA polymerase, buffer, and 5 mM dNTP mix for PCR
6. Kit to purify DNA from agarose gel fragments

2.6 Sequence Overlap Extension PCR (SOE-PCR)

1. Commercial DNA marker to estimate DNA concentration of the purified PCR fragments
2. Heat stable DNA polymerase, buffer, and 5 mM dNTP mix for PCR
3. Kit for cloning of PCR fragments

3 Methods

3.1 Isolation of Genomic DNA as Template for Intron Amplification

1. Add mercaptoethanol to the 2X CTAB buffer
2. Grind 100 mg plant material in a 1.5-ml microfuge tube immersed in liquid nitrogen. Use safety glasses
3. Add 500 µL 2X CTAB buffer, mix, and incubate at 65°C for 30–60 min
4. Centrifuge at 5000g for 10 min at room temperature

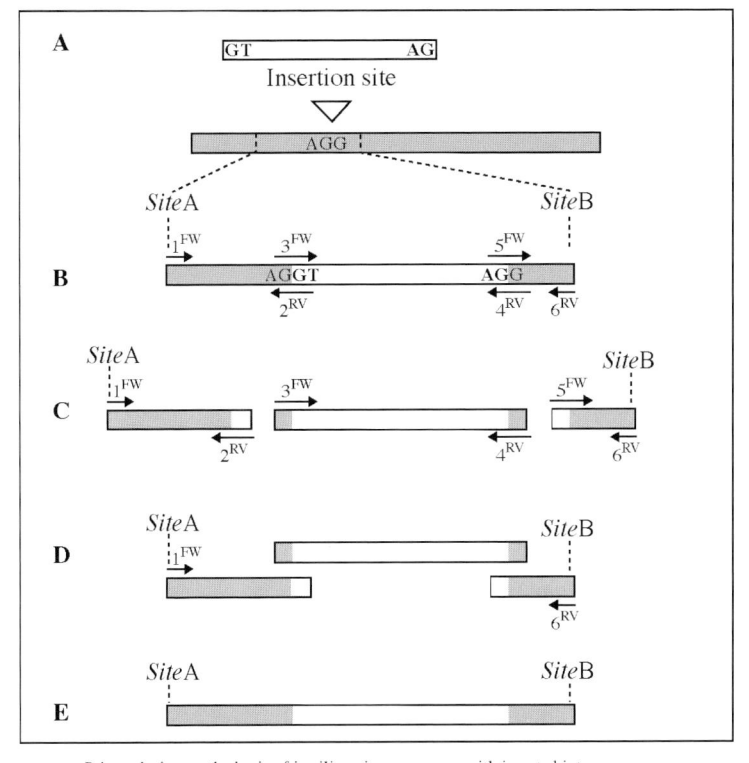

Primer design on the basis of in silico virus sequence with inserted intron
Primers 2^{RV} and 3^{FW} are complementary as are primers 4^{RV} and 5^{FW}

Fig. 2 Intron insertion using sequence overlap extension PCR (SOE-PCR). (**a**) Selection of an intron and intron insertion site matching the consensus. The insertion site should be situated between restriction sites *Site*A and *Site*B, which are suitable for reinserting the intron containing fragment in the complete virus sequence. (**b**) Primers are designed on the basis of an in silico construct. Primers indicated as *arrows* above the construct are virus sense, while primers below are antisense. Primers 1^{FW} and 6^{RV} contain the sequence of *Site*A and *Site*B, respectively. Complementary primers 3^{FW} and 2^{RV} at the 5′ exon-intron border and 5^{FW} and 4^{RV} at the 3′ intron–exon border are approximately 50 nucleotides in length containing approximately 20–25 nucleotides of virus sequence and 20–25 nucleotides of intron sequence. (**c**) Three separate PCR reactions are performed to amplify overlapping fragments of the region 5′ to the intron (primers 1^{FW} and 2^{RV}), the intron (primers 3^{FW} and 4^{RV}), and the region 3′ to the intron (primers 5^{FW} and 6^{RV}). (**d**) SOE-PCR is performed on a mixture of the three purified PCR fragments and primers 1^{FW} and 6^{RV}. (**e**) The PCR fragment of the virus sequence with intron. This fragment is separated from primers and unspecific fragment by gel electrophoresis, cloned, and checked by sequencing before it is reinserted into the complete virus sequence

5. Transfer the supernatant to a clean microfuge tube
6. Add an equal volume of chloroform/isoamylalcohol, shake the tube for 10 min
7. Centrifuge at 5000g for 2 min at room temperature
8. Repeat steps 5–7
9. Add 0.8 volume cold isopropanol and 0.1 volume 3 M sodium acetate

10. Centrifuge at 12000g for 15 min at room temperature
11. Remove the supernatant and add 1 mL 70% cold EtOH
12. Centrifuge at 12000g for 2 min at room temperature
13. Remove the supernatant and dry the DNA
14. Dissolve in 100 µL sterile H_2O
15. Or follow the instructions provided with the plant DNA purification kit

3.2   Identification of Suitable Introns

1. Identify introns from complete gene sequences, for example, in the NCBI database http://www.ncbi.nlm.nih.gov/entrez/query.fcgi
2. Select small introns (100–300 basepairs) with a sequence matching the splice site consensus as shown in Fig. 1
3. We have used the second intron (189 basepairs) of the ST-LS1 gene from *S. tuberosum* (accession X04753 nucleotide 2648–2836) and intron 2 (221 basepairs) from the NiR gene from *Phaseolus vulgaris* (accession U10419 nucleotide 3536–3756) successfully in *N. benthamiana*, *P. sativum*, and *C. annuum*.

3.3   In silico Test of the Selected Intron Inserted in the virus Sequence

1. If introns are inserted into the virus sequence in order to reduce toxicity, the intron insertion site should be placed in a region, which has been difficult to clone or amplify in *E. coli*. In addition the insertion site should match the AG/G exon border sequences as indicated in Fig. 1 and Fig. 2a (see note 1)
2. Paste in the intron sequence and run the virus sequence with the intron through a program, which predicts intron splice sites (e.g., NetPlantGene Server).
3. The output will display potential donor sites, acceptor sites, and branchpoints. Only the transcribed virus sense strand is relevant.
4. Check that the intron is recognized in the virus sequence. If the intron is not recognized, then try another insertion site and/or another intron sequence.
5. Also inspect the output for cryptic introns in the virus sequence, which may interfere with correct processing of the virus sequence (see note 1).

3.4   Primers and Design

1. Identify restriction sites, *Site*A and *Site*B, flanking the intron insertion site, which are suitable for reintroduction of the intron containing subclone into the full-length clone or a larger subclone (see note 7) Choose restriction sites that

are no more than 2 kb apart to avoid problems, which may arise with generation of long PCR products.

2. Six primers are needed as shown in Fig. 2b.

   a. Primers 1^{FW} and 6^{RV} are forward and reverse primers annealing to the regions with the restriction sites *Site*A and *Site*B (see note 2). Primers 1^{FW} and 6^{RV} are designed like normal PCR primers and it is advisable to check them for primer dimer formation and proper annealing the target sequence using for example vector NTI, which is available at http://www.invitrogen.com

   b. Primers 3^{FW} and 2^{RV} are complementary. Primer 3^{FW} is composed of 20–25 nucleotides matching the 3′ end of the virus sequence upstream the intron and 20–25 nucleotides matching the 5′ end of the intron. Primer 2^{RV} is composed of 20–25 nucleotides complementary to the 5′ end of the intron and 20–25 nucleotides complementary to the virus sequence upstream the intron.

   c. Primers 5^{FW} and 4^{RV} are complementary. Primer 5^{FW} is of composed 20–25 nucleotides matching the 3′ end of the intron sequence and 20–25 nucleotides matching the 5′ end of the virus downstream the intron. Primer 4^{RV} is composed of 20–25 nucleotides complementary to the virus sequence downstream the intron and 20–25 nucleotides complementary to the 3′ end of the intron (see note 3).

3. Make a 10 μM dilution of each primer in H_2O, mix well, and store at −20°C.

3.5 Amplification of Introns and Virus Sequences Flanking the Introns

1. Three separate PCR reactions are set up to amplify overlapping fragments: The virus region 5′ to the intron, the intron and the region 3′ to the intron (Fig. 2c). The virus regions 5′ and 3′ to the intron are amplified from a virus plasmid cDNA clone with primer pairs $1^{FW}/2^{RV}$ and $5^{FW}/6^{RV}$, respectively. The intron is amplified with primers 3^{FW} and 4^{RV} from purified plant DNA. Optimized heat-stable high-fidelity DNA polymerases with proofreading are recommended for the PCR. For example, Expand or Phusion, whereas pure *Pwo* or *Pfu* are less efficient.

2. The following two protocols are based on the instructions provided by the suppliers. Please refer to these instructions for further information and troubleshooting. Note also that other products can be used.

   2.1. PCR reaction in 50 μL with Expand™ (Roche). Thaw, mix, and centrifuge all solutions except the enzyme before use. Place on ice and assemble reactions on ice. Add in the following order: 38.25 μL H_2O; 5 μL 10X Expand High Fidelity with 15 mM $MgCl_2$ buffer; 2 μL 5 mM dNTPs; 1.5 μL 10 μM FW primer; 1.5 μL 10 μM RV primer; 1 μL template DNA (plasmid cDNA clone or genomic DNA); 0.75 μL Expand High Fidelity enzyme mix, mix, and centrifuge. Run the following PCR scheme: Initial denaturation 94°C for 2 min; 30 cycles of denaturation at 94°C for 15 s, annealing at 55°C for 30 s (see note 4), extension at 72°C for 60 s per 1,500 bases; final extension at 72°C for 7 min; 4°C. Go to 3.

2.2. PCR reaction in 50 μL with Phusion™ (Finnzymes). Thaw, mix, and centrifuge all solutions except the enzyme before use. Place on ice and assemble reactions on ice. Add in the following order: 31.5 μL H_2O (30 μL if DMSO is included); 10 μL 5X Phusion HF buffer; 2 μL 5 mM dNTPs; 2.5 μL 10 μM FW primer; 2.5 μL 10 μM RV primer; 1 μL template DNA (plasmid cDNA clone or genomic DNA); 1.5 μL DMSO (optional); 0.5 μL Phusion DNA polymerase, mix, and centrifuge. Run the following PCR scheme: Initial denaturation 98°C for 30 s; 30 cycles of denaturation at 98°C for 10 s, annealing at 55°C for 30 s (see note 5), extension at 72°C for 30 s per 1,000 bases; final extension at 72°C for 5–10 min; 4°C. Go to 3.

3. A small sample (1–5 μL) of each PCR reaction is analyzed by agarose gel electrophoresis to check that PCR products of the expected sizes have been amplified.

4. To purify the PCR products from primers and possible unspecific products, samples of the PCR reactions containing ≥ 200 ng of the correct PCR product are separated by electrophoresis on an agarose gel. The bands containing the PCR products are cut out of the gel and the DNA is purified from the gel using a gel band purification kit. Spin columns are very efficient and are provided by several commercial suppliers, for example GFX columns (Amersham Biosciences), MinElute and QIAquick (Qiagen), GenElute™ (Sigma), and SpinPrep™ (Novagen). Follow the supplier's instructions and elute the DNA in 20–25 μL sterile H_2O. Check again, by agarose gel electrophoresis of 1–2 μL of the eluted sample, that the PCR products have been recovered and estimate the DNA concentration by comparison to the bands in the DNA size marker.

3.6 Splicing by Overlap Extension PCR (SOE-PCR)

1. SOE-PCR reaction in 50 μL with Expand High Fidelity PCR system (Roche). Thaw, mix, and centrifuge all solutions except the enzyme before use. Place on ice and assemble reactions on ice. Add in the following order: H_2O to a total reaction volume of 50 μL; 5 μL 10X Expand High Fidelity with 15 mM $MgCl_2$ buffer; 2 μL 5 mM dNTPs; 1.5 μL 10 μM primer 1^{FW}; 1.5 μL 10 μM primer 6^{RV}; approximately 50 ng of each PCR product from the initial reactions; 0.75 μL Expand High Fidelity enzyme mix, mix, and centrifuge. Run the following PCR scheme: Initial denaturation 94°C for 2 min; 30 cycles of denaturation at 94°C for 15 s, annealing at 55°C for 30 s, extension at 72°C for 60 s per 1,500 bases;

final extension at 72°C for 7 min; 4°C. Note that Expand High Fidelity enzyme mix produces PCR products with 3′ dA overhangs. Go to step 2.

SOE-PCR reaction in 50 μL with Phusion™ (Finnzymes). Thaw, mix, and centrifuge all solutions except the enzyme before use. Place on ice and assemble reactions on ice. Add in the following order: H_2O to a total reaction volume of 50 μL; 10 μL 5X Phusion HF buffer; 2 μL 5 mM dNTPs; 2.5 μL 10 μM primer 1^{FW}; 2.5 μL 10 μM primer 6^{RV}; approximately 50 ng of each PCR product from the initial reactions; 1.5 μL DMSO (optional); 0.5 μL Phusion DNA polymerase, mix, and centrifuge. Run the following PCR scheme: Initial denaturation 98°C for 30 s; 30 cycles of denaturation at 98°C for 10 s, annealing at 55°C for 30 s, extension at 72°C for 30 s per 1,000 bases; final extension at 72°C for 5–10 min; 4°C. Note that Phusion™ produces blunt end PCR products. Go to step 2.

2. A small sample (1–5 μL) is analyzed by agarose gel electrophoresis to check that a PCR product of the expected size has been amplified.
3. The PCR product of the expected size is gel purified and the concentration determined. It is recommended to clone the fragment cloned using a T/A PCR cloning cloning kit for products with 3′dA overhangs (Expand) or blunt ends (Phusion™) (see note 6). The cloned fragment should be sequenced to check for errors before it is reinserted into the virus cDNA using restriction sites *Site*A and *Site*B (see note 7).

4 Notes

1. It is likely that sequences surrounding the intron may affect splicing efficiency, and the virus may contain cryptic 5′ and or 3′ splice sites, which can interfere with splicing of the true intron. The presence of cryptic splice sites in the virus sequence can be identified using, for example, the 'NetPlantGene Server' at http://www.cbs.dtu.dk/services/NetPGene/. To prevent cryptic splicing it may be necessary to eliminate either the 5′ or the 3′ splice site. This may be accomplished by silent mutations (Marillonnet et al., 2005). In our experience, cryptic sites do not seem to be a major problem if the goal is to generate an infectious clone because only a fraction of the transcripts need to be correctly spliced to initiate infection. However, insertion of true introns can improve the specific infectivity as demonstrated by Marillonnet et al. (2005). It is also known that sequences in the exon, known as exon splicing enhancers (ESE), are likely to play an important role in plant intron splicing. However, no splicing event in *Arabidopsis* has yet been shown to be enhancer dependent. For more information see http://www.tigr.org/software/SeeEse/background.html
2. Primers 1^{FW} and 6^{RV} should contain the sequence of the restriction sites *Site*A and *Site*B, or flank the sites so the restriction sites are contained within the PCR fragments. If the primers contain the restriction sites, the restriction site consensus should be placed at least three nucleotides from the 5′ end of the primer.
3. The length of the annealing sequence will depend on the GC content. It is advisable to use a longer primer for regions with low GC content.
4. Optimal annealing temperature depends on the melting temperature of the primers and the system used. Annealing at 55°C is usually a good starting point unless one of the primers anneal to a short and very AT-rich region. In this case, the annealing temperature should be lower.

5. The optimal annealing temperature will depend on the primer sequences, but in our experience 55°C is a good starting point.
6. PCR fragments are most conveniently cloned using PCR cloning kits adapted for cloning PCR products with dA overhangs or blunt ends. Although the PCR products described here contain restriction sites at least three nucleotides from the ends of the fragment, it is sometimes difficult to clone PCR fragments with the aid of such sites.
7. Multiple fragment ligations can facilitate reinsertion of the fragment containing the intron if the restriction sites *Site*A and *Site*B are not unique in the full-length clone. We have assembled potyvirus clones from up to five fragments with sticky ends in a single cloning step. The amounts of DNA of each fragment are adjusted when the fragments are cut from the agarose gel and can be purified together on a single spin column

References

1. Johansen, I.E. (1996) Intron insertion facilitates amplification of cloned virus cDNA in *Escherichia coli* while biological activity is reestablished after transcription *in vivo*. *Proc. Natl. Acad. Sci. USA* **93**, 12400–12405.
2. Satyanarayana, T., Gowda, S., Ayllon, M.A., and Dawson, W.O. (2003) Frameshift mutations in infectious cDNA clones of Citrus tristeza virus: a strategy to minimize the toxicity of viral sequences to Escherichia coli. *Virology* **313**, 481–491.
3. Gonzalez, J.M., Penzes, Z., Almazan, F., Calvo, E., and Enjuanes L. (2002) Stabilization of a full-length infectious cDNA clone of transmissible gastroenteritis coronavirus by insertion of an intron. *J. Virol.* **76**, 4655–4661.
4. Lopez-Moya, J.J. and Garcia, J.A. (2000) Construction of a stable and highly infectious intron-containing cDNA clone of plum pox potyvirus and its use to infect plants by particle bombardment. *Virus Res.* **68**, 99–107.
5. Yamshchikov, V., Mishin, V., and Cominelli, F. (2001) A new strategy in design of (+)RNA virus infectious clones enabling their stable propagation in E. coli. *Virology* **281**, 272–280.
6. Moury, B., Morel, C., Johansen, E., Guilbaud, L., Souche, S., Ayme, V., Caranta, C., Palloix, A., and Jacquemond, M. (2004) Mutations in *Potato virus Y* genome-linked protein determine virulence toward recessive resistances in *Capsicum annuum* and *Lycopersicon hirsutum*. *Mol. Plant Microbe Interact.* **17**, 322–329.
7. Constantin, G.D., Krath, B.N., MacFarlane, S.A., Nicolaisen, M., Johansen, I.E., and Lund, O.S. (2004) Virus-induced gene silencing as a tool for functional genomics in a legume species. *Plant J.* **40**, 622–631
8. Ratcliff, F., Martin-Hernandez, A.M., and Baulcombe, D. (2001) Tobacco rattle virus as a vector for analysis of gene function by silencing. *Plant J.* **25**, 237–245.
9. Brown, J.W.S. and Simpson, C.G. (1998) Splice site selection in plant pre-mRNA splicing. *Annu. Rev. Plant Physiol. Mol. Biol.* **49**, 77–95.
10. Simpson, C.G., Leader, D.J., and Brown, J.W.S. (1993) Characteristics of plant pre-mRNA introns. In *Plant Molecular Biology Labfax*, (Croy, R. R. D., ed.), BIOS Sci., Oxford, pp. 183–251.
11. Marillonnet, S., Thoeringer, C., Kandzia, R., Klimyuk, V., and Gleba, Y. (2005) Systemic *Agrobacterium tumefaciens*-mediated transfection of viral replicons for efficient transient expression in plants. *Nat. Biotechnol.* **23**, 718–723.
12. Horton, R.M., Hunt, H.D., Ho, S.N., Pullen, J.K., and Pease, L.R. (1989) Engineering hybrid genes without the use of restriction enzymes: Gene splicing by overlap extension. *Gene* **77**, 61–68.
13. Hebsgaard, S.M., Korning, P.G., Tolstrup, N. Engelbrecht, J., Rouze, P., and Brunak, S. (1996) Splice site prediction in Arabidopsis thaliana DNA by combining local and global sequence information. *Nucleic Acids Res.* **24**, 3439–3452.

Chapter 37

Analysis of Cell-to-Cell and Long-Distance Movement of *Apple Latent Spherical Virus* in Infected Plants Using Green, Cyan, and Yellow Fluorescent Proteins

Tsubasa Takahashi and Nobuyuki Yoshikawa

Abstract *Apple latent spherical virus* (ALSV) expressing green, cyan, and yellow fluorescent proteins (GFP, CFP, and YFP) was constructed and used to analyze the local and systemic movement of the virus in infected plants. In *Chenopodium quinoa* plants inoculated with GFP-ALSV, the infection foci first appeared as small fluorescent spots 2–3 days post inoculation (dpi). The GFP spots expanded as rings from 5 dpi, then fused to each other, and most fluorescence faded out at 10–12 dpi. In upper uninoculated leaves, GFP fluorescence was first observed 6–7 dpi on the basal area of mature leaves and on the entire area of young developing leaves. The appearance of fluorescent flecks on young leaves was first found on and near the class III and IV veins. ALSV labeled with two different fluorescent proteins (CFP-ALSV and YFP-ALSV) were used to investigate the distribution of identical, but differently labeled viruses in mixed infection. Fluorescence from CFP and YFP was in each case observed in separate areas in both inoculated and upper uninoculated leaves, indicating that populations of identical, but differently labeled viruses were replicated and distributed in discrete areas of infected leaves.

Keywords Fluorescent protein; GFP; CFP; YFP; Plant virus; Cell-to-cell movement; Long-distance movement; Mixed infection

1 Introduction

Green fluorescent protein (GFP) from jellyfish *Aequorea Victoria*, a 27 kDa monomer protein (1), has been widely used for monitoring gene expression and protein localization because it is nontoxic to plants, brightly fluorescent, very stable in cells, and nondestructive (2). Recombinant GFP-tagged viruses have been developed and used for study of viral cell-to-cell and long-distance movement in infected plants (2–7). Because the detection of GFP fluorescence may be the most simple and sensitive way to monitor the timing of viral movement, this tool would be of special importance in the study of plant virus infection dynamics in infected plant cells.

From: *Methods in Molecular Biology, Vol. 451, Plant Virology Protocols: From Viral Sequence to Protein Function*
Edited by G.D. Foster, I.E. Johansen, Y. Hong, and P.D. Nagy © Humana Press, Totowa, NJ

We have recently constructed infectious cDNA clones of *Apple latent spherical virus* (ALSV)-RNA1 and ALSV-RNA2 under the control of enhanced *Cauliflower mosaic virus* (CaMV) 35S promoter and developed an ALSV-RNA2-based vector for transient expression of foreign genes in host plants (8). The ALSV vector which expresses GFP has been used for the tracing of cell-to-cell movement of ALSV in infected plant tissues and for demonstrating that MP and three capsid proteins are all indispensable for the cell-to-cell movement of the virus (9). ALSV labeled with yellow and cyan fluorescent variants of GFP were also used to investigate the distribution of identical but differently labeled viruses in mixed infected plants (10).

2   Materials

2.1   Plants

Seeds of *Chenopodium quinoa* are planted in clay pots and grown in a greenhouse until eight-true leaf stage.

2.2   Infectious cDNA Clones of ALSV

Infectious cDNA clones of ALSV (pEALSR1 and pEALSR2L5R5) are used for construction of ALSV expressing GFP, YFP, and CFP (see note 1 and Fig. 1).

2.3   Fluorescent Proteins

Plasmids pEGFP-1, pECFP-1, and pEYFP-1 (BD Biosciences, CLONTECH) are used for amplification of GFP, CFP, and YFP sequence.

2.4   Purification of cDNA Clones

1. Competent cells for DNA cloning: *Escherichia coli* DH5αcells (STRATAGENE).
2. LB medium: 10 g bacto-teypepton, 5 g bacto-yeast extract, 10 g NaCl. Adjust the pH 7.5 with 5N NaOH. Adjust final volume to 1 L with deionized water. Sterilize by autoclave. Allow the LB medium to cool to 60°C or less, and add ampicillin. Ampicillin is made by dissolving in deionized water and sterilized by filtration through a 0.22-μ filter.
3. QIAGEN Plasmid Midi Kits (QIAGEN, Tokyo, Japan)
4. TE buffer: 10 mM Tris–HCl, 1 mM EDTA, pH 8.0. Shake until the solutes have dissolved and sterilize by autoclave.

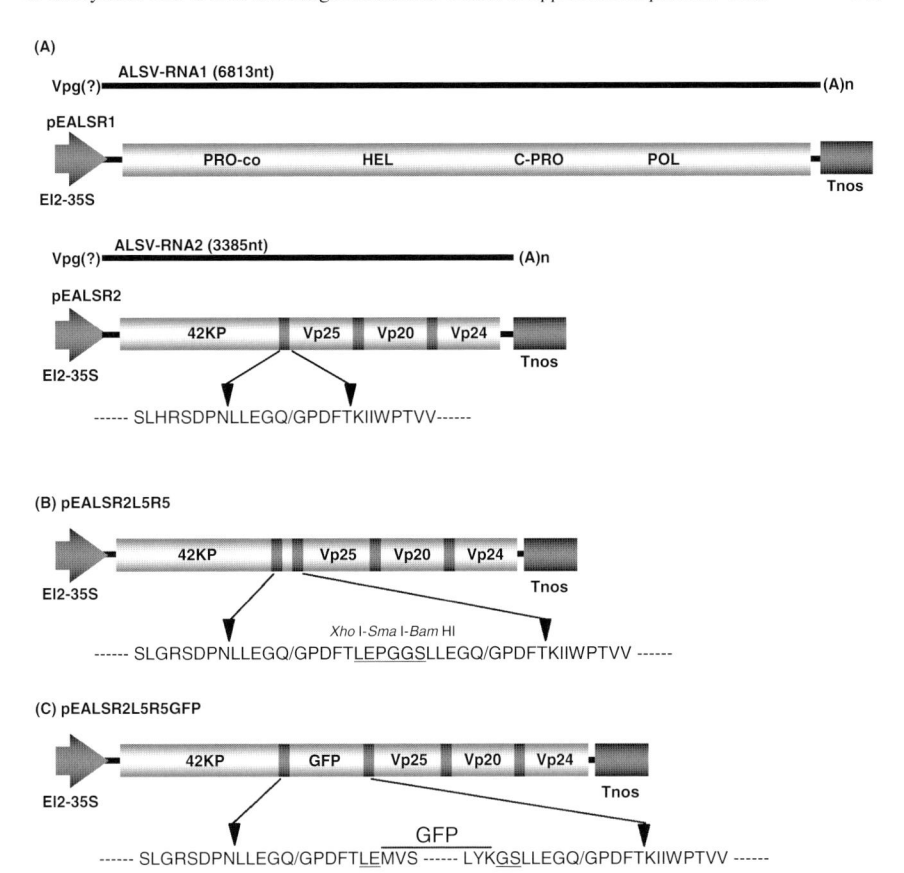

Fig. 1 Construction of infectious cDNA clones of *Apple latent spherical virus* (ALSV) and ALSV-RNA2 vectors. (**A**) Infectious cDNA clones of ALSV-RNA1 (pEALSR1) and ALSV-RNA2 (pEALSR2). An amino acid sequence near a cleavage site (/) between 42K and Vp25 was shown. (**B**) ALSV-RNA2 vector (pEALSR2L5R5). An amino acid sequence near two cleavage sites (/), and cloning sites *Xho*I, *Sma*I, and *Bam*HI (*underlined*) for foreign genes between 42K and Vp25 were shown. (**C**) pEALSR2L5R5GFP. An amino acid sequence near two cleavage sites (/), and GFP gene between 42K and Vp25 were shown. Amino acids derived from cloning sites were *underlined*

2.5   *Inoculation*

1. Purified infectious cDNA clones (pEALSR1 and pEALSR2L5R5GFP)
2. Infected *C. quinoa* leaves
3. 600-mesh carborundum (NACALAI TESQUE INC., Kyoto, Japan). Use carborundum after sterilization in a bottle by autoclave.
4. Latex finger cots
5. Inoculation buffer: 0.1 M Tris–HCl (pH 7.5), 0.1 M NaCl, 0.05 M $MgCl_2 \cdot 6H_2O$. Sterilize by autoclave

6. Mortar and pestle
7. Cotton swab
8. Water

2.6 Fluorescent Microscope

1. Fluorescent stereomicroscope: Fluorescent stereoscopic microscope VB-G25 (KEYENCE, Osaka, Japan) with the following filter sets: GFP-B filter sets (excitation filter 470/40 nm, and barrier filter 535/50 nm), CFP-B filter sets, (excitation filter 436/20 nm, and barrier filter 480/30 nm), and YFP-B filter sets (excitation filter 490/20 nm, and barrier filter 540/40 nm)
2. Digital camera: Olympus DP70 camera (OLYMPUS, Japan)
3. Personal computer for compilation of digital images

3 Methods

3.1 Construction of ALSV Vectors Expressing GFP, YFP, and CFP

1. ALSV expressing GFP, YFP, and CFP are constructed using ALSV-RNA2 vectors (pEALSR1 and pEALSR2L5R5) (see Fig. 1 and note 2)
2. The cDNA fragments containing GFP, CFP, and YFP genes are amplified from pEGFP-1 pEYFP-1 and pECFP-1, respectively, by polymerase chain reaction (PCR) using ExTaq DNA polymerase (TAKARA) and two primer pairs, XhoGFP [5′-CC<u>CTCGAG</u>ATGGTGAGCAAGGGCGAGGA-3′, containing a *Xho*I site (underlined)] and BamGFP [5′-CG<u>GGATCC</u>CTTGTACAGCTCGTC CA-3′, containing a *Bam*HI site (underlined)].
3. The amplified cDNA fragments of GFP, CFP, and YFP genes are double digested with *Xho*I and *Bam*HI, and ligated to pEALSR2L5R5 restricted with the same enzymes. Competent *E. coli* DH5α cells are transformed by the plasmids.
4. The bacteria colonies are selected by antibiotic sensitivity and cultured in a 3 mL of LB medium at 37°C overnight.
5. The plasmid DNAs are isolated from the cultured cells by the alkaline lysis method.
6. To identify the inserts, plasmid DNAs are restricted with *Xho*I and *Bam*HI and electrophoresed on a 1% agarose gel. The plasmids having an insert with the desired length are then sequenced.
7. The resulting pEALSR2-based vectors containing GFP, CFP, or YFP genes are designated pEALSR2L5R5GFP, pEALSR2L5R5CFP, and pEALSR2L5R5YFP, respectively.

3.2   Purification of Infectious cDNA Clones

1. Plasmids, pEALSR1, pEALSR2L5R5GFP, pEALSR2L5R5CFP, and pEAL-SR2L5R5YFP are propagated in *E. coli* DH5α cells.
2. *E. coli* DH5α cells transformed with pEALSR1, pEALSR2L5R5GFP, pEAL-SR2L5R5CFP, or pEALSR2L5R5YFP are prepared on a freshly streaked LB agar medium containing 50 μg/mL of ampicillin.
3. A single colony of transformed *E. coli* is inoculated in a 100 mL of LB medium containing 50 μg/mL of ampicillin at 37°C for overnight with vigorous shaking.
4. Plasmid DNAs are purified by the QIAGEN Plasmid Midi Kits (QIAGEN, Japan) according to the manufacturer's protocols, dissolved in TE buffer at a concentration of 1 μg/μL and stored at −80°C until use for inoculation.

3.3   Inoculation to Plants

3.3.1   Inoculation of Infectious cDNA Clones

1. Both pEALSR1 and pEALSR2L5R5GFP purified from large scale cultures of *E. coli* are mixed at a concentration of 1 μg/μL each.
2. A mixture of pEALSR1 and pEALSR2L5R5GFP (10 μL) is dropped onto the tip of a finger wrapped with latex finger cots and mechanically inoculated onto a leaf of *C. quinoa* plant (six true-leaf stage) (see note 3). The inoculation of infectious cDNA clones was conducted onto three leaves (third to fifth true leaves) per plant.
3. A surface of inoculated leaves is flushed with a sufficient amount of water to remove carborundum.
4. The symptoms consisting of chlorotic spots and mosaic appeared in the upper uninoculated leaves 9–12 days post inoculation (dpi).
5. Leaves with symptoms are collected and stored at −80°C until use for inoculation (see note 4).

3.3.2   Inoculation of GFP-ALSV onto Plants

1. Infected leaves with symptoms (0.1 g) stored at −80°C are homogenized in 0.4 mL of inoculation buffer in a mortar and pestle.
2. To investigate the local movement of ALSV on inoculated leaves and systemic movement into upper uninoculated leaves, *C. quinoa* plants (eight true leaf stage) were used for inoculation.
3. A cotton swab is soaked with leaf extracts and rubbed firmly but gently on the surface of the leaves (third to sixth true leaves) of *C. quinoa* plants (see note 3).
4. The surface of inoculated leaves is flushed with a sufficient amount of water to remove carborundum.
5. Inoculated plants were grown in a greenhouse condition.

3.4 Observation of Infected Plants by Fluorescent Microscope

1. Inoculated leaves were observed at 1–10 dpi using a fluorescent stereoscopic microscope equipped with an Olympus DP70 CCD camera, and digital images were stored and compiled in a personal computer. Figure 2 shows that the infection foci first appeared as fluorescent small spots (hereafter referred to as GFP lesions) on inoculated leaves at 2–3 dpi (Fig. 2a). As GFP lesions increased in size, the fluorescence in the center of lesions started to disappear and the GFP lesions expanded as rings from 5 dpi (Fig. 2c). The GFP lesions then fused to each other, and only the outline of the fused rings showed intense fluorescence (Fig. 2d, e). After 8 dpi, the fused rings were spread through the entire area of inoculated leaves, and most GFP fluorescence faded out (Fig. 2f). Digital graphs of GFP fluorescence were printed using a personal digital printer.

2. In upper uninoculated leaves, GFP fluorescence was first observed at 6–7 dpi on the basal area of mature leaves (eighth and ninth leaves) (Fig. 3c, d) and on the entire area of young developing leaves (10th and 11th leaves) (Fig. 3a, b). The area showing GFP fluorescence coincided with those showing vein clearing and chlorotic spots characteristic of ALSV infection. The appearance of fluorescent flecks on the 10th and 11th leaves was first found on and near the class III and IV veins (Fig. 4a, b), indicating that unloading of GFP-ALSV may occur predominantly from these minor veins in young developing leaves.

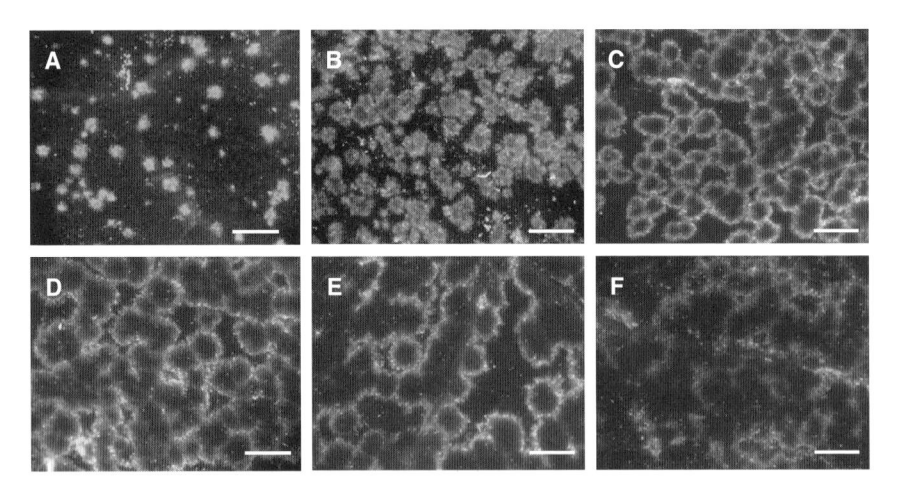

Fig. 2 GFP fluorescence on inoculated leaves of *C. quinoa* plant infected with GFP-tagged ALSV at (**A**) 3 dpi, (**B**) 4 dpi, (**C**) 5 dpi, (**D**) 6 dpi, (**E**) 7 dpi, and (**F**) 8 dpi. Bar = 2 mm

Fig. 3 GFP fluorescence on upper uninoculated leaves of *C. quinoa* inoculated with GFP-ALSV at 7 dpi. GFP-ALSV was inoculated onto third to sixth true leaves of *C. quinoa* (eight true leaf stage). GFP fluorescence on the entire area of young developing leaves (**A,B**) and on the basal area of mature leaves (**C,D**). (**A**) 11th leaf, (**B**) 10th leaf, (**C**) 9th leaf, (**D**) 8th leaf. Bar = 10 mm

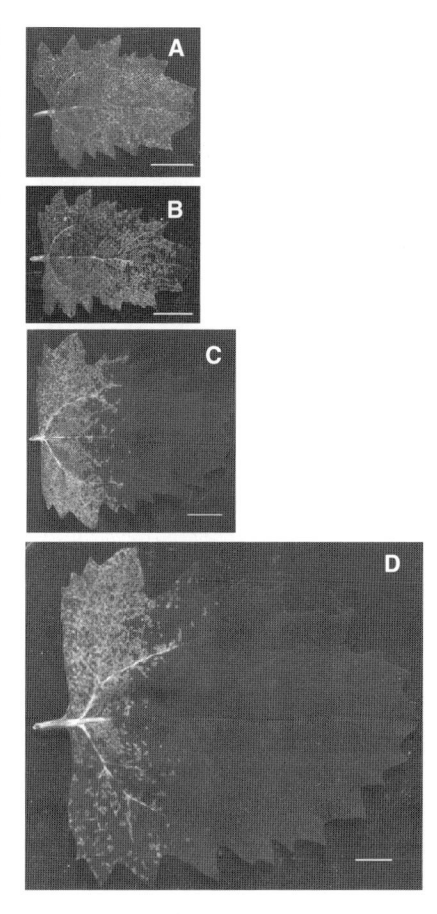

3.5 Analysis of Distribution of YFP-ALSV and CFP-ALSV in Mixed Infected Leaves

1. Inoculation of infectious cDNA clones (pEALSR1 plus pEALSR2L5R5CFP or pEALSR2L5R5YFP) onto *C. quinoa* plants were conducted as described above (see note 5).
2. Leaf tissue (0.1 g) infected with CFP-ALSV or YFP-ALSV was homogenized in 0.4 mL of inoculation buffer in a mortar and pestle. The crude sap was mixed (1:1) and mechanically inoculated onto four leaves (third to sixth true leaves) of *C. quinoa* plants (eight true leaf stage) as described above.
3. Fluorescent Images of CFP and YFP on identical leaves were acquired sequentially by a fluorescent stereoscopic microscope with settings optimal for CFP

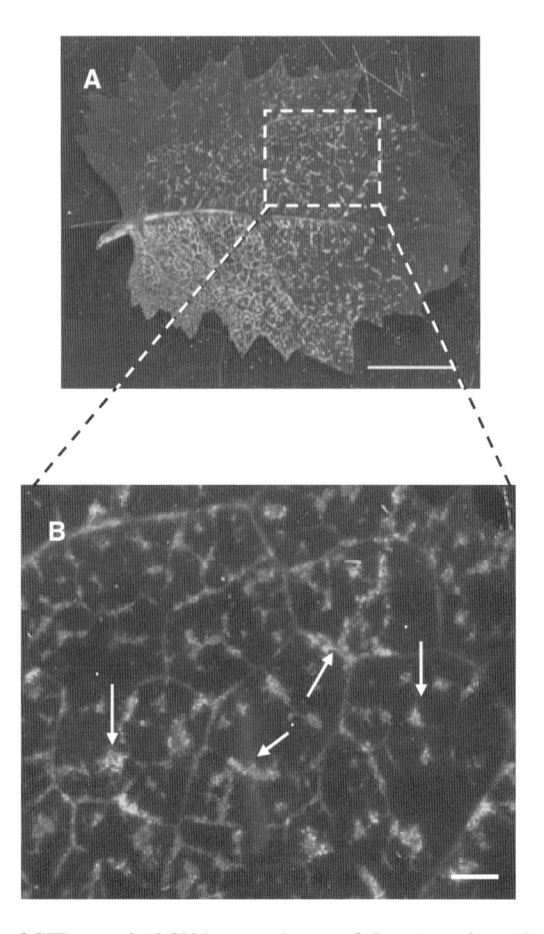

Fig. 4 Unloading of GFP-tagged ALSV in upper leaves of *C. quinoa* plant. (**A**) GFP fluorescence on minor veins of an entire area of young developing leaf. Bar = 10 mm. (**B**) Detail of the *boxed* region in (**A**). Fluorescence is unloading from class III and IV veins in upper sink leaves. Bar = 1 mm

(CFP-B filter sets; excitation filter 436/20 nm, and barrier filter 480/30 nm), followed by settings optimal for YFP (YFP-B filter sets; excitation filter 490/20 nm, and barrier filter 540/40 nm). Images of CFP and YFP were merged using Photoshop (Adobe Systems).

4. Both CFP and YFP fluorescence were found as small spots on coinoculated leaves at 3 dpi (Fig. 5a) (see note 6), and the spots expanded as rings from 5 dpi (Fig. 5b, c). When CFP and YFP rings expanded and met together, both rings fused to each other, and the fused rings showed an outline which consisted of both kinds of fluorescence (Fig. 5c). However, CFP and YFP fluorescence never overlapped and was always distributed separately. The separation of CFP and YFP was also found in upper uninoculated leaves (Fig. 5d–f).

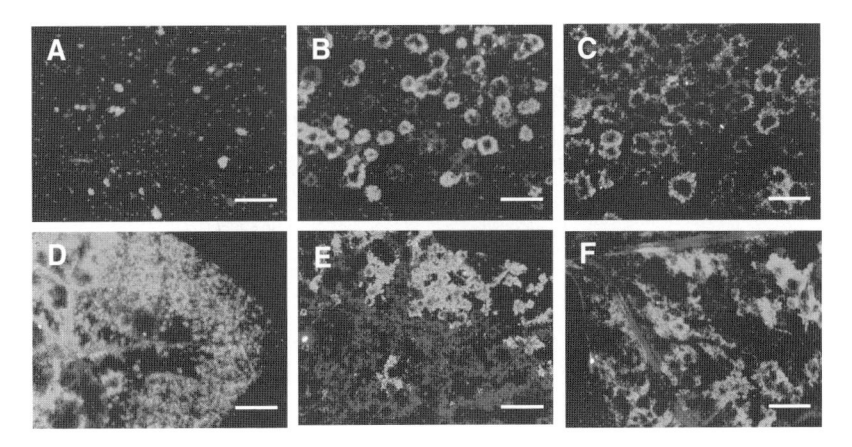

Fig. 5 The distribution of CFP-ALSV and YFP-ALSV on inoculated leaves (**A–C**) and upper uninoculated leaves (**D–F**) of *C. quinoa* plants co-inoculated with both viruses. CFP and YFP fluorescence on inoculated leaves at (**A**) 3 dpi, (**B**) 5 dpi, and (**c**) 6 dpi and fluorescence on upper uninoculated leaves at (**D**) 10 dpi, (**E**) 11 dpi, and (**F**) 13 dpi. Bar = 2 mm

4   Notes

1. ALSV is recently classified as a species in the newly established genus *Cheravirus* (11). Virus particles are isometric, c. 25 nm in diameter, and contain two single-stranded RNA species (RNA1 and RNA2) and three capsid proteins (Vp25, Vp20, and Vp24) (12). Infectious cDNA clones of ALSV-RNA1 (pEALSR1) and ALSV-RNA2 (pEALSR2) were successfully constructed using enhanced CaMV 35S promoter, and modified into viral vectors for stable expression of foreign genes in plants by molecular recombination technique (8). ALSV-RNA2 vector (pEAL-SR2L5R5) was constructed from pEALSR2 by creating artificial protease processing sites by duplicating the Q/G protease cleavage site between 42KP and Vp25 (8). The foreign gene fragments must be inserted in-frame at the *Xho*I, *Sma*I, and/or *Bam*HI sites, and they must not contain a stop codon in the foreign gene sequence.
2. General molecular techniques according to the standard protocol (13) were conducted for construction of viruses expressing fluorescent proteins.
3. Excessive rubbing will result in lethal damage to the leaf tissues from the carborundum. However, too little rubbing will not establish the infection of infectious cDNA clones or virus particles.
4. GFP expression in infected leaves was checked by fluorescent microscope before storage at −80°C. This is an important step because virus replication resulted in loss of GFP sequence on rare occasions. The level of viral coat protein could be checked by immunoblot analysis using specific antibodies against ALSV coat protein. The level of coat proteins of GFP-tagged virus was indistinguishable from that of wild type ALSV.
5. CFP and YFP variants are favorable as a pair for double-tagging research. Double-tagged plant viruses could be used for studying interactions between virus strains in coinfected plants (10).
6. In fluorescent microscopy, fluorescence of YFP could be generally observed as brighter than that of CFP (Fig. 5).

Acknowledgments We thank N. Sasaki and T. Sugawara for the photorecording of the fluorescence microscopy. We also thank T. Yamatsuta for assistance in construction of ALSV-based vectors. This work was supported in part by the 21st Century Center of Excellence Program from the Ministry of Education, Culture, Sports, Science and Technology of Japan.

References

1. Prasher, D. C., Eckenrode, V. K., Ward, W. W., Prendergast, F. G. and Cormier, M. J. (1992) Primary structure of the Aquorea Victoria green-fluorescent protein. *Gene* **111**, 229–233.
2. Oparka, K. J., Roberts, A. G., Santa Cruz, S., Boevink, P., Prior, D. and Smallcombe, A. (1997) Using GFP to study virus invasion and spread in plant tissues. *Nature* **388**, 401–402.
3. Baulcombe, D. C., Chapman, S. and Cruz, S. S. (1995) Jellyfish green fluorescent protein as a reporter for virus infection. *Plant J.* **7**, 1045–1053.
4. Oparka, K. J., Roberts, A. G., Prior D. A. M., Chapman, D., Baulcombe, D. S. and Santa Cruz, S. (1995) New ideas in cell biology. *Protoplasma* **189**, 133–141.
5. Casper, S. J. and Holt, C. A. (1996) Expression of the green fluorescent protein-encoding gene from a tobacco mosaic virus-based vector. *Gene* **173**, 69–73.
6. Roberts, A. G., Santa Cruz, S., Roberts I. M., Prior D. A. M., Turgeon, R. and Oparka, K. J. (1997) Phloem unloading in sink leaves of *Nicotiana benthamiana*: comparison of a fluorescent solute with a fluorescent virus. *Plant Cell* **9**, 1381–1396.
7. Wang, H. L., Sudarshana, M. R., Gilbertson, R. L. and Lucas, W. J. (1999) Analysis of cell to cell and long-distance movement of Bean dwarf mosaic geminivirus-green fluorescent protein reporter in host and non host species: identification of site of resistance. *Mol. Plant Microbe Interact.* **12**, 345–355.
8. Li, C., Sasaki, N., Isogai, M. and Yoshikawa, N. (2004) Stable expression of foreign proteins in herbaceous and apple plants using Apple latent spherical virus RNA2 vectors. *Arch. Virol.* **149**, 1541–1558.
9. Yoshikawa, N., Okada, K., Asanuma, K., Watanabe, K., Igarashi, A., Li. C. and Isogai, M. (2006) A movement protein and three capsid proteins are all necessary for cell-to-cell movement of apple latent spherical cheravirus. *Arch. Virol.* **151**, 837–848.
10. Takahashi, T., Sugawara, T., Yamatsuta, T., Isogai, M., Natsuaki, T. and Yoshikawa, N. (2007) Analysis of spatial distribution of identical and two distinct virus populations differently labeled with cyan and yellow fluorescent proteins in coinfected plants. Phytopathology **97**, 1200–1206.
11. Le Gall, O., Iwanami, T., Karasev, A. V., Jones, T., Letho, K., Sanfacon, H., Wellink, J., Wetzel, T. and Yoshikawa, N. (2005) Genus *Cheravirus*. In *Virus Taxonomy: 8th Report of ICTV* (Fauquet C. M., et al., eds.) pp. 803–805, Academic Press, San Diego, CA.
12. Li, C., Yoshikawa, N., Takahashi, T., Ito, T., Yoshida, K. and Koganezawa, H. (2000) Nucleotide sequence and genome organization of apple latent spherical virus: a new virus classified into the family Comoviridae. *J. Gen. Virol.* **81**, 541–547.
13. Sambrook, J. and Russell, D. W. (2001) *Molecular Cloning: A Laboratory Manual*, 3rd edn. Cold Spring Harbor Laboratory Press, Cold Spring Harbor, New York, NY.

Chapter 38
Agroinoculation: A Simple Procedure for Systemic Infection of Plants with Viruses

Zarir E. Vaghchhipawala and Kirankumar S. Mysore

Abstract Plant–virus interaction studies, for long, plagued by asynchronous/failed infections, have improved since the usage of *Agrobacterium* as a delivery agent for viral genomes. Popularly known as "agroinoculation," this method has revolutionized plant virology studies, leading to identification of viruses as casual agents of disease, viral genome mutagenesis and recombination analyses, and virus-induced gene silencing (VIGS) applications. We present here a brief overview of the recent applications of this method and a detailed protocol for agroinoculation and VIGS used in our laboratory.

Keywords *Agrobacterium*; Agroinoculation; Agroinfiltration; Agrodrench; VIGS; Tobacco rattle virus (TRV) vector

1 Introduction

The potential of using *Agrobacterium* as a vector to generate transgenic plants with genes of interest has been known since the early 1980s. This attribute of *Agrobacterium* has been recognized and duly exploited routinely by researchers seeking to understand the functions of their favorite gene(s). Plant virologists, who struggled with asynchronous/failed infections while using virus particles, naked RNA or cloned viral DNA as inoculum via mechanical means or using insect vectors, discovered the advantages of using *Agrobacterium* as a delivery agent. The first reports were published by Grimsley and associates in 1986 (1) and 1987 (2) where they were able to clone the genomes of Cauliflower Mosaic Virus (CaMV) and the maize streak virus into *Agrobacterium* binary vectors and showed disease onset and symptoms upon "agroinfection." Since then the agroinfection method (popularly known as agroinoculation) has been utilized extensively to deliver viruses inside plants for the following purposes: to validate both mono- and bipartite viruses as causal agents of disease, characterize novel viruses and their genomes via mutagenesis, recombinatorial analyses between related viruses, and for transient RNA interference (RNAi), and virus-induced gene silencing (VIGS) studies. Agroinoculation has also been invaluable in

From: *Methods in Molecular Biology, Vol. 451, Plant Virology Protocols: From Viral Sequence to Protein Function*
Edited by G.D. Foster, I.E. Johansen, Y. Hong, and P.D. Nagy © Humana Press, Totowa, NJ

identification of plant lines resistant to virus infection. A brief review of the various applications is discussed below.

1.1   Isolation and Identification of New Viral agents

A simple, short, efficient, and reproducible infectivity assay for insect transmitted, split genome geminivirus, the tomato golden mosaic virus (TGMV), was developed using agroinoculation (3). The first report of a whitefly transmitted geminivirus, the tomato yellow leaf curl virus (TYLCV), showed that whiteflies were able to transmit the virus from agroinoculated plants (with dimeric copies of cloned virus genome), exhibiting severe disease symptoms, to test plants (4). A leaf disc agroinoculation system was developed to differentiate between susceptible and resistant tomato genotypes to TYLCV infection (5). Leaf discs from agroinoculated susceptible genotypes (but not resistant genotypes) allowed for whitefly transmission to healthy plants and also allowed virus passage into regenerating plant tissue. The causal agent for a wheat dwarf outbreak, a wheat dwarf geminivirus (WDV), was isolated and shown by agroinoculation to be spread by leafhopper insects (6). Agroinoculation was also the preferred method by which the potato yellow mosaic geminivirus (PYMV) was shown to be infectious in *Nicotiana benthamiana*, tomato, and potato (7). Similarly, the mungbean yellow mosaic virus (MYMV) and the strawberry mild yellow edge potexvirus have been shown, by agroinoculation, to be the causal agent of disease in blackgram and strawberry, respectively (8, 9).

1.2   Mutational Analysis

The agroinoculation method has also been implemented to do mutational analysis of viral genomes. Klinkenberg et al. (1989) showed that coat protein deletion mutants of African cassava mosaic virus (ACMV) and TGMV have different fates upon agroinoculation in *N. benthamiana* (10). Mutational analysis of the various ORFs of beet curly top virus (BCTV) by agroinoculation into *N. benthamiana* and *Beta vulgaris* showed that ORF V2 was required for efficient systemic infection and symptom development while mutations in ORF C2/C3/C4 did not affect insect transmission (11). Mutational analysis in potato yellow mosaic virus DNA also identified the ORFs necessary for viral DNA replication, symptom appearance and severity in *N. benthamiana* (12). Sadowy and associates showed by mutational analysis using agroinoculation that ORF 0 was absolutely essential for viral accumulation while ORFs 1 and 2 encoded the proteinase and replicase domains in the potato leaf roll virus (13, 14).

1.3   Functional Studies on Viral ORFs

Agroinoculation has been the method of choice for researchers studying the functional attributes of various ORFs in viral isolates. Boulton et al. (1991) used an in vitro hybrid genome construction approach to distinguish between a narrow host range, mild symptom maize streak virus strain, and a broad host range severe disease causing strain (15). This approach coupled with agroinoculation, identified the primary determinants for host range, symptom severity, streak length, and timing of symptom appearance (15). Wild type infection caused by a mildly symptomatic tomato leaf curl virus (TLCV) ORF C4 mutant upon agroinoculation revealed a reversion of the mutation, implicating involvement of ORF C4 in symptom development (16). Genetic interchangeability of ORFs was also demonstrated in common host *N. benthamiana* between PYMV/ACMV and TGMV Gemini viruses via complementation using agroinoculation (17, 18). Recombination between CaMV and CaMV sequences present in transgenic *Nicotiana* plants translated to a change from a necrotic infection in inoculated leaves to a chlorotic mosaic infection pattern in the systemic leaves (19).

1.4   Germplasm Screening

The flexibility of agroinoculation method has allowed the screening of germplasm for varieties resistant to viral challenge. Accession lines of wild tomato species, which were resistant to TYLCV in the field, when tested by agroinoculation for resistance were found to be susceptible, indicating that agroinoculation lead to a breakdown of natural resistance (20). Resistance to rice tungro bacilliform virus (RTBV) could be distinguished from resistance to the leafhopper vector by agroinoculation during screening of rice germplasm and transgenic rice lines (21). Pathogenicity of maize streak virus (MSV) isolates and resistance of maize lines to MSV was also studied using an agroinoculation-based screen (22, 23). Several good sources of resistance to tomato leaf curl geminivirus (ToLCV) were identified in a screen of 90 isolates of *Lycopersicon* species by agroinoculation (24).

1.5   RNA Silencing

The agroinoculation assay has found a new application recently in the field of RNA silencing. Using the marker protein GFP, it was shown through an agroinoculation assay that the 126 kDa protein of tobacco mosaic virus (TMV) was able to delay GFP silencing, that is, it functioned as a suppressor of gene silencing (25). Several aspects of cytoplasmic RNA silencing were studied by agroinfiltration using an endogenous gene, VirP1 (26). A high titer of siRNAs was observed in the agroinfiltrated zone than one normally seen during a systemic silencing and the antisense–sense construct orientation performed better for silencing than its opposite orientation (26). A turnip crinkle virus (TCV)-based system was used to discriminate between cell-to-cell and systemic long-distance spread of RNA silencing in plants and only the agroinfiltration method, but not mechanical inoculation, was able to distinguish between the two modes of spread (27).

The dynamic nature of this method has thus lead to many modifications and many applications. Agroinoculation is popularly used for VIGS in plants (28).

This method involves the use of *Agrobacterium* as a vehicle to deliver a tobacco rattle virus (TRV)-derived VIGS vector that can express a fragment of a plant gene. This will result in transient silencing, through mRNA degradation, of a plant gene that has nucleotide sequence homology to the gene fragment used in the VIGS vector (29). TRV contains a bipartite positive-sense RNA genome, RNA1 and RNA2, with RNA2 being the component modified for cloning of genes of interest. TRV produces mild symptoms on the host and has a wide host range (30, 31). Our laboratory has also developed and published a modified agroinoculation procedure called "agrodrench" which provides certain advantages in certain situations (32). Leaf infiltrations are laborious for large-scale screenings, plus certain plants like soybean and maize are difficult to infiltrate. Leaf infiltration also does not induce efficient silencing in roots and also in leaves of some plants, including some varieties of tomato. Agrodrench protocol therefore gives efficient silencing in roots and is useful to study RNA silencing in several solanaceous species. A detailed protocol involving steps in agroinoculation of leaves and the agrodrench procedure currently used for VIGS is given below. The same procedure can also be used for nonVIGS related experiments to achieve systemic infection in plants with other viruses.

2   Materials

1. *Agrobacterium tumefaciens* strain GV2260 containing the TRV–VIGS constructs is used for VIGS experiments (note 1)
2. Three-week-old *N. benthamiana* plants (note 2) in the greenhouse operating at $23 \pm 3°C$ and 70% humidity and 16-h/8-h day/night cycle. Light intensities should be kept around $50–100\,\mu Es^{-1}m^{-2}$
3. Luria–Bertani (LB) broth/agar medium for growing the *Agrobacterium* cultures supplemented with the appropriate antibiotics
4. *Agrobacterium* induction medium: 10 mM $MgCl_2$, 10 mM MES buffer pH5.6, 150 μM Acetosyringone (made from 150 mM stock in DMSO)
5. 1-mL tuberculin syringes without needle
6. Shaker/incubators set to 28°C. High-speed table top cooled centrifuges
7. Spectrophotometer for measuring culture optical density at 600 nm
8. 10 mM MES buffer pH 5.6

3   Method

1. *N. benthamiana* seeds were germinated in flats with a soilless potting mixture, BM7 (Berger Co., Quebec, Canada). Two-week-old seedlings (one plant per pot) were transplanted to 10 cm diameter pots containing BM7 with application of fertilizer (20-10-20) and soluble trace element mix (The Scotts Co., Marysville,

OH, USA). Greenhouse conditions and light regimes were as mentioned above. Three-week-old seedlings (four-leaf stage) arranged on flood benches in the greenhouse is used as the starting material.

2. Three days before infiltration of constructs, streak *Agrobacterium* strains containing *pTRV1*, *pTRV2-GFP* (note 3), and/or *pTRV2-GOI* (gene of interest) (note 4) cultures from frozen stock on LB plates containing appropriate antibiotics. The TRV vectors used in our laboratory have kanamycin resistance marker while the *Agrobacterium* strain has a rifampicin resistance marker (note 5). Incubate plates at 29°C for 2 days. Inoculate a single *Agrobacterium* colony on day 3 in LB broth supplemented with appropriate antibiotics. Grow the culture overnight in a shaker incubator set to 29°C and 250 rpm.

3. On the day of infiltration, spin down the liquid cultures (5,000 rpm for 10 min) in 50-mL screw cap conical tubes (# 62.547.004 PP, Sarstedt, Newton NC) in a Sorvall biofuge Primo R table top centrifuge (rotor #7588) at room temperature.

4. Discard the supernatant carefully and resuspend pellet gently with *Agrobacterium* inoculation buffer (note 6), to a final OD_{600} of 1.0. OD (Eppendorf biophotometer, Eppendorf AG, Hamburg, Germany). Incubate at room temperature for 4–6 h with gentle shaking (50–100 rpm).

5. Pellet the culture(s) by centrifugation at 5,000 rpm for 10 min and resuspend the cultures in 10 mM MES buffer and adjust to $OD_{600} = 1.0$

6. For leaf infiltrations in *N. benthamiana*, mix the two cultures (adjusted to $OD_{600} = 1.0$) of *Agrobacterium* strains containing *pTRV1* and *pTRV2-GOI* in a 1:1 volumetric ratio in a separate tube. As a standard procedure for gene silencing studies in our laboratory, we infiltrate 10–12 plants per construct per experiment to account for variation. Two lower leaves on each plant are fully infiltrated on the abaxial surface with a needleless 1-mL tuberculin syringe, taking care not to damage the leaf surface excessively (note 7). Each experiment is repeated three times.

7. For the Agrodrench procedure, a 1:1 mixture of *Agrobacterium* strains carrying the *pTRV1* and *pTRV2-GOI* constructs should be drenched (3–5 mL) into the soil at the crown part of each plant using a 10 mL syringe.

8. Systemic leaf samples are collected at least 1 week later to measure the virus titer. For gene silencing experiments, leaf samples are collected approximately 3 weeks later for RNA isolation to assess for decrease in specific transcripts via quantitative RT-PCR.

4   Notes

1. A different *Agrobacterium* strain and a different viral genome can be used based on the experiment. The VIGS vectors *pTRV1* and *pTRV2* are gateway compatible vectors, obtained from Dr. Dinesh-Kumar, Yale University (30).

2. A different plant species can be used based on the experiment.

3. Vector control: A 451-bp *GFP* fragment was amplified using primers, gfpattB1: 5′-ggggacaagtttgtacaaaaaagcaggctCTTTTCACTGGAGTTGTCCC-3′ and gfpattB2: 5′-ggggaccactttgtacaa-

gaaagctgggtGCTTGTCGGCCATGATGTA-3′, from *GFP* gene and cloned into *pTRV2*. Since there is no *GFP* homolog in plants, this vector when inoculated will not cause any gene silencing effect in plants.

4. Genes of interest: As a visual marker gene for gene silencing, a 409 bp *NbPDS* (phytoene desaturase) fragment was cloned in *pTRV2* using the primers 5′-GGGGACAAGTTTTGTACAAAAAAGCAGG CCGGTCTAGAGG-CACTCAACTTTATAAACC-3′ and 5′-GGGGACCACTTTGTACAAGAAA-GCTGG GCGGGGATCCCTTCAGTTTTCTGTCAAACC-3′.

5. All plasmid constructs were introduced into electrocompetent *Agrobacterium* strain GV2260 cells by electroporation and the transformants were selected on media containing kanamycin (50 μg mL⁻¹) and rifampicin (50 μg mL⁻¹).

6. Acetosyringone is always added fresh to the *Agrobacterium* inoculation buffer at a final concentration of 150 μM from a 150 mM stock made in DMSO.

7. Care should be taken to expel any air bubbles in the tuberculin syringe prior to infiltration.

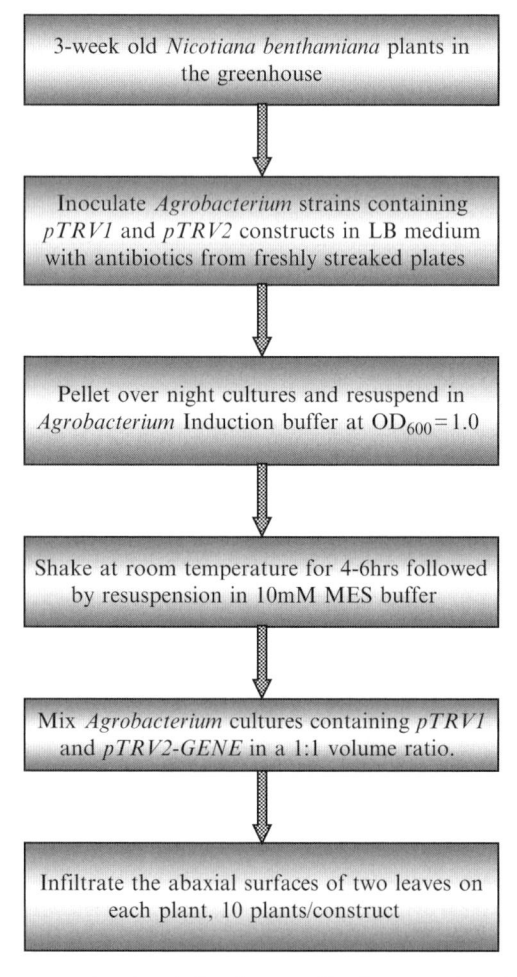

3-week old *Nicotiana benthamiana* plants in the greenhouse

↓

Inoculate *Agrobacterium* strains containing *pTRV1* and *pTRV2* constructs in LB medium with antibiotics from freshly streaked plates

↓

Pellet over night cultures and resuspend in *Agrobacterium* Induction buffer at OD₆₀₀ = 1.0

↓

Shake at room temperature for 4-6hrs followed by resuspension in 10mM MES buffer

↓

Mix *Agrobacterium* cultures containing *pTRV1* and *pTRV2-GENE* in a 1:1 volume ratio.

↓

Infiltrate the abaxial surfaces of two leaves on each plant, 10 plants/construct

Flowsheet indicating steps in virus-induced gene silencing procedure by means of Agroinoculation

References

1. Grimsley, N., Hohn, B., Hohn, T., and Walden, R. (1986) "Agroinfection," an alternative route for viral infection of plants by using the Ti plasmid. *Proc Natl Acad Sci* **83**, 3282–86.
2. Grimsley, N., Hohn, T., Davies, J. W., and Hohn, B. (1987) Agrobacterium-mediated delivery of infectious maize streak virus into maize plants. *Nature* **325**, 177–79.
3. Elmer, J. S., Sunter, G., Gardiner, W. E., Brand, L., Browning, C. K., Bisaro, D. M., and Rogers, S. G. (1988) Agrobacterium-mediated inoculation of plants with tomato golden mosaic virus DNAs. *Plant Mol Biol* **10**, 225–34.
4. Navot, N., Pichersky, E., Zeidan, M., Zamir, D., and Czosnek, H. (1991) Tomato yellow leaf curl virus: A whitefly-transmitted geminivirus with a single genomic component. *Virology* **185**, 151–61.
5. Czosnek, H., Kheyr-Pour, A., Gronenborn, B., Remetz, E., Zeidan, M., Altman, A., Rabinowitch, H. D., Vidavsky, S., Kedar, N., Gafni, Y., and Zamir, D. (1993) Replication of tomato yellow leaf curl virus (TYLCV) DNA in agroinoculated leaf discs from selected tomato genotypes. *Plant Mol Biol* **22**, 995–1005.
6. Bendahmane, M., Schalk, H.-J., and Gronenborn, B. (1995) Identification and characterization of wheat dwarf virus from France using a rapid method for geminiviurs DNA preparation. *Mol Plant Pathol* **85**, 1449–55.
7. Buragohain, A., Sung, Y., Coffin, R., and Coutts, R. (1994) The infectivity of dimeric potato yellow mosaic geminivirus clones in different hosts. *J Gen Virol* **75**, 2857–61.
8. Mandal, B., Varma, A., and Malathi, V. G. (1997) Systemic infection of *Vigna mungo* using the cloned DNAs of the blackgram isolate of mungbean yellow mosaic geminivirus through agroinoculation and transmission of the progeny virus by whiteflies. *J Phytopathol* **145**, 505–10.
9. Lamprecht, S., and Jelkmann, W. (1997) Infectious cDNA clone used to identify strawberry mild yellow edge-associated potexvirus as causal agent of the disease. *J Gen Virol* **78**, 2347–53.
10. Klinkenberg, F. A., Ellwood, S., and Stanley, J. (1989) Fate of African cassava Mosaic virus coat protein deletion mutants after agroinoculation. *J Gen Virol* **70**, 1837–44.
11. Stanley, J., Latham, J. R., Pinner, M. S., Bedford, I., and Markham, P. G. (1992) Mutational analysis of the monopartite geminivirus beet curly top virus. *Virology* **191**, 396–405.
12. Sung, Y., and Coutts, R. (1995) Mutational analysis of potato yellow mosaic geminivirus. *J Gen Virol* **76**, 1773–80.
13. Sadowy, E., Maasen, A., Juszczuk, M., David, C., Zagorski-Ostoja, W., Gronenborn, B., and Hulanicka, M. D. (2001) The ORF0 product of Potato leafroll virus is indispensable for virus accumulation. *J Gen Virol* **82**, 1529–32.
14. Sadowy, E., Juszczuk, M., David, C., Gronenborn, B., and Hulanicka, M. D. (2001) Mutational analysis of the proteinase function of Potato leafroll virus. *J Gen Virol* **82**, 1517–27.
15. Boulton, M. I., King, D. I., Markham, P. G., Pinner, M. S., and Davies, J. W. (1991) Host range and symptoms are determined by specific domains of the maize streak virus genome. *Virology* **181**, 312–18.
16. Rigden, J. E., Krake, L. R., Rezaian, M. A., and Dry, I. B. (1994) ORF C4 of tomato leaf curl geminivirus is a determinant of symptom severity. *Virology* **204**, 847–50.
17. Saunders, K., and Stanley, J. (1995) Complementation of African cassava mosaic virus AC2 gene function in a mixed bipartite geminivirus infection. *J Gen Virol* **76**, 2287–92.
18. Sung, Y., and Coutts, R. (1995) Pseudorecombination and complementation between potato yellow mosaic geminivirus and tomato golden mosaic geminivirus. *J Gen Virol* **76**, 2809–15.
19. Király, L., Bourque, J. E., and Schoelz, J. E. (1998) Temporal and spatial appearance of recombinant viruses formed between cauliflower mosaic virus (CaMV) and CaMV sequences present in transgenic *Nicotiana bigelovii*. *Mol Plant Microbe Interact* **11**, 309–16.

20. Kheyr-Pour, A., Gronenborn, B., and Czosnek, H. (1994) Agroinoculation of tomato yellow leaf curl virus (TYLCV) overcomes the virus resistance of wild *Lycopersicon* species. *Plant Breed* **112**, 228–33.
21. Cruz, F. C. S., Boulton, M. I., Hull, R., and Azzam, O. (1999) International Rice Research Institute, Los Banos, Philippines. *J Phytopathol* **147**, 653–59.
22. Martin, B. P., Willment, J. A., and Rybicki, E. P. (1999) Evaluation of maize streak virus pathogenicity in differentially resistant *Zea mays* genotypes. *Virology* **89**, 695–700.
23. Martin, B. P., and Rybicki, E. P. (2000) Improved efficiency of *Zea mays* agroinoculation with Maize streak virus. *Plant Dis* **84**, 1096–98.
24. Tripathi, S., and Varma, A. (2003) Identification of sources of resistance in *Lycopersicon* species to Tomato leaf curl geminivirus (ToLCV) by agroinoculation. *Euphytica* **129**, 43–52.
25. Ding, X. S., Liu, J., Chen, N.-H., Folimonov, A., Hou, Y.-M., Bao, Y., Katagi, C., Carter, S. A., and Nelson, R. S. (2004) The *Tobacco mosaic virus* 126-kDa protein associated with virus replication and movement suppresses RNA silencing. *Mol Plant Microbe Interact* **17**, 583–92.
26. Kościańska, E., Kalantidis, K., Wypijewski, K., Sadowski, J., and Tabler, M. (2005) Analysis of RNA Silencing in Agroinfiltrated Leaves of *Nicotiana benthamiana* and *Nicotiana tabacum*. *Plant Mol Biol* **59**, 647–61.
27. Ryabov, E. V., van Wezel, R., Walsh, J., and Hong, Y. (2004) Cell-to-cell, but not long-distance, spread of RNA silencing that is induced in individual epidermal cells. *J Virol* **78**, 3149–54.
28. Xie, Q., and Guo, H.-S. (2006) Systemic antiviral silencing in plants. *Virus Res* **118**, 1–6.
29. Herr, A. J., and Baulcombe, D. C. (2004) RNA silencing pathways in plants. *Cold Spring Harb Symp Quant Biol* **69**, 363–70.
30. Liu, Y., Schiff, M., and Dinesh-Kumar, S. P. (2002) Virus-induced gene silencing in tomato. *Plant J* **31**, 777–86.
31. Dinesh-Kumar, S. P., Anandalakshmi, R., Marathe, R., Schiff, M., and Liu, Y. (2003) *in* "Plant Functional Genomics" (Grotewold, E., Ed.), Vol. 236, pp. 287–93, Humana Press, Inc., Totowa, NJ.
32. Ryu, C.-M., Anand, A., Kang, L., and Mysore, K. S. (2004) Agrodrench: A novel and effective agroinoculation method for virus-induced gene silencing in roots and diverse Solanaceous species. *Plant J* **40**, 322–31.

Chapter 39
Geminivirus: Biolistic Inoculation and Molecular Diagnosis

Anésia A. Santos, Lilian H. Florentino, Acássia B.L. Pires, and Elizabeth P.B. Fontes

Abstract The *Geminiviridae* family is a large family of plant viruses that has single-stranded DNA genomes and infects a large variety of crop species. In this chapter, we describe a biolistic inoculation protocol that has been successfully used to propagate new species of geminivirus in permissive hosts with total DNA extracted from infected plants. This allows us to directly investigate the biological properties of uncloned and not sap-transmissible geminiviruses.

Keywords Geminiviruses; Infectivity assay; Diagnosis; Biolistic inoculation

1 Introduction

The geminiviruses are a large and diverse family of plant viruses that are packed as circular, single-stranded DNA genomes and are characterized by their unique morphological structure of paired-icosahedral capsides from which the family name was derived. The Geminiviridae family is taxonomically divided in four genera according to their host range, insect vector, phylogenetic relatedness, and genome organization, which may be in either single or double-component configuration (for review, see ref. 1). The names of the geminivirus genera were derived from the type members: *Begomovirus, Curtovirus, Mastrevirus*, and *Topocuvirus* (Table 1). Figure 1 illustrates the genomic organization of geminiviruses and exemplifies some species of each genus. The viral DNA is organized into divergent transcription units separated by an intergenic region (IR) of about 200 bp which contains the replication origin and two divergent promoters.

Begomovirus is the largest genus of this family and comprises the whitefly transmitted geminiviruses that infect dicotyledonous plants. The begomoviruses found in the western hemisphere typically have bipartite genomes (2), whereas several monopartite begomoviruses have been identified in the eastern hemisphere, such as Tomato yellow leaf curl virus (TYLCV) (3, 4) and Tomato leaf curl virus (ToLCV) (5), among others. Collectively, they are considered one of the most successful groups of plant viruses that infect a variety of economically important crops, such

From: *Methods in Molecular Biology, Vol. 451, Plant Virology Protocols: From Viral Sequence to Protein Function*
Edited by G.D. Foster, I.E. Johansen, Y. Hong, and P.D. Nagy © Humana Press, Totowa, NJ

Table 1 Genera of the *Geminiviridae* family, their type members and biological properties

Genus	Type member	Genome configuration	Vector	Host range
Mastrevirus	Maize streak virus (MSV)	Monopartite	Leafhoppers	Monocots and a few dicots
Curtovirus	Beet curly top virus (BCTV)	Monopartite	Leafhoppers	Dicots
Begomovirus	Bean golden mosaic virus (BGMV)	Mono- or bipartite	Whiteflies	Dicots
Topocuvirus	Tomato pseudo-curly top virus (TPCTV)	Monopartite	Treehoppers	Dicots

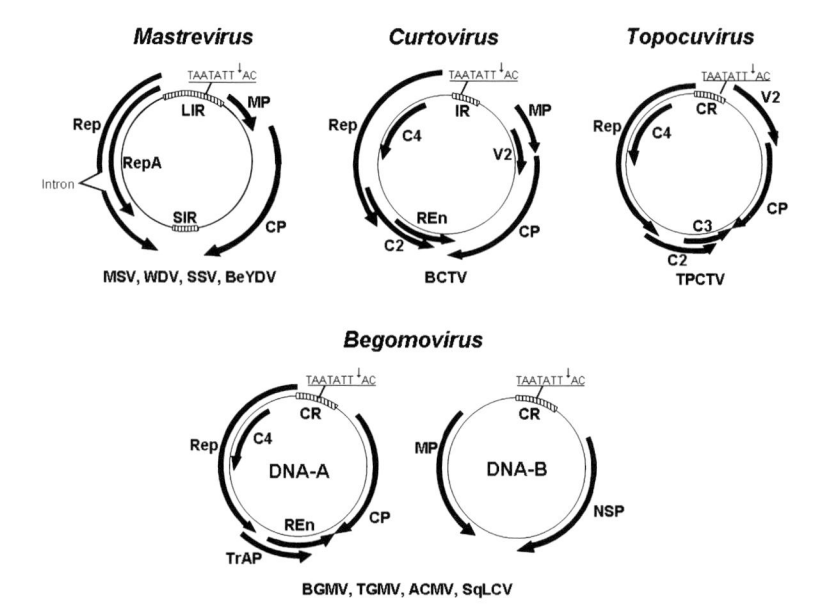

Fig. 1 The geminivirus genomic organization. The genomic organization of the four genera of the Geminiviridae family is show in the figure and some of most frequently studied members of each genus are given as examples. Mastrevirus: MSV, *Maize streak virus*; WDV, *Wheat dwarf virus*; SSV, *Sugarcane streak virus*; BeYDV, *Bean yellow dwarf virus*. Curtovirus: BCTV, *Beet curly top virus*. Topocuvirus: TPCTV, *Tomato pseudo-curly top virus*. Begomovirus: BGMV, *Bean golden mosaic virus*; TGMV, *Tomato golden mosaic virus*; ACMV, *African cassava mosaic virus*; SqLCV, *Squash leaf curl virus*. The viral proteins are named according to their functions as: RepA, replication-associated protein interacting with retinoblastoma; Rep, replication initiation protein; REn, replication enhancer protein; TrAP, transcriptional activator protein; CP, coat protein; MP, movement protein; NSP, nuclear shuttle protein. Proteins with less well-defined functions are named according to their positions in the genetic map, in which C stands for complementary strand and V, for virion sense. The non-coding regions are the large intergenic region (LIR) and the small intergenic region (SIR) in mastreviruses, the intergenic region in curtoviruses (IR) and the common region (CR) for topocuviruses and begomoviruses. The ORFs and the directions of transcription are designated by arrows. The invariant TAATATTAC sequence, located in the intergenic region is indicated together with the initiation site (↓) for rolling-circle DNA replication

as cassava, cotton, bean, pepper, and tomato (1). A current consensus prediction for the extent of begomovirus diversity holds that a high frequency of interspecies recombination resulted in the recent emergence of highly pathogenic virus geno-types causing a variety of serious begomovirus diseases (6–8). Accordingly, gemi-nivirus-associated epidemics have currently become an even greater threat to agriculture worldwide and protocols for geminivirus detection, cloning, inocula-tion, and species identification have been progressively developed and improved.

The geminiviruses replicate their genome by a rolling-circle mechanism via double-stranded DNA intermediates in the nuclei of infected cells (9). The conver-sion of ssDNA into double-stranded replicative form (RF) of the virus within the nucleus of infected cells relies totally on the host replication machinery. The syn-thesis of the RF provides the template for both transcription of viral genes and amplification of viral genomes, which is initiated by the viral Rep protein. As a replication initiator protein for the rolling circle mediated replication, Rep is a DNA-binding protein that exhibits sequence-specific endonuclease activity. Upon binding to the origin of replication (10, 11), Rep specifically nicks in a conserved nonanucleotide sequence (TAATATT-AC) and hence releases a 3'OH to initiate the rolling-circle replication by the host replication machinery. A round of replication generates the dsDNA intermediate and the virion-sense ssDNA, which can either be converted to dsDNA to serve another round of replication or be encapsidated as the geminate particle. As consequence, both ssDNA and dsDNA accumulated in infected plants and can be detected by Southern blot analysis (Fig. 2) or by the polymerase chain reaction (PCR) with specific primers (Table 2, Fig. 3). In this chapter, we describe the most used molecular methods for geminiviruses detection which rely on the detection of the viral DNA forms in infected plants. We also describe an efficient biolistic inoculation protocol that has been extensively used in our laboratory (7, 12) and can utilize either cloned infectious viral DNA or total DNA from infected plants as the inoculum.

Although serology is still a widely used technique for viral identification and diagnosis, the utilization of acid nucleic-based assays has been progressively increased. Hybridization of nucleic acid and PCR are the most common molecular methods for diagnostic of viral infection. While these molecular assays target the viral genome for diagnosis, serology relies on the coat protein accumulation for detection and viral identification. However, for the *Begomovirus* genus, the coat protein, encoded by the V1 or AV1 gene, is the most conserved protein among the begomoviruses (87% identity) and, thus, its identification per se does not allow us to distinguish among the different species of the genus. Furthermore, in some cases, upon geminivirus infection the viral DNA accumulates at very low concentration or, in some hosts, the infected tissue is obtained in such low amounts that the utili-zation of protocols based on PCR is crucial to yield large amounts of viral DNA. In this case, the amplification of segments of the viral genome with degenerate prim-ers that target highly conserved region of the begomovirus DNA (13) has proven to be the most efficient diagnostic technique for begomovirus infection. The associa-tion of the PCR-based diagnostic assay with high-fidelity DNA-dependent DNA polymerase can provide appropriate amounts of viral DNA for cloning, sequencing,

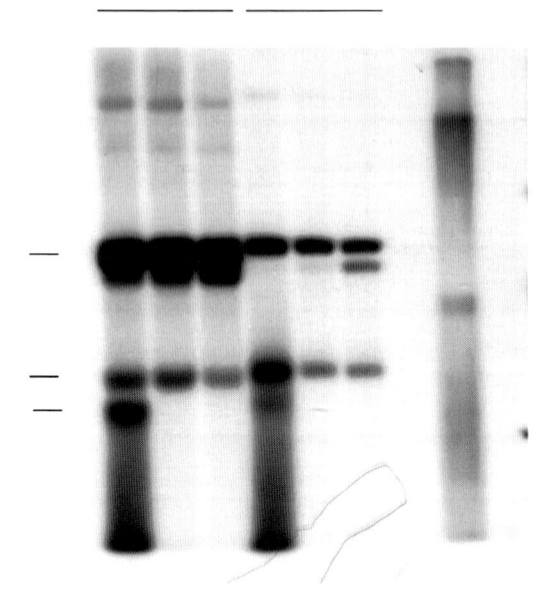

Fig. 2 Accumulation of viral DNA in infected plants analyzed by Southern hybridizations. Total DNA was isolated from infected tomato leaves [TCrLYV, Tomato crinkle leaf yellows virus, (7)] and from infected Sida rhombifolia leaves (SiMoV, Sida mottle virus, accession number AY090555), resolved on an agarose gel, transferred to nylon membranes, fixed on the membrane by UV, and hybridized with a ^{32}P-labeled probe specific for begomovirus DNA-A that corresponds to the TGMV-A fragment, position 870 a 2140. The positions of open circular (oc) and supercoiled (cc) ds DNA as well as single-stranded (ss) viral DNA forms are indicated on the left. The identity of the ssDNA was confirmed by the disappearance of the corresponding band upon digestion with Mung bean nuclease (Mgb) or S1 nuclease (S1N). ND refers to non-digested total DNA. TGMV-A refers to pTG1.3A which contains 1½ copies of TGMV-DNA-A, whereas TGMV-B is pTG1.4B that harbors partial tandem copies of TGMV-DNA-B. Healthy refers to total leaf DNA from asymptomatic plants (see notes 10–13)

Table 2 Degenerate primers for begomoviruses PCR-based diagnostic

Name	Sequence	Amplicon
PAL1v1978[a]	GCATCTGCAGGCCCACTYGTCTTYCCNGT	1.1-kb DNA-A fragment encompassing 5′ end of AL1, common region and 5′ end of AR1
PAR1c496[a]	AATACTGCAGGGCTTYCTRTACATRGG	
PCRc1[a]	CTAGCTGCAGCATATTTACRARWATGCCA	0.5-kb DNA-B fragment from the 5′ end of BL1 up to the first nucleotide of the CR
PBL12040[a]	GCCTCTGCAGCARTGRGRTCKATCTTC	
AR1Fwd[b]	ATACACTTTAATTCYAYATGCCTAAGCGGG	0.8-kb DNA-A fragment, AR1 coding region
AR1Rvs[b]	GAGCTGTTCGRRTCYCAACAGACAG	

[a]Rojas et al., 1983
[b]This work

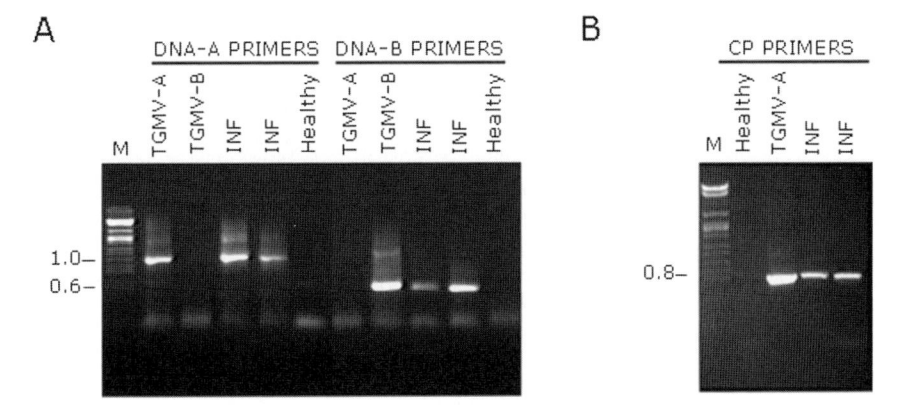

Fig. 3 PCR-based diagnostic of begomovirus infection. *N. benthamiana* seedlings were inoculated with total DNA from symptomatic tomato leaves by biolistic delivery (see notes 10–13). Total nucleic acid was extracted from systemically infected leaves (INF) and viral DNA was detected by PCR with DNA-A and DNA-B begomovirus specific primers (**a**) or with the coat protein flanking primers (**b**) at 7 days post-infection. Total DNA from healthy, mock-inoculated plants was used as control. DNA-A and DNA-B from TGMV were used as positive control. The numbers to the left correspond to the sizes in kb and positions of digested lambda DNA markers (M). The sequences of the primers are given in Table 2

and species identification based on nucleic acid conservation. The partial sequence of a geminivirus genome allows the development of specific protocols for cloning the entire circular genome from DNA of infected plants and the subsequent construction of biologically active geminivirus particles (14, 15).

The cloning of biologically active, infectious viral DNA requires the construction of tandemly repeated copies of viral DNA into a plasmid. Within plant cells, this recombinant plasmid containing 1½ copies of viral DNA directs by recombination the release of double-stranded circular viral DNA recovering the intact viral genome. Alternatively, and more efficiently, the release of viral sequences as in nature viral DNA within infected cells occurs through rolling-circle replication if the viral DNA is strategically cloned between two origins of replication. The initial step toward the construction of partial tandem repeats of viral DNA-A or DNA-B consists in the extraction of total DNA from infected tissues to be used as template in PCR with degenerate primers. Subsequently, the amplified fragment is cloned and sequenced. Based on this partial DNA sequence of the viral genome, partially overlapping and specific primers are strategically designed to amplify and to clone into a bacterial plasmid the entire circular DNA-A and/or DNA-B genomic components (14, 15). This provides the means to yield bacterial plasmid-derived infectious clones containing 1½ copies of the viral genome.

For viral infectivity assays from cloned DNA as inoculum, the host plants must be challenged with biologically active viral DNA through agroinoculation (16–18) or biolistic delivery (12, 19). The biolistic method of inoculation has been used by our laboratory with success in studies of begomovirus infectivity, when both DNA-A and DNA-B have been cloned, as well as for propagation of unknown begomovirus into permissive hosts through microprojectile bombardment with viral RF-enriched DNA fraction from infected plants.

If cloned viral DNA is available, the DNA constructions harboring 1½ copies of viral genome are maintained and proliferated in compatible *Escherichia coli* strains (e.g., JM 109 or DH5α), the recombinant plasmid is isolated as described (20) and the DNA concentration is determined by spectrophotometry at 260 nm prior to the precipitation into tungsten or gold particles. In the case of uncloned begomovirus species, the viral RF-enriched DNA fraction can be obtained from infected tissues, as described below, and directly precipitated into gold particles. Subsequently, the host plants are bombarded with the DNA containing-gold particles by gene gun and kept either in green house under standardized conditions or in growth chamber depending on the growth requirement of the plant species used as host. In the case of *Arabidopsis thaliana*, the plants are grown in a growth chamber at 22°C under long-day conditions (16-h light/8-h dark).

At regular interval periods, the inoculated plants must be evaluated for the development of typical geminivirus-associated symptoms, such as foliar necrosis, foliar chlorosis, mosaic, mottle, yellowing of the veins, stunted growth, epinasty, young leaf curly, young leaf death. Concomitantly, the confirmation of viral replication and viral DNA accumulation in infected host by the PCR using total DNA from inoculated plants as template and begomovirus-specific primers is necessary. Alternatively, the virus detection can be performed by Southern blot analysis, using fragments of DNA-A and/or DNA-B as probes (21, 22).

The PCR consists of the amplification, in geometric progression, of a certain DNA sequence, which is delimited by annealing sequences of rationally designed primers. For begomoviruses (13) and mastreviruses (23), universal and degenerate primers which are specific for DNA-A or DNA-B have been shown to be very effective in PCR-based diagnostic assays (Table 1). The degenerate primers consist of a mixture of oligonucleotides with variant nucleotides at certain positions that are targeted to conserved sequences and hence they are capable of covering efficiently the common genomic variability of the viruses from a determined genus. Species-specific primers can also be designed if the nucleotide sequence of the viral genome of interest is known.

A PCR assay is constituted by cycles with three major steps: (i) denaturation of the double strand DNA template by heating at 94°C for 30 s. This promotes the separation of the DNA strands, (ii) the reaction is incubated at 40–60°C for 1–2 min to promote the annealing of the primers to their target sequences on the template, and (iii) the reaction is incubated at 72°C for 1–3 min to allow synthesis of the new DNA molecules by the thermal stable DNA-dependent DNA polymerase (24). The major advantage of the PCR-based diagnosis is the extreme sensitivity of the assay which allows the detection of nucleic acids present in the infected tissues in amounts as low as picograms (10^{-12} g). The Southern analysis is based on the DNA hybridizations using labeled DNA-A or DNA-B fragments as probes and it is especially useful when the identification of the accumulated viral DNA forms is required.

2 Materials

2.1 Common Materials Needed for all Procedures Described in this Chapter

1. Liquid nitrogen
2. Ice
3. Micropipettes
4. Tips
5. 1.7-mL microcentrifuge tubes
6. Vortex
7. Microcentrifuge

2.2 Phenol–Chloroform Based Method of Plant Genomic DNA Extraction

1. Lysis buffer: 50 mM Tris–HCl pH 7.6, 100 mM NaCl, 50 mM EDTA, 0.5% (w/v) SDS, 10 mM β-mercaptoetanol. β-mercaptoetanol must be added immediately before use.
2. Buffered Phenol
3. Phenol:Chloroform 1:1 (v/v)
4. Chloroform
5. Isopropanol
6. Absolute ethanol
7. 70% (v/v) ethanol
8. TNE buffer: 10 mM Tris–HCl pH 7.6, 100 mM NaCl, 1 mM EDTA, 100 μg mL^{-1} RNase. It is important to add RNase immediately before use.
9. Sterile distilled water

2.3 Determination of DNA Concentration

1. Quartz cuvettes
2. Autoclaved distilled H$_2$O
3. Isolated plasmid DNAs or infected plant DNA
4. Spectrophotometer

2.4 Preparation of the Tungsten Particles

1. Tungsten M10 particles (0.7 μm diameter, BIORAD) or 1.0 μm gold particles
2. Sterile distilled H$_2$O
3. 50% (v/v) glycerol

2.5 Precipitation of Viral DNA on the Tungsten or Gold Particles

1. Tungsten particles or gold particles
2. 1 mg mL^{-1} plasmid DNA or 1 mg mL^{-1} RF-enriched fraction from infected tissue
3. 2.5 M CaCl$_2$
4. 0.1 M Spermidine
5. 100% (v/v) Ethanol
6. 20 mm tungsten or gold macrocarrier discs imbibed in 100% (v/v) ethanol
7. Petri plates – 160-mm diameter to store the discs with the DNA-coated tungsten or DNA-coated gold

2.6 Microprojectile Bombardment

1. PDS-1000/He accelerator (gene gun)
2. Vacuum pump
3. 450 psi (and/or 900 psi) rupture disk in isopropanol
4. Macrocarrier discs in ethanol
5. 0.7-μm Tungsten M-10 microcarriers or 1.0-μm Gold microcarriers
6. Holders
7. Metal mesh
8. 70% (v/v) ethanol
9. Host seedlings at the recommended developmental stage

2.7 Total DNA Extraction from Bombarded Plants

2.7.1 Quick Protocol of Total Plant DNA Extraction for PCR: Method I

1. Shaker with controlled speed and temperature at 37°C
2. Extraction buffer: 200 mM Tris–HCl pH 7.5, 250 mM NaCl, 25 mM EDTA, 0.5% (w/v) SDS.
3. Isopropanol
4. 70% (v/v) Ethanol
5. Sterile distilled H$_2$O

2.7.2 Plant Genomic DNA Extraction for PCR: Method II

1. Shaker with controlled speed and temperature at 37°C
2. Water bath

3. DNA extraction buffer: 0.1 M Tris–HCl pH 8.0, 0.05 M EDTA, 0.5 M NaCl, 0.01 M β-mercaptoethanol
4. 2-β-mercaptoethanol
5. 20% (w/v) SDS
6. 5 M potassium acetate
7. Isopropanol
8. 70% (v/v) Ethanol
9. Sterile distilled H_2O or TE buffer

2.8 Polymerase Chain Reaction (PCR)

1. Sterile 0.5 or 0.2-mL microcentrifuge tubes
2. Template DNA: 0.01–1 ng plasmid DNA or 0.1–1 μg infected plant DNA for a 50-μL reaction
3. Primers (10 μM): 18 a 25 nucleotides
4. 25 mM $MgCl_2$ (1–4 mM per reaction)
5. 10 mM dNTPs (0.2 mM per reaction)
6. Taq DNA polymerase (1–3 U)
7. Thermocycler
8. 10X Taq buffer: 200 mM Tris–HCl, pH 8.3, 500 mM KCl

2.9 Electrophoresis in Agarose Gel

1. Electrophoresis apparatus
2. Power supply
3. Autoclaved Milli-Q H_2O
4. Agarose
5. Sample buffer: 0.01 M Tris–HCl, pH 8.0, 50% (v/v) glycerol, 0.1% (w/v) bromophenol blue
6. 5X TBE: 0.45 M Tris–borate, pH 8.0, 0.01 M EDTA

3 Methods

3.1 Isolation of Total DNA from Infected Plants to be used as Inoculum

3.1.1 Phenol–Chloroform Based Method of Plant Genomic DNA Extraction

This protocol can be used for a large variety of plant species yielding a higher quality DNA sample with a final concentration sufficiently high for microprojectile bombardment and/or Southern blots.

1. Harvest young leaves (500 mg) from infected plants and immediately freeze them in liquid nitrogen.
2. Grind them to a fine powder in liquid nitrogen with mortar and pestle.
3. Transfer to a microtube and add 500 μL of lysis buffer and suspend the frozen powdered leaf tissue.
4. Incubate for 15 min at room temperature.
5. Add 250 μL of Tris-buffered phenol (pH 5.6–8.0). Vortex vigorously for 1 min. Let it sit for 2 min at room temperature.
6. Add 250 μL of chloroform. Vortex briefly and centrifuge the mixture at 16,000g for 2 min.
7. Transfer the supernatant to a clean, properly labeled microtube.
8. Repeat phenol/chloroform extraction.
9. Add 500 μL of chloroform. Vortex briefly.
10. Centrifuge the mixture at 16,000g for 2 min.
11. Transfer the supernatant to a clean, properly labeled microtube.
12. Precipitate the nucleic acids by adding 500 μL of isopropanol. Incubate for 5 min at room temperature.
13. Centrifuge at 16,000g for 5 min and discard the supernatant using pipette.
14. Add 500 μL of 70% (v/v) ethanol. Vortex briefly and centrifuge at 16,000g for 2 min.
15. Invert the microtube on absorbent paper and allow the pellet to air-dry for about 2–5 min.
16. Resuspend the pellet into 200 μL of TNE buffer, containing 0.1 mg mL^{-1} RNase.
17. Incubate 30 min at 37°C.
18. Extract contaminant proteins by adding equal volume of buffered phenol: chloroform (1:1). Vortex well.
19. Centrifuge at 16,000g for 7 min.
20. Transfer the aqueous phase to a clean tube and add equal volume of chloroform.
21. Vortex briefly.
22. Centrifuge at 16,000g for 2 min.
23. Transfer the aqueous phase to a clean microtube.
24. Add to the aqueous phase 2 volumes of absolute ethanol and mix gently.
25. Incubate at room temperature for 20 min or at −20°C for 10 min. A white DNA precipitated is visible at this stage.
26. Pellet the DNA at 16,000g for 5 min. Discard the supernatant and invert the tube on absorbent paper to remove excess of ethanol.
27. Wash the pellet with 500 μL of 70% (v/v) ethanol. Centrifuge and dry the pellet.
28. Resuspend the pellet in 30–50 μL of sterile H$_2$O and store at −20°C.

3.1.2 Buffering Phenol (see note 1)

1. Melt the phenol in H$_2$O bath in 1 L beaker at 45°C with loosened lid. It takes 1–2 h to melt.
2. Test melted phenol for degradation. Use a Pasteur pipette to draw up some phenol, if colorless, the phenol is okay to continue.

3. Add 0.5 g L^{-1} 8-OH quinoline (0.25 g per 500 mL phenol) (see note 2).
4. Work in fume hood. Add 100–200 mL TE$_{50:2}$, cap and shake vigorously with magnetic bar. Let the phases separate and discard the aqueous phase into a phenol-only waste bottle.
5. Repeat Step 4 twice.
6. With an indicator paper measure the pH of the aqueous phase, if acidic repeat above.
7. If neutral, add 100 mL TE$_{50:2}$.
8. Store at 4°C under TE.

3.1.3 Determination of DNA Concentration

1. For cloned bipartite geminivirus, separate three clean and properly labeled microtubes. Add to each one 499 µL of water.
2. Add to the control tube 1 µL of water.
3. Add to the other tubes 1 µL of the plant DNA extraction or 1 µL of the respective plasmid DNA extractions (viral DNA-A or DNA-B).
4. Calibrate the spectrophotometer for 260 and 280 nm readings and adjust the absorbance to zero for the control H$_2$O.
5. Replace the water in the cuvette with the first DNA sample.
6. Take the absorbance at 260 and 280 nm.
7. Repeat the Steps 6 and 7 for the other samples.
8. Evaluate the homogeneity of the preparations by the A$_{260}$/A$_{280}$ ratio. The quality of your purification is considered good if the observed A$_{260}$/A$_{280}$ ratio is between 1.8 and 2.0.
9. Calculate DNA concentration using the equation

$$C = \frac{A_{260} \times D \times E}{1.000},$$

where C = concentration (µg µL^{-1}), A_{260} = Absorbance at 260nm, D = dilution factor, E = Extinction coefficient: dsDNA = 50, ssDNA = 32, ssRNA = 40.

3.2 Biolistic Inoculation

3.2.1 Preparation of Tungsten Particles

1. Weigh 60 mg of tungsten M10 particles (0.7-µm diameter, BIORAD) or 1.0-µm gold particles (BIORAD) in a microcentritfuge tube and add 1 mL of 70% (v/v) ethanol. Vortex vigorously and keep under constant agitation for 15 min with the vortex.
2. Centrifuge at 15,000g for 5 min.
3. Remove the supernatant by pipetting carefully. Pay attention not to suck the particles that will not be adhered to the tube.

4. Add 1 mL of sterile distilled H_2O and vortex vigorously. Centrifuge at 7,500g for 8 min.
5. Repeat Step 4 twice
6. Suspend the particles in 1 mL of 50% (v/v) glycerol.
7. Store the labeled microtube at room temperature.
8. Every time pipetting particles from the tube, vortex the tube well.

3.2.2 DNA Precipitation on the Tungsten or Gold Microparticles

1. Transfer 50 μL of the microparticles suspension to a microtube and homogenize well the suspension by pipetting up and down.
2. Add 5–10 μL of DNA-A and DNA-B (concentration 1 mg mL^{-1}) or 25 μg of total DNA isolated from infected plants, homogenize well by pipetting.
3. Add 50 μL of 2.5 M $CaCl_2$ and homogenize quickly.
4. Add 20 μL of 0.1 M spermidine and homogenize quickly (see note 3).
5. Incubate at room temperature for 10 min under gentle agitation.
6. Centrifuge for 20 s and remove the supernatant carefully.
7. Resuspend the pellet in 150 μL of 100% (v/v) ethanol. Shake the tubes gently, centrifuge at 14,000g for 10 s and remove the supernatant.
8. Repeat Step 7.
9. Add 24 μL of 100% (v/v) ethanol. Homogenize vigorously.
10. Sonicate 1–3 s and spread the DNA-coated gold (tungsten) particles onto six macrocarrier discs (see note 4).
11. Set the macrocarrier discs in the holders, still wet with 70% (v/v) ethanol, and dry them. Place the macrocarriers in a vacuum dessiccator or let the particles to be air-dried on the bench.
12. Wash the metal mesh in 70% (v/v) ethanol and autoclave.

3.2.3 Microprojectile Bombardment

1. Turn on the P DS-1000/He accelerator (gene gun).
2. Open the main valve of the He tank completely. Before this, check if the other system valves are closed.
3. Open the second valve by turning it clockwise until the pressure gauge of the first chamber hits 450 psi for inoculation of *A. thaliana*, 900 psi for tomato, and 900 psi for *Nicotiana benthamiana*.
4. Set a rupture disk (wash in isopropanol) using the torque wrench. Put a metal mesh and a macrocarrier holder with a macrocarrier (see note 5).
5. Place the plants of *A. thaliana* at the seven-leaf developmental stage in the chamber, align the plants with the center of the macrocarrier holder, and keep an average distance of 3 cm between the macrocarrier and the plant to avoid destruction of the vegetal tissue. Tomato is often bombarded at the six-leaf stage, *N. benthamiana* at the four-leaf stage.

6. Close the door and start the vacuum by turning on the vacuum pump. When the gauge hits 20 in. Hg for *Arabidopsis* or 25 in. Hg for tomato and *N. benthamina*, push the fire switch (see note 6).
7. Check the pressure gauge in the chamber and release it, if necessary.
8. Release the vacuum immediately, but very carefully, after the particles are fired. When the vacuum is totally released, open the door. Remove the plant from the chamber.
9. Discard the used rupture disk, metal mesh, and macrocarrier.
10. Clean up the shooting chamber with 70% (v/v) ethanol.
11. Repeat Steps 4–10 for the next microprojectile bombardment (see note 7).
12. Close the main valve of the He tank. Close the door and start vacuum. Push the fire switch to release the pressure of the first chamber.
13. Close the second valve of the tank.
14. Release the vacuum and open the door. Clean up the chamber and store the accessories of the biolistic system in ethanol.
15. Turn off the gun.
16. Maintain the inoculated plants in growth chamber under appropriate conditions. In the case of tomato and *N. benthamiana*, after 48-h postinoculation, transfer the bombarded plants to large pots.

3.3 Polymerase Chain Reaction-Based Diagnostic Assay

3.3.1 Quick Protocol of Total Plant DNA Extraction for PCR: Protocol I

1. Harvest a young leaf (as low as 20 mg) of the host plant and immediately freeze in liquid nitrogen
2. Add 200 µL of extraction buffer and suspend the frozen leaf tissue with a plastic pistil
3. Leave on ice for 10 min at least.
4. Vortex for 5 s
5. Centrifuge twice at 14,000g for 5 min at room temperature
6. Transfer 100 µL of the supernatant into a microtube containing 100 µL of isopropanol (same volume) to precipitate the nucleic acids
7. Vortex briefly
8. Incubate at room temperature for 5–10 min
9. Centrifuge at 14,000g for 5 min at room temperature
10. Carefully remove the supernatant using a pipette (see note 8)
11. Wash the pellet by adding 100 µL 75% (v/v) ethanol and vortex briefly
12. Centrifuge at 14,000g for 5 min at room temperature
13. Discard the supernatant as Step 10
14. Repeat Steps 11–13
15. Invert the microtube on an absorbent paper and allow the pellet to air-dry for about 10 min at room temperature or vacuum-dry the pellet for 5 min

16. Suspend the pellet in 50 μL of sterile H_2O
17. Vortex briefly and leave for 30 min at room temperature to dissolve the pellet completely
18. Storage at −20°C (see note 9)

3.3.2 Plant Genomic DNA Extraction for PCR: Protocol II [adapted from Dellaporta et al. (25)]

1. Homogenize freshly harvested leaf tissue (as low as 50 mg) with 500 μL of extraction buffer in a microtube using a plastic pistil
2. Add 33 μL of 20% (w/v) SDS, vortex for 2 min and incubate at 65°C for 10 min
3. Add 160 μL of 5 M potassium acetate, vortex for 2 min, and spin for 10 min in microcentrifuge
4. Remove 450 μL of supernatant to a clean microtube avoiding the tissue debris. Repeat centrifugation if any debris is left in the supernatant
5. Add 0.5 volume of isopropanol, vortex, and centrifuge for 10 min at 14,000 rpm in microcentrifuge
6. Carefully remove with a pipette and discard the supernatant (see note 8)
7. Add to the pellet 500 μL of 70% (v/v) ethanol, vortex, centrifuge for 5 min, and carefully remove as much as supernatant as possible with a pipette
8. Vaccum-dry the pellet for 5 min
9. Resuspend the pellet in 100 μL of sterile H_2O or TE buffer. Use 1–5 μL for PCR

3.3.3 Polymerase Chain Reaction (PCR)

1. Add the following reagents to a microtube (0.2–0.5 mL): 3 μL of DNA, 5 μL of 10X reaction buffer, 5 μL of 25 mM $MgCl_2$, 1 μL of 10 mM dNTPs, 0.5 μL of each 10 μM primer solution (either the begomoviruses DNA-A or DNA-B degenerate primers or CP-flanking annealing primers, Table 2), 1 μL of Taq polymerase, and 34.5 μL of Milli-Q H_2O (final volume: 50 μL)
2. Keep the mixture on ice until taking it to the thermocycler
3. Program the thermocycler for the following conditions of cycle: Initial step at 94°C for 3 min, 30 cycles with the following parameters: denaturation at 94°C for 1 min, annealing at 50°C for 2 min, DNA extension at 72°C for 1 min per kb synthesized, an additional step at 72°C for 10 min to ensure that the nascent, amplified DNA fragments are blunted end at 5′ and 3′ ends.

3.3.4 Eletrophoresis in Agarose Gel

1. Heat and dissolve 0.8 g of agarose in 100 mL 0.5X TBE buffer
2. Add the 0.8% agarose solution into the electrophoresis apparatus cube using the proper combs to make the wells. Allow the agarose to solidify

3. Transfer 5 or 10 μL of each PCR reaction to a clean microtube
4. Add 1 or 2 μL of 6X loading sample buffer
5. After gel solidification, immerse the gel in 1X TBE within the electrophoresis apparatus
6. Load the samples into the wells
7. Connect the apparatus to the power supply and separate the amplified DNA fragments applying $15\,V\,cm^{-1}$
8. After separation, stain the agarose gel with ethidium bromide ($5\,\mu g\,mL^{-1}$) for 10 min. Alternatively, the ethidium bromide ($1\,\mu g\,mL^{-1}$) can be directly added to the agarose gel buffer or to the running buffer ($5\,\mu g\,mL^{-1}$) and the DNA is stained during electrophoresis
9. Visualize the DNA under UV

4 Notes

1. Wear gloves, coat, and glasses and work in fume hood. A solution of $TE_{50:2}$ 50 mM Tris–HCl (pH 7.2–8.0) and 2 mM EDTA/NaOH (pH 7.2–8.0) is needed. Use sterile stocks and sterile H_2O (800 mL for bottle of phenol). Buffer the phenol in its original jar (500 g). For simplicity, assume $1\,g\,mL^{-1}$ (500 mL phenol/bottle). Buffered phenol is available commercially and ready to use.
2. 8-OH quinoline is a nasty compound, do not breathe the dust (the concentration is not critical).
3. Spermidine solution is stored at 4°C in the dark.
4. Each DNA precipitation is for six macrocarrier discs, 2–3 μL aliquots/macrocarrier disc.
5. The distance between the rupture disk and the macrocarrier holder should be 1 cm.
6. At this moment, the rupture disk will be broken and the particles will be fired against the plant.
7. The chamber must be totally cleaned when different genomic components or different viral particles are used.
8. Beware that large nucleic acid pellets can be dislodged and be aspirated. To avoid this, it may be necessary to leave some (15–20 μL) of the supernatant behind.
9. This protocol yields DNA at low concentration and, thus, it is hardly visualized in ethidium bromide stained agarose gels. For PCR, use 1–3 μL per reaction.
10. In general, both Sida and tomato-infecting begomoviruses that have been found in the Brazilian territory are not sap transmissible. To overcome this problem, we have used the biolistic method to successfully propagate unknown and uncloned geminiviruses in permissive hosts by microprojectile bombardment with total DNA extracted from symptomatic *Sida rhombifolia* or tomato plants. This allows us to directly investigate the biological properties of the virus without having to clone it. In at least two cases, we found the symptoms to be the result of multiple infections, as the virus complex, which was obtained from the DNA of a single infected plant, segregated into distinct virus and developed different symptoms in bombarded *N. benthamiana* plants. The subsequent cloning and sequencing of the viral DNA-A's isolated from infected *N. benthamiana* confirmed the identity of the viruses. This was the case, for example, of SiMoV-[BR] (Sida mottle virus, accession number AY090555) and SYMV-[BR] (Sida yellow mosaic virus, accession number AY090558).
11. The biolistic inoculation of the permissive hosts *A. thaliana, N. benthamiana* and *Solanum lycopersicum* using our total DNA preparation from infected plants as the inoculum often results in 50% efficiency of infection. When infectious cloned viral DNAs are used as the inocula the efficiency of infection raises up to 100% for *Arabidopsis*, 90% for *N. benthamiana*, and 85% for tomato.

12. The Southern blot is also an efficient method for geminiviruses detection. In this case, we use the phenol–chloroform based method (Sect. 3.1.1) to obtain the total DNA from symptomatic plants and we follow exactly the protocols described in Sambrook et al. (20) for preparation of the probe, electrophoresis of total DNA, transfer to nylon membranes, and hybridization. The probes are DNA restricted fragments from cloned viral DNA-A and/or DNA-B. If DNA-A and/or DNA-B-specific probes are required, the common region must be absent (11, 21, 22). For begomovirus general probes, the coat protein coding region is the best choice, as it corresponds to the most conserved region among begomoviruses.

13. Agroinoculation has also proved to be an efficient method for begomovirus inoculation. The protocol is based on *Agrobacterium tumefaciens* mediated viral DNA inoculation. Thus, it absolutely requires that 1½ copies of DNA-A and DNA-B are cloned into binary vector for plant transformation, in case of bipartite begomoviruses. This precludes the use of total DNA from infected plants as the inoculum.

References

1. Rojas, M. R., Hagenm, C., Lucas, W. J., and Gilbertson, R. L. (2005) Exploiting chinks in the plant's armor: evolution and emergence of geminiviruses. *Annu. Rev. Phytopathol.* **43**, 361–394.

2. Hanley-Bowdoin, L., Settlage, S. B., Orozco., B. M., Nagar, S., and Robertson, D. (1999) Geminiviruses: models for plant DNA replication, transcription, and cell cycle regulation. *Crit. Rev. Plant Sci.* **18**, 71–106.

3. Kheyr-Pour, A., Bendahmane, M., Matzeit, V., Accotto, G. P., Crespi, S., and Groenenborn, B. (1991) Tomato yellow leaf curl virus from Sardinia is a whitefly-transmitted monopartite geminivirus. *Nucleic Acids Res.* **19**, 6763–6769.

4. Navot, N., Pichersky, E., Zeidan, M., Zamir, D., and Czosnek, H. (1991) Tomato yellow leaf curl virus: a whitefly-transmitted geminivirus with a single genomic component. *Virology* **185**, 151–161.

5. Dry, I. B., Rigden, J. E., Krake, L. R., Mullineaux, P. M., and Rezaian, M. A. (1993) Nucleotide sequence and genome organization of Tomato leaf curl geminivirus. *J. Gen. Virol.* **74**, 147–151.

6. Fondong, V. N., Pita, J. S., Rey, M. E. C., de Kochko, A., Beachy, R. N., and Fauquet, C. M. (2000) Evidence of synergism between African cassava mosaic virus and a new double-recombinant geminivirus infecting cassava in Cameroon. *J. Gen. Virol.* **81**, 287–297.

7. Galvão, R. M., Mariano, A. C., Luz, D. F., Alfenas, P. F., Andrade, E. C., Zerbini, F. M., Almeida, M. R., and Fontes, E. P. B. (2003) A naturally occurring recombinant DNA-A of a typical bipartite begomovirus does not require the cognate DNA-B to infect *Nicotiana benthamiana* systemically. *J. Gen. Virol.* **84**, 715–726.

8. Zhou, X., Liu, Y., Calvert, L., Munoz, C., Otim-Nape, G. W., Robinson, D. J., and Harrison, B. D. (1997) Evidence that DNA-A of a geminivirus associated with severe cassava mosaic disease in Uganda has arisen by interspecific recombination. *J. Gen. Virol.* **78**, 2101–2111.

9. Hanley-Bowdoin, L., Settlage, S. B., and Robertson, D. (2004) Reprogramming plant gene expression: a prerequisite to geminivirus DNA replication. *Mol. Plant Pathol.* **5**, 149–156.

10. Fontes, E. P. B., Luckowand, V. A., and Hanley-Bowdoin, L. (1992) A geminivirus replication protein is a sequence-specific DNA binding protein. *Plant Cell* **4**, 597–608.

11. Fontes, E. P. B., Eagle, P. A., Sipe, P. S., Luckow, V. A., and Hanley-Bowdoin, L. (1994) Interaction between a geminivirus replication protein and origin DNA is essential for viral replication. *J. Biol. Chem.* **269**, 8459–8465.

12. Fontes, E. P. B., Santos, A. A., Luz, D. F., Waclawovsky, A. J., and Chory, J. (2004) The geminivirus nuclear shuttle protein is a virulence factor that suppresses transmembrane receptor kinase activity. *Genes Dev.* **18**, 2545–2556.

13. Rojas, M. R., Gilbertson, R. L., Russel, D. R., and Maxwell, D. P. (1993) Use of degenerate primers in the polymerase chain reaction to detect whitefly-transmitted geminiviruses. *Plant Dis.* **77**, 340–347.
14. Briddon, R. W., Prescott, A. G., Luness, P., Chamberlin, L. C. L., and Markhan, P. G. (1993) Rapid production of full-length infectious geminivirus clones by abutting primer PCR (AbP-PCR). *J. Virol. Methods* **43**, 7–20.
15. Patel, V. P., Rojas, M. R., Paplomatas, E. J., and Gilbertson, R. L. (1993) Cloning biologically active geminivirus DNA using PCR and partially overlapping primers. *Nucleic Acids Res.* **21**, 1325–1326.
16. Girish, K. R. and Usha, R. (2005) Molecular characterization of two soybean-infecting begomoviruses from India and evidence for recombination among legume-infecting begomoviruses from South-East Asia. *Virus Res.* **108**, 167–176.
17. Rochester, D. E., Kositratana, W., and Beachy, R. N. (1990) Systemic movement and symptom production following agroinoculation with a single DNA of tomato yellow leaf curl geminivirus (Thailand). *Virology* **178**, 520–526.
18. Sunter, G., Hartitz, M. D., Hormuzdi, S. G., Brough, C. L., and Bisaro, D. M. (1990) Genetic analysis of Tomato golden mosaic virus: ORF AL2 is required for coat protein accumulation while ORF AL3 is necessary for efficient DNA replication. *Virology* **179**, 69–77.
19. Schaffer, R. L., Miller, C. G., and Petty, I. T. D. (1995) Virus and host-specific adaptations in the BL1 and BR1 genes of bipartite geminiviruses. *Virology* **214**, 330–338.
20. Sambrook, J., Fritsch, E. F., and Maniatis, T. (1989) *Molecular Cloning: A laboratory manual.* Cold Spring Harbor, NY: Cold Spring Harbor Laboratory Press.
21. Fontes, E. P. B., Gladfelter, H. J., Schaffer, R. L., Petty, I. T. D., and Hanley-Bowdoin, L. (1994) Geminivirus replication origins have a modular organization. *Plant Cell* **6**, 405–416.
22. Gladfelter, H. J., Eagle, P. A., Fontes, E. P. B., and Hanley-Bowdoin, L. (1997) Two domains of the AL1 protein mediate geminivirus origin recognition. *Virology* **239**, 186–197.
23. Rybicki, E. P. and Hughes, F. L. (1990) Detection and typing of Maize streak virus and other distantly related viruses of grasses by polymerase chain reaction amplification of a conserved viral sequence. *J. Gen. Virol.* **71**, 2519–2526.
24. Hartley, J. L. and Rashtchian, A. (1993) Dealing with contamination: enzymatic control of carryover contamination in PCR. *PCR Methods Appl.* **3**(2), S10–S14.
25. Dellaporta, S. L., Wood, J., and Hicks, J. B. (1983) A plant DNA minipreparation: version II. *Plant Mol. Biol. Rep.* **1**, 19–21.

Section 5
Genomics, Host Factors and Plant Based Studies

Chapter 40
Expression Microarrays in Plant–Virus Interaction

Kristina Gruden, Maruša Pompe-Novak, Špela Baebler, Hana Krečič-Stres, Nataša Toplak, Matjaž Hren, Polona Kogovšek, Lisa Gow, Gary D. Foster, Neil Boonham, and Maja Ravnikar

Since their conception in the late 1990s, microarray techniques have become a tool of choice for monitoring pangenomic gene expression. Although there are a large number of variations on the basic methodology the general approach remains standard and involves the comparison of a "test" RNA with a "control" RNA; in this case "healthy" and "virus-infected" plants. The protocol itself can be broken down into five main parts: RNA extraction, cDNA synthesis, hybridization, array scanning, and data analysis. The method presented is optimized for use with arrays based on glass slides spotted with cDNA, in this case 15,264 cDNAs from *Solanum tuberosum*. The labeling technique presented involves two steps: hybridization of cDNA produced using oligo-dT linker primers to the array and hybridization with a DNA dendrimer reagent comprising sequence complementary to the linker sequence bound to a fluorescent dye. We also present the use of the R environment for data analysis, generating statistical support for differential gene expression observed.

Keywords Virus infection; *Solanum tuberosum*; RNA extraction; Dendrimer labeling; cDNA microarray; Hybridization; Image analysis; Data analysis; R programming environment

1 Introduction

Plant responses to plant pathogens are complex, involving a range of signaling pathways (1) and show a broad spectrum of physiological and histological changes. Studying single components of the response in isolation can lead to limited conclusions or results, which fail to take into account the complex interactions between the different pathways of the response. Omics technologies are a major step forward in understanding plant–pathogen interactions as they offer a more holistic view of the processes involved. Expression microarrays are currently the most established technique for studying the transcriptome. The majority of transcriptomics work in the field of plant–pathogen interactions has been carried out on plant–bacterial and plant–fungal interactions, while plant responses to viruses are less studied and thus less well understood. The first microarray study on plant–virus interactions was published in 2003, reporting alteration in the gene expression profile of *Arabidopsis thaliana* after infection with *Tobacco Mosaic Virus* (2). In later research, responses to other viruses in *Arabidopsis* (3–5), as well as in poplar (6), maize (7), *Nicotiana benthamiana* (8), and potato (9) were monitored. These results still comprise only separate elements of processes in the hosts after pathogen attack. Still further functional studies on a wide range of hosts and viruses, in combination with other omics approaches that can lead to identification of general features and complete response cascades in plant–virus interactions, are needed.

Several platforms are currently available in the area of DNA microarray technology. They differ in the mode of preparation, type of solid support used and, most importantly, in the type of DNA printed. Briefly, some technologies are based on the use of short (25 nt) oligonucleotides (Affimetrix), and others on 50–70 mer oligonucleotides (Agilent, Qiagen). The third possibility is the printing of longer PCR products (10). In this chapter, the protocol presented is that optimized for cDNA arrays, specifically the potato arrays from The Institute for Genomic Research (TIGR) (http://www.tigr.org/tdb/potato/). In the section on bioinformatics (Sect. 3.9) we will introduce basic statistical approaches toward identifying differentially expressed genes. All further analysis is highly dependent on the individual experimental design and tools need to be adapted. R environment is widely considered to be the most suitable tool for use in microarray analysis, mainly due its high adaptability to the specificities of the experiment (11).

2 Materials

2.1 Sample and Total RNA Preparation

1. Inoculation buffer (20 mM sodium phosphate buffer, pH 7.6) (see note 1). Store at 4°C
2. RNeasy Plant Mini Kit (Qiagen). Store at room temperature
3. DNaseI, Amplification grade (Invitrogen). Store at −20°C
4. RNeasy MinElute Cleanup Kit (Qiagen). Store at room temperature
5. Absolute ethanol, molecular biology grade (Merck). Store at room temperature
6. 80% ethanol, molecular biology grade (see note 2). Store at room temperature
7. Agarose, molecular biology grade (Sigma). Store at room temperature
8. Loading dye solution (200 μL of 6X Loading Dye (Fermentas), 500 μL of glycerol (molecular grade), and 500 μL of nuclease-free water), stored at 4°C
9. 100-bp DNA ladder (Fermentas). Store at 4°C
10. 1X TAE buffer for agarose gel preparation: 40 mM Tris–acetate, 1 mM EDTA, pH 8.0. Store at room temperature
11. Ethidium bromide (1 μg μL^{-1}) (Sigma) (see note 3). Store at room temperature

2.2   cDNA Synthesis

1. 3DNA Array 900 Kit (Genisphere). Some components are light sensitive. Store at −20°C
2. Luciferase Control RNA (20 mg mL⁻¹) (Promega). Store in aliquots at −80°C
3. Superscript II Reverse Transcriptase (Invitrogen). Store at −20°C
4. 0.5 M NaOH/50 mM EDTA. Store in aliquots at 4°C for up to 6 months
5. 1 M Tris–HCl (pH = 7.5). Store at 4°C for up to 6 months
6. 10 mM Tris (pH = 8)/1 mM EDTA. Store at 4°C for up to 6 months
7. Microcon YM-30 Centrifugal Filter Devices (Millipore). Store at room temperature
8. 1X TE Buffer. Store at 4°C for up to 6 months

2.3   Microarray Prehybridization, cDNA, and 3DNA Hybridizations

1. TIGR Potato cDNA Arrays (http://www.tigr.org/tdb/potato/)
2. Prehybridization solution: 30% BSA (Sigma), 20X SSC, 10% SDS. Store at 4°C for up to 14 days
3. 3DNA Array 900 Kit (Genisphere). Store at −20°C
4. Salmon Testis DNA (Sigma). Store in aliquots at −20°C
5. LifterSlips (25x601-2-4789) (Erie Scientific Co.)
6. 0.1 M Dithiotreitol (DTT) (Sigma) solution. Store in aliquots at −20°C for up to 1 month
7. Post cDNA hybridization wash – first washing solution: 1X SSC, 0.1% SDS. Prepare fresh at room temperature
8. Post cDNA hybridization wash – second washing solution: 0.1X SSC, 0.1% SDS. Prepare fresh at room temperature
9. Post cDNA hybridization wash – third washing solution: 0.1X SSC. Prepare fresh at room temperature
10. Post 3DNA hybridization wash – first washing solution: 1X SSC, 0.1% SDS, 0.1 M DTT. Prepare fresh at room temperature
11. Post 3DNA hybridization wash – second washing solution: 0.1X SSC, 0.1% SDS, 0.1 M DTT. Prepare fresh at room temperature
12. Post 3DNA hybridization wash – third washing solution: 0.1X SSC. Prepare fresh at room temperature

2.4   Microarray Scanning, Image, and Data Analysis

1. High-resolution scanner (pixel size 5 or 10 μm), capable of producing quality images of scanned microarrays (e.g., LS200, TECAN)
2. Image-analysis software (e.g., ArrayPro Analyzer®, Media Cybernetics or other)

3. R software (language and environment for statistical computing and graphics, available free at http://www.r-project.org)

3 Methods

An overview of the methods used is shown in Fig. 1. In the described experiments, gene expression in virus–inoculated plants and mock-inoculated plants are compared. Plants are mechanically inoculated with sap of either healthy or virus infected plants. Plant material is collected at appropriate time after inoculation and snap frozen in liquid nitrogen. After RNA isolation, the residual genomic DNA in RNA samples is degraded. RNA concentrations are equilibrated between virus and mock-inoculated samples prior to cDNA synthesis. In contrast to commonly used direct and indirect labeling for microarray probe preparation, dendrimer labeling (12) is used in this case. cDNA synthesis is carried out with oligo-dT primers that include a "capture sequence" that is complementary to a DNA dendrimer labeled with either Cy3 or Cy5. The hybridization is carried out in two steps. First the cDNA is hybridized to the array, followed by the fluorescent dendrimer reagent. The appropriate fluorescent reagent (Cy3 or Cy5) will bind to the cDNA molecules according to which "capture sequence" they contain.

After final washing and drying, microarrays are scanned, and images (one for Cy3 and one for Cy5 dye) are processed by an image analysis program. During this step, a grid is placed over the images, low-quality spots are excluded and signal and background values for each spot are exported.

Subsequent data manipulation and analysis is carried out in an environment for statistical computing and graphics R (see note 4). R is similar to the S system, which was developed at Bell Laboratories by John Chambers and coworkers. It provides a wide variety of statistical and graphical techniques (linear and nonlinear modeling, statistical tests, time series analysis, classification, clustering, etc.). In the experiments, expression data from each microarray is imported to R, where it is combined with spotted cDNA information. All data is background corrected and normalized to correct for systematic and spatial biases. The effect of normalization is monitored by various pre- and postnormalization plots. Before searching for differentially expressed genes, expression values of duplicate spots are averaged and genes that show little variation after virus infection are filtered out. Genes that are significantly differentially expressed in virus-infected plants are selected on the basis of linear modeling and use of empirical Bayes method. Finally and most importantly, biological relevance of the results should be critically assessed. It depends on experimental design, but in order to obtain relevant data usually several biological repetitions are necessary.

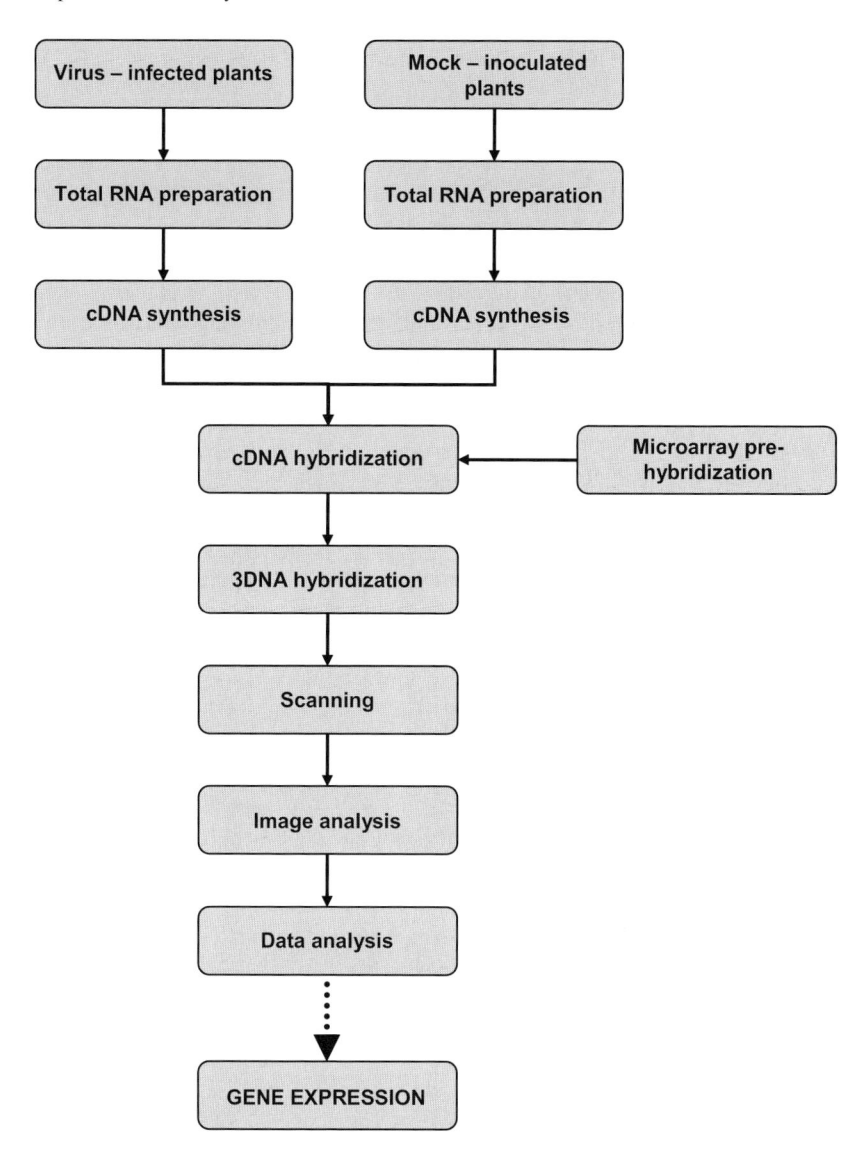

Fig. 1 Workflow of methods to study gene expression in plant–virus interaction by microarrays: plant material and total RNA preparation is followed by cDNA synthesis, microarray hybridization, and data analysis

3.1 Sample Preparation

3.1.1 Plant Material

1. Potato plants are propagated by stem-node segmentation. They are grown in modified MS medium and kept in a growing chamber at $19 \pm 2°C$ in the light and at $17 \pm 2°C$ in the dark, with 70–90 μmol m^{-2}s^{-1} radiation (Osram L36W/77 lamp) and a 16-h photoperiod.
2. After 2 weeks of cultivation, they are planted into soil and kept in a growing chamber for 4 weeks at $20 \pm 2°C$ in the light and at $18 \pm 1°C$ in the dark, at a relative humidity of $75 \pm 2\%$, with 120–150 μmol m^{-2}s^{-1} radiation (Osram L36W/77 lamp) and a 16-h photoperiod. Plants are watered every day with tap water.

3.1.2 Plant Inoculation

1. In each microarray experiment, the expression profile of mock and virus-inoculated potato leaves are compared. Inoculum is prepared from healthy and virus-infected potato plants, respectively; these plants are maintained in tissue culture.
2. To prepare inoculum, 1 g of plant material is ground thoroughly in 3 mL inoculation buffer (see note 1) in a pestle and mortar. The inoculum is incubated for 5 min at room temperature prior to use.
3. Potato plants to be inoculated are marked according to which inoculum they will receive (mock or virus). Lower leaves (3–4) from each plant are selected for inoculation and their petioles marked with permanent marker pen. Marked leaves are lightly dusted with carborundum powder and inoculum is gently spread over the leaf with a finger (see note 5).
4. Plants are incubated for 10 min, and afterward all inoculated leaves are rinsed with tap water. Plants are transferred back into the growth chamber where they are kept until harvesting (see note 6).

3.1.3 Sampling

1. Samples are collected 1-h postinoculation (see note 7).
2. Mock and virus-inoculated leaves are collected separately; leaves are torn off at the stem and immediately placed in liquid nitrogen (see note 7). Leaves from all plants in each inoculum group are pooled. The samples are ground to a fine powder in liquid nitrogen (see note 8) and approximately 300 mg of material is transferred to a 2-mL tube using a precooled spatula. Samples are kept in liquid nitrogen until subsequent analyses or stored at −80°C until further use.

3.2 Total RNA Preparation

3.2.1 Total RNA Isolation

Total RNA (totRNA) extraction from samples is carried out with an RNeasy Plant Mini kit (Qiagen) (see note 9). The kit is used according to the manufacturer's protocol (RNeasy Mini Handbook), with the following modifications:

1. All centrifugation steps are carried out at 18,000g
2. The addition of mercaptoethanol to the extraction buffer (RLT) is omitted without any loss in quantity or quality of yield.
3. Buffer RLT is preheated to 65°C prior to use
4. The amount of plant material for one extraction is increased to 300 mg. For this reason, the amount of extraction buffer (RLT) is increased to 800 µL. The larger samples volumes exceed the capacity of the columns provided in the kit. Liquid is therefore applied to the column in two aliquots with a short spin between each addition.
5. Optional step 9a in the protocol, to eliminate possible EtOH carryover is carried out
6. RNase-free water is preheated to 80°C prior to elution
7. Preheated RNase-free water (30 µL) is pipetted directly onto the RNeasy membrane (see notes 11 and 12). The column is incubated for 10 min at room temperature and then centrifuged for 2 min to elute the RNA (see note 13).

3.2.2 Genomic DNA Digestion

Following RNA extraction, RNA is treated with DNaseI (Amplification Grade, Invitrogen) to remove any contaminating genomic DNA. Genomic DNA can compete with cDNA in the microarray hybridization step and produce false results. The manufacturer's protocol is modified to reduce the amount of DNase I enzyme, without affecting efficiency of the reaction.

1. To 30 µL of eluted totRNA (approximate concentration 1.3 µg µL^{-1}), 4 µL of buffer, 4.7 µL of nuclease-free water, and 1.3 µL of DNase I enzyme are added.
2. The solution is gently mixed by pipetting, spun down, and incubated at room temperature for 15 min.
3. The reaction is stopped by adding 4 µL of EDTA (a component from the kit) to the sample and incubating at 65°C for 10 min.
4. Proceed immediately with totRNA cleanup.

3.2.3 Total RNA Cleanup

Following treatment with DNase, the RNA is concentrated and further purified to remove contaminating enzymes and buffers with an RNeasy MinElute Cleanup Kit

(Qiagen). The kit is used according to manufacturer's instructions with the following modifications:

1. All centrifugation steps are carried out at 18,000g
2. RNase-free water is preheated to 80°C
3. Preheated RNase-free water (14 µL) is pipetted directly onto the centre of the membrane (see note 11). The column is incubated for 10 min at room temperature (see note 12) and then centrifuged for 2 min to elute the RNA (see note 16).

3.2.4 Total RNA Equilibration

In order to get a relevant hybridization result, it is necessary to hybridize cDNA that corresponds to equal amounts of totRNA from the virus infected and control plants. RNA amounts are equalized by comparing the intensity of the bands from electroferograms following agarose gel electrophoresis (see note 17). Before preparing the gel, it is strongly recommended to rinse the cell base, gel tray, and comb (see note 18) with solutions in following order: 0.1 M NaOH, distilled water, 3% H_2O_2, and bi-distilled water to prevent degradation of RNA.

RNA samples are visualized on a 5-mm thick, 1.5% agarose gel. RNA (1 µL) is diluted 10 times in distilled water, 1 µL diluted RNA is added to 6 µL loading dye and 5 µL RNase-free water (see note 19). 100-bp DNA ladder (0.7 µL) is added to 6 µL loading dye solution and 5 µL RNase-free water. The gel is run at 80 V for 20 min, using POWER/PAC 1000 (BIO-RAD) (see note 20).

A digital image of the gel (electropherograme) is made with GelDoc Mega system in combination with UVI Photo MW software (Biosystematica, USA) (see note 21). Relative totRNA concentration is estimated by visual observation of the image. If the concentrations of totRNA from the mock and virus-inoculated plants are unequal, it should be equalized by adjusting the totRNA volume prior to proceeding with cDNA synthesis.

3.3   cDNA Synthesis

3.3.1   Preparation of RNA-RT Primer Mix

RNA-RT primer mix is prepared on ice by combining 8 µL of totRNA (see note 22) with 1 µL of RT Primer (incorporating the linker corresponding to either Cy3 or Cy5) (see note 23) and 2 µL *luciferase* mRNA (0.5 ng µL⁻¹) (external control). RNA-RT primer mix is gently mixed, briefly centrifuged and incubated at 80°C for 10 min. Primer mix is put on ice for at least 2–3 min, centrifuged, and kept on ice.

3.3.2   Preparation of Reaction Master Mix

Reaction Master Mix is prepared on ice, together for both RNA samples (control and infected). SuperScript II First Strand Buffer (8 µL of 5X) (Invitrogen), 4 µL of 0.1 M DTT, 2 µL of 10 mM dNTP, 2 µL of Superase-In, and 2 µL of Superscript II RNase H⁻ (Invitrogen) are combined and gently mixed (not vortexed), centrifuged, and kept on ice until further use.

3.3.3   cDNA Synthesis

Reaction Master Mix (9 µL) is added to each RNA-RT primer mix sample, gently mixed, and incubated at 42°C for 120 min. The reaction is stopped by adding 3.5 µL of 0.5 M NaOH/50 mM EDTA. Samples are incubated at 65°C for 10 min to denature the DNA–RNA hybrids and degrade the rest of the RNA. The reaction is neutralized with 5 µL of 1 M Tris–HCl. Control and infected sample are combined into one mixture, which will be hybridized together to one microarray. Empty microtubes are rinsed with 73 µL of 10 mM Tris (pH = 8)/1 mM EDTA, which is added to previously combined mixture.

3.3.4   cDNA Concentration

cDNA is concentrated with Millipore Microcon YM-30 Centrifugal Filter Devices, used according to manufacturer's instructions (see notes 11 and 24). Following elution, the total volume of cDNA is made up to 10 µL with water; the cDNA is stored at −20°C.

3.4   *Microarray prehybridization*

A prehybridization step is recommended in the TIGR microarray protocol (13) for reducing some types of nonspecific binding, a common cause of high background (see note 25).

1. The microarray is dipped in prewarmed (42°C, 30 min) prehybridization solution in a Coplin jar and incubated at 42°C for 45 min.
2. After the incubation, the array is washed for 3 s in the first Coplin jar filled with distilled water, transferred to distilled water in the second Coplin jar for 3 s, and finally washed for 3 s in a Coplin jar containing 100% isopropanol.
3. The array is dried by centrifugation at 1,000g for 1 min in an array holder. If a smear or dust is observed on the array, it is recommended to repeat washing in isopropanol before drying. Before further use, the dried slide and hybridization chamber are preheated to 50–55°C.

3.5 cDNA Hybridization

3.5.1 Hybridization Mix

Before starting, all the chemicals and samples needed are prepared as follows:

1. Formamide-Based Hybridization Buffer (2X) is thawed and resuspended by heating to 70°C for at least 10 min, thoroughly mixed, and centrifuged for 1 min at maximum speed.
2. For additional reduction of none specific hybridization, a competitor nucleic acid is required. For this purpose 1 μL (1 μg μL^{-1}) of Salmon testis DNA is denaturated by incubation at 98°C for 10 min and then immediately put on ice until further use.
3. For dimer breakdown, cDNA needs to be incubated at 65°C for 10 min and then immediately put on ice until further use.
4. LNA dT Blocker is thawed at room temperature (see note 26).
5. Hybridization mix (80 μL) is prepared by combining 1 μL of denatured Salmon testis DNA, 40 μL of preheated Formamide-Based Hybridization Buffer (2X), 2 μL of LNA dT Blocker, 28 μL of dH$_2$O, and 9 μL of preheated cDNA. The hybridization mix is combined by gently tapping the tube and then briefly centrifuged at 14,000g to ensure the mix is bubble free. The tube is incubated at 75–80°C for 10 min, and then kept at 45°C until loaded onto the microarray.

3.5.2 Loading the Microarray

1. The prewarmed microarray, placed on a previously cleaned bench, is covered by a LifterSlip. The rubber strips of the LifterSlip should be in contact with the surface of the slide. If necessary, use compressed air for removing fibers and dust from the LifterSlip.
2. Prewarmed hybridization mix is spun down and immediately carefully pipetted onto the microarray along the edge of the LifterSlip (see note 27). Bubbles in the pipette tip should be avoided.
3. The prewarmed hybridization chamber is prepared for incubation by pipetting up to 50 μL of 1X SSC into the channels. The microarray is placed into the chamber and the lid is firmly closed. The chamber is incubated in a water bath at 45°C for at least 16 h (overnight).

3.5.3 Post cDNA Hybridization Wash

1. Approximately 800 mL of each washing solution (Sect. 2.3) is prepared in 2-L glass beakers.
2. The microarray is carefully taken out of the hybridization chamber and the LifterSlip is removed by washing the array in the first washing solution (see note 25).

3. The microarray is placed in an array holder and washed at room temperature in the first washing solution for 5 min (see note 28). The array holder is then transferred to the second and third washing solutions, for 5 min each.
4. The microarray is dried by immediate centrifugation at 1,000g for 2 min in an array holder (see note 29). Before further use, the dried slide and hybridization chamber are preheated on 50–55°C.

3.6 3DNA Hybridization

3.6.1 Hybridization Mix

Before starting, all the chemicals and samples needed are prepared as follows:

1. 3DNA Array 900 Capture reagents are thawed in the dark at room temperature for 20 min. To break up aggregates that may form during the freezing process (see note 30), capture reagents are thoroughly vortexed for 3 s and after 10 min of incubation at 55°C, again vortexed at maximum setting for 3–5 s and briefly spun. Until further use, Capture reagents are kept in dark at room temperature.
2. Formamide-Based Hybridization Buffer (2X) is heated to 70°C, vortexed, and centrifuged to thaw and resuspend.
3. Anti-Fade Reagent, used to reduce fading of fluorescent dyes posthybridization, is thawed and 1 μL is resuspended in 100 μL 2X Formamide-Based Hybridization Buffer (see note 31).
4. Salmon testis DNA (1 μL of 1 μg μL^{-1}) is incubated at 98°C for 10 min and immediately put on ice until further use.
5. To reduce degradation of fluorescent dyes, further steps are performed in darkness.

3DNA hybridization mix (80 μL) is prepared by adding 40 μL of 2X Formamide-Based Hybridization Buffer (with added Anti-Fade Reagent), 2.5 μL of Cy3 3DNA Capture reagent, 2.5 μL of Cy5 3DNA Capture reagent, and 34 μL of dH$_2$O to denatured Salmon testis DNA. Hybridization mix is mixed by gentle tapping the tube and briefly centrifuged at 14,000g to ensure that the mix is free from bubbles. The tube is incubated at 80°C for 10 min and held at 45°C until loading on the microarray.

3.6.2 Loading on the Microarray

1. The hybridization chamber and work bench are cleaned with 70% ethanol. The prewarmed microarray is covered by a LifterSlip, with the rubber strips facing down, in contact with the surface of the slide. If necessary, compressed air is used to remove fibers and dust from the LifterSlip.

2. The prewarmed hybridization mix is spun down and immediately carefully pipetted onto the microarray along the edge of the LifterSlip (see note 27). Bubbles in the pipette tip are avoided.
3. Up to 50 μL of 1X SSC is pipetted into the channels of the hybridization chamber. The microarray is placed inside and the lid is firmly closed. The chamber is incubated in a water bath at 45°C for at least 4 h.

3.6.3   Post 3DNA Hybridization Wash

Post 3DNA hybridization washes are performed in the dark to avoid photobleaching and fading of fluorescent dyes. To reduce fading of the Cy5 dye, fresh DTT is included in the first and the second washing solutions.

1. Approximately 800 mL of each washing solution (Sect. 2.3) is prepared in 2-L glass beakers. The second and third washing solutions are prewarmed to 50–60°C.
2. The microarray is carefully taken out of the hybridization chamber and the LifterSlip is removed by washing the array in the first washing solution (see note 25).
3. The microarray is placed in an array holder and washed at room temperature in the first washing solution for 5 min (see note 28). The array holder is then transferred to the second and third washing solutions, for 5 min each.
4. The microarray is dried by immediate centrifugation at 1,000g for 2 min in an array holder (see note 29).
5. The dried microarray is stored in a black box to prevent entry of light and kept at room temperature until scanning. Scanning is carried out as soon as possible as the fluorescent dyes fade over time (see note 32).

3.7   Microarray Scanning

The protocol for scanning covers general steps since each scanning platform has its individual specificities. Microarrays can be scanned with any scanner that is capable of scanning at two channels. Any image-analysis software able to read signal and background intensities can be used. The use of software that has additional options of identifying, tagging, or filtering of spots based on personalized or default criteria (such as simple statistical calculations like signal to noise ratio) is recommended. This protocol for scanning and image analysis describes steps that are performed using LS200 scanner (TECAN) and ArrayPro Analyzer® software (Media Cybernetics) for two-color microarray experiments.

Excitation light induces fluorescence: Cy3 emits light in the green part of the visible spectrum (573–613 nm) and Cy5 emits light in the red part of the visible spectrum (672–712 nm), the emitted light is then detected by the scanner. Two images per microarray are obtained, one for each channel. Scanning parameters (PMT gain, oversampling (lines to average), resolution, etc.) are set in such a way that the brightest spots on the resulting image are not saturated (this is tested during a prescan). Note that some scanners are able to scan both channels at the same time and produce one composite image (see notes 33 and 34).

3.8   Image Analysis

3.8.1   Flipping the Image

Depending on the orientation of the slide during the scanning, an image rotation or flip might be needed (many microarrays have distinct spot pattern therefore making it easy to check the correct orientation of the image). The images are saved in the correct orientation.

3.8.2   Grid-Finding

Both images from one microarray slide are opened with image-analysis software and a grid is created on both (either loading a pre-existing grid template or starting a new one). The grid defines the spot area (feature) from which the signal is extracted by the image-analysis software. Creation of grids requires the slide layout information (number of columns and rows of blocks and the same for number of columns and rows of spots in each block). Slide layout is read by some image-analysis software directly from the GAL (gene array list) file or is entered manually. Make sure that the grid is overlaid perfectly over the spot pattern on the image and that no dust particles are recognized as a spot (if necessary, manually check and align the grid).

3.8.3   Loading the GAL File

The GAL file is loaded, where, beside the previously mentioned slide layout, the information on the spotted cDNA (clone name, putative annotation, validation information, sequence etc.) is assigned to every spot. GAL files can be stored as tab delimited .txt files.

3.8.4   Data Extraction

The basic data that is extracted from the image is

1. Signal (raw intensity): for example, trimmed mean signal (3% of the most and the least bright pixels in each spot are discarded and the mean is calculated

Fig. 2 Schematic representation of local corners background. This option calculates background values from the area between the cells outside of the ellipse inscribed within a bounding rectangular grid (black area on the picture)

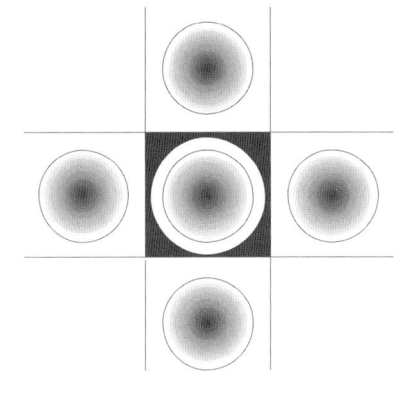

from the rest). Variable grid size can be used although this should not affect the signal data.

2. Background: because the background is usually not uniform throughout the image (at least small variations), background correction is advised (e.g., local background, local corners, see Fig. 2).
3. Normalization: no normalization is performed at the image analysis step. It is performed later (see Sect. 3.9.5).
4. Data extraction is automatically done for both channels (images) (see note 35).

3.8.5   Quality Control

First quality control is performed in image-analysis software. Spots that fail certain conditions are flagged (weighted) and excluded from further analysis. These conditions are

1. Nonvalidated spots (data from the microarray manufacturer)
2. Missing spots on the microarray
3. Uneven spots (smeared, doughnut shaped, etc.). A statistical approach is used, although the usage of morphological parameters is also possible. A spot is uneven if

$$\frac{signal(Cy3)}{STDEV(signal(Cy3))} < 1, \tag{1}$$

$$\frac{signal(Cy5)}{STDEV(signal(Cy5))} < 1. \tag{2}$$

This means that the standard deviation of signal in the spot area is high due to donut-shaped spots or light or dark dust particles that landed on the spot.

4. Uneven background (stains in background)

$$\frac{signal(Cy3)}{STDEV\,(background(Cy3))} < 3,$$ (3)

$$\frac{signal(Cy5)}{STDEV\,(background(Cy5))} < 3.$$ (4)

This is also known as signal-to-noise ratio. Spots that have uneven background (dust particles, shifted neighbor spot) that might affect the signal are selected and flagged.

5. Low-intensity signal

$$\frac{signal(Cy3)}{background(Cy3)} < 1.5,$$ (5)

$$\frac{signal(Cy5)}{background(Cy5)} < 1.5.$$ (6)

The spots that have signal lower than 1.5 times its local background are considered to have too faint a signal and are flagged.

In the end, a spot is flagged if it fails on any of the listed conditions (on any of the channels) (see note 36).

3.8.6 Data Export

Data are exported into a tab delimited txt file. The following parameters are needed for further analysis in R, but other parameters can be exported as well:

1. Name (putative annotation, from .gal file)
2. Clone_name: (from .gal file)
3. Gf (green foreground): Cy3 signal
4. Rf (red foreground): Cy5 signal
5. Gb (green background): Cy3 background
6. Rb (red background): Cy5 background
7. Ignore Filter: for each spot an Ignore Filter value is assigned. Flagged spots (determined by quality control described above) get a value of 0, and the others value 1. These values are used as weights for further analyses.

3.9 Data Analysis in R

3.9.1 Packages in R

There are several packages available to use with microarray data. Packages contain a number of functions and data sets that are used to work on a specific problem.

Packages in R cover most of the steps in microarray data analysis. They can be divided into several main groups, which follow the steps of microarray data processing:

1. Packages for quality control: These are designed to help assess the quality of spotted array experiments. They usually produce several plots and statistical measures that help to determine whether hybridizations and slides are of good quality and to visualize any problems that might occur during the preparation of microarrays (spotting) or during hybridizations. One such package is arrayQuality.
2. Packages for normalization that also search for differentially expressed genes: These are designed to perform various background correction procedures, within-array and between-array normalizations of data and functions which help to find differentially expressed genes. Such packages include marray, limma, and vsn.
3. Packages for clustering and annotation

Sometimes functions of different packages complement each other, sometimes they are very similar. Therefore, the next sections follow the steps in microarray data analysis and packages used in these steps are discussed.

All commands that are executable in R are printed in different font (`Courier new`). They can be typed directly in R GUI.

Sometimes commands are repeated many times. These commands are grouped together in scripts. In that case, only a script has to be run and all the commands contained in the script are executed (e.g., script for creating MA plots for all microarrays in the experiment and saving them to separate files). Scripts that are used in this protocol can be found in Sect. 5.

3.9.2 Installing and Loading of Packages

The following packages are needed for data manipulation described in this protocol: `arrayQuality`, `marray`, and `limma`. First they are downloaded (R GUI: `Packages` → `Install package(s) from Bioconductor...` → packages are selected one by one from the list). Packages are also downloaded from http://www.bioconductor.org/download as. zip files (the latest BioC released is selected) and manually installed (R GUI: `Packages` → `Install package(s) from local zip file(s)...` → packages are selected one by one).

All packages used in R session, are loaded every time you start the session with the command: `library(packageName)` where `packageName` is the actual name of the package to be loaded. Packages can be loaded automatically if `library` calls are included in the. `First` function, which is executed upon the session start-up. To load `marray`, `arrayQuality`, and `limma` packages at next start-up the command is typed:

```
First <- function(){
library(arrayQuality)
library(marray)
```

```
library(limma)
}
```

3.9.3   Importing Microarray Data into R

The following files are needed for successful loading of microarray data into R (all files are tab-delimited text files):

1. Data file: image-analysis output files created by ArrayPro Analyzer or any other image-analysis software (exported as the tab-delimited. txt files).
2. Targets file or phenoData file: contains information about which RNA sample was hybridized to each channel of each array, array numbers, file names of. txt data files and other important information about hybridization, samples, remarks etc. It is created in Microsoft Excel and saved as a tab delimited.txt file. Table 1 shows how a simple phenoData file can look like. Columns Cy5, Cy3 and fileName are necessary for data import and current application, but other information about hybridizations can be added as additional columns.
3. GAL file: file containing information about the microarray layout, names and IDs of spotted genes/clones, explained above.

The working directory in R GUI is set to *C:/Microarrays*. In this way all the files produced by R are saved in this output folder (R GUI: File → Change dir… C:/Microarrays).

To read the microarray data the following commands are executed:

```
library(limma)
```

#loads limma package

```
Targets <-readTargets("C:/Microarrays/phenoDataCulti-
var.txt")
```

#creates targets object

```
RG <- read.maimages(targets$fileName, path="C:/Micro-
arrays", columns = list(Rf="Rf", Gf="Gf", Rb="Rb",
Gb="Gb"), wt.fun = wtIgnore.Filter)
```

Table 1 An example of a phenoData file: list of microarrays (rows) used and their description

No.	Sample names	Cy5	Cy3	Slide no.	File name
1	1M, 1V	M	V	12902231	I_1_12902231.txt
2	2M, 2V	V	M	13016970	I_2_13016970.txt
3	3M, 3V	M	V	12902164	I_3_12902164.txt
4	4M, 4V	M	V	12902163	I_4_12902163.txt
5	5M, 5V	V	M	12902162	I_5_12902162.txt
…					

V viral inoculation; *M* mock inoculation
Note that columns 3, 4, and 6 have to be present with exact names: Cy5, Cy3, and File name

#reads intensity data and weights (flags for ignored spots) for all the genes for all arrays that are listed in the targets file (R object) into a RG object (see note 37).

```
RG$genes <- readGAL("C:/Microarrays/TIGRgalV3.txt")
```

reads the probe IDs and other probe-specific annotation stored in the GAL file into the *genes* component of RGList object. This command presumes that the GAL file is stored as TIGRgalV3.txt tab delimited. txt file in the C:/Microarray folder and contains columns named Block, Row and Column as well as cDNA (oligonucleotide) info (Clone_Name, annotations, etc.).

Once the gene array list is available, the print layout of the arrays is extracted from it with the command:

```
RG$printer <-getLayout(RG$genes)
```

#RG object now contains all the data needed.

```
RG$status <-RG$weights
```

#reads *weights* into *status* component of RG object (see note 38)

3.9.4 Quality Control Prior to Normalization

Packages marray and convert are loaded with commands:

```
library(marray)
library(convert)
```

Separate files with images of signal and background intensities of the slides are produced:

```
imageplot3by2(RG, z="Rb", prefix="SurfaceRb")
imageplot3by2(RG, z="Gb", prefix="SurfaceGb")
imageplot3by2(RG, z="G", prefix="SurfaceG")
imageplot3by2(RG, z="R", prefix="SurfaceR")
```

This is also done with one step by running the script *image.plots.R*.

An image plot for a single array is produced by (e.g. for array 2 in the experiment) (see note 39):

```
imageplot3by2(RG[,2], z="Rb")
```

3.9.5 Background Correction and Normalization

To subtract the background from the signal and to normalize data on all arrays in the experiment using global Loess normalization the following command is used:

```
MA <- normalizeWithinArrays(RG, method="loess",
bc.method="subtract")
```

Object MA (class MA List) that contains normalized data (see note 40) is created. To ignore background subtraction, simply `bc.method=` "`none`" is used. In order to see the effect of normalization (later during the visualizations of data on different plots) another MA object without normalization is created:

```
MAun <- normalizeWithinArrays(RG, method="none",
bc.method="subtract")
```

3.9.6   Various Pre- and Postnormalization Plots

3.9.6.1   Box Plots for Whole Arrays

Comparison of box plots of M-values for whole arrays prior to normalization gives some insight into variability between microarray hybridizations (Fig. 3). Boxplots for all microarrays in the experiment are plotted by running the script *big.boxplotUnNorm.R*. By running the script on normalized data (*big.boxplotNorm.R*) the effect of normalization is observed (see note 41). Alternatively simple command lines are run:

```
boxplot(MA$M~col(MA$M),names=colnames(MA$M),
par(cex=0.6))
boxplot(MAun$M~col(MAun$M),names=colnames(MAun$M),
par(cex=0.6))
```

3.9.6.2   Print-Tip Box Plots

By running the scripts *PrintTipPlots.pre.R* and *PrintTipPlots.post.R*, box plots for each print tip group (subgrid) for all arrays in experiment for unnormalized and

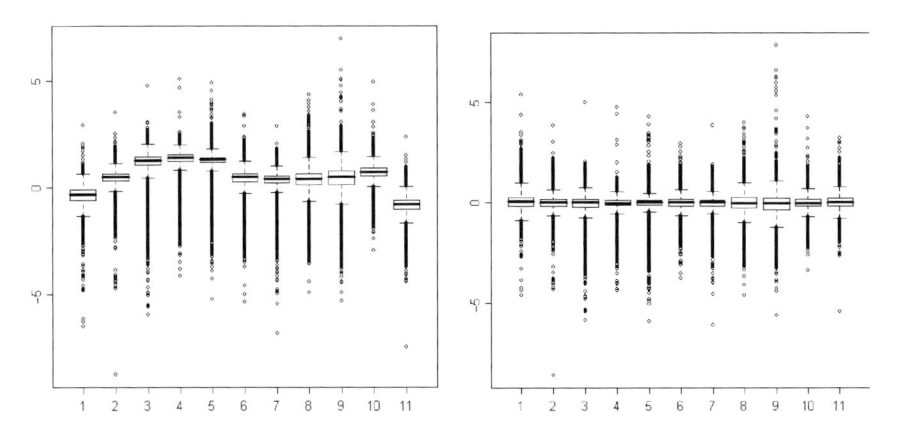

Fig. 3 Box-plot of M values on 11 microarrays before (*left*) and after loess normalization (*right*)

normalized data, respectively, are produced. Plots are saved into the working directory. The effect of normalization and the spatial variation of M-values are observed
across the arrays.

3.9.6.3   MA Plots

The following commands/scripts plot the MA plots for each array in the experiment
(Fig. 4). Using the MA$status part of the MA object which contains the information
about the weights. Spots with weight »0« are plotted as grey and spots with weight
»1« as black spots (see note 42):

```
plotMA(MA[,2], status=MA$status, values=c("0"","1""),
col=c("grey","black"), zero.weights=TRUE, cex=0.2)
```

Using this basic command an MA plot for microarray 2 (MA[,2]) in the experiment is plotted. MA plots for the rest of arrays are produced and saved. Alternatively
scripts *MAplots.all.R* and *MAplotsUnNorm.all.R* are run. All the plots in the experiment for normalized and unnormalized data respectively are produced and saved
separately as.png files in the working directory.

3.9.6.4   MA Plots for Print-Tip Groups

Viewing MA plots for each print tip group on the microarray enables observation
of the spatial variability of data on each microarray (similar to box plots for
print-tip groups created in section 3.9.6.2). The following command is used
for observation of all print-tip MA plots of one microarray together in one composite plot (e.g. plots for microarray 3 in the experiment):

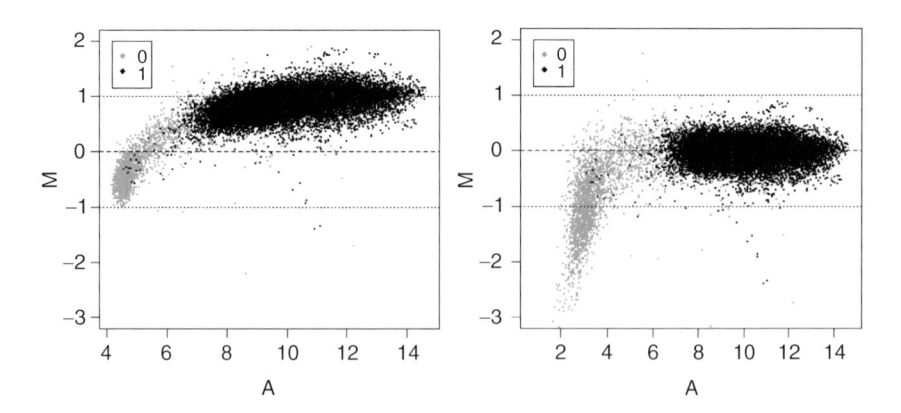

Fig. 4 MA plot of one microarray before (*left*) and after normalization (*right*).0 – bad quality
spots, excluded in image analysis, 1 – good quality spots

```
plotPrintTipLoess(MA[,3])
```

This can also be done by running the script *PrintTipLoessMAplotsNorm.R* for normalized or *PrintTipLoessMAplotsUnNorm.R* for unnormalized data.

3.9.7 Search for Differentially Expressed Genes with the Limma Package

3.9.7.1 Averaging Duplicates

On most microarrays spots are printed in duplicates (at least). In the following step the duplicate spots are averaged (for other options on how to take into account the information from duplicate spots in limma, see note 43). The following command is run by a function called my.mean (see note 44) which averages the M and A values for within-array duplicate spots and creates a new object, MAme, with averaged values:

```
MAme=my.mean(MA)
```

The function is viewed by the command:

```
page(my.mean)
```

3.9.7.2 Design Matrix

In the next step, a design matrix, which provides a representation of the different RNA targets that have been hybridized to the arrays, is created. In our case, the design matrix represents dye swapping, which is indicated in the *phenoData* file. The design object is created by the command:

```
design <- modelMatrix(targets, ref="M")
```

In a microarray experiment containing for instance five microarrays (Table 2), the design matrix is a quite simple vector $(-1,1,-1,-1,1)$.

Table 2 Design matrix: microarrays where the mock inoculated sample (M) is labeled with Cy3 and virus-inoculated sample (V) with Cy5 are marked with 1

No.	Sample names	Cy5	Cy3	Design
1	1M, 1V	M	V	−1
2	2M, 2V	V	M	1
3	3M, 3V	M	V	−1
4	4M, 4V	M	V	−1
5	5M, 5V	V	M	1

Dye swaps are marked with −1

3.9.7.3   M-Dependent Gene Filtering

To find differentially expressed genes with higher confidence, many genes that show low or no differential expression (genes that have M-values around zero) are filtered out. This is done by running the script *Mfilter.R*. The function is started from the MA object; an MAme object is created and a weight value of 0 is ascribed to all genes that have absolute M-values below 0.2 in (in this case) 3 out of 5 arrays. As a guide, comments on all commands for the process of filtering are inserted in the script (see note 45).

3.9.7.4   Significance Testing

For analyzing microarray experiments with the package limma, an approach named linear models is used. First the linear model is fitted to data that fully models the systematic part of microarray data (lmFit command). For statistical analysis and assessment of differential expression, empirical Bayes method is used to moderate the standard errors of the estimated log-fold changes (eBayes command). The commands used to execute this procedure are:

```
fit <- lmFit(MAfilter, design)
fit2 <- eBayes(fit)
```

3.9.7.5   Exporting Significant Data

To extract a table of the top ranked genes (top-table) from a linear model fit based on the calculated p-values the following function is used (see note 46):

```
Topt<-topTable(fit2, adjust.method="fdr", n=600)
```

To export the table into a tab-delimited.txt file use (see note 47):

```
write.table(topt,file="topTableCultivar.txt",sep="t")
```

4   Notes

1. Inoculation buffer (20 mM sodium phosphate buffer, pH 7.6) is made specifically for PVY[NTN] inoculation of potato plants. If other viruses or plants are used, inoculation buffer should be adapted accordingly.
2. 80% ethanol is prepared with nuclease-free water.
3. Ethidium bromide is a powerful mutagen. All the appropriate safety measures should be taken.
4. For every function used in R, a help document can be looked up which contains a short description of the function, its arguments as well as examples of its use (?functionName). A more structured help is available as a browsable .html file. It can be opened with:

`help.start()`. Most of the packages that have been loaded in R (library(packageName)) have a .pdf help file (called a Vignette) available with a more or less detailed description of the package and its functions. Vignettes can be opened in R GUI: Vignettes → package. More general help files are also available as .pdf files in R GUI: Help → Manuals (in PDF).

5. To minimize the chances of cross contamination, firstly mock inoculation is carried out followed by virus inoculation.

6. To study the effects of virus inoculation at different time points after inoculation, plants are incubated at chosen time points.

7. The most important thing during the sample-preparing procedure is that the leaves, once detached from the stems, are put in liquid nitrogen as quickly as possible and once they are frozen, they do not thaw. In case of thawing, RNA is rapidly degraded. Therefore, all the materials (mortars, pestles, tubes, spatulas, etc.) are precooled with liquid nitrogen and all the procedures until storing the material at −80°C are done as quickly as possible.

8. The yield of totRNA extraction depends on the quality of the grinding procedure. The finer the powder, the more totRNA is obtained from it.

9. RNA molecules are very unstable. All the precautions to avoid RNA degradation are carried out: the leaf tissue should not thaw prior to putting it into extraction buffer, RNase-free reagents should be used, work should be done fast, on clean surfaces, and gloves should be worn at all times. It is recommended to prepare all the necessary material for RNA isolation prior to starting the procedure.

10. The sample should be immediately transferred to RNeasy mini column.

11. Silica-gel membrane should not be damaged with the tip.

12. Use of preheated water (80°C) and 10-min incubation time prior to final centrifugation step results in higher yield of eluted totRNA.

13. The eluted totRNA is kept on ice or at −20°C until use in subsequent analysis. For longer storage, it is kept at −80°C.

14. totRNA cleanup is carried out immediately.

15. Sample is immediately transferred to RNeasy MinElute Spin Column.

16. The dead volume of the RNeasy MinElute Spin Column is 2 μL; elution with 14 μL of RNase-free water results in 12 μL elute.

17. Alternative methods can be used: Instead of preparing a gel, commercial ones are used like E-gels (Invitrogen, USA), Flash gels (Cambrex, USA). Alternatively, to measure absolute totRNA concentrations, NanoDrop (NanoDrop Technologies, USA) or Bioanalyzer (Agilent, USA) is used.

18. The comb with appropriate well volume is selected.

19. Original and diluted totRNA is kept on ice.

20. Can be easily adapted to other gel-run systems.

21. It can be done with any other detection system software.

22. If the volume of RNA after equilibration is not 8 μL, it needs to be adjusted to 8 μL with nuclease-free water.

23. For example, Cy3 is used for control and Cy5 for infected plant RNA.

24. It is important that the centrifugation steps do not exceed 14,000g. Otherwise, the reservoir membrane can detach from the reservoir and the sample is lost.

25. It is of great importance that the arrays do not dry out during the washing steps. Any delay in these steps may result in high background.

26. It is modified nucleotide that stabilizes the hybridization between complementary strands of nucleic acid by blocking all poly A sequences present in microarray features.

27. Capillary force will pull the hybridization solution under the LifterSlip. Be sure that the microarray and hybridization mix are still warm.

28. Shake the holder manually or put the glass jar on shaker and use the magnet and mix at 1,000g but lift the microarray holder above the magnet.

29. If smear or dust is observed on the array, rewash with strong agitation in third washing solution and redry.

30. The 3DNA Array 900 Capture reagents contain light-sensitive dyes; therefore, try to keep them as much in the dark as possible.
31. The stock solution can be stored at −20°C up to 2 weeks.
32. Keep microarrays at room temperature to prevent the condensation of water on their surface. It is advised to scan microarray the same day the hybridization was completed or at least in the next few days. Cy3 and Cy5 dyes degrade in time and are also very reactive with ozone. As a consequence the signal reduces with time.
33. Many scanning software packages have features for increasing the fluorescent dye signal by oversampling (multiple passes with the laser over the same area and adding up the fluorescence from each pass) and PMT gain (photo amplification). It is important to keep in mind that by increasing the signal in such way, noise is increased as well. It is also important to know that each pass of laser beam over the same microarray bleaches the fluorescent dyes. Most scanners are also able to scan with different resolutions (5 or 10 μm). Usually 10 μm is enough.
34. Make sure that the whole spotted area of the microarray is scanned. It is advised that every slide is entered in the scanner at exactly the same position (e.g., locked at the top left corner of the loading tray) and that the scan area for the same microarray type is always the same. In this way, the image-analysis software will have less problems locating the spots on microarray images (grid finding protocol).
35. Other settings can be used for these parameters:

    a. Signal: instead of trimmed mean, median can also be used.
    b. Background: there are many possibilities like global background (averaged over the entire image), signal from background cells, local background (e.g., local ring or corners around each spot). From our experiences, we find that local background is better than global.

36. All the parameters needed for calculation of these conditions (standard deviations) have to be calculated in the image-analysis software. If the same software is not able to flag the cells using these conditions, then this can be done in any other software (e.g., MS Excel).
37. RG object is an object of class RGList, a class used to store raw intensities as they are read from an image analysis output file (by command `read.maimages`). It also contains additional information about microarrays (find out more about RGList in limma Users guide or using help command? `read.maimages`).
38. By running R script *read.dataCultivar.R* (R GUI: `File → Source R code... → C:/ Microarrays/read.dataCultivar.R`) all these commands are executed automatically, RG object is created and its preview is shown. R scripts are available in Sect. 5. To create such a script simply open a new script in R (R GUI: `File → New script`) enter the text from the Sect. 5 into the empty window and save the script to *C:/Microarray* folder (R GUI: `File → Save as...`). If your files are located in a different folder system, please make sure to correct the file pathways in all R commands and all R scripts. The same procedure should be carried out with the rest of R scripts mentioned later in the protocol.
39. The function `imageplot3by2` will write image plots to files, six plots to a page. Images are rotated 90° counterclockwise (the first subgrid of the array is located in the bottom left corner). Files will be named in the following way: e.g. image1–3.png and saved into the working directory. The function `imageplot3by2` produces a spatial image plot of background or foreground data (from RG object). In that way, one can observe backgrounds and foregrounds for both channels for all arrays in the experiment and possibly see buffer smears, bubbles present in hybridization mix, etc. that were created during washing or hybridization of the slides.
40. The function `normalizeWithinArrays` allows a choice of a range of ormalization methods for microarray data such as Print-tip Loess normalization (default in `limma`), Global Loess normalization, Composite Loess normalization, and Shrunk Robust Splines normalization. Since global Loess normalization includes all spots in the normalization process not just spots within print-tip groups (Print-tip Loess) it will be used in this protocol. Background can also be

subtracted in different ways. More information on normalization methods, on MA List class of objects and on background subtraction can be found in the R help or in `limma` Users guide.

41. R creates plots in the same plotting window, one over the other. For several plots to be displayed at the same time in R in separate windows, a new blank plotting window has to be opened with command `x11()`. In this way, graphs will open in a separate active window in R GUI. It is possible to plot the graphs within the same graphics window if the `graphics.record` option is set with the following command:

```
options(graphics.record=T)
```

In this case PageUp and PageDown keys can be used to scroll through the graphs plotted within the same graphics window. The current graph can also be saved as an image:

```
R GUI: File → Save as → Png/Bmp/Jpeg.
```

42. To plot only the spots which have weight "1," the argument `zero.weights` in `plotMA` function has to be changed to `FALSE`.

43. `Limma` package can take into account correlation between within-array spot duplicates (pooled correlation method to make full use of the duplicate spots). But this feature can only be used if duplicate spots are printed equally throughout the microarray. TIGR potato arrays have duplicate spots but they are printed randomly throughout the array therefore average of M and A values was chosen. To make use of duplicate spot correlation a few other arguments of the `lmFit` function will have to be used (`ndups`, `spacing`, and `correlation`). See help on the `lmFit` function in R GUI and in the limma Users guide.

44. To insert the `my.mean` function into R workspace use the command `fix(my.mean)` which opens an R editor window. Clear the contents and enter the my.mean function text as printed in appendix as for all other R scripts. Save the function (R GUI: `File → Save`). Now the function can be run.

45. The filtering in the script `Mfilter.R` works in two steps, each is carried out by one function. Here are the two critical commands which can be modified:

```
Mfilt1=as.numeric(abs((MAmeMtemp))>0.2)
```

The cutoff value for M-values is entered here (…`>0.2`). The genes that have M-values below this threshold are good candidates to be filtered out. Simply change the `>0.2` value to whatever threshold suits you. However, this is only the first part of the filtering. The next critical step is defined by

```
Mfilt4=as.numeric(Mfilt3<3)
```

Remember that the experiment is composed of a few microarrays (e.g., five). Therefore, each gene has five M-values. If M-values are lower than 0.5 in only one experiment, we do not want to filter out this gene. If there is more than one M-value lower that 0.5 (e.g., three or more out of five), this gene can be filtered out. This is exactly what this command does. Simply change the `<3` in `Mfilt4` to whatever number you want to use as the criterion.

46. The `topTable` function also includes a possibility of adjusting p-values (argument `adjust.method`). One can choose from a variety of methods from the more strict methods like Bonferroni correction ("`bonferroni`") where p-values are multiplied by the number of comparisons to less conservative methods like Holm (1979) ("`holm`"), Hochberg (1988) ("`hochberg`"), Hommel (1988) ("`hommel`"), and Benjamini and Hochberg method ("`fdr`"). p-Value correction can be turned off by using the following argument `adjust.method="none"`. See `limma` Users guide for more detailed references.

The genes in the table are sorted by default by increasing p-values from top to bottom of the table. This means that the genes with the lowest p-values are statistically more likely to be differentially expressed than those with higher p-value.

It is possible that all the adjusted p-values will be large, meaning there is no good evidence for differential expression in the experiment. Sometimes all the p-values can be equal to one. That might happen if the lowest p-value is lower than $1/N$ (where N is the number of genes with nonmissing p-values). Lack of evidence for differentially expressed genes does not mean, however, that there are no differences in gene expression between arrays compared. The genes in the table are still listed according to their importance which means that the highest ranking genes may be differentially expressed – you can still use them as the result.

47. Tab-delimited .txt file will be saved to the working directly. Note that when opening any tables exported in R with Excel the row with names of the columns has to be shifted one cell to the right.

5   Appendix: R scripts

5.1   read.data.R

```
#reads the phenodata, signal data and gal file, creates
RG object and shows its summary.
library(limma)
targets <-readTargets("F:/ENSS/RRR/phenoDataCultivar.txt")
RG <- read.maimages(targets$fileName, path="F:/ENSS/
Data",columns = list(Rf="Rf",Gf="Gf",Rb="Rb",Gb="Gb"),
wt.fun=wtIgnore.Filter)
RG$genes <- readGAL("F:/ENSS/RRR/TIGRgalV3.txt")
RG$printer <-getLayout(RG$genes)
RG$status <-RG$weights
show(RG)
```

5.2   image.plots.R

```
#draws quality control images and saves them in the
working directory.
imageplot3by2(RG, z="Rb", prefix="ImagePlotRb")
imageplot3by2(RG, z="Gb", prefix="ImagePlotGb")
imageplot3by2(RG, z="G", prefix="ImagePlotG")
imageplot3by2(RG, z="R", prefix="ImagePlotR")
```

5.3   big.boxplotUnNorm.R

```
boxPlotAll <- function(x) {
{
png(filename=paste("bigBoxPlotUnNorm", ".png", sep=""))
```

```
boxplot(MAun$M~col(MAun$M),names=colnames(MAun$M))
dev.off()
}
}
boxPlotAll(MAun)
```

5.4 big.boxplotNorm.R

```
boxPlotAll <- function(x) {
{
png(filename=paste("bigBoxPlotNorm", ".png", sep=""))
boxplot(MA$M~col(MA$M),names=colnames(MA$M))
dev.off()
}
}
boxPlotAll(MA)
```

5.5 Print.TipPlots.pre.R

```
library(convert)
library(marray)
maRaw <- as(RG, "marrayRaw")
boxPlotRaw <- function(x) {
for(i in 1:6) {
png(filename=paste("PrintTipPre", i, ".png", sep=""))
boxplot(maRaw[,i], xvar="maPrintTip", yvar="maM",
main=paste("pre-normalization, chip", i), par(cex=0.6))
dev.off()
}
}
boxPlotRaw(maRaw)
```

5.6 Print.TipPlots.post.R

```
library(convert)
library(marray)
maNorm <- as(MA, "marrayNorm")
boxPlotNorm <- function(x) {
for(i in 1:6) {
png(filename=paste("PrintTipPost", i, ".png", sep=""))
boxplot(maNorm[,i], xvar="maPrintTip", yvar="maM",
```

```
main=paste("post-normalization, chip", i),
par(cex=0.6))
dev.off()
}
}
boxPlotNorm(maNorm)
```

5.7 MAplotsUnNorm.all.R

```
for(i in 1:5) {
png(filename=paste("MAplotUn", i, ".png", sep=""))
plotMA(MAun[,i], status=MA$status, values=c("0"","1""),
col=c("grey","black"), zero.weights=TRUE, cex=0.2)
dev.off()
}
```

5.8 MAplotsNorm.all.R

```
for(i in 1:5) {
png(filename=paste("MAplotNorm", i, ".png", sep=""))
plotMA(MA[,i], status=MA$status, values=c("0"","1""),
col=c("grey","black"), zero.weights=TRUE, cex=0.2)
dev.off()
}
```

5.9 PrintTipLoessMAplotsUnNorm.R

```
for(i in 1:5) {
png(filename=paste("PrintTipLoessUn", i, ".png",
sep=""))
plotPrintTipLoess(MAun[,i])
dev.off()
}
```

5.10 PrintTipLoessMAplotsNorm.R

```
for(i in 1:5) {
png(filename=paste("PrintTipLoessNorm", i, ".png",
```

```
sep=""))
plotPrintTipLoess(MA[,i])
dev.off()
}
```

5.11   my.mean.R

```
#function that averages the A and M values of duplicate
spots
#sorts MA object by Clone_name
"my.mean" <-
function (MA)
{
ind=order(MA$genes$Clone_name)
MA=MA[ind,]
M=MA$M
A=MA$A
w=MA$weights
#divides the rows for M-values and A-values into odds
and evens.
n=nrow(w)
M1=M[seq(1,n,2),]
M2=M[seq(2,n,2),]
w1=w[seq(1,n,2),]
w2=w[seq(2,n,2),]
A1=A[seq(1,n,2),]
A2=A[seq(2,n,2),]
#sums weights, averages M and A values, deletes every
second line subject MA$genes.
w=w1+w2
M=(M1*w1+M2*w2)/w
M[is.na(M)]=0
A=(A1*w1+A2*w2)/w
A[is.na(A)]=0
genes=MA$genes[seq(1,n,2),]
#Creates new MA object
MA$weights=w
MA$M=M
MA$A=A
MA$genes=genes
MA
}
```

5.12 MFilter.R

```
#Filtering by M-values on MA object (contains M-values
averaged by my.mean function)
MAme <-my.mean(MA)
# first creates MAmean object. If you created it already
it will rewrite it
Design <- c(1,-1,1,-1,1)
#enters the dye-swaps design
desOK <-matrix(rep(design,dim(MAmeMtemp)[1]),dim(MAmeMt
emp)[1],byrow=TRUE)
#creates a matrix that contains 15600 equal rows (design
vector)
MAmeMtemp <-MAme$M*desOK
#takes into account the dye-swaps and corrects the M-
values so that they can be averaged in the next step
Mfilt1=as.numeric(abs((MAmeMtemp))>0.2)
Mfilt2=matrix(Mfilt1,dim(MAmeMtemp)[1],dim(MAmeMtemp)[2])
#creates a matrix with values (0 if abs(M<0.2)
Mfilt3=rowSums(Mfilt2)
#sums the above calculated values across all arrays
Mfilt4=as.numeric(Mfilt3<3)
Mfilt5=matrix(rep(Mfilt1,dim(MAmeMtemp)[2]),dim(MAmeMtemp)
[1],dim(MAmeMtemp)[2])
#creates a matrix of new weights
MAnew_weights <- MAme$weights*Mfilt5
#multiplies filter values with weights
MAnew_weights[is.na(MAnew_weights)]=0
#the following three commands builds up new MA object
(MAfilter) on the basis of the MAme object - of course
it replaces the old weights with the new ones
MAfilter <-MAme
MAfilter$weights <-NULL
#removes the old weights
MAfilter$weights <-MAnew_weights
# enters the new, corrected weights, into the new object
```

Acknowledgments We thank Dr. Andrej Blejec for his support in R programming and Mrs. Lidija Matičič for excellent technical assistance.

References

1. Birch, P.R.J. and Kamoun, S. (2000) Studying interaction transcriptomes: coordinated analyses of gene expression during plant-microorganism interactions. In R. Wood (ed.), New technologies for life sciences: a trends guide. Elsevier Science, New York, N.Y. s77–s82.
2. Golem, S. and Culver, J.N. (2003) *Tobacco mosaic virus* induced alterations in the gene expression profile of *Arabidopsis thaliana*. *Mol. Plant Microbe In.* **16**, 681–688.
3. Whitham, S.A., Quan, S., Chang, H.S., Cooper, B., Estes, B., Zhu, T., Wang, X. and Hou, Y.M. (2003) Diverse RNA viruses elicit the expression of common sets of genes in susceptible *Arabidopsis thaliana* plants. *Plant J.* **33**, 271–283.
4. Huang, Z., Yeakley, J.M., Garcia, E.W., Holdridge, J.D., Fan, J.B. and Whitham, S.A. (2005) Salicylic acid-dependent expression of host genes in compatible *Arabidopsis*-virus interactions. *Plant Physiol.* **137**, 1147–1159.
5. Marathe, R., Guan, Z., Anandalakshmi, R., Zhao, H. and Dinesh-Kumar, S.P. (2004) Study of *Arabidopsis thaliana* resistome in response to cucumber mosaic virus infection using whole genome microarray. *Plant Mol. Biol.* **55**, 501–520.
6. Smith, C.M., Rodriguez-Buey, M., Karlsson, J. and Campbell, M.M. (2004) The response of the poplar transcriptome to wounding and subsequent infection by a viral pathogen. *New Phytol.* **164**, 123–136.
7. Shi, C., Ingvardsen, C., Thummler, F., Melchinger, A.E., Wenzel, G. and Lubberstedt, T. (2005) Identification by suppression subtractive hybridization of genes that are differentially expressed between near-isogenic maize lines in association with sugarcane mosaic virus resistance. *Mol. Gen. Genomics* **273**, 450–461.
8. Senthil, G., Liu, H., Puram, V.G., Clark, A., Stromberg, A. and Goodin, M.M. (2005) Specific and common changes in *Nicotiana benthamiana* gene expression in response to infection by enveloped viruses. *J. Gen. Virol.* **86**, 2615–2625.
9. Pompe-Novak, M., Gruden, K., Baebler, Š., Krečič-Stres, H., Kovač, M., Jongsma, M., and Ravnikar, M. (2006) Potato virus Y induced changes in the gene expression of potato (*Solanum tuberosum* L.). *Physiol. Mol. Plant P.* **67**, 237–247.
10. Barrett, J.C. and Kawasaki, E.S. (2003) Microarrays: the use of oligonucleotides and cDNA for the analysis of gene expression. *Drug Discov. Today* **8**, 134–141.
11. Gentleman, R.C., Carey, V.J., Bates, D.M., Bolstad, B., Dettling, M., Dudoit, S., Ellis, B., Gautier, L., Ge, Y., Gentry, J., et al. (2004) Bioconductor: open software development for computational biology and bioinformatics. *Genome Biol.* **5**, R80
12. Stears, R.L., Getts, R.C. and Gullans, S.R. (2000) A novel, sensitive detection system for high-density microarrays using dendrimer technology. *Physiol. Genomics* **3**, 93–99.
13. Hegde, P., Qi, R., Abernathy, K., Gay, C., Dharap, S., Gaspard, R., Hughes, J.E., Snesrud, E., Lee, N. and Quackenbush, J. (2000) A concise guide to cDNA microarray analysis. *Biotechniques* **29**, 548–554, 556.

Chapter 41
Genome-Wide Screens for Identification of Host Factors in Viral Replication

Tadas Panavas, Elena Serviene, Judit Pogany, and Peter D. Nagy

Abstract The central step in virus infection cycle is replication, which depends on viral and host factors. Model hosts, such as yeast, can be very valuable to identify host factors and study the functional interactions of host factors with viral proteins and/or the virus nucleic acids. The advantages of using yeast include the availability of (i) single gene-deletion library, (ii) the essential gene library (yTHC), (iii) the controllable small or large-scale expression of viral proteins and nucleic acids, and (iv) the rapid growth of yeast strains. Here, we describe procedures, which facilitate high-throughput analysis of tombusvirus replication in yeast.

Keywords Yeast; High-throughput transformation; High-throughput viral RNA extraction; Northern blotting; Tombusviruses; Host factors

1 Introduction

Viruses are intracellular pathogens that are a threat to all living organisms. The central step in virus infection cycle is replication, which depends on viral and host factors. Model hosts, such as yeast, can be very valuable to identify the host factors and study their functional interactions with viral proteins and nucleic acids (1–5). The advantages of using yeast include the availability of (i) single gene-deletion library (YKO), (ii) the regulatable essential gene library (yTHC), (iii) the controllable expression of viral proteins and nucleic acids, and (iv) the rapid growth of yeast strains. The high-throughput approach described here is suitable to conduct genome-wide screens to identify all the host proteins affecting tombusvirus replication.

Tombusvirus replication is launched by coexpression of three viral components in yeast cells. These include p33 and p92 replication proteins expressed constitutively via the *ADH1* promoter (plasmids pHisGBK-His33 and pGAD-His92), and the replicon RNA, which is expressed from the galactose/glucose regulatable *GAL1* promoter (plasmid pYC/DI-72) (6–8). After induction with galactose, the replicon RNA starts robust replication that can reach up to ribosomal RNA levels (6–8). Also, the tombusvirus replicase complex can be purified from these yeast cells for biochemical studies (7, 9).

From: *Methods in Molecular Biology, Vol. 451, Plant Virology Protocols: From Viral Sequence to Protein Function*
Edited by G.D. Foster, I.E. Johansen, Y. Hong, and P.D. Nagy © Humana Press, Totowa, NJ

This unit describes high-throughput approaches for analysis of tombusvirus replication in yeast, including cotransformation of yeast strains with three expression plasmids, followed by culturing of the transformants and analysis of viral RNA replication. Each step is performed in 96-well plates using a multichannel pipetman or liquid handling robot. The methods described here can easily be scaled up or down as needed.

2   Materials

2.1   *Yeast High-Throughput Transformation*

1. The yeast single gene deletion library (YKO) and the essential gene (yTHC) library (Open Biosystems, Huntsville, AL). These libraries consist of ~4,800 and 800 strains, respectively, that can be cultured in 96-deep-well plates.
2. Plasmids pGAD-His92, pHisGBK-His33 and pYC/DI-72 (Fig. 1), which can produce the viral-coded p92 and p33 replication proteins and the replication competent DI-72 repRNA, respectively, in yeast (7, 8). These plasmids carry *LEU2*, *HIS3*, and *URA3* genes that allow the selection for transformants on growth media lacking leucine, histidine, and uracil (Fig. 1).
3. 1 M Lithium acetate (LiAc). Filter sterilize and store at room temperature (RT).
4. YPD containing 200 mg/L G418. YPD medium (1 L) contains 10 g Yeast extract, 20 g peptone, 20 g Dextrose in 1 L sterile, deionized water. Autoclave media and then add 1 mL of 200 mg mL^{-1} G418. Store it at 4°C in the dark.
5. SC-ULH/glucose medium, which contains 6.7 g Yeast nitrogen base without amino acids, 1.4 g Dropout mixture without tryptophan, leucine, uracil, histidine, 0.1 g Tryptophan, 20.0 g Dextrose (glucose) in 1 L sterile H$_2$0. Autoclave media and then add 1 mL of 200 mg mL^{-1} G418. Store it at 4°C in the dark.
6. SC-ULH/glucose medium with 2% agar. Add 10 g Difco agar to 500 mL SC-ULH/glucose medium (see above). Dry plates in the hood for 1–2 h without lid. Store the plates at 4°C in the dark until use.
7. 50% w/v PEG$_{3350}$. Filter sterilize it and store it in the dark at RT.
8. 96 well, 2 mL round bottom plates
9. A 96-pin replicator
10. Breathable membranes for covering plates
11. Storage mats
12. Incubators equipped with plate holders
13. Table top or floor centrifuge equipped with rotor that can accommodate deep-well plates
14. Spectrophotometer, which is preferably read in plate format
15. Multichannel pipetmans (20 and 300 µL) or a liquid handling robotic instrument

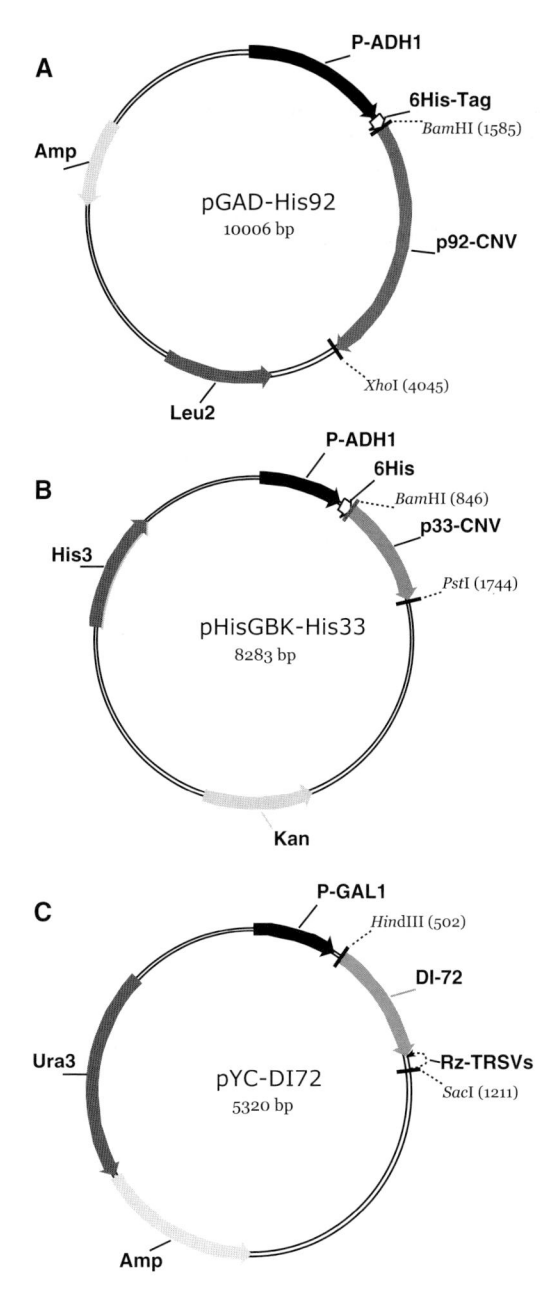

Fig. 1 Schematic representation of plasmids used to launch TBSV RNA replication in yeast. All three plasmids are cotransfected to the same yeast cells. (**a**) p92, derived from the closely related Cucumber necrosis virus (CNV, which supports as efficient replication of TBSV RNA as the TBSV replication proteins), is expressed constitutively from the *ADH1* promoter. There is a six-histidine tag (6His-Tag) at the N terminus of p92 to facilitate affinity purification. (**b**) The CNV p33 carrying a 6His-Tag is expressed constitutively from the *ADH1* promoter. (**c**) TBSV defective interfering (DI-72) replicon RNA is transcribed from the galactose/glucose inducible/repressible *GAL1* promoter. There is a ribozyme (Rz-TRSVs) at the 3' end of DI-72, which generates the authentic 3' end sequence, via a self-cleavage that occurs with ~50% efficiency in yeast

16. Transformation mix (10 mL) contains 6.7 mL PEG$_{3350}$ (50% w/v), 1.0 mL 1.0 M LiAC, 0.5 mL single-stranded-DNA, sonicated (10 mg mL^{-1}) (10), 0.1 mL pGAD-His92 plasmid (~0.3 mg mL^{-1}), 0.1 mL pHisGBK-His33 plasmid (~0.3 mg mL^{-1}), 0.1 mL pYC/DI-72 plasmid (~0.3 mg mL^{-1}), 1.5 mL sterile H$_2$0. Vortex transformation mix thoroughly before use!

2.2 High-Throughput Viral RNA Extraction from Yeast

1. SC-ULH/galactose medium
2. Phenol, water saturated
3. Chloroform
4. Absolute Ethanol
5. 3 M Sodium Acetate (NaAc), pH 5.3
6. 0.5 M EDTA, pH 8.0
7. 10% SDS
8. Doxycycline. 10 mg mL^{-1} stock solution in 50% ethanol
9. 20X TBE: 1 L contains 216 g Tris, 110 g Boric acid, 14.88 g EDTA, and sterile dH$_2$O
10. Ethidium bromide (10 mg mL^{-1})
11. RNase-free water
12. 1X Yeast RNA extraction buffer: 100 mL contains 86 mL RNase-free dH$_2$O, 1.66 mL 3 M NaAc pH 5.3, 2 mL 0.5 M EDTA pH 8.0, and 10 mL 10% SDS
13. Centrifuge with a rotor that can hold 96-deep-well plates

2.3 Viral RNA Analysis with Agarose Gel Electrophoresis

1. 20X TBE buffer: 1 L contains 216 g Tris, 110 g Boric acid, 14.88 g EDTA, and sterile dH$_2$O.
2. 2X RNA loading dye: 20 mL contains 1 mL bromophenol blue/xylene cyanol dye mixture (Sigma, B3269) and 19 mL formamide. Store it at 4°C in the dark.
3. Ultrawide agarose gel electrophoresis unit (with 96 or 2 × 96 wells), multichannel pipetman compatible.

2.4 Viral RNA Analysis with Northern Blotting

1. 1 L of 20X SSC contains 88.23 g Tri-sodium citrate and 175.32 g NaCl, pH 7.0. Autoclave it. Store at RT.
2. Northern wash solution I: 1 L contains 890 mL sterile dH$_2$O, 100 mL sterile 20X SSC, and 10 mL 10% SDS.

3. Northern wash solution II: 1 L contains 985 mL sterile dH_2O, 5 mL sterile 20X SSC, and 10 mL 10% SDS.
4. Nylon membrane (Hybond X-L, Amersham-Pharmacia)
5. UltraHyb hybridization buffer (Ambion, Inc)
6. Transblot apparatus (Biorad Transblot Semi-dry)
7. Hybridization chamber with sealable tubes
8. PhosphoImager or FluoroImager (depending on the probe used)

3   Methods

3.1   Yeast High-Throughput Transformation

1. Aliquot 0.3-mL YPD containing 200 mg L^{-1} G418 into a 96-deep-well plate, using a multichannel pipetman.
2. Inoculate each well with a different yeast colony from one of the yeast libraries using a 96-pin replicator. Cover plate with breathable membrane.
3. Grow yeast overnight with shaking at 300 rpm in a 30°C incubator shaker.
4. Next morning, transfer ~5–10 μL of the overnight yeast cultures into a new plate that contains fresh 0.25 mL YPD per well with 200 mg L^{-1} G418. The final OD should be ~0.3 at 600 nm wavelength. For most yeast strains, this will take only ~5–10 μL of overnight culture.
5. After covering the plate with breathable membrane, grow diluted cultures for another 4 h at 300 rpm in a 30°C incubator shaker.
6. Centrifuge cultures for 2 min at 2,000 rpm (750 g) at RT and decant the medium. Turn the plate upside down, then tap it gently, upside down, on a paper towel for few seconds to remove as much supernatant as possible.
7. Add 0.2 mL of sterile water to each well, then cover the plate with a storage mat and resuspend yeast by vortexing.
8. Centrifuge the plate for 2 min at 2,000 rpm at RT and decant the supernatant.
9. Add 200 μL of 100 mM LiAc to each well, cover the plate with a storage mat and resuspend the pellet by vortexing. Incubate yeast for 15 min at RT.
10. Prepare the transformation mix during the incubation.
11. Centrifuge the plate for 2 min at 2,000 rpm and pipette off the LiAc solution with a multichannel pipetteman.
12. Add 95 μL of transformation mix to each well, cover the plate with breathable membrane and resuspend the yeast by vortexing.
13. Incubate yeast at 30°C for 30 min. There is no need for shaking during incubation.
14. Apply heat shock to yeast by placing the plate into a 42°C water bath for 40 min.
15. Centrifuge the plate for 2 min at 2,000 rpm and discard the transformation mix by pipetting it off with a multichannel pipetteman.
16. Add 100 μL of sterile water to each well, cover the plate with a storage mat, and resuspend the yeast by vortexing.

17. Plate each yeast strain on a prelabeled agar plate containing SC-ULH/glucose minimal medium. Thoroughly resuspend each yeast strain separately by pipetting it up and down just before plating, because yeast cells are heavy and sediment to the bottom very quickly.
18. Incubate the plates at 30°C until colonies appear (2–4 days).
19. Store the plates at 4°C in the dark. Before storage, streak individual yeast colonies on fresh agar plates containing SC-ULH/glucose minimal medium, followed by incubation at 30°C. Store the plates at 4°C in the dark.

3.2 High-Throughput Viral RNA Extraction from Yeast

Total RNA could be efficiently extracted from yeast cultures grown in 96-deep-well plates. The RNA obtained is used for agarose gel electrophoresis and Northern blotting.

1. Dispense 0.5 mL SC-ULH medium with 2% galactose+200 mg L^{-1} G418 into a 96-deep-well plate. Add 0–10 mg L^{-1} doxycycline to the growth medium for yeast transformants obtained from the yTHC library (11, 12). The amount of doxycycline needed should be tested for each strain. Doxycycline regulates the expression of target genes in the yTHC library (Open Biosystems, www.openbiosystems. com).
2. Inoculate each well in a 96-deep-well plate with a separate yeast transformant (see above) using 0.3-mL multichannel pipette tips. Transfer only a little amount of cultured yeast. After the transfer, pipette the mixture up and down a few times using a multichannel pipetteman. It is important to transfer approximately the same amount of cells into each well to get even growth for each yeast strain.
3. Cover the plate with breathable membrane and grow cultures at 23°C at 300 rpm for ~48 h.
4. Pellet yeast at 2,500 rpm (1,200g) for 5 min at RT and decant the media. Turn the plate upside down, then, tap it gently, upside down, on a paper towel for few seconds to remove as much fluid as possible.
5. Add 300 μL of 1X yeast RNA extraction buffer and 300 μL of phenol to each well, followed by covering the plate with a 96-well storage mat.
6. Vortex plate and incubate it for 4 min in a 65°C water bath.
7. Place plate into an ice slurry for 2 min and then centrifuge it at 2,500 rpm (1,200g) for 5 min at RT.
8. Transfer 220 μL of the aqueous (upper) phase into a new plate already containing 220 μL phenol/chloroform (mixed in 1:1 ratio) in each well. Cover the plate with a 96-well storage mat and mix contents by vortexing.
9. Centrifuge the plate at 2,500 rpm for 5 min at RT.
10. Transfer 150 μL of the aqueous phase into a new plate which already contains 7 μL per well of 3 M NaAc and 350 μL per well of 100% EtOH. Cover the plate with a fresh 96-well storage mat and mix contents by turning the plate upside down a few times.

11. Incubate the plate for at least 1.5–2 h at −20°C.
12. Centrifuge the plate at 5,700 rpm (6,100g) for 30 min at 4°C and then decant the ethanol by turning the plate upside down, then tapping it gently on a paper towel for a few seconds.
13. Add 300 µL per well of 70% EtOH to pellet, cover the plate with a 96-well storage mat and mix the solutions by turning the plate upside down a few times.
14. Centrifuge the plate at 5,700 rpm for 15 min at 4°C.
15. Decant the 70% ethanol wash solution by turning the plate upside down, then tapping it gently on a paper towel for few seconds.
16. Dry the pellet in a speedVac for ~1 h with medium heating. The pellet contains total RNA from yeast, including ribosomal and viral RNAs in the largest amounts. The RNA samples can be stored at −80°C until use.

3.3 Viral RNA Analysis with Agarose Gel Electrophoresis

17. Add 30 µL 1X RNA dye mixture to pellet, then cover the plate with a 96-well storage mat and dissolve the pellet by vortexing.
18. Load 10 µL of sample on a 1–1.5% agarose gel (containing 0.5X TBE + 100 mg L^{-1} ethidium bromide) using a multichannel pipette. If you are planning to do Northern blot after electrophoresis, then heat your samples at 85°C for 5 min, then put the plate immediately on ice before loading the samples onto the gel.
19. Perform agarose gel electrophoresis for 75 min at 200 V.
20. Take a digital image of the agarose gel under UV light (305 nm) using a gel documentation system.

3.4 Viral RNA Analysis with Northern Blotting

1. Precut extra-thick filter paper and nylon membrane to the exact size of the agarose gel. Handle the nylon membrane and the filter paper with clean gloves. Mark the membrane with a marker to know the active side, which will have the blotted RNA.
2. Equilibrate the agarose gel, the filter papers, and the nylon membrane in sterile 0.5X TBE buffer for 10–15 min in RNase-free trays at RT with gentle shaking.
3. Prepare gel sandwich on a semidry transblotting instrument as follows. First place the prewetted extra-thick filter paper onto the anode surface (bottom), followed by the nylon membrane; then the equilibrated gel (the wells side of the gel facing up), followed by another extra-thick filter paper on the top. Carefully remove air bubbles between the papers, gel, and the membrane during the preparation of the sandwich.

4. Secure the top cathode and place the cover on.

5. Set power supply for 15 V constant voltage and run the blotting for 30 min.

6. After blotting, wash the nylon membrane briefly with sterile 2X SSC at RT in an RNase-free tray. Put membranes on clean filter paper to remove excess liquid.

7. Place the membrane with the RNA samples facing up on a filter paper and put it into a UV chamber, followed by cross-linking (70 mJ). After removal of the nylon membrane from the UV chamber, it can be used immediately for Northern hybridization or stored between two sheets of regular filter paper at RT.

8. Place the membrane into 2X SSC buffer for 5 min to wet it evenly and then place the membrane into a hybridization tube (the side of the membrane with the RNA sample should face toward the solution). Make sure that no air bubbles are trapped between the membrane and the hybridization tube.

9. Add 15 mL of Ultrahyb (Stratagene) hybridization buffer to the hybridization tube, and then seal it with a cap. Place tubes into hybridization oven.

10. Prehybridize the membranes for ~1 h at 68°C by constantly rotating the tube.

11. Meantime, denature the probe with 50% formamide (final concentration) at 85°C for 5 min. Preparation of single-stranded RNA or oligo DNA probe can be found in (13).

12. Add 75 μL of denatured probe into the hybridization tube containing the prehybridization solution and the membrane and then seal it with a cap.

13. Hybridize the probe to the viral RNA on the membrane overnight at 68°C by constantly rotating the tube.

14. At the end of hybridization, discard hybridization solution. Collect all the radioactive liquid waste in a labeled container if radioactive probe was used.

15. Add ~50 mL of Northern wash solution I to the hybridization tube, seal the tube and then rotate the tube in the hybridization oven for 5 min at 68°C.

16. Repeat the above washing step twice with ~100 mL Northern wash solution I for 30 min at 68°C.

17. Then repeat the washing step twice with ~100 mL Northern wash solution II for 15 min at 68°C.

18. At the end of the washing steps, remove the membrane from the hybridization tube, dry it on a clean filter paper, and then wrap it with Saran wrap.

19. Place the wrapped membrane into the phosphorimager cassette in such a way that the side carrying the viral RNA faces the phosphorimager screen.

20. Scan the phosphorimager screen with a PhosphoImager after sufficient amount of exposure of the membrane to the phosphorimager screen (~2–48 h).

4 Notes

1. It is critical to avoid cross-contamination of yeast strains during the entire manipulation! However, each strain carries a barcode that can be verified using predesigned PCR primer sequences, which are available at *http://www-sequence.stanford.edu/group/yeast_deletion_project/PCR_strategy.html/*

2. G418 is necessary only if you work with the YKO library or the yTHC library. But keep in mind that only the mutant strains are resistant to G418, whereas the parental strains are G418 sensitive.

3. Requirement for shaking: Because yeast cells sediment quickly, it is important to shake the plates vigorously (~300 rpm) for fast growth.

4. Sealing of plates: Tight closing of the plates during centrifugation and culturing is important to avoid cross-contamination.

5. Accurate pipetting: The interphase between phenol/chloroform and the aqueous phase should not be pipetted to new plates during phenol/chloroform extraction.

6. Reagents purity: RNase-free reagents and instruments are critical to obtain high-quality viral RNA samples. Each step should be performed with RNase-free solutions and equipment. Use of clean RNase-free gloves is recommended. RNase contamination from contaminated equipment or surfaces can be removed by using hot 10% SDS or RNase AWAY from Ambion, Inc.

7. Trouble shooting:

Absence of yeast colonies after transformation: Use of inadequate amount of yeast cells and/or not enough plasmids for transformation.

Lack of yeast growth: Incorrect choice of media/antibiotics. Carefully match selection markers in the plasmids and the composition of the minimal media.

Absence of viral RNA in control samples: Contamination of samples and/or instruments with RNase.

8. Time considerations: Transformation of 96 or 2 × 96 yeast strains can be performed in a single day. Incubation of transformed yeast takes 2–4 days. Growing yeast strains in plates for virus RNA extraction takes 2 days, whereas RNA extraction from 96 or 2 × 96 yeast strains, followed by agarose gel electrophoresis and semidry blotting could be done in a single day. Northern blotting and phosphoimager-based analysis require an additional day or two depending on (i) the amount of target RNA, which can vary between low and high depending on the accumulation of viral RNA and (ii) the labeling efficiency of the probe. Each step can be scaled down or up as needed to optimize the screening process based on the capacity of the given laboratory.

5   Anticipated Results

The level of viral RNA replication in individual yeast strains should be compared with the amount of virus RNA accumulation in the parental yeast strain. The ribosomal RNA levels, which are detectable on the agarose gel, could be used as internal standards to reduce sample-to-sample variations. The amount of yeast cells can also be measured using a spectrophotometer with plate-reading capability.

Altogether, the presented high-throughput methods are suitable to analyze genome-wide libraries, such as the yeast YKO and the yTHC, for the effect of single genes on tombusvirus replication (4, 5, 11, 12). These methods are also expected to be suitable for analyzing other viruses, which are engineered to replicate in yeast.

Acknowledgments   This work was supported by NIH-NIAID.

References

1. Kushner, D. B., Lindenbach, B. D., Grdzelishvili, V. Z., Noueiry, A. O., Paul, S. M., and Ahlquist, P. (2003) *Proc Natl Acad Sci USA* **100,** 15764–9.
2. Ahlquist, P., Noueiry, A. O., Lee, W. M., Kushner, D. B., and Dye, B. T. (2003) *J Virol* **77**, 8181–6.
3. Nagy, P. D., and Pogany, J. (2006) *Virology* **344**, 211–20.
4. Serviene, E., Shapka, N., Cheng, C. P., Panavas, T., Phuangrat, B., Baker, J., and Nagy, P. D. (2005) *Proc Natl Acad Sci USA* **102**, 10545–50.
5. Panavas, T., Serviene, E., Brasher, J., and Nagy, P. D. (2005) *Proc Natl Acad Sci USA* **102**, 7326–31.
6. Panaviene, Z., Panavas, T., and Nagy, P. D. (2005) *J Virol* **79**, 10608–18.
7. Panaviene, Z., Panavas, T., Serva, S., and Nagy, P. D. (2004) *J Virol* **78**, 8254–63.
8. Panavas, T., and Nagy, P. D. (2003) *Virology* **314**, 315–25.
9. Serva, S., and Nagy, P. D. (2006) *J Virol* **80**, 2162–9.
10. Gietz, R. D., and Woods, R. A. (2002) *Methods Enzymol* **350**, 87–96.
11. Serviene, E., Jiang, Y., Cheng, C. P., Baker, J., and Nagy, P. D. (2006) *J Virol* **80**, 1231–41.
12. Jiang, Y., Serviene, E., Gal, J., Panavas, T., and Nagy, P. D. (2006) *J Virol* **80**, 7394–404.
13. Sambrook, J., Maniatis, T., and Fritsch, E.F. (1989). Molecular Cloning: A Laboratory Manual, 2nd edition, Cold Spring Harbor Laboratory Press, Cold Spring Harbor, NY.

Chapter 42
Phosphorylation of Movement Proteins by the Plasmodesmal-Associated Protein Kinase

Jung-Youn Lee

Abstract Plant viruses encode movement proteins (MPs) which play important roles in spreading their infectious materials throughout host plants. This infection is facilitated by cell-to-cell trafficking of MPs through specialized channels termed plasmodesmata, which involves specific interactions between MPs and host factors. Recently, we have reported the identification of a host protein kinase named plasmodesmal-associated protein kinase (PAPK) which specifically phosphorylates a subset of noncell autonomous proteins in vitro, including MPs of *Tobacco mosaic virus* (TMV) and *Bean dwarf mosaic virus* (BDMV). Biochemical purification of PAPK was achieved by developing a method in which a series of liquid chromatographic separations of plasmodesmal-enriched subcellular fractions was coupled with phosphorylation assays using TMV MP as a substrate. Application of this approach may prove useful in isolating other host kinases that interact with various viral components.

Keywords Protein kinase; Movement protein; Phosphorylation; Posttranslational modification; TMV; Plasmodesmata; Casein Kinase 1

1 Introduction

Protein phosphorylation provides a central regulatory mechanism in signal transduction and cell signaling pathways in many organisms. It appears that plant viruses exploit this host mechanism during the infection process in which a host protein kinase(s) phosphorylates various viral components (1–6). Among these, phosphorylation of the 30-kD movement protein (MP) of *Tobacco mosaic virus* (TMV) within the infected host cells has been shown to play an important role in regulating cell-to-cell movement of this protein (7–10). In order to isolate the host protein kinase responsible for this MP phosphorylation, we set out a biochemical approach in which TMV MP was used as a specific substrate and plasmodesmal-enriched cell wall proteins (PECP) as a source for the protein kinase activity (11). The purified enzyme was named plasmodesmal-associated protein kinase (PAPK) as its plasmodesmal localization was presumed and subsequently demonstrated. Substrate specificity test showed that PAPK could phosphorylate other non-cell autonomous proteins including a MP of *Bean dwarf mosaic virus* (BDMV) in addition to TMV MP. These results and previous studies on phosphorylation of TMV MP support a model that PAPK has broad substrate specificity and may be involved in modulating the activity of MPs of various types of plant viruses (12).

An overall scheme for PAPK purification from tobacco cell culture is depicted in Fig. 1. PECP fractions were prepared from tobacco cells by eluting proteins from clean cell wall pellets. It is important to note that purification of PAPK was successful, despite its low expression level, chiefly due to a protocol that we have developed to isolate PECP fraction on a large scale by utilizing a plant cell suspension culture system (13). Enrichment of PAPK activity within PECP subfractions was tested by in vitro phosphorylation of TMV MP which was prepared by denaturation and renaturation of inclusion bodies formed during its expression in *Escherichia coli*. Subsequently, a series of liquid chromatographic protein separations coupled with in-solution phosphorylation assays were performed to purify PAPK (Fig. 2). Finally, in-gel phosphorylation assays were employed to identify a specific protein band as the PAPK within a partially purified fraction that contained few protein bands.

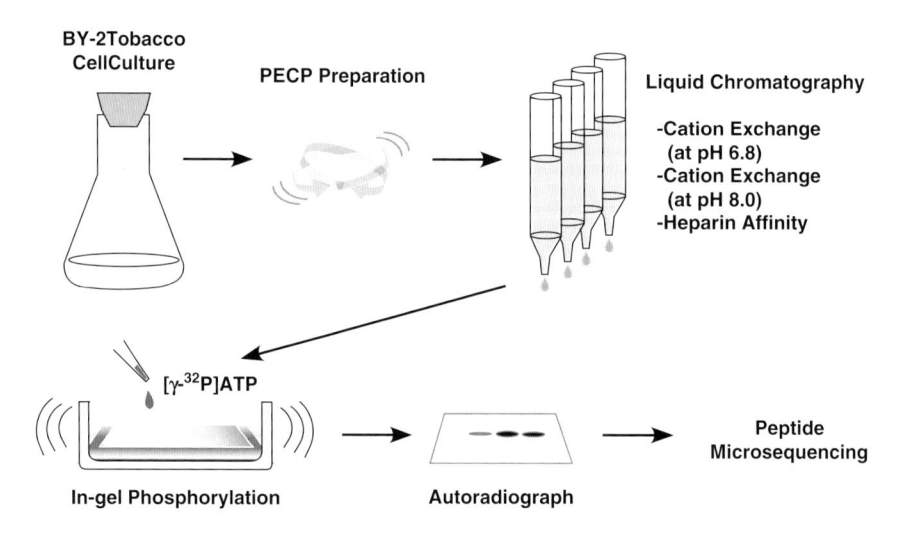

Fig. 1 Overall scheme of PAPK purification. Suspension cultured tobacco cells (BY-2) were harvested and homogenized to prepare plasmodesmal-enriched cell wall proteins (PECP). PAPK was purified from PECP by further fractionation through a series of liquid chromatography involving ion exchange and heparin columns. PAPK activity was followed by phosphorylation assays using TMV MP as a substrate. After PAPK was sufficiently purified as a discrete protein band in a SDS-PAGE gel and confirmed by in-gel phosphorylation, its molecular identity was revealed by peptide microsequencing

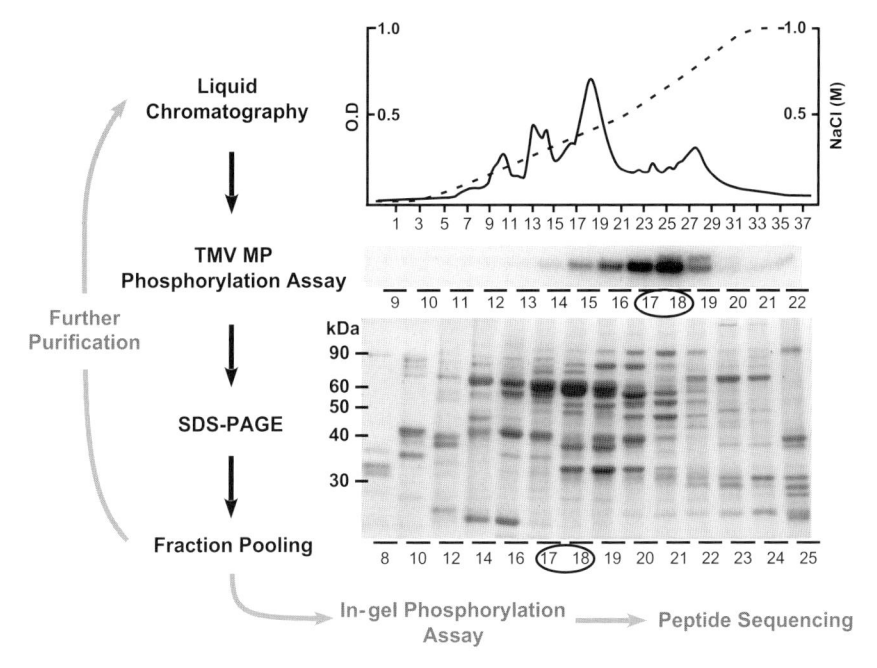

Fig. 2 An example exhibiting PAPK purification steps. The PECP subfraction was loaded onto a HiTrap-SP column equilibrated with buffer A and fractionated by salt gradient. The collected fractions were then tested for PAPK activity by phosphorylation of TMV MP and for protein profiles by SDS-PAGE. Finally, fractions with high PAPK activity and low protein complexity (i.e., fractions #17 and 18) were pooled for further purification. Following several rounds of liquid chromatography, the protein band corresponding to PAPK was confirmed by in-gel phosphorylation and subjected to protein microsequencing

2 Materials

2.1 Tobacco Cell Suspension Culture

1. Culture media: 4.3 g L^{-1} MS salt (Sigma), 3% (w/v) sucrose, 0.5 g L^{-1} MES, 1 mg L^{-1} thiamine, 0.1 g L^{-1} myo-inositol, 200 mg L^{-1} KH$_2$PO$_4$. Adjust pH of media to 5.7 with 1–5 N KOH before autoclaving. Store sterile media at room temperature (see note 1).
2. Suspension culture and inoculation: wide-bore pipettes (sterile and disposable), culture flasks, foam plugs, aluminum foil sheets, and shaker. Cell inoculation needs to be performed in a laminar-flow clean bench.
3. Fernbach flasks for large-scale cultures

2.2   Expression and Purification of MPs

1. Transformed expression host cells (e.g., BL21 pLys [DE3]) (see note 2)
2. Protease inhibitors: Leupeptin, Aprotinin, and PMSF. Dissolve Leupeptin and Aprotinin (USBiochemical) in nano-pure water to $10\,\text{mg mL}^{-1}$ (1000X). Store in aliquots at -20°C (see note 3). Prepare $100\,\text{mM}$ PMSF (100X) in 100% EtOH. Store at -20°C.
3. Lysis Buffer: B-Per (Pierce) supplemented with protease inhibitors (1X), $10\,\text{mM}$ EDTA, $0.2\,\text{mg mL}^{-1}$ lysozyme. Make the buffer just before use.
4. IPTG (1 M): store in aliquots at -20°C.
5. Denaturation buffer stock solutions (10X each): 20% (w/v) SDS and $1\,\text{M}$ $NaHCO_3$, pH 9.0. Store each stock at room temperature.
6. Renaturation buffer (5X): $50\,\text{mM}$ Tris–HCl, pH 8.0, $5\,\text{mM}$ EDTA, $500\,\text{mM}$ NaCl, 50% glycerol. Store at room temperature.
7. DTT: weigh DTT powder as needed each time before use and add directly to the dialysis buffer. Dissolve DTT by stirring the buffer on a magnetic stirrer.
8. Dialysis apparatus: Dialyser Cassette (Pierce) with MMCO 10,000

2.3   PECP Preparation

1. Homogenization buffer (10X): $400\,\text{mM}$ HEPES, pH 6.8, $200\,\text{mM}$ KCl, $50\,\text{mM}$ KH_2PO_4, $10\,\text{mM}$ EDTA. Store at 4°C. Add glycerol to 10% when preparing 1X homogenization buffer (buffer H).
2. Detergent wash buffer (10X): 10% (w/v) CHAPS in 1X buffer H
3. BeadBeater homogenizer (BioSpec) equipped with 350-mL chambers and 1-mm glass beads
4. Protein concentrators: Amicon Centricon Plus 80 with MWCO 10,000 (Millipore); Microcon with MWCO 10,000 (Millipore)

2.4   In-solution Phosphorylation Assays

1. [γ-^{32}P] ATP ($3000\,\text{Ci mmol}^{-1}$ [Amersham Biosciences]): Radioactive materials are hazardous. Please follow institutional guidelines and instructions for the handling and use of radioactive materials and for the proper disposal of the wastes (see note 4).
2. Phosphorylation assay buffer (4X): $200\,\text{mM}$ Hepes, pH 7.5, $40\,\text{mM}$ $MgCl_2$, $8\,\text{mM}$ EGTA. Store at 4°C.
3. DTT ($20\,\text{mM}$): store at -20°C.
4. Protein dilution buffer (1X): $5\,\text{mM}$ Hepes, pH 7.5

5. Prepare 3 mM ATP (6X stock) in nano-pure water and store in aliquots at −20°C. Make some working stocks (0.5 mM) out of these and store in small (typically 0.1–0.2 mL) aliquots at −20°C (see note 5).
6. Hot ATP mix (5X): 90 µL of 500 µM ATP mixed with 2 µCi of [γ-^{32}P] ATP
7. X-ray film, cassettes, and X-ray film developer
8. Gel air-drying kit: frames and cellophane sheets (Fisher). Gels can be dried in a frame inside a hood overnight or for 2–3 h using a gel drier (Bio-Rad) inside a hood (see note 6).

2.5 SDS-Polyacrylamide Gel Electrophoresis (PAGE)

1. Gel running buffer (10X): 250 mM Tris–HCl, pH 8.3, 1.92 M glycine, 1% SDS, stored at room temperature.
2. Resolving buffer (4X): 1.5 M Tris–HCl, pH 8.8, 0.4% SDS, stored at room temperature.
3. Stacking buffer (4X): 0.5 M Tris–HCl, pH 6.8, 0.4% SDS, stored at room temperature.
4. Acrylamide mix (40T:3C): 40% acrylamide/bis solution, stored at 4°C.
5. Ammonium persulfate (10%): prepare a small quantity (e.g., 1 mL) of the solution in water when needed. Store at room temperature up to ~10 days.
6. N,N,N,N'-tetramethyl-ethylenediamine (TEMED), stored at 4°C.
7. SDS sample-loading buffer (6X): 10% SDS, 60% (v/v) glycerol, 360 mM Tris–HCl, pH 6.8, 6 mM DTT, 0.025% (w/v) bromophenol blue. Store in aliquots at −20°C.

2.6 Protein Purification using Liquid Chromatography

1. Fast Performance Liquid Chromatography (FPLC)
2. Prepare all buffers in cold nano-pure water, filter through a membrane (0.22-µ pore-size), and de-gas on ice. Store at 4°C.
3. Filtered, de-gased, nano-pure water; store at 4°C.
4. HiTrap (Amersham-Pharmacia) columns (1 mL size): e.g., HiTrap-SP (cation exchanger) and HiTrap-Heparin
5. HiTrap-SP chromatography: equilibration buffer contains 40 mM HEPES, pH 6.8, 2 mM EDTA, 10% glycerol (buffer A). Elution buffer contains 40 mM HEPES, pH 6.8, 2 mM EDTA, 10% glycerol, 1 M NaCl (buffer B). Also, prepare the same buffers in pH 8.0.
6. HiTrap Heparin chromatography: equilibration buffer contains 40 mM HEPES, pH 6.8, 2 mM EDTA (buffer C). Elution buffer contains equilibration buffer plus 1 M NaCl (buffer D).

2.7 In-gel Protein Phosphorylation

1. [γ-^{32}P] ATP (3000 Ci mmol^{-1} [Amersham Biosciences]): Radioactive materials are hazardous. Please follow institutional guidelines and instructions for the handling and use of radioactive materials and for the proper disposal of their wastes (see note 4).
2. Prepare all the following buffers fresh except 0.5 M HEPES, pH 7.5, which can be stored at room temperature.
3. Gel wash buffer: 20% isopropanol, 50 mM HEPES, pH 7.5
4. Gel incubation buffer (buffer P): 50 mM HEPES, pH 7.5, 2 mM β-mercaptoethanol
5. Denaturation buffer: buffer P containing 6 M guanidine–HCl
6. Renaturation buffers: R1, buffer P containing 3 M guanidine–HCl; R2, buffer P containing 1 M guanidine–HCl; R3, buffer P containing 0.04% Tween 20
7. Equilibration buffer: 50 mM HEPES, pH 7.5, 10 mM MgCl$_2$, 2 mM EGTA
8. Phosphorylation assay buffer (1X): equilibration buffer supplemented with 2 mM DTT and ATP mix (10 μCi mL^{-1} [γ-^{32}P] ATP)
9. Stop solution: 5% Trichloro acetic acid (TCA), 1% sodium pyrophosphate
10. Proteins: partially purified PAPK fraction for in-gel auto and substrate phosphorylation and TMV MP for in-gel substrate phosphorylation
11. Large plastic containers and square Petri dishes

3 Methods

Isolation of a host kinase that phosphorylates a certain viral movement protein involves preparation of protein materials, including a specific MP as a substrate and a large quantity of plant cell extract as a source for the host kinase activity. Viral movement proteins can be expressed in *E. coli* but usually form inclusion bodies; however, a fairly pure inclusion body preparation can be achieved as described below, eliminating the need for affinity tags which might compromise the activity of viral movement proteins. It is important to note that the availability, on a large scale, of a subcellular fraction in which the target kinase activity is enriched is essential before setting up a purification scheme.

In order to aid PAPK purification, PECP was subfractionated in a manner that the proteins were eluted sequentially from clean cell walls by applying step gradients of calcium. Tobacco cells were used to optimize this protocol; however, it can be applied to culture cells derived from other plant species or to plant tissues. The advantages of using cell cultures for PECP preparation are that they are easily obtainable in large scale in a short period and small space, which can provide an analytical amount of PECP fraction, and that effective cell disruption method, on a large scale, is available.

For the purification of PAPK from a PECP subfraction, a series of liquid chromatographic separations using FPLC was set up, which was coupled with phosphorylation assays of the resulting fractions to probe enzyme activity (see Fig. 2). Finally, in-gel auto and substrate phosphorylation assays (14) were critical steps in determining which protein band corresponded to the PAPK in a partially purified fraction containing several discrete protein bands when resolved by SDS-PAGE. This approach also helped eliminate further purification steps which would be required for purifying proteins to near homogeneity.

3.1 Maintenance of Cell Suspension Culture

1. BY-2 tobacco suspension-culture cells can be grown at 23–25°C (without the need for a growth chamber) with agitation on a shaker (~150 rpm). Maintain the culture by transferring 2-mL inoculum to 100 mL fresh media every 7 days (see note 1).
2. For a large-scale culture, transfer 50 mL of inoculum to 750 mL of media (made in each Fernbach flask) and culture for 5–7 days before harvesting the cells. Usually, four Fernback flasks (total of 3.2 L culture) are sufficient to collect ~1 kg cells.

3.2 Expression and Purification of Viral Movement Proteins

1. Inoculate transformed BL21 cells into 2-mL LB medium supplemented with antibiotics (LB$^+$) and culture overnight at 37°C while shaking at 250 rpm.
2. On the following day, start the second culture in fresh LB$^+$ medium at a dilution of 1:200 (cell volume: medium volume) and grow cells to OD ~0.7 allowing about 3–4 h of growth for a second culture volume of 250 mL. The rest of the procedure assumes a 250-mL culture volume.
3. Induce protein production by adding IPTG to 0.5 mM and continue the culture for an additional 3 h at 37°C.
4. Collect the bacterial cells by centrifugation at 5,000g for 15 min at 4°C. Decant the medium and remove the liquid as much as possible. Store cell pellets at −80°C until cell lysis (see note 7).
5. Resuspend the frozen cell pellet in 12-mL bacterial cell lysis buffer. Pipet up and down using wide-bore 25-mL glass pipettes to loosen up and resuspend the pellet (see note 8).
6. Incubate the resuspended cells at room temperature with gentle shaking for 20 min until the resuspended cells become fully lysed.
7. Sonicate the samples to complete cell lysis by repeating 20–30 s pulses (see note 9).
8. Spin samples down for 5 min at 13,000g. Discard the supernatant.
9. Resuspend the pellets with 0.1X B-Per and resuspend the pellets by brief sonication.
10. Repeat steps 8–9 to wash the pellets. Additional washes may be performed.
11. Spin the samples down for 5 min at 13,000g. Discard the supernatant.
12. Resuspend the pellets (inclusion bodies) in 3 mL of 1X denaturation buffer containing 2% SDS and 0.1 M NaHCO$_3$, pH 9.0, and pulse sonicate into solution.

13. Heat the resuspended inclusion body pellets in boiling water for 4 min.
14. Cool samples on ice for 2 min (see note 10).
15. Perform renaturation of proteins by a stepwise dialysis in which the ratio between renaturation and denaturation buffers is gradually increased from 20:80 to 100:0 (see note 11). Set up and run the dialysis at room temperature. Change buffers 2–3 times a day with each dialysis lasting at least 4 h.
16. Repeat the last dialysis against 100% renaturation buffer for 4–16 h at room temperature.
17. Centrifuge dialyzed samples at 20,000g for 15 min to remove insoluble proteins.
18. Check protein purity and concentration by running various volumes (i.e., 1–5 μL) of samples in SDS-PAGE (see Sect. 3.5). Estimate the concentration of samples by comparing their band intensities with those of BSA standards run in the same gel and stained with GelCode Blue (Pierce).
19. Concentrate proteins, if necessary, by using Amicon protein concentrator units.

3.3   Cell Homogenization and PECP Preparation

1. Harvest BY-2 cells (~1 kg) grown in Fernbach flasks by filtering 3.2 L of the culture through Miracloth (Calbiochem).
2. Incubate the cell paste in 3 L of a 75 mM calcium solution by gentle stirring at room temperature for 1 h. This helps to remove some noncovalently bound cell wall proteins. Meanwhile, set up for the cell homogenization in a cold room and make sure that all the buffers and homogenization apparatuses are cooled to 4°C before use.
3. Drain the buffer and collect the cells by filtering through large funnels of which are layered with a sheet of Miracloth. Rinse the cell paste by pouring water over it and draining.
4. Weigh the collected cell paste and transfer to the cold room for the homogenization process.
5. Add ~150 mL of cold buffer H (1X) containing a mixture of 1X protease inhibitors to each BeadBeater (BioSpec Products) chamber which is half filled with cooled 1-mm glass beads. Transfer ~90 g cell paste to each chamber and fill the chambers to the top with buffer H. Carefully close the chambers, excluding as much air as possible (see note 12).
6. Mount the chamber onto the BeadBeater motor. Homogenize the cells by running the BeadBeater for 1 min and cooling at least 5 min before another run (see note 13).
7. Check the homogenate under a microscope to determine whether additional runs are needed to achieve >90% cell disruption.
8. Transfer the cell homogenate to precooled 500-mL centrifuge bottles by decanting the supernatant and leaving the beads in the chamber. Recover as much homogenate as possible by rinsing the beads a few times with 1X buffer H and collecting all the suspended homogenate.

9. Collect the cell wall pellets by centrifuging the cell homogenate at low speed (5 min at 300–500g). Remove the supernatant by carefully and slowly decanting so as not to disturb the loose cell wall pellets.

10. Wash the cell wall pellets by resuspending them in 1X buffer H followed by low-speed centrifugation. Collect the supernatant and store temporarily at 4°C for the subsequent analyses (protein quantification, SDS-PAGE, and phosphorylation assays).

11. Repeat the wash step several times until white-colored pellets are obtained.

12. Estimate the volume of the cell wall pellet in each centrifuge bottle and add an equal volume of 2% CHAPS buffer. Incubate the cell wall suspension for 1 h with vigorous agitation at room temperature. Brief sonication (3–5 times of a 30-s pulse) of the solution helps dispersion of the pellet.

13. Centrifuge the cell wall suspensions at 3,000g for 15 min and save the supernatant as the Detergent-extract; store temporarily at 4°C for further analyses.

14. Wash the residual pellets by resuspension in 1X buffer H. Collect washed pellets by centrifugation at 3000g for 15 min.

15. Extract PECP subfractions by incubating the cell wall pellets with a step gradient of 30, 75, 100, and 200 mM of calcium. Add an equal volume of solution to the cell wall pellet and resuspend by brief sonication.

16. Extract proteins by vigorously shaking the resuspended pellet for 1 h at room temperature or overnight at 4°C.

17. Centrifuge the mix at 3,000g for 15 min and save the supernatant as a PECP subfraction and save the pellet for the next round of calcium extraction.

18. Repeat steps 15–16 until the 200-mM calcium extract is recovered.

19. Filter each subfraction (supernatant, detergent-extract, and PECP) through filter paper (Q2 paper, Fisher Scientific) to get rid of insoluble particles. Concentrate the proteins by ultrafiltration using Centricon Plus-80 (Millipore).

20. Dialyze the concentrated protein samples against a buffer containing 50 mM HEPES, pH 7.5, 2 mM EGTA at 4°C. Centrifuge the dialyzed samples at 20,000g for 15 min to remove any aggregates.

21. Determine protein concentration by Bradford assay (Bio-Rad). If necessary, concentrate samples further by using Centricon or Microcon concentrators until a desired concentration (2–3 mg mL^{-1}) is achieved.

22. Check the protein profile of each subcellular fraction by SDS-PAGE followed by staining with GelCode Blue (Pierce) (see Sect. 3.5).

3.4   In-solution Phosphorylation Assays

1. Label 1.5-mL microfuge tubes and precool them on ice. Place all the buffers, proteins, and other reagents on ice. Assemble and keep the assays on ice until incubation at 30°C in a water bath.

2. The following phosphorylation assays are performed in a 50-μL reaction volume (see note 14). Prepare an enzyme mix containing: 10 μL 5X assay buffer; 15 μL nano-pure water; 5 μL of 20 mM DTT; 1 μg of substrate proteins (TMV MP, BDMV BC1, etc.) in which the volumes are adjusted to 5 μL; and 5–10 μg of PECP subfractions with the volumes adjusted to 5 μL. In the case of FPLC fractions, take 5 μL from each fraction. Mix the components well by very gentle and brief vortexing as each component is added to the tubes (see note 15).

3. Centrifuge the tubes briefly at 4°C to collect the liquid at the bottom of each tube.

4. All the following procedures involving [γ-^{32}P] ATP should be performed behind a radiation-safe shield in a properly equipped radiation work area (see note 4). Prepare the hot ATP mix containing 100 μL of 500 mM cold ATP and 2 μCi of [γ-^{32}P] ATP (3,000 Ci mmol^{-1} [Amersham Biosciences]). Vortex gently and centrifuge the mix briefly at room temperature. Dispose contaminated pipette tips, tubes, and gloves in a dry-radioisotope waste container.

5. Preincubate the tubes containing the enzyme mix at 30°C for 10 min. Transfer each tube to a water bath at 30-s intervals.

6. Add 10 μL hot ATP mix to each enzyme mix at 30-s intervals, mix well by gentle vortexing, and incubate at 30°C for 10 min for the phosphorylation reactions.

7. Stop the phosphorylation reactions by adding 10 μL 6X SDS sample loading buffer and vortexing.

8. Resolve the reactions in 10 or 12% SDS-PAGE (see Sect. 3.5). Load 15–30 μL of each reaction mixture in the gel.

9. Rinse the gel with water and discard appropriately in a liquid-radioisotope waste container after checking the radioisotope contamination with a Geiger counter. Stain the gel with GelCode Blue (Pierce). Dry the gel for exposure to X-ray film.

10. Assemble the exposure cassette and store at −80°C for 3–16 h before developing.

3.5 SDS-PAGE Analysis of Purified Proteins

1. These instructions are for mini-gel systems purchased from CBS Scientific, but are adaptable to those from other suppliers.

2. For the gel-cassette assembly, prepare a clean set of glass plates, a gasket, two spacers (1-mm thick), a 10-well comb (1-mm thick), and three clips. Clean the glass plates first with water, and then with 70% ethanol, followed by acetone using Kim Wipes (Fisher) before assembling the cassette.

3. Prepare the resolving gel solutions. For a 12% resolving gel, make a 10-mL solution containing 2.5 mL of 4X resolving buffer, 3 mL acrylamide mix, 4.5 mL

water, 50 μL ammonium persulfate, and 5 μL TEMED. Add the TEMED just before pouring the gel.

4. Pour the gel leaving a space (up to ~1 cm below the tip of a comb) for the stacking gel. Slowly overlay the gel solution with water-saturated butanol and allow the gel to polymerize at room temperature. Remove the butanol and rinse the top surface of the gel gently with deionized water. Blot away excess liquid.

5. Prepare the stacking gel solution containing 1.25 mL of 4X stacking buffer, 0.5 mL acrylamide mix, 3.2 mL water, 50 μL ammonium persulfate, and 2 μL TEMED. Insert a comb such that one end of the comb is plunged in the stacking gel space and the other end is sticking out in the air. Pipette in the gel solution through the open side of the gel assembly and slowly lower the comb with care not to trap any air bubbles. Allow the gel to polymerize at room temperature.

6. Remove the clips, gasket, and comb and rinse the wells of the gel with water. Make sure there are no bits of acrylamide left in any of the wells.

7. Clip the gel, front glass side inward, to the SDS-PAGE gel running unit. If only one gel is being run, clip a back glass plate to the opposite side of the unit to prevent the buffer from running out. Fill the top of the unit with 1X gel running buffer and make sure that there is no leakage. Fill the bottom reservoir with running buffer.

8. Cover the unit with a lid and connect to a power supply. Run the gel at 20 mA through the stacking gel and at 25 mA through the resolving gel until the dye front reaches the bottom of the gel.

3.6 Protein Purification using FPLC

1. These instructions assume the use of and work experience with the AKTA FPLC system (Amersham-Pharmacia). Handling, temporary storage, and purification of protein samples need to be carried out at 4°C.

2. Prepare protein samples by dialyzing against buffer A, pH 6.8, overnight for the fractionation through a cation exchange column (e.g., HiTrap SP column).

3. Centrifuge the sample at 20,000g for 20 min and transfer the supernatant to a new tube and store until loading onto a FPLC column (HiTrap SP).

4. Wash FPLC pumps thoroughly with water and fill the pumps with equilibration buffer A and B, pH 6.8.

5. Attach a 1-mL HiTrap-SP column to the FPLC. Wash the column, at a flow rate of 1 mL min^{-1}, first with 10 mL of buffer A, pH 6.8, and then with 10 mL of buffer B, pH 6.8.

6. Equilibrate the column with 10 mL of buffer A, pH 6.8, or until the absorbance recording becomes stabilized at a base line.

7. Load the column with a protein sample (or a pool of selected fractions) at a flow rate of 0.5 mL min^{-1} (see note 16). Collect 1-mL fractions throughout the chromatography except the flowthrough which can be collected in larger fractions, depending on the loading size.

8. Wash the column with 10–20 mL of equilibration buffer. Elute proteins with a linear salt gradient made by mixing buffers A and B in a gradient volume of 20 mL (see Fig. 2, top panel).

9. Upon completed elution, regenerate the column by washing with 5 mL of buffer B followed by 10 mL of buffer A. The column can be stored at 4°C and reused.

10. Pick every other fraction for phosphorylation assays (see Sect. 3.4) as demonstrated in Fig. 2. Examine the protein profiles of these fractions by resolving the proteins in SDS-PAGE (see Sect. 3.5).

11. Pool the peak fractions with TMV MP-phosphorylating activity and dialyze against buffer A, pH 8.0, for the second round of purification using the cation exchange at higher pH than the first round. Prepare FPLC with a new HiTrap-SP column equilibrated with buffer A at pH 8.0. Repeat purification steps 3–10.

12. Pool the peak fractions and dialyze the sample against buffer C to perform further purification by employing a HiTrap Heparin column. Prepare FPLC and a Heparin column for chromatography with buffers C and D by basically repeating the procedures described in steps 3–10 except collecting fractions of smaller size (0.5 mL).

13. Choose the peak fractions for further analysis to identify a discernable protein band as the PAPK by in-gel phosphorylation assays (see Sect. 3.7).

3.7 In-gel Phosphorylation Assays

1. During in-gel phosphorylation assays, incubation of the gel in a container with various buffers involves gentle agitation on a shaker at room temperature. It is critical to perform the assays with freshly made buffers. In-gel substrate phosphorylation, in which TMV MP was polymerized to a final concentration of ~300 μg mL^{-1} polyacrylamide gel, was performed basically following the procedure described below for in-gel autophosphorylation assays (see note 4).

2. Resolve a peak fraction selected from the Heparin chromatography and phosphorylation analyses by SDS-PAGE (see Sect. 3.5).

3. Following the SDS-PAGE, transfer the gel into a large plastic container filled with 200 mL of gel wash buffer. Cover the container and incubate the gel with gentle shaking for 1 h to remove SDS from the gel. Decant the buffer and repeat this gel wash with fresh buffer.

4. Transfer the gel into a container filled with 200 mL of buffer P. Incubate the gel with gentle shaking for 1 h to remove isopropanol.

5. Transfer the gel into a container filled with 100 mL of denaturation buffer. Incubate the gel with gentle shaking for 1 h.

6. Transfer the gel into a new container. For protein renaturation, incubate the gel with gentle shaking for 3 h in 100 mL of buffer R1 followed by 3-h incubation in 100 mL of buffer R2. Finally, transfer the gel into 100 mL of buffer R3 and incubate with gentle shaking for an additional 3 h (see note 17).

7. Repeat the incubation of the gel in 100 mL of fresh buffer R3.

8. Transfer the gel into a container filled with 100 mL of equilibration buffer. Incubate the gel with gentle shaking for 15 min. Change the buffer and equilibrate the gel for an additional 15 min.

9. Perform the following steps involving the use of [γ-^{32}P] ATP behind a radiation-safe shield in a properly equipped radiation work area (see note 4). Transfer the gel into a small square Petri dish and add 6 mL of phosphorylation assay buffer. Cover the dish and incubate the gel with gentle shaking for 1 h.

10. Remove the assay solution and discard appropriately as radiation waste. Rinse the gel briefly with water and discard the solution as radiation waste after checking the contamination with a Geiger counter.

11. Transfer the gel into a large plastic container filled with 500 mL of phosphorylation stop buffer. Incubate the gel with gentle shaking for 1 h. Discard the solution as radiation waste after checking the contamination with a Geiger counter.

12. Repeat the gel treatment with fresh stop buffer three additional times. Discard the solution as radiation waste after checking the contamination with a Geiger counter.

13. Stain the gel with GelCode Blue (Pierce). Dry the gel for exposure to X-ray film. Assemble the exposure cassette and store at −80°C for 3–16 h before developing.

14. Determine the protein band corresponding to the PAPK based on the autoradiograph and the protein profile.

4 Notes

1. Make fresh media before depleting old media. When using a fresh batch of media for the first time, include an additional inoculation using a flask of old media from the batch that has previously been confirmed to be free of contamination and for normal cell growth. This will provide a backup culture in case the fresh media has any problems.

2. We use bacterial host BL21 pLys [DE3] for expression of movement proteins because it worked well for production of inclusion body. Also, this host cell is lysed by B-Per (Pierce) fairly efficiently.

3. Multiple freeze–thawing is allowed but make small aliquots. Refreeze the stock if not depleted.

4. A radiation-safe work area is essential for kinase assays involving [γ-^{32}P] ATP. The typical work area for the assays will need to be equipped with shields, a Geiger counter, a water bath set at 30°C with a microfuge tube rack or equivalent, a vortex, a centrifuge, dry and liquid-radioisotope waste containers, and a timer. Monitoring contamination by wipe-test and scintillation counting before and after the use of radioisotope and proper disposal of contaminated materials are required. Use two pairs of gloves when handling radioisotope and performing phosphorylation assays. This will provide an additional protection against direct contamination of bare hands in case the outer gloves are punctured or torn. Check hands frequently with a Geiger counter for contamination and change the outer gloves if inner gloves are free of contamination.

5. ATP working stocks should not be refrozen but disposed if thawed once.

6. Overdrying tends to crack the gel. Do not leave the gel inside the hood for too long.
7. Freezing helps cell lysis.
8. It is important to resuspend the cell pellet well for maximum protein recovery. Use a spatula to break frozen cell pellets in lysis buffer prior to pipetting. The cell lysate will turn milky when a large quantity of inclusion bodies is present.
9. Sonication in 50-mL centrifuge or Falcon tubes is acceptable. If dealing with a smaller cell culture, use several 1.5-mL eppendorf tubes instead.
10. Do not let samples sit on ice longer than 2 min, as this will precipitate SDS and proteins. If this happens, the protein will be of no use.
11. For example, we perform six steps of dialysis with an increasing ratio of renaturation buffer to denaturation buffer (20:80, 40:60, 60:40; 80:20, and two changes at 100:0).
12. Minimizing air bubbles in the chamber is very important for efficient homogenization. If the color of the homogenate changes from bright yellow to brown, this might indicate too much air and poor homogenization.
13. We usually find at least five runs are needed for effective cell disruption. Make sure that the homogenates are kept cool during and between processing. Use of wet ice helps.
14. This reaction volume can be proportionally adjusted as needed. We also perform 25-μL assays routinely when the protein concentration is high enough to allow for a smaller assay volume to be used.
15. Set the vortex at low to medium speed. Avoid vigorous mixing in any case when dealing with protein samples as it may result in denaturation of the proteins.
16. The flow rate can be adjusted up to 1 mL min^{-1} depending on the back-pressure.
17. At this point, the gel could be left in buffer R3 overnight at room temperature.

Acknowledgments The author would like to thank Byung-Chun Yoo for helpful discussion, Dong-Jin Kim for technical assistance, and Jessica Habashi for proofreading the manuscript. This work was supported by National Science Foundation grant (MCB0445626) and National Institute of Health grant (P20 RR-15588) from the National Center for Research Resources.

References

1. Atabekov, J. G., Rodionova, N. P., Karpova, O. V., Kozlovsky, S. V., Novikov, V. K., and Arkhipenko, M. V. (2001) Translational activation of encapsidated potato virus X RNA by coat protein phosphorylation. *Virology* **286**, 466–474.
2. Kim, S. H., Palukaitis, P., and Park, Y. I. (2002) Phosphorylation of cucumber mosaic virus RNA polymerase 2a protein inhibits formation of replicase complex. *EMBO J.* **21**, 2292–2300.
3. Matsushita, Y., Yoshioka, K., Shigyo, T., Takahashi, H., and Nyunoya, H. (2002) Phosphorylation of the movement protein of Cucumber mosaic virus in transgenic tobacco plants. *Virus Genes* **24**, 231–234.
4. Citovsky, V., McLean, B. G., Zupan, J. R., and Zambryski, P. (1993) Phosphorylation of tobacco mosaic virus cell-to-cell movement protein by a developmentally regulated plant cell wall-associated protein kinase. *Genes Dev.* **7**, 904–910.
5. Ivanov, K. I., Puustinen, P., Gabrenaite, R., Vihinen, H., Ronnstrand, L., Valmu, L., Kalkkinen, N., and Makinen, K. (2003) Phosphorylation of the potyvirus capsid protein by protein kinase CK2 and its relevance for virus infection. *Plant Cell* **15**, 2124–2139.
6. Shapka, N., Stork, J., and Nagy, P. D. (2005) Phosphorylation of the p33 replication protein of Cucumber necrosis tombusvirus adjacent to the RNA binding site affects viral RNA replication. *Virology* **343**, 65–78.

7. Waigmann, E., Chen, M. H., Bachmaier, R., Ghoshroy, S., and Citovsky, V. (2000) Regulation of plasmodesmal transport by phosphorylation of tobacco mosaic virus cell-to-cell movement protein. *EMBO J.* **19**, 4875–4884.
8. Trutnyeva, K., Bachmaier, R., and Waigmann, E. (2005) Mimicking carboxyterminal phosphorylation differentially effects subcellular distribution and cell-to-cell movement of Tobacco mosaic virus movement protein. *Virology* **332**, 563–577.
9. Karpova, O. V., Ivanov, K. I., Rodionova, N. P., Dorokhov, Y. L., and Atabekov, J. G. (1997) Nontranslatability and dissimilar behavior in plants and protoplasts of viral RNA and movement protein complexes formed in vitro. *Virology* **230**, 11–21.
10. Karpova, O. V., Rodionova, N. P., Ivanov, K. I., Kozlovsky, S. V., Dorokhov, Y. L., and Atabekov, J. G. (1999) Phosphorylation of tobacco mosaic virus movement protein abolishes its translation repressing ability. *Virology* **261**, 20–24.
11. Lee, J. Y., Taoka, K., Yoo, B. C., Ben-Nissan, G., Kim, D. J., and Lucas, W. J. (2005) Plasmodesmal-associated protein kinase in tobacco and Arabidopsis recognizes a subset of non-cell-autonomous proteins. *Plant Cell* **17**, 2817–2831.
12. Lee, J. Y., and Lucas, W. J. (2001) Phosphorylation of viral movement proteins - regulation of cell-to-cell trafficking. *Trends Microbiol.* **9**, 5–8.
13. Lee, J. Y., Yoo, B. C., Rojas, M. R., Gomez-Ospina, N., Staehelin, L. A., and Lucas, W. J. (2003) Selective trafficking of non-cell-autonomous proteins mediated by NtNCAPP1. *Science* **299**, 392–396.
14. Yoo, B. C., Lee, J. Y., and Lucas, W. J. (2002) Analysis of the complexity of protein kinases within the phloem sieve tube system – characterization of Cucurbita maxima calmodulin-like domain protein kinase 1. *J. Biol. Chem.* **277**, 15325–15332.

Chapter 43

Virus-Induced Gene Silencing as a Tool to Identify Host Genes Affecting Viral Pathogenicity

Xiaohong Zhu and S.P. Dinesh-Kumar

Abstract Host factors are crucial determinants of viral pathogenicity. Identifying host factors and their contributions to virus infections may lead to the development of novel antiviral strategies. The recently developed virus-induced gene silencing (VIGS) approach offers a rapid means to knock down expression of a given gene in plants. VIGS can be used to determine biological function of candidate genes or to discover new genes that play a role in a given biological pathway. Here, we describe genome-wide Tobacco rattle virus (TRV)-based VIGS screening methods to identify host factors involved in viral pathogenicity.

Keywords Virus-induced gene silencing; Host factors; Viral pathogenicity; RNAi; TRV-based VIGS; Functional genomics; TMV; *Nicotiana benthamiana*

1 Introduction

A successful virus invasion of the host results from a compatible host–virus interaction (1–4). Viruses with relatively small genomes start their infection cycle by replicating within the infected cells, then spread locally from initially infected cells to neighboring cells through plasmodesmata, and finally move into the vascular system to systemically infect the plant (2). Although the concept that host proteins are involved in most steps of virus infection is well established, the identification and functional characterization of host factors with roles in virus replication, cell-to-cell movement and long distance movement are still difficult to accomplish.

In yeast, a high-throughput approach has been developed to identify host factors that affect brome mosaic virus (BMV) and tomato bushy stunt virus (TBSV) replication by using a yeast single-gene deletion library (5, 6). However, this systematic screen is limited by the potential underestimation of host factors due to functional redundancy and artificially high viral replication levels resulting in reduced dependence on host factors. Genetic screens in the model plant *Arabidopsis thaliana* have uncovered several loci that affect viral replication (7–9) or systemic movement (10–12).

From: *Methods in Molecular Biology, Vol. 451, Plant Virology Protocols:*
From Viral Sequence to Protein Function
Edited by G.D. Foster, I.E. Johansen, Y. Hong, and P.D. Nagy © Humana Press, Totowa, NJ

Biochemical approaches using viral replicase proteins and movement proteins (MP) as probes have isolated several cellular interacting proteins from affinity purifications, gel overlay assays, or cDNA expression library screens. Subunits of host translation initiation factor eIF3 in barley and tomato were copurified with BMV (13) and TMV (14) replication proteins, respectively. A tobacco cell wall enzyme pectin methylesterase (PME) (15) and microtubule-associated protein MPB2C (16) that interacts with TMV MP were isolated using gel bolt overlay binding assay and a membrane-based yeast interaction screen respectively. In a cDNA expression library screen, two putative transcriptional coactivators KELP and MBF1 were isolated as interactors of tobamovirus (ToMV) MP (17, 18). Although most of the identified cellular interacting partners described above were confirmed by other biochemical approaches, their biological roles in viral pathogenicity have not yet been determined.

Virus-induced gene silencing (VIGS) or RNA interference (RNAi) approaches provide a useful method for identification of host factors involved in viral pathogenicity. RNAi is a posttranscriptional gene silencing (PTGS) mechanism in plants (19). Double-stranded RNA (dsRNA) or hairpin RNA introduced into plants transiently, or by transgenes, can induce sequence-specific degradation of host mRNA homologous to the introduced sequence that effectively suppresses the target gene expression. RNAi circumvents the limitations of other reverse-genetic tools including insertional mutagenesis and antisense approaches (20, 21). VIGS is an RNAi approach that is based on the fact that RNA viruses are both inducers and targets of PTGS (22). A virus genome is engineered to carry exogenous sequences within a T-DNA expression cassette without loss of infectivity. The recombinant virus can trigger RNAi response against virus and foreign sequences when delivered into plants by Agroinfiltration (21, 23, 24).

We have developed an improved tobacco rattle virus (TRV)-based VIGS system to study gene function in plants (25, 26). VIGS has been used successfully to study biological functions of candidate genes and also to perform genome-wide screens in a high-throughput manner (21, 24, 27, 28). Therefore, a reverse genetic approach based on VIGS offers the opportunity to validate biological functions of candidate host factors and also will aid in identification of new host factors that play a role in a viral pathogenicity. Below, we describe a TRV-based VIGS procedure for identification of host factors in *Nicotiana*–TMV model system. However, this method could be adapted for other host–virus interaction system to identify host factors involved viral pathogenicity.

2   Materials

1. *Agrobacterium tumefaciens* strain C58C1 (GV2260)
2. 200 mM 3′–5′ Dimethoxy 4′-hydroxy acetophenone (acetosyringone)
3. Dimethyl formamide (DMF)
4. 1 M 2-[*n*-Morpholino] ethane sulfonic acid (MES)
5. 1 M Magnesium chloride ($MgCl_2$)

6. Infiltration medium (10 mM $MgCl_2$, 10 mM MES, and 200 μM acetosyringone)
7. TRV-based VIGS vectors: pTRV1 and *Nicotiana benthamiana* normalized cDNA library constructed in pTRV2
8. Positive control, pTRV2-*NbPDS* (*Phytoene desaturase*)
9. *A. tumefaciens* strain C58C1 harboring TMV-GFP plasmid
10. 96-well plate and 96-deep-well box
11. 96-pronger
12. 1-mL Syringe
13. Mortar and pestle (sterile)
14. Cheesecloth
15. Sponge (sterile)
16. *N. benthamiana* seeds
17. Watering tray, 90 mm square pots and propagation domes (Myers Industries, Inc. OH, USA)
18. Lite Carts (Indoor Gardening Supplies, MI, USA)
19. Water-soluble fertilizer (Peat-Lite®, The Scotts Company OH, USA)
20. UV lamp (B100AP/R, Mineralogical Research Co. CA, USA)
21. Razor blade

3 Method

TRV is a bipartite positive sense RNA virus. The TRV genome consists of two RNA molecules: RNA1, which encodes RNA-dependent RNA polymerase, the movement protein, and cysteine-rich 16K protein; and RNA2, which encodes the coat protein and two nonstructural proteins of 40K and 30K. Since the 40K and 30K sequences are not required for virus infection by Agroinfiltration, we have removed these sequences and inserted a multiple cloning site (MCS) to clone target gene sequence for silencing (25). cDNA clones of RNA1 and modified RNA2 were inserted into a T-DNA expression cassette to generate T-DNA plasmids of RNA1 (pTRV1) and RNA2 (pTRV2). To initiate VIGS, *A. tumefaciens* cultures harboring pTRV1 and pTRV2 carrying a normalized cDNA library clone inserted into the MCS are mixed in a 1:1 ratio and infiltrated into leaves of *N. benthamiana*. The VIGS of a gene corresponding to the cDNA sequence in the library will normally occur around 10 days postinfiltration (dpi). To monitor the effect of silenced gene on TMV infection in *N. benthamiana*, a GFP tagged TMV (TMV-GFP) is used to allow one to track virus replication, cell-to-cell movement, and long-distance movement.

3.1 Plant Preparation

1. Germinate seeds in pots covered by propagation domes at 23–25°C.
2. Transplant 2-week-old seedlings into individual pots inserted in a tray and cover with a propagation dome. Uncover the trays 3 days after transplanting and fertilize seedlings once with water-soluble fertilizer (see note 1).

3.2   Agrobacterium Preparation

3.2.1   Preparation of pTRV1 Containing *Agrobacterium*

Transform pTRV1 to *Agrobacterium* strain GV2260 and select transformants on LB plates with Kanamycin (50 μg/mL), Rifampicin (25 μg/mL), Streptomycin (50 μg/mL), and Carbenicillin (50 μg/mL) (hereinafter referred to as KRSC). Inoculate pTRV1 *Agrobacterium* clone into 2 mL of KRSC-LB and allow cultures to grow 48 h at 26°C. From this, prepare glycerol stock and store at −80°C.

3.2.2   Preparation of pTRV2 Normalized cDNA Library Containing *Agrobacterium*

1. Transform pTRV-cDNA library into *Agrobacterium* strain GV2260 by electroporation (Bio-Rad, CA, USA). Select transformants on LB plates with KRSC.
2. Pick *Agrobacterium* clones individually using toothpick and array into 96-deep-well box containing 0.6 mL of KRSC-LB and allow cells to grow for 48 h at 26°C. Prepare two glycerol stocks in 96-well box format from each culture using an automatic, liquid-handling system (Biomek® FX Beckman Coulter, Inc., CA, USA) or automatic multichannel pipette and store these stock boxes at −80°C.

3.2.3   Preparation of Cultures Containing pTRV1

1. Streak out *Agrobacterium* glycerol stocks containing pTRV1 onto KRSC-LB plate and incubate at 26°C for 48 h
2. Inoculate pTRV1 *Agrobacterium* into 100 mL KRSC-LB and allow it to grow overnight at 26°C with vigorous shaking (200 rpm)
3. Harvest cells by spinning at 3,000 rpm for 30 min. Resuspend cells in the infiltration media at OD_{600}~1.0 (see note 2)

3.2.4   Preparation of Cultures Containing pTRV2 Normalized cDNA Library

1. Inoculate pTRV2 normalized cDNA library *Agrobacterium* glycerol stocks in 96-well boxes onto KRSC-LB plates using a 96 pronger (see note 3). Allow plates to incubate for 48 h at 26°C.
2. Inoculate *Agrobacterium* clones from plates, using a 96 pronger into a 96 deep-well box where each well contains 1.2 mL of KRSC-LB and a 2-mm diameter glass bead (see note 4). Allow the culture to grow overnight at 26°C with vigorous shaking (200 rpm).

3. Harvest cells by spinning at 3,000 rpm for 30 min. Resuspend cells in 0.1 mL infiltration media by vortexing (see note 5). Randomly choose six samples from 96 wells and check optical density (OD) at a wavelength of 600 nm (OD_{600}). Adjust the final volume and OD_{600} of each well to 0.6 mL and ~1.0, respectively.

3.2.5 Mixing of Cultures Containing pTRV1 and pTRV2-cDNA Library

Add 0.6 mL of OD_{600} = ~1.0 pTRV1 culture prepared in Sect. 3.2.1 to each well containing 0.6 mL pTRV2-cDNA library culture prepared in Sect. 3.2.2. Incubate this final infiltration mixture at room temperature for 3 h (see note 6).

3.2.6 Preparation of Controls

1. Transform pTRV2 without any insert and pTRV2-*NbPDS* into *Agrobacterium* strain GV2260 and incubate at 26°C for 48 h. From this, prepare glycerol stock and store at −80°C.
2. Streak out *Agrobacterium* glycerol stocks containing pTRV2 alone and pTRV2-*NbPDS* onto KRSC-LB plate and incubate at 26°C for 48 h.
3. Inoculate pTRV2 alone and pTRV2-*NbPDS* Agro separately into 5 mL KRSC-LB and allow it to grow overnight at 26°C with vigorous shaking (200 rpm).
4. Harvest cells by spinning at 3,000 rpm for 30 min. Resuspend cells in the infiltration media at OD_{600}~1.0.
5. Mix equal volume of pTRV1 containing *Agrobacterium* culture prepared in Sect. 3.2.1 and cultures containing pTRV2 alone or pTRV2-*NbPDS*. Incubate this final infiltration mixture at room temperature for 3 h.

3.3 Infiltration of Agrobacterium into Plants

1. Fill a 1-mL needle-less syringe with the mixture of *Agrobacterium*
2. Make a small 0.05–0.1-mm slit using the corner of razor blade in the two lower leaves of four-leaf stage *N. benthamiana* plants (see note 7). Place the opening of the 1-mL needle-less syringe containing the *Agrobacterium* mixture over the slit and place your finger from other hand on the opposite face of the leaf. Then slowly squeeze the syringe to infiltrate the *Agrobacterium* mixture and allow it to spread through the entire leaf. Infiltrate two plants for each TRV2-cDNA clone during initial screening (see note 8).
3. Infiltrate two plants with pTRV2 alone and pTRV2-*NbPDS* Agro mixture as controls prepared in Sect. 3.2.6.
4. Maintain infiltrated plants at 23–25°C in light carts or growth chamber (see note 9).
5. The VIGS of a gene corresponding to your cDNA library sequence in TRV2 vector occurs between 6 and 10 days postinfiltration of *Agrobacterium* (see note 10).

3.4 TMV-GFP Virus Preparation

1. Grow *Agrobacterium* containing TMV-GFP plasmid in KRSC-LB overnight at 26°C and resuspend cells in infiltration media. Adjust OD_{600} = ~1.0 and incubate the infiltration culture at room temperature for 3 h (see note 11).
2. Infiltrate two leaves of six-leaf stage *N. benthamiana* plants using 1-mL needleless syringe as described in Sect. 3.3.
3. Monitor upper leaves for TMV-GFP spread under UV light and collect the leaves, which are almost completely filled with virus, approximately 10–12 dpi. Grind leaf tissue in sterile water on ice and filter the homogenate through four layers of cheesecloth. Keep the filtered homogenate as TMV-GFP virus inoculum on ice until used or store at −80°C for later use (see note 12).

3.5 TMV-GFP Inoculation

1. Infect the upper leaves of the silenced plants around 10–12 dpi of pTRV2-cDNA library by rubbing TMV-GFP inoculum onto the surfaces of the leaves using a sterile sponge.
2. Infect TRV2 alone infiltrated plants and nonsilenced plants with TMV-GFP in a similar manner.
3. Start to monitor plants for the presence of GFP on the inoculated leaf around 2 days after infection of TMV-GFP by observing fluorescence under UV illumination. Continue to follow spread of virus from initial infection sites, if any are observed into upper leaves.

If the silenced host gene has no effect on TMV infection, then GFP fluorescence in the inoculated leaves and upper leaves will be similar to that of nonsilenced or TRV alone control plants infected with TMV-GFP. If a silenced host gene is required for virus replication and/or cell-to-cell movement, it is anticipated that no or less GFP fluorescence is observed on the TMV-GFP inoculated leaves compared to the control. If the silenced host genes are essential for systemic movement of virus, the small infection foci are expected to appear only on the inoculated leaves and no or delayed GFP fluorescence is anticipated on the systemic upper leaves (see note 13).

3.6 Confirmation of Candidate Genes

1. To confirm VIGS of candidate cDNA clones that shows aberrant virus replication or movement, retest these for the effect on TMV-GFP pathogenicity using at least six independent plants. Only those clones that show VIGS phenotype in all six independent plants tested are carried for further analyses.
2. Rescue plasmids from *Agrobacterium* containing cDNA library clone (see note 14) and sequence the candidate genes.

3. Conduct RT-PCR to determine the degree and specificity of silencing by using a primer pair, upstream or downstream flanking the region of the targeted genes.

4 Notes

1. Watering tray provides consistent reservoir of water to each plant. Seedlings are fertilized once using water-soluble fertilizer after being uncovered to achieve easy infiltration.
2. pTRV1 Agro culture has to be prepared on the same day as pTRV2-cDNA library.
3. Two Agro plates are prepared from each glycerol stock plate. The 96 pronger is sterilized by soaking in 70% ethanol for 2 min and dried with paper towels.
4. Glass beads are used to facilitate aeration within wells and accelerate uniform growth. It is easy to resuspend cells in infiltration media by leaving glass beads inside wells when vortexing box.
5. Acetosyringone can be dissolved in DMF or DMSO and stored as aliquots of 200 mM stock solution at −20°C.
6. Minimum time, which is required to activate *vir* genes of Agro, is 3 h.
7. Four-leaf stage plants are optimal for infiltration in terms of silencing efficiency.
8. To prevent cross contamination, caution must be taken by carefully handling the culture and changing gloves and syringe between infiltrations.
9. Temperature affects efficiency of VIGS. Reproducible silencing results are normally obtained at 23–25°C. If multiple layers of cart are used, the temperature of each layer may vary due to the heat released from the lamps on the lower layer.
10. Control *NbPDS* silencing phenotype that is photobleaching of green leaves resulting from silencing of *phytoene desaturase* gene should be visible between 6 and 10 dpi of *Agrobacterium*.
11. Infiltration of TMV-GFP is normally done on the fourth day after the infiltration of pTRV2-cDNA library so that virus inoculums will be prepared to infect the silenced plants 7 days later.
12. Virus preparation can be stored at −80°C. However, fresh virus inoculums are recommended for robust infection.
13. In addition to monitoring fluorescence, it is suggested that the transcription or expression level of replicase protein or coat protein or movement protein is documented by a northern blot or western blot on the confirmation stage.
14. Plasmids can be rescued by extracting DNA from Agro cells using DNA extraction kit (QIAGEN Inc. CA, USA) and transforming plasmid DNA into *E. coli* cells.

Acknowledgments We thank Tessa Burch-Smith for critical reading of the manuscript and Yule Liu for helpful comments. VIGS work in S.P.D.-K. laboratory is supported by National Science Foundation Plant Genome Grant DBI-0211872.

References

1. Nelson, R. S., and Citovsky, V. (2005) *Plant Physiol 138*, 1809–1814.
2. Waigmann, E., Ueki, S., Trutnyeva, K., and Citovsky, V. (2004) *Crit Rev Plant Sci 23*, 195–250.

3. Whitham, S. A., and Wang, Y. (2004) *Curr Opin Plant Biol 7*, 365–371.
4. Ahlquist, P., Noueiry, A. O., Lee, W. M., Kushner, D. B., and Dye, B. T. (2003) *J Virol 77*, 8181–8186.
5. Kushner, D. B., Lindenbach, B. D., Grdzelishvili, V. Z., Noueiry, A. O., Paul, S. M., and Ahlquist, P. (2003) *Proc Natl Acad Sci USA 100*, 15764–15769.
6. Panavas, T., Serviene, E., Brasher, J., and Nagy, P. D. (2005) *Proc Natl Acad Sci USA 102*, 7326–7331.
7. Lellis, A. D., Kasschau, K. D., Whitham, S. A., and Carrington, J. C. (2002) *Curr Biol 12*, 1046–1051.
8. Tsujimoto, Y., Numaga, T., Ohshima, K., Yano, M. A., Ohsawa, R., Goto, D. B., Naito, S., and Ishikawa, M. (2003) *EMBO J 22*, 335–343.
9. Yamanaka, T., Ohta, T., Takahashi, M., Meshi, T., Schmidt, R., Dean, C., Naito, S., and Ishikawa, M. (2000) *Proc Natl Acad Sci USA 97*, 10107–10112.
10. Lartey, R. T., Ghoshroy, S., and Citovsky, V. (1998) *Mol Plant Microbe Interact 11*, 706–709.
11. Whitham, S. A., Anderberg, R. J., Chisholm, S. T., and Carrington, J. C. (2000) *Plant Cell 12*, 569–582.
12. Chisholm, S. T., Parra, M. A., Anderberg, R. J., and Carrington, J. C. (2001) *Plant Physiol 127*, 1667–1675.
13. Quadt, R., Kao, C. C., Browning, K. S., Hershberger, R. P., and Ahlquist, P. (1993) *Proc Natl Acad Sci USA 90*, 1498–1502.
14. Osman, T. A. M., and Buck, K. W. (1997) *J Virol 71*, 6075–6082.
15. Dorokhov, Y. L., Makinen, K., Frolova, O. Y., Merits, A., Saarinen, J., Kalkkinen, N., Atabekov, J. G., and Saarma, M. (1999) *FEBS Lett 461*, 223–228.
16. Kragler, F., Curin, M., Trutnyeva, K., Gansch, A., and Waigmann, E. (2003) *Plant Physiol 132*, 1870–1883.
17. Matsushita, Y., Deguchi, M., Youda, M., Nishiguchi, M., and Nyunoya, H. (2001) *Mol Cells 12*, 57–66.
18. Matsushita, Y., Miyakawa, O., Deguchi, M., Nishiguchi, M., and Nyunoya, H. (2002) *J Exp Bot 53*, 1531–1532.
19. Baulcombe, D. (2004) *Nature 431*, 356–363.
20. Waterhouse, P. M., and Helliwell, C. A. (2003) *Nat Rev Genet 4*, 29–38.
21. Burch-Smith, T. M., Anderson, J. C., Martin, G. B., and Dinesh-Kumar, S. P. (2004) *Plant J 39*, 734–746.
22. Baulcombe, D. C. (1999) *Curr Opin Plant Biol 2*, 109–113.
23. Robertson, D. (2004) *Annu Rev Plant Biol 55*, 495–519.
24. Lu, R., Martin-Hernandez, A. M., Peart, J. R., Malcuit, I., and Baulcombe, D. C. (2003) *Methods 30*, 296–303.
25. Liu, Y., Schiff, M., Marathe, R., and Dinesh-Kumar, S. P. (2002) *Plant J 30*, 415–429.
26. Liu, Y., Schiff, M., and Dinesh-Kumar, S. P. (2002) *Plant J 31*, 777–786.
27. Lu, R., Malcuit, I., Moffett, P., Ruiz, M. T., Peart, J., Wu, A. J., Rathjen, J. P., Bendahmane, A., Day, L., and Baulcombe, D. C. (2003) *EMBO J 22*, 5690–5699.
28. Liu, Y., Schiff, M., Czymmek, K., Talloczy, Z., Levine, B., and Dinesh-Kumar, S. P. (2005) *Cell 121*, 567–577.

Chapter 44
Yeast Two-Hybrid Assay to Identify Host–Virus Interactions

Stuart A. MacFarlane and Joachim F. Uhrig

Abstract The small size of most plant virus genomes and their very limited coding capacities requires that plant viruses are dependent on proteins expressed by the host plant for all stages of their life cycle. Identification of these host proteins is essential if we are to understand in any meaningful way the interactions that exist between virus and plant. A variety of methods are now available to isolate and study interacting proteins, however, the yeast two-hybrid (Y2H) assay system, which was one of the earliest mass analysis methods to be developed [Nature 340:245–246, 1989] remains one of the most popular and amenable approaches in current use.

The Y2H method works by expressing two candidate interacting proteins together in the yeast cell. The (bait and prey) proteins under study are fused either to a promoter-specific DNA-binding domain or to a transcription activation domain. Interaction in the yeast nucleus between the bait and prey proteins brings the transcription activation and DNA-binding domains together so that they can initiate expression of a reporter gene. The reporter may be nonselective, such as the β-galactosidase (LacZ) protein, or be selective by complementing a chromosomal mutation in a metabolic pathway for, for example, leucine or histidine biosynthesis. Individual bait proteins can be screened for interaction against a library of prey proteins, with any yeast colonies that grow on selective plates containing potential interacting partners.

Using the Y2H system, a number of plant proteins interacting with viral proteins have been identified, recently, increasing our knowledge of the molecular basis of viral infection and host defense mechanisms.

Keywords GAL4; LexA; Interaction mating; 3-amino-1,2,4-triazole (3AT); Yeast colony PCR; Gap-repair cloning

From: *Methods in Molecular Biology, Vol. 451, Plant Virology Protocols:*
From Viral Sequence to Protein Function
Edited by G.D. Foster, I.E. Johansen, Y. Hong, and P.D. Nagy © Humana Press, Totowa, NJ

1   Introduction

In the yeast two-hybrid (Y2H) system, the interaction of any pair of (bait and prey) proteins is linked to the expression (transcription) of a number of reporter genes that either allow the growth of the yeast cell on a particular selective medium or can be monitored by some colorimetric test (for reviews on recent methodological developments see refs. 2–4). The system relies on the finding that many transcriptional activator proteins are comprised of modular domains for promoter binding and recruitment of RNA polymerase, and that these domains remain functional when separated from one another. By fusing them to two other proteins that interact with each other, the DNA-binding domain (BD) and transcription activation domain (AD) are brought into close proximity and are able to promote gene expression.

Y2H systems use two different combinations of BD and AD modules. In one, both the BD and AD are from GAL4, a yeast transcription enhancing factor that activates genes involved in galactose metabolism. In the presence of galactose, GAL4 activates transcription by binding to a galactose-specific upstream activating sequence (UAS$_G$) in the promoter regions of these genes. In yeast strains that have been constructed for the Y2H system, both the chromosomal GAL4 gene and the Gal80 gene (which inhibits GAL4 activity) have been deleted. This enables the GAL4 that is reconstituted by bait and prey fusion protein interaction to be active in the absence of galactose and in the presence of glucose.

An alternative system uses the DNA-BD of the *Escherichia coli* lexA protein, which is a repressor protein that regulates expression of genes in the SOS pathway that respond to stresses such as radiation damage to DNA. The lexA protein binds to a specific sequence known as the LexA operator, and single or multiple copies of this sequence are introduced into the promoter regions of the reporter genes that are used for Y2H analysis using this protein. The lexA BD is combined with a variety of AD constructs including, the GAL4 AD, the Herpes Simplex Virus VP16 protein (which is a very strong transcriptional activator) and a protein encoded by intergenic *E. coli* sequences, the B42 "acid blob," which is a weaker activator than the GAL4 AD.

The bait/prey interaction needs to take place in the yeast nucleus to activate reporter gene expression. The GAL4 protein naturally locates to the yeast nucleus, whereas the lexA protein does not. This may be overcome either by increasing the level of expression of the lexA-bait fusion protein, or by including a nuclear localization signal, e.g., SV40 large T antigen NLS, into the fusion protein. All the plasmids used in both Y2H systems contain both yeast and bacterial replication origins and so can be grown in bacteria for all cloning steps and transferred to yeast for the screening steps.

The lexA "interaction trap" Y2H system was designed by researchers in the laboratory of Roger Brent, and a large amount of background information as well as detailed protocols are posted on the Web pages of the Russ Finley and Erica Golemis laboratories (http://proteome.wayne.edu/Update.html, http://www.fccc.edu/research/labs/golemis/InteractionTrapInWork.html). A yeast strain appropriate for this system is EGY48, in which the UAS sequence of the chromosomal LEU2 gene is replaced with three copies of the dimeric LexA operator. This re-engineered gene is not expressed under normal conditions and the yeast does not grow in the absence of exogenously supplied leucine. The bait plasmid (e.g., pEG202) uses the constitutive ADH promoter to express the bait:lexA fusion protein and carries the HIS3 gene which complements the his3 mutation in EGY48. Growing the yeast without added histidine ensures that the bait plasmid is retained in the cell. The prey plasmid (e.g., pJG4-5) uses the galactose-inducible GAL1 promoter and fuses the prey protein to the SV40 nuclear localization peptide and the B42 acid blob protein. This plasmid carries the TRP1 gene which complements the trp1 mutation in EGY48 and allows it to be selected by growing the cells without added tryptophan. Both plasmids carry the high copy 2-μ origin of replication and can be maintained simultaneously in yeast, unlike the situation that occurs in most bacteria where plasmids with the same replication origin cannot coreplicate in the same cell. When yeast cells are transformed with plasmids expressing a pair of interacting bait and prey proteins, the lexA BD and acid blob AD come together and are competent to bind to the lexA operator sites upstream of the LEU2 gene and allow growth of the yeast in the absence of added leucine (and histidine and tryptophan). As expression of the prey protein is induced by galactose (supplemented with raffinose) and repressed by glucose, growth on media containing either of these carbon sources can be tested in order to detect nonspecific activation of the LEU2 gene. An alternative bait plasmid, pGilda, expresses the bait:lexA fusion protein from the inducible GAL1 promoter. This is useful if constitutive expression of the bait protein causes toxicity problems. As an additional assay for bait:prey interaction, the yeast cells can be transformed with a third, reporter plasmid (e.g., pSH18-34) which carries the URA3 gene for selection of transformants and the lacZ gene downstream of four dimeric LexA operators to monitor bait:prey interaction. This construct is very sensitive and able to detect weak bait:prey interactions. Other lacZ reporters exist having fewer lexA operators, these are less sensitive and useful for analyzing strongly interacting proteins. Commercial sources of lexA-based Y2H systems include OriGene (DupLEX-A Yeast Two-Hybrid System) and Invitrogen (Hybrid Hunter Two-Hybrid System).

Appropriate yeast strains for the GAL4 Y2H system are AH109, PJ69-4A, Y190, and Y187. Here the bait plasmid (e.g., pAS1, pGBKT7) expresses the bait: GAL4 BD fusion from the ADH1 promoter and carries the TRP1 gene which allows it to be selected for by growing the cells without added tryptophan. The prey plasmid (e.g., pACT2, pGADT7) expresses the prey:GAL4 AD fusion from the ADH1 promoter and carries the LEU2 gene which allows it to be selected for by growing the cells without added leucine. Interaction between the bait:GAL4 BD and prey:GAL4 AD proteins forms a functional GAL4 protein that can activate expression of reporter genes that are integrated into the chromosome of this yeast strain. These are the HIS3 gene (for growth without added histidine), ADE2 (for growth without added adenine), the endogenously GAL4-regulated MEL1 gene, and the lacZ gene from *E. coli*. MEL1 and LacZ encode for an α- and a ß-galactosidase enzyme, respectively, that can be used as Y2H reporters using either qualitative or semiquantitative colorimetric assays.

The different steps involved in Y2H analysis are similar regardless of which system is chosen. A first consideration is whether the system is to be used to analyze the (possible) interaction between two known proteins or whether the aim is to screen a library of prey proteins to find those that interact with a particular bait protein. The first scenario is a straightforward, simple, and rapid process, whereas, the latter is complicated by the need to obtain or produce a library of suitable quality and by the associated increase in scale of all the necessary "hands-on" activities.

However, recent advances in "interaction mating," a method making use of yeast cell conjugation to combine two plasmids in a yeast cell, simplify the screening of Y2H libraries.

For the LexA system, the library is transformed into a MATa yeast strain such as EGY48 or RFY231, a compatible (MATα) bait strain would be RFY206 (5). For GAL4 interaction mating, a suitable combination of strains would be, for example, Y187 (MATα) and PJ69-4A (MATa). The next step is to plate the yeast onto SC media lacking three amino acids (one to select each plasmid and the third, leucine (lexA) or histidine (GAL4), to select for bait:prey interaction. In some Y2H strains for the GAL4 system, the GAL1-UAS regulating the HIS3 gene is not completely silent. Here, 3-amino-1,2,4-triazole (3-AT), a competitive inhibitor of HIS3 activity, can be added to the media to suppress background activity or to adjust assay conditions according to potential autoactivating properties of individual bait proteins. Colonies start to appear on these selective plates after 2–3 days and the fastest growing colonies may contain the most strongly interacting proteins, nevertheless, *bona fide* (weaker) interactors may continue to appear as long as 7–10 days after plating. These colonies are transferred to fresh plates for further testing and archiving. Subsequent assays for growth of the colonies on media without adenine (for GAL4) and activation of lacZ expression identify them as true positives, although, control experiments must be done to show that neither the bait nor prey proteins by themselves activate reporter gene expression.

If no colonies of potential interactors are obtained when screening a library, there may be several explanations. Firstly, the library may not contain any true prey clones, perhaps it was constructed from RNA isolated from a tissue type lacking interactors or collected at the wrong time point in the plant development. The library also may not be sufficiently large or well represented to include all possible clones, and the yeast transformation may not have been efficient enough to obtain clones covering the entire transcriptome. Attention to detail and the sourcing of appropriate RNA populations may overcome these problems. Another possibility is that the bait protein is not expressed adequately and/or is excluded from the nucleus. Immunoblot analysis can be used to check that the bait protein is expressed, and truncation of the protein gene to remove inhibitory domains, for example, transmembrane domains, may improve expression sufficiently to allow Y2H screening.

If colonies with probable interactors are isolated, the task now is to confirm their identity. The prey plasmid DNA can be isolated from yeast (mixed with the bait plasmid) and retransformed into *E. coli* strain KC8. This bacterium has mutations in tryptophan, leucine, and histidine biosynthesis, which can be complemented by the TRP1, LEU2, or HIS3 genes carried on the different Y2H plasmids, allowing the bacterium to grow on minimal media lacking any of these essential amino acids. Once isolated from bacteria, the plasmid can be sequenced using appropriate primers to determine the identity of the prey clone.

Alternatively, using specific primers, the prey sequence can be PCR amplified directly from the yeast cells for cloning and/or sequencing. It is possible that more than one prey plasmid could be present and maintained in a yeast colony that has come through the screening process, and so it is important to retest each clone after isolation in *E. coli* or reconstruction using the PCR-amplified fragment. Most cDNA clones in Y2H libraries will not be full-length and further investigation of interacting protein pairs could require cloning complete prey protein genes into yeast plasmids and rescreening them for specific interaction with the bait protein. However, as was discussed above for bait proteins, inhibitory domains may be present in the full-length prey protein that means it does not behave in yeast exactly as does the truncated protein.

The field of Plant Virology has been slow to utilize the Y2H technique to study plant:virus interactions. The technique has been criticized in that virus and plant proteins may not be expressed correctly (with the necessary posttranslational modifications) in yeast and that the yeast nucleus is not the correct environment in which to analyze virus:plant protein interactions. There is also a feeling that the Y2H system always produces a large amount of false positives that devalue the results obtained by this approach. It is certainly correct that not all interactions will be identified by this approach but the technique includes many controls that can add confidence to the interactions that are detected. Also, it is important that all interactions that are identified in yeast must be corroborated by alternative approaches such as gel overlay or immune precipitation techniques. In actual fact, all available techniques to analyze protein:protein interactions do so under artificial conditions with little or no recognition of the multiprotein complexes that must continually form and reform in the living plant cell.

There is a growing body of publications showing the utility of the Y2H system to the investigation of virus:plant interactions, and these include many functionally different virus proteins: The coat protein (CP) of *Turnip crinkle virus* was found to interact with TIP, a member of the NAC family of proteins that are involved in plant development and defense (6). Further Y2H analysis of truncated CP identified a NAC-interacting domain, and single site substitutions in the CP, that abolished interaction with TIP also allowed the virus to overcome HR-mediated resistance in *Arabidopsis*. In another study, a different NAC protein from tomato was found to interact with the replication enhancer protein (Ren) from the geminivirus *Tomato leaf curl virus* (7). Expression of this protein was increased by virus infection, both the NAC protein and Ren protein colocalized to the plant cell nucleus, and overexpression of the NAC protein increased replication of the virus. A different geminivirus protein, the AL2 protein of *Tomato golden mosaic virus* (as well as orthologous proteins in other geminiviruses), was found to interact with SNF1-kinase, and knockdown of SNF1-kinase expression made plants more susceptible to virus infection (8). AL2 also interacts in Y2H with adenosine kinase (ADK), and virus infection reduces ADK expression in plants, and further work links targeting of

ADK with suppression of (antiviral) RNA silencing (9, 10). Other targets of virus silencing suppressor proteins have also been identified. The *Tobacco etch virus* HC-Pro silencing suppressor interacted in yeast with rgs-CaM, a calmodulin-related protein that itself functions as a silencing suppressor when expressed in plants from a virus vector (11). The Tomato bushy stunt virus P19 silencing suppressor protein interacts with plant ALY proteins that are likely to be involved in transcriptional activation and export of RNAs from the nucleus (12). Coexpression with P19 led to a relocalization of some of the ALY proteins from the nucleus to the cytoplasm, and overexpression of some of the ALY proteins inhibited P19 silencing suppression activity (T. Canto, S. MacFarlane, and J. Uhrig, (13). The RNA helicase domain of the *Tobacco mosaic virus* (TMV) replicase protein interacted in a Y2H screen with an ATPase protein and a subunit of the oxygen-evolving complex of photosystem II (14). Silencing of each of these plant genes by VIGS gave different results, targeting the ATPase reduced TMV accumulation two-fold whereas targeting the photosystem protein increased TMV accumulation 10-fold. In another study, the same domain of the TMV replicase protein interacted with PAP1, a regulator of auxin response genes involved in plant development (15). Infection of plants with TMV reduced PAP1 expression preventing its localization to the nucleus, and virus in which the replicase protein was mutated to abolish interaction with PAP1, could replicate and move normally but produced attenuated disease symptoms. For virus movement proteins (MPs), the *Tomato spotted wilt virus* NSm protein was found to bind to DnaJ-domain proteins, which are regulators of the Hsp-70 chaperone protein that is also known to be involved in closterovirus movement (16). The movement protein of TMV was found to interact with pectin methylesterase, a cell wall protein that was isolated by affinity chromatography of cell wall extracts and immobilized MP (17). As a final example, one of the most widely studied virus: plant protein interactions is that between the small, genome-linked viral protein (VPg) of potyviruses and various host translation initiation factors (eIF4E and eIF(iso)4E) (18, 19). The importance of this interaction has been demonstrated by showing that natural resistance of some plants to various potyviruses is caused by mutations to the eIF4E proteins that make them unable to interact with virus VPg (20, 21).

2 Materials

2.1 *Yeast Strains and Vectors*

LexA-System:

EGY48 (*MATα ura3his3 leu2::3LexAop-LEU2 trp1::hisG LYS2*)
RFY231 (*the same as EGY48 but without the trp1-1 allele*)
RFY206 (*MATa ura3-52 his3&Aelig;200 leu2-3 lys2&Aelig;201 trp1::hisG*)
L40 (*MATa trp1 leu2 his3 LYS2::lexA-HIS3 URA3::LexA-lacZ*)

GAL4-system:

AH109 (*MATa, trp1, leu2, ura3, his3, gal4Δ, gal80Δ, LYS2::GAL1$_{UAS}$-GAL1$_{TATA}$-HIS3, GAL2$_{UAS}$-GAL2$_{TATA}$-ADE2, ura3::MEL1$_{UAS}$-MEL1$_{TATA}$-lacZ*)
PJ69-4A (*MATa trp1-901 leu2-3112 ura3-52 his3-200 gal4. Δ. gal80. Δ. LYS2:: GAL1-HIS3 GAL2-ADE2. met2::GAL7-lacZ.*)
Y187 (*MATA, ura3-52, his3-200, ade2-101, trp1-901, leu2-3,112, gal4Δ, met-, gal80Δ, URA3::GAL1$_{UAS}$-GAL1$_{TATA}$-lacZ*)
Y190 (*MATa, ura3-52; trp1-901; ade2-101; leu2-3; 112 his3-200r; gal4D; gal 80D; URA3:GAL1-lacZ; LYS2::GAL1-HIS3; cyhr*)

Vectors

LexA-system bait: e.g., pEG202
LexA-system prey: e.g., pJG4-5
GAL4-system bait: e.g., pAS2, pGBKT7
GAL4-system prey: e.g., pACT2, pGGADT7

2.2   Media

1. **YPAD**, for 1 L: 20 g Difco peptone, 10 g yeast extract, 100 mg adenine (Sigma A9126), 18 g Agar. Adjust pH to 5.8, autoclave, cool to ca. 55°C, add 50 mL filter-sterilized 40% glucose.
2. **YCM**, for 1 L: 10 g Yeast Extract, 10 g Difco Peptone. Adjust to pH 3.5, autoclave at 121°C for 20 min (the media may appear turbid), cool to ca. 55°C and add 50 mL filter-sterilized 40% Glucose.
3. For YCM agar plates, dissolve 10 g yeast extract and 10 g Difco Peptone in 475 mL H$_2$O, adjust to pH 4.5 and autoclave at 121°C for 20 min. Separately autoclave 475 mL H$_2$O + 15 g Agar. Cool YCM media and Agar to ca. 55°C, mix well and add 50 mL filter-sterilized 40% Glucose.
4. Synthetic complete dropout (SC dropout) medium for 1 L: 6.7 g Difco (or Sigma) Yeast Nitrogen Base (without amino acids). Add the following amino acids, leaving out Leucine and/or Tryptophan and/or Histidine to prepare the respective dropout media:

Alternatively use commercially available amino acid dropout mixtures (e.g., BD Biosciences). Adjust the pH to 5.6. For solid medium, add 15% Agar. Autoclave at 121°C for 15 min, cool to ca. 55°C. Add 50 mL filter-sterilized 40% glucose (final concentration 2%) and, if required add 3–50 mL of a filter-sterilized 1 M 3-amino-1,2,4-triazole (3-AT) solution. Store the 3-AT stock solution at 4°C in the dark.

2.3   Yeast Transformation

1. 50% PEG 4000 (filter sterilized)
2. 1 M lithium acetate (LiAc) (filter sterilized)

Arginine	50 mg L^{-1}
Aspartic Acid	80 mg L^{-1}
Histidine	20 mg L^{-1}
Isoleucine	50 mg L^{-1}
Leucine	100 mg L^{-1}
Lysine	50 mg L^{-1}
Methionine	20 mg L^{-1}
Phenylalanine	50 mg L^{-1}
Threonine	100 mg L^{-1}
Tryptophan	50 mg L^{-1}
Tyrosine	50 mg L^{-1}
Uracil	20 mg L^{-1}
Valine	140 mg L^{-1}
Serine	20 mg L^{-1}
Adenine	120 mg L

3. 100 mM LiAc (filter sterilized)
4. 2 mg mL^{-1} herring sperm DNA, sheared by repeated passage through a hypodermic needle to reduce the viscosity of the solution, denatured by boiling for 10 min and rapid cooling on ice, stored at −20°C.

2.4   Protein Preparation from Yeast Cultures for Western Blotting

1. Glass beads (425–600 µm; Sigma)
2. Protease Inhibitor Stock Solution (prepare fresh): Pepstatin A 0.1 mg mL^{-1}, Leupeptin 0.03 mM, Benzamidine 145 mM, Aprotinin 0.37 mg mL^{-1}
3. PMSF (phenylmethyl-sulfonyl fluoride) stock solution (100X): Dissolve 0.1742 g PMSF (Sigma #P7626) in 10 mL isopropanol. Store at room temperature in the dark
4. Cracking buffer stock solution (for 100 mL): 8 M Urea (48 g), 5% w/v SDS (5 g), 40 mM Tris–HCl (pH 6.8) (4 mL of a 1 M stock solution), 0.1 mM EDTA (20 µL of a 0.5-M stock solution), 0.4 mg mL^{-1} Bromophenol blue (40 mg), deionized H$_2$O (to a final volume of 100 mL)
5. Cracking buffer (sufficient for one protein preparation): Cracking buffer stock solution 1 mL (recipe above), β-mercaptoethanol 10 µL, Protease inhibitor solution 70 µL, prechilled (recipe above), PMSF 50 µL of 100X stock solution

2.5   Y2H Test to Detect Interactions between Individual Bait and Prey Proteins

1. ß-Galactosidase Assay to monitor expression of the LacZ gene

   a) Liquid nitrogen

b) Sterile Whatman #5 or VWR grade 410 paper filter discs
c) Z-Buffer: Dissolve 16.1 g $Na_2HPO_4.7H_2O$, 5.5 g $NaH_2PO_4.2H_2O$, 0.75 g KCl, and 0.264 g $MgSO_4.7H_2O$ in 1 L H_2O, adjust to pH 7.0 and autoclave. Z-buffer can be stored at room temperature for up to 1 year.
d) ß-Mercaptoethanol (100%)
e) X-Gal stock solution: Dissolve 5-bromo-4-chloro-3-indolyl-β-D-galactopyranoside in *N,N*-dimethylformamide (DMF) at a concentration of 20 mg mL^{-1}. Store in the dark at −20°C

2. α-Galactosidase Assay to monitor expression of the MEL1 gene

a) PNP-α-Gal Solution: 100 mM, *p*-nitrophenyl α-D-galactopyranoside (Sigma #N0877) in H_2O. For 10 mL, dissolve 30.1 mg of PNP-α-Gal in 10 mL of H_2O. Filter sterilize. Prepare solution fresh before each use
b) 10X Stop solution: 1 M Na_2CO_3 in deionized H_2O (Sigma #S7795)
c) 1X NaOAc. 0.5 M sodium acetate, pH 4.5 (Sigma #S7545)
d) Assay Buffer: Prepare Assay Buffer fresh, before each use, by combining 2 volumes 1X NaOAc Buffer with 1 volume PNP-α-Gal Solution [2:1 (v/v) ratio]. Mix well

2.6 Screening Libraries with the Y2H System

1. Hemocytometer

2.6.1 Library Screening by Interaction Mating

1. YPAD medium supplemented with 10% PEG6000, filter sterilized (for simplified interaction mating method)

2.7 Analysis of Positive Colonies from Y2H Screenings

1. Plasmid isolation

a) Glass beads (425–600 μm; Sigma)
b) Resuspension Buffer (P1): 50 mM Tris-Cl, pH 8.0, 10 mM EDTA, 100 μg mL^{-1} RNase A
c) Lysis buffer (P2): 200 mM NaOH, 1% SDS (w/v)
d) Neutralization buffer (N3): 3.0 M potassium acetate pH 5.5

2. Yeast colony PCR

a) 10 N NaOH stock solution

3 Methods

The Y2H system is an *in vivo* method that relies on the activation of reporter genes
by two fusion protein constructs in living yeast cells. This means that there are a
number of requirements that the proteins used in this assay have to fulfill. First of
all, the proteins have to be expressed properly in the yeast cell and they have to be
targeted to the nucleus where the interaction takes place. This may be a problem,
if the proteins under investigation themselves contain structural features that target
them to different subcellular compartments. Membrane proteins, for example, are
generally not suitable for Y2H analyses. Furthermore, it is important that both the
bait and the prey proteins do not interfere with reporter gene function and do not
have a negative influence on yeast metabolism and viability in general. In particular,
therefore, the bait proteins should be tested for suitability in advance (see Sect.
3.2). In the following sections, methods for the use of the GAL4-based Y2H
system are described in detail. With minor modifications taking into account the
different selection markers, these methods can also be used with the LexA-based
Y2H system.

3.1 Small-Scale Single and Double Transformation of Yeast Cells

For Y2H experiments involving single bait and prey constructs, and similarly for
the preparation of bait-expressing yeast strains for interaction mating, high efficiency of the transformation is not required. Therefore, standard yeast transformation protocols can be applied. A simple and widely used procedure suitable for a
wide range of yeast strains is the LiAc method

1. Pick a well-growing yeast colony of your favorite yeast strain from a freshly
   streaked plate into 5 mL YPAD medium (pH 5.8) and incubate overnight at
   30°C with shaking (ca. 250 rpm). After 16–18 h, this gives a stationary culture
   of $OD_{600} > 1.5$
2. Subculture 1 mL into 50 mL YPAD and grow at 30°C with shaking for 3½–4 h.
   The OD_{600} should be between 0.6 and 1.2
3. Collect cells by low-speed centrifugation (4,000 rpm, 5 min, room temperature)
4. Discard the medium and resuspend the cells by vortexing in 25 mL of sterile
   dH_2O
5. Respin cells (4,000 rpm, 5 min, room temperature), discard supernatant and
   resuspend in 1 mL of sterile dH_2O. Transfer to a microcentrifuge tube and respin
   at top speed for 10 s to pellet cells
6. Resuspend in 550 µL 100 mM LiAc pH 7.5 and transfer 50-µL aliquots to 11
   sterile microcentrifuge tubes
7. Centrifuge for 10 s at top speed to pellet cells, and remove the supernatant

8. To each tube add, in order

   240 µL 50% PEG 4000
   36 µL 1 M LiAc
   25 µL single-stranded DNA
   50 µL plasmid DNA (250–500 ng of each plasmid)

9. Resuspend and mix thoroughly by pipetting or vigorous vortexing. Incubate at 30°C for 25 min with occasional shaking/mixing, and then incubate at 42°C for 25 min without shaking

10. Centrifuge cells at low speed (4,000 rpm, 10 s), remove medium and resuspend in 200 µL sterile dH$_2$O

11. Spread aliquots onto SC dropout medium (spread gently with spreading bar or using sterile glass beads). Allow to air-dry and incubate at 30°C for 2–3 days for colonies to develop (note 1).

3.2 Pre-Testing of Bait and Prey Constructs

The Y2H system requires the correct expression of two fusion proteins in the yeast cell. Therefore, especially with regard to the interpretation of negative results, the presence and correct size of bait and prey proteins in the yeast should be confirmed. This can be done by western blot analysis using commercially available monoclonal antibodies to the GAL4-BD or -AD, respectively. Furthermore, owing to the fact that the Y2H system is an *in vivo* method, making use of the transcription and translation machineries of living yeast cells, it is obvious that the analysis of bait or prey proteins that interfere with these processes or are toxic for the yeast cell is not possible. Toxicity, which might be a problem particularly when using viral proteins as bait, becomes apparent by the failure to obtain transformants or by low growth rates of transformed cells. While toxic proteins cannot be used in cases where yeast growth is the signal for monitoring protein interactions (e.g., library screenings), the use of inducible promoters and the LacZ reporter gene might allow the testing of individual toxic bait and prey proteins for interaction.

1. Protein preparation from yeast cultures for western blotting

   (a) Pick a well-growing yeast colony of bait and/or prey-containing yeast strains into 5 mL of the appropriate SC dropout media and incubate overnight at 30°C with shaking (ca. 250 rpm). Use a single colony from a freshly streaked plate not older than 3–4 days. After 16–18 h, this gives a stationary culture of OD$_{600}$ > 1.5. Similarly prepare an overnight culture of an untransformed yeast colony as a negative control.

   (b) Subculture 1 mL into 50 mL YPAD and grow at 30°C with shaking for 31/2–4 h. The OD$_{600}$ should be between 0.4 and 0.6 (note 2). Calculate the total number of OD_{600} units for each sample by multiplying the OD_{600} (of a 1-mL sample) by the culture volume.

(c) Collect cells by low-speed centrifugation (4,000 rpm) for 5 min at 4°C

(d) Discard the supernatant and resuspend the cells in 50 mL dH_2O precooled on ice

(e) Centrifuge cells as in step (b), discard supernatant, freeze cells in liquid nitrogen, and store at −70°C until use

(f) Prewarm cracking buffer to 60°C

(g) Quickly thaw cell samples by separately resuspending each one in prewarmed cracking buffer. Use 100 µL of cracking buffer per 7.5 OD600 units of cells (note 3)

(h) Transfer the cell suspensions to microcentrifuge tubes and add ca. 80 µL of glass beads per 7.5 OD600 units of cells

(i) Vortex vigorously for 1 min

(j) Centrifuge at top speed for 5 min at 4°C to pellet debris and unbroken cells

(k) Transfer the supernatants to fresh microcentrifuge tubes and place on ice

(l) Re-extract the pellets by adding 50 µL cracking buffer, incubating the tubes in a boiling water bath for 5 min, vortexing vigorously for 1 min and centrifuging at top speed for 5 min at 4°C

(m) Combine the second supernatant with the corresponding first supernatant (from step (k). Samples may be stored at −70°C or on dry ice

(n) Proceed with standard SDS-PAGE protocols followed by western blotting and detection of the fusion proteins using commercially available monoclonal antibodies to the Gal4 DNA-binding domain (bait proteins) or the Gal4 activation domain (prey proteins) (note 4).

2. Test for autoactivity of bait proteins

The Y2H system relies on the transcriptional activation of reporter genes. Therefore, one major source of false-positive results is an intrinsic potential of the bait protein to activate transcription when directed to the promoters of the reporter genes by the fused Gal4 DNA-binding domain. A necessary control prior to using particular proteins as baits is to test whether reporter genes are activated by bait constructs without any prey proteins.

This can be done simply by plating the yeast strains transformed with the bait construct onto SC double dropout plates lacking Trp and His, supplemented with an appropriate amount of 3-AT. Usually 3–5 mM of 3-AT is sufficient to suppress the background leakage of the Gal4-dependent promoters used in many Y2H strains.

In case of growth on these control plates, the 3-AT concentration can be increased up to 50–100 mM. If bait proteins still autoactivate the reporter genes, these constructs are not suitable for use with the Y2H system and should be modified by, for example, the deletion of domains to get rid of the transcription activating potential of the bait protein (note 5).

3.3 Y2H Test to Detect Interactions between Individual Bait and Prey Proteins

The yeast strains commonly used for Gal4-based Y2H assays offer several different reporter genes to monitor protein interactions. There are the growth reporters HIS3 (yeast strains AH109, PJ69-4A, Y190) and ADE2 (AH109, PJ69-4A), allowing the selection for prototrophic growth, and there are two possibilities to monitor protein interactions by measuring enzymatic activities of either β-galactosidase or the secreted α-galactosidase MEL1 (AH109, PJ69-4A, Y190, Y187).

After direct or successive double transformation of the yeast strains with the respective bait and prey plasmids, the growth reporters are monitored by plating the cells onto the respective SC dropout media supplemented with the appropriate amount of 3-AT (see step (2) in Sect. 3.2) to suppress any residual potential of the bait protein to activate transcription (note 6).

1. ß-Galactosidase Assay to monitor expression of the LacZ gene

   There are several different protocols to assay β-galactosidase activity in yeast cells, but the most simple and sensitive method is the colony-lift assay.

   (a) Use agar plates (SC dropout media appropriate for the yeast strains to be assayed) with freshly transformed yeast colonies or restreak the yeast strain of interest, and grow for 2–4 days at 30°C
   (b) Freshly prepare Z-buffer/X-Gal solution by adding 0.27 mL β-mercaptoethanol and 1.67 mL X-Gal stock solution to 100 mL of Z-buffer
   (c) For each assay, place a sterile Whatman #5 or VWR grade 410 paper filter disc in a clean Petri dish (90-mm diameter) and soak it with 2 mL of Z-buffer/XGal solution
   (d) Layer a sterile filter disc (paper, nitrocellulose, or nylon) onto the plate of transformants. Leave the filter on the plate for ca. 30 s to wet completely and pick up some yeast from each colony (note 7)
   (e) Using forceps remove the filter from the plate and freeze it by immersing in liquid nitrogen for 5–10 s. This causes rupture of the yeast cells and releases the ß-galactosidase enzyme
   (f) Remove the filter from the liquid nitrogen and thaw it at room temperature
   (g) Taking care to avoid air bubbles, layer the filter colony-side up onto the X-Gal-containing filters prepared in step (c)
   (h) Incubate the filters at 30°C until blue staining appears (note 8)
   (i) Let the filters dry in a fume hood

2. α-Galactosidase Assay to monitor expression of the MEL1 gene

   The Mel1 gene encodes a secreted α-galactosidase that can be assayed spectrophotometrically as a Y2H reporter directly from the culture media without the need to lyse the yeast cells. The assay can be performed in microtiter plates and might be used as a semiquantitative measure to assess relative interaction strengths (note 9).

The following protocol describes the method using microtiter plates. However, the assay can be scaled-up and performed similarly in conventional microcentrifuge tubes.

a) Inoculate 2–5 mL of the appropriate SC dropout media with a well-growing yeast colony from a freshly streaked plate and grow overnight (16–18 h) at 30°C with shaking
b) Freshly prepare the Assay Buffer and equilibrate at room temperature
c) Measure the OD_{600} of each sample and centrifuge 1 mL of the culture at top speed for 3 min
d) Carefully transfer the supernatant (containing the secreted α-galactosidase) into a clean microcentrifuge tube and store at room temperature
e) Transfer 16 µL of the supernatant to a well of a clear microtiter plate
f) Add 48 µL of Assay Buffer to each sample
g) Seal the plate to avoid evaporation and incubate at 30°C for 60 min
h) Terminate the reaction by adding 136 µL of 10X Stop Solution
i) Measure OD_{410} of each sample
j) Calculate α-galactosidase units according to Lazo et al. (22)

3.4   Screening libraries with the Y2H system

The Y2H system is not just a technique to investigate the potential protein interactions of individual bait or prey proteins. An additional major and powerful application of the system is the screening of cDNA or genomic libraries, with the aim of identifying novel interaction partners, potentially providing information about cellular functions and mechanisms.

To screen a complex library, there is the need to assay large numbers (usually several millions) of doubly transformed yeast cells in parallel, each expressing a particular combination of bait and prey constructs. Selection is achieved by activation of reporter genes promoting prototrophic growth of the yeast cell that can be detected by growth of colonies on suitable selective media. There are two approaches to obtain the required numbers of cells expressing the bait and prey constructs: on the one hand by successive double transformation and on the other hand by mating compatible haploid yeast strains, each individually transformed with either the bait or the prey construct to obtain diploid zygotes containing both plasmids.

3.4.1   Library Screening by Double Transformation

In contrast to the single or double transformation of individual bait and/or prey constructs, the transformation of cDNA or genomic libraries requires high transformation efficiencies to obtain several million transformants in order not to decrease the library's complexity.

3.4.2 High-Efficiency Protocol for Yeast Library Transformation

Prior to the transformation of a library, small-scale test transformations should be carried out to determine the actual transformation efficiency of the particular yeast strain used. The method can then be scaled-up to obtain the desired number of transformants.

1. Pick a well-growing yeast colony of your bait-expressing yeast strain from a freshly streaked plate into 5 mL of the respective SC dropout media selecting for presence of the plasmid. Incubate overnight at 30°C with shaking (ca. 250 rpm). Determine the cell titer by counting the cells using a hemocytometer (note 10)
2. Transfer the appropriate culture volume that yields 2.5×10^8 cells to a centrifugation tube. Collect cells by centrifugation at 4,000 rpm for 5 min at room temperature
3. Resuspend cell pellet in 50 mL of prewarmed (30°C) YPAD media and incubate at 30°C with shaking until cell density is 2×10^7 cells per mL, which usually takes 3–5 h (note 11)
4. Collect cells by low speed centrifugation (4,000 rpm, 5 min, room temperature)
5. Discard medium and resuspend the cell pellet in 1/2 volume of sterile dH$_2$O
6. Respin cells (4,000 rpm, 5 min, room temperature), discard supernatant and resuspend in 1 mL of sterile 100 mM LiAc pH 7.5 and transfer to a microcentrifuge tube
7. Centrifuge for 10 s at top speed to pellet cells, and remove the supernatant
8. Resuspend in 550 μL 100 mM LiAc pH 7.5 and transfer 50 μL aliquots to 11 sterile microcentrifuge tubes
9. Collect cells by centrifugation, discard supernatant and add the components of the transformation mix in order from top to bottom

   240 μL 50% PEG 4000
   36 μL 1M LiAc
   25 μL single-stranded DNA (2 mg mL^{-1}, freshly denatured by boiling for 10 min and rapid cooling on ice)
   50 μL library plasmid DNA (use 0.1, 0.2, 0.5, 1, and 2 μg of the library DNA)

10. Resuspend and mix thoroughly by pipetting or vigorous vortexing. Incubate at 30°C for 30 min with occasional shaking/mixing, and then incubate at 42°C for 20 min
11. Centrifuge cells at low speed (4,000 rpm, 3 min), remove transformation mix and resuspend the cells in 1 mL of SC dropout media. Prepare 1:10, 1:50, 1:100, 1:500, and 1:1000 dilutions and spread 100 μL of each dilution onto SC dropout plates

12. Allow plates to air-dry and incubate at 30°C for 2–3 days to select for transformants
13. Count colonies to determine transformation rate and calculate the scale-up to obtain a sufficient number of transformants to completely represent the library. This number is dependent on the complexity of the primary cDNA library. For an average library containing 10^6 independent clones, three to five million yeast transformants should be sufficient.
14. For transformation of the library modify the protocol according to the following table:

		10X	20X	30X	50X	70X
Step 3	YPAD	50 mL	100 mL	150 mL	250 mL	350 mL
Step 8	100 mM LiAc	3 mL	3 mL	3 mL	5 mL	7 mL
Step 9	50% PEG$_{3350}$	2.4 mL	4.8 mL	7.2 mL	12 mL	16.8 mL
	1 M LiAC	350 μL	700 μL	1.05 mL	1.75 mL	2.45 mL
	SS-DNA (2 mg mL^{-1})	500 μL	1 mL	1.5 mL	2.5 mL	3.5 mL
	Plasmid DNA + H$_2$O	250 μL	500 μL	750 μL	1.25 mL	1.75 μL
	Incubation at 42°C	30 min	35 min	40 min	45 min	60 min
Step 11	SC dropout	20 mL	40 mL	40 mL	40 mL	40 mL

Do not merely add more library DNA to the same number of competent yeast cells when trying to obtain higher numbers of transformants, as this can result in the yeast cells containing several different prey plasmids, which will confuse subsequent analysis. For a transformation scale of 20X and higher, cells should be plated onto 100 large (12 × 12 cm) Petri dishes. Gently spread ca. 400 μL per plate using a spreading bar or sterile glass beads. In parallel, plate 5, 2, 1, 0.5, and 0.1 μL of the cell suspension onto SC-Leu/Trp plates to assess transformation efficiency and to calculate the total number of double transformants assayed in the screening. Incubate the plates at 30°C for 3–5 days or, in the case of slow growing cells, up to 14 days, until colonies develop.

3.4.3 Library Screening by Interaction Mating

Despite the high-efficiency transformation protocols, transforming yeast cells with a library is rather laborious. Y2H screening by successive double transformation as described above requires up to 100 plates per screening. A far more efficient way of obtaining doubly transformed yeast cells is the so-called interaction mating. Interaction mating makes use of the yeast life cycle with its haploid and diploid phases. If the nutritional status is appropriate, haploid cells of opposite mating type (MATa and MATα) can conjugate and form diploid zygotes. Under the right conditions, this is a rather efficient process and can be used to combine two plasmids transformed individually into haploid yeast strains into the same diploid zygote.

Not only is the mating rather efficient, in contrast to a chemical transformation procedure that inevitably poses much stress to the cell, but also zygotes are fit and viable, resulting in faster growth and therefore shorter incubation times for the positive colonies to develop. An additional advantage is that the library is transformed into yeast in advance and can be stored in aliquots at −70°C. In that way, a single library transformation can yield enough material for more than 50 screenings.

As a convention, usually the bait constructs are transformed into yeast strains of mating type "a" (MATa, e.g., AH109, PJ69–4A, Y190) and the prey constructs into yeast strains of mating type "α" (MATα, e.g., Y187).

3.4.4   Preparation of the Library for Interaction Mating

1. Use the high-efficiency transformation protocol described above to transform a MATα yeast strain, e.g., Y187 with the library of choice
2. Incubate the plates at 30°C until small colonies are visible, which usually takes approximately 2 days (note 12)
3. Collect the cells from the plates by using 10 mL YPAD per plate and a spreading bar to wash off the colonies and measure the OD_{600} of the suspension
4. Incubate on ice for 1 h
5. Collect the cells by centrifugation at 4,000 rpm for 5 min at 4°C, discard the supernatant and resuspend the pellet in ice-cold YPAD to a final cell density of 40 OD_{600} mL^{-1}
6. Add an equal volume of sterile ice-cold 50% glycerol, mix thoroughly, and distribute into convenient aliquots. A 1-mL aliquot containing a cell number equaling 20 OD_{600} units is sufficient for one Y2H library screening
7. Incubate at −20°C for 30 min, then transfer to −70°C (note 13)
8. Next day determine the titer of viable cells by quickly thawing one aliquot, counting the cells with a hemocytometer and plating serial dilutions onto the appropriate selection media. The titer should not be below 20%.

3.4.5   Library Screening

1. Inoculate 50 mL of the appropriate SC dropout media (4% Glucose) with an amount of bait cells equaling approximately 5–10 well-grown colonies and grow with shaking at 30°C for 16–20 h until cell density is 2×10^7 cells per mL (count cells using a hemocytometer). Always use freshly streaked plates. The growth rate of bait cultures can vary considerably and should be tested in advance.
2. Quickly thaw the required number of aliquots of the frozen library by adding them straight to an at least 20-fold volume of prewarmed (30°C) YPAD medium. Incubate at 30°C for 1 h with shaking (ca. 200 rpm)
3. Determine the cell density of the library suspension by counting the cells using a hemocytometer and mix in a sterile centrifuge tube 10 mL of the bait culture (2×10^8 cells) with an aliquot of the library suspension similarly equaling 2×10^8 cells

4. Collect cells by centrifugation (room temperature, 4,000 rpm, 5 min) and discard the supernatant
5. Resuspend the cell pellet in 4 mL of YCM pH 3.5, transfer to a 100 mL Erlenmeyer flask and incubate at 30°C for 105 min with shaking
6. Spin down the cells by low-speed centrifugation (4,000 rpm, 5 min, room temperature). Discard medium and resuspend the cells by vortexing or pipetting in a small volume of sterile dH_2O, then dilute in 500 mL sterile dH_2O.
7. Collect cells on a 47-mm membrane filter (0.45-μm pore size) by vacuum filtration using a 500-mL filter funnel (effective filtration area: 13.1 cm^2) and incubate filters for 4.5 h at 30°C on YCM (pH 4.5) agar plates
8. Put filters into sterile 50-mL disposable plastic tube with cap, add 10 mL of SC triple dropout media (or sterile water), vortex to completely resuspend cells from filter into solution
9. Plate 5, 2, 1, 0.5, and 0.1 μL onto SC-Leu/Trp plates to assess mating efficiency and to calculate the total number double transformants (zygotes) assayed in the screening. While this mating protocol can yield 20 million zygotes or more, many bait proteins seem somehow to interfere with mating, decreasing this number. However, five million or more zygotes should be obtained routinely which for most libraries would be sufficient.
10. Plate the suspension onto 10–20 agar plates (12 × 12 cm) containing the appropriate SC triple dropout media. Incubate for 5–14 days until colonies have formed (note 14)

3.5   Analysis of Positive Colonies from Y2H Screenings

Positive colonies should be restreaked on selection media to confirm growth independent of the high cell density in the library screening. When handling large numbers of colonies, positive colonies from the screening should be picked into microtiter plates containing ca. 50 μL of sterile dH_2O in each well. Resuspended cells can then be conveniently transferred to selection plates using a 96-pin tool. In that way, different growth markers (HIS3 or ADE2) as well as different stringencies achieved by different concentrations of 3-AT, can be tested in parallel.

A necessary further control to exclude potential false-positive results is to isolate the candidate prey construct and retransform it into yeast together with the original bait construct and/or with control constructs, to confirm the specific activation of the Y2H reporter genes (note 15).

3.5.1   Isolation and Characterization of Plasmid DNA
        from Positive Colonies

Preparing plasmid DNA from yeast cultures is in principle similar to plasmid preparations from *E. coli*. However, there are two factors making it a little more difficult: the yeast cell wall is hard to disrupt and the yield of plasmid DNA usually is very low.

A simple protocol includes the following steps:

1. Pick a yeast colony into 5 mL of triple dropout media (SC lacking Leu/Trp/His + 3-AT). Incubate overnight (20–24 h) at 30°C with shaking (ca. 250 rpm)
2. Collect cells by centrifugation (room temperature, 4,000 rpm, 5 min) and discard the supernatant
3. Resuspend the pellet in 250 µL of buffer P1
4. Add ca. 0.4 g glass beads and break the cells by vortexing vigorously for 5 min
5. Add 250 µL of buffer P2 and mix immediately
6. Add 350 µL of buffer N3 and mix thoroughly
7. Purify the plasmid either by phenol/chloroform extraction and ethanol precipitation or by using plasmid purification columns
8. Transform *E. coli* KC8 by electroporation and select transformants on dropout media lacking leucine to select for the prey plasmid
9. Isolate the plasmid from *E. coli* by conventional plasmid preparation methods

An important control to confirm the screening result is now to cotransform an appropriate yeast strain (e.g., AH109) with the isolated prey plasmid together with the original bait or together with control plasmids (empty bait vector, unrelated bait constructs). Transformants should then be assayed for activation of different Y2H reporter genes by growth on SC dropout media lacking Leu/Trp/His or lacking Leu/Trp/Ade, respectively, and by monitoring the enzymatic activities of the LacZ and MEL1 gene products.

3.5.2 Yeast Colony PCR

Plasmid recovery from yeast cultures is a rather time-consuming and laborious method that might not be suitable when handling large numbers of positive colonies. Yeast colony PCR, amplifying specifically the insert of the selected prey constructs might be a more simple alternative, allowing the researcher quickly to proceed with the control experiments to exclude false-positive results and to determine the identity of the candidate interacting protein by sequencing.

1. Pick a well-growing colony into 25 µL 0.02 N NaOH (freshly prepared) and incubate ~5 min at room temperature
2. Set up PCR reaction as follows (final concentrations):

dNTPs	0.1 mM each
Template	2 µL
Primers	10 pmol each
Taq-buffer	1X
MgCl$_2$	2.5 mM
Taq-Polymerase	1.25 U
H$_2$O	to 50 µL

PCR primers should anneal to the prey plasmids from the library at least 150 base pairs upstream and downstream of the cDNA insert. Flanking sequences of sufficient length are necessary for reconstitution of the prey plasmid by recombination cloning (see below). In our hands, the best results are achieved using rather long PCR primers, combining the annealing and elongation steps, and with the following PCR conditions (note 16):

	94°C (2 min)
40 cycles	94°C (45 s)–72°C (2 min)
	94°C (45 s)
	72°C (5 min)

3. Analyze 5 µL of the PCR reaction by agarose gel electrophoresis. Even if only a rather weak band is visible, usually the amount of PCR product is sufficient for the following gap-repair cloning steps to be used to confirm the screening results.

3.5.3 Analysis of Prey Candidates by Recombination Cloning

Yeast is able to carry out homologous recombination. Gap-repair cloning takes advantage of this efficient repair mechanism, allowing an insert to be cloned into a vector directly in yeast without restriction/ligation and without a cloning step in *E. coli* (23) (Fig. 1).

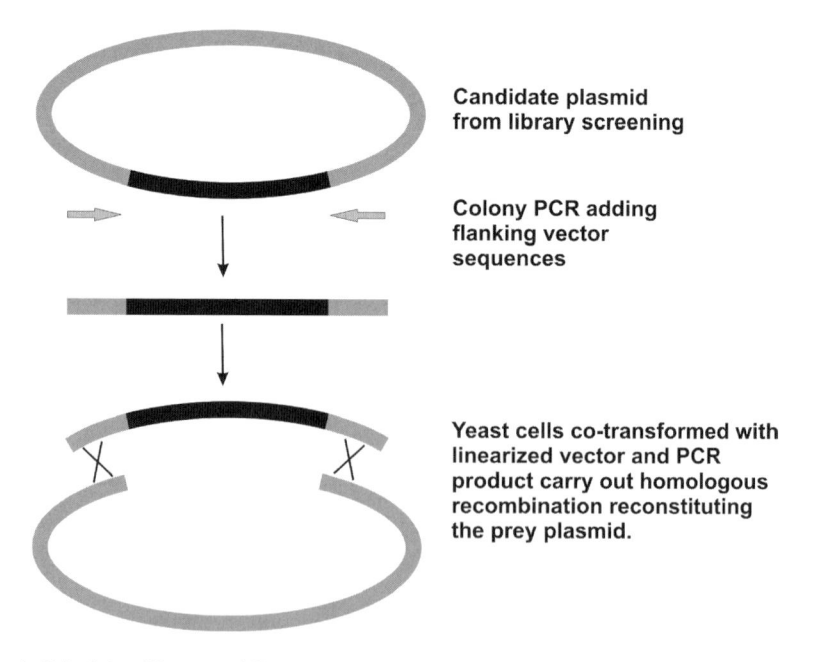

Candidate plasmid
from library screening

Colony PCR adding
flanking vector
sequences

Yeast cells co-transformed with
linearized vector and PCR
product carry out homologous
recombination reconstituting
the prey plasmid.

Fig. 1 Principle of "gap-repair" recombination cloning in yeast

1. Linearize the library activation domain vector by cutting it to completion within the multiple cloning site using two or three different restriction endonucleases. Purify the linearized vector by agarose gel electrophoresis and extraction of the respective band.

2. With the regular or high-efficiency transformation method, transform an appropriate Y2H strain with 5, 10, 20, and 50 ng of the purified linear AD vector to confirm absence of (undigested) background plasmid (note 17).

3. Cotransform your bait-expressing yeast strain with 2 μL of PCR product from the colony PCR and 10–20 ng of linearized AD vector. In parallel, similarly transform control strains pretransformed with empty BD vector or expressing an unrelated bait protein.

4. Incubate on SC-Leu/Trp plates for 2–3 days until colonies appear indicating a successful reconstitution of the prey plasmid.

5. Analyze transformants for expression of the Y2H reporter genes by plating on the respective selection media and/or by assaying MEL1 or LacZ activity.

4 Notes

1. For a "quick-and-dirty" method, the protocol can be modified as follows: Step 2 can be omitted, instead, spin down 1 mL of the overnight culture, discard the supernatant, and proceed directly to step 8. After mixing thoroughly, incubate the tube at 42°C for 40–60 min, and proceed with step 9 using 1 mL of sterile water instead of 200 μL. When transforming a single plasmid, this quick method usually results in a sufficient number of transformants, however, the lower efficiency of a double transformation might necessitate use of the regular transformation protocol.

2. Do not grow yeast cultures to higher density because in late logarithmic growth phase expression of the fusion proteins driven by the ADH promoter is downregulated. Furthermore, in the late growth phase endogenous proteases accumulate.

3. If the cell pellet does not thaw quickly, place the sample briefly at 60°C. PMSF is thermolabile; therefore, add an additional aliquot of the 100X PMSF stock solution to the samples every 10 min throughout the protein preparation process.

4. Western blotting and antibody detection reveals only the presence and the correct size of bait and prey fusion proteins, but not their correct nuclear localization. While with the conventional GAL4 Y2H system there is no simple way to show that the fusion proteins are indeed directed to the nucleus, the LexA system offers a possibility to check the functionality of the bait fusion protein directly.

    In some instances we observed differences in the potential of bait proteins to autoactivate in single or double transformed cells. Here, the proper control would be to cotransform yeast strains with the bait construct together with an empty prey vector or with prey constructs expressing proteins that do not interact with the respective bait protein (we routinely use pSNF4 expressing the yeast SNF4 protein, which so far has not been reported to interact with any viral proteins).

5. To test prey proteins for auto-activation is usually not necessary. However, from Y2H library screenings we have isolated a few prey constructs that promote yeast growth without the presence of any bait protein, but these are rare events and are easily identified by the regular controls that should be performed while testing the specificity of Y2H results (see Sect. 3.5).

6. 30°C might not in all cases be the optimal temperature to assay protein interactions if using baits and/or prey proteins from organisms that need different growth conditions. Weak protein interactions especially might be detected only at lower temperatures.

7. Take care to avoid air bubbles. Label the filter asymmetrically, for example, by poking holes through the filter into the agar to allow association of the results to the colonies on the plate.

8. Staining of strong positives may be visible already after a few minutes whereas weaker positives may take overnight or longer to turn blue. For longer incubation, seal the plates to avoid drying of the filters.

9. The assay is not very robust. Therefore, care should be taken to use cells from similar growth stages (preferably late logarithmic phase) and exactly adjusted cell numbers. Samples should be assayed at least in triplicate.

10. Accuracy at this step seems to be quite important to obtain maximal transformation efficiency. However, instead of counting the cells, cell density can be determined by measuring the OD_{600}. For many yeast strains, an OD_{600} of 1 equals $1-2 \times 10^7$ cells per mL.

11. Alternatively, directly inoculate 50 mL of the appropriate SC dropout media (4% Glucose) with an amount of bait cells equaling approximately 5–10 well-grown colonies and grow with shaking at 30°C for 16–20 h until cell density is 2×107 cells per mL. However, the growth rate of bait cultures can vary considerably and should therefore be tested in advance.

12. The colonies should grow separately and should be just visible. Although longer incubation would yield material for far more screenings, older colonies contain significant fractions of dead cells decreasing the mating efficiency.

13. The optimal way to freeze yeast cells to obtain the maximal survival rate would be to slowly decrease the temperature by 1°C min−1. However, the technical equipment needed might not be available in a standard molecular biological laboratory and the described method should yield 20–30% of viable cells after thawing, which is sufficient for Y2H screenings.

14. For a simplified version of an interaction mating library, screening the protocol can be modified as follows: Prepare the bait preculture and the library as described and measure the OD600. Mix a volume of the bait culture corresponding to 20 OD600 with a volume of the library culture equaling 20 OD600, spin down the cells, discard supernatant, and resuspend in 20 mL of YPAD containing 10% polyethylene glycol 6000. Transfer the suspension to a 100-mL Erlenmeyer flask and incubate at 30°C for 5 h with slow shaking (80 rpm). Collect the cells by centrifugation, discard supernatant, resuspend the pellet in ca. 10 mL of sterile dH_2O and plate onto 10–20 agar plates (12 × 12 cm) containing the appropriate selection media.

15. While there are numerous reasons for obtaining a false-negative result with the Y2H system, false-positives are more easily recognized and excluded by the appropriate control experiments. There are basically two classes of common false-positive clones occurring in Y2H screenings. There are the so-called "random" false positives, including cells or colonies that grow only under the particular conditions of a library screening (large cell numbers, high cell density). These are easily identified either by their inability to grow after restreaking onto selective media (see Sect. 3.5) or by the fact that the positive result cannot be reproduced by isolation of the prey construct and cotransformation with the bait plasmid (see Sect. 3.5.1 or 3.5.3). The second class comprises prey constructs that activate the Y2H reporters nonspecifically and independent of the particular bait used for the screening. These false-positives are identified by cotransformation with empty bait vector or with control plasmids expressing unrelated bait proteins. A list of common false positives from Y2H screenings can be found at http://www.fccc.edu/research/labs/golemis/InteractionTrapInWork.html.

Our own experience from more than 500 Y2H library screenings shows that every cDNA library seems to contain its own specific false-positives. In our hands, common (reproducible) false-positive plant proteins include several ribosomal proteins (At1g43170, At1g56045, At2g27720), photosystem I subunit III precursor (At1g31330), an ids4-like protein (AT5g20150), and JAB1/CSN5 (At1g22920), which has been shown recently to actually interact with the Gal4 DNA-BD (24).

16. For many common Gal4 AD vectors (pACT2, pGAD10, pGADT7, pGAD_GH) the same oligonucleotides can be used for colony PCR. We obtained good results using the primers AD5 (5′-GGACGGACCAAACTGCGTATAACGCGTTTGGAATCACTACAGGGATG-3′) and AD3 (5′-GCGACCTCATGCTATACCTGAGAAAGCAACCTGACCTACAGGAAAGAG-3′) annealing in the activation domain and in the ADH terminator, respectively. These primers

allow a PCR with a combined annealing and elongation step at 72°C. This not only shortens the PCR cycling times but also seems to give more reliable and specific results.

17. The gap-repair cloning method often is not very efficient, yielding sometimes only a few colonies. It is therefore important that the linearized vector does not produce any background. If there are problems with background, further restriction with additional enzymes, dephosphorylation and purification may be necessary.

References

1. Fields, S. and Song, O. (1989) A novel genetic system to detect protein-protein interactions. *Nature* **340**, 245–246.
2. Cusick, M.E., Klitgord, N., Vidal, M. and Hill, D.E. (2005) Interactome: gateway into systems biology. *Hum. Mol. Genet.* **14**, R171–R181.
3. Ito, T., Chiba, T. and Yoshida, M. (2001) Exploring the protein interactome using comprehensive two-hybrid projects. Trends Biotechnol. 19, S23–S27.
4. Fields, S. (2005) High-throughput two-hybrid analysis. *FEBS J.* **272**, 5391–5399.
5. Finley Jr., R.L. and Brent, R. (1994). Interaction mating reveals binary and ternary connections between Drosophila cell cycle regulators. *Proc. Natl. Acad. Sci. USA* **91**, 12980–12984.
6. Ren, T., Qu, F. and Morris, T.J. (2000) HRT gene function requires interaction between a NAC protein and viral capsid protein to confer resistance to Turnip crinkle virus. *Plant Cell* **12**, 1917–1925.
7. Selth, L.A., Dogra, S.C., Rasheed, M.S., Healy, H., Randles, J.W. and Rezaian, M.A. (2005) A NAC domain protein interacts with tomato leaf curl virus replication accessory protein and enhances viral replication. *Plant Cell* **17**, 311–325.
8. Hao, L., Wang, H., Sunter, G. and Bisaro, D.M. (2003) Geminivirus AL2 and L2 proteins interact with and inactivate SNF1 kinase. *Plant Cell* **15**, 1034–1048.
9. Wang, H., Hao, L., Shung, C-Y., Sunter, G. and Bisaro, D.M. (2003) Adenosine kinase is inactivated by geminivirus AL2 and L2 proteins. *Plant Cell* **15**, 3020–3032.
10. Wang, H., Buckley, K.J., Yang, X., Buchmann, R.C. and Bisaro, D.M. (2005) Adenosine kinase inhibition and suppression of RNA silencing by geminivirus AL2 and L2 proteins. *J. Virol.* **79**, 7410–7418.
11. Anandalakshmi, R., Marathe, R., Ge, X., Herr Jr., J.M., Mau, C., Mallory, A., Pruss, G., Bowman, L. and Vance, V.B. (2000) A calmodulin-related protein that suppresses posttranscriptional gene silencing in plants. *Science* **290**, 142–144.
12. Uhrig, J., Canto, T., Marshall, D. and MacFarlane, S.A. (2004) Relocalization of nuclear ALY proteins to the cytoplasm by the Tomato bushy stunt virus p19 pathogenicity protein. *Plant Physiol.* **135**, 2411–2423.
13. Canto, T., Uhrig, J.F., Swanson, M., Wright, K.M. and MacFarlane, S.A. (2006). Translocation of the TBSV P19 protein into the nucleus by ALY compromises its silencing suppressor activity. *J. of Virol.* **80**, 9064–9072.
14. Abbink, T.E.M., Peart, J.R., Mos, T.N.M., Baulcombe, D.C., Bol, J.F. and Linthorst, H.J.M. (2002) Silencing of a gene encoding a protein component of the oxygen-evolving complex of photosystem II enhances virus replication in plants. *Virology* **295**, 307–319.
15. Padmanabhan, M.S., Goregaoker, S.P., Golem, S., Shiferaw, H. and Culver, J.N. (2005) Interaction of the Tobacco mosaic virus replicase protein with the Aux/IAA protein PAP1/IAA26 is associated with disease development. *J. Virol.* **79**, 2549–2558.
16. Soellick, T.R., Uhrig, J.F., Bucher, G.L., Kellman, J.W. and Schreier, P.H. (2000) The movement protein NSm of Tomato spotted wilt tospovirus (TSWV): RNA binding, interaction with the TSWV N protein, and identification with interacting plant proteins. *Proc. Natl. Acad. Sci. USA* **97**, 2373–2378.

17. Chen, M.H., Sheng, J., Hind, G., Handa, A.K. and Citovsky, V. (2000) Interaction between the tobacco mosaic virus movement protein and host cell pectin methylesterases is required for viral cell-to-cell movement. *EMBO J.* **19**, 913–920.
18. Wittman, S., Chatel, H., Fortin, M.G. and Laliberte, J.F. (1997) Interaction of the viral protein genome linked of turnip mosaic potyvirus with the translational eukaryotic initiation factor (iso) 4E of Arabidopsis thaliana using the yeast two-hybrid system. *Virology* **234**, 84–92.
19. Schaad, M.C., Anderberg, R.J. and Carrington, J.C. (2000) Strain-specific interaction of the tobacco etch virus Nia protein with the translational initiation factor eIF4E in the yeast two-hybrid system. *Virology* **273**, 300–306.
20. Ruffel, S., Dussault, M.H., Palloix, A., Moury, B., Bendahmane, A., Robaglia, C. and Caranta, C. (2002) A natural recessive gene against potato virus Y in pepper corresponds to the eukaryotic initiation factor 4E (eIF4E). *Plant J.* **32**, 1067–0175.
21. Gao, Z., Johansen, E., Eyers, S., Thomas, C.L., Ellis, N. and Maule, A.J. (2004) The potyvirus recessive resistance gene, sbm1, identifies a novel role for translation initiation factor eIF4E in cell-to-cell trafficking. *Plant J.* **40**, 376–385.
22. Lazo, P.S., Ochoa, A.G. and Gascón, S. (1978) α-galactosidase (melibiase) from Saccharomyces carlsbergenesis: structural and kinetic properties. *Arch. Biochem. Biophys.* **191**:316–24.
23. Ma, H., Kunes, S., Schatz, P.J. and Botstein, D. (1987) Plasmid construction by homologous recombination in yeast. *Gene* **58**, 201–21.
24. Nordgard, O., Dahle, O., Andersen, T. O. and Gabrielsen, O. S. (2001) JAB1/CSN5 interacts with the GAL4 DNA binding domain: a note of caution about two-hybrid interactions. *Biochimie* **83**, 969–71.

Index

Printed in the United States of America